OXFORD MEDICAL PUBLICATIONS

Genetic Biochemical Disorders

OXFORD MONOGRAPHS ON MEDICAL GENETICS

General Editors:

M. BOBROW
P. S. HARPER
A. G. MOTULSKY
C. R. SCRIVER

Former Editors

J. A. FRASER ROBERTS
C. O. CARTER

OXFORD MONOGRAPHS ON MEDICAL GENETICS No 12

Genetic Biochemical Disorders

PHILIP F. BENSON, M.D., Ph.D., M.Sc., F.R.C.P., F.R.C.Path., D.C.H.

Consultant Paediatrician
Al Qassimi Hospital
Sharjah
United Arab Emirates

Formerly
Reader in Clinical Biochemical Genetics
Paediatric Research Unit
The Prince Philip Research Laboratories
Guy's Hospital Medical School
and
Honorary Consultant Clinical Biochemical Geneticist and
Honorary Consultant Paediatrician to Guy's Hospital

ANTHONY H. FENSOM, B.Sc., Ph.D., C.Chem., F.R.S.C., M.R.C.Path.

Director, Supraregional Laboratory for Genetic Enzyme Defects
Paediatric Research Unit
The Prince Philip Research Laboratories
Guy's Hospital Medical School

OXFORD   NEW YORK   TORONTO   TOKYO
OXFORD UNIVERSITY PRESS

Oxford University Press, Walton Street, Oxford OX2 6DP

Oxford   New York   Toronto
Delhi   Bombay   Calcutta   Madras   Karachi
Petaling Jaya   Singapore   Hong Kong   Tokyo
Nairobi   Dar es Salaam   Cape Town
Melbourne   Auckland

and associated companies in
Beirut   Berlin   Ibadan   Nicosia

Oxford is a trademark of Oxford University Press

First published, 1985
Reprinted, 1986
First published in paperback, 1986

British Library Cataloguing in Publication Data

Benson, P.F.
  Genetic biochemical disorders.——(Oxford
monographs on medical genetics; no. 12)
  1. Metabolism, Inborn errors of
  2. Biochemical genetics
  I. Title   II. Fensom, Anthony H.
  616.3'9042   RC627.8

  ISBN 0–19–261193–3
      0–19–261642–0 (pbk)

Library of Congress Cataloging in Publication Data

Benson, P.F. (Philip F.)
  Genetic biochemical disorders.
  (Oxford monographs on medical genetics; no. 12)
  (Oxford medical publications)
  Includes bibliographies and index.
  1. Metabolism, Inborn errors of.   I. Fensom, Anthony H.
  II. Title.   III. Series.   IV. Series:
  Oxford medical publications. [DNLM: 1. Metabolism,
  Inborn Errors. WD 205 B474g]
  RC627.8.B46   1985   616.3'9   84–25485

  ISBN 0–19–261193–3
      0–19–261642–0 (pbk)

Set by Cambrian Typesetters, Frimley, Surrey
Printed in Great Britain by
St Edmundsbury Press Ltd,
Bury St Edmunds, Suffolk

Preface

During the last ten years, the authors have supervised the Supraregional Laboratory for Genetic Enzyme Defects. In this capacity we have investigated several thousand children suspected of having inborn errors of metabolism, and their families. We have carried out prenatal diagnosis in about 200 pregnancies at risk for metabolic diseases. For prevention and treatment of metabolic disorders, prompt and precise diagnosis are often essential. However, many of these conditions are rare, and clinicians may not have seen individuals affected with many preventable or treatable metabolic diseases. Our experience leads us to believe that there is a need for a book summarizing the salient facts on the clinical, genetical, pathological, and biochemical features of inborn errors, reviewing therapy where this is possible, or methods of prenatal diagnosis when it is not. This book is aimed, therefore, at a wide readership, but especially practising and student paediatricians, neurologists, geneticists, obstetricians, pathologists, medical biochemists, dieticians, and others involved in diagnosis, prevention and treatment of patients with metabolic diseases.

Sharjah and   P. F. B.
London   A. H. F.
September 1984

Acknowledgements

We are most grateful to our colleague Dr. J. A. Fraser Roberts for his constant support and encouragement during the preparation of this book. We would like to thank several of our colleagues for allowing us to publish illustrations, in particular Dr. K. Hugh-Jones for Fig. 2.2(c) showing the face of a young boy with Hurler's disease; Professor R. G. Spector for his photograph of the brain of a boy with Hurler's disease (Fig. 2.4(b)); Mrs. M. Jackson for the electrophoresis of urinary and amniotic fluid glycosaminoglycans (Fig. 2.6(a)–(d)); Dr. W. Schutt for Fig. 2.7 showing two patients with Scheie's disease; Dr. G. Snodgrass and Dr. K. Hugh-Jones for Figs. 2.18 and 2.19 showing clinical and radiological features of Morquio's disease; Mr. D. E. Mutton for Fig. 3.3 showing the appearance on phase microscopy of a fibroblast from a patient with I-cell disease; Professor C. Wood for Fig. 3.5 showing a patient with mannosidosis; Dr. S. R. Aparico for Fig. 4.14 showing the electron micrograph of renal epithelium from a patient with Fabry's disease; Dr. B. R. G. Neville for Fig. 4.16 showing arthropathy in Farber's disease; Professor D. R. Turner for Fig. 4.17 showing the electronmicrograph of a neurone from a patient with Farber's disease; Professor A. Niederwieser for advice on the planning of Fig. 5.2 showing the phenylalanine hydroxylation system; Professor I. A. Magnus for Fig. 7.3(a)–(c) showing the progressive facial changes in a patient with xeroderma pigmentosum; Dr. F. Giannelli and William Heinemann Medical Books Ltd. for Fig. 7.4 showing a patient with xeroderma pigmentosum; Dr. F. Giannelli for Fig. 7.5 showing the scheme of excision repair of DNA; and Spastics International Medical Publications for permission to reproduce Figs. 2.1, 2.2(e) and 5.1.

We are greatly indebted to Miss C. Beech and Miss J. Garner for typing the manuscript and to Miss E. Manners for assisting with the references; to Mrs B. Merchant and the staff of the Paediatric Research Unit Laboratory for their continued help; and to Miss P. M. Archer, Miss K. Montague and the Department of Medical Illustration and Photography of Guy's Hospital, and to Mr. L. Kelberman of the Paediatric Research Unit for their help in preparing illustrations.

Contents

1. Introduction

PATTERNS OF INHERITANCE

Since the basic principles of genetic inheritance are described clearly elsewhere (e.g. Fraser Roberts and Pembrey 1985), only those aspects with direct relevance to genetic counselling and prenatal diagnosis of metabolic disorders will be considered here.

Dominant inheritance, that is, where there is full expression of the clinical phenotype in at least some heterozygotes for enzyme defects, appears to be limited to the porphyrias, familial hypercholesterolaemia and possibly Crigler–Najjar syndrome type II. Only one parent is usually the carrier and is often affected (but to a variable degree). The risk of transmission to their children is one in two, irrespective of sex. Partial clinical expression, however, occurs in some heterozygous female carriers for recessive X-linked disorders, including Fabry's and Menkes' diseases and Lowe's syndrome. This may be due, in theory, to the existence in these subjects of a high proportion of cells in which (as discussed below) there is inactivation of the wild gene carried on 'Lyonized' X-chromosomes.

Autosomal recessive inheritance

In autosomal recessive inheritance the abnormal gene is carried on a non-sex chromosome. Characteristically, both parents are carriers but have no clinical expression of the disease. However, there are exceptions; for example, cataracts have been reported in probable heterozygotes for galactokinase deficiency. When both parents are heterozygotes for the same mutant gene, the risk of transmission to their children, irrespective of sex, is one in four. The carrier state will be transmitted, on average, to two in four children, while one in four children will be homozygous for the normal gene.

The carrier frequency may be calculated as follows:

Assuming random mating, if the frequency of homozygotes is p^2 for a recessive gene of frequency p; and q^2 for homozygotes for the normal allele of frequency q (that is, $1-p$), the heterozygote frequency will be $2pq$. For practical purposes, in view of the low value for p, one may equate q to 1 (instead of $1-p$) and calculate the frequency of carriers by doubling the gene frequency, which is calculated as the square root of the frequency of the homozygotes for the mutant gene (affected individuals). For example, if the frequency of the homozygotes is 1 in 10 000, the gene frequency will be 1 in 100, and the frequency of the heterozygote 1 in 50.

Consanguinity is more common among parents of children with autosomal recessive disorders than in others (Fig. 1.1).

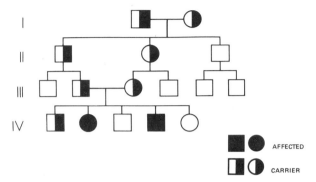

Fig. 1.1. Pedigree illustrating the transmission of autosomal recessive genes through heterozygotes (carriers), to homozygotes (affected), arising from first cousin matings.

The rarer the disease, the lower is the frequency of carriers in the general population. In mutations of extreme rarity, the genes may be limited to only a few families so that the disease will only occur in children of parents belonging to these families. This results in a higher than average incidence of consanguinity amongst parents of affected children. First cousins have a 1 in 8 chance of carrying the same gene. For example, it may be calculated that the frequency of first cousin marriages, which in the United Kingdom is approximately 0.5 per cent, increases to 3 per cent in parents of children with a recessive disease with a frequency of 1 in 10 000; and to 9 per cent when the frequency is 1 in 100 000.

Although heterozygote carriers of autosomal recessive disorders are nearly always normal clinically, there is often a gene dosage effect on the biochemical phenotype which can be demonstrated either as reduced enzyme activity or by an abnormal response after loading with the enzyme substrate. This is an important phenomenon in prenatal diagnosis, since it may allow characterization of a disease for which a family is at risk when the propositus has died without precise enzyme diagnosis.

The children of an affected individual have a 1 in 2 chance of themselves being carriers, irrespective of sex.

X-linked (sex-linked) inheritance

In X-linked inheritance, the abnormal gene is carried on an X-chromosome.

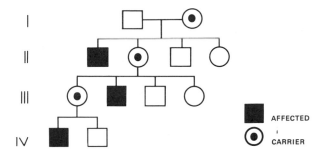

Fig. 1.2. Pedigree illustrating the transmission of X-linked genes to carrier daughters and affected hemizygote males.

In the majority of cases, the mother is the carrier and (with a few exceptions as mentioned above) will be clinically normal. The risk of transmitting the disease to her sons and the carrier state to her daughter will be 1 in 2. Male inheritance is irrespective of the paternal genotype or of consanguinity. X-linked recessive disease occurs almost exclusively in hemizygous males (Fig. 1.2). Exceptionally, the disease occurs in female homozygotes as a result of mating between a carrier mother and an affected father, or when the mutant gene is present in only a single dose, but the other X-chromosome is either absent (e.g. in gonadal dysgenesis or Turner's syndrome) or partially inactivated as a result of a translocation.

In X-linked dominant inheritance, however, e.g. in hypophosphataemic rickets, or ornithine carbamyl transferase deficiency (qv), the clinical phenotype may be expressed as fully as in the hemizygote male.

According to the Lyon hypothesis, in the early female embryo there is random inactivation of one of the two X chromosomes in each cell. This results in a cellular mosaicism with two populations of cells, in which either the maternal or paternal X-chromosome is expressed. The hypothesis implies that the same X-chromosome which is functionally inactivated in one cell continues to be inactivated in daughter cells derived from it, either in the female or in cell culture clones. The hypothesis has been proven for many X-linked loci by demonstration that some cells express one allele and others the other allele. Some X-linked loci, for example those for steroid sulphatase, or the Xg^a blood group, which are on the short arm of the X-chromosome, escape inactivation.

The phenomenon is useful for detection of carriers of X-linked genes by demonstrating two populations of cells, one with normal, and the other deficient enzyme activity. In practice, such cells are usually cultured fibroblasts derived from a skin biopsy a few millimetres in diameter. Hair root analysis is an interesting alternative, since there is a reasonable chance that the half-dozen or so cells attached to plucked hairs will be derived mainly from one cell type. Enzyme assay of hair root cells has the theoretical advantage that the cells may be collected from separate parts of the scalp, thus increasing the chance of detecting mosaicism; the technique has detected female carriers of Fabry's and Hunter's diseases. However, failure to detect two populations of cells can never exclude the existence of mosaicism.

When affected hemizygous males reproduce, all the sons will escape the disease and all the daughters will be carriers.

Prenatal diagnosis

Early prenatal diagnosis followed by selective termination of pregnancy is the only way at present by which families at risk for certain genetic diseases can choose to have only unaffected children. The introduction of prenatal diagnostic procedures in recent years has revolutionised management of genetic diseases which can be detected in the foetus. Thus, at counselling interviews, genetic and procedural risks are explained and discussed with the parents, who are also made aware of the possibility of certain diagnostic errors. With these reservations clearly in mind, the parents can consider whether, if the tests indicate that the foetus is affected, they would choose to have the pregnancy terminated. Prenatal diagnosis is usually only undertaken if the parents have decided that they would request termination should the foetus be affected.

Indications for prenatal diagnosis

In general, prenatal diagnosis is indicated where a pregnancy is at risk for a serious genetic disease for which there is no satisfactory treatment. There are large variations in the degree of risk acceptable to different families. The great majority of parents without religious, ethical or other objections will request prenatal diagnosis for genetic risks of 1 in 4, which apply to most cases of metabolic disorders, or for risks in the order of 1 in 50, which applies for trisomy-21 for mothers aged about forty. It has been suggested that prenatal diagnosis should be offered to parents when the risk is 1 in 100 or more, that is, for approximately 8 per cent of all pregnancies (Clinical Genetics Society 1978).

Usually the family is known to be at risk because there has been a previous affected child. In this event it is essential that a precise enzymic diagnosis be established so that specific enzyme testing is possible for monitoring subsequent pregnancies. If a child is dying with an undiagnosed illness, which is suspected of being an inborn error of metabolism, it is important that skin fibroblast cultures be initiated and either assayed for suspected enzyme deficiencies or stored under liquid nitrogen so that they may be tested later, and a liver sample is collected as soon as possible after death. This may allow post mortem diagnosis and prenatal diagnosis should the mother become pregnant again.

In sex-linked disorders the mother may be found to be a carrier because she has an affected brother or soral nephew and may be offered amniocentesis even if she has not had affected children.

Genetic screening of ethnic populations known to have a high incidence of carriers is another method for establishing that both spouses are carriers of deleterious genes. The best example is the screening of Ashkenazi Jews in whom Tay–Sachs disease is about one hundred times more common than in others, and the carrier frequency is approximately 1 in 27. Over a quarter of a million were screened during 1969–1979, identifying 251 at-risk couples in whom 623 pregnancies were monitored, yielding 148 affected foetuses.

This approach may gain wider acceptability for other disorders in the future if mass screening procedures become practicable for other genetic disorders.

PROCEDURES

Amniocentesis

Currently, the most reliable methods for prenatal diagnosis of most metabolic disorders in the foetus depend on direct enzyme assay of cultured amniotic cells collected at amniocentesis. Uncultured amniotic cells have been used for prenatal diagnosis of Tay-Sachs disease, but since approximately four-fifths of amniotic cells are non-viable, they are often unreliable for assay of unstable enzymes. Cell-free amniotic fluid receives contributions both from the mother and the foetus, so that demonstration of enzymic activity in the fluid may not rule out the existence of genetic deficiency in the foetus; this has been demonstrated for α-glucosidase in a foetus with Pompe's disease (glycogenosis type 2).

Other metabolic disorders, for example the GM_2-gangliosidoses and Hunter's disease (MPS II), have been detected by amniotic fluid enzyme assay but more experience is required before the degree of reliability of these assays can be established.

Accumulation of metabolites or storage material in the amniotic fluid can be detected in some pregnancies where the foetus is affected. Examples include methylmalonic acidaemia, the mucopolysaccharidoses, and galactosaemia.

Complications may arise in twin pregnancies, which can be missed even with ultrasound. Moreover, sometimes only one sac can be tapped, and an additional problem is that during amniocentesis, fluid can be transferred from one sac to the other (Campbell *et al.* 1976). When only one pair of twins is affected, the parents may elect to allow the pregnancy to continue. Cardiac puncture has been used to cause death of a Hurler foetus, allowing the normal twin to be delivered (Aberg *et al.* 1978).

Contamination of amniotic cells with maternal cells has given rise to diagnostic errors in 0.10 to 0.51 per cent of samples (Mossman *et al.* 1981). When cultured amniotic cells have an XX chromosome complement, it may be possible to establish whether they are foetal rather than maternal by studying marker chromosomes (Buchanan *et al.* 1980) or the HLA antigens of amniotic cells and of parental leucocytes (Niazi *et al.* 1979). Whenever possible, when cells are female, in addition to assaying cultured cells, a chemical test should be carried out on the amniotic fluid supernatant, which is independent of maternal-cell contamination. We have reported correct prenatal diagnosis of a Hurler foetus by demonstrating elevated amniotic fluid heparan and dermatan sulphates when amniotic fluid cells were of maternal origin (Mossman *et al.* 1981).

A further problem is that mutant foetal cells lacking specific enzyme activity may become contaminated with mycoplasma during culture, and on testing may be found to have acquired enzyme activity. We have reported the appearance of argininosuccinic acid lyase activity in mutant cells from a patient with argininosuccinic aciduria which became infected with *Mycoplasma hyorhinis* and we stress that all cells should be screened for mycoplasma contamination (Fensom *et al.* 1980).

Foetoscopy

Foetoscopy with real-time ultrasound guidance offers the opportunity of direct examination of the foetus and for collection of foetal blood, skin, or liver. The former is the method of choice for detection of the majority of the haemoglobinopathies and the haemophilias. It is also useful for direct enzyme assay for mothers presenting in advanced gestation, when amniotic cell culture would delay assay to a stage of pregnancy when the foetus may be viable. We have used foetal blood for prenatal testing of pregnancies at risk for galactosaemia (Fensom *et al.* 1979*a*), Hurler's disease (Rodeck *et al.* 1983*a*), homocystinuria (Fensom *et al.* 1983), metachromatic leucodystrophy (Rodeck *et al.* 1983*b*), and propionic acidaemia (Fensom *et al.* 1984*a*). However, in a pregnancy terminated because the foetus had cystinosis, foetal leucocytes isolated immediately after termination did not have elevated intracellular free cystine concentration. In this pregnancy the diagnosis was confirmed on cultured foetal fibroblasts. In skilled hands, the procedure has a foetal loss of only 2–3 per cent (Rodeck and Campbell 1978; Rodeck 1980).

Chorionic villi biopsy

Chorionic villi biopsy under foetoscopic and ultrasound localization has allowed accurate foetal sexing in 40 cases at 6–12 weeks' gestation (Kazy *et al.* 1982). In this study, no abortions followed 124 procedures carried out without monitoring. These authors found villi to have measurable activity of ß-glucosidase, ß-galactosidase, ß-hexosaminidase, ß-glucuronidase, α-fucosidase, sphingomyelinase, and arylsulphatase A. More recently, activities of many other enzymes have been demonstrated in villi and cultured villi (Simoni *et al.* 1983; Fensom *et al.* 1984*b*). Already the technique has been used for prenatal diagnosis of the haemoglobinopathies (Williamson *et al.* 1981; Old *et al.* 1982), Hunter's disease (Kleijer *et al.* 1984*a*), atypical Tay–Sachs disease (Besançon *et al.* 1984), Lesch–Nyhan syndrome (Gibbs *et al.* 1984), citrullinaemia (Kleijer *et al.* 1984*b*), and argininosuccinic aciduria (Vimal *et al.* 1984).

If found to be reliable and safe for prenatal diagnosis, this method will be preferable to amniocentesis since it will allow *in utero* diagnosis in the first trimester when termination of pregnancy can be by aspiration rather than by inducing labour by prostaglandins.

DNA recombinant techniques

The human haploid genome consists of about 3×10^9 nucleotides of which only about 10 per cent code for structural or regulatory genes (Bishop 1974).

A study of DNA polymorphisms of genetic loci linked to inherited disease may prove useful for diagnosis (Botstein *et al.* 1980). For example, using a DNA polymorphism for restriction enzyme sites linked to the sickle cell gene (Kan and Dozy 1978; Phillips *et al.* 1980), prenatal diagnosis of sickle cell anaemia is possible in a high proportion of foetuses. However, more recently, the sickle cell gene itself has been detected by a DNA recombinant technique (Geever *et al.* 1981).

DNA polymorphisms may also allow prenatal diagnosis of ß-thalassaemia (Kan *et al.* 1980; Little *et al.* 1980).

Mapping of the human genome is a realistic possibility and should allow diagnosis of genetic disease by DNA analysis rather than by identifying mutant enzymes or other proteins in the future. Potentially, this new methodology is one of the most powerful approaches to diagnosis of individuals and foetuses affected by genetic diseases.

2. The mucopolysaccharidoses

The mucopolysaccharidoses (MPS) are a group of inherited lysosomal storage disorders of connective tissue each with distinctive phenotypes and a progressive course due to a severe deficiency of an enzyme which usually catalyses a step in the degradation of glycosaminoglycans (GAG).

The first descriptions were by Hunter (1917) of two brothers with the only sex-linked variety, and by Hurler (1920) of two patients with the most severe form. The diseases have both eponymous and numerical (McKusick 1972) designations (Table 2.1).

Excessive quantities of GAG were reported in liver (Brante 1952) and urine (Dorfman and Lorincz 1957). Later GAG in urine were identified as dermatan sulphate and heparan sulphate in Hurler (MPS IH), Scheie (MPS IS), and Hunter (MPS II) diseases (Kaplan 1969); as predominantly heparan sulphate in the Sanfilippo diseases (MPS IIIA-D) (Sanfilippo *et al.* 1963); keratan sulphate and chondroitin sulphate in classical Morquio's disease (MPS IVA) (Pedrini, *et al.* 1962); dermatan sulphate in Maroteaux–Lamy disease (MPS VI) (Maroteaux *et al.* 1963); and chondroitin sulphate in some cases of ß-glucuronidase deficiency (Sly *et al.* 1973) or rarely in Hurler's disease (Benson *et al.* 1972; Babarik *et al.* 1974; Spranger *et al.* 1974*a*).

Lysosomes distended with storage material in cells from patients with Hurler's disease, described as metachromatic granules in cultured fibroblasts (Danes and Bearn 1965), were identified in liver by Van Hoof and Hers (1964) by electron microscopy.

Fratantoni *et al.* (1968*a*, 1969*a*) demonstrated that cultured fibroblasts from patients with Hurler's and Hunter's diseases incorporated radiosulphate (^{35}S-sulphate) into GAG more rapidly than those from normal individuals because of defective GAG degradation rather than because of accelerated synthesis. Correction of this defect by the products of cells of other genotypes was also demonstrated (Fratantoni *et al.* 1968 *b*). These 'corrective factors' were shown to be specific for different types of MPS. For example, the abnormal GAG degradation of Hurler cells could be corrected by growing them in medium which had been used to culture Hunter cells, i.e. which contained the 'Hurler factor', later identified as α-L-iduronidase (Shapiro *et al.* 1976*a*). Such complementation studies were used to show that cells from patients with Hurler's disease could not correct the defect in Scheie cells, indicating that the two diseases were probably allelic (Wiesmann and Neufeld 1970).

Table 2.1 *Classification of the mucopolysaccharidoses (MPS)*

Disease	Excessive urinary GAG	Enzyme deficiencies	Main clinical features
MPS IH Hurler	DS, HS	α-iduronidase	Early presentation, often in first year; lumbar gibbus and corneal clouding; stiff joints; dwarfism; hepatosplenomegaly; coarse facies; scaphocephaly; hydrocephalus; progressive mental deterioration; death in first decade; AR
MPS IS Scheie	DS, HS	α-iduronidase	Stiff joints, especially of hands, with carpal tunnel syndrome; corneal clouding; normal intelligence; aortic incompetence; survival to adult life; AR
MPS IH/S (double heterozygote) Hurler/Scheie	DS, HS	α-iduronidase	Clinical features intermediate between MPS IH and MPS IS; AR
MPS IIA Hunter (severe)	DS, HS	Sulphoiduronate sulphatase	Similar to MPS IH but milder. Kyphosis and corneal clouding very rare; X-linked
MPS IIB Hunter (mild)	DS, HS	Sulphoiduronate sulphatase	Visceral and skeletal changes similar to MPS IIA but intelligence normal. Survival to adult life; X-linked
MPS IIIA Sanfilippo A	HS	Heparan N-sulphatase	Severe mental handicap noticeable first two years; no growth retardation; mild skeletal and visceral changes; survival into 2nd and 3rd decades; AR
MPS IIIB Sanfilippo B	HS	N-acetyl- α -glucosaminidase	As in MPS IIIA; AR
MPS IIIC Sanfilippo C	HS	Acetyl CoA: α -glucosaminide N-acetyltransferase	As in MPS IIIA; AR
MPS IIID Sanfilippo D	HS	N-acetylglucosamine 6-sulphate sulphatase (substrate prepared from heparan sulphate)	As in MPS IIIA; AR

Table 2.1 (*continued*)

Disease	Excessive urinary GAG	Enzyme deficiencies	Main clinical features
MPS IVA Classical Morquio	KS, CS	N-acetylgalactosamine 6-sulphate sulphatase	Severe skeletal deformities, including sternal protrusion, dwarfism, dorsal kyphosis, odontoid hypoplasia, compression of cervical cord, and dysostosis multiplex; corneal clouding; aortic valve disease; AR
MPS IVB Variant Morquio (MPS V renamed MPS IS)	KS	ß-galactosidase	Similar to classical form but milder. Heart murmurs not observed; AR
MPS VI Maroteaux-Lamy (severe and mild variants)	DS	N-acetylgalactosamine 4-sulphate sulphatase (arylsulphatase B)	Similar to MPS IH, but normal intelligence; facies not so markedly abnormal; compression of cervical cord; AR
MPS VII Sly	HS, DS or CS, DS or CS	ß-glucuronidase	Variable, e.g. Sly *et al.* 1973; Gehler, *et al.* 1974; Beaudet *et al.* 1975; psychomotor retardation; dysostosis multiplex with gibbus; hepatosplenomegaly; AR
Matalon, Wappner, Brandt, and Korwitz 1982	KS, HS	N-acetylglucosamine 6-sulphate sulphatase (substrate prepared from keratan sulphate)	Rapid psychomotor regression; petit mal; blindness; dysostosis multiplex; inheritance unknown

Abbreviations: DS = dermatan sulphate
HS = heparan sulphate
KS = keratan sulphate
CS = chondroitin sulphate
AR = autosomal recessive

Moreover, fibroblasts from certain patients with Sanfilippo's disease could correct each other and were therefore presumably due to different enzyme defects (Kresse *et al.* 1971). The accuracy of these lines of reasoning was subsequently confirmed by the elucidation of specific enzyme deficiencies.

Hurler's disease (MPS IH); α-L-iduronidase deficiency

Clinical features

Hurler's disease is the most severe of the MPS. As in the other types, infants appear normal at birth. During the first year there is progressive corneal clouding and the formation of a characteristic angular kyphosis or gibbus at the lower dorsal or upper lumbar vertebrae (Fig. 2.1(a)). Growth is usually normal during the first year, but subsequently becomes stunted. The characteristic coarse facies (Fig. 2.2) responsible for the early label of gargoylism become progressively more obvious after the first year, together with distension of the abdomen, hepatosplenomegaly, umbilical and often inguinal hernias. Arthropathy causes stiffness of joints with limitation in range of movement and a characteristic 'claw-hand' appearance. There is usually a persistent nasal discharge. Deafness occurs in the majority of patients due to 'glue-ears' and sensory impairment. The hair is coarse and the skin feels tough. The skull becomes scaphocephalic. Papilloedema is common after a few years. Heart murmurs are common.

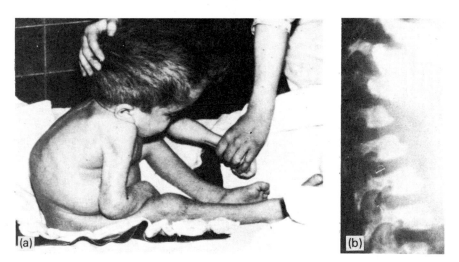

Fig. 2.1. (a) Patient aged five years with Hurler's disease; note dorso-lumbar gibbus. (b) X-ray showing characteristic 'beaking' of upper lumbar vertebrae in Hurler's disease.

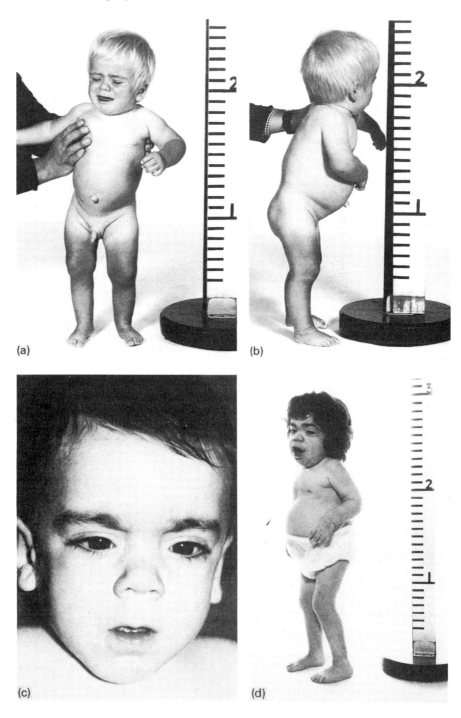

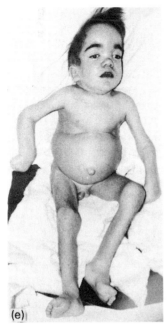
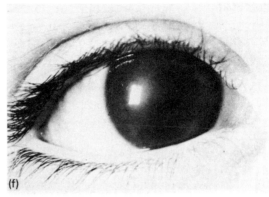

Fig 2.2. Characteristic clinical features of Hurler's disease. (a) 11 months of age, early facial changes and umbilical hernia. (b) Same patient showing dorso-lumbar gibbus, abdominal distension. (c) Facies of another male patient at 11 months of age. (d) A girl at three years of age; typical facies, abdominal distension and arthropathy. (e) An unrelated patient aged five years; typical facies, abdominal distension, umbilical hernia and arthropathy. (f) Corneal clouding at 12 months of age.

Cardiovascular manifestations were reviewed by Krovetz *et al.* (1965), who studied 15 patients and by Schrieken *et al.* (1975). The salient findings were increased pulmonary resistances. Systolic and diastolic murmurs are also common. Radiological examination reveals generalized skeletal involvement or dysostosis multiplex. By 5–18 months of age, lateral views of the spine show characteristic 'beaking' of the D12–L2 vertebrae (Fig. 2.1(b)) caused by atrophy of the upper and anterior aspect of the body of these vertebrae, at the apex of the gibbus. The ribs are widened or spatulate, but narrow abruptly close to the vertebral attachments. In the skull premature closure of the sagittal suture results in scaphocephaly. The sella turcica is enlarged, creating the J-shaped deformity. Detailed description of the skeletal deformity is given by McKusick (1972).

There is a progressive mental retardation until the patient becomes bedridden. Death from cardiac failure and pulmonary infections usually occurs towards the end of the first decade.

Pathological findings

Characteristically, there is accumulation of storage material in a wide variety of cells (Tondeur and Neufeld 1974), including connective tissue, the lining of arteries (Renteria *et al.* 1976), heart valves (Renteria and Ferrans 1976), cornea, hepatocytes (Fig. 2.3) and Kupffer cells (McKusick 1972). By light microscopy, cells distended by storage material have been described as clear, Hurler or balloon cells. Electron microscopy reveals that the material is largely membrane-bound and probably lysosomal (Van Hoof 1973; Goldfischer *et al.* 1975). In neurones, lysosomal inclusions resemble lamellar or zebra bodies seen in some sphingolipidoses and may contain lipids in addition to heparan and dermatan sulphates (Suzuki 1972; Constantopoulos *et al.* 1976).

In cultured fibroblasts, metachromatic granules represent lysosomes distended by storage material, stained by cationic dyes (Alcian blue, toluidine blue), which on analysis has been shown to consist mainly of dermatan sulphate, with smaller quantities of heparan and chondroitin sulphates (Matalon and Dorfman 1969). Demonstration of metachromasia in circulating lymphocytes was a useful diagnostic test (Muir *et al.* 1963) before the development of a specific enzyme assay.

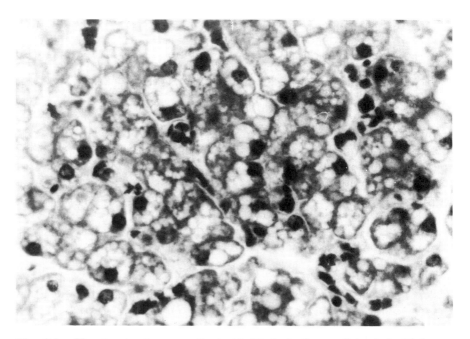

Fig. 2.3.   Hepatocytes from a patient with Hurler's disease distended with large vesicles, identified as lysosomes containing GAG.

In 58 necropsies reviewed by Krovetz *et al.* (1965), the main cardio-vascular findings were nodular enlargement of the mitral and tricuspid valve edges, and endocardial fibroelastosis.

Hydrocephalus (Fig. 2.4(a) and (b)) is common, usually communicating, and may result from obstruction of the subarachnoid space by meningeal thickening and adhesion in the region of the tentorium.

Molecular defects

There is a severe deficiency of the lysosomal hydrolase α-L-iduronidase (iduronidase) (Bach *et al.* 1972; Matalon and Dorfman 1972) required for stepwise degradation of dermatan sulphate and heparan sulphate (Fig. 2.5, reaction 2; Fig. 2.17, reaction 7). Iduronidase has been identified with the Hurler corrective factor (Shapiro *et al.* 1976*a*). It may be reasoned, therefore, that correction involves active pinocytosis of iduronidase which catalyses degradation of stored GAG in phagosomes into which it becomes incorporated. Human kidney iduronidase has a molecular weight of about 60 000 and consists of two chains of 30 000 which are probably products of a single polypeptide (Rome *et al.* 1978; Myerowtiz and Neufeld 1981).

Patients with Hurler's and Hunter's disease have a striking reduction of liver but not fibroblast (Benson *et al.* 1971) ß-galactosidase (Öckerman 1968; Ho and O'Brien 1969) which is not directly involved in degradation of heparan and dermatan sulphates. Kint (1973, 1974) has shown that ß-galactosidase activity in Hurler and Hunter livers can be restored by addition of cetylpyridinium chloride (CPC) which precipitates GAG, and considers that this is evidence that deficiency of ß-galactosidase results from formation of a complex between the enzyme and stored GAG. Furthermore the isoenzyme pattern of five hydrolases was abnormal. This pattern could be reproduced *in vitro* by the addition of chondroitin sulphate to a homogenate of normal liver (Kint *et al.* 1973). Kint (1974) considers that secondary ß-galactosidase deficiency may be responsible for the accumulation of sphingolipids in brain neurones. Dermatan sulphate and heparan sulphate are stored in the viscera and excreted in the urine in excessive quantities in the ratio of about 7:3.

Diagnosis

The definitive diagnosis is the demonstration of severe deficiency of iduronidase in cultured fibroblasts (Matalon and Dorfman 1972; Bach *et al.* 1972), leucocytes (Liem and Hooghwinkel 1975) or serum (unpublished observations). Several substrates have been used, including phenyl-α-L-iduronide (Bach *et al.* 1972), α-L-iduronosyl-anhydromannose (Matalon and Dorfman 1972), iduronosyl-anhydro (1-^{3}H)-mannitol 6-sulphate

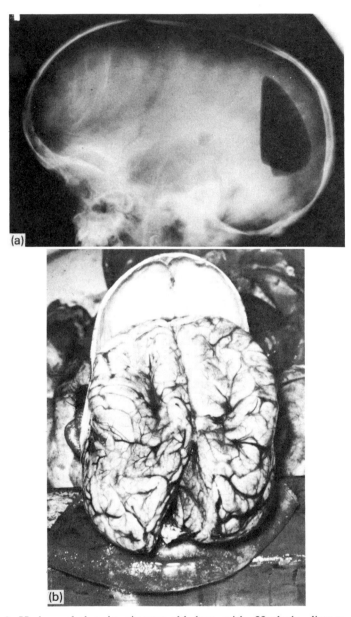

Fig. 2.4. Hydrocephalus in six-year-old boy with Hurler's disease. (a) Air ventriculogram in prone position. Distension of the posterior horns of the lateral ventricles; some air is in the third ventricle, but not enough to confirm distension. The presence of air in the posterior fossa indicates that there is a communicating hydrocephalus with a block probably at the level of the tentorium. (b) Appearance of brain at necropsy showing collapse of cortex into drained lateral ventricles.

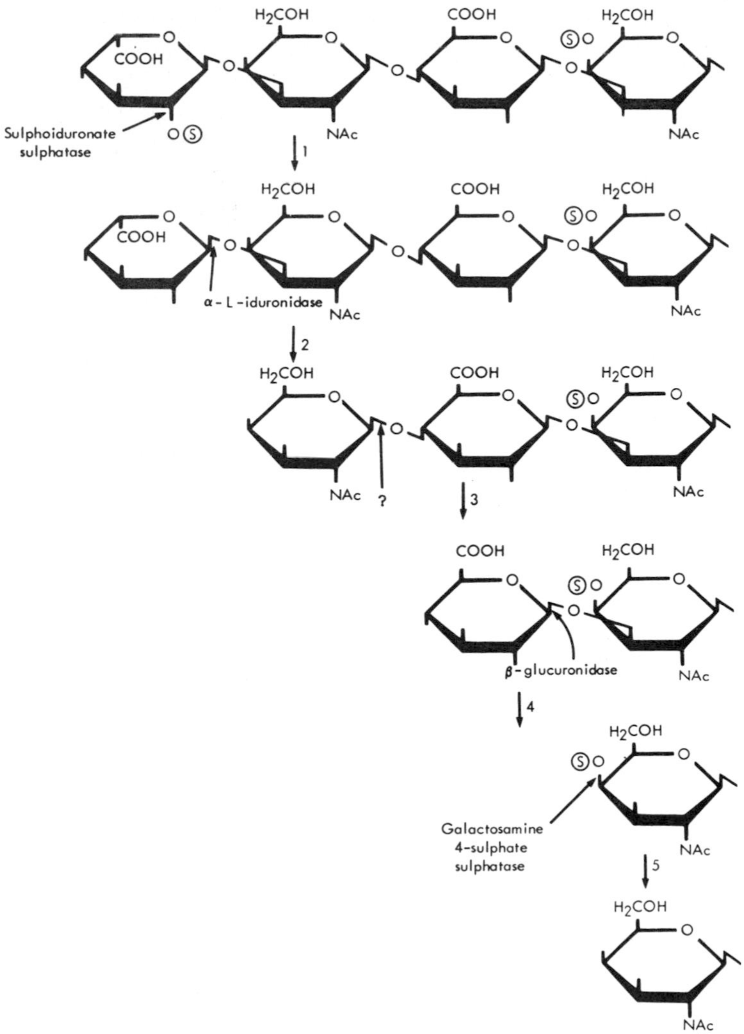

Fig. 2.5. Pathway of dermatan sulphate degradation.

(Hopwood 1979) and 4-methylumbelliferyl- α -L-iduronide (MU-iduronide) (Stirling *et al.* 1978, 1979). The last-named is used routinely by most centres because the assay is sensitive and simple.

Often before iduronidase assay the urine will be tested to demonstrate an excess of dermatan and heparan sulphates. The demonstration of excessive urinary excretion of GAG is commonly used as a screening test for all the MPS. It is important that urine should not be allowed to stand at room temperature since resulting degradation of polymeric GAG may lead to false negative results. Rapid screening procedures often produce errors, possibly because they tend to be carried out by laboratories with limited experience. They are based on the production of a precipitate with CPC (Pennock 1970; Pennock *et al.* 1976) or a metachromatic spot on paper impregnated with Alcian blue (Berry and Spinanger 1960).

More reliable methods (Matalon and Dorfman 1972) involve demonstration of elevated urinary polymeric GAG excretion after precipitation with 5-aminoacridine (Muir and Jacobs 1967; Dean *et al.* 1971) or CPC (Di Ferrante 1967) and assaying for hexuronic acid (Bitter and Muir 1962); or by measuring optical absorbance of Alcian blue-precipitated GAG (Whiteman 1973). Where only random urine samples are available, correction may be made by reference to creatinine excretion. Methods for assay of dermatan and heparan sulphates for the diagnosis of Hurler's disease include the determination of galactosamine/glucosamine ratio, since the main amino sugar of heparan sulphate is glucosamine, and of dermatan and chondroitin sulphates galactosamine. Chondroitin sulphates can be removed by digestion with hyaluronidase (Dean *et al.* 1971). Alternatively, the GAG can be separated by cellulose acetate electrophoresis (Whiteman 1973) or by elution from a Dowex 1 column with increasing sodium chloride concentrations (Schmidt 1974). For diagnostic purposes, we use two-dimensional electrophoresis by the method of Whiteman (1973) (Fig. 2.6 (a) and (b)).

Genetic counselling and prenatal diagnosis

Hurler's disease has an autosomal recessive form of inheritance. The incidence has been estimated in British Columbia and Westfalen as about 1 per 100 000 births (Lowry and Renwick 1971; Spranger 1972), giving a carrier frequency of about 1 in 150 persons. Assuming an annual birth rate of 700 000 in the United Kingdom, and an average 10-year survival for affected children, it may be calculated that seven babies with Hurler's disease would be born annually, that there would be about 70 affected surviving children, and that 7 per cent would have consanguineous parents.

Carrier detection, which was unreliable using phenyl-α-L-iduronide as substrate (Bach *et al.* 1973*a*), is now possible using the 4-methylumbelliferyl substrate (Stirling *et al.* 1978, 1979) for assay of cultured fibroblast iduronidase; however, a larger number of controls and obligate heterozygotes need to be assayed before reliability can be assessed.

The method of choice for prenatal diagnosis is direct determination of cultured amniotic cell MU-iduronidase (Stirling *et al.* 1979). The less sensitive assay using the phenyl substrate may fail to discriminate between affected and heterozygous foetuses (Bach *et al.* 1973*a*). Chemical assay of

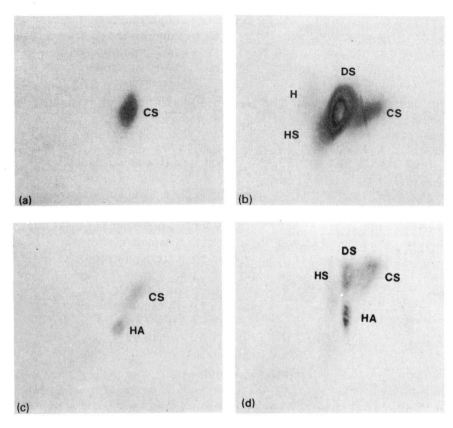

Fig. 2.6. Two-dimensional electrophoresis of Alcian Blue-precipitated GAG. (a) Normal urine (b) Hurler urine (c) Amniotic fluid from normal pregnancy (d) Amniotic fluid where the foetus had MPS IH.CS, chondroitin sulphate; DS, dermatan sulphate; HS, heparan sulphate; HA, hyaluronic acid; H, heparin-like component.

amniotic fluid GAG in a pregnancy at 18 weeks' gestation (Brock *et al.* 1971) and in two further pregnancies at 14 and 16 weeks' gestation respectively (Matalon *et al.* 1972*a*) failed to detect a Hurler foetus. However, two-dimensional electrophoresis of Alcian blue-precipitated GAG allows discrete separation of urinary-derived dermatan sulphate, heparan sulphate and chondroitin sulphate and allows comparison of their concentrations (Fig. 2.6(c) and (d)). This method (Whiteman and Henderson 1977; Mossman *et al.* 1981; Mossman and Patrick 1982) is more sensitive than earlier methods used for amniotic fluid GAG assay and has been used for prenatal diagnosis in 39 pregnancies for accurate detection of foetuses with Hunter, Hurler, and Sanfilippo diseases. In one of these pregnancies a Hurler foetus was accurately predicted because of excessive amniotic fluid dermatan sulphate concentration, even though amniotic cell cultures were overgrown by maternal cells with iduronidase activity in the heterozygote range (Mossman *et al.* 1981).

It is recommended therefore that analysis of amniotic fluid GAG using the methods of Whiteman and Henderson (1975) be used in addition to direct iduronidase assay of cultured amniotic cells. In pregnancies presenting in advanced gestation, or when there is amniotic cell culture failure, GAG analysis may be particularly useful. Alternatively, we have excluded Hurler's disease by assay of foetal leucocytes collected at foetoscopy (Rodeck *et al.* 1983*a*).

Demonstration of accelerated radiosulphate incorporation into cultured amniotic cell GAG was the first method used for accurate *in utero* prediction of a Hurler foetus (Fratantoni *et al.* 1969*b*), but this method has been replaced by iduronidase assay.

Aberg *et al.* (1979) performed lethal cardiac puncture on a twin affected with Hurler's disease, avoiding abortion of the unaffected co-twin.

Scheie's disease (MPS IS); α-L-iduronidase deficiency

Clinical features

Stiffness and 'claw' deformity of the hands (Fig. 2.7(a) and (b)) presenting between two and seven years of age; progressive severe corneal clouding, noticed between six and 20 years of age; and aortic valve disease, with normal intelligence, are the cardinal features first clearly described as a 'forme fruste of Hurler's disease' by Scheie, *et al.* (1962).

Carpal tunnel syndrome with loss of sensation of the fingertips is a common complication. Cervical cord compression by thickened meninges has been reported (Kennedy, *et al.* 1973). Stiffness of other joints, pes cavus, and genu valgum also occur. The facies are coarse and a broad mouth is typical. Aortic valve stenosis and regurgitation occur commonly. In contrast to Hurler's disease, height is usually normal. Glaucoma and deafness have been reported. Since survival of untreated patients into mature adult life is usual, operative procedures for relief of aortic valve disease, glaucoma, and nerve entrapments should be carried out where indicated.

Pathology

In a necropsy of a 35-year-old man who died with cardiac failure (Dekaban *et al.* 1976), lesions similar to those found in Hurler's disease were found in the viscera, including thickening of all heart valves. Histologically, numerous 'gargoyle cells' were found in hepatocytes, Kupffer cells, spleen, and connective tissue. In contrast, however, the cortical neurones appeared histologically normal, but on electron microscopy a few lipofuscin inclusions were demonstrated.

Molecular defect

Failure of genetic complementation in heterokaryons after hybridization of fibroblasts from patients with Hurler and Scheie diseases (Fortuin and Keijer 1980, Galjaard 1980a) is consistent with allelism at the iduronidase locus. This is supported by biochemical demonstration that the residual enzyme activities in cultured fibroblasts from patients with the two diseases have different kinetics.

Thus, using a sensitive iduronidase assay with iduronosyl anhydro-(1-^{3}H)-mannitol 6-sulphate as substrate, Hopwood and Muller (1979) have shown K_m values for Hurler, Scheie, and control cells to be 656, 50, and 53

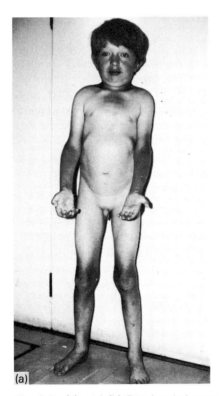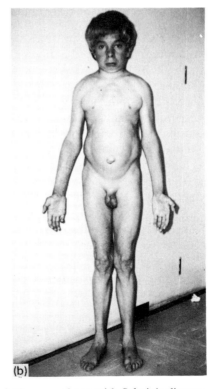

Fig. 2.7. (a) and (b) Brothers of seven and 13 years of age with Scheie's disease, showing characteristic 'claw' hand deformities.

μmol/l, respectively. Smaller differences were found in pH optima and thermal stability.

Mucopolysacchariduria is quantitatively and qualitatively similar to that described for MPS IH.

Diagnosis

As for Hurler's disease, the diagnosis is confirmed by demonstrating severe deficiency of α-L-iduronidase in leucocytes (Liem and Hooghwinkel 1975) or cultured fibroblasts (Stirling *et al.* 1978, 1979); and excessive urinary excretion of dermatan sulphate and heparan sulphate.

Discrimination from Hurler's disease is usually clear from the clinical features, but differences in kinetics of residual iduronidase activities in MPS IH and MPS IS have been reported (Hopwood and Muller 1979). Residual iduronidase activity in Scheie fibroblasts has been reported by Matalon and Deanching (1977).

Genetic counselling and prenatal diagnosis

Scheie's disease has an autosomal recessive mode of inheritance. The incidence in British Columbia and Westfalen has been established as about 1 per 500 000 births (Lowry and Renwick 1971), giving a carrier frequency of about 1 in 350 persons. Assuming an annual birth rate of 700 000 in the United Kingdom, only one or two affected neonates would be expected to be born annually. It may be calculated that the parents would be first cousins in approximately 18 per cent of cases. At the time of writing, no attempts at prenatal diagnosis have been reported. The methods would be as for Hurler's disease.

Animal model

Deficiency of α-L-iduronidase has been discovered in a short-haired cat (Haskins *et al.* 1979*a,b*; 1981) and the Plott hound (Shull *et al.* 1982).

Hurler–Scheie double heterozygote or genetic compound (MPS IH/S); α-L-iduronidase deficiency

Simultaneous heterozygosis for mutant allelic genes was first discussed by J.B.S. Haldane (1937–1938). In view of the evidence summarized earlier suggesting that Hurler's and Scheie's diseases are due to distinct allelic mutations, it may be predicted that double heterozygotes (or genetic compounds) for these alleles should exist. Several possible examples have been reported (McKusick *et al.* 1972; Stevenson *et al.* 1976).

Clinical features

As predicted, the features of double heterozygotes are intermediate to those observed in patients who are homozygotes for either allele. Thus features are first noticed between one and two years of age; mental retardation is absent or moderate; there is moderate limitation of joint mobility, with hepatosplenomegaly, dysostosis multiplex, usually without gibbus, stunting of growth, and corneal clouding. The course is slowly progressive.

Diagnosis

As in patients with Hurler's and Scheie's diseases, there is excessive excretion of urinary dermatan and heparan sulphates. The activities of fibroblast iduronidase are deficient. Cultured fibroblasts show impaired degradation of GAG which is correctable by Hurler factor.

Genetic counselling and prenatal diagnosis

The disorder has an autosomal recessive form of inheritance. Assuming homozygote frequency for Hurler's and Scheie's diseases to be 1 in 100 000 and 1 in 500 000, respectively, and using the adapted Hardy–Weinburg equation (McKusick *et al.* 1972):

$$p^2 + 2pq_1 + 2pq_2 + 2q_1q_2 + q^2_1 + q^2_2 = 1$$

where p = the frequency of the wild gene, q_1 of the Hurler gene and q_2 of the Scheie gene, the frequency of the double heterozygote compound $2q_1q_2$ will be approximately 1:112 000.

The risk of recurrence in sibs of an affected individual is 1 in 4.

The incidence of first cousin parents should be the same as in the general population, that is, about 0.5 per cent in the United Kingdom. However, reports of parental consanguinity (Jensen *et al.* 1978; Kaibara *et al.* 1973) suggest that the phenotype MPS IH/S described above may also result from homozygosity of a third mutant allele.

Prenatal diagnosis has not been reported, but the methods would be the same as those for Hurler's disease.

It should be noted that the existence of double heterozygotes is based on sound genetic principles, but proof that individual subjects with intermediate clinical phenotypes have such genotypes is lacking. At present the possibility has not been excluded that there may be several variant iduronidase alleles producing various degrees of iduronidase-deficient phenotypes. This is further supported by the report of a patient with iduronidase deficiency, but without the Hurler or Scheie phenotype (Gardner and Hay 1974).

Hunter's disease (MPS II); sulphoiduronate sulphatase deficiency

Based upon analysis of 52 severely and 31 mildly affected patients with Hunter's disease, Young *et al.* (1982) concluded that patients can be assigned clearly to these two groups. However, McKusick and Neufeld (1983) considered that a sharp differentiation into two forms is probably not justified.

Clinical features

(a) Severe type

In general, the clinical features of Hunter's disease are similar to those of Hurler's disease. Differences include absence of clinically detectable corneal clouding, absence of gibbus, and marked preponderance of male patients, owing to sex-linked inheritance. However, corneal clouding (Spranger *et al.* 1978), gibbus (Benson *et al.* 1979*a*), and female patients (Milunsky and Neufeld 1973; Neufeld *et al.* 1977) have been reported. The presence of nodular skin lesions over the scapulae and upper arms tends to occur in Hunter's rather than Hurler's disease. The common form of Hunter's disease often presents between two and five years of age with partial deafness, recurrent respiratory infections, diarrhoea, hernias, or stunted growth. Convulsions occur in about two-thirds of patients over 10 years of age (Young *et al.* 1982). On examination, the characteristic coarse facies (Fig. 2.8(a)–(f)) are usually present, together with hepato-splenomegaly, joint stiffness (Fig. 2.9(a) and (b)) with characteristic limitation in extension of interphalangeal joints (Fig. 2.10) and clawing of hands (Fig. 2.11(a) and (b)). Skeletal changes of dysostosis multiplex including 'beaking' of lower dorsal and upper lumbar vertebrae (Benson *et al.* 1979*a*) – but usually without kyphosis – are similar to those described for Hurler's disease. Dentigenous cysts appear as radio-translucent areas around unerupted teeth (Lustmann *et al.* 1975).

As in Hurler's disease, there are often diastolic and systolic murmurs. Progressive mental and physical deterioration (Fig. 2.12(a)–(d)), accompanied by failing vision due to papilloedema (Fig. 2.13) and sometimes retinitis pigmentosa, usually lead to death by 15 years. Mean age of death in 53 patients was 11.8 years (Young *et al.* 1982).

(b) Milder type

Patients with the milder forms (Fig. 2.14(a) and (b)) who survive into their twenties or thirties, tend to suffer from compression of the median and ulnar nerves at the carpal and cubital tunnels respectively (Karpati *et al.* 1974; Swift and McDonald 1976). In a family reported by McKusick

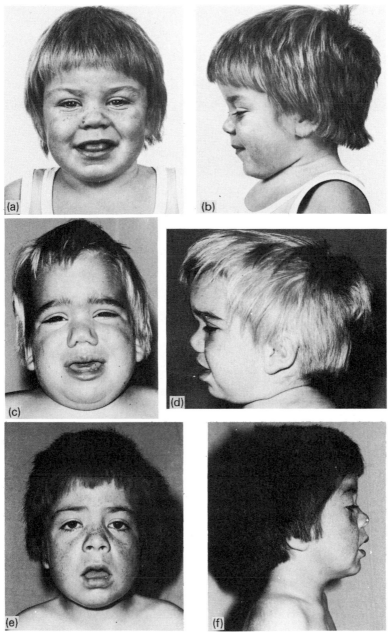

Fig. 2.8. (a)–(f) Coarse facies of boys with severe form of Hunter's disease:
(a)(b) Boy aged five years three months
(c)(d) Unrelated boy aged four years eight months
(e)(f) Unrelated boy aged nine years two months.

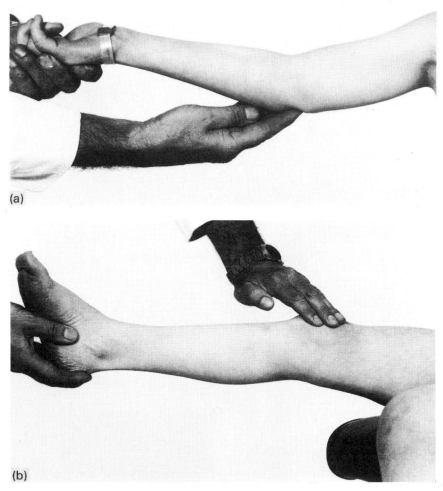

Fig. 2.9. Limitation of extension of (a) elbow and (b) knee joints of a ten-year-old boy with Hunter's disease.

(1972), three affected males had died aged 58, 60, and 61, and a fourth survived aged 61. An unrelated patient survived to 87 years (Hobolth and Pedersen 1978). An affected male with a low sperm count had a daughter who gave birth to two affected sons (Di Ferrante and Nichols 1972). Three other males with progeny have been reported (Hobolth and Pedersen 1978). Young *et al.* (1982) found that patients with the mild form of Hunter's disease had a lower incidence of behavioural disorders, diarrhoea, convulsions, and terminal cachexia than patients with the severe form.

Pathology

As in the severe (Hurler) and mild (Scheie) forms of iduronidase deficiency, more storage material has been reported in cerebral neurones of the severe than in the mild forms of Hunter's disease (Spranger 1972).

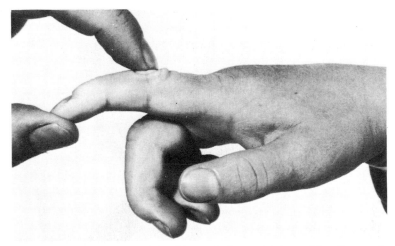

Fig. 2.10. Limitation of extension of interphalangeal joints of forefinger of a ten-year-old boy with Hunter's disease.

Storage material has also been reported in corneal cells both by histological and electron microscopical studies. As for Hurler's disease, there is evidence of intralysosomal storage of both dermatan and heparan sulphates.

Molecular defects

The disease is due to deficiency of sulphoiduronate sulphatase (Fig. 2.5 reaction 1; Fig. 2.17, reaction 6), leading to intralysosomal accumulation of dermatan and heparan sulphates since sulphated iduronate residues occur in both molecules. No differences have been demonstrated in the enzyme defects of the mild or severe forms (Bach *et al.* 1973*b*; Liebaers and Neufeld 1976), residual enzyme activity being minimal or undetectable in both.

Human plasma sulphoiduronate sulphatase consists of a single polypeptide chain with molecular weight about 80 000 (Liebaers and Neufeld 1976).

Heparan sulphate and dermatan sulphate are excreted in excessive quantities in the urine, usually in about equal quantities, but sometimes heparan sulphate predominates.

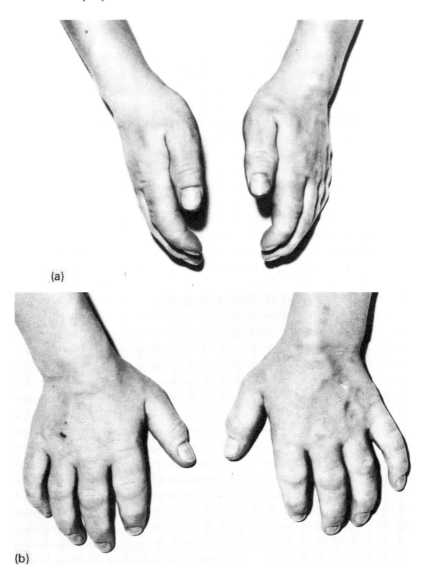

(a)

(b)

Fig. 2.11. (a) and (b) Characteristic 'claw' hands of an 11-year-old boy with Hunter's disease.

The forms, of varying severity, may be allelic, but there is no support for this either by cell hybridization or enzyme kinetic studies. Since sulpho-iduronate sulphatase activity is markedly deficient in multiple sulphatase deficiency, this differential diagnosis must always be considered.

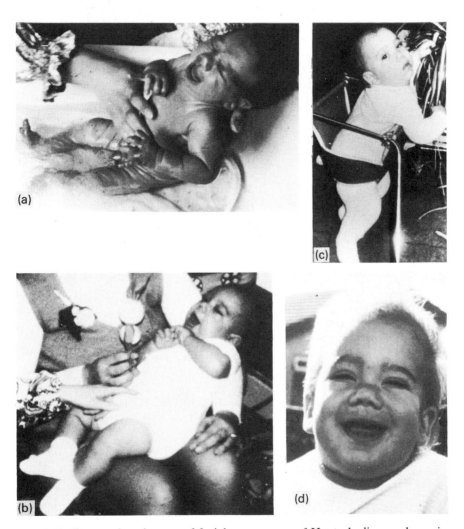

Fig. 2.12. Progressive changes of facial appearance of Hunter's disease shown in family photographs. See Fig. 2.8 (c) and (d) for facial appearance when aged four years eight months.
(a) aged three weeks
(b) aged five months
(c) aged seven months
(d) aged 12 months.

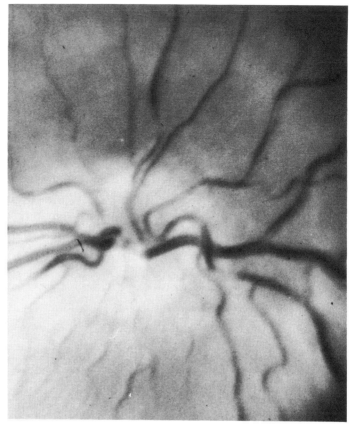

Fig. 2.13. Papilloedema in a boy with Hunter's disease aged 11 years.

As in Hurler's disease, there is a secondary deficiency of ß-galactosidase demonstrable in liver and spleen.

Diagnosis

Excessive urinary excretion of GAG is detected either by screening tests or by more reliable methods as described for Hurler's disease. The proportion of heparan sulphate is usually higher than in Hurler's disease, and is usually about the same as that of dermatan sulphate, but in some cases is considerably higher, giving rise to suspicion that the patient may have a Sanfilippo syndrome. Specific diagnosis, however, is established by demonstrating a severe deficiency of sulphoiduronate sulphatase in serum, lymphocytes or cultured fibroblasts (Liebaers and Neufeld 1976).

Demonstration that accelerated radiosulphate incorporation of cultured fibroblasts can be corrected by addition to the culture medium of the Hunter factor (Cantz *et al.* 1972) was a useful test before the specific enzyme assay was available.

Metachromatic leucocyte inclusions are common (Muir *et al.* 1963) but their demonstration is rarely needed as a diagnostic test.

Genetic counselling and prenatal diagnosis

The disease is the only MPS known to have a sex-linked recessive mode of inheritance. Affected patients are therefore usually males but two girls have been reported (Milunsky and Neufeld 1973; Neufeld *et al.* 1977) and it has been estimated that two per cent of Hunter families include a female with Hunter's disease (Neufeld *et al.* 1977). The incidence has been estimated as being less than 1 per 100 000 liveborn (Lowry and Renwick

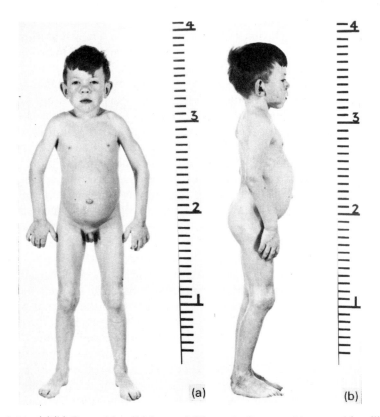

Fig. 2.14. (a)(b) Boy with mild form of Hunter's disease with normal intelligence at 12 years.

1971; Spranger 1972), but in Israel, Schaap and Bach (1980) estimated the incidence as 1 per 67 500 births.

According to the Lyon hypothesis (Lyon 1962), it has been shown, by cloning fibroblasts (Migeon *et al.* 1977), that carrier females have two populations of cells, one with normal and one with severely deficient sulphoiduronate sulphatase activities. The occurrence of females with Hunter's disease could theoretically be due to differential inactivation (Lyonization) of the X-chromosome carrying the paternal (wild) gene, rather than the maternal (Hunter) gene for sulphoiduronate sulphatase. Preferential survival of cells with the Hunter gene has been described in cells after prolonged cultivation, after freezing in liquid nitrogen (Booth and Nadler 1974; Donnelly and Di Ferrante 1975) but not confirmed (Migeon *et al.* 1977).

Another theoretical possibility is that there is an autosomal recessive form of Hunter's disease (Epstein *et al.* 1976; Neufeld *et al.* 1977).

A further explanation is that the female Hunter patients have an inherited mutation on their maternal X-chromosome, and either absence or inactivation due to chromosomal translocation or a new mutation of the sulphoiduronate sulphatase locus on the paternal X-chromosome. The most important practical application is the inadvisability of relying only on foetal sexing for prenatal diagnosis of Hunter's disease; reliable prenatal diagnosis can be carried out by direct enzyme assay of cultured amniotic cells and possibly of amniotic fluid (Liebaers *et al.* 1977), but the latter requires confirmation. As in other types of MPS, confirmation should also be sought by determination of elevated amniotic fluid GAG by two-dimensional electrophoresis (Whiteman and Henderson 1977).

Demonstration of accelerated radiosulphate incorporation into cultured amniotic cells, which is corrected by the Hunter factor (Fratantoni *et al.* 1968*b*), can be used for further confirmation if considered necessary.

As in other types of X-linked disorders, sisters and maternal aunts of Hunter patients and other females at risk for the carrier state should receive genetic counselling when they approach child-bearing age. Proof that a female is a carrier can currently be achieved by demonstrating two populations of cells, one with normal and one with severely deficient sulphoiduronate sulphatase activity. Since cells with the normal paternal chromosome can correct the maternal Hunter cells in culture, it is necessary to clone cells to prove the existence of two populations (Migeon *et al.* 1977). Most fibroblast cultures are derived from skin biopsies only a few millimetres in width. A much larger area of skin can be monitored by plucking hairs from widely separated locations on the scalp and analysing hair roots with about six adhering cells for sulphoiduronate sulphatase activity (Yutaka *et al.* 1978; Nwokoro and Neufeld 1979).

Inability to demonstrate the carrier state by any of the above-mentioned methods cannot guarantee that heterozygosis has been excluded. At present, therefore, potential high-risk carriers should receive genetic counselling and where appropriate should be offered diagnostic amniocentesis.

The Sanfilippo diseases (MPS III) A–D

Clinical features

This group of disorders was identified because the predominant GAG in the urine is heparan sulphate (Harris 1961; Sanfilippo *et al.* 1963). Typically, there is severe mental defect with mild somatic or skeletal involvement, somewhat coarse facies (Figs 2.15 and 2.16), no gibbus, and

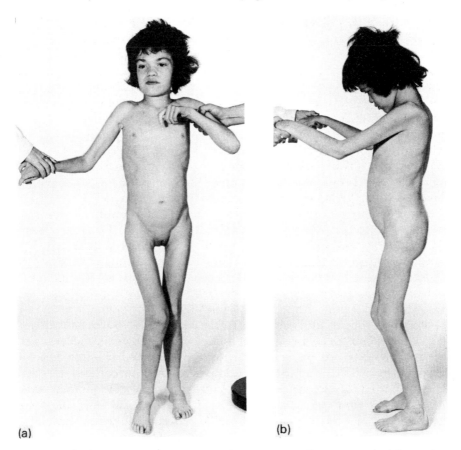

(a)   (b)

Fig. 2.15. (a)(b) Patient aged 14 years with Sanfilippo disease type A; (a) anterior and (b) lateral views of same patient.

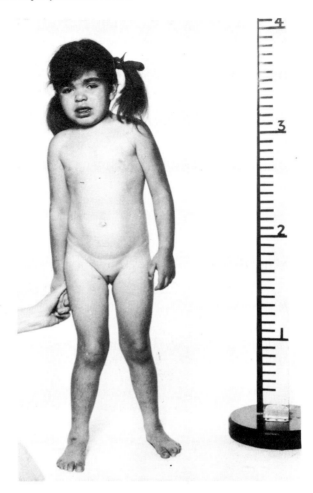

Fig. 2.16. Patient aged five years with Sanfilippo disease type B.

absent or very slight corneal clouding. Behaviour disturbances and delayed
development are the commonest presenting symptoms and usually are first
noticed between two and six years of age. Affected children become
hyperactive, irritable, aggressive, and destructive. In contrast to other
types of MPS, only one-third are considered to have normal early
development (Van de Kamp 1979), the others being late in motor
development and talking. Rampini (1976) reported 40 out of 55 children
from the literature were mentally retarded before three years of age.
However, there is considerable variation in the age of onset and clinical
course, some patients becoming bedridden before the age of six years,

while others are still walking in their twenties. Hepatosplenomegaly is usually slight, while height is usually normal. Joint involvement is absent or minimal. Radiologically, there is only mild dysostosis multiplex, the vertebral bodies usually having a biconvex shape, and the posterior parts of the calvaria being unusually thick. The cardiovascular system is usually unaffected (Schrieken *et al.* 1975) but severe mitral stenosis has been reported in a patient with Sanfilippo B disease aged three years (Herd *et al.* 1973).

From an analysis of 75 patients (26 of whom died) and from data in the literature, Van de Kamp (1979) considered that patients with Sanfilippo A disease (average age at death 14 years) die earlier than those with either Sanfilippo B (25 years) or Sanfilippo C (21 years). However, McKusick (1972) considers that types A and B are indistinguishable clinically. Intertype and intratype variability was evident in six patients with type A, 23 with type B, and 14 with type C. Moreover, the existence of mild and severe variants with type B suggests that there may be allelic variations (Van de Kamp *et al.* 1981).

Pathology

The cerebral ventricles are enlarged. Histologically there is a deficiency of neurones, with ballooning of those remaining. Cytoplasmic inclusions of granular, lamellar types or zebra bodies have been described (Wallace *et al.* 1966; Cain *et al.* 1977).

Abnormalities of the mitral valves have been reported in two patients (Cain *et al.* 1977).

Molecular defects

The Sanfilippo diseases are due to deficiency of lysosomal enzymes required for the complete exclusive degradation of heparan sulphate (Fig. 2.17) and heparin. Partial degradation may be catalysed by endo-glycosidases (Roden 1980). In Sanfilippo A there is deficiency of heparan *N*-sulphatase which catalyses release of sulphate from the nitrogen atom of glucosamine *N*-sulphate residues (reaction 4). In Sanfilippo B, there is deficiency of the enzyme α -*N*-acetylglucosaminidase (reaction 2). Since there appears to be no lysosomal α -glucosaminidase, the α -linked glucosamine residues left at the non-reducing ends of the GAG chains after action of the *N*-sulphatase must be *N*-acetylated before they can be cleaved. The enzyme catalysing this acetylation, acetyl CoA: α -gluco-saminide *N*-acetyltransferase, is the enzyme deficient in Sanfilippo C (reaction 5). This is the only known lysosomal enzyme which is not a hydrolase and which requires a co-substrate (acetyl CoA). In the most

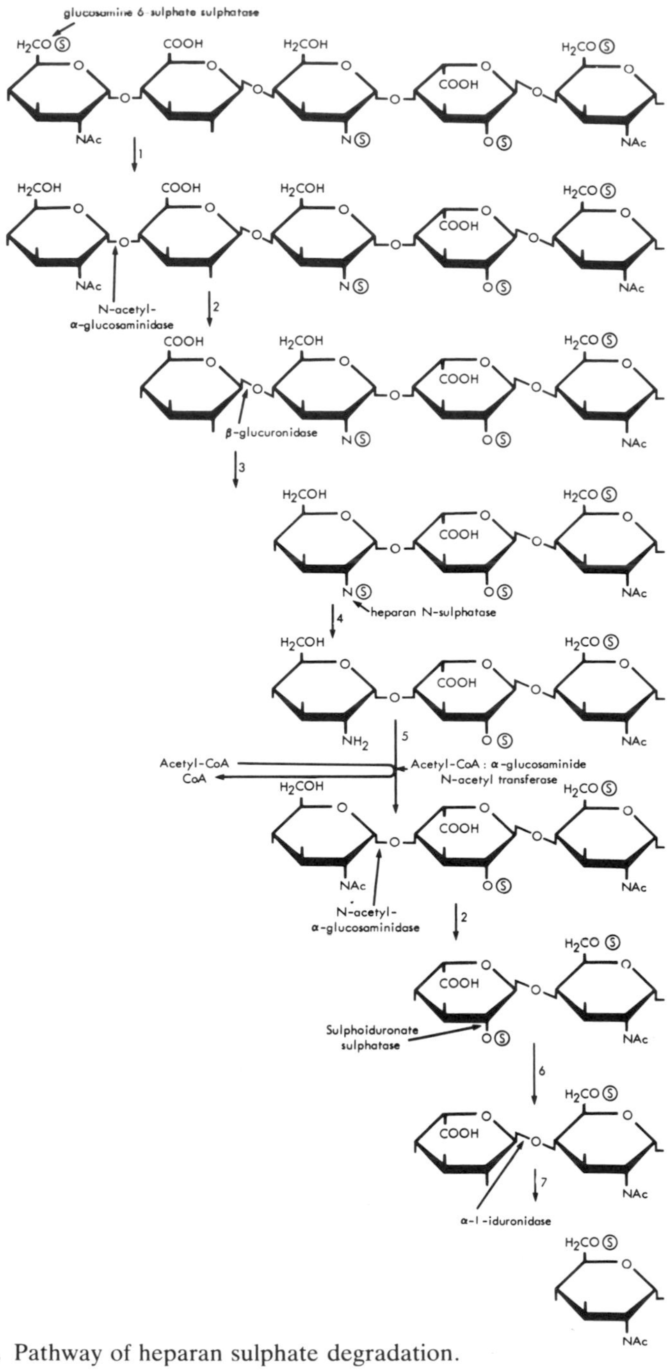

Fig. 2.17. Pathway of heparan sulphate degradation.

recently described Sanfilippo D there is deficiency of the enzyme *N*-acetylglucosamine 6-sulphate sulphatase (reaction 1) acting on 6-sulphated *N*-acetylglucosamine residues of heparan sulphate but not those in keratan sulphate (Kresse *et al*. 1980).

As for Hurler's and Hunter's diseases, there is a secondary deficiency of liver ß-galactosidase (Ho and O'Brien 1969).

In Sanfilippo B disease, the presence of material cross-reacting with antibodies against α-*N*-acetylglucosaminidase indicates the presence of an inactive enzyme, probably because of a structural gene mutation (von Figura and Kresse 1976). The excessive urinary GAG is almost exclusively heparan sulphate (Sanfilippo *et al*. 1963; Whiteman and Young 1977).

Diagnosis

Heparan sulphaturia will lead to a strong suspicion that the patient has a form of Sanfilippo disease, but as stated earlier, this may also occur in Hunter's disease. The methods for screening of urines and for GAG assay are as described for Hurler's disease. Sanfilippo types A, B and C are readily diagnosed by enzyme assay. In MPS IIIA deficiency of heparan *N*-sulphatase (heparan sulphate sulphaminohydrolase; heparin sulphamidase) may be demonstrated in cultured skin fibroblasts and leucocytes (Kresse 1973; Matalon and Dorfman 1974). In MPS IIIB, deficiency of α-*N*-acetylglucosaminidase may be demonstrated in cultured skin fibroblasts (O'Brien 1972*a*), plasma or leucocytes (Liem *et al*. 1976). In MPS IIIC, deficiency of acetyl-CoA: α-glucosaminide *N*-acetyltransferase has been demonstrated as diminished conversion of a heparan sulphate-derived unacetylated trisaccharide to the acetylated trisaccharide (Klein *et al*. 1978) or, more conveniently for routine use, by diminished conversion of glucosamine to *N*-acetylglucosamine (Hopwood and Elliott 1981). In MPS IIID, deficiency of *N*-acetylglucosamine 6-sulphate sulphatase has been demonstrated as an inability by cultured fibroblasts to release sulphate from *N*-acetylglucosamine 6-sulphate residues in oligosaccharides derived from heparan sulphate (Kresse *et al*. 1980).

Genetic counselling and prenatal diagnosis

Inheritance is autosomal recessive for Sanfilippo A–C and possibly also for D, as one of the two patients described was a girl (Kresse *et al*. 1980). The incidence is probably much more common than generally believed since it is clear that many cases are undiagnosed. Thus of the cases investigated at the Paediatric Research Unit, approximately half had been seen regularly by paediatricians and treated undiagnosed for behaviour disorders and mental handicap. McKusick (1972) wrote: 'The disorder may recur appreciably more often then in one out of every 100 000 to 200 000 people as estimated by Terry and Linker (1964)'. Van de Kamp (1979), from his study of 75 patients in Holland, calculated the incidence to be 1:24 000.

In Britain, the great majority of cases of Sanfilippo are type A. Whiteman and Young (1977) reported 28 who were type A and only one patient, from Poland, who was type B. In Holland, the distribution seems to be different, since of 66 patients (Van de Kamp 1979) only 32 (48 per cent) were type A while 20 (30 per cent) were type B and 14 (21 per cent) type C. So far, only four patients have been reported with type D (Gatti *et al.* 1982).

Schmidt *et al.* (1977) found that normal lymphocyte-rich preparations isolated in a Ficol gradient had 6–15 times more sulphamidase activity than granulocytes, but used mixed leucocytes for carrier testing. Mean activities for normal leucocytes were 66.8 ± 24.0 (SD) pmol sulphate released/h/mg protein; for six Sanfilippo patients 0.0–2.0 units, and for 8 parents, 28.7 units ± 10.9 (SD) (43 per cent of control mean). Although there was some overlap between activities of normal and obligate heterozygotes, the probability of an individual being a carrier of the Sanfilippo A gene could be calculated.

In Sanfilippo B, serum N-acetyl-α-D-glucosaminidase was assayed in three families with affected individuals using three different substrates, namely p-nitrophenyl N-acetyl-α-D-glucosaminide, phenyl-N-acetyl-α-D-glucosaminide, or uridine 5'-diphospho-N-acetylglucosamine (von Figura *et al.* 1973). All three methods were satisfactory, but the first is less time-consuming and gives lower residual activities in affected patients, and was therefore considered the most suitable for clinical purposes. Using the p-nitrophenyl substrate, homozygous patients had residual activities of only 1–3 per cent of normal control mean. Heterozygous parents had activities 25–46 per cent of control mean (four parents tested using the p-nitrophenyl substrate, but 16 per cent using the other two substrates).

Correct prenatal diagnosis of a foetus affected with Sanfilippo A was described by Harper *et al.* (1974), while monitoring of four pregnancies at risk is reported by Kleijer *et al.* (1978) in their series of prenatal diagnosis in 118 pregnancies at risk for metabolic disease. Tests of choice are the sulphamidase assay (Kresse 1973; Schmidt *et al.* 1977) combined with two-dimensional electrophoresis of Alcian Blue-precipitated GAG (Whiteman and Henderson 1977; Mossman and Patrick 1982).

Sanfilippo type B has been diagnosed prenatally by enzyme assay of cultured amniotic cells, and by demonstrating heparan sulphate in the amniotic fluid (Mossman *et al.* 1983).

No prenatal monitoring of Sanfilippo types C or D has been reported.

Morquio's disease (classical MPS IVA); keratan sulphate excretors and non-keratan sulphate excretors; *N*-acetylgalactosamine 6-sulphate sulphatase deficiency; MPS IVB; β-galactosidase deficiency

The salient clinical features are severe skeletal deformity (including dwarfism), corneal clouding, and aortic valve disease. The first descriptions of the bony abnormalities were by Morquio (1929) and by Brailsford (1929). Before the recognition that Morquio's disease, as now defined, has specific extraskeletal, biochemical, and genetic features, numerous non-specific cases of 'short trunk' dwarfism were reported in the literature as 'Morquio's disease'. However, even cases with corneal clouding, aortic disease, and leucocyte inclusion are heterogenous. Some mild cases have been found to be non-keratan sulphate excretors (McKusick 1972), but since in patients with Morquio's disease keratan sulphate excretion decreases with age (Campailla and Martinelli 1969; Linker *et al.* 1970) and since non-keratan excretors tend to be post-pubertal, it is possible they excreted excess keratan sulphate when younger. The clinical features resembling those of Morquio's disease have also been described in so-called MPS IVB, which differs from 'classical' Morquio's disease both in the deficient enzyme and type of mucopolysacchariduria (O'Brien *et al.* 1976; Arbisser *et al.* 1977; Groebe *et al.* 1980; Trojak *et al.* 1980).

Clinical features

The clinical phenotype is predominantly one of progressive skeletal deformity, usually leading, ultimately, to cervical cord compression. The earliest deformities may be noticed towards the end of the first year, when knockknees, short neck and sternal protruberance (pectus carinatum) (Fig. 2.18) may cause parents to seek medical advice. Linear growth is initially normal, but slows and becomes arrested during the end of the first decade, in severe cases leading to dwarfism. A detailed description of skeletal abnormalities is given by McKusick (1972). The main features, in addition to those mentioned above, are flat vertebrae (platyspondyly), deformity and fragmentation of the femoral heads, and severe hypoplasia or even absence of the odontoid process (Fig. 2.19). The last-named, combined with laxity of ligaments, usually leads to compression of cervical cord, leading to either acute quadriplegia or myelopathy of slower onset, and eventually death, usually by the age of 20, although survival into the 50s is recorded (McKusick 1972). Cardiorespiratory failure in some cases results from aortic incompetence and reduced respiratory function, owing to thoracic deformity, and sometimes respiratory tract obstruction (Fig 2.18). Corneal clouding is usually slight and may not occur until the end of the first decade. Nerve deafness is common, and dental enamel is poorly

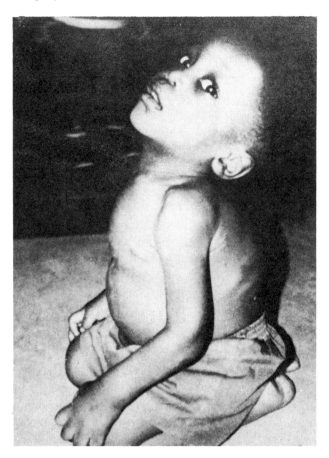

Fig. 2.18. Boy aged five years with Morquio's disease. Note sternal protruberance (pectus carinatum) and other chest deformities. The neck is hyperextended to overcome respiratory obstruction.

formed. In contrast to some other forms of MPS, joints are excessively mobile and intelligence is normal.

Pathology

Marked hypertrophy of the anterior longitudinal ligament of C1 and C2, which could accentuate cord compression, was reported in a necropsy by Einhorn *et al.* (1946). Microscopically, the architecture of bone and cartilage is distorted. Chondrocytes are swollen and packed with membrane-bound vacuoles. In the skin, similar vacuoles are present in basal and Malphigian layers, but not in other cells (Hollister *et al.* 1975). Cytoplasmic inclusions have also been described in liver (Tondeur and Locb 1969) and cerebral neurones (Gilles and Deuel 1971).

Molecular defects

Patients with classical Morquio's disease have a defect in catabolism of keratan sulphate (Fig. 2.20) and chondroitin 6-sulphate (Fig. 2.21) due to severe deficiency of *N*-acetylgalactosamine 6-sulphate sulphatase (Matalon *et al.* 1974*a*; Singh *et al.* 1976; Horwitz and Dorfman 1978). It seems likely that this enzyme catalyses the desulphation of galactose 6-

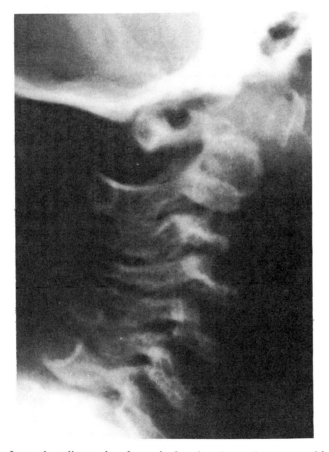

Fig. 2.19. Lateral radiograph of cervical spine in a three-year-old boy with Morquio's disease. Note hypoplasia of the odontoid process and irregularity and flattening of cervical vertebrae.

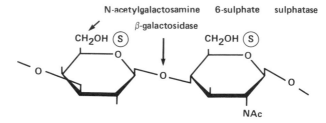

Fig. 2.20. Enzyme defects in the degradation of keratan sulphate.

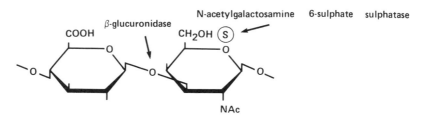

Fig. 2.21. Enzyme defects in the degradation of chondroitin 6-sulphate.

sulphate residues in keratan sulphate (which does not contain *N*-acetylgalactosamine 6-sulphate residues) because of structural similarities between the two sugars. A claim for *in vitro* confirmation of this based on studies using galactitol 6-sulphate as substrate (Di Ferrante *et al.* 1978) was disputed by Glössl *et al.* (1979) who found that neither crude nor highly purified preparations of the enzyme would desulphate this substrate. More recently, however, by using keratan (^{35}S) sulphate prepared from bovine cornea as substrate, Glössl and Kresse (1982) obtained evidence that the enzyme does indeed possess galactose 6-sulphate sulphatase activity.

The residual enzymes in fibroblasts from two patients with milder forms of MPS IVA have been reported to have different kinetic properties to that in the classic severe form (Glössl *et al.* 1981), suggesting the possibility of allelic variants. Despite the evidence for residual enzyme activity in severe MPS IVA, no cross-reacting material was obtained in five patients with the severe disease (Glössl *et al.* 1980).

The deficient enzyme in MPS IVB is ß-galactosidase which hydrolyses the ß-galactosyl residues of keratan sulphate and probably of the ganglioside GM₁ (O'Brien *et al.* 1976; Arbisser *et al.* 1977; Groebe *et al.* 1980; Trojak *et al.* 1980). The mutations causing either MPS IVB or generalized GM₁-gangliosidosis may therefore involve the same enzyme and be allelic. This might explain several features in common such as bone deformities, storage of keratan sulphate, and corneal clouding (Babarik *et al.* 1976). A gene coding for ß-galactosidase has been assigned to chromosome 3 (Shows *et al.* 1979).

A patient with excessive excretion in the urine of both heparan sulphate and keratan sulphate has been reported (Matalon *et al.* 1978) whose cultured fibroblasts exhibited deficiency of *N*-acetylglucosamine 6-sulphate suphatase, assayed using a sulphated disaccharide derived from keratan sulphate as substrate (Matalon *et al.* 1982). This enzyme is distinct from the *N*-acetylglucosamine 6-sulphate sulphatase which is deficient in MPS IIID (q.v.) and which is assayed using a monosulphated trisaccharide derived from heparan sulphate as substrate. The excessive excretion of heparan sulphate in the patient of Matalon *et al.* (1978) awaits explanation.

A report of *N*-acetylglucosamine 6-sulphate sulphatase deficiency in another patient (Di Ferrante *et al.* 1978) was later withdrawn (Di Ferrante 1980).

At the time of writing, the exact role of *N*-acetylglucosamine 6-sulphate sulphatase in the degradation of keratan sulphate awaits elucidation since Kresse *et al.* (1981) have shown that *N*-acetylglucosamine 6-sulphate may be liberated intact from keratan sulphate by ß-hexosaminidase A, the enzyme which is deficient in Tay–Sachs disease (q.v.).

Diagnosis

Children with classical Morquio's disease before the age of 10 years have had an excess of urinary keratan sulphate (Pedrini *et al.* 1962; Dean *et al.* 1971) and chondroitin 6-sulphate (Kaplan *et al.* 1968; Kaplan 1969; Dean *et al.* 1971), but in older patients keratan sulphate excretion may be within normal limits (Campailla and Martinelli 1969; Linker *et al.* 1970). Since keratan sulphate does not contain a uronic acid, screening tests described for urinary GAG in MPS do not detect keratan sulphaturia, although a borderline high normal, or even doubtfully elevated result is not uncommon owing to the excessive urinary excretion of uronic acid-containing chondroitin sulphate.

Tests for keratan sulphate in the urine usually depend on the demonstration of an elevated galactose content in a hydrolysate of CPC-precipitated GAG. A simpler test which appears reliable is the demonstration of material with characteristic mobility in two-dimensional electrophoresis of Alcian Blue-precipitated material.

The definitive test for diagnosis of classical Morquio's disease is the demonstration of severe deficiency of cultured fibroblast or leucocyte *N*-acetylgalactosamine 6-sulphate sulphatase (Matalon *et al.* 1974*a*; Singh *et al.* 1976; Horwitz and Dorfman 1978; Kresse *et al.* 1981). A sensitive assay (Glössl and Kresse 1978) using as substrate the trisaccharide 6-sulpho-*N*-acetylgalactosamine-glucuronic acid-6-sulpho-*N*-acetyl (1-^{3}H) galactosaminitol, depends on recovery and radioassay of the desulphated compounds from an ion exchange column after elution with 0.4M sodium chloride.

Morquio's cultured fibroblasts do not exhibit accelerated incorporation of radiosulphate (Hollister *et al*. 1975), a phenomenon which is characteristic of cells derived from patients with MPS I–III, VI, and some cases of VII.

Patients with MPS IVB can readily be diagnosed by demonstration of severe deficiency of leucocyte or cultured fibroblast ß-galactosidase activity using the synthetic 4-methylumbelliferyl or *p*-nitrophenyl ß-galactosides as substrates (O'Brien *et al*. 1976; Arbisser *et al*. 1977; Groebe *et al*. 1980; Trojak *et al*. 1980). These patients have keratan sulphaturia without chondroitin sulphaturia (Arbisser *et al*. 1977; Groebe *et al*. 1980).

Genetic counselling

Both MPS IVA and IVB have an autosomal recessive mode of inheritance. In view of considerable heterogeneity in 'short trunk dwarfism', the Morquio's diseases, and spondyloepiphyseal dysplasia, which may be due to MPS IVB, or to uncharacterized genetic disease with autosomal recessive, sex-linked, or dominant inheritance, it is essential to attempt a precise biochemical diagnosis by enzyme assay and characterization of urinary GAG for genetic counselling. Estimates of the incidence have varied from 1:40 000 to less than 1:200 000 (Lowry and Renwick 1971; McKusick 1972). There appear to be no data on carrier detection. Prenatal diagnosis of MPS IVA has been reported by assay of the cultured amniotic cell *N*-acetylgalactosamine 6-sulphate sulphatase (von Figura *et al*. 1982). It would be interesting to see if an excess of amniotic fluid keratan sulphate could be demonstrated by two-dimensional electrophoresis (Mossman *et al*. 1981; Mossman and Patrick 1982).

Maroteaux–Lamy disease (MPS VI); *N*-acetylgalactosamine 4-sulphate sulphatase deficiency

The co-existence of striking skeletal abnormalities and corneal clouding resembling those of Hurler's disease, but with preservation of normal intelligence combined with mucopolysacchariduria which is predominantly or exclusively dermatan sulphaturia, was first recognized by Maroteaux *et al*. (1963).

Clinical features

Classical and milder forms have been described (McKusick 1972). In the classical form, early symptoms are growth retardation after the first year and progressive skeletal deformities such as dorso-lumbar kyphosis or sternal protrusion after the second year. Coarse facial features, severe corneal clouding and restriction of joint mobility all resemble those of Hurler's disease (Fig. 2.22 (a) and (b)). The hips are severely affected with contractures and deformity of the femoral heads. Neurological complications include cord compression due to atlantoaxial subluxation or thickening of the cervical dura (Peterson *et al*. 1975), carpal tunnel syndrome and hydrocephalus. Pregnancy was accompanied by exacerbation of myelopathy in one patient (Peterson *et al*. 1975). Hepatosplenomegaly and hernias are common. Dwarfism allows differential diagnosis from patients with Scheie's disease who are normal in height. Cardiac abnormalities are common, and include aortic incompetence or stenosis and mitral incompetence, which may lead to cardiac failure and death towards the end of the second decade (Schrieken *et al*. 1975). Few patients with the classical disease survive into their twenties, but mild (Quigley and Kenyon 1974; Di Ferrante *et al*. 1974) and intermediate (Peterson *et al*. 1975) cases have longer survival.

Radiological abnormalities are reviewed by McKusick (1972). They can be detected by the age of two years. The appearance of beaking of the lumbar spine and later of kyphosis are similar to that of Hurler's disease (Fig. 2.23). Fragmentation of the femoral epiphysis and hypoplasia of the odontoid peg are typical.

Pathology

We have been unable to find a necropsy reported in the literature. Metachromatic inclusions in leucocytes (Mittwoch 1963; Muir *et al*. 1963) are striking and more marked than in other types of MPS, and have been described as Alder bodies. We know of one patient in whom the diagnosis was suggested by the histopathologist (Dr B. Lake) on the appearance of leucocytes alone. Cytoplasmic, probably intralysosomal inclusions have also been described in cornea, conjunctiva, skin (Quigley and Kenyon 1974), Kupffer cells and hepatocytes (Tondeur and Neufeld 1974).

The molecular defect

In Maroteaux–Lamy disease there is severe deficiency of the lysosomal enzyme *N*-acetylgalactosamine 4-sulphate sulphatase (Fig. 2.5, reaction 5) (Matalon *et al*. 1974*b*; O'Brien *et al*. 1974), leading to accumulation in the urine and (by analogy to the other MPS) probably in tissues, of dermatan sulphate with 4-sulphated *N*-acetylgalactosamine residues at the non-reducing end of their chains. Some accumulation of chondroitin 4-sulphate, which also contains these residues, can also be anticipated.

The enzyme defect was originally identified as deficiency of aryl-

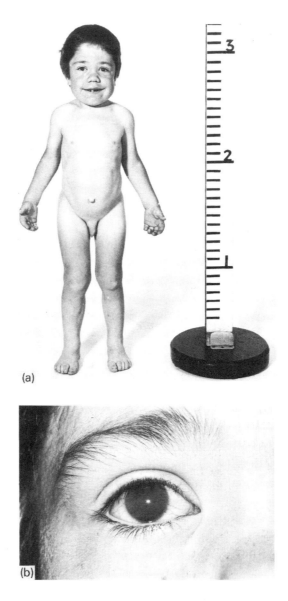

Fig. 2.22. (a) Seven-year-old boy with Maroteaux–Lamy disease. (b) Close up of eye showing corneal clouding.

sulphatase B (Stumpf *et al.* 1973). This enzyme has been shown to be identical to *N*-acetylgalactosamine 4-sulphate sulphatase (Matalon *et al.* 1974*b*; O'Brien *et al.* 1974).

Diagnosis

The simplest method of diagnosis is the demonstration of arylsulphatase B deficiency in leucocytes (Beratis *et al.* 1975*a*) or cultured fibroblasts (Fluharty *et al.* 1974*a*) using as substrates either *p*-nitrocatechol sulphate (inhibiting the A isoenzyme with barium acetate); or 4-methylumbelliferyl sulphate, after separation from the A isoenzyme by DEAE cellulose chromatography (Fluharty *et al.* 1974*a*). The isoenzymes arylsulphatase A and B are distinct both immunologically (Neuwelt *et al.* 1971) and genetically (see molecular defect of metachromatic leucodystrophy). The demonstration that the mucopolysacchariduria is predominantly (80–95 per cent) or exclusively dermatan sulphate is a useful screening procedure. The methodology is as described for Hurler's disease.

Genetic counselling and prenatal diagnosis

The disorder has an autosomal recessive mode of inheritance. The disease is probably as rare as Scheie's disease, but the frequency is unknown. Classic, mild and intermediate forms all have the same enzyme deficiency, and are probably allelic, with the last-named possibly being a double heterozygote of the genes which are homozygous in the classic and severe forms respectively.

Carrier detection has been attempted by assay of leucocyte aryl-sulphatase B (Beratis *et al.* 1975*a*) but more individuals need to be tested to assess reliability.

Prenatal diagnosis has been reported in three affected foetuses (Kleijer *et al.* 1976*a*; Van Dyke *et al.* 1981) by demonstrating arylsulphatase B deficiency in cultured amniotic cells.

Animal model

Deficiency of arylsulphatase B has been found in Siamese cats (Haskins *et al.* 1980).

ß-glucuronidase deficiency (MPS VII)

There have been such marked differences in the clinical phenotypes of patients with ß-glucuronidase deficiency (Sly *et al.* 1973; Danes and Degnan 1974; Gehler *et al.* 1974; Beaudet *et al.* 1975; Gitzelmann *et al.*

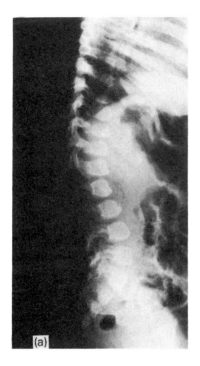

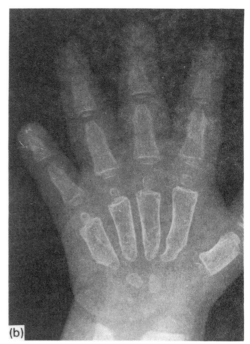

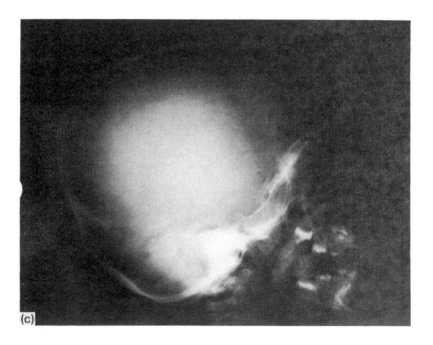

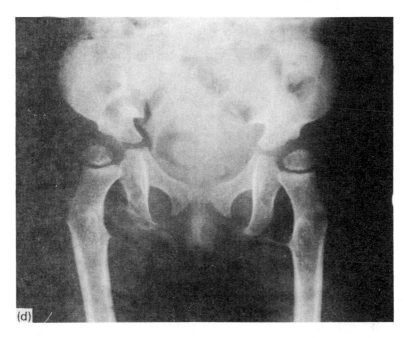

Fig. 2.23. Radiological abnormalities in a two-year-old girl with Maroteaux–Lamy disease. (a) Spine: There is prominent inferior beaking involving D12, L1, L2 and, to a lesser extent, L3, L4 and L5. There is posterior scalloping, involving particularly L3, L4 and L5. (b) Hand: There is constriction of the proximal metaphyses of the digits, metacarpals and the distal metaphyses of the proximal phalanges. (c) Skull: There is splaying of the sutures. The skull is enlarged and a 'J'-shaped sella is present. (d) Pelvis: The superior acetabular portions are rather constricted and there are shallow acetabular fossae. Rather vertical ischia are present. There is bilateral coxa valga.

1978; Guibaud *et al.* 1979) that genetic heterogeneity seems likely. The spectrum spans from a severe progressive mucopolysaccharidosis leading to death at 2.75 years of age (Beaudet *et al.* 1975) to a normal phenotype, including intellect, at 17 years (Danes and Degnan 1974).

Clinical features

In view of the diversity of clinical presentation, the disease should be suspected in individuals with undiagnosed disorders manifesting with features of the MPS. In the most severe form (Patient K.B. of Beaudet *et al.* 1975), hepatosplenomegaly was present at birth; metachromatic granules (Alder–Reilly anomaly) were present in bone marrow granulocytes by 13 days. She developed ascites and massive inguinal hernias.

Radiologically, dysostosis multiplex, including wedge deformity of the lumbar vertebrae, and broadening of the ribs was present at three months. Corneal clouding was noted at seven months. She had numerous episodes of otitis media and bronchopneumonia; there was growth retardation and progressive skeletal changes until death at 2.75 years.

The least severely affected were the 17-year-old female detected by a positive spot test for urinary GAG (performed for unrecorded reasons), said to be physically and intellectually normal (except for genu valgum and club foot at birth), but who radiologically appears to have mild scoliosis (see radiograph in Danes and Degnan 1974); and two brothers (Gitzelmann *et al.* 1978) who up to 15.5 and 19 years of age, respectively, had normal intelligence. There was non-symptomatic, mild kyphoscoliosis in both and hypertelorism in one. Other patients have had intermediate phenotypes. First described was a black male who at seven weeks of age had unusual facies, metatarsus adductus, heptatosplenomegaly, umbilical hernia and dorsolumbar gibbus. Short stature was apparent by 18 months. In his third year, pigeon breast deformity and bilateral inguinal hernias developed. During the first three months he had had three hospitalizations for pneumonia. The corneas were clear at 30 months.

The patient of Gehler *et al.* (1974) had similar features but had corneal clouding.

The patient T.Y. of Beaudet *et al.* (1975) had hypertension, coarctation of the aorta, and aortic incompetence at eight years of age, but at 13 years of age he had normal intelligence and height, no hepatosplenomegaly or gross corneal clouding, but ophthalmological examination revealed fine superficial stromal changes in the corneal periphery, compatible with a mucopolysaccharidosis.

Of two affected North African sisters whose parents were first cousins (Guibaud *et al.* 1979), one was noticed to have facial dysmorphism, club feet, and swollen extremities soon after birth. At 18 months, growth and psychomotor development were normal, but she had dorsolumbar kyphosis; the other at four years of age had facial dysmorphism and mild skeletal abnormalities. Neither had features of a storage disease.

Radiological features have been as varied as the clinical phenotypes. Appearances have ranged from those of mild kyphoscoliosis (Gitzelmann *et al.* 1978) to those 'found in several of the classic mucopolysaccharidoses' (Sly *et al.* 1973).

Pathology

Circulating leucocytes have shown metachromatic inclusions in most, but not all (e.g. Guibaud *et al.* 1979) patients. Two grafts were inserted in the abdominal aorta of case T.Y. of Beaudet *et al.* (1975) at 13 years of age.

Histology of an aortic biopsy revealed fibromuscular dysplasia with abundant intercellular mucopolysaccharides and large vacuolated cells.

The molecular defect

In all patients reported, there has been severe deficiency of ß-glucuronidase (Fig. 2.5, reaction 4; Fig. 2.17, reaction 3). This has been demonstrated in leucocytes (Sly *et al.* 1973), cultured fibroblasts (Hall *et al.* 1973) and serum (Beaudet *et al.* 1975). The demonstration of antigenically cross-reacting material to goat ß-glucuronidase antiserum in four unrelated patients (Bell *et al.* 1977) suggests the presence of mutant ß-glucuronidase in these patients. ß-glucuronidase deficiency might be expected to cause a block in catabolism of polymers containing ß-linked glucuronic acid residues including dermatan, heparan, and chondroitin sulphates, and hyaluronic acid. However, although about half the carbohydrate chains of chondroitin sulphates and hyaluronic acid consist of ß-linked glucuronic acid residues, considerable degradation of these GAG would be expected by tissue hyaluronidases, yielding tetramers which escape detection by methods in which only precipitable material is analyzed.

Diagnosis

The diagnosis is confirmed by demonstrating severe deficiency of ß-glucuronidase in leucocytes (Sly *et al.* 1973), cultured fibroblasts (Hall *et al.* 1973) or serum (Beaudet *et al.* 1975).

When comparing activities of ß-glucuronidase in cultured fibroblasts, care should be observed in standardizing cell population density since enzyme activity rises considerably as cultures become confluent (Russell *et al.* 1971). In the interpretation of results, it should be noted that ß-glucuronidase in bovine serum is taken up by mutant cells and may cause some rise in intracellular activity (Glaser and Sly 1973). Both phenyl-ß-D-glucuronide (Robins *et al.* 1968) and 4-methylumbelliferyl-ß-D-glucuronide (Glaser and Sly 1973) can be used as substrates for the enzyme assay.

Reports on the degree and type of mucopolysacchariduria have varied. In the first case reported, there was only mild excess excretion of urinary GAG (8.1–12.5 mg/24h) between 24 and 31.5 months. Initially these were identified as chondroitin 4- and 6-sulphates, but according to Beaudet *et al.* (1975), this patient excreted primarily dermatan and heparan sulphates. Excessive excretion of dermatan and heparan sulphates has been found in the majority of subsequently reported patients, but chondroitin sulphaturia was reported in two affected sisters (Guibaud *et al.* 1979).

Although most patients have had cellular evidence of impaired GAG catabolism such as leucocyte metachromatic inclusion bodies and accelerated radiosulphate incorporation into cultured fibroblast GAG, these have been absent respectively in patients reported by Guibaud *et al.* (1979) and Beaudet *et al.* (1975).

Genetic counselling and prenatal diagnosis

The disease has an autosomal recessive mode of inheritance as evidenced by its occurrence in sisters with consanguineous parents, and by the demonstration of partial ß-glucuronidase deficiency in leucocytes (Sly *et al.* 1973; Beaudet *et al.* 1975; Guibaud *et al.* 1979), and fibroblasts and serum (Guibaud *et al.* 1979; Beaudet *et al.* 1975) of parents. The disease is one of the rarest forms of MPS. Only about 12 patients have been reported in eight years. Two pregnancies at risk have been monitored by assay of ß-glucuronidase in cultured amniotic cells. Unaffected and heterozygous foetuses were accurately predicted by Guibaud *et al.* (1979) and Maire *et al.* (1979), respectively.

When cultured amniotic cells are assayed for ß-glucuronidase activity, cell population densities should be standardized, and possible uptake of enzyme by mutant cells from the culture medium considered (as for the fibroblast assay — see 'Diagnosis').

The structural locus for ß-glucuronidase has been assigned to chromosome 7 (Chern and Croce 1976; Francke 1976).

The occurrence of genetic heterogeneity, including possibly double heterozygotes for different mutations at the ß-glucuronidase locus, is strongly suggested by the diversity of the clinical phenotype. This is supported by the report of differences in titration patterns of cross-reactive ß-glucuronidase antigen of fibroblasts from four unrelated patients with ß-glucuronidase deficiency (Bell *et al.* 1977).

Animal model

Mice homozygous for the *Gus^h* allele have marked ß-glucuronidase deficiency, but defective GAG catabolism has not been demonstrated. However, they have been used to study enzyme replacement (Hudson *et al.* 1980).

In mice, Paigen *et al.* (1979) have reported that the *Gur* locus exerts a regulatory effect on transcription of ß-glucuronidase messenger RNA.

In rat liver, ß-glucuronidase is distributed both in the lysosomes and in endoplasmic reticulum, while in the mouse the enzyme in both sites is coded by the same structural gene (Ganschow and Paigen 1967).

Treatment of the mucopolysaccharidoses

Enzyme replacement

The demonstration by Fratantoni *et al.* (1968*b*) that the defect in GAG metabolism could be corrected in cultured fibroblasts from patients with either MPS IH or MPS II by the products of cells with other genotypes immediately suggested the possibility that this could be achieved in patients. As discussed above, these cell products or corrective factors were identified as forms of the enzymes which were deficient in the MPS and selectively taken up by fibroblasts (Shapiro *et al.* 1976*a*). Since patients with the MPS appear normal physically and mentally in the first few months or years after birth, enzyme replacement before the onset of regression appeared to be a reasonable possibility.

The practical difficulties, however, are numerous. Firstly, the blood–brain barrier: plasma enzymes pass poorly into the CSF. Moreover, it is virtually unknown whether neurones can take up enzymes even if they are presented to the neurone surface. Modification of the blood–brain barrier by intra-arterial mannitol or arabinose allows about one per cent of infused enzymes to enter the brain (Neuwelt *et al.* 1981). Theoretically, the situation may seem less difficult in MPS IV, IS, or VI, in which connective tissues and the skeletal system are involved rather than the central nervous system.

Secondly, if there is absence of cross-reacting material to enzyme antibodies, it is probable that the immune system of patients receiving such enzymes would produce antibodies to the 'foreign' protein. Thus although cross-reacting material has been detected for MPS IIIB and MPS VII, none could be detected for MPS IVA. It is theoretically likely therefore that attempts at enzyme replacement for classical Morquio's disease would result in production of antibodies against the enzyme. Patients with MPSIH and IS have sufficient residual iduronidase activity to demonstrate altered enzyme kinetics (Hopwood and Muller 1979) and are therefore likely to have mutant enzymes.

Thirdly, most lysosomal enzymes have short half-lives intracellularly (O'Brien *et al.* 1973; Von Figura and Kresse 1974; Achord *et al.* 1977) so that frequent, even daily, administration would be necessary.

Fourthly, if the enzyme is purified prior to administration, it is possible that oligosaccharide residues acting as recognition markers for cell surface receptors may be damaged. In an attempt to introduce a recognition marker, Murray *et al.* (1983) have shown recently that treating gluco-cerebrosidase by sequential enzymatic deglycosylation so as to expose mannose residues considerably increases its uptake by rat Kupffer cells.

Infusion of fresh frozen plasma was reported to have caused transient clinical and biochemical improvements in Hurler's and Hunter's diseases

(Di Ferrante *et al*. 1971) and in Sanfilippo diseases types A and B (Dean *et al*. 1973) or other forms of MPS (McKusick *et al*. 1972), but this was not confirmed after infusion of plasma or blood by others (Dekaban *et al*. 1972; Erickson *et al*. 1972; Danes *et al*. 1972). A superior response was found after infusion of fresh lymphocytes (Knudson *et al*. 1971) or leucocytes (Moser *et al*. 1974).

Attempts were made to overcome the problems of short enzyme lives by providing a source of enzyme by skin (Dean *et al*. 1975) or cultured fibroblast (Dean *et al*. 1979, 1981) transplant, but although these procedures were followed by biochemical evidence of enzyme replacement and GAG catabolism, no definite clinical improvement followed.

The only treatment which has reversed the clinical and biochemical features of MPS has been bone marrow transplantation (BMT) for MPS IH (Hobbs *et al*. 1981). Still to overcome, however, are the problems of graft-versus-host disease and other complications of BMT. It is still too early to judge whether cerebral damage can be averted by this procedure.

Surgical treatment

Other forms of treatment are lamellar corneal grafts to overcome severe corneal clouding in MPS IH/IS (Kajii *et al*. 1974) and in other types of MPS, and carpal tunnel decompression in MPS II (Swift and McDonald 1976), MPS VI and other types of MPS where carpal tunnel syndrome occurs. Orthopaedic procedures (Kopits 1976) are often required for MPS IV and VI, such as posterior spinal fusion to prevent atlantoaxial dislocation (Kopits *et al*. 1972) and correction of genu valgum by osteotomy. Correction of glue ears and provision of hearing aids can cause dramatic improvement in speech development in various types of MPS. Aortic valve replacement for aortic valve disease may be necessary in some types of MPS (Wilson *et al*. 1980). Repair of inguinal hernias is indicated in several of the MPS.

3. Disorders of glycoprotein metabolism — the oligosaccharidoses

Under these headings we describe a group of progressive disorders of glycoprotein metabolism in which there is accumulation of oligosaccharides in tissues and urine. In some there is also accumulation of sphingolipids, glycolipids, and glycosaminoglycans in the tissues, but in contrast to the mucopolysaccharidoses, there is no excess of urinary mucopolysaccharides.

Some of these disorders were designated by Spranger and Wiedemann (1970) as the genetic mucolipidoses (ML), and identified by numbers I–III. Merin *et al.* (1975) suggested the inclusion of ML IV. The numerical classification was useful during a period when the molecular defects were poorly understood, and is retained for ML III (pseudo-Hurler polydystrophy) and ML IV which is described as a sphingolipidosis (ganglioside sialidase deficiency). However, since the discovery that in ML I there is a deficiency in neuraminidase (sialidase) activity, the name of 'sialidosis' (Durand *et al.* 1977) has been suggested and adopted widely, although it is recognized that there are clinical and genetic variants (Lowden and O'Brien 1979). Moreover, ML II is now usually called 'I-cell disease' because cultured fibroblasts contain cytoplasmic inclusions (De Mars and Leroy 1967; Leroy and De Mars 1967; Tondeur *et al.* 1971). These disorders of glycoprotein metabolism can be classified as either disorders of glycoprotein synthesis (or more specifically of post-translational modification) or as disorders of glycoprotein catabolism.

DEFECTS IN POST-TRANSLATIONAL MODIFICATION OF LYSOSOMAL ENZYMES AND OTHER GLYCOPROTEINS

I-cell disease and pseudo-Hurler polydystrophy

I-cell disease (mucolipidosis II or ML II) and pseudo-Hurler polydystrophy (mucolipidosis III, or ML III) have similar biochemical defects, though they may be less pronounced in the latter. Clinically, patients with I-cell disease resemble patients with a severe form of Hurler's disease without mucopolysacchariduria, leading to death by five years of age; while patients with ML III resemble individuals with Scheie's, or Maroteaux–Lamy disease, also without mucopolysacchariduria, surviving to adult life, but with moderate mental handicap.

In both disorders several lysosomal hydrolases are moderately or severely deficient within cultured fibroblasts, but elevated in the culture medium. Stable lysosomal enzymes are also present in excess in plasma.

The basic defect in both disorders is in the phosphorylation of mannose residues of lysosomal enzymes, which therefore lack mannose 6-phosphate recognition markers and fail to reach lysosomes by a receptor-mediated process. This implies that at least some of the phosphorylated enzyme precursors are secreted into the extracellular space before entering the cell and the lysosomes, where they join phosphorylated enzyme precursors which have not left the cell, before 'trimming' by limited proteolysis to become 'mature' enzymes.

Clinical features of I-cell disease; mucolipidosis II (ML II)

Skeletal abnormalities may be present at, or before, birth. Congenital dislocation of the hips is common. Within a few months patients develop a Hurler-like face (Fig. 3.1), hernias, thoracic deformity (Fig. 3.2),

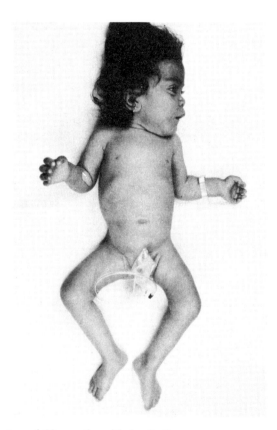

Fig. 3.1. Patient aged 18 months with I-cell disease.

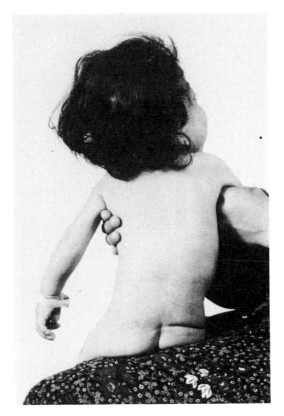

Fig. 3.2. Gibbus in patient with I-cell disease.

hyperplasia of the gums, restriction of joint mobility and hepato-splenomegaly. There is progressive psychomotor retardation, but no corneal clouding. Survival beyond five years of age is unusual. There are marked radiological skeletal changes resembling those seen in Hurler's disease. Widening of the diaphysis and periostial thickening, especially of the radius and humerus, are typical (Patriquin *et al.* 1977; Lemaitre *et al.* 1979). Cardiac murmurs due to aortic and mitral valve pathology are common.

Pathology of I-cell disease

Phase microscopy of cultured fibroblasts reveals numerous intracellular inclusions (Fig. 3.3) (Leroy and De Mars 1967; De Mars and Leroy 1967) which on electron microscopy have been identified as large lysosomes. They contain heterogenous storage material with the ultrastructural appearance of membranous whorls (Tondeur *et al.* 1971; Gilbert *et al.*

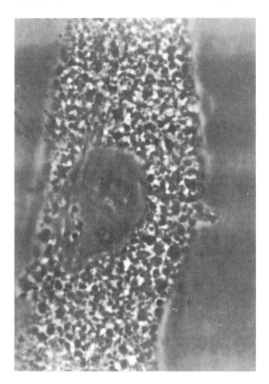

Fig.3.3. Phase microscopy of cultured fibroblasts from patient with I-cell disease. Note numerous large cytoplasmic inclusions.

1973). There is a tissue- or organ-specific distribution of cells with distended lysosomes. Particularly affected are chondrocytes, fibroblasts, glomerular epithelial cells, Schwann cells, vascular perithelial cells, peripheral neurones, and hepatic histiocytes. Variably affected or unaffected are hepatocytes, Kupffer cells, leucocytes, and cortical neurones (Kenyon and Sensenbrenner 1971; Tondeur *et al.* 1971; Gilbert *et al.* 1973; Scot *et al.* 1973; Martin *et al.* 1975). Brain and liver cells, which are not affected histologically, have been found to have a normal content of glycolipid. This distribution is also in accord with that of multiple enzyme deficiencies, which involve connective tissue rather than liver, brain, and leucocytes (see 'Molecular defects', p. 59). Histochemical studies of the storage material have yielded conflicting results. Some workers have reported positive staining with Alcian blue (metachromasia), periodic acid–Schiff (PAS), and Sudan black, indicating, respectively, the presence of negatively charged, carbohydrate, and lipid-containing material (Tondeur *et al.* 1971; Gilbert *et al.* 1973), but Martin *et al.* (1975) could not find evidence of mucopolysaccharide storage. Accumulation of sulphated mucopolysaccharides in cultured fibroblasts was reported by Schmickel *et al.* (1975).

Clinical features of mucolipidosis III (pseudo-Hurler polydystrophy)

The major clinical features of ML III were defined by Spranger and Wiedemann (1970). Later Thomas *et al.* (1973), Taylor *et al.* (1973), and Kelly *et al.* (1975) reported enzymatic and cytological characteristics similar to those described for I-cell disease.

Stiffness of the hands and shoulders, between two and four years of age, is a common presenting symptom. The stiffness of the hands progresses to a claw hand deformity by six years of age. Severe stunting of growth is evident by six to eight years in males and about two years later in females. Coarsening of the face is present by six years of age. Radiologically there is dysostosis multiplex, usually causing suspicion that the patient may have a mucopolysaccharidosis, but mucopolysaccharide excretion is always normal. Spranger *et al.* (1974*b*) considered that severe pelvic and bizarre vertebral changes were specific for ML III. They described low iliac wings, with hypoplastic bodies, flattened and irregular femoral epiphyses, valgus deformity of the femoral necks, underdevelopment of the posterior parts of the dorsal vertebral bodies, and hypoplasia of the anterior third of the lumbar vertebral bodies. Gibbus was present in four of 12 patients described by Kelly *et al.* (1975). Slit lamp examination reveals mild peripheral corneal clouding which increases with age, but is never as severe as in Hurler's or Maroteaux–Lamy diseases.

In about two-thirds of patients there is carpal tunnel syndrome with loss of sensation of the fingers and wasting of the thenar muscles. Nearly all patients have aortic and/or mitral valvular disease, but cardiac failure is rare under the age of twenty (Kelly *et al.* 1975). Intelligence is usually moderately to severely reduced but can be normal (Kelly *et al.* 1975). The disease is usually more severe in males than in females. The life-span is not well established. Kelly *et al.* (1975) report that of 12 proven cases, the oldest three were 28, 26, and 22 years of age.

Pathology of mucolipidosis III

We have been unable to find any reports of necropsies in ML III patients. Phase microscopy of cultured fibroblasts reveals abundant intracellular granular inclusions similar to those seen in I-cell disease, which on electron microscopy are identified as single-membrane limited lysosomes containing amorphous electron-dense and electron-translucent areas, together with membraneous lamellar bodies (Taylor *et al.* 1973).

Molecular defects

Cultured fibroblasts from patients with I-cell disease show decreased intracellular activities and increased extracellular activities of several lysosomal enzymes (Wiesmann *et al.* 1971*a*, 1974; Leroy *et al.* 1972; Table 3.1). The intracellular activities are decreased to a greater extent for some enzymes (α -iduronidase, arylsulphatase A) than for others (ß-glucuronidase, α-mannosidase), while acid phosphatase and ß-glucosidase are distributed normally.

In general, enzymes which have decreased activities inside the cell have increased activities outside the cell if they are stable in the culture medium, plasma or other extracellular fluid (Wiesmann *et al.* 1971*a*; Wiesmann and Herschkowitz 1974; Berman *et al.* 1974*a*). Normal enzyme activities are present in cells and organs with normal histology (such as liver, brain and leucocytes) (Tondeur *et al.* 1971; Leroy *et al.* 1972; Gilbert *et al.* 1973).

The explanation of why the clinical course of I-cell disease resembles that of the mucopolysaccharidoses, rather than that of deficiencies of other affected enzymes, may be that the defect is confined mainly to connective tissue cells.

As for I-cell disease, in ML III some intracellular lysosomal enzymes have considerably reduced activities (Thomas *et al.* 1973). When compared with values of control means (100 per cent), cultured fibroblast residual activities (Kelly *et al.* 1975) for *N*-acetyl-ß-glucosaminidase were 12–13

Table 3.1 *Distribution of lysosomal hydrolases in cultured I-cell fibroblasts*

Decreased intracellular activities	Increased extracellular activities	Normal distribution
Arylsulphatase A	Arylsulphatase A	Acid phosphatase
N-acetyl-ß-glucos-aminidase	*N*-acetyl-ß-glucos-aminidase	ß-glucosidase
N-acetyl-ß-galacto-saminidase	ß-glucuronidase	
ß-glucuronidase α -galactosidase ß-galactosidase	α -galactosidase	
α -iduronidase	ß-galactosidase	
Sulphoiduronate sulphatase	α -mannosidase	
α-mannosidase α-fucosidase Galactosyl ceramide ß-galactosidase		
Sphingomyelinase		

per cent; for ß-galactosidase 6–15 per cent; α-fucosidase 16–42 per cent; α-mannosidase, 12–72 per cent; arylsulphatase A, 6–24 per cent. A corresponding increase in activity was observed in serum of patients when compared with activities in serum of controls, including *N*-acetyl-ß-glucosaminidase (7–35 times that of controls), ß-galactosidase (6–24 times), α-fucosidase (3–12 times) and arylsulphatase A (4–28 times).

Radiosulphate incorporation into cultured fibroblast glycosamino-glycans in all 11 patients with ML III studied was 2.1–6.1 times higher than the control mean (Kelly *et al.* 1975).

The first clue of the basic defect came from the experiments of Hickman and Neufeld (1972) who showed that I-cells would take up, and retain normally, enzymes secreted into the culture medium by normal cells. However enzymes secreted by I-cells could not be taken up by other cell strains. These workers suggested that the hydrolases were abnormal in that they lacked a recognition marker required to enter the cells and reach the lysosomes. A similar explanation was suggested for the defect in ML III cells (Hickman *et al.* 1974; Glaser *et al.* 1974).

The identification of this marker in normal lysosomal enzymes, and the demonstration of its absence in I-cell enzymes and reduction in ML III enzymes, were achieved by a series of brilliant experiments. Since endocytosis by normal cultured fibroblasts of ß-hexosaminidase was reduced by periodate treatment it was reasoned that the marker was carbohydrate (Hickman *et al.* 1974). It was argued that since the kinetics of uptake were consistent with binding to cell surface receptors, these could be saturated by competitive inhibitors. Of the inhibitors tested, mannose 6-phosphate was the most effective (Sando and Neufeld 1977; Kaplan *et al.* 1977; Ullrich *et al.* 1978). Furthermore, treatment of lysosomal enzymes by phosphatase diminished their uptake by normal cells. It was therefore proposed, (Kaplan *et al.* 1977) and later demonstrated, that lysosomal enzymes contained mannose 6-phosphate (Natowizc *et al.* 1979; Distler *et al.* 1979; von Figura and Klein 1979; Hasilik and Neufeld 1980*a*) which could act as the recognition marker. The demonstration that mannose 6-phosphate was absent in lysosomal enzymes secreted by the I-cells (Bach *et al.* 1979*a*) confirmed the hypothesis.

The primary enzyme defect in I-cell disease, therefore, involves the mechanism of phosphorylation of the mannose at the non-reducing end of the oligosaccharide comprising six to eight mannose residues linked to newly synthesized lysosomal enzymes (Tabas and Kornfeld 1980; Varki and Kornfeld 1980; Hasilik *et al.* 1980). This phosphorylation is thought to proceed by two steps (Waheed *et al.* 1981*a*): firstly, the transfer of *N*-acetylglucosamine 1-phosphate from UDP-*N*-acetylglucosamine to the C-6 of mannose to form a phosphate diester; and secondly, the release of the *N*-acetylglucosamine by a phosphodiesterase (Waheed *et al.* 1981*b*). The absence of the first reaction in I-cells (Hasilik *et al.* 1981; Reitman and Kornfeld 1981; Reitman *et al.* 1981; Waheed *et al.* 1982) and the considerable reduction in ML III cells (Reitman *et al.* 1981) has recently been demonstrated, thus establishing the defect in both disorders as deficiency of uridine 5′-diphosphate-*N*-acetylglucosamine: glycoprotein *N*-acetylglucosaminylphosphotransferase (*N*-acetylglucosamine 1-phosphotransferase).

Defective phosphorylation of lysosomal enzymes in I-cell disease and ML III prevents or reduces their re-entry into the cell, and therefore their arrival at the lysosomes. This may explain defective 'trimming' by proteolysis (Frisch and Neufeld 1981) of these enzymes, which normally probably occurs in lysosomes, and results in enzymes with higher than normal molecular weights (Hasilik and Neufeld 1980*b*). Moreover the abnormal extracellular lysosomal enzymes are over-sialated, and are abnormally anionic (Vladutio and Rattazzi 1975; Miller 1978). Urinary (Strecker *et al.* 1977) and fibroblast (Thomas *et al.* 1976) glycoproteins are also over-sialated. In fact Strecker *et al.* (1976*b*) and Strecker and Michalski (1978) suggested that neuraminidase deficiency was the primary defect in I-cell disease. It seems likely however that the disproportionately greater deficiency of neuraminidase than other enzymes in I-cell disease is due to greater instability.

The deficiency of *N*-acetylglucosamine 1-phosphotransferase in I-cell disease was originally demonstrated in cultured fibroblasts, but later studies showed that the enzyme is also absent in liver (Owada and Neufeld 1982; Waheed *et al.* 1982), spleen, kidney, and brain (Waheed *et al.* 1982). Since these organs have normal activities of the lysosomal enzymes (ß-galactosidase excepted) which are deficient in cultured fibroblasts and other connective tissue cells in I-cell disease, it seems likely that they possess an additional mechanism for introducing acid hydrolases into lysosomes which is independent of the mannose 6-phosphate recognition markers (Owada and Neufeld 1982; Waheed *et al.* 1982).

Diagnosis

For both I-cell disease and ML III the diagnosis is most conveniently made by demonstrating elevated activities of lysosomal enzymes (Table 3.1) in the serum (Wiesmann *et al.* 1971*a*, 1974; Leroy *et al.* 1972). Absence of elevated mucopolysacchariduria will exclude the diagnosis of Hurler's and Hunter's diseases.

In cultured fibroblasts the diagnosis is supported by demonstration of abundant coarse intracellular granules on phase microscopy (Fig 3.3); and confirmed by demonstrating the abnormal distribution of lysosomal enzyme activities shown in Table 3.1, and deficiency of *N*-acetylglucosamine 1-phosphotransferase activity.

Genetic counselling and prenatal diagnosis

I-cell disease. Since both sexes are affected in similar proportions the disorder probably has an autosomal recessive form of inheritance. This is supported by the report that some parents have a partial defect of serum *N*-acetyl-ß-D-glucosaminidase on DEAE column chromatography (Van Elsen *et al*. 1976), and birth of an affected child to consanguineous parents (personal observations). The occurrence of the disease in siblings (foetuses) of either sex is also consistent with autosomal recessive inheritance. Heterozygote detection by assay of fibroblast *N*-acetylgluco-samine 1-phosphotransferase activity has been reported (Varki *et al*. 1982).

Prenatal diagnosis has been reported by demonstrating elevated activities of amniotic fluid supernatant lysosomal enzymes (Huijing *et al*. 1973; Warren *et al*. 1973), or altered properties of α-mannosidase (Owada *et al*. 1980). However, the diagnosis should be confirmed by demonstrating intracellular lysosomal enzyme deficiencies (Table 3.1) of cultured amniotic cells (Aula *et al*. 1975*a*; Matsuda *et al*. 1975; Gehler *et al*. 1976; Kleijer *et al*. 1978). It should be noted that as discussed for ß-glucuronidase, activities of several lysosomal enzymes are sensitive to cell population densities, so it is important to standardize the conditions of control and test cells before assay (Butterworth *et al*. 1973; Heukels-Dully and Niermeijer 1976).

ML III. Autosomal recessive inheritance is indicated by reports of the disorder in affected brothers and sisters (Kelly *et al*. 1975; Stein *et al*. 1974) and parental consanguinity (Stein *et al*. 1974).

It has been reported that electrophoresis of extracellular enzymes (Vladutiu and Rattazzi 1975), assay of serum ß-hexosaminidase (Van Elsen *et al*. 1976; Vidgoff and Buist 1977; Jolly and Desnick 1979) or assay of cultured fibroblast or leucocyte *N*-acetylglucosamine 1-phospho-transferase activities (Varki *et al*. 1982) are useful methods for carrier detection.

Prenatal diagnosis should be possible by testing cultured amniotic cells by methods described under diagnosis for cultured fibroblasts. As demonstrated for I-cell disease it will be interesting to test the amniotic fluid supernatant for elevated activities of lysosomal enzymes.

DEFECTS OF GLYCOPROTEIN DEGRADATION

Catabolism of glycoproteins

Glycoprotein catabolism is thought to proceed almost exclusively inside the lysosomes by sequential hydrolysis. The first step is a partial degradation of the protein core by proteolytic enzymes resulting in the release of glycopeptides. The second step is cleavage of the asparagine residue from *N*-acetylglucosamine catalysed by *N*-aspartyl-ß-glucosaminidase, which is deficient in aspartylglycosaminuria. This releases oligosaccharides with the following structure:

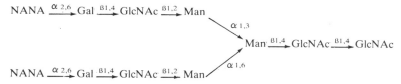

The oligosaccharides are then catabolized first by ß-endo-*N*-acetylglucosaminidase acting at the reducing end of the chains and then from the non-reducing ends in stepwise reactions catalysed by neuraminidase (deficient in sialidosis), ß-galactoisidase (deficient in GM$_1$-gangliosidoses), hexosaminidiase A and B (deficient in Sandhoff's disease), α-mannosidase (deficient in mannosidosis), and ß-mannosidase (for which no storage disease has been described in humans, although a caprine form is known).

Sialidosis; mucolipidosis I; type 2 GM$_1$-gangliosidosis variant; neuraminidase deficiency; the cherry-red-spot-myoclonus syndrome

Several clinical phenotypes have been reported in patients with deficiency of sialidase (neuraminidase; *N*-acetylneuraminic acid hydrolase). Classification according to the absence (Type 1) or presence (Type 2) of dysmorphic features has been suggested by Lowden and O'Brien (1979) with type 2 subdivided according to the age of onset into infantile and juvenile types, some of type 2 also having deficiency of ß-galactosidase. The designation of sialidosis was suggested by Durand *et al.* (1977).

This essentially clinical classification has some predictive value with regard to prognosis and appears to be upheld by ethnic distribution. However, at the molecular level it is challenged by the report (Hoogeveen *et al.* 1980) that if cultured fibroblasts derived from patients with combined neuramindase/ß-galactosidase deficiency are fused with either type 1 or type 2 sialidosis cells (without ß-galactosidase deficiency), there is an increase in neuraminidase activity (3–9 times the values after parental fusion). On the other hand, there is no increase in neuraminidase activity when type 1 cells are fused with type 2 cells (without ß-galactosidase deficiency). These experiments suggest that type 1 and type 2 patients (without ß-galactosidase deficiency) share the same genetic defect, which is different to that in patients with the combined neuraminidase/ß-galactosidase defect.

Fibroblasts from patients with I-cell disease also have deficiency of intracellular neuraminidase as part of a multiple lysosomal enzyme deficiency which is corrected by fusion with cells from patients with sialidosis types 1 or 2 with or without ß-galactosidase deficiency, indicating distinct genetic defects. (Hoogeveen *et al.* 1980). Ganglioside sialidase deficiency or ML IV is discussed under the sphingolipidoses.

Clinical features

Sialidosis type 1. This type is the normosomatic group of Lowden and O'Brien (1979) and was described by Rapin *et al.* (1978) as the 'cherry-red-spot-myoclonus syndrome'. Decreasing visual acuity and myoclonus at 8–15 years of age are characteristic presenting symptoms. Cherry-red spots in the macular region of the fundus are the only constant sign, but punctate lens opacities have been reported (Durand *et al.* 1977).

Myoclonus is of the action type, i.e. it is induced by movement and emotion. Spasms begin in the limbs and become progressively more severe, eventually preventing purposive movements; the spasms are resistant to anticonvulsants. Grand mal occurs in about half the patients and ataxia in a quarter. Intelligence is either normal or mildly retarted. Visual acuity becomes progressively impaired. Some patients die in the third decade, but survival into the fourth has been reported (Thomas *et al.* 1978; Rapin *et al.* 1978). Some patients have experienced burning pain in their extremities, exacerbated by heat, as is characteristic in Fabry's disease (Rapin *et al.* 1978). Of ten patients in the literature whose ethnic origin was recorded, seven were of Italian ancestry (Lowden and O'Brien 1979).

Sialidosis type 2. This is the dysmorphic group of Lowden and O'Brien (1979). The presenting symptom is the abnormal facial appearance (Fig. 3.4) which is reminiscent of the coarse Hurler's disease facies. Typically, there is hypertelorism, the nasal bridge is flat, and the lips are thickened. The head circumference is increased and height decreased. Macular cherry-red spots occurred in 16 of 17 cases reviewed by Lowden and O'Brien (1979). The exception was described by Kelly and Graetz (1977). Myoclonus and ataxia are present in about 75 per cent of patients. The majority have moderate mental retardation.

Other observed features include cataracts, corneal clouding, hepatomegaly (especially if onset is early) joint stiffness, deafness, and hernias. Lowden and O'Brien (1979) further subdivided sialidosis type 2 according to infantile or juvenile onset. Of 14 juvenile onset cases, 10 were Japanese. In this group, onset was between eight and 15 years of age, and the earliest symptoms were joint stiffness, coarse features, or decreased visual acuity. The oldest survivor is 25 years old. Hepatomegaly (without splenomegaly) occurred in all five patients in the infantile group, but in none of the juvenile group.

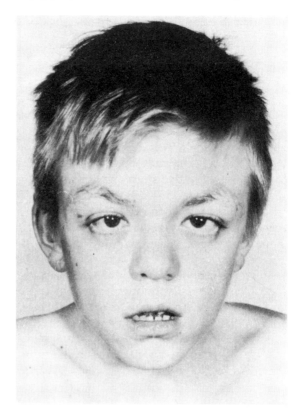

Fig. 3.4. Patient with sialidosis type 2, aged 13 years. Note dysmorphic features.

Radiologically changes resemble but are milder than those of Hurler's disease, and include beaking of the first or second lumbar vertebra, wide ribs, thickening of the skull, and a coarse texture of the bones.

Combined neuraminidase and β-galactosidase deficiency. Combined neuraminidase and ß-galactosidase deficiency was first described by Wenger *et al.* (1978*a*) in a patient previously reported as a new variant of GM₁-gangliosidosis (Justice *et al.* 1977). Two other patients initially diagnosed as GM₁-gangliosidosis variants were later recognized as having the combined deficiency (Lowden and O'Brien 1979).

According to Lowden and O'Brien (1979), ß-galactosidase activity was low in tissues, but normal in plasma of all juvenile onset sialidosis type 2 patients studied, normal in two infantile onset type 2 patients, but absent in one. The clinical features are as described for sialidosis type 2. Several patients with type 2 sialidosis have had angiokeratoma corporis diffusum characteristic of Fabry's disease (Loonen *et al.* 1974; Suzuki *et al.* 1977; Miyatake *et al.* 1979; Kobayashi *et al.* 1979).

Mixed types. It is clear from the literature that assignation into types 1 and 2 is not always simple, since several cases of sialidosis have features of both types. Thus, of patients designated as type 1 (Lowden and O'Brien 1979), two of 18 had mental retardation, and one (Durand *et al.* 1977) had hepatomegaly; a patient who presented with myoclonus at 18 years of age, coarsening of the face and dysostosis multiplex, lacked mental retardation until some dementia was noted at 48 years of age (Miyatake *et al.* 1979); a patient who died at 22 years of age, designated as type 2 because of coarse facies, mental retardation and dysostosis multiplex, had features intermediate between the juvenile (absence of hepatomegaly) and infantile (onset before one year of age, normal ß-galactosidase activity) types (Winter *et al.* 1980).

Pathology

Vacuolated or foam cells have been found in the minority of patients with sialidosis type 1 and consistently in patients with type 2, in lymphocytes (Durand *et al.* 1977; Goldstein *et al.* 1974*a*), liver (Justice *et al.* 1977; Durand *et al.* 1977), and bone marrow cells (Durand *et al.* 1977). Neuronal lipidosis with enlarged neuronal lysosomes has been reported (Gonatas *et al.* 1963) in the patient later studied by Rapin *et al.* (1978).

Molecular defects

In all types of sialidosis there is marked deficiency of neuraminidase in leucocytes and cultured fibroblasts (O'Brien 1978*a*, 1981; O'Brien and Warner 1980; Thomas *et al.* 1978). Sialidosis cells show defective cleavage of both 2,3- and 2,6-linked substrates. Urinary oligosaccharides are more than 100 times normal. Twelve different types with similar structure have been characterized by Spranger *et al.* (1977), Strecker *et al.* (1977) and Durand *et al.* (1977). All contain sialyl residues at the non-reducing end in either $2 \rightarrow 3$ (20 per cent) or $2 \rightarrow 6$ (80 per cent) linkage with a ß-galactosyl residue. It has been suggested (Lowden and O'Brien 1979) that the mutant enzyme retains higher residual activity for $2 \rightarrow 3$ linkages, which results in a preponderance of $2 \rightarrow 6$-linked oligosaccharides in storage material and urine. Some sialic acid remains bound to glycoprotein. This was demonstrated in both sialidosis types 1 and 2 as abnormal electrophoretic mobility due to a greater negative charge (accelerated anodal mobility) of several glycosylated enzymes, including acid phosphatase, *N*-acetyl-ß-hexosaminidase, α-glucosidase, and adenosine deaminase. Treatment with neuraminidase restored normal electrophoretic mobilities (Swallow *et al.* 1981).

An understanding of the molecular basis for the combined neuraminidase/ß-galactosidase deficiency has been provided by the work of Galjaard and colleagues. Complementation of the neuraminidase deficiency of cultured fibroblast derived from patients with combined neuraminidase/ ß-galactosidase deficiency by sialidosis type 1 or 2 cells (without ß-galactosidase deficiency) indicated that the combined deficiency is genetically distinct from isolated neuraminidase deficiency (Hoogeveen *et al.* 1980). Measurement of the turnover time of ß-galactosidase in different mutant fibroblast strains showed that the enzyme in combined deficiency cells has a half-life of about one tenth that in normal or GM_1-gangliosidosis cells (Van Diggelen *et al.* 1980, 1981) due to an enhanced rate of degradation (Van Diggelen *et al.* 1982). Since both ß-galactosidase and neuraminidase activity are found to be enhanced by a factor secreted from normal, sialidosis (without ß-galactosidase deficiency) or GM_1-gangliosidosis cells (Hoogeveen *et al.* 1980, 1981), it was suggested that the cells lack a 'protection factor' which has the role of protecting ß-galactosidase and neuraminidase against premature intralysosomal degradation. Evidence for this proposal came from immunoprecipitation studies (D'Azzo *et al.* 1982) which demonstrated complete absence of a 32 000 dalton enzymatically inactive glycoprotein component in combined deficiency cells which was present in normal cells; and absence of a 54 000 dalton precursor of the 32 000 dalton component from the culture medium. The authors proposed that the 32 000 dalton component may normally be required to unite ß-galactosidase and neuraminidase in a complex within the lysosomal membrane.

Diagnosis

The diagnosis is by demonstration of neuraminidase deficiency in leucocytes or cultured fibroblasts. The methylumbelliferyl substrate (Warner and O'Brien 1979) is now preferred to neuraminlactose (Cantz *et al.* 1977). Prolonged sonication of cells leads to marked loss of neuraminidase activity and should be avoided (Warner and O'Brien 1979). The ß-galactosidase deficiency can be detected by using either the *p*-nitrophenyl or methylumbelliferyl-ß-D-galactopyranosides as substrate in leucocytes or fibroblasts. Thin-layer chromatography of urine is a useful screening procedure to demonstrate an increase of total oligosaccharides after staining with orcinol, or specifically of sialated oligosaccharides after staining with resorcinol (Holmes and O'Brien 1979).

Genetic counselling and prenatal diagnosis

The reports of affected siblings, high frequency of parental consanguinity, and equal sex-ratio (Lowden and O'Brien 1979) attest to autosomal recessive inheritance in sialidosis types 1 and 2. This is confirmed by the demonstration of intermediate deficiency of neuraminidase activity in parents of both type 1 (Thomas *et al*. 1978; Johnson *et al*. 1980*a*) and type 2 (Kelly and Graetz 1977) patients. Parents of patients with combined neuraminidase and ß-galactosidase deficiency show a partial defect of the former enzyme, but normal activity of the latter (Wenger *et al*. 1978*a*).

Prenatal diagnosis has been reported in several pregnancies (Johnson *et al*. 1980*a*; Steinman *et al*. 1980), including one where the foetus had the combined deficiency, by demonstration that cultured amniotic cells had diminished neuraminidase activity towards the methylumbelliferyl substrate (Kleijer *et al*. 1979).

Although the number of reported cases is still low, at present there appear to be possible ethnic predilections, with an increased incidence of type 1 in Italians, and of type 2 in Japanese (Lowden and O'Brien 1979).

Mannosidosis

The first reported patient with mannosidosis (Öckerman 1967*a*, *b*) had Hurler-like features, cloudy areas in the lens capsule, storage of mannose-rich oligosaccharides and deficiency in the activity of lysosomal α-mannosidase. Subsequent reports have revealed heterogeneity in the clinical phenotype and differences in the properties of residual mannosidase activities.

Clinical features
Severe and mild forms have been described. It is not yet clear whether a sharp distinction between a severe infantile phenotype, or type I, and a milder juvenile/adult phenotype, or type II, is justified (Desnick *et al*. 1976*a*). Slight facial dysmorphism, mental retardation, and dysostosis multiplex occur in all patients as the disease advances, but in the mild form these features are less marked.

In the severe form there is hepatosplenomegaly with kyphoscoliosis (Fig. 3.5) and gibbus (Aylsworth *et al*. 1976), deafness, severe dysostosis multiplex, and recurrent severe infections, sometimes leading to early death, as in the patient reported by Desnick *et al*. (1976*a*) who died aged three and a half years, and the patient described by Öckerman who died at four and a half years.

In the mild form, partial deafness and mental retardation are usually the presenting features, with mild dysostosis multiplex being evident radiologically. Mental retardation is progressive in the first decade, but in some patients there is little evidence of progression after the first decade. Survival to 26 years of age (Booth *et al*. 1976; Kistler *et al*. 1977) and mild mental retardation (Bach *et al*. 1978) have been reported.

Cataracts occur in most, and hernias in about half the patients. Height is usually normal. Radiologically all patients have had dysostosis multiplex,

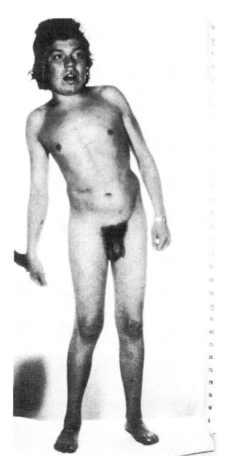

Fig. 3.5. Patient with mannosidosis aged 16 years. Note kyphoscoliosis.

including beaking of upper lumbar vertebrae, widening of the long bones and of the ribs, and coarse trabeculation of metacarpals and phalanges with increased density of the skull.

Pathology

Necropsies have been reported in three patients (Kjellman *et al.* 1969; Desnick *et al.* 1976*a*; Sung *et al.* 1977). Neurones of cerebrum, brain stem, and spinal cord are distended with periodic acid–Schiff (PAS)-positive (i.e. carbohydrate) material which on electron microscopy appears to be intralysosomal. There is also neuronal loss and some demyelination.

Liver biopsies show storage vacuoles in hepatocytes and Kupffer cells (Autio *et al.* 1973). The hepatic storage material does not stain with Alcian blue (i.e. is not negatively charged) or PAS.

Lymphocytes (Kjelman *et al.* 1969), histiocytes in connective tissue, and bone marrow (Kistler *et al.* 1977) are also vacuolated.

Liver and brain have a considerably increased mannose content (Öckerman 1969, 1973; Autio *et al.* 1973).

Neutrophils have depressed chemotactic responsiveness and impaired phagocytosis of bacteria (Desnick *et al.* 1976*a*).

Molecular defects

The disease is caused by severe deficiency of acidic α-mannosidase (Öckerman 1967*b*, 1973). Both lysosomal acidic isoenzymes (A and B), described initially by Marsh and Gourlay (1971) in rat liver and then in human liver, are affected (Carroll *et al.* 1972; Desnick *et al.* 1976*a*) but the neutral enzyme (α-mannosidase C) can be shown to retain normal activity when separated by either ion-exchange chromatography or isoelectric focusing (Carroll *et al.* 1972), or cellulose acetate electrophoresis (Poenaru and Dreyfus 1973).

More recently the liver α-mannosidase B component has been subdivided into B_1 and B_2 (Chester *et al.* 1975); while a second component of α-mannosidase C has been identified (Phillips *et al.* 1974).

Similar isoenzymes exist in cultured fibroblasts (Burditt *et al.* 1980*a*; Burton and Nadler 1978; Gordon *et al.* 1980), cultured amniotic cells (Hultberg *et al.* 1976), and in leucocytes (Masson *et al.* 1974; Nakagawa *et al.* 1980).

Residual activity in cultured fibroblasts has not exceeded 10 per cent of normal values in patients described by Aylsworth *et al.* (1976); Beaudet and Nichols (1976); Booth *et al.* (1976) and Desnick *et al.* (1976*a*) but was 15–20 per cent of normal values in a brother and sister described by Bach *et al.* (1978) and 5–16 per cent in three unrelated patients studied by Burditt *et al.* (1980*a*). Mannosidase deficiency has also been demonstrated in liver, spleen, and brain (Öckerman 1967*b*; Autio *et al.* 1973), and in serum and leucocytes (Masson *et al.* 1974; Desnick *et al.* 1976*a*; Booth *et al.* 1976; Kistler *et al.* 1977).

The effect of divalent metallic ions on the residual enzyme has shown differences in different patients with mannosidosis. Thus in some patients residual acidic mannosidase, like the enzyme from normal tissues, is stimulated by zinc ions (Hultberg and Masson 1975; Desnick *et al.* 1976*a*; Kistler *et al.* 1977) but not in others (Burton and Nadler 1978). Moreover, cobalt ions, which inhibit the activity of normal mannosidase, stimulate residual activity of some mannosidosis patients (Hultberg and Masson 1975; Desnick *et al.* 1976*a*) but not others (Button, Fensom and Benson 1980, unpublished). Cross-reacting material against antibodies to both types of normal acid α-mannosidase isoenzymes has been demonstrated in mannosidosis fibroblasts, indicating the presence of mutant enzyme, probably due to a structural gene mutation (Mersmann and Buddecke 1977), but could not be demonstrated in three unrelated patients (Burditt *et al.* 1980*a*).

The urine contains a considerable increase in mannose and glucosamine-containing oligosaccharides (Kistler *et al.* 1977; Yamashita *et al.* 1980; Matsuura *et al.* 1981). Seven types of mannose-rich oligosaccharides from five patients with mannosidosis were characterized by Strecker *et al.* (1976*a*). They all possess an α-mannose residue at their non-reducing terminal, and an *N*-acetylglucosamine residue in the terminal reducing position and are related to glycans present in numerous glycoproteins.

The most abundant compound is the trisaccharide Man-(1 → 3)-α-(1 → 4)-ß-GlcNAc (Norden *et al.* 1973; Strecker *et al.* 1976*a*; Yamashita *et al.* 1980; Matsuura *et al.* 1981). However the structure of four pentasaccharides suggested by Strecker *et al.* could not be confirmed by Yamashita *et al.* (1980) or Matsuura *et al.* (1981) who respectively characterized 13 and 14 higher molecular weight oligosaccharides from mannosidosis urine. Cultured mannosidosis fibroblasts show defective degradation of ^{3}H-mannose-labelled material. This defect is corrected by adding purified pig kidney α-mannosidase to the culture medium (Mersmann *et al.* 1976).

Diagnosis

The diagnosis is readily established by demonstrating severe reduction in the activity of acidic α-mannosidase in serum, leucocytes, or cultured fibroblasts (Beaudet and Nichols 1976; Desnick *et al.* 1976*a*; Kistler *et al.* 1977).

There is no excessive mucopolysacchariduria. The demonstration of mannose-rich oligosaccharides in the urine by thin-layer chromatography is a useful screening procedure. In specialized laboratories these may be further characterized by gas–liquid chromatography or nuclear magnetic resonance (Van Halbeek *et al.* 1980; Matsuura *et al.* 1981).

Genetic counselling and prenatal diagnosis

Since the original description 15 years ago, at least 67 cases have been reported (Chester *et al.* 1982). The disease has an autosomal recessive mode of inheritance. Parental α-mannosidase activity may be reduced in leucocytes (Booth *et al.* 1976; Maire *et al.* 1978; Bach *et al.* 1978), fibroblasts (Aylsworth *et al.* 1976; Booth *et al.* 1976) and serum (Masson *et al.* 1974; Bach *et al.* 1978) but there is overlap with the normal range. In all four heterozygotes reported by Desnick *et al.* (1976*a*), mannosidase activities for leucocytes and serum fell to within the normal range. Masson *et al.* (1974) found that heterozygotes could be distinguished from normal individuals, and patients with mannosidosis, when mannosidase activities were expressed as ratios to ß-hexosaminidase activities either for fibroblasts (six heterozygotes) or leucocytes (four heterozygotes) but not when results were expressed as total or specific activities. Hirani *et al.* (1977) found that plasma acid mannosidase activities in four parents of children with mannosidosis overlapped with the normal range when expressed in units per plasma volume, but that a differential assay based on the difference in thermal stability of acidic and true intermediate α-mannosidase (calculated from a formula by substituting activities of heated and unheated enzyme at pH 4.2 and 5.2) allowed complete discrimination when expressed as the ratio of intermediate to acid α-mannosidase activities.

Prenatal diagnosis of an affected foetus was reported by Maire *et al.* (1978) using the methylumbelliferyl substrate for α-mannosidase assay of cultured amniotic cells three weeks after amniocentesis at 15–16 weeks' gestation. The authors report that at high substrate concentration, in cultured fibroblasts, mannosidosis cells have similar mannosidosis activity to control cells. They consequently assayed amniotic cells using a range of substrate concentrations and demonstrated that residual α-mannosidase in the amniotic cells from the affected foetus had lower affinity for the substrate (K_m 20mM) than α-mannosidase from normal cells (K_m 0.9mM). Prenatal diagnosis in three pregnancies at risk for mannosidosis was reported by Poenaru *et al.* (1979).

We have monitored a pregnancy at risk for mannosidosis by conventional α-mannosidase assay of cultured amniotic cells three weeks after amniocentesis performed at 18 weeks' gestation. Amniotic cell mannosidase activity was approximately 50 per cent of control values, and a heterozygous foetus was predicted. An unaffected foetus was confirmed by cord blood leucocyte assay, and a heterozygous mannosidosis genotype was diagnosed by demonstrating approximately 50 per cent α-mannosidase activity in the newborn cultured skin fibroblasts (Tansley, Fensom, and Benson 1981, unpublished observation). The gene for acidic α-mannosidase has been assigned to chromosome 19 by somatic cell hybridization studies (Champion *et al.* 1978).

Animal model

In Angus cattle, mannosidosis is relatively common. As in humans, there is a deficiency of acidic α-mannosidase and storage of mannose and *N*-acetylglucosamine-containing oligosaccharides (Hocking *et al.* 1972) in homozygotes, and a partial enzyme defect in heterozygotes (Jolly *et al.* 1977). The effects of zinc supplementation, and of a natural 'lymphocyte transplant' are described under 'Treatment' below.

Mannosidosis has recently also been described in a domestic cat (Burditt *et al.* 1980*b*; Walkey *et al.* 1981).

In a neurodegenerative disorder in goats a major storage material in brain was Man-(1 → 4)-ß-GlcNAc-(1 → 4)-ß-GlcNAc (Jones and Laine

1981). As expected from this structure the animals were found to have a deficiency of ß-mannosidase (Jones and Dawson 1981) which had not previously been encountered.

Treatment

Attempts at stimulation of the residual acidic α-mannosidase activities by oral zinc sulphate supplementation in two patients with mannosidosis resulted in decreases in urinary oligosaccharide concentration, but not in detectable activation of residual enzyme in plasma, leucocytes or tears (Grabowski *et al*. 1977, 1980).

In a 12-month-old calf with mannosidosis, zinc sulphate supplementation by rumen tube over a four-day period was accompanied by a small increase in liver and leucocyte mannosidase activities. In longer term experiments three affected calves were fed zinc-treated milk. In liver and kidney there was a modest increase in α-mannosidase activity and decrease of storage-type oligosaccharides. On histopathological examination, no differences attributable to mannosidosis were found in brain, lymph node, or liver between zinc-supplemented and control mannosidosis calves (Jolly *et al*. 1980). In a chimaeric male calf with mannosidosis which had obtained a natural lymphocyte transplant from its female twin as a result of shared placentas, and which had 70 per cent female peripheral blood lymphocytes, there was considerable reduction in degree of pathological changes but the progressive neurological course was not arrested (Jolly *et al*. 1976).

Fucosidosis

Two siblings with dysmorphic Hurler-like features were described by Durand *et al*. (1966), and first shown to have α-fucosidase deficiency by Van Hoof and Hers (1968). Type 1 (infantile) and type 2 (juvenile and adult) forms have been reported, but more genetic heterogeneity seems likely from the diverse clinical phenotypes. About one third of approximately 60 cases reported have been Italian, eight of whom came from two families in Calabria (Durand *et al*. 1982).

Clinical features

In the type 1 or infantile form, (Durand *et al*. 1969; Larbrisseu *et al*. 1978; Alhadeff *et al*. 1978*b*) progressive psychomotor retardation is usually noted in the first eight months, commencing with hypotonia and progressing to spasticity, decerebrate rigidity, and death by the sixth year. Convulsions are common. Clinical features include Hurler-like facies, hepatosplenomegaly, cardiomegaly, and dorsolumbar kyphosis. Recurrent respiratory infections and muscular hypotonia may occur in the first year. Other patients with infantile onset have had coarser Hurler-like features, including the facial appearance and dysostosis multiplex (Lee *et al.* 1977), cardiomegaly (Durand *et al.* 1982), but no corneal clouding. Radiologically, beaking of upper lumbar vertebrae and dorsolumbar kyphosis are similar to those of Hurler's disease.

In type 2, onset has been reported from five years (Borrone *et al.* 1974; McPhee *et al.* 1975) to 18 years of age (Patel and Zeman 1976). The salient features are psychomotor retardation, often presenting with difficulty in walking progressing to inability to stand, coarse facial features, dysostosis multiplex, and angiokeratoma corporis diffusum (as in patients with Fabry's disease), with juvenile or adult onset (Kornfeld *et al.* 1977).

Other clinical findings have included tortous conjunctival vessels (Snyder *et al.* 1976) and minimal hepatosplenomegaly. Clinical heterogeneity has been reported in the same family (Romeo *et al.* 1977; Durand *et al.* 1979). Radiographs show thickness of the skull, flattening and irregularity of vertebral bodies, spatulate ribs, shallow acetabular cavities, fragmented femoral head epiphyses (Borrone *et al.* 1974), beaking of upper lumbar vertebrae, and hypoplasia of the odontoid peg (Landing *et al.* 1976).

Pathology

A characteristic feature is the widespread occurrence of cytoplasmic intralysosomal inclusions and vacuoles. They are common in peripheral blood lymphocytes and bone marrow histiocytes (Durand *et al.* 1967; Loeb *et al.* 1969) and hepatocytes. They are intralysosomal (Durand *et al.* 1967) and contain both carbohydrates (PAS-positive) and lamellated lipids (osmic acid positive) but not mucopolysaccharides (i.e. do not produce metachromasia with Alcian blue). The brain has scarce neurones with membrane-bound vesicles containing granular or lamellar material. Oligodendrocytes and astrocytes also contain cytoplasmic vacuoles or granules which stain positively for carbohydrate and lipids (Loeb *et al.* 1969; Durand *et al.* 1967, 1969).

The ultrastructure of angiokeratoma lesions has been studied by Kornfeld *et al.* (1977) who found storage vesicles which were clear, or contained granular and lamellar structures in various cell types including fibroblasts, sweat gland cells, and endothelial cells of blood vessels.

Molecular defects

There is severe deficiency of α-L-fucosidase in leucocytes (Landing *et al.* 1976; Borrone *et al.* 1974; McPhee *et al.* 1975), fibroblasts (McPhee *et al.* 1975; Landing *et al.* 1976; Beratis *et al.* 1977), serum (Borrone *et al.* 1974), brain, liver, and kidney (Van Hoof and Hers 1968; Loeb *et al.* 1969). Residual activity has been reported as varying from absent in fibroblasts and leucocytes (McPhee *et al.* 1975) to about 10 per cent of normal in leucocytes (Borrone *et al.* 1974).

No increase in fucosidase activity was found after hybridization of type 1 and type 2 fibroblasts (Beratis *et al.* 1977). No cross-reacting material to anti- α -fucosidase serum was detected in fucosidosis liver (Alhadeff *et al.* 1975, 1978*a*; Thorpe and Robinson 1978) suggesting either absence of mutant enzyme or an alteration of its immunologically specific site. The storage material has been characterized as L-fucose-containing glyco-peptides in brain (Loeb *et al.* 1969) and as pentaglycosylceramide in liver (Dawson and Spranger 1971). Since this has a similar structure to two blood-group-active substances, namely H antigen and Lewis A antigen, Dawson (1972) suggested that the accumulating sphingolipid has a structure similar to the H antigen, which is a common catabolite of the A and B antigens. This would explain why patients with fucosidosis have increased antigenicity of blood group substances (Borrone *et al.* 1974). A fucose-containing decasaccharide, and to a lesser extent a disaccharide, are the main storage molecules in brain (Tsay and Dawson 1975). L-fucose is most commonly found in the non-reducing terminal of glycoproteins, oligosaccharides, and glycolipids (Kornfeld and Kornfeld 1973).

Cultured fucosidosis fibroblasts take up purified human placental α -fucosidase from culture medium (Turner *et al.* 1979).

An unexplained observation is that some patients with type 1 have a considerable increase in sweat sodium chloride concentration which returns to normal terminally (Durand *et al.* 1969; Alhadeff *et al.* 1978*b*), and a more modest increase in tears (Durand *et al.* 1969). Rarely, sweat sodium chloride is increased in type 2 (Landing *et al.* 1976).

Diagnosis

The diagnosis can be established by demonstrating severe deficiency of α -fucosidase in leucocytes using the methylumbelliferyl derivative as substrate (Van Hoof and Hers 1968; Borrone *et al.* 1974; Di Matteo *et al.* 1976; Patel and Zeman 1976) or in cultured fibroblasts (Borrone *et al.* 1974; Di Matteo *et al.* 1976). In the latter, tissue α-fucosidase must be assayed under carefully controlled conditions, since activity is sensitive to population cell density (Heukels-Dully and Niermeijer 1976). Serum fucosidase activities have a wide range in normal adults and should not be relied upon for diagnosis (Ramage and Cunningham 1975; Ng *et al.* 1976), but reduced values have been reported in patients (Patel and Zeman 1976; Borrone *et al.* 1974; Di Matteo *et al.* 1976).

Genetic counselling and prenatal diagnosis

These disorders have an autosomal recessive mode of inheritance, the parents having lower average α-fucosidase activities than controls in leucocytes (Beratis *et al.* 1975*b*; Ng *et al.* 1976), lymphocytes (Bugiani 1982) and in cultured fibroblasts (Beratis *et al.* 1977), but discrimination is incomplete (Gatti *et al.* 1980). This group of disorders is rare except in Calabria in southern Italy (Durand *et al.* 1982). The incidence has not been established. Prenatal diagnosis for fucosidosis is complicated by the marked variation in enzyme activity according to the conditions of culture. Thus immediately after sub-culturing of confluent cultures, normal amniotic cells have high activities that fall after a few days, to rise again as cells become confluent. Moreover, epithelial cells, which initially may represent a high proportion of cultured amniotic cells, have been reported to have higher fucosidase activity than fibroblast-like cells which are more typical of established amniotic cell cultures (Butterworth and Guy 1977).

Prenatal diagnosis in a pregnancy at risk for fucosidosis was reported by Matsuda *et al.* (1975), who found cultured amniotic epithelial cell fucosidase to be 30 per cent of two control cultures and predicted a heterozygous foetus. The pregnancy however ended with the birth of undiagnosed twins, both of whom were affected. It may be speculated that the error arose by assay of epithelial-like cells which have α-fucosidase activity four to six times higher in the initial culture than the fifth passage (Poenaru *et al.* 1976) from the pregnancy at risk, or by assay of non-confluent control cells. Correct prenatal diagnosis has been reported by Poenaru *et al.* (1976), Durand *et al.* (1979), and Bugiani (1982).

Human α-fucosidase shows genetic polymorphism with three common phenotypes, Fu1, Fu2, and Fu2–1, representing homozygosis or heterozygosis for two autosomal co-dominant alleles, Fu^1 and Fu^2 with respective frequencies in New York whites of 0.753 and 0.247, and among blacks of 0.926 and 0.074 (Turner *et al.* 1975; Harris 1980). Up to six discrete isozymes have been detected in leucocytes and liver. Since in fucosidosis there is deficiency of α-fucosidase activity in all electrophoretic isozymes (Alhadeff *et al.* 1974), they must be determined by a single gene locus. This has been assigned to chromosome 1 by analysis of somatic cell hybrids (Turner *et al.* 1978).

The α-L-fucosidase (α-FUC) locus is closely linked to the Rhesus blood group on band 32 –band 34 of the short arm of chromosome 1. Based on familial segregation studies, Cook *et al.* (1977) suggested that the map of 1p is 1 pter-PGD-(6-phosphogluconate dehydrogenase)-E-1-(ellipto-cytosis)-Rh-α-FUC-PGM-1 (phosphoglucomutase-1)-centromere. The Goss and Harris (1977) method of studying recombination in families, and synteny tests in hybrid cells suggests the sequence of loci to be PGM-1-UMPK (uridine monophosphate kinase)-AK-2(adenylate kinase-2)-α-FUC-ENO-1-(enolase-1).

Aspartylglycosaminuria

The first report of aspartylglycosaminuria (AGU) was in a severely retarded adult brother and sister in whose urine Pollitt *et al.* (1968) found an unknown ninhydrin-positive compound. On hydrolysis it gave aspartic acid, ammonia, and glucosamine, in equimolar amounts. Identification of the compound as 2-acetamido-1-(β^1-L-aspartamido)-1,2-dideoxy-ß-D-glucose (AADG; Fig. 3.6) was achieved by infra-red spectrophotometric studies and chromatographic and elemental analysis (Jenner and Pollitt 1967). During the four years following the first report, 34 Finnish cases were reported by Autio (1972).

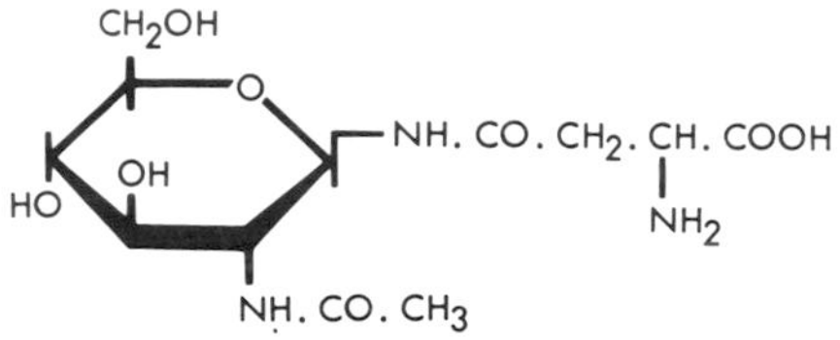

Fig. 3.6. The structure of 2-acetamido-1-(ß-L-aspartamido)-1,2-dideoxy-ß-D-glucose (AADG).

Clinical features

During the first three months the affected infants appear normal but then develop recurrent respiratory infection, diarrhoea, and hernias. Delayed speech development is a common early symptom, in addition to pro-gressive global mental retardation which is mild initially but severe by school age. Mild dysmorphic facial coarseness is noticeable during the first few years, and becomes progressively more marked, with wrinkled facial skin. A hoarse deep voice first noticed in the first decade is characteristic. Adult speech is severely defective or lacking. Growth is initially normal but falls below the 10th percentile by 10 years. Other features are systolic murmurs, macroglossia, crystalline lens opacities, hypotonia, and joint hypermobility. Radiologically, there are progressive changes which become more obvious at seven to ten years. The skull is thickened, the vertebrae have irregular outlines and the long bones show cortical thinning. Fractures occur with slight trauma.

   The oldest known patient is 52 years. Ten of the Finnish patients have died aged 27–46 years, eight from pulmonary infections or empyema, one from drowning, and one from peritonitis due to a perforated ulcer (Aula *et al.* 1982*a*).

Pathology

Peripheral blood lymphocytes show numerous cytoplasmic vacuoles, which on electron microscopy are identified as lysosomes distended with storage material (Aula *et al.* 1975*b*).

In brain, liver, and kidney there are distended lysosomes containing storage material (Arstila *et al.* 1972). However frozen sections of liver do not usually stain with histochemical reagents such as oil red-O, Sudan black-B, Alcian blue or PAS (Aula *et al.* 1982*a*). The endoplasmic reticulum appears dilated, and the Golgi apparatus small (Arstila *et al.* 1972; Isenberg and Sharp 1975).

Molecular defects

Severe deficiency of 1-aspartamindo-ß-*N*-acetylglucosamine amidohydrolase (AADGase), an enzyme which catalyses the release of *N*-acetylglucosamine from AADG, was first reported in seminal fluid (Pollitt *et al.* 1968) and plasma (Pollitt and Jenner 1969). This enzyme normally cleaves the linkage between *N*-acetylglucosamine and asparagine, and therefore between carbohydrate and protein of glycoproteins, and probably initiates catabolism of the oligosaccharide chain of glycoproteins. Substantial residual activity was found in brain, liver, and spleen, and normal activity in kidney by Palo *et al.* (1972), but reduced activity in kidney was reported by Dugal (1977). AADGase deficiency can be demonstrated in cultured fibroblasts (Aula *et al.* 1974), lymphocytes, and serum (Aula *et al.* 1976). The storage material in liver and brain is mainly AADG (Fig. 3.6) (Palo and Savolainen 1973). In lymphocytes also the main storage material has been shown to be AADG using a highly sensitive gas-chromatographic-mass-spectrometric technique (Maury and Palo 1980).

In addition to AADG, several minor aspartyloligosaccharides accumulate in liver (Maury 1979), neural tissue, spleen, kidney, and thyroid (Maury 1980), with the following structure

$$\text{ß-Gal-}(1 \rightarrow 4)\text{-ß-GlnNAc-Asn} \quad (1)$$
$$\alpha\text{-NANA-}(2 \rightarrow 3)\text{-ß-Gal-}(1 \rightarrow 4)\text{-ß-GlcNAc-Asn} \quad (2)$$
$$\alpha\text{-NANA-}(2 \rightarrow 4)\text{-ß-Gal-}(1 \rightarrow 4)\text{-ß-GlcNAc-Asn} \quad (3)$$
$$\alpha\text{-Man-}(1 \rightarrow 6)\text{-ß-Man-}(1 \rightarrow 4)\text{-ß-GlcNAc-}(1 \rightarrow 4)\text{-ß-GlcNAc-Asn} \quad (4)$$

Diagnosis

The diagnosis can be made by demonstrating an excess of AADG and higher molecular weight glycoasparagines in the urine by high voltage electrophoresis (Pollitt *et al.* 1968; Autio 1972) or by a combination of Sephadex and ion-exchange chromatography. Aula *et al.* (1982*a*) have found two-dimensional thin-layer chromatography most convenient. Confirmation should be sought by demonstrating reduced activity of

AADGase in lymphocytes (Aula *et al.* 1976) or cultured fibroblasts (Aula *et al.* 1974) using AADG as substrate and measuring liberated *N*-acetylyglucosamine by Ehrlich's reaction.

Genetic counselling and prenatal diagnosis

The disease has an autosomal recessive mode of inheritance. There appears to be good discrimination between carriers and either normal or affected individuals by measuring AADGase activity in lymphocytes (Aula *et al.* 1976) or cultured fibroblasts (Aula *et al.* 1974). There is a high incidence in Finland where by 1982 138 patients had been diagnosed from 108 families (Aula *et al.* 1982*a*). Prenatal diagnosis of an affected foetus by enzyme assay of cultured amniotic cells has been reported by Aula *et al.* (1984). The foetus was aborted at 20 weeks' gestation and the diagnosis confirmed by demonstration of AADGase deficiency in cord blood foetal lymphocytes and cultured foetal skin fibroblasts and chorionic villus cells.

4. The sphingolipidoses and other lipid storage disorders

THE GANGLIOSIDOSES

Tay–Sachs disease; GM$_2$-gangliosidosis type 1

First described by a British ophthalmologist, Tay (1881) and an American neurologist, Sachs (1896), this rapidly progressive disease leads to decerebrate rigidity, blindness and death, usually by three years of age. It is the commonest inborn error of metabolism diagnosed prenatally, and at the time of writing the only one in which detection of carriers by mass population screening allows identification of couples at risk, and by subsequent prenatal diagnosis, prevention of even one affected child.

Clinical features

Affected infants appear normal at birth, but by six months of age have become unusually lethargic, floppy, and poor feeders. A characteristic sign is an abnormally accentuated startle reflex to sudden sharp noises. Motor development, which is initially within normal limits, becomes progressively retarded. Thus head raising in prone and supine positions occurs normally, but sitting and especially standing are delayed, while walking is rarely achieved. Cherry-red spots in the centre of the macula surrounded by pale, almost white areas develop in the great majority of, but not all, patients. After about 12 months the head becomes abnormally large, there is blindness, progressive deafness, spasticity, and decerebrate rigidity. Feeding becomes increasingly difficult. Death occurs usually by three years of age from pulmonary infection.

Pathology

Specific pathological changes are limited to the nervous system. Neuronal lipidosis with ballooning of cytoplasm with osmiophilic materials occurs in the cerebrum, cerebellum, and the autonomic nervous system including the rectal mucosa. There is demyelination and decrease in the number of axons both in the cerebrum and cerebellum. Cerebral cortical gliosis is widespread and is the cause of macrocephaly. Neuronal storage material is localized in distended lysosomes and exhibits acid phosphatase activity (Wallace *et al.* 1964). In electronmicrographs the storage material appears as closely packed concentric layers of membrane.

Molecular defects

In Tay–Sachs disease there is a severe deficiency of ß-*N*-acetylhexo-saminidase A, which catalyses cleavage of the terminal *N*-acetyl-ß-D-galactosamine from the ganglioside GM_2 (Okada and O'Brien 1969) (Fig. 4.1). This defect was predicted by Svennerholm (1962) and Brady (1966) from the structure and catabolism of the storage material GM_2. The discovery of the enzyme defect was made possible by the demonstration of two hexosaminidase components in human spleen (Robinson and Stirling 1968). The two hexosaminidase isoenzymes designated A (hex A) (which is defective in Tay–Sachs disease) and B (hex B) are both active against molecules containing ß-D-*N*-acetylglucosamine and ß-D-*N*-acetylgalac-tosamine moieties, and are both lysosomal hydrolases. They can be separated by ion-exchange column chromatography or electrophoresis (Fig. 4.2). A third weakly-acting brain isoenzyme, hexosaminidase C, is not lysosomal, is active only as a *N*-acetylglucosaminidase, has a neutral pH optimum (Braidman *et al.* 1974), and is not implicated in the sphingolipidoses.

Three other forms of acidic hexosaminidases have been identified. Two have electrophoretic forms intermediate to the major forms hex A and hex B, and have been designated I_1 and I_2 (Price and Dance 1972; Ellis *et al.* 1975) and may represent heterogeneity in the carbohydrate component of hex B (Beutler and Kuhl 1975). Form I_2 is present in serum and increases in activity during pregnancy, and has also been called the P-form (Stirling 1972; Swallow *et al.* 1974). The fifth acidic form hex S has a faster anodal electrophoretic mobility than hex A and cross-reacts with anti-hex A serum (Geiger *et al.* 1977). It is inactive in Tay–Sachs disease (Reuser and Galjaard 1976), but active in Sandhoff's disease (see below).

Antibodies against purified hex B cross-react against hex A (Carroll and Robinson 1973; Srivastava *et al.* 1974), indicating a common subunit designated ß. Geiger and Arnon (1976) and Lee and Yoshida (1976) have purified hex A and hex B from human placenta. Each isoenzyme could be

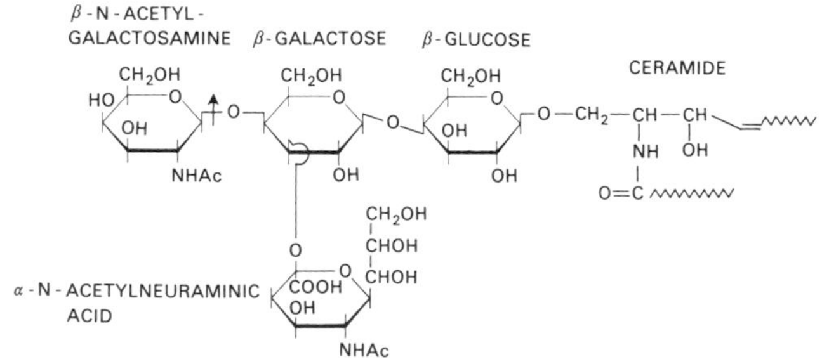

Fig. 4.1. Ganglioside GM_2.

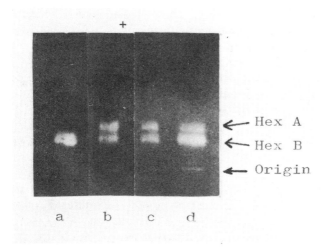

Fig. 4.2. Cultured fibroblast and amniotic cell ß-hexosaminidase A and B separated by cellulose acetate electrophoresis. Gels were stained with 4MU ß-*N*-acetylglucosamine at pH 4.4. (a) Tay–Sachs disease fibroblasts (b) and (c) Normal amniotic cells (d) At-risk amniotic cells where the foetus was a heterozygote.

dissociated into four subunits with a molecular weight of approximately 25 000. Hex B had a structure β_4 while hex A had an equal proportion of ß and another polypeptide designated α, consistent with the structure $\alpha_2\beta_2$. Thus Tay–Sachs disease can be explained by a mutation of the α polypeptide. More recently the number of subunits in hex A has been questioned, and a structure of $\alpha\beta_2$ (Mahuran and Lowden 1980) or $\alpha_n\beta_2$ (Warner and O'Brien 1982) has been suggested.

In Tay–Sachs disease cross-reacting material has been demonstrated to anti-A and anti-B, possibly representing the presence of common antigen, but no hex A antigen has been demonstrated in liver (Carroll and Robinson 1973; Srivastava and Beutler 1974; Bartholomew and Rattazzi 1974).

RNA from cultured fibroblasts from four patients with Tay–Sachs disease (three of Ashkenazi descent, and one non-Jewish), did not direct the translation of immunoprecipitable α-chain in a rabbit reticulocyte system, in contrast to RNA from control cells. However, RNA from the fibroblasts of another non-Jewish Tay–Sachs patient did synthesize the α-chain. This chain was abnormal, however, since it was relatively insoluble, and could be labelled with ^{3}H-mannose (normally occurring in the endoplasmic reticulum), but not with ^{32}P-phosphate (normally occurring in the Golgi apparatus). The authors speculate that this insoluble α-chain is not transported to the Golgi apparatus, nor to lysosomes or the exterior of the cell (Proia and Neufeld 1982). These elegant studies, therefore, suggest the existence of genetic heterogeneity among non-Ashkenazi patients with Tay–Sachs disease.

Using hamster–human hybrids, Hoeksema *et al.* (1977) and Gilbert *et al.* (1975a) have assigned the gene for the α subunit to chromosome 15 and the ß subunit to chromosome 5.

The observations cited above − that hex S activity is absent in Tay–Sachs disease, and that hex S (which has a similar molecular weight to hex A and hex B of 100 000) cross-reacts with anti-hex A, but not anti-hex B − are all consistent with an α_4 structure for hex S.

In addition, a low molecular weight glycoprotein stimulatory factor or activator is necessary for GM_2-ß-hexosaminidase activity but not for hexosaminidase activity against synthetic substrates (Conzelman and Sandhoff 1978).

The storage material in Tay–Sachs disease brain is ganglioside GM_2 which is increased 100–300 times normal, and the asialo derivative GA_2 which is increased by about 20 times the normal value. In viscera the total ganglioside content is not significantly increased, but the proportion of GM_2 is higher than normal (O'Brien 1983).

Diagnosis

The diagnosis of Tay–Sachs disease is established by demonstration of hex A deficiency in serum (O'Brien *et al.* 1970), leucocytes, cultured fibroblasts (Okada and O'Brien 1969; Okada *et al.* 1971), saliva, or tears (Singer *et al.* 1973; Goldberg *et al.* 1977). Since hex B is more heat stable than hex A, the latter can be demonstrated as the proportion of heat labile ß-hexosaminidase after a standardized heat inactivation (e.g. 3h at 50°C) using 4-methylumbelliferyl-2-acetamido-2-deoxy-ß-D-glucopyranoside as substrate. Alternatively, hex A can be separated prior to assay using electrophoresis on starch gel (Swallow *et al.* 1976) or cellulose acetate (Galjaard *et al.* 1974a) or by DEAE-cellulose column chromatography (Dance *et al.* 1970).

Genetic counselling and prenatal diagnosis

Tay–Sachs disease has an autosomal recessive mode of inheritance. Carriers can be detected by demonstrating reduced serum hex A activity. In borderline subjects the leucocyte enzyme is also assayed. The probability that the test would miss a carrier has been estimated as less than 1 in 30 000 (Kaback *et al.* 1973a, 1974).

In women taking oral contraceptives (Kaback *et al.* 1973a) or who are pregnant (Lowden and La Ramee 1972), the serum hex A activity falls to heterozygote values, but the leucocyte activities are not reduced and can be used for carrier detection.

The disease occurs approximately one hundred times more commonly in Ashkenazi Jews from the Balkan provinces than in other populations. In the United States, carrier frequencies have been estimated as 0.026 for Ashkenazis and 0.0029 for non-Jews, and homozygote frequencies as 0.0027 and 0.000028, respectively (Kozinn *et al.* 1957; Aronson and Volk 1962; Aronson 1964; Myrianthopoulos and Aronson 1966). Estimates for homozygotes from non-Jewish populations may be up to 100 per cent too high since they include at least some cases of Sandhoff's disease. The reason for the high gene frequency in Ashkenazi Jews is not known; two mechanisms have been suggested. Firstly, resistance to pulmonary tuberculosis among ancestor carriers has been proposed as a heterozygote advantage (Myrianthopoulos and Aronson 1966). Such a mechanism should cause a continued increase in the gene frequency only in populations exposed to tuberculosis. Secondly, a founder effect (Chase and McKusick 1972) has been suggested. This implies that the present high gene frequency amongst Ashkenazis is a transient phenomenon due to chance occurrence of the mutant gene later in history from a founder originating from an Ashkenazi community. This model predicts that there will be a decrease in the gene frequency in the future.

Mass screening for Tay–Sachs disease couples amongst Ashkenazis has been organized (Kaback 1981). Up to June 1980, 312 214 individuals were screened, 12 763 carriers detected, and 268 couples at risk identified. In these 252 pregnancies were monitored by diagnostic amniocentesis while a further 562 pregnancies were monitored because of the birth of a previous child with Tay–Sachs disease. A total of 175 affected foetuses were detected, 168 of which were aborted. Of 150 newborn or abortuses examined, all predictions were accurate, except for one false negative.

Tay–Sachs disease is the commonest genetic enzyme defect diagnosed prenatally (Milunsky 1973; Galjaard 1980*b*) but considerable care is required to avoid errors in view of possible variant alleles which have been reported and because the enzyme is moderately unstable. Two incorrect diagnoses, neither by assay of cultured cells, have been a false negative (Ellis *et al.* 1973) and a false positive (Rattazzi and Davidson 1972). The former may have been due to loss of hex B activity in dilute homogenates. Addition of albumin can diminish such loss (Okada *et al.* 1971).

Prior to prenatal diagnosis enzyme assay should be carried out on cultured fibroblasts from patients in order to determine that they are carriers of classical Tay–Sachs disease, rather than of a variant (see below). Assay should then be carried out on amniotic fluid, and repeated on cultured amniotic cells (O'Brien 1977, 1983). In addition to determination of hex A by the heat denaturation method we routinely test for the presence of the 'A' component by cellulose acetate electrophoresis (Fig. 4.2).

Sandhoff's disease; GM$_2$-gangliosidosis type 2

Clinical and pathological features

The clinical phenotype, course, and pathological changes in the nervous system are identical to those of Tay–Sachs disease. In Sandhoff's disease, however, there is some extraneurological involvement in that there are a few vacuolated lymphocytes, and vacuolated histiocytes in bone marrow, spleen, lymph nodes, and lungs; and sometimes prominent vacuolation of the epithelial cells of the loops of Henle. On electron microscopy, these cells are shown to contain cytoplasmic, lamellar inclusions (Sandhoff and Harzer 1973).

Molecular defects

In Sandhoff's disease both hex A and hex B are deficient in all tissues examined. As described under 'Molecular defects' for Tay–Sachs disease, immunological and biochemical studies indicate that hex A and hex B have a common peptide subunit. Sandhoff's disease therefore can be explained by a mutation affecting the ß gene, causing deficiency of both hex A ($\alpha_2\beta_2$) and hex B (β_4). Immunological studies in tissues of patients with Sandhoff's disease suggest genetic heterogeneity. Thus, in several patients material has been demonstrated which cross-reacts with anti-hex B-antibody and which has similar electrophoretic mobility to hex A and B (Srivastava and Beutler 1974), but in other single patients either hex A or hex B antigens have been demonstrated (Carroll and Robinson 1973). RNA from cultured fibroblasts from two patients with Sandhoff's disease, in contrast to RNA from control cells, did not direct the translation of immunoprecipitable ß chain in a rabbit reticulocyte system (Proia and Neufeld 1982). The gene has been assigned to chromosome 5 (Hoeksma et al. 1977).

Confirmation that Tay–Sachs and Sandhoff's diseases result from different gene mutations was obtained by demonstrating the appearance of hex A in Sandhoff–Tay–Sachs fibroblast heterokaryons, which was absent in either cell type before fusion (Thomas et al. 1974; Rattazzi et al. 1976; Galjaard et al. 1974a).

In Sandhoff's disease the storage material in brain contains more GA$_2$ and less GM$_2$ than in Tay–Sachs disease and also contains globoside (Fig 4.3) (Sandhoff and Harzer 1973). In contrast to Tay–Sachs disease however, GA$_2$, GM$_2$, and globoside are also increased in liver, spleen, kidney, and plasma (Suzuki et al. 1971a; Krivit et al. 1972) and in urine (Krivit et al. 1972).

Diagnosis

The diagnosis of Sandhoff's disease can be made by demonstrating

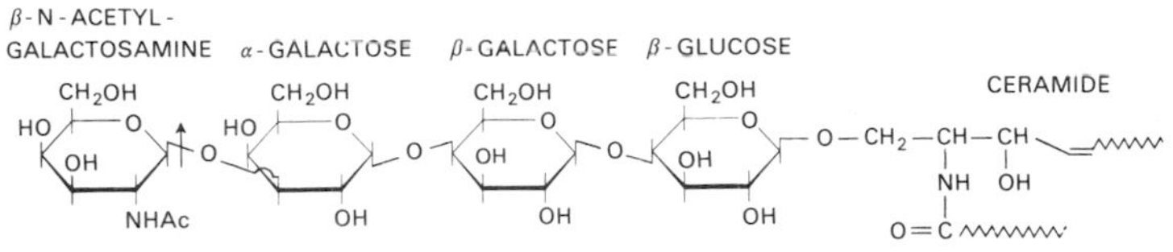

β-N-ACETYL-
GALACTOSAMINE   α-GALACTOSE   β-GALACTOSE   β-GLUCOSE

Fig. 4.3. Globoside.

deficiency of total ß-hexosaminidase in serum (Sandhoff and Harzer 1973; Okada *et al.* 1972) or cultured fibroblasts (Okada *et al.* 1972) using a methylumbelliferyl substrate (Okada *et al.* 1972). A low level of residual activity is due to the presence of hex S (Swallow *et al.* 1976).

Genetic counselling and prenatal diagnosis

Sandhoff's disease has an autosomal recessive form of inheritance. Reduced ß-hexosaminidase activities have been demonstrated in serum or leucocytes of parents of index cases (Desnick *et al.* 1973*a*; Harzer *et al.* 1975). Prenatal diagnosis has been reported in several pregnancies (Desnick *et al.* 1973*a*; Harzer *et al.* 1975*a*; Kleijer *et al.* 1978). The most reliable methods involve total ß-hexosaminidase assay of cultured amniotic cells. Amniotic fluid activity is also reduced but this finding should always be confirmed by subsequent assay of cultured cells. In the first report of an affected pregnancy, amniotic fluid globoside was 18 times higher than in controls (Desnick *et al.* 1973*a*).

Tay–Sachs disease variant AB(+); deficiency of a non-enzymic activator protein

Several patients have been reported (Kolodny *et al.* 1973; De Baecque *et al.* 1975; Conselmann and Sandhoff, 1978) with the clinical phenotype of Tay–Sachs disease and cerebral accumulation of GM_2, who had normal hex A and hex B activities when tested with synthetic substrates, but no activity when GM_2 was used as the substrate. Kidney extracts from the patient reported by Conselmann and Sandhoff (1978) showed absence of an activator protein. This protein can be demonstrated in tissue extracts from normal individuals or from patients with classical Tay–Sachs or Sandhoff's disease and activates purified hex A-catalysed degradation of GM_2 (Li *et al.* 1981). The activator consists of two components of molecular weights 25 000 and 60 000. The smaller component was absent in the patient reported by Conselmann and Sandhoff (1978).

Healthy adults with deficiency of hex A

In three families there was deficiency of hex A in fibroblasts (Navon *et al.*

1973; Vidgoff *et al.* 1973; O'Brien *et al.* 1978), when measured with synthetic substrates, in healthy adults. However, using GM_2 as substrate there was about 50 per cent of normal hex A activity in leucocytes (Tallman *et al.* 1974) and in fibroblasts (O'Brien *et al.* 1978; O'Brien and Geiger 1979).

In all these families, women who had hex A deficiency had given birth to children with Tay–Sachs disease. It has been suggested that the adults with normal phenotypes and hex A deficiency are double heterozygotes at the Tay–Sachs (i.e. α-peptide) locus, one allele carrying the classical Tay–Sachs mutation, and the other a mutation which leads to a product that lacks activity against synthetic substrates, but retains activity against GM_2. This explanation is supported by enzyme activity studies in a Pennsylvania Dutch family where genes for Tay–Sachs disease and of a mutant allele show allelic segregation as predicted by the hypothesis (Kelly *et al.* 1976).

A 43-year-old woman from this family, and an unrelated six-month-old healthy male from an Ashkenazi family were studied in considerable detail by Thomas *et al.* (1982). They concluded that these individuals probably have a physiologically normal genetic condition manifested by absence of hex A activity in serum, and a relative preservation of activity against both GM_2 ganglioside and 4-methylumbelliferyl-N-acetyl-ß-glucosaminide. Antibodies to the α-chain of hex A precipitate the protein which has GM_2-N-acetylgalactosaminidase activity but is without activity against synthetic substrates (O'Brien and Geiger 1979).

Juvenile GM_2-gangliosidosis; GM_2-gangliosidosis type 3

Clinical features

Ataxia and progressive spasticity are the common presenting features with onset between two and six years of age. There is intellectual deterioration, sometimes with psychotic changes. As the disease progresses, there is increasing motor deterioration, and sometimes there are epileptic fits. Flexion deformities develop, but visual acuity, unlike in Tay–Sachs and Sandhoff's diseases, is retained until late. Death between five and 15 years is usually due to pulmonary infections (Suzuki *et al.* 1970*a*; Menkes *et al.* 1971; Brett *et al.* 1973).

Pathology

Specific pathological changes are limited to the nervous system. Cytoplasmic inclusions (Suzuki *et al.* 1970*a*; Menkes *et al.* 1971) and lamellar bodies in neurones resemble those of Tay–Sachs and Sandhoff's diseases.

Molecular defects

Deficiency of hex A in juvenile GM_2-gangliosidosis has varied from partial

(Suzuki *et al.* 1970*a*; Okada *et al.* 1970) to almost complete (Brett *et al.* 1973). In one patient activity was much higher against synthetic substrate (41–43 per cent of normal) than against GM_2, where it was undetectable (Zerfowski and Sandhoff 1974).

The fundamental defect is probably a mutation of the α-subunit allele different from that causing Tay–Sachs disease.

Diagnosis

In view of the residual activity of hex A, which can be as high as 30–40 per cent of normal in serum, leucocytes, and cultured fibroblasts, the most reliable assay for diagnosis employs GM_2 as substrate (Zerfowski and Sandhoff 1974).

Genetic counselling and prenatal diagnosis

Autosomal recessive inheritance is suggested by the demonstration of a partial hex A defect in parents (Brett *et al.* 1973; Okada *et al.* 1970). Carrier testing and prenatal diagnosis should be carried out by enzyme assays using GM_2 as substrate.

Adult GM_2-gangliosidosis

Two adult Ashkenazi sibs with GM_2-gangliosidosis have been reported (Rapin *et al.* 1976). Onset was in childhood with a slowly progressive spinocerebellar degeneration with impairment of walking and posture. There was atrophy of voluntary muscles, foot drop, pes cavus, dysarthria, and ataxia. Vision was normal and there were no cherry-red spots. Intelligence was normal. There was severe deficiency of hex A in serum and leucocytes. Neuronal lipidosis and elevated GM_2 levels were found in an affected sister who had died at 16 years of age. A third adult patient with absent serum and fibroblast hex A activities was reported by Kaback *et al.* (1978). He developed leg cramps at 16 years of age, followed by ataxia, dysarthria, spasticity, fibrillation of voluntary muscles, and proximal muscle wasting at 18 years. There were no cherry-red spots. Both parents had partial deficiency of serum and fibroblast hex A activities.

GM_1-gangliosidosis type 1; generalized gangliosidosis

Clinical features

Apathy, oedema of the extremities, laboured, noisy respiration, and feeding difficulties within a few weeks or months of birth are common presenting features. Motor development is manifestly retarded with delay in supine head raising. Four cases known to the authors failed to sit unsupported or to crawl. Hepatosplenomegaly occurs early. The facies become coarse, and the gums hypertrophy. Dorsolumbar kyphosis occurs in some cases. Macular cherry-red spots are present in about half the patients. As in Tay–Sachs disease, the startle reflex to sudden noises may be accentuated and there may be macrocephaly. Faint clouding of the cornea and cardiomegaly without valvular involvement occurs in some patients (Benson *et al.* 1976*a*; Babarik *et al.* 1976). There is progressive neurological deterioration with blindness, deafness, convulsions, and spasticity, leading to death, usually in the first two years (O'Brien 1983). Radiologically, there is progressive generalized dysostosis multiplex with beaking of the upper lumbar vertebrae, periostial thickening at the shafts of the long bones, and spatulate ribs (Spranger *et al.* 1974*b*).

Pathology

There is neuronal lipidosis in the central and autonomic nervous systems. Neurones are distended by intralysosomal storage material which is sudanophilic (lipid) and periodic acid–Schiff positive (carbohydrate). Glial cells are metachromatic (Suzuki 1968; Petrelli and Blair 1975). Lysosomal storage occurs in renal epithelium, hepatocytes and Kupffer cells. On electron microscopy both amorphous and lamellar structures are seen (Gonatas and Gonatas 1965; Roels *et al.* 1970). Cytoplasmic vacuolation and lysosomal storage occurs in visceral histiocytes of the reticulo-endothelial system, and in peripheral blood monocytes.

Molecular defects

There is severe deficiency of GM_1-ß-galactosidase activity, the enzyme which catalyses conversion of ganglioside GM_1 to GM_2. This has been demonstrated in brain, liver (Norden and O'Brien 1973), cultured fibroblasts, and leucocytes (Sloan *et al.* 1969*a*; Singer and Schafer 1970; Benson *et al.* 1976*a*). Residual activity is less than one per cent of normal when assayed against GM_1. There is also ß-galactosidase deficiency against galactose-containing glycoproteins (McBrinn *et al.* 1969). The ß-galactosidase which is deficient in GM_1-gangliosidosis is distinct from the ß-galactosidase catalysing cleavage of the terminal galactose from galactosyl ceramide, which is deficient in Krabbe's disease, or from lactosyl ceramide (Wenger *et al.* 1974*a*). The ß-galactosidase deficiency in GM_1-gangliosidosis can be demonstrated by using the artificial substrates *p*-nitrophenyl-ß-D-galactopyranoside (Gatt and Rapport 1966; Benson *et al.* 1976*a*) or 4-methylumbelliferyl-ß-D-galactopyranoside (Öckerman 1968; Benson *et al.* 1976*a*).

   Human liver ß-galactosidase consists of four main isoenzymes (Ho and O'Brien 1969; Robinson 1974; Cheetham *et al.* 1975). Three are lysosomal isoenzymes designated A_1, A_2 (formerly ß-galactosidase A) and A_3 (formerly B), and are deficient in GM_1-gangliosidosis, while the fourth is neutral. Antibodies against the main isoenzyme A precipitates A_1, A_2, and A_3 thus demonstrating that they contain a common peptide (Norden *et al.* 1974; Frost *et al.* 1978) but do not precipitate neutral ß-galactosidase (Norden *et al.* 1974) or galactocerebrosidase, confirming that they are different proteins. The observation that there is competitive inhibition between GM_1 and the methylumbelliferyl substrate suggests that they are cleaved at the same catalytic site as ß-galactosidase A (Norden *et al.* 1974).

The demonstration that in cells from affected patients there are approximately normal quantities of material which cross-react with anti-ß-galactosidase antibodies suggests the presence of a mutant enzyme (Meisler and Rattazzi 1974; Ben-Yoseph *et al.* 1977).

Chemical characterization of the storage substance has revealed about 30 types of galactose-containing glycoproteins (Strecker and Montreuil 1979), the commonest being an octasaccharide with the structure:

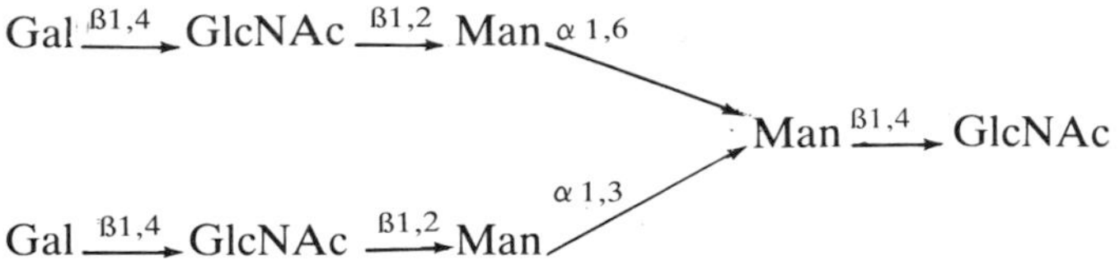

$$\text{Gal} \xrightarrow{\text{ß1,4}} \text{GlcNAc} \xrightarrow{\text{ß1,2}} \text{Man} \xrightarrow{\alpha 1,6}$$

$$\text{Man} \xrightarrow{\text{ß1,4}} \text{GlcNAc}$$

$$\text{Gal} \xrightarrow{\text{ß1,4}} \text{GlcNAc} \xrightarrow{\text{ß1,2}} \text{Man} \xrightarrow{\alpha 1,3}$$

(Wolfe *et al.* 1974), which may appear in urine (Ng and Wolfe 1975).

Storage material in the nervous system is mainly GM_1-ganglioside (Fig 4.4) and its asialated derivative GA_1 (Suzuki *et al.* 1971c; Berra *et al.* 1974). These gangliosides are increased to about ten times their normal levels in the cerebral cortex. GM_1 is 20–50 times normal in liver (O'Brien 1983).

Diagnosis

As stated above, in infants with the characteristic clinical phenotype the diagnosis can be made by demonstrating severe deficiency of ß-galactosidase in leucocytes or fibroblasts (Singer and Schafer 1970; Sloan *et al.*

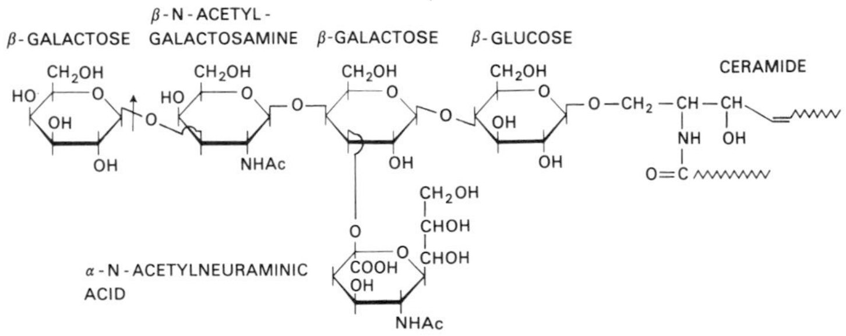

Fig. 4.4. Ganglioside GM_1.

1969*a*; Benson *et al.* 1976*a*) using the artificial methylumbelliferyl substrate. It should be noted, however, that in addition to generalized gangliosidosis several distinct disorders have been reported in patients with leucocyte and fibroblast ß-galactosidase deficiency. These include juvenile and adult types of GM_1-gangliosidosis, ß-galactosidase deficiency variants, Morquio's disease type B (MPS IVB), and the combined neuraminidase/ß-galactosidase deficiency. The theoretical relation of these phenotypes to different mutations of ß-galactosidase differentially affecting activities against varied substrates is discussed by O'Brien (1983) and below. It is important, therefore, when making a diagnosis to consider the enzyme defect in relation to clinical features (see p. 94).

Genetic counselling and prenatal diagnosis

The disorder has an autosomal recessive mode of inheritance. Partial enzyme deficiencies have been reported in carriers, but discrimination is incomplete in leucocytes (Singer and Schaffer 1970) and fibroblasts (Babarik *et al.* 1976; Suzuki *et al.* 1977). Prenatal monitoring for GM_1-gangliosidosis with correct prediction of unaffected foetuses was reported by Kaback *et al.* (1973*b*), and in three pregnancies by Kleijer *et al.* (1976*b*). An affected foetus was correctly diagnosed by Lowden *et al.* (1973) three weeks after diagnostic amniocentesis at 14 weeks' gestation. Cultured amniotic cell ß-galactosidase was one per cent of normal using the methylumbelliferyl substrate. At 17 weeks' gestation there was already marked evidence of abnormal lysosomal storage in spinal cord neurones. Two pregnancies, one affected and one unaffected, were reported by Kaback *et al.* (1973*b*).

The structural gene for the acid ß-galactosidases A_1, A_2, and A_3 has been assigned to the short arm of chromosome 3 (Naylor *et al.* 1980).

Juvenile GM_1-gangliosidosis; GM_1-gangliosidosis type 2

The juvenile form of GM_1-gangliosidosis differs from generalized GM_1-gangliosidosis by a later onset, a more slowly progressive course, and absence of hepatosplenomegaly, macular cherry-red spot, facial dysmorphism, and clinically evident skeletal involvement.

Clinical features

Normal development during the first year, followed by the appearance of locomotor ataxia and loss of manipulative skill is the usual presentation. The disorder was first reported by Derry *et al.* (1968) and reviewed by O'Brien (1972*b*, 1983) and Gilbert *et al.* (1975*b*). However, the age of onset has been variable and has been as late as 5–6 years in some cases (Patton and Dekaban 1971; Gilbert *et al.* 1975*b*).

The clinical course is one of slowly progressive neurological deterioration with loss of speech, internal strabismus, spasticity progressing to decerebrate rigidity and anticonvulsant-resistant fits. Recurrent pulmonary infections lead to death, usually between three and 10 years of age (Wolfe *et al.* 1970; O'Brien 1972*b*; Gilbert *et al.* 1975*b*).

Although the skeletal system is not clinically abnormal, radiologically there may be beaking of the lumbar vertebrae and deformity of the metacarpal and pelvic bones, which in one case were present at seven months of age, before the onset of neurological symptoms (O'Brien 1972*b*).

Pathology

There is lysosomal storage in neurones which is similar to that found in generalized GM_1-gangliosidosis. Involvement of renal glomerular epithelium is also similar to that seen in the infantile type, but visceral histiocytosis is less marked. Lymphocytes and bone marrow histiocytes are vacuolated. (Suzuki 1968; Wolfe *et al.* 1970; Mamelle *et al.* 1975).

Molecular defects

As for type 1, there is severe deficiency of ß-galactosidase and a normal quantity of cross-reacting material to anti-ß-galactosidase antibodies (Meisler and Rattazzi 1974; Ben-Yoseph *et al.* 1977). In most type 2 patients residual ß-galactosidase activity against GM_1-ganglioside in liver is two to five times higher than in those with type 1 (Norden and O'Brien 1975); and in cultured skin fibroblasts five to 30 times higher (Pinsky *et al.* 1974; O'Brien and Norden 1977).

Storage of GM_1 and GA_1 in cerebral cortex in type 2, as in type 1, is about 10 times higher than normal, but in contrast to type 1 is not raised in liver.

Genetic counselling and prenatal diagnosis

The disorder has an autosomal recessive mode of inheritance. A partial enzyme deficiency occurred in leucocytes and fibroblasts of both parents in one family (Benson and Babarik, unpublished observations) but insufficient data are available to assess the reliability of carrier detection. Failure to restore ß-galactosidase activity in heterokaryons produced by fusion of type 1 and type 2 cells suggests that the two disorders are due to non-complementing allelic mutations (Galjaard *et al.* 1975). Prenatal diagnosis of an affected foetus was reported by Booth *et al.* (1973) by assay of 4-methylumbelliferyl-ß-D-galactosidase activity of amniotic cells which had been cultured for four weeks following amniocentesis at 16 weeks' gestation.

GM₁-gangliosidosis in adults; GM₁-gangliosidosis type 3

A few patients have been reported with clinical course similar to type 2 (juvenile form) but with later onset, and usually a more slowly progressive course. There has been cerebellar ataxia with dysarthria and later spasticity and intellectual impairment, but without fits (Stevenson *et al.* 1978; Wenger *et al.* 1980*a*).

Other ß-galactosidase deficient variants

These include:
(a) patients with moderate skeletal involvement and cherry-red spots with combined neuraminidase deficiency, now classified as a form of sialidosis (q.v.) (Lowden and O'Brien 1979);
(b) four patients with severe skeletal dysplasia without neurological involvement, classified as Morquio's disease type B (MPS IVB) (q.v.) with ß-galactosidase deficiency (O'Brien *et al.* 1976; Arbisser *et al.* 1977; Groebe *et al.* 1980).

Relation of ß-galactosidase deficiency to clinical phenotype

O'Brien (1983) suggests that the diverse effects of a single gene mutation involving acid ß-galactosidase may be explained by a one gene–one polypeptide–many substrate model. If, for example, there is a significant residual catalytic activity for galactoglycoproteins but not for GM₁, there will be a neurological deterioration because of GM₁ storage. Alternatively, if the converse applies, there will be galactoglycoprotein storage resulting in skeletal deformity without cerebral involvement. In the later onset patients considerably higher residual activities of ß-galactosidase in liver and fibroblasts (Norden and O'Brien 1975; O'Brien and Norden 1977) might explain the slower course.

Animal models

GM₁-gangliosidosis has been reported in Friesian cattle (Donnelly *et al.* 1973), beagles (Read *et al.* 1976; Rittmann *et al.* 1980) and (with ß-galactosidase deficiency) in Siamese and other domestic cats (Baker *et al.* 1971; Holmes and O'Brien 1978; Blakemore 1972).

GM₃-gangliosidosis; N-acetylgalactosamine transferase deficiency

A single male patient has been described with a possible defect in ganglioside biosynthesis (Max *et al.* 1974; Tanaka *et al.* 1975). The son of unrelated Ashkenazi Jews, the patient died at 14 weeks. The clinical phenotype included motor weakness, failure to thrive, gingival hypertrophy, macroglossia, inguinal hernias, and coarse facies. There were no storage vacuoles or cytoplasmic inclusions, but there was spongy degeneration of the white matter, and swollen astrocyte processes. In the brain there was an increase in the concentration of GM_3 with a deficiency of GM_1, GD_{1a} and GD_{1b}, and a reduction to 10 per cent of normal in the activity of GM_3-UDP-N-acetylgalactosaminyl transferase (Max *et al.* 1974). Enzyme activities in parents were not reported. The similar phenotype in a maternal uncle raises the possibility of X-linked inheritance. However, O'Brien (1983) considers that at present there is no evidence to include GM_3-gangliosidosis or N-acetylgalactosamine transferase deficiency among the hereditary diseases of human beings.

Mucolipidosis IV (MLIV)

Corneal clouding from the first few weeks after birth and a slowly progressive neurological deterioration during the first few months, without visceral or skeletal involvement, are characteristic of this disorder.

The first seven patients reported were all Ashkenazi Jews (Berman *et al.* 1974*b*; Merin *et al.* 1975; Newell *et al.* 1975; Tellez-Nagel *et al.* 1976; Kohn *et al.* 1977) but subsequently the disorder has been reported in a Caucasian, and in Algerian and Moroccan Arabs (Goutières *et al.* 1979).

Clinical features

Corneal clouding occurs as early as six weeks of age (Berman *et al.* 1974*b*). In one child, corneal opacities decreased between 10 and 24 months (Tellez-Nagel *et al.* 1976) in another they reappeared a few months after corneal transplant (Kohn *et al.* 1977).

Neurological regression may be slow during the first year but becomes more pronounced during the second year. There is delayed motor development, hypotonia, and bilateral pyramidal tract signs. Later there is very severe mental and motor deterioration and microcephaly. Abnormal retinograms and optic atrophy have been reported (Merin *et al.* 1975; Newell *et al.* 1975; Goutières *et al.* 1979). The oldest patient was 23 years of age (Newell *et al.* 1975).

Pathology

We are not aware of a report on a necropsy. Cytoplasmic inclusions in biopsies of skin, conjunctiva, and brain, and in histiocytes of bone marrow aspirates are characteristic. In skin and conjunctival biopsies, both lamellar fibrillogranular inclusions have been reported in fibroblasts, endothelial cells, Schwann, and perineural cells (Goutières *et al.* 1979). Cultured fibroblasts contain complex membrane-bound bodies with concentric lamellae (Goutières *et al.* 1979). Bone marrow aspirates contained histiocytes with cytoplasm distended by sudanophilic, weak periodic acid-Schiff (PAS)-positive material (Berman *et al.* 1974*b*; Merin *et al.* 1975). A brain biopsy (Tellez-Nagel *et al.* 1976) showed PAS-and sudan black B-positive granular material in neurones and glial cells. Peripheral blood leucocytes are usually normal, but Goutières *et al.* (1979) reported that electron microscopy revealed some lamellar cytoplasmic inclusions in a few lymphocytes.

Molecular defects

Cultured fibroblasts have shown accumulation of both gangliosides and glycosaminoglycans. Bach *et al.* (1975) demonstrated that fibroblast GD_3 is elevated to 1.5–2 times normal levels, while GM_3 is raised to a lesser extent. By direct chemical analysis, and by using radioactive precursors, Bach *et al.* (1977) reported a 10-fold increase of hyaluronic acid, and a threefold increase of sulphated glycosaminoglycans in cultured fibroblasts. Total ganglioside showed a 50 per cent elevation in brain biopsy grey matter and a threefold elevation in white matter (Tellez-Nagel *et al.* 1976).

The urine of reported patients has not contained excess of either mucopolysaccharides or oligosaccharides. A partial deficiency of ganglioside neuraminidase has been reported (Bach *et al.* 1979*b*; Hahn *et al.* 1980).

More recently a severe deficiency of fibroblast neuraminidase against GD_{1a} and GM_3 ganglioside was reported in a 22-year-old Italian female with a typical clinical phenotype of ML IV (Caimi *et al.* 1982).

Genetic counselling and prenatal diagnosis

The report of consanguineous parents in both a Caucasian and an Arab Moroccan patient is consistent with autosomal recessive inheritance, as is the occurrence of the disorder in siblings from two families (Goutières *et al.* 1979).

Prenatal diagnosis was attempted in two pregnancies by electron microscopy of cultured amniotic fluid cells. Abnormal membranous cytoplasmic inclusions were demonstrated in the cells from both pregnancies. Both were terminated but only one foetus was examined. Numerous inclusions in the brain, spinal cord, skin, and conjunctiva were considered to confirm that the foetus was affected (Kohn *et al.* 1977). Tellez-Nagel *et al.* (1976), however, reported cellular inclusions in skin fibroblasts from the mother of a patient, so that the possibility cannot be excluded that the foetuses studied by Kohn *et al.* (1977) were heterozygotes.

Prenatal diagnosis in three other pregnancies at risk for MLIV in the same mother was reported by Bach *et al.* (1980). In two pregnancies, amniotic cells were collected at amniocentesis at 15–16 weeks' gestation and cultured. Both foetuses were considered to be affected because amniotic cells contained a threefold increase of GD_3, a twofold increase of GM_3, and an approximately fourfold increase of dermatan and heparan sulphates. Ganglioside sialidase activity was 0.29 nmol substrate/mg protein/h (units) in one foetus (simultaneous controls 0.45 and 0.98 units), and a mean of 0.16 units in the other (simultaneous controls 0.29–0.52 units). In the third pregnancy normal values were obtained and the foetus was considered to be unaffected.

OTHER SPHINGOLIPIDOSES

Gaucher's disease

Gaucher's disease is the commonest sphingolipid storage disease. The adult type (type 1), which is characterized by hepatosplenomegaly, hypersplenism, skeletal involvement and lipid-containing foam cells, is about 30 times more common in Ashkenazi Jews than in other populations. The infantile type (type 2) is characterized in addition by progressive psychomotor regression, leading to death by two years of age and the juvenile type (type 3), which is the rarest form, has a later onset, but is also characterized by neurological deterioration. A few cases with similar course to type 3 have been reported in adults.

Clinical features

Type 1: adult type; chronic non-neuronopathic Gaucher's disease. Enlargement of the spleen is usually the first clinical feature. In spite of the designation 'adult type', this form of Gaucher's disease can present from infancy to the sixth or seventh decade (Mathoth *et al.* 1974; Hodson *et al.* 1979). Exceptionally, splenomegaly is absent (Brinn and Glaubman 1962). The liver is enlarged later, and to a lesser degree. Bruising of the skin on relatively minor trauma is characteristic. After some years the skin has a light yellowish colour. Febrile episodes are accompanied by generalized pain in trunk and limbs. Pains in bones and joints are also common. The abdomen is sometimes enlarged by distension of the colon, in addition to splenomegaly. There may be mild microcytic anaemia, pathological fractures, recurrent pulmonary infections, and occasionally pulmonary hypertension. Arthritis of the hip joints and degeneration of vertebral bodies are common.

In a few patients with the adult phenotype, neurological abnormalities, including intellectual impairment, fits, and ataxia have occurred (Miller *et al.* 1973; Soffer *et al.* 1980). Survival into late adult life is the rule, but some patients die younger because of pulmonary infections. Radiologically, the lower third of the femora are expanded, giving an Erlenmeyer flask appearance.

Type 2: infantile type; acute neuronopathic Gaucher's disease. Early symptoms include protruberant abdomen and failure to thrive between three and six months of age, when hepatosplenomegaly is usually present. After six months of age, neurological regression and developmental arrest are the outstanding features. The voluntary muscles become hypertonic and spastic with exaggerated reflexes. Involvement of the extrapyramidal systems is common. There is laryngeal spasm, swallowing difficulty and strabismus. Some patients have fits. Psychomotor deterioration is rapid and leads to recurrent pulmonary infection and death. Average survival in 67 patients was nine months (Fredrickson and Sloan 1972*a*). The literature has been reviewed by De Berranger *et al.* (1975).

Type 3: juvenile type; subacute neuronopathic Gaucher's disease. This is the rarest form and is distinguished from type 2 by its later onset and slower course. Hepatosplenomegaly, psychomotor retardation and convulsions are the principal features. There may be spasticity, strabismus and ataxia (Dreborg *et al.* 1980; Neil *et al.* 1979; Sack 1980; Nishimura and Barranger 1980). Red cell glucocerebroside was three times normal in a patient with type 3 (Brady *et al.* 1974). Foam cells occur in visceral reticuloendothelial cells. Glucocerebroside content was 23 times normal in a liver biopsy from a 15½-year-old affected boy (Brady *et al.* 1974).

Cases of Gaucher's disease from two districts in northern Sweden have been found to have characteristic clinical and biochemical features, distinguishing them from classical types 1 and 2 and placing them within the type 3 group. Because these patients are more clearly defined than others of type 3, and since they probably all belong to the same family tree, they have been termed the Norrbottnian type (Dreborg *et al.* 1980; Svennerholm *et al.* 1982).

Pathology

Type 1. Before discovery of the enzyme defect, identification of the characteristic morphology of Gaucher cells, usually in the bone marrow, was the only method for confirming the diagnosis. These cells are enlarged histiocytes, swollen with accumulated lipid. They occur in the reticulo-endothelial system including the spleen, medullary region of the lymph nodes, Peyer's patches, liver sinusoids, and bone marrow. Gaucher cells also occur in alveolar capillaries and pulmonary lymphatics and may be related to the clinical features of frequent pulmonary infection and occasionally pulmonary hypertension. They are also found in the adrenal cortex, pancreas, and thyroid. Cytochemically they stain positively with periodic acid–Schiff reagent, indicating the presence of carbohydrate, and with *p*-nitrophenylphosphate at acid pH, indicating acid phosphatase activity. By electron microscopy, Gaucher cells have been shown to contain enlarged lysosomes (Fisher and Reidbord 1962; Jordan 1964), which contain microtubules composed of twisted fibrils (Lee *et al.* 1973). The cells have altered surface infrastructure (Djaldetti *et al.* 1979) and surface markers (Burns *et al.* 1977).

Hypersplenism occurs in the great majority, leading to thrombocytopenic purpura and other forms of haemorrhage, anaemia, and leucopenia. Serum acid phosphatase activity is usually increased.

Type 2. In the brain, Gaucher cells (see type 1) are typically in clusters in cerebral perivascular spaces. Neuronal storage cannot usually be demonstrated (Espinas and Faris 1969). More commonly, there is demyelination and diffuse astrocytosis (Levine and Hoenig 1972; Verity and Montasir 1976). Infiltrations of Gaucher cells in the extraneurological systems are similar to those described for type 1.

Molecular defects

There is a severe deficiency of ß-glucocerebrosidase, the ß-glucosidase which catalyses cleavage of glucose from glucocerebroside (glucosyl ceramide, Fig. 4.5). This was demonstrated independently in spleen by Brady *et al.* (1965, 1966a) using (^{14}C)-glucosyl ceramide as substrate; and by Patrick (1965), who showed that extracts of Gaucher spleen had deficient ß-glucosidase activity towards both glucocerebroside and the artificial substrate *p*-nitrophenyl-ß-D-glucoside. Subsequently the ß-glucosidase deficiency was demonstrated using 4-methylumbelliferyl-ß-D-glucopyranoside as substrate in liver (Öckerman and Kohlin 1968), leucocytes (Beutler and Kuhl 1970; Hultberg and Öckerman 1970; Wenger *et al.* 1978b) and fibroblasts (Beutler *et al.* 1971; Ho *et al.* 1972a; Hultberg *et al.* 1973).

There are at least two isoenzymes of ß-glucosidase in human leucocytes, as demonstrated by two pH optima of 4.0 and 5.3 (Beutler and Kuhl 1970).

Subcellular fractionation of various tissues reveals a tightly membrane-bound lysosomal ß-glucosidase and a cytosolic form (Barrett and Heath 1977). Both forms are active against synthetic substrates (4-methylumbelli-

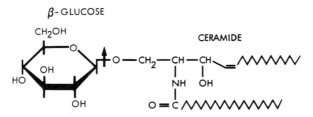

Fig. 4.5. Glucocerebroside.

feryl and *p*-nitrophenyl derivatives) but only the lysosomal form hydrolyses glucocerebroside.

Ho and O'Brien (1971) and Ho and Light (1973) have demonstrated that acid ß-glucosidase and glucocerebrosidase activities depend on the presence of a soluble acidic glycoprotein (Factor P) and a heat-labile membrane-bound protein (Factor C) but the physiological role of these factors is not yet accepted (Peters *et al.* 1977*a*, *b*).

There is appreciable residual activity of glucocerebrosidase in Gaucher's disease in the range of 3–45 per cent of normal (Beutler *et al.* 1971; Ho *et al.* 1972*b*; Hultberg *et al.* 1973; Peters *et al.* 1977*b*; Brady 1978; Fensom and Benson, unpublished observations), but no correlation with the clinical features was observed by Mathoth *et al.* (1974). Investigation of possible correlation in Gaucher's disease of deficiencies among newly discovered isozymes of ß-glucocerebrosidase (Ginns *et al.* 1980; Maret *et al.* 1980) will be of interest.

Pentchev *et al.* (1978) purified glucocerebrosidase from normal spleen and adult Gaucher spleen, until it was enriched by about 26 000 times. The main difference in kinetics was considerably increased K_m for Gaucher enzyme. There was no appreciable difference between the two enzymes in heat stability and pH optimum.

Liver cerebroside in biopsies is elevated 23- to 389-fold above normal (Brady 1978). Accumulation of glucocerebroside in histiocytes is most marked in spleen. The source is thought to include sphingolipid membrane components of mammalian cells including kidney (Martensson 1966), muscle (Max *et al.* 1970), liver (Kwiterovitch *et al.* 1970), leucocytes (Miras *et al.* 1966), blood platelets (Snyder *et al.* 1972), senescent red cells (Yamakawa and Nagai 1978) and mast cell membranes (Sweeley 1980).

Only a small fraction of the glucocerebroside catabolized each day accumulates in Gaucher's disease; moreover, the few milligrams of cerebroside which are stored in Gaucher's disease account for only a small proportion of visceral enlargement which in the case of liver can represent an extra kilogram in weight. It has been suggested that the remaining enlargement could represent a compensatory provision of residual cerebrosidase (Brady 1978).

In type 2, glucocerebrosidase deficiency is as described for type 1. The disease may be due to different allelic mutations. This is unproven but consistent with failure to produce genetic complementation by somatic cell hybridization (Turner and Hirschhorn 1977), the difference in ethnic distribution, and the fact that affected siblings with one exception always have the same form of the disease (but see 'Genetic counselling and prenatal diagnosis', p. 101). There appears to be a normal total brain cerebroside concentration. However, Svennerholm (1967) reported that while cerebroside hexose was 100 per cent glucose in age-matched controls, it was 30 per cent galactose and 70 per cent glucose in brain from type 2 children. However, no such difference was observed by Philippart and Menkes (1967) or French *et al.* (1969).

The recent work of Nilsson and Svennerholm (1982) has shed some light on the molecular basis for the different phenotypic expression of the neuronopathic and non-neuronopathic forms of Gaucher's disease. These workers found that brains of five patients with the infantile neuronopathic type had considerable elevation (3.8–8.8 μmol/kg of cerebrum; controls, undetectable) of glucosylsphingosine (psychosine), that is the deacylated product of glucocerebroside, while patients with juvenile Gaucher's disease had lower, but still elevated concentrations (0.8–4.6 μmol/kg). Moreover the brain of a patient with presumed genetic compound (double heterozygote) of the adult, non-neuronopathic form and the juvenile form had the lowest concentration (0.7 μmol/kg) of glucosylsphingosine. Since bases such as galactosylsphyingosine (psychosine), which accumulates in Krabbe's disease, and glucosylsphingosine, are known to be highly toxic to the brain, the accumulation of the latter may explain the extensive neuronal cell loss and psychomotor regression in the neuronopathic forms of Gaucher's disease.

It is tempting to speculate that the mutation in the non-neuronopathic form of Gaucher's disease leads to the synthesis of a form of glucocerebrosidase which lacks catalytic activity for glucocerebroside but retains activity for glucosylspingosine while allelic mutations in the neuronopathic forms lead to enzymes lacking the activities to varying degrees against both substrates.

Diagnosis

The simplest methods for confirming the diagnosis are by demonstrating severe depression of ß-glucosidase activity in leucocytes (Beutler and Kuhl 1970; Wenger *et al.* 1978*b*) or fibroblasts (Beutler *et al.* 1971; Ho *et al.* 1972*b*; Hultberg *et al.* 1973; Turner and Hirschhorn 1978) using 4-methylumbelliferyl-ß-D-glucoside as substrate. The fluorogenic assay is more specific in the presence of sodium taurocholate (Peters *et al.* 1976; Wenger *et al.* 1978*b*; Choy and Davidson 1980*a*). Others have used more difficult assays employing (^{14}C)-glucosyl ceramide (Kampine *et al.* 1967; Svennerholm *et al.* 1979), (^{14}C)-stearoylglucocerebroside (Ho and O'Brien 1971), or glucocerebroside in which the ceramide double bonds had been catalytically tritiated (Harzer 1980). Results with a chromogenic substrate, 2-hexadecanoyl-4-nitrophenyl-ß-D-glucopyranoside have also been reported (Gal *et al.* 1976*a*; Johnson *et al.* 1980).

Genetic counselling and prenatal diagnosis

Type 1 has an autosomal recessive mode of inheritance. Parent–offspring inheritance has been reported but the possibility that the unaffected parent was a carrier has not been excluded. Dominant inheritance therefore remains unproven. Type 1 is about 30 times more frequent in Ashkenazi Jews than in other populations. The incidence in Ashkenazis has been estimated as high as 1 in 2500 births (Fried *et al.* 1963; Grocn 1964; Brady 1978). Estimates of carrier frequency amongst Ashkenazis have varied from 1:100 to 1:30.

In only one family has more than one type of Gaucher disease been reported in full siblings (Wenger *et al.* 1982). It was suggested that the mother, who was clinically normal, and had leucocyte ß-glucosidase activity in the range of patients with Gaucher's disease was a double heterozygote for two mutant alleles (g^1/g^2) at the ß-glucosidase locus and that the father was a carrier of a third mutant allele, plus the normal wild type gene (g^3/G). It was further suggested that the two children with the acute neuronopathic forms had genotype g^1/g^3, and that a 25-year-old son with adult Gaucher's disease inherited a g^2/g^3 genotype. This suggestion does not explain why the mother was clinically normal.

Cultured skin fibroblast ß-glucosidase activities are reduced in carriers but there is some overlap with normal activities (Brady *et al.* 1971; Beutler *et al.* 1971; Ho *et al.* 1972a). Beutler *et al.* (1976) have found that assay of monocyte activities gives good discrimination. Wenger *et al.* (1978b) were able to discriminate between 12 carriers and eight controls by leucocyte ß-glucosidase assays using the 4-methylumbelliferyl substrate. Good discrimination was also achieved by Raghavan *et al.* (1980) assaying leucocyte ß-glucosidase activity.

Although the use of the natural substrate is still advocated by some workers, it appears that the addition of sodium taurocholate to the incubation mixture inhibits non-specific ß-glucosidase activity allowing glucocerebrosidase activity to be assayed reliably using the 4-methylumbelliferyl substrate (Peters *et al.* 1976; Wenger *et al.* 1978b; Choy and Davidson 1980a,b).

The first prenatal diagnosis of Gaucher's disease (type 2) was by Schneider *et al.* (1972b). Six pregnancies at risk were monitored using both natural and synthetic substrates by Wenger (1978). Håkansson and Svennerholm (1979) reported prenatal monitoring of 13 pregnancies for Gaucher's disease. Kitagawa *et al.* (1978) have reported prenatal monitoring of pregnancies at risk for the juvenile and adult forms of Gaucher's disease.

As for some other lysosomal enzymes, cell population densities must be standardized before assay (Heukels-Dully and Niermeijer 1976).

Treatment

Type 1 patients usually develop hypersplenism, which is relieved by splenectomy. Symptomatic treatment includes analgesics for bone pains, blood transfusion for anaemia and orthopaedic treatment for fractures and hip joint destruction. Intravenous administration of purified glucocerebrosidase from human placenta caused significant decrease in liver and erythrocyte cerebroside (Brady *et al.* 1974, 1982; Pentchev *et al.* 1975).

In the neuronopathic types of Gaucher's diseases there is the additional problem of the blood–brain barrier. Modification of the blood–brain barrier by intravenous infusion of mannitol or arabinose (Barranger *et al.* 1979) allows about one per cent of infused enzymes to enter the brain (Neuwelt *et al.* 1981). Recently Murray *et al.* (1983) have shown that exposing mannose residues of glucocerebrosidase, by sequential enzymatic deglycosylation, considerably increases its uptake by rat Kupffer cells. Transplantation of spleen (Groth *et al.* 1971) or kidney (Groth *et al.* 1980) was unsuccessful.

Animal model

Canine Gaucher's disease has been reported by Vandwater *et al.* (1979).

Krabbe's disease; globoid cell leucodystrophy

First described in five Danish patients by Krabbe (1916), further characterized by the report of 32 Swedish cases by Hagberg *et al.* (1970) and of the specific enzyme defect (Suzuki and Suzuki 1970), this leucodystrophy is usually fatal by two years of age.

Clinical features

The onset of progressive retardation in development is usually at three to six months of age, and although there is steady retrogression, the course of the disease was divided into three stages by Hagberg (1963a). In stage 1 there is generalized hyperirritability, hyperaesthesia, episodic fever of unknown origin, and some joint stiffness. The child becomes hypersensitive to slight stimuli and there may be fits. Regression of motor development is noticeable. Cerebrospinal fluid protein is increased. In stage 2 there is rapid and severe motor and mental deterioration, marked hypertonicity with extended crossed legs and backward-bent head. Tendon reflexes are exaggerated and there are tonic and clonic fits, optic atrophy, and sluggish light reflexes. In stage 3, which is achieved in a few weeks or months, children develop decerebrate rigidity, blindness, deafness, and lack contact with the surroundings. This stage lasts until death, which usually occurs at around two years of age.

The head is usually, but not always, small. The tendon reflexes become depressed after about six months and may be lost because of progressive peripheral neuropathy (Dunn *et al.* 1969; Lieberman *et al.* 1980). CAT scans show atrophy of white matter.

Late-onset globoid cell leucodystrophy

Some patients with later onset and slower course have been reviewed by Crome *et al.* (1973). In cases reported before 1970, enzymological confirmation was not possible, the diagnosis being established by demonstration of globoid cells in the white matter. However, some late onset cases have been confirmed by enzyme assay (Young *et al.* 1972; Kolodny *et al.* 1980). In these cases the concentration of cerebrospinal fluid protein has usually been normal (Crome *et al.* 1973), but occasionally elevated (Malone *et al.* 1975; Kolodny *et al.* 1980), and peripheral neuropathy has not been reported.

Pathology

The brain is proportionately reduced in size. The gyri are shrunken and the sulci widened. On section, the white matter is greyish-white and has a rubbery feel owing to demyelination. Histologically, changes are much more marked in the white than the grey matter. The outstanding abnormalities are the presence of numerous globoid cells, demyelination, and astrocytic gliosis.

Two types of globoid cells have been distinguished: large (20–50 μm) cells with 15–20 nuclei situated near the plasma membrane (globoid bodies); and epithelioid, or globoid cells which are smaller, round or oval mononucleated cells. It is probable that these types of cells are of similar origin (Austin and Lehfeldt 1965; Olsson *et al.* 1966). The cycoplasm stains moderately with periodic acid–Schiff, Sudan black B and Sudan IV, does not exhibit metachromasia with toluidine blue, and is strongly positive for acid phosphatase (Wallace *et al.* 1963; Allen and De Veyra 1967; Oehmichen 1980; Oehmichen *et al.* 1980). The mononuclear globoid cells are widespread in the white matter, but are concentrated in clusters in perivascular spaces.

The origin of globoid cells has been widely discussed. They can be produced by experimental implantation of galactocerebroside in rat brains (Austin *et al.* 1961; Austin and Lehfeldt 1965; Olsson *et al.* 1966). The effect is specific since implantation of sulphatide, glucocerebroside, ceramide, ganglioside, or acid mucopolysaccharides does not produce globoid cell infiltration. These experiments suggested that globoid cells were of mesodermal origin. This was supported by the work of Oehmichen and Gruninger (1974), who labelled cell DNA with thymidine and followed the transformation of mesodermal cells into globoid cells following galactocerebroside implantation into animal brain. Evidence now suggests that these mesodermal cells or histiocytes enter the central nervous system from the blood stream (Rosemann and Friede 1967).

Demyelination of white matter of the brain is severe, and occurs in the cerebrum, cerebellum and in the spinal cord and is associated with axonal loss and depletion of oligodendroglial cells. Areas of demyelination are occupied by a dense astrocytic population. The grey matter is only occasionally affected with focal areas of necrosis.

Electron micrographs of globoid cells reveal numerous inclusions which are moderately electron dense, straight, curved, or crystaloid, or take the form of straight or twisted tubules (Yunis and Lee 1970). The ultra-structure of experimentally produced globoid cells by injection of galactocerebroside into rat brains showed similar inclusions (Suzuki 1970; Andrews and Menkes 1970).

Dunn *et al.* (1969) found degenerative changes in axons and myelin sheaths in all of seven patients. The optic nerve shows similar changes to those of cerebral white matter (Harcourt and Ashton 1973).

Molecular defects

There is deficiency of the enzyme galactosylceramide ß-galactosidase in white and grey matter, liver, kidney, spleen (Suzuki and Suzuki 1970; Austin *et al.* 1970), leucocytes, serum and fibroblasts (Suzuki and Suzuki 1971; Farrell *et al.* 1973; Svennerholm *et al.* 1979), but appreciable residual activities have been reported in cultured fibroblasts (Suzuki and Suzuki 1971; Farrell *et al.* 1973; Besley and Bain 1976). It should be noted that the various types of GM_1-gangliosidosis are also due to deficiency of a lysosomal ß-galactosidase. However, there are wide differences in the substrate specificities; for example, in GM_1-gangliosidosis, but not in Krabbe's disease, there is deficiency in hydrolysis of the artificial substrates 4-methylumbelliferyl-ß-D-galactoside and *p*-nitrophenyl-ß-D-galactoside, and of gangliosides GM_1 and GA_1. Conversely, in Krabbe's disease but not in GM_1-gangliosidosis, there is deficiency in hydrolysis of galactocere-broside (Fig. 4.6), galactosylsphingosine (psychosine) (Fig. 4.7) (Miyatake and Suzuki 1972), lactosylceramide (Wenger *et al.* 1974*a*) and mono-galactosyl diglyceride (Wenger *et al.* 1973). However, cleavage of lactosylceramide is catalysed by both types of ß-galactosidases under different assay conditions (Tanaka and Suzuki 1975; Wenger *et al.* 1975). The considerable differences between the clinical phenotypes of the GM_1-gangliosidoses and Krabbe's disease can be explained on the basis of these

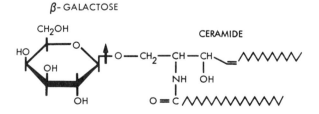

Fig. 4.6. Galactocerebroside.

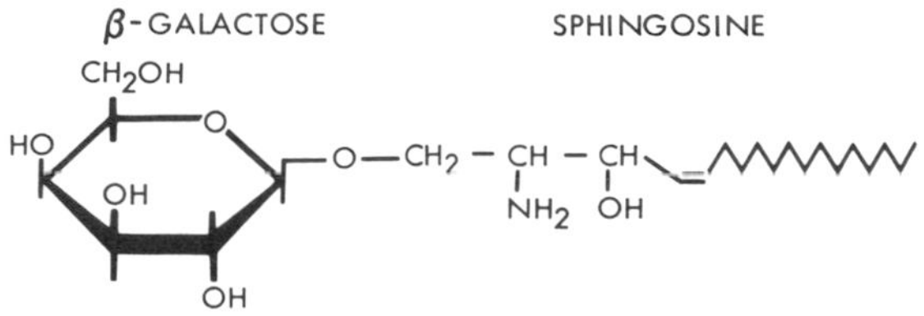

Fig. 4.7. Psychosine.

wide differences in the substrate specificities of the two affected ß-galactosidases. Furthermore, lack of accumulation of lactosylceramide in either group of disorders is consistent with the fact that it is a substrate for both types of ß-galactosidase.

In normal brain, cerebroside is almost exclusively a component of myelin and oligodendroglia. This is consistent with the observations that it exists in very low concentrations before myelination (Wells and Dittmer 1967; Cuzner and Davison 1968) and in brains of individuals with severe demyelinating disorders (e.g. Schilder's disease, metachromatic leuco-dystrophy).

The activity of galactocerebroside ß-galactosidase is low in newborn brain (Bowen and Radin 1969). With the acceleration of myelination shortly after birth there is a rapid rise in the activity of this enzyme (Bowen and Radin 1969) and in the turnover of myelin. Suzuki and Suzuki (1970) have suggested that in the brain of patients with Krabbe's disease, galactocerebroside cannot be catabolized, and causes globoid cell infiltration. As myelination proceeds, the increasing number of globoid cells leads to massive death of oligodendroglial cells and their myelin extensions. Alternatively, Miyatake and Suzuki (1972) suggested that psychosine synthesized within oligodendroglia during the period of active myelination might reach toxic levels and kill the cells.

In brains of patients affected by Krabbe's disease there is deficiency rather than accumulation of galactocerebroside. Thus the total yield of myelin is reduced from about 1000 mg/10 g net weight to only 3.8 mg/10 g. The percentage composition of galactolipids in myelin was, for normal brain: cerebroside 13.6 per cent, sulphatide 3.8 per cent; and for Krabbe's brain: 12.4 per cent and 4.6 per cent respectively (Eto *et al.* 1970). However, more recently, Vanier and Svennerholm (1975) and Svennerholm *et al.* (1980) reported a 100-fold increase in galactosylsphingosine (deacylated galactocerebroside or psychosine) in white matter of Krabbe's brains. This substance is highly toxic, as shown by the experiment in which injection of a mixture of 3 per cent psychosine and 97 per cent galactocerebroside into rat brain caused more tissue degeneration than when galactocerebroside was injected alone (Suzuki *et al.* 1976). Since psychosine is a substrate of galactocerebroside ß-galactosidase its accumulation in Krabbe's disease can be understood. However, it appears to be synthesized from UDP-galactose and sphingosine, since its formation by deacylation of galactocerebroside could not be demonstrated (Lin and Radin 1973). Psychosine could not be detected in normal brain (Svennerholm *et al.* 1980). Vanier and Svennerholm (1976) suggested that psychosine is formed from UDP-galactose and sphingosine, and rapidly hydrolysed under normal conditions but accumulates in Krabbe's disease. In their study of brains from 18 patients who had died of Krabbe's disease, Svennerholm *et al.* (1980) reported accumulation of galactose ß-1 → 4 galactosylceramide and suggested that this is another lipid which in normal brain has a rapid turnover, but accumulates in Krabbe's disease. These authors suggest that psychosine, rather than galactocerebroside, is the primary storage substance of white matter of Krabbe's disease brains which accumulates in oligodendroglial cells, leading to their death.

Diagnosis

The disorder should be suspected clinically in infants with progressive psychomotor degeneration, and elevated cerebrospinal fluid protein, especially if the tendon reflexes are sluggish or absent. The diagnosis can be confirmed by the demonstration of severe deficiency of galactocerebroside ß-galactosidase in serum, leucocytes or fibroblasts (Suzuki 1971; Farrell *et al.* 1973; Besley and Brain 1976; Svennerholm *et al.* 1979). Since activity in normal serum is low and unstable, leucocytes are normally preferred for diagnostic assay.

Enzyme assay using natural substrate requires preparation of the substrate galactocerebroside which is tritiated in the position of galactose by oxidation with galactose oxidase followed by borotritide reduction. In the enzyme assay, after incubation, galactose released is recovered by lipid solvent partition and assayed for radioactivity (Suzuki 1977).

A chromogenic procedure has been described (Gal *et al.* 1977) for assay of galactocerebroside ß-galactosidase using an analogue of galactocerebroside: 2-hexadecanoyl-amino-4-nitrophenyl-ß-D-galactopyranoside. However, Besley and Bain (1978) have found this chromogenic assay to be less sensitive and less specific than the assay using the natural substrate.

More recently, Besley and Gatt (1981) have described the use of a further chromogenic (ω-2,4,6-trinitrophenyl aminolauroyl galactosyl sphingosine) and a fluorogenic (11-(9-anthroyloxy)-undecanoyl galactosyl sphingosine) substrate for reliable enzymatic diagnosis of Krabbe's disease in leucocytes, fibroblasts, and cultured amniotic cells.

Genetic counselling and prenatal diagnosis

Hagberg *et al.* (1970) calculated the incidence in Sweden to be 1.9 per 100 000 births, but this would seem to be higher than in non-Scandinavian countries. In Japan the incidence has been calculated roughly as 0.2–0.3 per 100 000 births (Suzuki and Suzuki 1983). The disorder has an autosomal recessive form of inheritance. Several consanguineous parents have been reported. Partial deficiency of galactocerebroside ß-galactosidase has been reported in serum (Suzuki and Suzuki 1971), leucocytes and cultured fibroblasts from carriers (Suzuki and Suzuki 1971, 1973; Farrell *et al.* 1973; Besley and Bain 1976) but there is some overlap with control values (Svennerholm *et al.* 1975; Suzuki 1977; Wenger *et al.* 1974*b*).

Prenatal diagnosis for Krabbe's disease was first reported by Suzuki *et al.* (1971*b*). Amniocentesis was at 14 weeks' gestation. Cultured amniotic cell galactocerebroside ß-galactosidase activity after five weeks' culture was five per cent of control values. The pregnancy was terminated and the diagnosis was confirmed in the abortus. Subsequently there have been several series of reports of prenatal monitoring for Krabbe's disease (Patrick 1978; Galjaard 1979; Vanier *et al.* 1981). Okada *et al.* (1979) diagnosed Krabbe's disease in one of twins.

Animal models

Globoid cell leucodystrophy has been reported in West Highland terriers (Frankhauser *et al.* 1963). Histological changes are identical to those of Krabbe's disease, including peripheral nerve lesions (Fletcher *et al.* 1971). Suzuki *et al.* (1970*b*) have demonstrated that as in humans, there is deficiency of galactocerebroside ß-galactosidase activity. On isoelectro-focussing, only the major peak of enzyme activity is decreased in canine serum, while all three peaks are almost absent in human Krabbe's disease (Suszuki *et al.* 1974). Austin *et al.* (1968) reported that affected canine brain had slightly increased cerebroside content, but reduced total lipid. The disease has been reported in a single sheep, with considerably reduced galactosylceramide ß-galactosidase activity (Pritchard *et al.* 1980).

The nearest animal model to the human disease is the twitcher mouse, which has galactosylceramide ß-galactosidase deficiency (Kobayashi *et al.* 1980*a*, *b*). Heterozygotes have a partial enzyme defect (Kobayashi *et al.* 1982).

Niemann–Pick disease

First reported in a Jewish child who died in his second year, by a German paediatrician, Niemann (1914) and characterized histologically by Pick (1927), disorders of this group have in common excessive accumulation of sphingomyelin (Fig. 4.8). Four types were designated A–D by Crocker

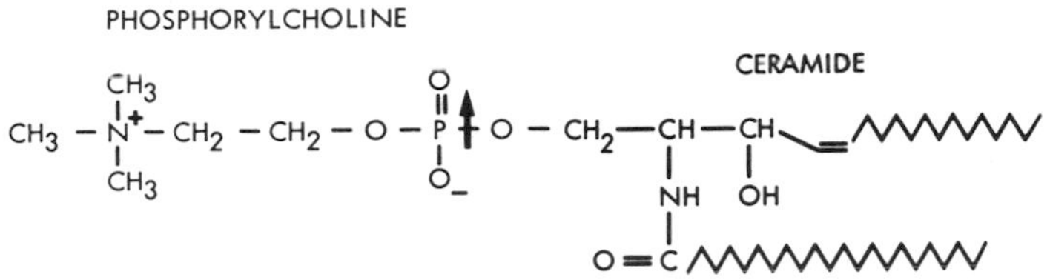

PHOSPHORYLCHOLINE

CERAMIDE

Fig. 4.8. Sphingomyelin.

(1961) according to the age of onset, speed of progression, ethnic distribution, and predominance of either visceromegaly or neurological regression. Type E was added by Fredrickson and Sloan (1972*b*). Neville *et al.* (1973) suggested a classification into types I–VI.

The existence of an infantile neuronopathic form (type A) and of a distinct non-neuronopathic visceral form (type B) is widely accepted. Furthermore, in these two types there is a severe deficiency of sphingomyelinase in addition to storage of sphingomyelin (Fig. 4.8), which is in liver, spleen, and brain in type A and in liver and spleen in type B.

In the chronic neuronopathic or juvenile forms (types C and D) there is both neurological regression and hepatosplenomegaly, but sphingomyelin storage is slight or absent and total sphingomyelinase activity is only modestly reduced, although in some cases an isoenzyme deficiency has been reported (Callahan *et al.* 1975). The designation type D has been reserved for patients of French origin in Nova Scotia with a type C-like illness. Type E includes adults with sphingomyelin accumulation in some tissues, without neurological abnormalities but with a possible isoenzyme deficiency in some cases (Callahan *et al.* 1975). Several patients have been reported with atypical phenotypes (Wenger *et al.* 1977; Schneider *et al.* 1978; Dewhurst *et al.* 1979).

Clinical features

Type A (acute neuronopathic form). This is the commonest form. There is mental and motor deterioration in the first year. Hepatosplenomegaly occurs early and has been detected by six months of age. Macular cherry-red spots occur in about half the patients. There is progressive severe wasting of subcutaneous fat and voluntary muscle mass, and abdominal distension. The skin often has a yellowish colour. Progressive psychomotor deterioration leads to death, usually due to bronchopneumonia, by the third year.

Type B (visceral form without neurological involvement). Splenic enlargement may be noticed in the first two years, while hepatic enlargement occurs later. Both viscera become considerably enlarged by four to six years of age. Life expectancy in this form of Niemann–Pick disease is not well recorded. Radiologically the lung fields show reticular or diffuse infiltration. In some cases the appearance of the chest X-ray is the first feature which attracts attention, hepatosplenomegaly being discovered on physical examination.

Type C chronic neuronopathic form. The onset of neurological regression is usually after the first or second year. There is loss of skills including speech. Patients become ataxic and develop convulsions. As the disorder progresses, there is hypertonia and psychomotor deterioration, leading to death between five and 15 years.

Type D (Nova Scotia variant). Patients who are of French origin from Nova Scotia, may present with jaundice shortly after birth and hepatospleno-megaly. After two to four years there is neurological regression, with progressive psychomotor deterioration, ataxia, and convulsions.

Type E (adult, non-neuronopathic form). This rare group includes adults with moderate hepatosplenomegaly and foam cells in the marrow.

Pathology

In type A there is massive enlargement of the spleen and moderate enlargement of the liver. Large (20–90 μm) vacuolated (foamy) histiocytes are found in the bone marrow, lymph nodes, spleen, pulmonary alveoli, kidneys, and adrenal medulla. Unstained cells under phase contrast microscopy have a 'mulberry-like' appearance owing to numerous lipid droplets. Cells may contain the brown pigment ceroid (lipofuscin). Staining of tissue sections with haematoxylin and eosin sometimes reveals blue-green material, which has inspired the name 'sea-blue histiocytes'. The cells are sudanophilic and stain moderately positive with periodic acid–Schiff reagent.

On electron microscopy foam cells are seen to contain cytoplasmic lipid inclusions which are concentrically laminated or granular.

Cytoplasmic inclusions have also been reported in peripheral blood monocytes (Lazarus *et al.* 1967), hepatocytes, Kupffer cells, and renal epithelial cells.

In Niemann–Pick disease types A and C, the brain is atrophic, firm, and weighs 50–90 per cent of normal. Histologically there is lipid storage in neurones, glia, and ganglion cells, marked loss of neurones, and proliferation of astrocytes and glia. There is also demyelination which may be secondary to neuronal loss (Ivemark *et al.* 1963; Van Bogaert *et al.* 1963; Phillipart *et al.* 1969).

In patients with type B, the nervous system is unaffected. Numerous foam cells are present in bone marrow and reticuloendothelial cells of spleen, liver, and lymphatic system (Miller and Reimann 1972).

In patients with type C, pathological changes are similar to those of type A, but are of later onset (Phillipart *et al*. 1969; Emery *et al*. 1972; Pellissier *et al*. 1976).

In patients with type D, numerous foam cells are present in bone marrow. Lipid inclusions have been reported in blood monocytes, but skin fibroblasts appear normal (Vethamany *et al*. 1972).

Molecular defects

Using sphingomyelin labelled with ^{14}C in the choline residue, Brady *et al*. (1966*b*) demonstrated that liver from patients with type A disease had 0–9 per cent of normal activity of sphingomyelinase, the enzyme which catalyses the reaction: sphingomyelin + H_2O → ceramide + phosphoryl-choline. Later it was shown that patients with type B disease also had severe sphingomyelinase deficiency (Schneider and Kennedy 1967). No enzyme defect was found in types C and D (Orii *et al*. 1975). In cultured fibroblasts there is marked deficiency of sphingomyelinase activity in types A and B (Sloan *et al*. 1969*b*; Callahan and Khalil 1975; Vanier 1978; Fensom *et al*. 1977) and partial deficiency in type C (Gal *et al*. 1975; Vanier 1978; Fensom *et al*. 1977; Besley 1978). Leucocytes show sphingomyelinase deficiency in types A and B, but not in type C (Vanier 1978). On isoelectric focussing two major forms (I and II) of liver sphingomyelinase can be identified. One form (isoenzyme II) was absent in liver from a patient with type C disease (Callahan *et al*. 1974). Brain has two major and three minor forms of sphingomyelinase. Callahan *et al*. (1975) were also able to show absence of a major form in the brain of another type C patient. Muller and Harzer (1980), found that a highly purified preparation from the brain of a type C patient had 59 per cent of the sphingomyelinase specific activity of a control brain. On isoelectric focussing, Besley (1977) demonstrated seven peaks of sphingomyelinase activity (isoenzymes) in normal fibroblasts and severe deficiency in the two most cathodic forms in fibroblasts from a type C patient. Fibroblasts from a type E patient showed absence of isoenzyme II, but increased activity of isoenzyme I (Callahan *et al*. 1975).

Besley *et al*. (1980) have reported somatic cell hybridization studies which show that complementation is produced by fusion of Niemann–Pick types A or B fibroblasts with type C, but not when type A and B are fused together. This indicates the involvement of at least two genes in the disease. The evidence of Christomanou (1980) for the absence of an activating factor in type C which normally stimulates degradation of sphingomyelin (and also glucocerebroside) correlates with the hybridization studies if it is assumed that the activating factor is a protein coded for by one gene, the second gene coding for sphingomyelinase.

Beaudet and Manshrek (1982) showed that cultured fibroblasts of normal humans took up, and hydrolysed in a 24 hour period, 79–92 per cent of [choline-*N*-methyl-^{14}C] sphingomyelin added to the culture medium. The equivalent values for type A cells were 2–7 per cent, and for type B 15–49 per cent, suggesting a greater residual sphingomyelinase activity in type B than in type A cells.

Sphingomyelin (Fig. 4.8) in spleen of patients with type A can increase from 15 mg/g dry weight to 370 mg/g dry weight (Fredrickson and Sloan 1972*b*), and in two patients with type B to 250 and 280 mg/g dry weight (Crocker 1961; Fredrickson and Sloan 1972*b*). In type C there is a wide range of accumulation from 25 to 255 mg/g dry weight, while only minimal increase was present in type D (Fredrickson and Sloan 1972*b*). There is a moderate increase in grey matter of type A brains but only a slight increase in white matter.

No marked increase of sphingomyelin occurs in brains of patients with type C. The sphingomyelin which accumulates in Niemann–Pick disease may originate from turnover of a wide range of cells or cell components. Sphingomyelin is a constituent of myelin sheaths, erythrocytic stroma, endoplasmic reticulum, mitochondria, and plasma membranes.

Diagnosis

In cultured fibroblasts, Niemann–Pick disease types A and B can be diagnosed by demonstrating severe deficiency of sphingomyelinase, using either the natural substrate (Sloan *et al.* 1969*b*; Callahan and Khalil 1975; Fensom *et al.* 1977), or chromogenic substrate 2-hexadecanoylamino-4-nitrophenylphosphorylcholine (NHP) (Gal *et al.* 1975), or a fluorescent substrate, bis-4-methylumbelliferyl pyrophosphate, to detect phosphodiesterase deficiency (Fensom *et al.* 1977).

In leucocytes types A and B can be diagnosed by demonstrating severe deficiency of total sphingomyelinase using the natural substrate (Kampine *et al.* 1967; Brady *et al.* 1971*a*; Wenger 1977; Vanier 1978) or with the chromogenic assay.

Deficiency of fibroblast sphingomyelinase isoenzymes has been demonstrated in only a few patients with type C (Callahan *et al.* 1975; Besley 1977) or type E (Callahan *et al.* 1975). A partial sphingomyelinase deficiency is found in the majority of type C patients (Vanier 1978). Besley (1978) found that when the neutral form of sphingomyelinase was removed from a fibroblast extract, reliable detection of type C was possible using the methylumbelliferyl substrate, when residual activity was 20–30 per cent of normal values.

Sphingomyelinase activity in cultured fibroblasts from patients with type D Niemann–Pick disease is within normal range (Sloan *et al.* 1969*b*).

Genetic counselling and prenatal diagnosis

Niemann–Pick disease types A–D have an autosomal recessive form of inheritance. Insufficient data are available for patterns of inheritance in the isolated patients with type E or other variants.

In carriers of types A or B, partial sphingomyelinase deficiencies have been demonstrated in leucocytes (Brady *et al.* 1971*a*; Zitman *et al.* 1978; Vanier 1978) and cultured skin fibroblasts using either natural (Fensom *et al.* 1977; Vanier 1978) or artificial (Gal *et al.* 1975; Fensom *et al.* 1977; Besley 1978) substrates.

Prenatal diagnosis of a type A foetus (Epstein *et al.* 1971) was established by the demonstration that amniotic cell sphingomyelinase was 4 per cent of two control amniotic cultures after five weeks of culture, following amniocentesis at 13 weeks' gestation. The diagnosis was confirmed after termination of pregnancy at 19 weeks' gestation. Foetal liver sphingomyelin concentration was 17 times higher than in age-matched normal foetal liver (Schneider *et al.* 1972*a, b*). Several other accurate prenatal predictions have been reported; the biggest series was by Wenger (1977, 1978), who reported two of six monitored pregnancies to be affected. Other prenatal diagnoses have been reported by Patrick *et al.* (1977), Chazan *et al.* (1978), Harzer (1979), and Vanier *et al.* (1979). The report by Patrick *et al.* (1977) was of particular interest since they used the chromogenic sphingomyelin analogue HNP (see 'Diagnosis') in addition to the natural substrate. Amniocentesis was at 14 weeks' gestation. After three and a half weeks of culture amniotic fluid cells hydrolysed HNP to only 4.5 per cent of the mean control value.

The partial defect found in cultured fibroblasts from type C patients makes this form of the disease particularly difficult to diagnose prenatally (Harzer *et al.* 1978).

Animal models

A disease which appears similar to type A has been reported in Siamese cats (Wenger *et al.* 1980*b*), and in a poodle (Bundza *et al.* 1979). A form with features of type C has been described in mice (Pentchev *et al.* 1980).

Sulphatase deficiencies

Metachromatic leucodystrophy (MLD) is a group of disorders characterized by progressive neurological regression, accumulation of lipids containing a galactosyl 3-sulphate residue, and a severe deficiency of the enzyme cerebroside sulphate sulphatase, or arylsulphatase A (ASA). They are subdivided into the late infantile, juvenile and adult forms. The first two are characterized by normal early development followed by loss of muscular co-ordination and gait difficulties, leading to mental retardation, quadriplegia, and early death. The adult form, which is rarer, differs not only by its later onset, but by its presentation, which usually is of mental illness with features of dementia and psychosis, followed some years later by motor disturbance.

Multiple sulphatase deficiency is a rare and distinct entity resembling late infantile MLD clinically, but with additional features typical of certain types of mucopolysaccharidoses. This disorder is considered separately since there is not only deficiency of ASA but also of arylsulphatase B, arylsulphatase C, and at least five other sulphatases.

A quite distinct disorder, placental or steroid sulphatase deficiency, is also described. It is sex-linked and leads to ichthyosis without intellectual or other neurological defect and is due to deficiency of the microsomal enzyme arylsulphatase C, which appears to be identical to placental or steroid sulphatase. Deficiencies of sulphatases implicated in MPS II, IIIA, IIID, IVA, VI, and against ß-linked (keratan sulphate-derived) *N*-acetylglucosamine 6-sulphate, are discussed in the chapter on the mucopolysaccharidoses.

Metachromatic leucodystrophy; sulphatide lipidosis

Metachromatic staining of the nervous system (see section on Pathology) was first described by Alzheimer (1910) in a patient with diffuse sclerosis; but was not reported by Scholz (1925) in his clinico-pathological study of progressive familial leucodystrophy. This was remedied by Peiffer (1959), who 34 years later demonstrated metachromasia in frozen sections from Scholz's patients. Sulphatide storage was identified by Jatzkewitz (1958) and Austin (1959), and ASA deficiency demonstrated by Austin *et al.* (1963, 1965*b*) and Mehl and Jatzkewitz (1965, 1968).

Clinical features

Late infantile form. Zlotogora *et al.* (1981) found that less than 60 per cent of children with the late infantile form of MLD walk by 16 months, and that 15 per cent never achieve independent walking; while 90 per cent of normal children walk independently by 15 months. MacFaul *et al.* (1982) reported that 24 children presented between six and 25 months (mean 17) with delay or deterioration in walking. However, the majority learn to walk unaided and to speak in short sentences. The first symptoms are loss of acquired motor skills, including walking, which becomes unsteady. At this stage there is hypotonia of the limbs, especially of the lower limbs, often genu recurvatum, and diminished or even absent tendon reflexes, giving rise to suspicion of myopathy or peripheral neuropathy (De Silva and Pearce 1973). Intermittent pains in the limbs may be severe (Hagberg 1963*b*). Rarely, there may be early pyramidal tract signs, including hypertonus and exaggerated tendon reflexes, but these are characteristics of the more advanced stages, after about 15 months from the onset.

As the disorder progresses, ataxia develops and speech becomes impaired. Eventually all locomotion is lost and patients develop decerebrate rigidity, weakness, and lack of co-ordination of pharyngeal musculature, with optic atrophy in the majority. A typical appearance of the fundi has been described in patients with late infantile MLD, namely a greyish discolouration of the retina and macula with a central red spot resembling that seen in Tay–Sachs disease (Cogan *et al.* 1970). Convulsions occur in about one third of the patients and tend to be a late feature (MacFaul *et al.* 1982). Finally, all voluntary actions are lost, including speech and swallowing. The patients are blind and helpless, but with care and tube feeding can survive several years before death, usually from broncho-pneumonia (Hagberg 1963*b*; Schutta *et al.* 1966).

Motor conduction velocity was slowed in all 18 children tested by MacFaul *et al.* (1982).

Computerized tomography (CT) scans show diagnostic symmetrical loss of white matter density. In advanced stages there is a little dilation of the ventricles (Buonnano *et al.* 1978).

Presymptomatic cases may have normal CT appearances (Procopis 1979; Benson P.F., Fensom A.H., and Neville B.G.R., unpublished observations).

Juvenile form. The age of onset is between three and 20 years and is remarkably similar in affected siblings, consistent with the occurrence of several distinct genetic entities. Younger patients tend to have an illness characterized by motor dysfunction, especially clumsiness of gait, with a course which is similar to, but occurring at a later age than, that in patients with the late infantile form. Older children tend to have clinical features with greater resemblance to the adult form, which is characterized by mental illness with features of psychosis, dementia and emotional disorders. Muscular rigidity, postural abnormalities and cerebellar mani-festations have been described. In an analysis of 38 children with MLD, MacFaul *et al.* (1982) assigned 24 to the late infantile group. One boy presented at 13 years, while the others were designated either as early juvenile or intermediate (seven children) presenting with a gait disorder between four and six years; or as juvenile (six children) presenting with educational or behavioural difficulties between six and ten years. Motor conduction velocity was slowed in four of eight patients investigated with the early juvenile, or juvenile forms.

Adult form. By definition, this rare form occurs when symptomatic onset is after 20 years of age (Müller *et al.* 1969; Pilz *et al.* 1977). More important than age of onset is the predominance of mental, rather than motor, symptoms which can also occur in patients with the juvenile age onset. Dementia manifests with poor memory, decreasing intelligence in previously able individuals, and emotional abnormalities. Psychotic changes take the form of schizophrenic thought disturbances. Motor defects may also occur and present with clumsiness and dysarthria. Pyramidal tract signs predominate, with hypertonus and brisk tendon reflexes. Neurological signs also indicate extrapyramidal affliction with Parkinsonian features; also cerebellar dysfunction with ataxia, including truncal ataxia, nystagmus and intention tremor; and peripheral neuropathy with diminished conduction velocity in peripheral nerves (Bosch and Hart 1978; Tagliavini *et al.* 1979).

In the later stages, convulsions occur, the patient deteriorates, loses speech and becomes immobilized and incontinent.

Pathology

Macroscopically the white matter appears grey-brown, may show cystic changes, and feels firm. Histologically there is loss of myelin sheaths, and abundant accumulation of granular lipid masses, both within and outside perivascular macrophages. These lipids are strongly metachromatic, that is they stain pink or red with cationic dyes such as Alcian blue or cresyl violet. Metachromasia is exhibited by high molecular weight or aggregated molecules which are polyanionic owing to repeating sulphate, phosphate or carboxyl residues. They include sulphates, glycosaminoglycans (mucopolysaccharides), sulphatides and gangliosides. The mechanism is unknown but is thought to depend on additional resonance energy, from interaction of the anion-bound cationic dye molecules, causing transmission of longer wavelength light (Dulaney and Moser 1978). A method for metachromatic staining of sulphatide in MLD using Alcian blue has been described by Lampert and Lewis (1975). There is cytochemical evidence that sulphatide is responsible for the metachromasia. For example, the material is identified as lipid since it stains with Sudan black, and is lipid solvent extractable. Suzuki *et al.* (1967) isolated the metachromatic bodies by differential centrifugation, and identified sulphatide as the predominant lipid component.

Besides the characteristic metachromatic lipid infiltration of white matter, there is oligodendroglial cell loss and distension of selected neurones with metachromatic material. Affected neurones are in cranial nerve nuclei, thalamus, hypothalamus, and anterior and dorsal root ganglion cells. Cerebral neurones are spared, though Betz cells may be affected. Peripheral nerves show interesting patterns of demyelination in addition to metachromasia of Schwann cells and phagocytes. Thus in teased-out preparations of sciatic and other peripheral nerves (Dayan 1967; Thomas *et al.* 1977) from cases with MLD, there is a pattern of segmental demyelination, i.e. demyelination between some adjacent nodes of Ranvier, but not others, along the same nerve fibre.

Metachromatic material is consistently seen in viscera and in several extraneural systems (Wolfe and Pietra 1964). For example, in kidney tubules, hepatic histiocytes, Kupffer cells, hepatocytes, and in endocrines including the adrenal cortex, anterior pituitary, and testis. The gall bladder is particularly affected. It tends to be small, fibrotic and is infiltrated with metachromatic macrophages. There are mucosal polyps containing metachromatic material (Wolfe and Pietra 1964; Kleinman *et al.* 1976).

In the eyes, there is metachromasia without distension of the retinal ganglion cells (Cogan *et al.* 1970). On electron microscopy, two types of inclusions were found by Gregoire *et al.* (1966) in oligodendroglia: one, a type showing a 'herringbone' pattern with a banding period of 100 Å, but which on cross section was hexagonal; and the other concentrically laminated 58 Å spheres. The herringbone inclusions are specific to MLD. They have been demonstrated in purified sulphatides. They have also been described in epithelium from parts of renal tubules and liver from MLD patients (Resibois 1971) and foetuses (Leroy *et al.* 1973; Wiesmann *et al.* 1975). Cultured skin fibroblasts do not show inclusions unless sulphatide is added to the culture medium; characteristic herringbone material then appears (Rutsaert *et al.* 1973).

The presence of abnormal inclusions in nearly all sural nerve Schwann cells, even when they surrounded fibres which were not yet myelinated, led Webster (1962) to speculate that inclusions were not formed from products of demyelination. This can also be deduced by the presence of abnormal inclusions in some Schwann cells in affected foetuses before the onset of myelination (Leroy *et al.* 1973; Wiesmann *et al.* 1975).

Abnormal cytoplasmic inclusions have been reported in oligodendroglia, astrocytes (Aurebeck *et al.* 1964) and anterior horn cells. The inclusions are membrane-bound and acid phosphatase positive, which indicates that they are intralysosomal (Resibois 1969).

Molecular defects

There is severe deficiency of ASA or cerebroside sulphatase, demonstrated by Austin *et al.* (1963, 1965*b*) and Mehl and Jatzkewitz (1965, 1968) in brain, liver, and kidney (Jatzkewitz and Mehl 1969); and in leucocytes (Percy and Brady 1968; Kihara *et al.* 1973*a*; Dubois *et al.* 1980), cultured fibroblasts (Porter *et al.* 1969, 1971*a*; Kihara *et al.* 1973*a*; Leroy *et al.* 1973), serum (Beratis *et al.* 1973; Singh *et al.* 1975; Hashimoto *et al.* 1978), tears (Libert *et al.* 1979), saliva (Den Tandt and Jaeken 1979) and urine (Thomas and Howell 1972).

The natural substrates of ASA are galactosyl (i.e. sulphatide, Fig. 4.9) and lactosyl sulphatide, (Harzer and Benz 1974) sulphogalactosyl-glycerolipid, sulphogalactosylsphingosine (Eto *et al.* 1974*a*), and ascorbic acid 2-sulphate (Manowitz *et al.* 1977). Deficient sulphohydrolase activity against galactosyl sulphatide in tissues from patients with MLD has already been discussed; deficient activity against each of the other substrates has also been demonstrated (Fluharty *et al.* 1976). Deficiency of sulpho-galactosylsphingosine sulphatase might lead to an accumulation of sulphogalactosyl sphingosine, which is highly toxic and which, as has been postulated for galactosyl sphingosine in Krabbe's disease, might lead to myelin degeneration.

Although there is cross-reacting material in MLD tissues against antibody to normal liver ASA (Stumpf *et al.* 1971; Luijten *et al.* 1978), any residual activity has been attributed to arylsulphatase B (Shapira and Nadler 1975).

Since there appears to be no difference in the residual ASA activity in late infantile, juvenile, or adult forms of MLD (Porter *et al.* 1971*a*; Percy *et al.* 1977) it is interesting that a correlation has been reported between the amount of accumulation of ^{35}S-cerebroside sulphate by cultured fibro-blasts, and the age of onset of MLD. Thus Porter *et al.* (1971*a*) found that cultured fibroblasts from patients with late infantile MLD accumulated the lipid more rapidly, and released ^{35}S-sulphate more slowly than control fibroblasts. Cells from patients with the juvenile type had intermediate values, while fibroblasts from an adult type patient gave values inter-mediate between those of the juvenile type and of controls.

Differences in the ability to hydrolyse sulphatide presented to fibroblasts in culture can also be visualized by microscopy. Thus, fibroblasts from the late infantile type develop visible lipid-laden lysosomes when cultured in a medium containing added sulphatide, while no detectable differences were observed under the same conditions between cells from the adult type and controls (Percy *et al.* 1977).

Hybridization of cultured fibroblasts from patients with the late infantile, and adult forms of MLD does not restore ASA activity, suggesting that the mutations in the two forms of the disease are allelic (Kaback *et al.* 1972).

Sulphatide (Fig. 4.9) storage was first identified by Jatzkewitz (1958) and Austin (1959). Although, in the late infantile type of MLD, the sulphatide content is increased by three to ten times normal, the total lipids of the white matter are reduced to approximately two-thirds of normal. Excess sulphatide is present in myelin isolated prenatally (Suzuki *et al.* 1967), and postnatally (Poduslo *et al.* 1976). The reduction of total lipids in white

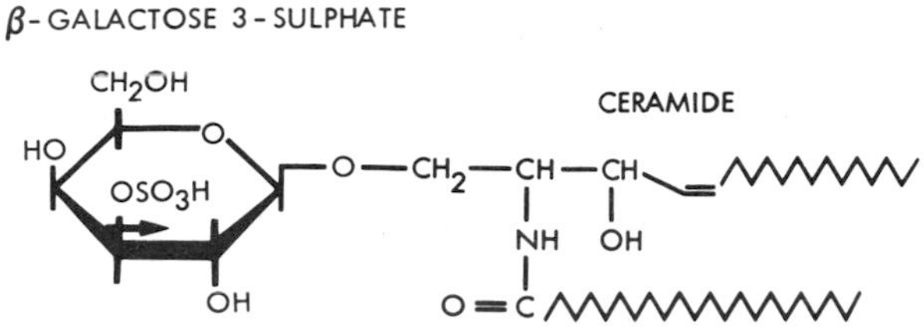

β- GALACTOSE 3 - SULPHATE

Fig. 4.9. Sulphatide.

matter is accounted for by loss of cholesterol, phospholipids, sphingo-myelin, cerebrosides, and other myelin lipids.

In brain from patients with the adult type, sulphatide concentration is less increased than in the late infantile type (Pilz and Heipertz 1974).

A deficiency of long-chain fatty acids has been found in MLD white matter sphingomyelin and cerebrosides (O'Brien and Sampson 1965). Long-chain (C21–26) unsubstituted fatty acids were reduced from 78 per cent in controls to 22 per cent in MLD. This has been interpreted to indicate a return to pre-myelination pattern (Malone and Stoffyn 1966) coinciding with demyelination, which is also seen in other demyelinating disorders (Stallberg-Stenhagen and Svennerholm 1965).

Cerebroside sulphate sulphatase activator deficiency as a cause of the juvenile type

Jatzkewitz and Stinshoff (1973) showed that an activator for cerebroside sulphate sulphatase was present in human tissue from both normal subjects and MLD patients. Later the factor was extensively purified and shown to be a protein with molecular weight about 21 500, and a model for the activation process was proposed (Fischer and Jatzkewitz 1975). A variant form of MLD which may be due to deficiency of the cerebroside sulphate sulphatase activator was reported by Stevens *et al.* (1981) in two siblings of consanguineous parents (Shapiro *et al.* 1979a). The clinical features were consistent with the juvenile type of the disease but the activities of cultured fibroblast ASA were about half the normal value. However, hydrolysis of cerebroside sulphate by growing cultured fibroblasts was very defective. Supplementation of the fibroblasts with cerebroside sulphatase activator, by the method of Fischer and Jatzkewitz (1975), normalized the response in the loading test.

Diagnosis

In most laboratories, the diagnosis of MLD is usually confirmed by the demonstration of severe deficiency of ASA activity in leucocytes or cultured fibroblasts (for references, see Molecular defects, p. 117). However in some heterozygotes, activities of ASA overlap with those of affected patients (Lott *et al.* 1976; Kihara *et al.* 1973*a*; Dubois *et al.* 1980). In contrast, cerebroside sulphate sulphatase activity is usually completely absent in patients (Raghavan *et al.* 1981).

An independent method for diagnosis of MLD in cultured fibroblasts is to demonstrate that they are unable to hydrolyse ^{35}S-sulphatide added to the culture medium. This method of sulphatide loading (Porter *et al.* 1970; Fluharty *et al.* 1978) is probably the most reliable laboratory technique to establish the diagnosis in patients, and foetuses, especially when one or both parents have severely reduced ASA activity (see 'Genetic counselling and prenatal diagnosis', p. 121). Furthermore, although cultural fibroblasts from MLD patients have severe deficiency of ASA and show abnormal accumulation of ^{35}S-sulphatide, they do not exhibit metachromatic granules unless sulphatide is added to the culture medium (Porter *et al.* 1970). Fibroblasts from normal individuals which can degrade sulphatide do not become metachromatic.

The excessive accumulation of sulphatide by MLD fibroblasts in culture can be largely corrected by the addition of ASA from normal human urine to the culture medium (Wiesmann *et al.* 1971*b*; Porter *et al.* 1971*b*).

It appears therefore that the study of sulphatide metabolism of growing intact cells is a better guide to the genotype than study of enzyme activity in disrupted cells.

Assay of ASA is usually carried out using *p*-nitrocatechol sulphate (*p*NS) as substrate. Since arylsulphatase B (ASB) also releases sulphate from *p*NS, conditions of the assay are adjusted to minimize ASB activity. Thus the procedure of Percy and Brady (1968) is carried out at pH5, rather than at pH6, which favour ASB activity, in the presence of chloride and pyrophosphate, which inhibit ASB more than ASA (Dulaney and Moser 1977*a*). Using these conditions, ASB contributes only 2 per cent of activity.

Methods where ^{35}S-labelled galactosyl sulphatide is used as substrate are specific for ASA (Booth *et al.* 1975) but the substrate is not available commercially and must be prepared biosynthetically (Fluharty *et al.* 1974*b*).

Methods for differential assay of ASA and ASB include partitioning before assay by electrophoresis on cellulose acetate gel strip and staining by incubation with 4-methylumbelliferyl sulphate (Rattazzi *et al.* 1973). Alternatively, a method has been developed which utilizes selective heat inactivation of ASA rather than ASB at 60°C (Kolodny and Mumford 1976).

Discrimination between homozygote affected individuals on the one hand and heterozygotes with very low ASA and cerebroside sulphatase activities on the other can sometimes be assisted by assay of urinary sulphatide excretion, peripheral nerve conduction velocity and sural nerve biopsy. Excessive excretion of sulphatide in the urine of patients with MLD has been demonstrated as metachromatic material, in renal epithelial cells or fragments (Austin 1957; Lake 1965; MacFaul *et al*. 1982) or after extraction by lipid solvents (Austin 1957). More precise techniques allowing quantitative assay of sulphatides have been reported by Philippart *et al*. (1971) and Pilz *et al*. (1973). The last-named group has demonstrated excessive urinary excretion of sulphatide in three individuals, before the onset of symptoms.

Genetic counselling and prenatal diagnosis

The late infantile, juvenile, and adult forms of MLD have an autosomal recessive mode of inheritance. Thus one in four siblings, irrespective of sex, are affected (Gustavson and Hagberg 1971; Schutta *et al*. 1966). Consanguinity was reported in three of 11 affected families by Gustavson and Hagberg (1971). The diagnosis of heterozygotes is discussed under 'Diagnosis', p. 119.

The incidence of the late infantile form has been estimated as 1:40 000 in Sweden. Estimates from reported series of patients (Gustavson and Hagberg 1971; Schutta *et al*. 1966) suggest that the incidence of the juvenile form may be four times rarer than the late infantile form, or about 1:160 000. An exceptionally high incidence of late infantile MLD (1 per 75 live births) has been reported in a small isolate of 1000–1200 Habbanite Jews (Zlotogora *et al*. 1980).

Accurate prenatal diagnosis of MLD has been reported by Leroy *et al*. (1973), Wiesmann *et al*. (1975) and van der Hagen *et al*. (1973) by demonstration of reduced ASA activity in cultured amniotic cells.

Reduced activity of ASA has also been reported in amniotic fluid supernatant from affected pregnancies (Borresen and van der Hagen 1973; Harzer *et al*. 1975*b*; Rattazzi and Davidson 1977) but until more data are available results obtained by this method should be confirmed by assay of cultured amniotic cells.

Arylsulphatase A activities of cultured amniotic cells from affected pregnancies using *p*NS as substrate (Dulaney and Moser 1977*a*; Percy and Brady 1968; see 'Diagnosis') were reduced to 7 per cent (Leroy *et al*. 1973), 3 per cent (Wiesmann *et al*. 1975) and 4 per cent (Van der Hagen *et al*. 1973) of control values.

Several studies have shown wide variations of ASA activities in cultured cells, according to cell population densities and the period following subcultivation. Care should be taken, therefore, in standardizing culture conditions before assay (Heukels-Dully and Niermeijer 1976).

It is important to assay the parents' fibroblast ASA activity before attempting diagnosis, since as discussed under 'Diagnosis', p. 119, in some cases heterozygous values are very low. Thus in a family reported by Dubois *et al.* (1977), cultured fibroblasts from the father of an affected child had cerebroside sulphatase activity of 10.7 per cent of the control mean, but ASA activity was only 5.9 per cent. In the family reported by Booth *et al.* (1975), cultured amniotic cells from a pregnancy at risk for MLD had cerebroside sulphatase activity of 50 per cent of control mean and ASA activities ranging from 20 per cent to 60 per cent of normal. Postnatal studies indicated that the infant was a heterozygote. In another family (Kihara *et al.* 1980), MLD was diagnosed in a girl where the mother was in the 10th week of pregnancy. The mother's leucocyte ASA was undetectable (control mean 106.1; father 28.1 nmoles/h/mg protein). Fibroblast cultures were initiated but not sufficiently advanced for assay when cultured amniotic cells were assayed at 17 weeks' gestation following amniocentesis at 13 weeks. Cultured amniotic fluid cell ASA activity was absent, but in view of the absence of activity in maternal leucocytes, the interpretation was questionable. Remaining amniotic cells were cultured further and examined by a modification of the sulphatide loading test (Porter *et al.* 1970) by growing cells in medium containing ^{35}S-labelled sulphatide. After incubation, cellular sulphatide was extracted and assayed for radioactivity. A clear result was obtained, with 82–95 per cent of incorporated sulphatide being hydrolysed by control cells, and none by amniotic cells. The foetus was aborted at 20 weeks' gestation and the prenatal diagnosis confirmed. Subsequently, maternal fibroblasts were shown to have ASA and cerebroside sulphatase activities of 9.1 per cent and 1.2 per cent of control values, respectively.

Treatment

Evidence of correction of the enzyme deficiency has been obtained in cultured fibroblasts (Porter *et al.* 1971*b*; Wiesmann *et al.* 1971*b*) or in foetal brain cells or explants (Poduslo *et al.* 1976) by incubation with exogenous enzyme *in vitro*.

Enzyme administration has been attempted by Greene *et al.* (1969*a*) by intravenous administration of purified brain ASA in a three-year-old boy with late juvenile MLD. High ASA levels were present in liver 6h and 18h after administration. The same workers also gave the enzyme intrathecally, and demonstrated activity in the cerebrospinal fluid 8h later.

In an attempt to overcome the 'blood–brain barrier', Austin (1973*a*) injected purified brain ASA in the subarachnoid space. A pyrexial reaction precluded further administration.

Since vitamin A appears to be needed for sulphatide synthesis, Melchior and Clausen (1968) restricted vitamin A intake to a three-year-old girl with late juvenile MLD, for four months. Although the clinical features were unchanged, the authors demonstrated reduced urinary glycolipid excretion during the period of treatment. More encouraging results were reported by Moosa and Dubowitz (1971), who noticed unequivocal clinical improvements in a three-year-old girl with MLD, six weeks after starting a low-vitamin A diet. No further deterioration occurred over the next two years when diet was maintained. Failure to prevent regression by low vitamin A diet, however, was reported by Warner (1975). A low sulphur diet in two patients with MLD produced a 90 per cent reduction in urinary inorganic sulphate excretion, but no reduction in sulphatide excretion, and no change in the rate of incorporation of intravenously administered ^{35}S-sulphate into urinary sulphatide (Moser *et al.* 1967). Further experience however has shown that neither form of diet prevents the clinical progression.

Multiple sulphatase deficiency; mucosulphatidosis

Multiple sulphatase deficiency (MSD) is a rare disorder characterized clinically by the combined features of late infantile MLD and of a mucopolysaccharidosis. There is deficient activity of at least seven sulphatases, two of which have structural genes on the X-chromosomes, and the rest on autosomes (Murphy *et al.* 1971; Austin 1973*b*; Eto *et al.* 1974*b*, *c*; Basner *et al.* 1979).

Clinical features

Early development is normal, but during the first two years there is a disturbance of motor function, as described for late infantile MLD, but progression is usually more severe. Affected children may learn to stand with support but do not usually reach the stage of walking, or acquire meaningful speech. Bilateral optic atrophy leads to blindness. In addition, there is lumbar kyphosis, hepatosplenomegaly, and coarsening of facial features. Ichthyosis, which occurs in some patients with MSD also occurs in isolated deficiency of steroid sulphatase (see X-linked ichthyosis, p. 126). The progressive disease leads to death usually between three and 12 years (Austin 1973*b*).

Radiologically there is dysostosis multiplex resembling that of Hurler's disease, including beaking of the first and second lumbar vertebra, thickening of phalangeal bones and a 'J-shaped' sella turcica (Austin 1973*b*).

A patient with MSD with onset in the neonatal period has been reported by Vamos *et al.* (1981), and we are aware of a further case. The patient of Vamos *et al.* (1981) was small for dates at birth. He had a disproportionate dwarfism with short neck and limbs, bilateral club feet, and stiff joints. A facial dysmorphism with coarse features, low set ears, and epicanthic folds were noted. The liver was enlarged 3 cm below the costal margin and bone

X-rays showed severe hypoplasia of all vertebral bodies and multiple epiphyseal dysplasia. At age two-and-a-half months hydrocephalus, hepatosplenomegaly, and poor somatic and psychomotor development were present, together with corneal clouding which had not previously been reported in MSD.

Pathology

As for MLD, the cerebral white matter feels firm. Histologically there is loss of myelin sheaths. Numerous macrophages with metachromatic granules are seen in perivascular areas. There is loss of oligodendroglia. Electron microscopy reveals lamellar and granular storage material (Raynaud *et al.* 1975). There are metachromatic cytoplasmic inclusions in peripheral blood and bone marrow leucocytes (Austin 1973*b*).

Molecular defects

The primary defect is not known but there is deficiency of arylsulphatase (AS) A, B, and C in tissues, leucocytes and cultured fibroblasts (Murphy *et al.* 1971; Eto *et al.* 1974*c*). Compared with control cells, Fiddler *et al.* (1979) reported that MSD fibroblasts from two patients had ASA activities of less than 1 per cent and about 7 per cent of normal, respectively; ASB of about 5 per cent and 10 per cent; and ASC of 40–60 per cent. Deficiency of cholesterol sulphate sulphatase and dehydroepiandrosterone sulphate sulphatase have been found in liver by Murphy *et al.* (1971), and in liver, kidney, and brain by Eto *et al.* (1974*b*), who also demonstrated decreased psychosine sulphate sulphatase and cerebroside sulphate sulphatase activities. Evidence for deficiency of sulphoiduronate sulphatase and heparan *N*-sulphatase was provided by the work of Eto *et al.* (1974*c*) who found that MSD fibroblasts did not correct abnormal mucopolysaccharide catabolism in fibroblasts from patients with Hunter or Sanfilippo A disease. Basner *et al.* (1979) demonstrated deficiency of these last-named sulphatases in four MSD fibroblast strains by direct assay of the enzymes with natural substrates, and also found deficiency of *N*-acetylgalacto-samine and *N*-acetyglucosamine 6-sulphate sulphatases, and of ASB (using UDP-*N*-acetylgalactosamine 4-sulphate as substrate) in all four strains.

Thus there is evidence of deficient activity of enzymes whose structural genes are sited either on the X-chromosome (sulphoiduronate sulphatase, steroid sulphatase) or on different autosomes (e.g. ASA on chromosome 22, ASB on chromosome 5), and which are either lysosomal (e.g. ASA, ASB, sulphoiduronate sulphatase, heparan *N*-sulphatase) or microsomal (steroid sulphatase).

Fiddler *et al.* (1979) examined cross-reacting material in cultured skin fibroblasts from individuals with MSD, MLD (and therefore ASA deficiency), Maroteaux–Lamy disease (and therefore ASB deficiency), and normal controls. They found that in MSD cells, cross-reacting materials to anti ASA and anti-ASB were present, but in reduced concentrations which were proportional to reduced enzyme activities. Since the two enzymes are structurally unrelated, the authors suggested that the disorder was the result of a defect in a process which coordinates the expression of several sulphohydrolases. This is compatible with the concept that deficient sulphatase activities in MSD are due to reduced amounts of otherwise normal enzymes. A similar conclusion was reached by Fluharty *et al.* (1978) and Basner *et al.* (1979). This is supported by the studies of Chang and Davidson (1980) who demonstrated that somatic hybrids produced by fusing cells from patients with MLD and MSD showed complementation with restoration of ASA activity. The complemented enzyme resembled normal ASA with respect to heat stability, pH optimum, K_m, electrophoretic mobility, and immunological reactivity.

Furthermore, Fluharty *et al.* (1979) have shown that MSD fibroblasts expressed ASA deficiency when cultured at low pH (less than 7); while in high pH media (7.4), cells produced increased enzyme activities. This phenomenon was shown to be reversible. Kresse *et al.* (1980) have shown that all the sulphatases with low activity (with the possible exception of heparan *N*-sulphamidase) are stimulated by the addition of sodium thiosulphate to the medium for one or two months. All these studies are consistent with the view that the structural genes for the affected sulphatases are normal, but that this monogenic disease is due to a defective regulatory gene, causing reduced amounts of structurally normal enzymes. It remains to be established if this is due to reduced synthesis, or accelerated degradation.

The urine in MSD contains a five- to ten-fold increase in the concentration of mucopolysaccharides (Austin 1973*b*), which consist of dermatan sulphate and heparan sulphate in approximately equal quantities (Murphy *et al.* 1971); and an increase in the concentration of cholesterol sulphate. A five-fold increase of mucopolysaccharides has been reported in kidney, while an increased concentration of cholesterol sulphate has been reported in liver, kidney, and plasma (Austin 1973*b*; Murphy *et al.* 1971). Cultured fibroblasts show defective degradation of radiosulphate-labelled glycosaminoglycans (Eto *et al.* 1974*c*).

Diagnosis

The diagnosis of MSD can be confirmed by demonstrating deficiency of ASA, ASB, and ASC in either leucocytes (Humbel 1976) or cultured fibroblasts (Eto *et al.* 1974*c*; Fiddler *et al.* 1979). Further confirmation includes demonstration of excess urinary glycosaminoglycans (Austin 1973*b*) and cholesterol sulphate (Murphy *et al.* 1971).

Genetic counselling and prenatal diagnosis

The occurrence of the disorder in sib pairs and in both sexes is consistent with autosomal recessive inheritance but neither decreased sulphatase activity in carriers nor prenatal diagnosis appear to have been established.

X-linked ichthyosis; steroid sulphatase deficiency

This is not a sphingolipidosis but is included here for convenience.

Ichthyosis is a disorder in which there are visible scales of the skin, the appearance of which has been compared to adherent fish-scales. Several clinical forms have been described, either occurring sporadically, or with various patterns of inheritance (Goldsmith 1976; Wells and Kerr 1965, 1966).

Shapiro *et al.* (1978) and Koppe *et al.* (1978) independently demonstrated that patients with X-linked ichthyosis have deficiency of steroid sulphatase, a disorder which was first recognized because pregnant women with affected male foetuses excreted reduced amounts of oestriol and had low serum oestriol concentrations (France and Liggins 1969; Shapiro *et al.* 1976*b*).

Clinical features

Ichthyosis is present at birth in about one-fifth of patients, becoming noticeable in the others by the end of the first year. Only affected males have been reported. None of 61 females considered to be heterozygotes on genetic grounds had any signs of ichthyosis (Kerr and Wells 1965). Ninety-one affected males from 31 families had similar clinical features. There was dry scaling on anterior and posterior surfaces of all limbs and the trunk in all cases. The scalp was involved in 63 per cent of cases and the neck in 57 per cent. The axillae, antecubital fossae, and popliteal fossae were affected in 20 per cent of cases, but the palms and soles appeared normal. In 70 per cent of cases the scales were dark (ichthyosis nigricans) and large (more than 4 mm in greatest width). In most cases the head and neck tend to clear with advancing age, while the abdomen and lower limbs become more markedly involved. Warm weather favours scale formation.

In primiparous women, steroid sulphatase deficiency has been associated with prolonged gestation and sometimes difficulty with induction of labour. Multiparous women, however, have usually had normal spontaneous labour and delivery (France *et al.* 1973; Oakey *et al.* 1974; Braunstein *et al.* 1976; Tabei and Heinrich 1976). The disorder does not appear to have an adverse effect on the foetus or newborn.

Pathology of X-linked ichthyosis

Histologically, affected males have an increased width of the stratum corneum (hyperkeratosis), and the stratum granulosum. A perivascular infiltration occurs in all cases. The rete ridges are abnormally prominent in some lesions.

Female heterozygotes have normal histology.

Molecular defects and diagnosis

Patients with X-linked ichthyosis have absent or considerably reduced cultured fibroblast steroid sulphatase activity (Shapiro *et al.* 1978; Kubilus *et al.* 1979). The assay involves incubation of cell homogenates with tritiated dehydroepiandrosterone (DHEA) sulphate at pH 7.2 for two hours. The DHEA is extracted into benzene or ether and assayed for radioactivity.

Placental steroid sulphatase hydrolyses steroid alcohol 3-sulphates and is distinct from arylsulphatase (AS)A and ASB, but is probably identical to ASC. Placental steroid sulphatase deficiency has been reported in several pregnancies (France *et al.* 1973; Oakey *et al.* 1974; Tabei and Heinrich 1976). A combined deficiency of placental sulphatase and placental total arylsulphatase has been reported (France *et al.* 1973; Oakey *et al.* 1974; Braunstein *et al.* 1976; Tabei and Heinrich 1976).

Steroid sulphatase in the placenta catalyses conversion of androgen sulphates such as dehydroepiandrosterone sulphate to free androgens, which can then be converted to oestrogens. In placental steroid sulphatase deficiency oestrogen excretion is about 1 mg/24h, compared with 40–60 mg/24h in normal pregnancies. Pregnandiol and 17-oxogenic steroid excretion is within the normal range. Placental sulphatase activity against pregenolone sulphate and oestrone sulphate is virtually undetectable (Oakey *et al.* 1974). Plasma concentrations of oestradiol-17ß progesterone, 17 α-hydroxyprogesterone, 11ß-hydroxycorticosteroids and corticosteroid-binding globulin are reduced (Oakey *et al.* 1974).

Differential diagnosis from foetal adrenal hypoplasia, which has a high mortality, and which is also associated with very low maternal oestriol levels (Fliegner *et al.* 1971) may be achieved by demonstrating steroid sulphatase deficiency as lack of rise in urinary oestrogens after infusion (Braunstein *et al.* 1976) or intra-amniotic instillation (Tabei and Heinrich 1976) of DHEA sulphate.

Immunological and complementation studies have shown no heterogeneity between patients from different geographical regions (Shapiro and Mohandes 1980).

Genetic counselling and prenatal diagnosis

Placental steroid sulphatase deficiency and X-linked ichthyosis have been described only in male foetuses and patients, respectively.

The frequency of placental steroid sulphatase deficiency has been estimated as less than 1 in 5000 pregnancies (Oakey *et al.* 1974). In Berkshire, England, the frequency of sex-linked ichthyosis was about 1:6000 males (Kerr and Wells 1965; Wells and Kerr 1966), while in the same study the frequency of the autosomal dominant type of ichthyosis was 1:2662 males. The observed high incidence of X-linked ichthyosis indicates that it is the most frequent known X-linked mutation after those affecting Xg blood groups, partial colour blindness, Duchenne's muscular dystrophy and, in certain populations, glucose 6-phosphate dehydrogenase deficiency.

The gene for steroid sulphatase in closely linked with the gene for the Xg^a locus and is situated on the tip of the short arm of the X-chromosome (Muller *et al.* 1980; Tiepolo *et al.* 1980). The steroid sulphatase locus, like the Xg^a locus (Fialkow 1970; Ducos *et al.* 1971) appears to escape Lyonization as shown by the presence of only one population of cells with normal enzyme activity in female heterozygotes (Shapiro *et al.* 1979*b*).

Late prenatal diagnosis of affected pregnancies is possible by demonstrating low urine oestriol. Early prenatal diagnosis has been reported by Hähnel *et al.* (1982) at 16 weeks' gestation by demonstration of low activity of steroid sulphatase in cultured amniotic cells and high levels of dehydroepiandrosterone sulphate in amniotic fluid, in addition to very low oestriol concentrations in maternal blood and urine, increased maternal plasma dehydroepiandrosterone sulphate, and very high maternal urinary 16 α-hydroxydehydroepiandrosterone sulphate.

The demonstration of high steroid sulphatase activity in normal chorionic villi (Lam *et al.* 1984) suggests the possibility for prenatal diagnosis during the first trimester of pregnancy.

Fabry's disease; angiokeratoma corporis diffusum; α-galactosidase A deficiency

This disease was described independently in the same year in Germany by Fabry (1898) and in England by Anderson (1898). Eponymous designation by general consent favours the former author, who erroneously considered the lesions to be purpura nodularis, rather than the latter, who accurately identified angiokeratoma and suggested that the involvement of the kidneys might also be due to a generalized abnormality of blood vessels.

Hemizygote males affected by this X-linked disorder suffer episodes of excruciating pain and develop the characteristic skin lesions usually in the second decade. Vascular disease and renal failure appear later, and characteristically are the cause of death in the fifth decade. A milder form of the disease occurs in some heterozygous females. The disease is due to accumulation of ceramide trihexoside Gal-Gal-Glc-Cer (Sweeley and

Klionsky 1964) because of deficient activity of a lysosomal trihexosidase (galactosyl hydrolase) (Brady *et al.* 1967), identified by Kint (1970) as α-galactosidase and later shown to be the predominant 'A' isoenzyme (Johnson and Brady 1972; Ho *et al.* 1972*b*).

Clinical features

Affected male hemizygotes are symptomless until the end of the first decade or the beginning of the second. The most distressing symptom is severe pain of the extremities, which characteristically occurs in crises which last from a few minutes to one or more days. The pain is excruciating, and burning in quality. Crises are often provoked by heat, exercise, or emotional upset, and are sometimes accompanied by fever. They may be interpreted as symptoms of rheumatic fever or neurosis (personal observations; Bagdale *et al.* 1968; Lockman *et al.* 1973). In between crises, a burning, sustained less severe discomfort is experienced by most patients (Wise *et al.* 1962; Lockman *et al.* 1973).

Angiokeratoma usually first appear about the time pain is experienced (Johnston *et al.* 1968). Usually they occur in clusters of sharply-defined dark-red lesions which do not fade on pressure, and are flat, or raised, smooth, or hyperkeratotic. The distribution of angiokeratoma is variable. They often occur on scrotum, penis (Fig. 4.10), buttocks (Fig. 4.11), back, thighs and abdomen (especially round the umbilicus (Fig. 4.12). There is a tendency to symmetry. Pains and angiokeratoma resembling those of Fabry's disease sometimes occur in sialidosis type 2, and in patients with fucosidosis. Rarely Fabry's disease has been reported to occur without

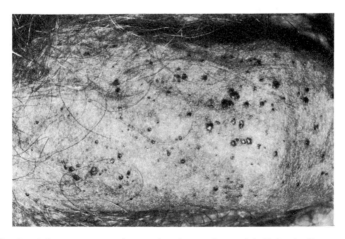

Fig. 4.10. Angiokeratoma on the penis of a patient with Fabry's disease.

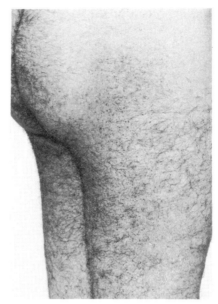

Fig. 4.11. Angiokeratoma on the buttocks of a patient with Fabry's disease.

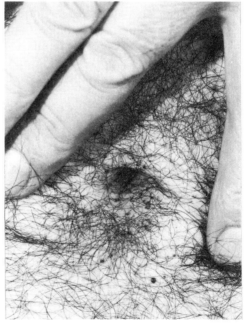

Fig. 4.12. Angiokeratoma around umbilicus of a patient with Fabry's disease.

angiokeratoma (Romeo *et al.* 1972, 1975; Clarke *et al.* 1971). Diminished sweating or anhydrosis is common.

During the third decade pains and crises subside and the patients tend to feel a general improvement until the onset of renal and cardiovascular disease, usually in the fourth decade. Frequent presenting symptoms in the thirties are due to hypertension and include cerebral infarction or haemorrhage, and myocardial infarction (Wise *et al.* 1962) leading to hemiplegia and angina. Patients often have valvular disease of the heart such as mitral regurgitation or aortic incompetence (Desnick *et al.* 1976*b*). Other complications include fits and regression of personality.

Progressive renal failure is accompanied by proteinuria, defective tubular function (Pabico *et al.* 1973), and azotaemia, and is often treated by dialysis and renal transplantation (Desnick *et al.* 1972; Phillipart *et al.* 1972). Death is usually in the early forties (Colombi *et al.* 1967), though survival into the sixties has been recorded (Johnston *et al.* 1968).

Ocular lesions of Fabry's disease are characteristic and may be diagnostic even if the disease is not suspected (personal observation). The earliest change is diffuse corneal haziness seen on slit lamp examination which is present before symptomatic onset, even during infancy (Spaeth and Frost 1965). With increasing age, corneal opacities appear as whorls from the centre of the cornea, extending to the periphery (Wise *et al.* 1962).

Tortuous conjunctival (Fig. 4.13) and retinal vessels are also early features and may be seen in childhood.

Other features include limitation in extension of terminal interphalangeal joints (Wise *et al.* 1962), lymphoedema, and varicose veins of the legs.

Heterozygote females

The majority of female heterozygotes experience attenuated symptoms (Wise *et al.* 1962), some are symptomless (Avila *et al.* 1973), and a few experience severe manifestations. The most common manifestation in heterozygote females are corneal opacities, similar to those described for hemizygote males (Wise *et al.* 1962; Johnston *et al.* 1966, Weingeist and Blodi 1973).

Skin lesions are less common but are similar although less severe than in hemizygous males. Pains in hands and feet may be provoked by high environmental temperature, or pyrexial illness, as in hemizygotes.

More serious features, including hypertension, proteinuria, and renal failure, occur rarely (personal observations; Wise *et al.* 1962; De Groot 1964; Colombi *et al.* 1967).

Pathology

At necropsy, the naked eye changes are those associated with renal failure

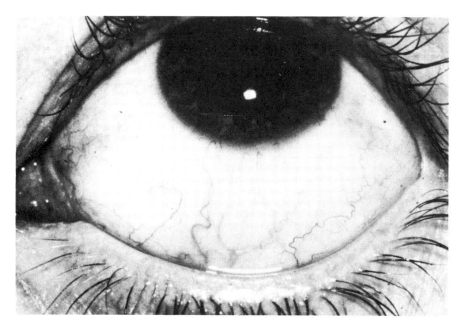

Fig. 4.13. Tortuous conjunctival vessels in a patient with Fabry's disease.

and hypertension, including left ventricular hypertrophy, myocardial ischaemia, and cerebral vascular thrombosis or haemorrhage.

Histologically, the punctate skin lesions contain dilated small blood vessels just below the epidermis in dermal papillae, with aneurysmal dilation in deeper layers. There is lipid storage in the endothelium, perithelium and muscle of small blood vessels (Pittelkow *et al.* 1957; Sagebiel and Parker 1968). Lipid storage also occurs in sweat gland epithelium and in erector pili muscle fibres (Witschel and Mayer 1968). There may be keratosis over the suface of protruding lesions. The finding of lipid deposits in biopsies of skin which appear normal clinically (Sagebiel and Parker 1968) suggests that these early changes may progress to vascular pathology.

Early changes in the kidney are deposits of lipid in epithelial cells (Fig. 4.14) of the glomeruli, distal tubules, and the loops of Henle, and in glomerular endothelium and Bowman's capsules. The proximal tubules are affected least, and probably later (McNary and Lowenstein 1965). Lipid-laden tubular epithelial cells become detached and appear in the urine (Desnick *et al.* 1970).

In the nervous system lipid deposits occur in neurones of the autonomic system, in the perineural sheaths of peripheral nerves, in basal ganglia, and in cranial nerve nuclei. Lipid deposits are also frequent in blood vessels of the nervous system (Ohnishi and Dyck 1974; Tabira *et al.* 1974).

As in the rest of the body, abnormal lipid deposits are widespread in the vascular system of the eyes. Deposits occur in endothelium, perithelium, and muscle of ocular blood vessels of the lens, cornea, conjunctiva, iris, and ciliary body (Weingeist and Blodi 1973).

Lipid infiltration of cardiac valves and muscle cells, causing cardiomegaly, is common (Desnick *et al.* 1976*b*).

Vacuolated lipid-laden cells are found generally in the reticuloendothelial system, including foam cells in bone marrow (Frost *et al.* 1966).

The stored material is often crystalline and shows birefringence with the appearance of Maltese crosses. It is lost if the tissue is treated with lipid solvents such as xylene. The lipid is retained on formalin fixation, and is periodic acid–Schiff positive, even after glycogen is removed by diastase digestion. It is not metachromatic and is positive to lipid stains (Frost *et al.* 1966; Van Mullem and Ruiter 1968).

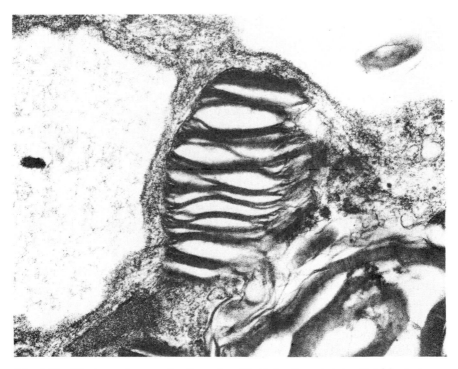

Fig. 4.14. Electronmicrograph of renal epithelial cell containing lipid inclusions, showing a membrane-like periodicity (we are grateful to Dr. S.R. Aparicio for this picture).

On electron microscopy, lipid deposits are seen in lysosomes which are identified by the single limiting membrane and positive staining for acid phosphatase (Hashimoto *et al.* 1965). Storage material typically has lamellar or concentric arrangements (Frost *et al.* 1966; Van Mullem and Ruiter 1970) and has been detected in skin, kidney, and most viscera.

Molecular defects

Fabry's disease is caused by severe deficiency of the lysosomal enzyme α-galactosidase A. Deficient ceramide trihexosidase activity in Fabry's disease was first demonstrated by Brady *et al.* (1967) with a radiolabelled natural substrate for the enzyme assay. Using the artificial substrates 4-methylumbelliferyl-α-D-galactoside and *p*-nitrophenyl-α-D-galactoside, Kint (1970) first showed that the deficient enzyme was an α-galactosyl hydrolase. This observation was soon followed by confirmation of the alpha anomeric configuration of the terminal galactosyl residue in the ceramide trihexoside which accumulates in Fabry's disease patients (Li and Li 1971). It was further shown that there were two forms of α-galactosidase (designated A and B) which have different thermostability (Desnick *et al.* 1973*b*; Kint 1971; Beutler and Kuhl 1972), and electrophoretic mobilities. The major isoenzyme was shown to be deficient in Fabry's disease (Ho *et al.* 1972*b*; Johnson and Brady 1972).

Deficiency of α-galactosidase A has been demonstrated in numerous tissues, including liver (Ho 1973), kidney (Kano and Yamakawa 1974), body fluids such as plasma (see 'Diagnosis' p. 136), tears (Johnson *et al.* 1975), leucocytes, and fibroblasts (see 'Diagnosis').

No immunological material cross-reacting with α-galactosidase A has been demonstrated in patients with Fabry's disease (Beutler and Kuhl 1973; Rietra *et al.* 1974; Hamers 1978), suggesting either absence of enzyme or, if an altered enzyme is present, a structural gene mutation affecting the immunologically reactive site. Hamers *et al.* (1977) conclude from their somatic cell hybridization experiments that absence of immunologically detectable α-galactosidase A is unlikely to be due to a regulatory gene mutation.

Residual α-galactosidase activity in Fabry's disease against artificial substrates is largely due to the presence of the α-galactosidase B isoenzyme. (Schram and Tager 1981). Immunologically there is no cross-reactivity between antibodies to α-galactosidase B and α-galactosidase A (Beutler and Kuhl 1972; Rietra *et al.* 1974; Schramm *et al.* 1977; Hamers 1978). α-galactosidase B is highly active as an *N*-acetyl-α-galactosaminidase (Dean *et al.* 1977), with which it is identified immunologically (Schramm *et al.* 1977, 1978). The gene coding for α-galactosidase A is on the X-chromosome (see 'Genetic counselling', p. 136); and for α-galactosidase B (*N*-acetyl- α-galactosaminidase) on chromosome 22 (De Groot *et al.* 1978).

There is progressive accumulation of the triglycosylceramide (ceramide trihexoside) Gal-Gal-Glc-Cer (Fig. 4.15), and other neutral glycosphingo-lipids with terminal α-galactosyl residues, but no accumulation of glycosaminoglycans or acidic sphingolipids (such as sialated gangliosides or sulphatides). Ceramide trihexoside predominates in extra-neuronal tissues, unlike gangliosides, which occur in both neural and extraneural tissues, but especially in the central nervous system. It is derived from globoside (GalNAc-Gal-Gal-Glc-Cer) which is found in kidney and erythroctes (Vance *et al.* 1969; Burkholder *et al.* 1980). The first step in the degradation of globoside is catalysed by ß-*N*-acetylhexosaminidase, yielding ceramide trihexoside, which in Fabry's disease accumulates because of deficiency of α-galactosidase, the exoglycosidase, which normally catalyses cleavage of the α-galactosyl residue. Lesser accumulation has been reported of digalactosyl ceramide (galabiosylceramide; Gal-Gal-Cer) in kidneys (Sweeley and Klionsky 1963; Christenson-Lou 1966) and pancreas (Schibanoff *et al.* 1969); and of blood group B-active glycosphingolipids in a patient of blood group B (Wherret and Hakamori 1973).

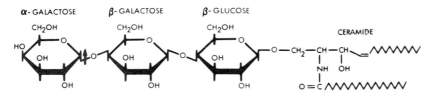

Fig. 4.15. Ceramide trihexoside.

In Fabry's disease concentrations of glycosphingolipids have been found to be 30- to 300-fold higher than normal (Schibanoff *et al.* 1969). Apart from erythrocytes, all tissues and cell types examined have had sphingo-lipid accumulation. The kidneys, lymphatic system and the prostate are especially affected (Christensen-Lou 1966; Miyatake 1969). Plasma ceramide is three- to four-fold higher than in normal individuals (Vance *et al.* 1969). Abnormally increased excretion of sphingolipid in the urine has also been reported (Desnick *et al.* 1970).

The relation of the symptoms to the basic defect is not known but some speculation may be made. There is a generalized disorder of the smaller blood-vessels characterized by narrowing, because of lipid deposition, and swelling and proliferation of the endothelium (Schibanoff *et al.* 1969). Such narrowing predisposes to ischaemia, infarction, and hypohydrosis and may contribute to pain. It has been suggested that the pain may result from involvement of peripheral nerve fibres or the dorsal root ganglia (Tabira *et al.* 1974; Tome *et al.* 1977; Sung 1979); and that febrile episodes may be due to involvement of hypothalmus (Rahman and Lindenberg 1963). Renal failure can be attributed to lipid deposition in specific sites of the nephrons and in small renal blood vessels (see 'Pathology', p. 131).

Diagnosis

The diagnosis of an affected male hemizygote is readily established by assay of total α -galactosidase activity in serum or plasma (Mapes *et al.* 1970; Brewster *et al.* 1973; Desnick *et al.* 1973*b*) but care should be taken to avoid loss of activity, since the enzyme is unstable (Mayes and Beutler 1977). The enzyme deficiency can also be demonstrated in leucocytes (Kint 1970; Desnick *et al.* 1973*b*), cultured fibroblasts (Romeo and Migeon 1970; Galjaard *et al.* 1974*b*; Fensom *et al.* 1979*b*), or tears (Johnson *et al.* 1975). However, since the fibroblast enzyme activity increases during culture, care should be taken in standardizing conditions of the patient and control cells (Heukels-Dully and Niermeijer 1976). Either the *p*-nitrophenyl or 4-methylumbelliferyl derivatives may be used as substrates (Kint 1970) but the latter is usually preferred because of the greater sensitivity of the fluorogenic assay (Dean and Sweeley 1977).

Genetic counselling and prenatal diagnosis

Fabry's disease has an intermediate X-linked (Fraser Roberts and Pembrey 1978) mode of inheritance (De Groot 1964; Johnston *et al.* 1969). Several methods have been used for detection of carrier females. This is important for sisters, maternal aunts, and other female relatives because of the possibility of prenatal diagnosis and of preventing the birth of any affected children.

The most reliable test for heterozygotes is the demonstration of two populations of cells, one with normal and the other with absent α -galactosidase A activity according to the Lyon hypothesis. This has been demonstrated in cloned skin fibroblasts (Romeo and Migeon 1970) and by demonstrating variable activity by hair root analysis (Spense *et al.* 1977; Beaudet *et al.* 1978; Vermorken *et al.* 1980). Reduced total α-galactosidase activity in cultured fibroblasts has been reported against the 4-methylumbelliferyl substrate in one heterozygote by Wood and Nadler (1972) and in four heterozygotes by Ho *et al.* (1972*b*); and against the *p*-nitrophenyl substrate in one heterozygote (Brady *et al.* 1971*b*). Fensom *et al.* (1979*b*) found that the best discrimination of carriers in uncloned fibroblasts was the ratio between α -galactosidase A and total ß-galactosidase. However, others have concluded that no single assay is completely discriminatory between carrier and non-carrier females. Thus, Rietra *et al.* (1976) recommended enzyme assay in plasma and leucocytes, together with determination of neutral glycolipids in urinary sediments. Spense *et al.* (1977) found overlap between controls and two carriers in plasma enzyme activity; and between controls and three carriers in leucocyte enzyme activity. They advocated determination of α-:ß-galactosidase ratios as a useful additional test.

It has been estimated that new mutations are responsible for one in six affected men (Wise *et al.* 1962).

Brady *et al.* (1971*b*) were the first to diagnose Fabry's disease *in utero*. Amniocentesis was carried out at 17 weeks' gestation. After four weeks of culture, amniotic cells had 10 per cent of normal α-galactosidase activity. The foetal sex was male. The pregnancy was terminated and the diagnosis of Fabry's disease was confirmed by enzyme assay of foetal tissues, and by demonstrating a ten-fold increase in ceramide trihexoside in foetal liver.

Galjaard *et al.* (1974*b*) have monitored a pregnancy at risk for Fabry's disease 10 days after amniocentesis at 15 weeks' gestation, using a micromethod. The demonstration of normal total α-galactosidase activity when expressed per cell number allowed accurate diagnosis of an unaffected foetus.

Using somatic cell hybridization studies the structural gene for α-galactosidase A has been localized to the long arm of the X-chromosome, Xq22–q24 (Shows *et al.* 1978; De la Chapelle and Miller 1979; Weil *et al.* 1979).

Treatment

Relief of pain. Numerous analgesics and other drugs have been tried to relieve the excruciating burning acroparaesthesia. Several reports suggest that diphenylhydantoin (Lockman *et al.* 1973; Dupperrat *et al.* 1975), carbamazepine (Lenoir *et al.* 1977) alone or both given together (Atzpodien *et al.* 1975) may be effective.

Transplantation of kidneys or foetal liver. Renal transplants have usually been undertaken, after a period of dialysis because of renal failure. It was hoped initially that the kidney might be a source of α-galactosidase A (Desnick *et al.* 1972; Philippart *et al.* 1972; Beutler *et al.* 1973) but subsequent experience failed to provide evidence for enzyme replacement (Spense *et al.* 1976; Grunfeld *et al.* 1975; Van den Bergh *et al.* 1976).

Foetal liver transplants in three male patients with Fabry's disease failed to produce a rise in enzyme or a decrease of ceramide trihexoside in plasma (Touraine *et al.* 1979*a, b*; Touraine and Malik 1980).

Enzyme replacement. As for the mucopolysaccharidoses, and metachromatic leucodystrophy, *in vitro* studies have shown that in culture small quantities of the deficient enzyme can be taken up from the medium by fibroblasts from hemizygous patients with Fabry's disease, and produce catabolism of accumulated substrate (Dawson *et al.* 1973). However, a clinical cure by administration of enzyme would be surprising because of the frequency of the Fabry phenotype in heterozygous females who on average would be expected to have approximately half the normal quantity of enzyme.

The administration of normal plasma produced a brief rise in the patient's plasma enzyme activity, and a decrease in plasma ceramide trihexoside (Mapes *et al.* 1970).

Administration of purified enzyme derived from human placenta (Brady *et al.* 1973), spleen and plasma (Desnick *et al.* 1979, 1980) succeeded in lowering circulating ceramide trihexoside.

Plasmapheresis. A reduction in plasma ceramide trihexoside to near normal values was achieved by three plasmaphereses at two-day intervals by Moser *et al.* (1980*a*). It is too early to judge whether this form of treatment will improve the long-term prognosis.

Farber's disease; lipogranulomatosis; ceramidase deficiency

Farber *et al.* (1957) reported three patients with lipogranulomatosis characterized by painful arthritis, subcutaneous periarticular nodules, hoarseness of voice, and a progressive course, usually leading to death in the first few years. Ceramide storage occurs in the subcutaneous nodules, and (in some cases) in viscera (Prensky *et al.* 1967; Sugita *et al.* 1973) because of severe deficiency of ceramidase activity (Sugita *et al.* 1972; Dulaney *et al.* 1976).

Clinical features

The clinical features of 27 patients are analysed by Moser and Chen (1983). The most characteristic features are painful swelling of the interphalangeal (Fig. 4.16), metacarpal, wrist, elbow, ankle, and knee joints, which together with a hoarse voice usually present between two weeks and four months. Later the painful joint swellings diminish and nodular thickening appears round the joints and tendon sheaths. Discrete subcutaneous nodules characteristically form around the affected joints and in areas liable to pressure, such as the occiput. Occasionally nodules have disappeared.

Joint contractures lead to flexion deformities of knees, wrists, and fingers. Nodules sometimes form in remote areas, and in the pharynx can give rise to dysphagia. Recurrent pulmonary infections are common and are often the cause of death.

Other features include a large tongue, cardiac involvement, and moderate hepatomegaly.

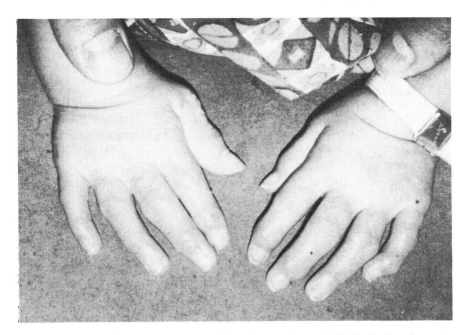

Fig. 4.16. Swelling of interphalangeal joints in a patient with Farber's disease aged two years.

The majority of patients have progressive and severe psychomotor regression. Salaam infantile spasms have been reported (Becker *et al.* 1976; Neville and Turner unpublished observations). Most patients have had evidence of lower motor neurone lesions including hypotonia, muscular atrophy, and diminished tendon reflexes with electronmyographic evidence of denervation. This has been attributed to peripheral neuropathy (Vital *et al.* 1976) and neuronal storage in anterior horn cells (Moser and Chen 1983) but in one published case to myopathy (Schonenberg and Lindenfelser 1974). Ophthalmological abnormalities include granulomatous lesions in the conjunctiva, a grey macula with a faint central cherry-red spot (Cogan *et al.* 1966) and corneal opacities (Ozaki *et al.* 1978; Amirhakimi *et al.* 1976).

Most patients die before four years of age, but some have survived with relatively mild symptoms into the second decade (Samuelsson and Zetterstrom 1971; Barrière and Gillot 1973; Pavone *et al.* 1980).

Cerebrospinal fluid protein concentration has been elevated in eight of ten patients investigated (Moser and Chen 1983).

Pathology

Histologically there is infiltration of macrophages or histiocytes in the subcutaneous periarticular, articular, and other tissues. Granulomata show macrophages, multinucleated cells, and lymphocytes around a central area infiltrated with foam cells. Other cells, for example chondrocytes of cartilage or endocardial cells of heart valves, (Abul-Haj *et al.* 1962) show mainly glycolipid storage.

Granulomata have been described at necropsy in all patients. They occur especially in the region of the joints, joint capsules, and adjacent bones. Granulomata occur on the parietal pleura. The lungs are consistently affected with areas of consolidation, which is shown histologically to be due to alveolar infiltration with numerous macrophages.

Heart valves have been affected in about one-third of patients examined at necropsy. There was thickening of the mitral valve and chordae tendinae, and thickening of aortic valves and nodular lesions.

Electron microscopy shows changes typical of sphingolipidoses with single-membrane, acid phosphatase-positive distended lysosomes containing curvilinear tubular structures. Evidence that this storage material is ceramide includes the production of tubular bodies in cultured fibroblasts from patients with Farber's disease by the addition of ceramides containing non-hydroxy fatty acids to the culture medium (Rutsaert *et al.* 1977). Lamellar storage material or Zebra bodies (Fig. 4.17) occurs in neurones and may represent storage of gangliosides especially GM$_3$ (Moser *et al.* 1969).

Molecular defects

Ceramides (Fig. 4.18) are degraded by ceramidases. Several forms have been described including acid, neutral, and alkaline ceramidases which catalyse the reaction ceramide + H$_2$O $\rightarrow$ sphingosine + fatty acid. A deficiency of acid ceramidase in Farber's disease was demonstrated in post mortem tissues (Sugita *et al.* 1972) and later in cultured fibroblasts (Dulaney *et al.* 1976) and peripheral blood leucocytes (Moser and Chen 1983). Alkaline ceramidases have normal activity (Sugita *et al.* 1975).

Ceramide accumulation occurs in subcutaneous nodules, kidney (Moser *et al.* 1969; Becker *et al.* 1976; Samuelsson and Zetterstrom 1971; Toppet *et al.* 1978) and variably in liver (Ozaki *et al.* 1978), other tissues, and plasma. Ceramide accumulation can be demonstrated by thin-layer chromatography (Moser *et al.* 1969), high performance liquid chromatography (Sugita *et al.* 1974; Yahara *et al.* 1980) or gas–liquid chromatography combined with mass spectometry (Samuelsson and Zetterstrom 1971).

Ceramides which accumulate in Farber's disease differ from normal ceramides in that they contain significant amounts of 2-hydroxy fatty acids (Moser *et al.* 1969; Sugita *et al.* 1973). However, no hydroxy fatty acid was demonstrated in ceramide from a subcutaneous nodule (Samuelsson and Zetterstrom 1971). The 2-hydroxy fatty acids in Farber's disease consist

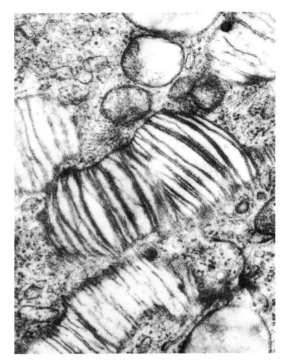

Fig. 4.17. Electronmicrograph showing intraneuronal lamellar lipid inclusions (zebra bodies) in Farber's disease.

mainly of the C24 acids with smaller proportions of C22, C20, and C18 (Samuelsson and Zetterstrom 1971; Sugita *et al.* 1973). Three- to ten-fold increase in ganglioside accumulation in liver, lymph nodes and subcutaneous nodules is consistent with PAS-positive material in foam cells from these tissues (Moser *et al.* 1969).

Ceramide has a key role in sphingolipid structure and metabolism. It participates in synthesis and degradation of gangliosides, myelin (galactosyl ceramide, sulphatide), membranes (sphingomyelin), and other glyco-sphingolipids.

Granuloma formation may be simply a response to ceramide, as can be shown by subcutaneous injection of ceramide into rats (Moser *et al.* 1969). The absence of lesions in bone marrow and reticuloendothelial cells in

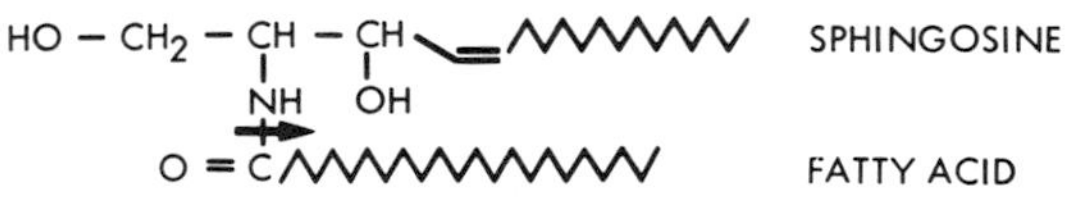

$$HO - CH_2 - CH - CH \diagdown\!\!=\!\!\diagup\!\!\!\bigwedge\!\!\bigwedge\!\!\bigwedge\!\!\bigwedge \quad \text{SPHINGOSINE}$$
$$\underset{NH}{\overset{|}{}} \quad \underset{OH}{\overset{|}{}}$$
$$O = C \bigwedge\!\!\bigwedge\!\!\bigwedge\!\!\bigwedge\!\!\bigwedge\!\!\bigwedge \quad \text{FATTY ACID}$$

Fig. 4.18. Ceramide.

Farber's disease may be due either to the activity of neutral alkaline ceramidases in these tissues, or to the re-utilization of undegraded ceramide for the synthesis of glycolipids.

Diagnosis

The disease can be diagnosed by demonstration of severe deficiency of acid ceramidase in cultured skin fibroblasts or leucocytes (Dulaney *et al.* 1976; Dulaney and Moser 1977*b*; Moser and Chen 1983). Additional confirmation includes histology of a subcutaneous nodule, which shows granuloma formation, and macrophages with lipid cytoplasmic inclusions (Farber *et al.* 1957). As described above, these granulomata have characteristic cytochemical properties and ultrastructural appearance.

Further evidence for the diagnosis of Farber's disease includes chemical evidence of ceramide accumulation in biopsies of subcutaneous nodules (Amirhakimi *et al.* 1976; Schmoeckel and Hohlfed 1979).

Genetic counselling and prenatal diagnosis

The disorder has an autosomal recessive mode of inheritance, as shown by the occurrence of affected sibs of both sexes with unaffected parents, and supported by the report of parental consanguinity (Pavone *et al.* 1980) and by the demonstration of partial ceramidase deficiency in cultured skin fibroblasts from both parents (Dulaney *et al.* 1976; Fensom *et al.* 1979*c*; Pavone *et al.* 1980).

Prenatal diagnosis has been carried out in four pregnancies in two mothers. In three of the pregnancies the foetus was accurately considered to be unaffected (Dulaney and Moser 1977; Fensom *et al.* 1979*c*). In the fourth, cultured amniotic cell ceramidase activity was 8 per cent of control, and the pregnancy was terminated. An affected foetus was confirmed by demonstration of increased ceramide concentration in kidney and liver; and reduced ceramidase activity in brain and cultured fibroblasts (Fensom *et al.* 1979*c*).

OTHER LIPID STORAGE DISEASES

Refsum's disease; phytanic acid storage disease; phytanic acid α-hydroxylase deficiency

A clinical syndrome characterized by the tetrad (sometimes incomplete) of retinitis pigmentosa, peripheral polyneuropathy, cerebellar ataxia, and a high cerebrospinal fluid protein concentration was described by Refsum (1945, 1946) in five patients from two Norwegian families. In liver and spleen from a seven-year-old girl who had died of the disease, Klenk and Kahlke (1963) found accumulation of a lipid which they characterized as

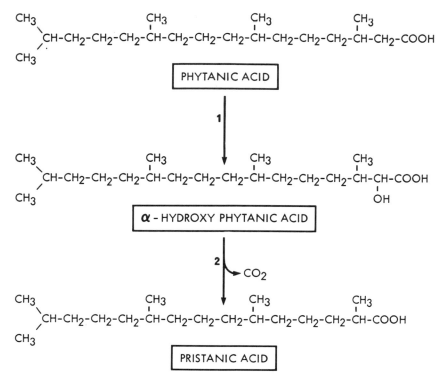

Fig. 4.19. Enzyme steps in the oxidation of phytanic acid: (1) Phytanic acid
α-hydroxylase (2) α-hydroxyphytanic acid oxidase.

being predominantly phytanic acid (Fig 4.19) using gas chromatography.
Subsequent studies established the pathway for metabolism of phytanic
acid (Steinberg *et al.* 1965). Moreover, it was shown that in patients with
Refsum's disease phytanic acid oxidation was less than 5 per cent of normal
(Mize *et al.* 1969). The primary defect was shown to be reduced activity of
phytanic acid α-hydroxylase, which normally catalyses the conversion of
phytanic acid to α-hydroxyphytanic acid.

These series of studies were interesting both theoretically because they
revealed the existence of a new metabolic pathway, and in practice because
they paved the way to dietary treatment by restriction of phytanic acid and
its precursors.

Clinical features

Progressive impairment of vision, especially in limited light (night
blindness), unsteadiness of gait and clumsiness are typical early symptoms.
There is constriction of the visual fields, and retinitis pigmentosa. The skin

is often dry, hyperkeratotic, and even ichthyotic. There is symmetrical peripheral neuropathy with loss or impairment of peripheral sensations and tendon reflexes. In older patients progressive nerve deafness is common. There is a wide variation in the age of onset which is from childhood to late adult life. Sudden aggravation sometimes accompanied by fever can be provoked by surgery or pregnancy, and is followed by gradual remission. Some patients have anosmia. Sudden death without a demonstrable specific cause has occurred in half the reported fatalities and has been attributed to cardiac arrhythmia in view of reports of impaired auriculo-ventricular conduction.

Pathology

Bronchopneumonia is usually the cause of death when this can be established. The nervous system appears normal to the naked eye.

Histologically, two main types of changes have been described in peripheral nerves: a proliferation of Schwann cells and fibroblasts causing an 'onion bulb' appearance, and infiltration of peripheral nerves by a homogenous material which separates individual nerve fibres and stains weakly with toluidine blue and PAS (Cammermeyer *et al.* 1954; Cammermeyer 1956; Flament-Durand *et al.* 1971). There is loss of neurones, axons, and myelin, and infiltration with lipid-laden macrophages in spinal cord, brain stem, and cerebellum.

On electron microscopy, the 'onion bulb' structures appear as concentric layers of Schwann cell processes (Fardeau *et al.* 1970; Flament-Durand *et al.* 1971).

Molecular defects

There is severe deficiency in the activity of phytanic acid α-hydroxylase, the enzyme which catalyses the conversion of phytanic acid to pristanic acid by α-oxidation (Fig. 4.19) (Mize *et al.* 1969). This can be demonstrated as defective release by cultured fibroblasts of $^{14}CO_2$ from universally-labelled ^{14}C-phytanic acid added to the culture medium (Herndon *et al.* 1969).

The deficiency of phytanic acid α-oxidation causes accumulation of phytanic acid in tissues and serum. In normal individuals, serum phytanic acid is just detectable, representing 0.1–0.5 per cent of total fatty acids (Avigan, 1966), but in Refsum's disease it may represent up to one-third of total fatty acids (Klenk and Kahlke 1963; Campbell and Williams 1967; Fryer *et al.* 1971).

Phytanic acid which accumulates in patients with Refsum's disease appears to come entirely from the diet and not from endogenous synthesis. Attempts to demonstrate biosynthesis of phytanic acid in a patient with Refsum's disease either from labelled mevalonic acid (Steinberg *et al.*

1965, 1967) or D_2O over a four- to five-month period (Steinberg *et al.* 1967, 1970) were unsuccessful. Although minimal incorporation of deuterium from D_2O was found, this could have resulted from the conversion of phytol to phytanic acid.

The largest sources of dietary phytanic acid in humans are cow's milk and its products, cheese, butter, and cream (Ackman and Hooper 1973). Daily intake of phytanic acid has been estimated as 56–89 mg (Steinberg 1983).

Another substance with a polyisoprenoid structure is phytol, which differs from phytanic acid in having a $\triangle^2$-double bond and a primary alcohol rather than a carboxylic acid group at C1. Phytol is readily converted into phytanic acid by oxidation of the alcohol residue, and reduction of the double bond mainly in this order (Mize *et al.* 1969; Baxter *et al.* 1967).

In the rat, phytol that is bound by ester linkage to a propionic acid side chain of one of the pyrrole rings of chlorophyll in green vegetables is largely unabsorbed and is therefore an unimportant source of phytanic acid (Baxter and Steinberg 1967). Dietary free phytol is a much smaller source of phytanic acid than the preformed acid (Steinberg *et al.* 1970). Smaller and less important sources of phytanic acid include the side chain of vitamin K_2 (phylloquinone) and possibly the very small component synthesized from bacterial flora.

Diagnosis

The diagnosis is established by the demonstration of abnormally high levels of serum phytanic acid, estimated by the gas–liquid chromatographic technique (Avigan 1966; Nevin *et al.* 1967; Try 1969) and by demonstrating a severe deficiency of cultured fibroblast phytanic acid α-hydroxylase (Herndon *et al.* 1969; Poulos 1981).

Genetic counselling and prenatal diagnosis

Refsum's disease has an autosomal recessive mode of inheritance. Cultured fibroblasts from eight parents of affected children oxidized phytanic acid at 46–59 per cent of the rate of normal individuals, there being no overlap with individuals who were homozygous normal or homozygous affected (Herndon *et al.* 1969).

Rarely, heterozygotes may have serum phytanic acid concentration elevated to values observed in affected homozygotes (Kahlke and Richterich 1965; Nevin *et al.* 1967). Moreover following oral phytol loading heterozygotes may have higher elevation of serum phytanic acid than controls (Gautier *et al.* 1973).

The disease is rare, but the incidence is not known. We are not aware of any attempts at prenatal diagnosis.

Treatment

The principle of treatment is to deplete the body of phytanic acid by administration of a diet low in phytanic acid. The success of treatment is monitored by assay of plasma phytanic acid concentrations (Quinlan and Martin 1970; Gibberd *et al.* 1979). The role of plasma exchange or plasmapheresis (Gibberd *et al.* 1979; Moser *et al.* 1980*a*) remains to be established, but is probably useful to reduce phytanic acid storage initially. Where there has been a good fall of plasma phytanic acid (Steinberg *et al.* 1970; Kark *et al.* 1971) by reducing the daily dietary intake of phytanic acid from about 60 mg to less than 3 mg, progression of the disorder has been arrested. Such drastic reduction in dietary intake involves avoidance of butter, margarine, cheese, milk, palm oil, herring, and mammalian liver. If plasma phytanic acid concentrations do not fall sufficiently it may be necessary to place the patient on a mainly artificial diet.

Rapid mobilization of adipose tissue, by low-calorie diets, should be avoided since the associated release of phytanic acid can cause severe relapse.

Wolman's disease; cholesteryl ester storage disease

Abramov *et al.* (1956) described a child with hepatosplenomegaly, calcification of the adrenals, and abdominal distension who died at two months of age. They reported accumulation of cholesterol and triglycerides in liver, adrenals, spleen, and lymph nodes and demonstrated that the cholesterol was in the esterified form. Two affected sibs from the same family were reported later (Wolman *et al.* 1961).

Patrick and Lake (1969*a*,*b*) demonstrated deficiency of acid lipase in liver and spleen. This deficiency was shown against both cholesteryl esters and triglycerides.

Clinical features

Persistent projectile vomiting and abdominal distension during the first few weeks are typical presenting features. Diarrhoea (Marshall *et al.* 1968; Lough *et al.* 1970) and mild pyrexia (Crocker *et al.* 1965) have been reported. Moderately severe anaemia usually occurs by the second month. Vacuolated lymphocytes are common (Crocker *et al.* 1965; Marshall *et al.* 1968). Massive hepatosplenomegaly is a constant feature and may be present in the first week after birth (Lough *et al.* 1970). A specific feature is bilateral calcification of the adrenals and has been present in the great majority, but not all patients (Marshall *et al.* 1968; Leclerc *et al.* 1971; Schaub *et al.* 1980).

Radiologically, the adrenals are considerably enlarged, stippled, and semilunar in shape, with the concave surfaces facing medially and inferiorly.

There is fat malabsorption (Eto and Kitagawa 1970; Lough *et al.* 1970) and depressed adrenal response to corticotrophin (Crocker *et al.* 1965; Eto and Kitagawa 1970).

Failure to thrive and malaise are progressive and become marked by the third month, although neurological development is usually normal. Death usually occurs by six months of age but may be delayed to 14 months.

Patients with a later onset and a more slowly progressive course (Patrick and Lake 1973) may be classified as cholesteryl ester storage disease (Beaudet *et al.* 1977; Michels *et al.* 1979). Onset may be as late as the second decade (Schiff *et al.* 1968). In this group of patients adrenal calcification has been reported in only one case (Beaudet *et al.* 1977; Michels *et al.* 1979).

Pathology

To the naked eye, the adrenal glands are bright yellow, normal in shape, and symmetrically enlarged, and weigh two to three times more than normal. They are firm and contain gritty areas of calcification. On section, the cortex is bright yellow, but the medulla whitish-grey.

Histologically, the cells of the adrenal glands are swollen, vacuolated and contain sudanophilic lipid, but the general architecture is retained. The calcium is in the form of fine granules, or lumps (Marshall *et al.* 1968). Electron microscopy reveals lipid in globular or crystalline deposits, which have been identified as cholesteryl esters and triglycerides, both by staining (Lough *et al.* 1970) and biochemical properties (Lake and Patrick 1970; Ellis and Patrick 1976).

Lipid-laden histiocytes, or 'foam cells' are widespread. They have been reported in the spleen, lymph nodes, bone marrow, thymus, and tonsils (Marshall *et al.* 1968; Patrick and Lake 1973).

The liver, which may weigh twice as much as normal, appears yellow and greasy at necropsy. Histologically the portal spaces may be infiltrated with lymphoid cells and the hepatic lobular architecture considerably distorted (Marshall *et al.* 1968; Lough *et al.* 1970).

Depletion of cortical neurones and infiltration of the meninges by foamy histiocytes have been reported (Crocker *et al.* 1965).

Molecular defects

Acid lipase deficiency in liver and spleen was demonstrated as absence of activity to hydrolyse cholesteryl oleate, glyceryl tridecanoate, and glyceryl tripalmitate at pH 4.6 by Patrick and Lake (1969*a,b*). They also showed that the lipase present in Wolman's disease is inhibited by diethyl *p*-nitrophenyl phosphate (E600) which inhibits alkaline microsomal esterase

(Lake and Patrick 1970). A similar deficiency was found in leucocytes (Young and Patrick 1970).

The cholesteryl ester content of liver, spleen, and adrenals is increased. That of liver has been seven to 161 times greater than normal (Patrick and Lake 1969*a*, *b*; Lake and Patrick 1970). They include the oxygenated esters 7 α-cholesterol, 7 ß-cholesterol, 7-ketocholesterol, and 5,6 α- and 5,6 ß-epoxycholesterol, which constitute about 10 per cent of the total cholesteryl esters (Assmann *et al.* 1975).

The triglyceride content of liver, spleen, and adrenals is also increased. In liver, the triglyceride content has been twice to 10 times greater than normal (Lough *et al.* 1970; Assmann *et al.* 1975).

Diagnosis

Calcification of the adrenals is almost always present, but is not diagnostic since it may occur in neoplasms or haemorrhages.

The diagnosis is established by demonstrating severe deficiency of acid ester hydrolase activity in cultured fibroblasts or leucocytes. A convenient assay utilizes *p*-nitrophenyl laurate as substrate (Beaudet *et al.* 1977). A sensitive fluorometric assay was described by Kelly and Bakhru-Kishore (1979).

Genetic counselling and prenatal diagnosis

The disorder has an autosomal recessive mode of inheritance as evidenced by its frequent occurrence in siblings of both sexes, with unaffected parents; parental consanguinity (Wallis *et al.* 1971; Raafat *et al.* 1973); and the demonstration of a partial enzyme defect in leucocytes (Patrick and Lake 1973; Schaub *et al.* 1980) and cultured fibroblasts (Kyriakides *et al.* 1972) in parents of children with either Wolman's disease or cholesteryl ester storage disease (Beaudet *et al.* 1977; Orme 1970).

Prenatal diagnosis was reported by Patrick *et al.* (1976) by assay of amniotic cells cultured for three weeks after amniocentesis at 15 weeks' gestation. Residual activities were 2 per cent of normal value against ^{14}C-tripalmitin, and 15 per cent against *p*-nitrophenyl palmitate. Prenatal monitoring of two pregnancies at risk for Wolman's disease was reported by Coates *et al.* (1978). Cultured amniotic cell acid lipase activity was normal in one pregnancy. Deficiency in the other pregnancy was demonstrated both as a reduction in total acid lipase activity, using ^{14}C-triolein as substrate, and as deficiency of the A isoenzyme after separation by electrophoresis and staining using the 4-methylumbelliferyl substrate.

Coates *et al.* (1978) found that acid lipase activity of the amniotic fluid supernatant was not reliable for prenatal diagnosis.

5. Disorders of amino acid metabolism

DISORDERS OF SULPHUR AMINO ACID METABOLISM

Homocystinuria; homocystinaemia

In the great majority of patients, homocystinuria and homocystinaemia are caused by deficiency in the activity of cystathionine ß-synthase. This enzyme catalyses the coupling of homocysteine with serine, in the presence of vitamin B_6 (pyridoxine), to form cystathionine, in the transulphuration pathway (Fig. 5.1). The existence of genetic heterogeneity is suggested by the different quantities of residual enzyme activity in individual patients, and by the response to administration of pyridoxine by about half the patients. Both these phenomena tend to be concurrent among sibs, and therefore probably depend on the existence of several different mutations. In this form of homocystinaemia the blood methionine concentration is elevated, and the excretion of urinary inorganic sulphate following methionine loading is decreased.

In a small proportion of patients with homocystinuria and homocystinaemia there is deficiency of remethylation of homocysteine to methionine. This may be due to a number of causes both genetic and non-genetic. These include deficiency of the cofactor of the 5-methyltetra-hydrofolate–homocysteine methyltransferase reaction, namely methyl-cobalamin (MeCbl). Such deficiency may arise from nutritional deficiency, or defective intestinal absorption of vitamin B_{12}; failure in the metabolic conversion of the hydroxycobalamin form of B_{12} to the active cofactor MeCbl; or failure to retain MeCbl. A further cause is defective synthesis of the cosubstrate of the remethylation reaction of homocysteine, namely 5-methyltetrahydrofolate, because of deficiency in the activity of 5,10-methylenetetrahydrofolate reductase.

In the various types of homocystinuria due to defective remethylation of homocysteine there is usually low serum methionine concentration, and normal excretion of urinary inorganic sulphate following methionine loading.

Other rare causes of homocystinuria include bacterial conversion of cystathionine in patients with cystathioninuria (Levy and Mudd 1973); and administration of 6-azauridine triacetate (Hyanek *et al.* 1969).

An alternative pathway for homocysteine remethylation utilizes betaine as the methyl donor (Fig. 5.1) and is catalysed by betaine–homocysteine

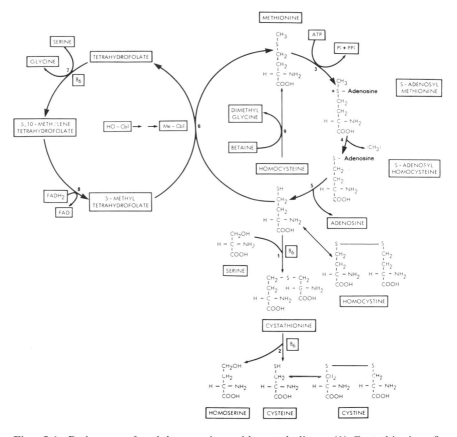

Fig. 5.1. Pathways of sulphur amino-acid metabolism: (1) Cystathionine ß-synthase (2) γ-cystathionase (3) Methionine adenosyl transferase (4) Trans-methylase (5) Adenosylhomocysteinase (6) 5-methyltetrahydrofolate homo-cysteine methyl transferase (7) Serine hydroxy-methyl transferase (8) 5,10-methyl-enetetrahydrofolate reductase (9) Betaine-homocysteine methyl transferase.

methyltransferase. However, this pathway appears unable to compensate for the blocked 5-methyltransferase pathway. No genetic defect of homocysteine remethylation in this pathway has yet been described.

Deficiency of cystathionine ß-synthase

Clinical features

There is considerable clinical heterogeneity; some patients have ectopia lentis but normal intelligence and no thrombo-embolic tendency (Ritchie and Carson 1973). Others have ectopia lentis and mild skeletal changes

(McKusick *et al.* 1971; Drayer *et al.* 1980), while others are mentally retarded and have marked skeletal involvement (Brenton *et al.* 1972).

Dislocation of the optic lens occurs in over 90 per cent of patients (Francois 1972) but is usually not present in the first three years, becoming increasingly more frequent during the first decade. The dislocation, which is usually downwards, is due to tearing of the zonular fibres which connect the lens to the ciliary body. Thickened zonular remnants may be seen as a fringe along the equatorial edge of the lens (Ramsey and Dickson 1975). The edge of the iris may be seen to quiver on movement of the eye or head (iridodonesis). Other ocular complications include myopia and astigmatism secondary to ectopia lentis (McKusick *et al.* 1971; Johnston 1978), anterior dislocation with painful pupillary obstruction and glaucoma (Lieberman *et al.* 1966; Elkington *et al.* 1973; Johnston 1978), cataracts (Schimke *et al.* 1965) and retinal artery occlusion (Wilson and Ruiz 1969).

The most serious complications result from arterial or venous thrombosis, which can occur even in the first year (Dunn *et al.* 1966). These may involve the carotid, coronary, and pulmonary arteries, leading to severe sequelae or sudden death (Dunn *et al.* 1966; Carey *et al.* 1968*a*; McKusick *et al.* 1971), and are sometimes precipitated by surgery (Komrower and Wilson 1963). Surgery, arteriography, and venography precipitate thrombosis, and should be avoided whenever possible.

Malar flush and livedo reticularis of the extremities are present in over half the patients, and may be secondary to thromboembolism.

Skeletal abnormalities are common (Schedewie *et al.* 1973). Radiologically, there is striking osteoporosis in many adult patients (Schimke *et al.* 1965; Morreels *et al.* 1968; Brill *et al.* 1974), and occasionally this occurs in the first decade (Morreels *et al.* 1968). The vertebrae are frequently biconcave especially posteriorly (Schimke *et al.* 1965; Brenton *et al.* 1972). Scoliosis is present (Brill *et al.* 1974) more often than kyphosis (Morreels *et al.* 1968). Arachnodactyly is present in less than half the patients (McKusick *et al.* 1971; Brill *et al.* 1974).

Mental retardation may be a presenting symptom and is usually detected as delay in motor milestones in the first two years. Intelligence quotients (IQ) vary according to method of ascertainment. Clearly they will be low if the patients are identified by screening retarded patients. On the other hand, when ascertainment is mainly by the presence of ectopia lentis, only about half the patients have below-normal intelligence (McKusick *et al.* 1971).

Progressive impairment of intellect is suggested by the observation that in some families older affected children have lower IQ than their younger affected sibs (Yoshida *et al.* 1968; Verma *et al.* 1974). However, in other families, IQ seems to be independent of age (Dunn *et al.* 1966; Turner *et al.* 1967; Carey *et al.* 1968*a*).

Generalized convulsions occur in about 10 per cent of cases (White *et al.* 1965; Carey *et al.* 1968*a*) while myoclonic fits and *petit mal* have been reported in a few cases. Hemiparesis occurs following cerebrovascular thrombosis (Dunn *et al.* 1966). Various electroencephalographic abnormalities have been reported, including an excess of slow waves and of sharp wave discharges (Dunn *et al.* 1966; Turner *et al.* 1967).

Affected males can father clinically normal children (McKusick *et al.* 1971), but affected mothers seem to have a high incidence of foetal loss (Ritchie and Carson 1973; Brenton *et al.* 1977; Kurczynski *et al.* 1980).

Pathology

In the eye, zonular remnants on the lens or ciliary body appear histologically to form a PAS positive amorphous layer (Henkind and Ashton 1965; Ramsey *et al.* 1972). On electron microscopy, the zonular remnants appear as tangled unorganized filaments (Ramsey and Dickson 1975). Vascular occlusion occurs in the brain (Gibson *et al.* 1964; Dunn *et al.* 1966; Hopkins *et al.* 1969*a*), coronary arteries (Schimke *et al.* 1965; Carey *et al.* 1968*a*) and kidneys (Dunn *et al.* 1966). Arterial walls are characteristically affected with internal fibrous thickening (White *et al.* 1965; Gibson *et al.* 1964). The medial muscle appears frayed and infiltrated with collagen (Gibson *et al.* 1964). Thrombophlebitis has produced pulmonary emboli (Schimke *et al.* 1965). Fatty infiltration of hepatocytes, especially in the centrilobular regions, has been reported (Gaull *et al.* 1974).

Molecular defects

Deficiency of cystathionine ß-synthase was first demonstrated by Mudd *et al.* (1964) in a liver biopsy. Subsequently, this has been confirmed in numerous patients in whom residual activity has been either undetectable or as high as 10 per cent of the control value (Gaull *et al.* 1974; Perry 1974). Deficiency of cystathionine ß-synthase has been demonstrated also in brain (Laster *et al.* 1965*a*; Mudd *et al.* 1966), cultured skin fibroblasts (Uhlendorff and Mudd 1968; Uhlendorff *et al.* 1973) and PHA-stimulated lymphocytes (Goldstein *et al.* 1973*a*).

The reason why only some patients respond to pyridoxine administration (Barber and Spaeth 1967) is not known, but the response is similar in sibs (Longhi *et al.* 1977; Wilcken and Turner 1978; Valle *et al.* 1980*a*; Curczynski *et al.* 1980).

Residual cystathionine ß-synthase activity has been found in all livers examined from B_6-responsive patients but not B_6-unresponsive patients. These data are consistent with the explanation that responsiveness to B_6 depends on stimulation of residual enzyme. Responsiveness to B_6 by the patient, however, is commonly, but not always correlated with the presence of residual activity in cultured fibroblasts. Conversely un-responsiveness to B_6 has been reported even when no residual cystathionine ß-synthase activity has been detected in fibroblasts (Uhlendorff *et al.* 1973; Fowler *et al.* 1978; Bittles and Carson 1981).

Responsiveness to B_6 *in vitro*, however, correlates poorly to responsiveness by the patient (Uhlendorff *et al.* 1973; Fowler *et al.* 1978). Kim and Rosenberg (1974) partially purified cystathionine ß-synthase from cultured fibroblasts from three pyridoxine-responsive patients and from normal individuals. The addition of pyridoxal phosphate stimulated a smaller relative increase of cystathionine ß-synthase activity in the enzyme from patients than from controls.

It seems clear that responsive patients do not have pyridoxine deficiency, as evidenced by absence of anaemia or convulsions or of biochemical abnormalities indicative of pyridoxine deficiency such as elevated urinary excretion of xanthurenic acid (Hollowell *et al.* 1968; Barber and Spaeth 1969), or decreased urinary excretion of pyridoxine metabolites (Morrow and Barnes 1972). Also, there is no evidence that pyridoxine stimulates an alternative metabolic pathway for homocysteine metabolism.

Characteristic alterations in plasma are an increase in the normal fasting value of methionine from less than 0.03 μmol/ml up to 2 μmol/ml, and the appearance of homocysteine (which is not normally detectable) to concentrations of about 0.2 μmol/ml (Brenton *et al.* 1965; Perry *et al.* 1967*a*). Although fresh blood and urine contain homocysteine, when analysed it is usually recovered as the disulphide homocystine. In patients, homocystine excretion in the urine may exceed 1 mmol per day (Brenton *et al.* 1965), though urinary methionine may be within normal limits owing to renal tubular reabsorption (Carson *et al.* 1963, 1965). There is evidence that remethylation of homocysteine contributes to the accumulation of methionine. Thus treatment with folic acid (Carey *et al.* 1968*b*; Morrow and Barnes 1972) or betaine (Perry 1974; Smolin *et al.* 1981) increases the methionine/homocystine ratio.

In both normal individuals and in patients with some residual cysta-thionine ß-synthase activity a substantial proportion of urinary sulphur is in the form of sulphate, either inorganic or esterified. This has been demonstrated by tracer studies using ^{35}S-L-methionine (Brenton and Cusworth 1966) and after methionine loading (Laster *et al.* 1965*b*). Since inorganic sulphate is formed from cysteine, it is not clear why patients who have a block in cysteine synthesis appear to be able to produce sulphate relatively efficiently, albeit at a reduced rate. Thus, tracer studies showed that two normal individuals given doses of ^{35}S-methionine excreted 76 and 86 per cent of their urinary radioactivity as total sulphate in 48 hours, as compared with a patient with cystathionine ß-synthase deficiency who excreted 45 per cent (Brenton and Cusworth 1966). There is evidence that production of cysteine from methionine in cystathionine ß-synthase deficient patients may occur via an alternative metabolic route involving enhanced activity of cysteine desulphydrase (Benson *et al.* 1969*a*).

Besides homocysteine and methionine, several derivatives occur in plasma and urine of patients with cystathionine ß-synthase deficiency. These include:

(a) the mixed disulphide of homocysteine and cysteine, which in untreated patients is present usually in a lower concentration than homocystine, but in patients on low methionine–high cysteine diets, persists when homocystine is barely detectable (Sardharwalla *et al.* 1968);

(b) *S*-adenosylhomocysteine in urine, but not plasma (Perry *et al.* 1966, 1967*a*);

(c) homolanthionine, possibly synthesised by γ-cystathionase-catalysed condensation of homocysteine and homoserine (Perry *et al.* 1966);

(d) oxidation products of homocysteine. It is not known whether these occur *in vivo* or are produced as artefacts during analysis. They include homocysteic and homocysteine sulphinic acids (Ohmori *et al.* 1972);

(e) derivatives of homocysteine, the mixed disulphide of homocystine or homolanthionine, including

S-(3-hydroxyl-3-carboxy- η-propylthio)-cysteine,

S-(2-hydroxyl-2-carboxyethylthio)-homocysteine,

S-(3-hydroxyl-3-carboxy- η-propylthio)-homocysteine

(Kodama 1971);

(f) methionine sulphoxide (Laster *et al.* 1965*b*);

(g) an unidentified form of bound methionine (Laster *et al.* 1965*b*).

In patients with deficiency of cystathionine ß-synthase activity there is deficiency of the enzyme reaction product cystathionine in brain (Gerritsen and Waisman 1964) and its metabolite cyst(e)ine in plasma and urine (Perry *et al.* 1968*a*). The inverse relation between plasma homocysteine and cysteine concentrations suggests that reduction in plasma cysteine may be due to removal by formation of the mixed disulphide with homocysteine (Perry *et al.* 1968*a*; Seashore *et al.* 1972). This is supported by the observation that cysteine administration is more effective in raising blood cysteine concentration when plasma concentration of homocysteine is lower.

Little is known of the relation between the molecular defects and the clinical phenotype. In rats, administration of methionine or homocystine (Brown and Allison 1948; Cohen *et al.* 1958) retards growth.

In two children who died with homocystinaemia, there was widespread vascular pathology consisting of fibrosis of the intima and media, proliferation of the perivascular connective tissue, and thickening of the media. However, in one (with a disorder of cobalamin metabolism) there was low plasma methionine and high plasma cystathionine, while in the other (with presumed cystathionine ß-synthase deficiency) there was presumably high plasma methionine and low plasma cystathionine. McCully (1969) therefore suggested that the vascular pathology resulted from accumulation of homocysteine (or its derivatives) in both children. This is supported by the report of extensive vascular thrombosis (Kanwar *et al.* 1976; Wong *et al.* 1977*a*) and vascular pathology (Kanwar *et al.* 1976; Baumgartner *et al.* 1980) in patients with 5,10-methylenetetrahydrofolate reductase deficiency who also have accumulation of homocysteine, but have low blood methionine concentration.

It has been shown that in patients with cystathionine ß-synthase deficiency there is a defect in cross linkage between the three polypeptide chains of collagen. This is evidenced by the greater proportion of soluble collagen (Harris and Sjoerdsma 1966*a*, *b*; Kang and Trelstad 1973) and a lower proportion of collagen aldehyde residues and of cross-linked compounds. Further, it was shown that the addition of homocysteine (but not of homocystine) to solutions of rat skin collagen caused failure of formation of fibrils; and a decrease in cross-links (Kang and Trelstad 1973). Siegel (1975) also found that the addition of homocysteine (but not of homocystine) to chick bone collagen in the presence of lysyl oxidase caused a decrease in cross linkage.

It is interesting that ß-aminopropionitrile-induced interference with collagen cross-linkage in rats (lathyrism) results in kyphoscoliosis in addition to other connective tissue abnormalities. Ectopia lentis is unlikely to be a manifestation of collagen abnormality since protein of the zonular fibres of the lens contains a high proportion of cyst(e)ine, not found in collagen. Since patients with sulphite oxidase deficiency have dislocated lenses (Irreverre *et al.* 1967; Shih *et al.* 1977; Johnson *et al.* 1980*d*; Beemer and Delleman 1980) it has been suggested (Irreverre *et al.* 1967) that in both disorders there might be disruption of disulphide bonds due either to sulphonation of sulphite or exchange with homocysteine.

There are two main theories to explain the thromboembolic tendency. In some but not all patients with cystathionine ß-synthase deficiency it has been related to increased platelet adhesiveness, which can be measured by the adherence to glass (McDonald *et al.* 1964; Cusworth and Dent 1969). This is supported by the experimental production of platelet adhesiveness by the addition of homocysteine to normal platelets (McDonald *et al.* 1964); and a change towards normality in platelet adhesiveness in some but not all pyridoxine-responsive patients when treated with pyridoxine (Carson and Carre 1969; Barber and Spaeth 1969). The other theory is that the thrombotic tendency is secondary to widespread vascular injury which may exist without thrombosis, and which, as discussed above, has been attributed to homocystinaemia.

Diagnosis

A simple method for establishing presumptive diagnosis when there are typical clinical features is the demonstration of excessive concentration of homocystine in the urine by the cyanide–nitroprusside test (Spaeth and Barber 1967). More specifically, amino-acid analysis of plasma will reveal elevated methionine and homocystine, reduced cystine concentrations and absent cystathionine. Plasma samples should be deproteinized within a few minutes of collection since on standing homocysteine forms disulphide bonds with protein with which it is removed on deproteinization.

Definitive diagnosis can be established by demonstrating severe deficiency of cultured fibroblast cystathionine ß-synthase activity (Uhlendorff *et al.* 1973; Fowler *et al.* 1978; Bittles and Carson 1981). It should be emphasized that demonstration of homocystinuria and homocystinaemia is not sufficient to establish a diagnosis of cystathionine ß-synthase deficiency. The diagnosis is strengthened by demonstrating elevation of blood methionine, but the definitive test is the fibroblast enzyme assay.

Treatment

Pyridoxine. Reversal of biochemical abnormalities by administration of pyridoxine (Barber and Spaeth 1967, 1969) has occurred in over half the reported patients. The degree of response to pyridoxine may be dose-dependent (Carson and Carre 1969). Average initial doses are 150 mg/day in the young infant, 300 mg/day in an older child and 500 mg/day in an adult, given in several divided doses. Although some patients respond to doses of 25 mg/day (Hollowell *et al.* 1968), doses of 1000 mg/day for several weeks may be needed to obtain some response (Perry 1974). In some patients, folate deficiency needs to be corrected before a pyridoxine response is observed (Morrow and Barnes 1972). The success of treatment is monitored by assay of plasma methionine, homocysteine and cystine at the same time each day, e.g. one hour after lunch. If there is no rise in the plasma cystine after about five days, the dose of pyridoxine can be doubled. Gaull *et al.* (1968) reported a patient in whom a dose of 1200 mg a day was needed in combination with dietary restriction of methionine before a response was obtained. A response is usually observed within three to six weeks.

Some patients who respond to pyridoxine still show methionine intolerance following methionine loading, and some dietary restriction of methionine is probably advisable (Mudd *et al.* 1970*a*).

Diet. Low methionine diets supplemented with cyst(e)ine and/or pyridoxine, when commenced in infancy before onset of irreversible pathology, control the biochemical abnormalities and appear to prevent thromboembolic phenomena (Komrower and Sardharwalla 1971; Carson 1975). Treatment in older children appears to improve school performance

(McKusick *et al.* 1971). However, three of 21 patients classified as non-B_6 responders had IQ 74–80, and two had developed ectopia lentis (Pullon 1980).

Low methionine diets can be maintained by feeding mixtures of amino acids (excluding methionine) and other dietary constituents. In infancy suitable semi-synthetic diets are Albumaid Methionine Low or Albumaid RVHB X methionine (Methionaid in USA) available from Scientific Hospital Supplies Ltd, Liverpool, UK. Ten g/kg body weight/day will provide approximately 3 g protein/kg body weight/day. Fat needs to be added either in the form of double cream or 50 per cent arachis oil in water (Prosparol, Duncan, Flockhard & Co. Ltd., Greenford, Middx., UK). Vitamins need to be added. A suitable preparation is Ketovite (tablets and liquid) (Paines & Byrne Ltd., Greenford, Middx., UK), which includes folic acid, vitamin B_{12}, and other B group vitamins. Essential requirements of methionine can be added in the form of cow's milk, which has a seasonal variation in the methionine content of 50–140 mg/100 ml. Initially about 50 ml of milk per day can be added, but the amount should be assessed by monitoring plasma concentrations of methionine and homocystine (Carson 1975; Committee on Nutrition 1976). Mixed feeding can be introduced at about two months in the form of low-protein rusks, Aminex (Cow & Gate Ltd., Trowbridge, Wilts., UK).

For older children foods are classified according to their methionine content and allowed accordingly (for details see Carson 1975).

Cystine supplements, e.g. soluble calcium cystinate are given in doses of 150–200 mg/kg body weight. It should be noted that plasma cystine may not rise until plasma homocystine and the mixed disulphide have been reduced (Komrower and Sardharwalla 1971; Perry 1974).

Enhancement of remethylation. Administration of betaine (Fig. 5.1) has caused a fall in plasma homocystine, and the mixed disulphide, but a marked rise in plasma methionine (Komrower and Sardharwalla 1971) and clinical improvement (Smolin *et al.* 1981).

A similar biochemical response was reported after feeding with choline, a betaine precursor (Perry *et al.* 1968*a*). It is not known, however, whether this redistribution is beneficial.

As noted earlier, folic acid deficiency may need to be corrected before pyridoxine therapy becomes effective. In two patients with folic acid deficiency and cystathionine ß-synthase deficiency, folate administration produced a fall in homocystine excretion but a rise in methionine excretion (Carey *et al.* 1968*b*). Treatment with pyridoxine and vitamin B_{12} in addition to folic acid in another patient produced a fall in homocystine concentration without methionine accumulation (Morrow and Barnes 1972).

Symptomatic treatment. Decreased platelet survival in cystathionine ß-synthase deficient patients has been reversed by administration of dipyridamole (100 mg four times a day). Freedom from thromboembolic episodes over a three-year period was also reported in two pyridoxine-unresponsive patients who previously had had several thromboses, following treatment with dipyridamole, 100 mg once daily, and acetyl-salicylic acid, 1 g once daily, or alternatively dipyridamole, 100 mg four times a day on its own (Harker *et al.* 1974). Surgical procedures can provoke thromboembolic complications and should be avoided whenever practical. Ocular surgery has been fatal in several patients (Carson *et al.* 1965; Henkind and Ashton 1965; Francois 1972). It has been suggested that oral contraceptives and pregnancy should be avoided (Perry 1974).

Genetic counselling and prenatal diagnosis

From mass screening of about 2.6 million infants, the incidence may be calculated to be about 1:200 000 (Benson and Polani 1978). McKusick (1972) suggested a higher incidence of 1:45 000. The disorder has an autosomal recessive mode of inheritance. Obligate heterozygotes on average have lower fibroblast cystathionine ß-synthase activities than normal but overlap with the control range occurred in eight of nine heterozygotes reported by Uhlendorff *et al.* (1973), and in four of six reported by Bittles and Carson (1973, 1981). However no overlap occurred in five heterozygotes reported by Fleisher *et al.* (1973). In PHA-stimulated lymphocytes, activities in three of 17 heterozygotes overlapped with controls (Goldstein *et al.* 1973a).

Liver biopsy will not be considered a justifiable procedure for carrier detection, but it is noteworthy that of eight heterozygotes studied (Finkelstein *et al.* 1964; Laster *et al.* 1965a; Gaull *et al.* 1974), only one value of liver cystathionine ß-synthase activity overlapped with controls (Gaull *et al.* 1974).

The possibility of detection of carriers by the pattern of excretion of metabolites in urine was studied by Sardharwalla *et al.* (1974), after the administration of methionine (0.1 g/kg body weight) to 12 heterozygotes and 12 controls who had fasted for 12h. Using sensitive assays utilizing an iodoplatinate system, complete discrimination was found between hetero-zygotes and controls for the following criteria:

1. Carriers but not controls had detectable homocystinuria 6 or 9h after loading.

2. The concentration of the mixed homocysteine–cysteine disulphide was consistently higher in carriers.

3. The ratios of either homocystine or the mixed disulphide of cystine or of homocysteic acid to cysteic acid of the oxidised urines were all higher in

carriers than in controls. More experience is required before this relatively simple method can be considered to be reliable.

Prenatal diagnosis of an affected foetus was reported by Fowler *et al.* (1982) after amniocentesis at 16 weeks' gestation. An unaffected foetus had also been correctly predicted in the same family. A pregnancy at risk was monitored by Fleisher *et al.* (1974), by assay of cystathionine ß-synthase activity of amniotic cells which had been cultured for five and a half weeks after amniocentesis at 16 weeks' gestation, and an unaffected foetus accurately predicted.

Bittles and Carson (1973) monitored a pregnancy in an affected mother whose foetus could be predicted to be a heterozygote, or if the father was a carrier, either a heterozygote or affected. A heterozygous foetus was predicted.

Cystathionine ß-synthase activity is very low or undetectable in chorionic villi, but increases after tissue culture to a level comparable with that in cultured amniotic cells (Benson *et al.* 1983). First trimester diagnosis of the disease would therefore depend on assay of the enzyme in cultured villi.

Defective remethylation of homocysteine to methionine

The cofactor for the main reaction by which homocysteine is remethylated to methionine, catalysed by 5-methyltetrahydrofolate: homocysteine methyltransferase (5-methyl THF:HC-MT) is MeCbl. The only other known reaction in which a cobalamin derivative acts as cofactor is the conversion of L-methylmalonyl-CoA to succinyl-CoA, catalysed by methylmalonyl-CoA mutase, in which the cofactor is adenosylcobalamin (AdoCbl). Deficiency or failure to retain cobalamin (Cbl) therefore usually gives rise to homocystinuria and methylmalonic acidaemia.

There are three groups of disorders in which homocystinuria is due to defective remethylation of homocysteine:

(a) Deficiency of Cbl due to inadequate intestinal absorption or dietary intake, also giving rise to methylmalonic acidaemia.

(b) Defective formation, or retention, of MeCl and/or AdoCbl also giving rise to methylmalonic acidaemia. At least two different mutations have been identified, which give rise to this combined defect, using complementation studies. These have been designated cblC and cblD. Homocystinuria in this group of disorders is caused by secondary deficiency of 5-methyl THF:HC-MT.

(c) Deficiency in the synthesis of cosubstrate of the 5-methyl THF:HC-MT reaction, namely 5-methyltetrahydrofolate (5-methyl THF), due to deficiency of 5,10-methylene THF reductase (Fig. 5.1). Since this reaction does not involve the metabolism of AdoCbl, this deficiency does not give rise to methylmalonic acidaemia.

(a) Malabsorption or dietary deficiency of vitamin B_{12}

Malabsorption of vitamin B_{12} in spite of normal gastric intrinsic factor glycoprotein and normal serum vitamin B_{12} binding capacity leads to deficiency of hydroxycobalamin in serum, and of MeCbl and AdoCbl in cells, and to the combination of homocystinuria, methylmalonic aciduria, and megaloblastic anaemia (Gräsbeck *et al.* 1960; Imerslund and Bjornsted 1963; Mohamed *et al.* 1965). The biochemical abnormalities may be readily reversed by parenteral administration of vitamin B_{12} (Hollowell *et al.* 1969).

Homocystinuria, cystathioninuria and methylmalonic aciduria due to nutritional deficiency of B_{12} occurred in a six-month-old infant who had been reared exclusively on breast milk of his vegetarian mother (Higginbottom *et al.* 1978).

(b) Combined deficiency of 5-methyltetrahydrofolate: homocysteine methyltransferase, and of methylmalonyl-CoA mutase

At least seven children with combined homocystinuria and methylmalonic acidaemia due to a defect in the metabolism or retention of Cbl have been reported.

Complementation studies have established that children with this disorder can be assigned to one of two complementation groups, designated cbl C and cbl D (Gravel *et al.* 1975; Willard *et al.* 1978).

Clinical features. Patients in the cbl C group have had severe – usually fatal – illnesses presenting in infancy. Death has occurred between four and seven years. Two survivors are severely retarded (Levy *et al.* 1970*a*; Dillon *et al.* 1974; Anthony and McLeay 1976; Baumgartner *et al.* 1979*a*; Carmel *et al.* 1980); while the two brothers in the cbl D group have had a much milder course presenting later (Goodman *et al.* 1970). Of the six children in the cbl C group, five presented in infancy with failure to thrive, convulsions, developmental retardation, and anaemia which was megaloblastic in four. Four had haemolytic crises, and three congestive cardiac failure. Three were male and three female.

The two brothers in complementation group D both had biochemical abnormalities, but the two-year-old was clinically normal, and the 14-year-old had mild mental handicap but severe behaviour problems. He is the only patient in either complementation group who has had thromboembolism.

In both complementation groups cbl C and cbl D the biochemical abnormalities included homocystinuria, methylmalonic acidaemia, hypomethionineaemia, and normal serum Cbl concentration.

Molecular defects and diagnosis. Absorption of Cbl from the gut, its transport in the plasma in a form bound to the transcobalamins (TC), the receptor-mediated absorptive endocytosis of the TCII-Cbl complex into the cell followed by the intralysosomal hydrolysis of the complex, with release of Cbl are described later under 'Transcobalamin I and II deficiency' (p. 163). As far as these mechanisms have been examined, they appear to function normally in patients with the cbl C and cbl D mutations (Youngdahl-Turner *et al.* 1978, 1979; Willard *et al.* 1978). The apoenzymes of 5-methylTHF:HC-MT and of methylmalonyl-CoA mutase appear to function normally in the presence of cofactors (Goodman *et al.* 1970; Willard *et al.* 1978; Willard and Rosenberg 1978*a, b*; Mellman *et al.* 1979*a*). Their deficient activity in cell culture can be normalized by hydroxycobalamin (OH-Cbl) (Willard *et al.* 1978, Mudd *et al.* 1970*b*).

Three main abnormalities have so far been demonstrated: first, considerably reduced Cbl content of fibroblasts, liver and kidney (Mudd *et al.* 1969; Dillon *et al.* 1974; Linnell *et al.* 1976; Baumgartner *et al.* 1979*b*); second, severely defective conversion by cultured fibroblasts of ^{57}Co-cyanocobalamin (CN-Cbl), or ^{57}Co-OHCbl to either MeCbl or AdoCbl; and third, severely defective retention of ^{57}Co-CN-Cbl by cultured fibroblasts (Mahoney *et al.* 1971; Willard *et al.* 1978). In accord with the greater clinical severity of patients with the cbl C than the cbl D mutation, cbl C cells have a more pronounced defect than cbl D cells. Furthermore the preferential utilization of CN-Cbl rather than OH-Cbl is more marked in cbl C than cbl D cells (Mudd *et al.* 1970*b*; Mellman *et al.* 1979*b*). Although the precise defects of synthesis and retention of cbl by cbl C and cbl D cells are not understood, the nature of the two mutations appear to be similar, but of differing severity.

The diagnosis is suspected when the characteristic biochemical abnormalities of homocystinuria, methylmalonic acidaemia, and hypomethionineaemia with normal serum Cbl are demonstrated. The diagnosis can be confirmed by demonstrating defective conversion of CN-Cbl or OH-Cbl to MeCbl or AdoCbl; or defective retention of CN-Cbl by cultured fibroblasts (see above).

Genetic counselling. The existence of equal numbers of males and females with the cbl C mutation is consistent with autosomal recessive inheritance. The nature of inheritance of the cbl D mutation is not yet known. It should be possible to diagnose both types prenatally by study of cultured amniotic cells.

Treatment. It is not yet known whether cobalamin therapy from birth will prevent clinical deterioration.

(c) 5,10-methylenetetrahydrofolate reductase deficiency

The synthesis of 5-methyltetrahydrofolic acid (5-methyl THF) (the methyl donor for the remethylation of homocysteine to methionine) from 5,10-methylene-tetrahydrofolic acid is catalysed by 5,10-methylene tetrahydrofolate reductase (methylene THF reductase) (Fig. 5.1). Deficiency of this enzyme has been demonstrated in at least 14 patients.

There is a wide spectrum of severity and of clinical features, related to the age of onset of symptoms. Onset in infancy is associated with apnoea, coma (Harpey *et al.* 1981), seizures, and death (Narisawa *et al.* 1977), while teenage onset has been characterized by psychosis, moderate mental handicap (Mudd *et al.* 1972; Freeman *et al.* 1975) and peripheral neuropathy (Erbe 1979), or proximal muscle wasting, normal intelligence (Mudd *et al.* 1972), and survival into adult life. Patients with onset at five to seven years of age, as in the four children of the same family described by Wong *et al.* (1977*a*), have had an intermediate phenotype with normal early development and subsequent regression, severe mental handicap, spastic paresis, and extensive vascular thrombosis. Three of the four had died by 13 years of age. Other patients with similar clinical phenotypes were reported by Wong *et al.* (1977*b*) Baumgartner *et al.* (1979*a*) and Erbe (1979).

Methylene THF reductase deficiency has been demonstrated in fibroblasts (Mudd *et al.* 1972; Wong *et al.* 1977*b*; Rosenblatt and Erbe 1977), leucocytes, liver, kidney, brain (Narisawa *et al.* 1977), and lymphocytes (Wong *et al.* 1977*b*). Since 5-methyl THF is the main form of folate in serum and cells, serum and fibroblast folate are reduced (Harpey *et al.* 1981; Rosenblatt *et al.* 1979). There is increased concentration of serum homocysteine, reduced or undetectable serum methionine (Harpey *et al.* 1981) and homocystinuria. In contrast to patients with homocystinuria due to 5-methyl THF:homocysteine methyltransferase deficiency, patients with 5,10-methylene THF reductase deficiency do not have methylmalonic acidaemia.

At post mortem there has been extensive thrombosis (Wong *et al.* 1977*a*; Kanwar *et al.* 1976), intimal hyperplasia, and disruption of elastic lamellae similar to the vascular changes seen in homocystinuria due to cystathionine ß-synthase deficiency.

Response to therapy has varied. There was a good response to folic acid therapy at 15 years of age (Mudd *et al.* 1972; Freeman *et al.* 1975) and very good response at seven and a half months in a comatose infant treated with pyridoxine, methionine, folic acid, and vitamin B_{12} (Harpey *et al.* 1981), but unresponsiveness to folic acid has also been reported (Wong *et al.* 1977*a*). Heterozygotes have reduced enzyme activities in lymphocytes (Wong *et al.* 1977*b*).

Transcobalmin I and II deficiency

Transcobalamin II deficiency

Absorption of cobalamin (Cbl) from the gut is dependent on its binding to the glycoprotein 'intrinsic factor' (IF) discovered by Castle and later characterized. Deficiency of IF causes pernicious anaemia. The IF component of the IF-Cbl complex binds with specific receptor sites in the ileum and releases the Cbl through the ileal membrane into the portal blood. Initial binding in the serum is to the ß-globulin, transcobalamin II (TCII) (Hall and Finkler 1967; Hom 1967) but over 90 per cent of Cbl in serum is bound to glycoprotein transcobalamin I (TCI) (Allen 1965; Ellenborgen 1979). The TCII–Cbl complex is recognized by, and binds to specific high-affinity cell surface receptors on cultured fibroblasts (Youngdahl-Turner *et al.* 1978) and is internalized by absorptive endocytosis (Youngdahl-Turner *et al.* 1979) before release of Cbl by proteolysis of the complex by lysosomal proteases. Free Cbl then leaves the lysosomes and either enters the mitochondria where it is converted to AdoCbl by reductions and adenosylation, or remains in the cytosol and is converted to MeCbl (Mellman *et al.* 1977; Kolhouse and Allen 1977) (Fig 6.2(b)).

Megaloblastic anaemia, which is characteristic of pernicious anaemia, has been reported in at least six children with an inherited deficiency or abnormality of TCII (Hakami *et al.* 1971; Hitzig *et al.* 1974; Burman *et al.* 1979; Linnell *et al.* 1980). Also a 32-year-old woman whose pancytopaenia at six weeks of age had responded to liver extract and blood transfusions subsequently developed a megaloblastic anaemia. She was found to have immunologically cross-reacting material to TCII which failed to bind cobalamin (Seligman *et al.* 1980). Full haematological response followed treatment with Cbl, 1 mg given intramuscularly each week. Deficiency of TCII is responsible for reduced absorption of Cbl from the intestine and intracellular depletion of Cbl. In addition to severe neonatal megaloblastic anaemia, there is stunting of growth. Since most of the circulating Cbl is bound to TCI, plasma total Cbl concentration is normal.

The disorder has an autosomal recessive mode of inheritance. A partial deficiency of TCII has been demonstrated in both parents of an affected child (Hakami *et al.* 1971).

Since there is no homocystinaemia or hypomethioninaemia, TCII does not appear to be important in supplying Cbl to synthesize methyl-Cbl in the 5-methyl THF-homocysteine methyltransferase reaction. Similarly, as there is no methylmalonicacidaemia, TCII does not seem to be essential for the transport of Cbl for the synthesis of sufficient adenosyl-Cbl required for the methylmalonyl-CoA mutase reaction (Scott *et al.* 1972). However, a reduced rate of synthesis of adenosyl-Cbl but not of methyl-

Cbl by cultured fibroblasts from a patient suggests a possible link between TCII-mediated Cbl transport and intracellular synthesis of adenosyl-Cbl (Linnell *et al*. 1980).

Complete clinical and haematological remission followed treatment with pharmacological doses of B_{12} in the patient described by Hitzig *et al*. (1974).

Transcobalamin I deficiency

Two brothers aged 46 and 47 years with deficiency of TCI were unable to maintain normal Cbl concentrations in their sera, but were without clinical or biochemical evidence of B_{12} deficiency (Carmel and Herbert 1969).

Hypermethioninaemia due to hepatic methionine adenosyltransferase deficiency

Screening of newborn for hypermethioninaemia in order to detect homocystinuria has identified at least five children, two males and three females, without homocystinuria, but with persistent hypermethioninaemia due to deficiency of hepatic methionine adenosyltransferase activity (Gaull 1977; Gaull and Tallan 1974; Finkelstein *et al*. 1975; Gaull *et al*. 1981). Liver biopsy extract showed residual activity of up to 18 per cent of normal at high methionine concentration, but up to 30 per cent of normal at low concentration (Finkelstein *et al*. 1975).

Isoenzymes in leucocytes, fibroblasts and lymphoid cell lines had normal activities (Gaull *et al*. 1981).

Up to seven years of age no clinical abnormality has been reported.

Inheritance is uncertain, but may be autosomal recessive as children of both sexes have been affected, with normal parents, two of whom have had decreased oral methionine tolerance, and one of whom (a father) had approximately half normal methionine adenosyltransferase activity in the liver (Gaull *et al*. 1981).

Hypermethioninaemia

Hypermethioninaemia has been reported in children with diverse clinical features and can probably arise from a number of different causes in addition to deficiencies of cystathionine ß-synthase and hepatic methionine adenosyltransferase. Hypermethioninaemia with other hyperamino-acidaemias was identified with an incidence of one in 7750 newborn tested between four and 10 weeks of age by Levy *et al*. (1969*a*). In the 22 infants detected in a screening programme by Komrower and Robins (1969) it was benign and transient. All were receiving artificial milk feeds and therefore a relatively high protein (and methionine) diet. In three other cases with more persistent hypermethioninaemia, there was associated tyrosinaemia, anaemia (Hb 7.0–9.2 g/dl) and elevation of serum alkaline phosphatase activities, but no evidence that the disorder produced permanent harm.

A relation seems to be established with hypermethioninaemia on the one hand and either benign transient tyrosinaemia (sometimes with anaemia) or severe acute tyrosinaemia on the other. The reports of elevated serum alkaline phosphatase and transaminase activities in benign hyper-methioninaemia raise the question of possible liver dysfunction (Iber *et al.* 1957*a*) and therefore of a closer relation to acute tyrosinaemia where severe liver damage is the rule (Gjessing and Halvorsen 1965; Scriver *et al.* 1966*a*).

Hypermethioninaemia and tyrosinaemia have also been reported in hereditary fructose intolerance (Halvorsen and Gjessing 1970; Rosen *et al.* 1977), and in a child with hepatic and renal dysfunction (Smith and Strang 1958). The urine of the last child had an unpleasant odour which suggested the name 'oast house disease'.

Methionine malabsorption syndrome

Only two patients have been reported with this disorder. Both patients had a peculiar unpleasant smell likened to 'burnt sugar' or 'an oast house'. Both patients had white hair.

The first patient (Jepson *et al.* 1958; Smith and Strang 1958) had recurrent febrile episodes accompanied by oedema and severe mental retardation. She was apathetic, hypotonic and had extensor spasms, dying at 10 months of age. She excreted excessive quantities of 2-hydroxybutyric acid as well as increased amounts of phenylalanine, tyrosine, methionine, phenylpyruvic acid, phenylacetic acid, indoleacetic acid, *p*-hydroxyphenyl-acetic acid and *p*-hydroxyphenyllactic acid.

The second patient (Hooft *et al.* 1965) was a girl born of non-consanguineous parents who had convulsions from two years of age, and suffered from episodes of diarrhoea. Her IQ was 27. She excreted excessive quantities of 2-hydroxybutyric acid (12–80 mg/day) and other 2-keto acids, in particular branched chain 2-ketoacids. Loading studies with methionine aggravated her diarrhoea. When treated by a methionine-low diet her convulsions and diarrhoea stopped, her EEG became normal and her urine became free of 2-hydroxybutyrate. After oral methionine loading the parents and three siblings excreted 2-hydroxybutyric acid in urine and faeces.

Cystathioninuria; γ-cystathionase deficiency

Harris *et al.* (1959) reported a mentally retarded adult woman who excreted excessive quantities of cystathionine in urine. They suggested that this could arise from deficiency in the activity of γ-cystathionase, the enzyme which cleaves cystathionine to cysteine and α-ketobutyric acid, the latter being formed from homoserine which is a reaction intermediate (Fig. 5.1). This was confirmed by demonstration of marked decrease in the activity of the liver enzyme in the absence of pyridoxal phosphate. Mental retardation is now considered to have been fortuitous and cystathioninuria to be probably benign.

Clinical features

Following the original report by Harris *et al.* (1959) of cystathioninuria in a mentally retarded female, screening of mentally handicapped individuals has revealed a few other patients. However, it is now clear that cystathioninuria can occur in normal individuals (Hooft *et al.* 1969; Perry *et al.* 1968*b*) or in patients with a wide variety of clinical features which are probably unrelated to the enzyme deficiency. These have included nephrogenic diabetes insipidus (Perry *et al.* 1967*b*), thrombocytopenic purpura, and renal calculi (Mongeau *et al.* 1966). In an 18-year-old male, cystathioninuria, goitre, retarded growth and partial hearing loss all responded to thyroxine (Nishikawa *et al.* 1970).

Molecular defects

Activity of γ-cystathionase in cultured fibroblasts from individuals with cystathioninuria has been reported as undetectable by Pascal *et al.* (1975*a*, *b*; 1978) but as 15 per cent of control mean by Bittles and Carson (1974). γ-cystathionase deficiency has also been demonstrated in liver (Frimpter 1965; Finkelstein *et al.* 1966; Kint and Carton 1971) and in cultured lymphocytes (Pascal *et al.* 1975*b*). Finkelstein *et al.* (1966) and Kint and Carton (1971) showed that human γ-cystathionase also had homoserine dehydratase activity, which was also deficient in the affected patients. γ-cystathionase deficiency in the great majority of cases is reversible by administration of large doses of the cofactor pyridoxine (Frimpter *et al.* 1963). As has been suggested for cystathionine ß-synthase deficiency, pyridoxine-responsiveness may depend on the existence of residual γ-cystathionase activity. Thus no such activity was detected in a non-responsive individual who, unlike pyridoxine-responsive subjects, had no cross-reactive material to γ-cystathionase antibodies (Pascal *et al.* 1975*b*).

Although deficiency of pyridoxine leads to cystathioninuria (Scriver and Hutchinson 1963; Linkswiler 1970), there is no evidence of pyridoxine deficiency in patients with cystathioninuria.

In addition to excessive quantities of cystathionine, the urine of patients has had excess of cystathionine sulphoxide (Kodama *et al.* 1975). Increased concentration of *N*-acetylcystathionine in the urine has been reported (Perry *et al.* 1968*c*), and also of smaller quantities of compounds formed by transamination followed by reduction or carboxylation of cystathionine or *N*-acetylcystathionine (Kodama *et al.* 1975).

Diagnosis

Cystathioninuria occurs as a transient phenomenon in newborn, clearing up usually by five months of age (Lyon *et al.* 1971) and has been reported in association with galactosaemia (Lieberman *et al.* 1967), neural tumours (Gjessing 1963), hepatoblastoma (Voute and Wadman 1968), vitamin B_6 deficiency (Park and Linkswiler 1970), and thyrotoxicosis (Gjessing 1964).

It may also occur in homocystinuria due to remethylation defects, for example 5,10-methylenetetrahydrofolate reductase deficiency (Shih *et al.* 1972).

Definitive diagnosis therefore depends on demonstration of γ-cysta-thionase deficiency (see Molecular defects', p. 166).

Treatment

The current view is that cystathioninuria is a benign condition and requires no treatment *per se*.

Genetic counselling

In Massachusetts an incidence of 1:70 370 has been found by screening of 633 331 newborns (Levy *et al.* 1980). It is clear, however, that the reported incidence will depend on the sensitivity of the test, and on the age of the infants screened. Thus in Australia no cases were found of cystathioninuria persisting to six months of age, by screening of 200 000 infants (Turner and Brown 1972), but using more sensitive methods 13 were detected in 35 809 infants, only two of whom had cystathioninuria persisting to five months (Lyon *et al.* 1971). If these two persistent excreters had γ-cystathionase deficiency, the incidence was about 1:18 000.

The disorder has an autosomal recessive mode of inheritance. Reduced γ-cystathionase activity in parents (presumably heterozygotes) of individuals with cystathioninuria has been demonstrated as increased cystathionine urinary excretion, either on normal diets in about 50 per cent, or after methionine loading in nearly 90 per cent (Mudd and Levy 1978). Heterozygotes may also have raised concentration of plasma cystathionine (Whelan and Scriver 1968*a*); and reduced γ-cystathionase activity of lymphoid cell lines (Pascal *et al.* 1978).

Evidence for genetic heterogeneity is provided by the degree of responsiveness of liver γ-cystathionase activity to added pyridoxal phosphate *in vitro* which has varied from almost none (Pascal *et al.* 1975*b*, 1978) to about 30-fold to 50-fold (Frimpter 1965; Pascal *et al.* 1978). As for patients with cystathionine ß-synthase deficiency there is a fairly good correlation between the degree of responsiveness to pyridoxine, and the degree of residual activity, both of which tend to be concurrent in siblings.

In view of the innocent nature of the condition, prenatal diagnosis has not been performed.

ß-mercaptolactate-cysteine disulphiduria

At least four patients have been reported with this rare disorder, two unrelated males who were mentally retarded and two sisters who were mentally and physically normal.

The first male was investigated at 46 years of age (Crawhall *et al.* 1968*a*; Ampola *et al.* 1969). He had generalized convulsions and an IQ of 36 (Crawhall 1969*a*). The parents of the second male were first cousins. He had obesity, hypogonadism, valgus deformity of the knees, bilateral cataracts, and an IQ of about 60 at 23 years of age (Law and Fowler 1976).

The two sisters were detected as part of a population screening programme in Switzerland. At 11 and 13 years they had IQ's of 114 and 104, respectively (Niederwieser *et al.* 1973). It is not possible at present to assert whether this is a benign condition or whether it may have been associated with the clinical phenotype of the two retarded patients. Crawhall has speculated that there may be defective metabolism of ß-mercaptopyruvate by ß-mercaptopyruvate sulphurtransferase, or of cysteine, for example, by cysteine oxidase, but proof for either of these mechanisms is lacking.

Sulphite oxidase deficiency

This rare disorder has been described in at least three children all of whom had severe neurological disorders. In one, it was fatal in the third year; in another the parents both had a partial sulphite oxidase deficiency in cultured fibroblasts, suggesting autosomal recessive inheritance and the potential for prenatal diagnosis; in a third patient 'combined' deficiency of sulphite oxidase and xanthine oxidase was associated with xanthine urolithiasis, and the possibility of molybdenum 'dependency'.

The first patient, reported at two and a half years, had severe brain damage with mental retardation, dislocated ocular lenses and excessive excretion of *S*-sulphocysteine, sulphite and thiosulphate in the urine (Irreverre *et al.* 1967). Since sulphite ions can react with cysteine to form *S*-sulphocysteine, which has a structure $HO_2S \cdot S \cdot CH_2 \cdot CH(NH_2)COOH$, the authors considered the possibility that because of a defect in the oxidation of sulphite to sulphate there was a deficiency in the activity of the enzyme sulphite oxidase. The patient died soon after the diagnosis had been established, but at necropsy liver sulphite oxidase deficiency was confirmed (Mudd *et al.* 1967). The parents had had seven other children, of whom four were normal and three had died within a few days of birth and may have had sulphite oxidase deficiency.

Sulphite oxidase deficiency was also reported in a four and a half year old boy with ectopia lentis, choreo-athetosis, and spastic bilateral hemiparesis (Shih *et al.* 1977). Altered behaviour had been noted at 17 months and episodes of 'thrashing about' for one hour at 17 and 19 months. At two years he had diarrhoea, vomiting, staggering, myoclonic jerks, occasional flinging movements of the limbs, and ataxic gait. Later he had generalized convulsions followed by aphasia and right hemiparesis. At 28 months he had right-sided choreiform movements, regression of speech, left-sided hemiparesis, and scissoring of both legs. Ophthalmological examination at two years of age showed no abnormality, but at four years revealed subluxation of both lenses. He excreted large quantities of sulphite, thiosulphate, and *S*-sulphocysteine in the urine. His urinary inorganic sulphate excretion was reduced. Sulphite oxidase in cultured skin fibroblasts was less than 5 per cent of normal values in the patient and about 50 per cent in the father and 30 per cent in the mother, consistent with an autosomal recessive mode of inheritance. A low sulphur–amino acid diet produced a good biochemical response.

Duran *et al.* (1978*a*) reported a female infant who presented at 10 days of age with feeding difficulties. She had dysmorphic features, bilateral dislocation of the lenses and generalized convulsions. At 14 months she passed small xanthine urinary calculi, and by two years she was hypertonic and severely retarded. She had excessive excretion of sulphocysteine, taurine, and thiosulphate, and low excretion of sulphate in the urine, characteristic of sulphite oxidase deficiency. She also had low blood uric acid, xanthinuria, and marked deficiency of xanthine oxidase activity in a jejunal biopsy. The authors speculate that since xanthine oxidase and sulphite oxidase are virtually the only two human enzymes requiring molybdenum as a cofactor (Johnson *et al.* 1980*c*), the combined deficiency might be due to a form of 'Molybdenum dependency'. Molybdenum administration, however, did not cause any response. Xanthine oxidase deficiency, first decribed by Dent and Philpot (1954), has been reported in about 40 patients, about half of whom have urolithiasis (Cartier and Perignon 1978).

Homocystinsulphonuria

Two brothers aged 10 and 12 years with severe mental handicap were found to have homocystine sulphone in their urine and plasma (Lemonnier *et al.* 1971).

The cause and mode of inheritance is unknown.

Cystinosis; Lignac–Fanconi syndrome; cystine storage disease

The characteristic feature in cystinosis is intracellular accumulation of free (i.e. non-protein) cystine. Cystine crystals are formed characteristically in the cornea, bone marrow, leucocytes, and other tissues and cells. The causal metabolic defect has not yet been identified, in spite of a considerable amount of research. Recently, however, defective efflux of cystine from lysosomes has been demonstrated. Clinically there is a wide spectrum of severity, but siblings from the same family have a similar course.

The most severe form, nephropathic cystinosis, causes impairment of renal tubular function and presents as the Fanconi syndrome, with thirst and polyuria soon after six months of age. Glomerular damage occurs later leading to uraemia and death, usually before 10 years of age.

Some patients have a 'late-onset' form, also called 'intermediate' or 'adolescent' cystinosis. Clinical onset is usually after five years of age and survival is often until the second decade.

Finally, there is a rare 'benign' variant where the occurrence of cystine crystals in the cornea, bone marrow, and leucocytes is consistent with absence of symptoms and normal life span.

It is well recognized today that cystinosis is a distinct entity from cystinuria, which is characterized by an increased excretion in the urine of the amino acids cystine, ornithine, arginine, and lysine because of defective renal tubular reabsorption (Dent and Harris 1951). Historically, the two disorders were confused since the first reported child with cystinosis (Abderhalden 1903) had four relatives with cystinuria. Moreover, cystine urinary calculi, which are characteristic of cystinuria, have been reported in a few patients with cystinosis. These have been attributed to the high cystine excretion as part of.generalized aminoaciduria due to tubular dysfunction in patients with nephropathic cystinosis (Bickel *et al.* 1952).

Clinical features

Nephropathic cystinosis. Typical first symptoms are the onset of thirst and polyuria shortly after six months of age, sometimes causing dehydration and pyrexia. In the second year, patients tend to fall behind in growth, and are usually below the third centile in height and weight. They may develop rickets if on an average intake of vitamin D.

Photophobia occurs in most patients. Slit lamp examination reveals numerous refractile opacities in the cornea and conjunctiva, which are pathognomonic of cystinosis but are not present at birth (Schneider *et al.* 1969). Characteristic retinal changes have been observed as early as five weeks of age when only a few crystals were present in the iris (Schneider *et al*. 1969). There is patchy depigmentation in the periphery of the retina interspersed with clumps of pigment or discrete pigment stippling (Wong *et al*. 1967*a*; Wong 1973). The majority of children have blond hair and fair skin.

As the disease progresses, there is evidence of glomerular involvement with casts and erythrocytes appearing in the urine and rising blood urea nitrogen and creatinine concentrations.

Patients are usually mobile until the last few months of terminal renal failure, but often develop hypothyroidism, which may be due to cystine deposits in the thyroid gland (Chan *et al*. 1970). The disease is usually fatal by 10 years of age.

Late-onset (intermediate; adolescent) cystinosis. The age of symptomatic onset has varied from 18 months (Jonas and Schneider 1983) to 17 years (Pittman *et al*. 1971), but in individual families the age of onset tends to be similar in affected siblings (Goldman *et al*. 1971; Hooft *et al*. 1971; Spear *et al*. 1971). The renal disease is milder than in nephropathic cystinosis, but all reported patients have had crystalline deposits in cornea, conjunctiva, and bone marrow. Whereas in patients with nephropathic cystinosis skin pigmentation is noticeably less than that of their parents, either fair (Aaron *et al*. 1971) or normal (Goldman *et al*. 1971) skin pigmentation has been reported in intermediate variants. Photophobia, peripheral retinopathy, and growth retardation occur in some, but not all, patients (Weber *et al*. 1979).

Benign cystinosis. Patients with benign cystinosis do not have renal dysfunction, retinopathy, or any symptoms attributable to cystinosis but have crystalline deposits in the cornea, bone marrow and leucocytes. The condition is usually diagnosed on routine examination (Cogan *et al*. 1957; Schneider *et al*. 1968*a*; Kroll and Lichte 1973; Dodd *et al*. 1978).

Pathology

The unique feature of nephropathic and intermediate forms of cystinosis is heavy deposition of cystine crystals in the reticuloendothelial cells of bone marrow, liver, spleen, and lymphatic system, without inflammatory reaction. Cystine crystals may be lost during histological preparation and are most consistently observed in unstained frozen or alcohol-fixed tissues by phase microscopy (Spear 1973, 1974). In the kidneys a 'swan neck' deformity of the proximal convuluted tubules may occur in the first year (Teree *et al*. 1970). Kidneys in terminal renal failure show scarred glomeruli, tubular degeneration, and interstitial fibrosis. Abundant cystine crystals occur in interstitial cells and, to a lesser extent, in glomerular and tubular cells. Electronmicrographs reveal membrane-enclosed crystalline deposits which are probably in lysosomes (Jackson *et al.* 1962; Goodman *et al.* 1973*a*; Spear 1974).

Goodman *et al.* (1973*a*) showed that cystine gradually accumulates in transplanted kidneys, even from a cadaveric donor who was unlikely to have been a heterozygote for cystinosis. Microscopically, the proximal renal tubules of transplanted kidneys have not shown 'swan neck' deformity or cystine deposits even after periods of over one year. Cystine deposits have appeared in the donor interstitial cells but not in the epithelial cells.

In the eye, cystine cells have been demonstrated in corneal epithelial cells, histiocytes, and keratocytes, and in the conjunctiva (Wong 1972; Francois *et al.* 1972). Electronmicrographs indicate that the deposits are intralysosomal (Wong *et al.* 1970; Kenyon and Sensenbrenner 1974). Ultrastructural studies of a cystinotic lymph node also showed membrane-limited crystalline bodies resembling lysosomes and associated with the lysosomal marker acid phosphatase (Patrick and Lake 1968).

Molecular defects

Free (non-protein) cystine in leucocytes has been estimated to be about 80 times normal in nephropathic cystinosis and 30 times normal in the benign variant (Schneider *et al.* 1967*a*). Free-cystine in leucocytes in intermediate cystinosis has an intermediate accumulation (Hooft *et al.* 1971; Kroll and Lichte 1973; Goldman *et al.* 1971). Cystine accumulation is predominantly in neutrophils, rather than monocytes or lymphocytes (Schulman *et al.* 1970*a*).

Cultured fibroblasts also have a considerable accumulation of free cystine (Schneider *et al.* 1967*b*) which is about 100 times greater than normal in cells derived from patients with the nephropathic form. A lower, but elevated, free cystine content has been reported in fibroblasts from patients with the intermediate form (Goldman *et al.* 1971; Hooft *et al.* 1971; Kroll *et al.* 1973). An accumulation of about 50 times normal has been reported in fibroblasts from patients with the benign form (Schneider *et al.* 1967*b*, 1968*a*).

In an aborted foetus the free cystine levels in the liver, spleen and kidney were considerably elevated (Schneider *et al.* 1974; States *et al.* 1975). Cystine crystals were probably present in the foetal cystinotic liver (Haynes *et al.* 1980).

Electron microscopical studies indicate that intracellular free cystine in cystinotic cells is within lysosomes. This suggests an abnormality of cystine metabolism or transport.

Cystine metabolism requires its reduction to cysteine, followed by (a) degradation to inorganic sulphate and taurine, (b) incorporation into protein, and (c) conversion into glutathione (Schneider *et al.* 1968*b*; Tietze

1973; Schulman 1973). Numerous studies on the activities of enzymes catalysing these metabolic pathways have failed to reveal any defects in cystinotic cells (Patrick 1962; Tietze *et al.* 1972; States *et al.* 1974; Waterson *et al.* 1974; Kaye and Nadler 1976).

Early studies on cystine transport in cystinotic cells were inconclusive. No abnormality was found in uptake of ^{35}S-cysteine in cultured fibroblasts (States *et al.* 1974) or leucocytes (Schneider *et al.* 1968*b*). Moreover no abnormality has been found in γ-glutamyl transpeptidase of cystinotic cells (Patrick *et al.* 1979) which according to Meister (1973) may play a role in amino acid transport.

More recently, however, it was found that when intact leucocytes (Steinherz *et al.* 1982*a*) or lysosome-rich fractions (Steinherz *et al.* 1982*b*) from cystinotic patients were loaded with cystine by incubation with cystine dimethylester, cystine efflux was considerably less than in preparations from normal individuals. Moreover Gahl *et al.* (1982) showed that the transport system in lysosomes from normal leucocytes is saturable, and is stimulated by ATP, suggesting that it is a carrier-mediated, active (energy-dependent) process; and that this process was deficient in preparations from cystinotic patients, and functioning at about half normal maximal rates in carriers.

These results are consistent with the hypothesis that the primary defect in cystinosis is in a transport mechanism of cystine efflux from the lysosomes, rather than from the cells.

In the earlier stages of the disease, biochemical changes in blood and urine are those of the Fanconi syndrome with defective renal tubular reabsorption of specific components of the glomerular filtrate leading to glycosuria which can be mild or severe (Bickel 1955), generalized aminoaciduria with increased cystine excretion (usually in proportion to that of other aminoacids), phosphaturia, and proteinuria, including the excessive excretion of the light chains of gamma globulin (Waldman *et al.* 1973). There is an increased excretion of urinary bicarbonate and ammonium ion. The urine is alkaline, in spite of a metabolic acidosis. There is hypophosphataemia and raised serum alkaline phosphatase activity, indicating subclinical or active rickets, which may also be associated with hypocalcaemia. Anaemia is common and may precede azotaemia. Hypokalaemia and acidosis are common, especially in the later stages of the disease.

Diagnosis

After the first year or so, the presence of refractile bodies in the cornea on slit lamp examination is pathognomonic.

The diagnosis is confirmed by demonstrating an excess of free intracellular cystine in leucocytes (Schneider *et al.* 1967*a*, 1968*b*) or cultured fibroblasts (Schneider *et al.* 1967*b*; Kroll *et al.* 1974; Willcox and Patrick 1974).

Cystinosis is one of the commonest causes of the Fanconi syndrome and should be suspected in children with vitamin D-resistant rickets and biochemical abnormalities of blood and urine, summarized under 'Molecular defects' above. Other causes of the Fanconi syndrome are reviewed by Schneider and Seegmiller (1972). Some are summarized in Table 5.1.

Table 5.1 *Causes of the Fanconi syndrome*

Genetic Causes
   Cystinosis (Bickel *et al.* 1952)
   Galactosaemia (Komrower *et al.* 1956)
   Tyrosinaemia, type 1 (Gentz *et al.* 1965)
   Hereditary fructose intolerance (Baerlocker *et al.* 1978)
   Familial nephrosis
   Lowe's syndrome (Abbassi *et al.* 1968)
   Wilson's disease (Morgan *et al.* 1962)

Toxic
   Lead, mercury
   Lysol, maleic acid, vitamin D

Other disorders
   Nephrotic syndrome
   Multiple myeloma
   Amyloidosis

Genetic counselling and prenatal diagnosis

All forms of cystinosis have an autosomal recessive mode of inheritance. Heterozygotes of the nephropathic form have accumulation of free cystine in leucocytes and cultured fibroblasts (Schneider *et al.* 1967*a*, *b*). Less information is available on intracellular free cystine in heterozygotes for the other types of cystinosis, but those for the adult form have significant accumulation (Goldman *et al.* 1971). A relatively high incidence of nephropathic cystinosis of one per 25 909 has been reported in Brittany, as compared with one per 326 440 in the rest of France (Bois *et al.* 1976). These incidences indicate gene frequencies of 0.0062 and 0.0018 respectively. The diversity in the severity of intermediate forms coupled with the similarity amongst siblings from the same family (Goldman *et al.* 1971; Hooft *et al.* 1971) suggests the existence of several distinct mutations, each with characteristic features.

Prenatal monitoring of a pregnancy at risk for cystinosis was first reported by Schulman *et al.* (1970*b*). Amniocentesis was at 27 weeks' gestation. Amniotic cells were cultured for seven to eight weeks before ^{35}S-cystine was added to the cultured medium. After four days, cells were sonicated in the presence of *N*-ethylmaleimide (NEM) and deproteinized. Glutathione-NEM, cysteine-NEM and cystine were separated from the supernatant by ion-exchange column chromatography and assayed for radioactivity. An unaffected, possibly heterozygous, foetus was accurately predicted.

Schneider *et al.* (1974) used a similar method for prenatal diagnosis. Amniotic cells were cultured for five weeks after amniocentesis at 18 weeks' gestation. Ion exchange chromatography was preferred to either high voltage electrophoresis or thin-layer chromatography for separation of products. Amniotic cells from the pregnancy at risk had considerably elevated ^{35}S-cystine/^{35}S-cysteine-NEM and ^{35}S-cystine/^{35}S-glutathione-NEM ratios. An affected foetus was predicted and the pregnancy terminated. The diagnosis was confirmed by the demonstration of 50–100 times elevation of free cystine in foetal liver, kidney, placenta, thymus, spleen and cultured fibroblasts. No cystine crystals were found in the viscera.

The method of choice for prenatal diagnosis, which is simple and sufficiently sensitive for results to be available after about three weeks of amniotic cell culture, was described by Willcox and Patrick (1974). Cells are grown in medium containing ^{35}S-cystine, sonicated in the presence of NEM and deproteinized. ^{35}S-cystine and ^{35}S-glutathione-NEM in the supernatant are separated by thin-layer chromatography and detected by autoradiography after about three days' exposure to X-ray film. The presence of a marked ^{35}S-cystine band in cystinotic extracts is clear by visual examination alone, but can be quantitated by densitometry of the X-ray; or by scintillation counting of the areas scraped from the thin layer.

A sensitive method for separating ^{35}S-cystine, after complexing with NEM, has been reported by States and Segal (1969), and States *et al.* (1975) using high voltage electrophoresis.

Treatment

Drugs. Treatment with dimercaptopropanol and DL-penicillamine in an attempt to reactivate thiol-dependent enzymes appeared hopeful initially (Clayton and Patrick 1961). Subsequent studies failed to confirm any beneficial effects (Crawhall *et al.* 1968*b*; Hambraeus and Broberger 1967).

The observations that intracellular free cystine of cultured cystinotic fibroblasts could be lowered by the addition to the culture medium of dithiothreitol (DTT) (Goldman *et al.* 1970), cysteamine (Thoene *et al.* 1976) or ascorbic acid (Kroll and Schneider 1974) suggested the use of these agents for treatment. Reduction of free cystine in leucocytes after treatment with DTT was reported in two patients by Depape-Brigger *et al.* (1977), but long term effects are not yet known.

Cysteamine administration has been effective in lowering free cystine of leucocytes (Schneider *et al.* 1981) and has been tried in several patients

(Thoene *et al.* 1976; Roy and Pollard 1978; Girardin *et al.* 1979; Yudkoff *et al.* 1981) but requires further evaluation.

Although cysteamine is well tolerated when administered in slowly increasing doses (Corden *et al.* 1981) it is unpalatable, and may be replaced by its derivative phosphocysteamine (Thoene and Lemon 1980) which appears equally effective, and is tasteless.

A multicentre trial on ascorbic acid therapy failed to show any benefit and was discontinued (Schneider *et al.* 1979).

Diet. No benefit has been derived by placing cystinotic patients on methionine- and cystine-low diets (Seegmiller *et al.* 1968; Bickel *et al.* 1973).

These results are consistent with the report by Thoene *et al.* (1977*a*) that cystinotic cells accumulate cystine mainly from products of intracellular protein degradation.

General. Rickets usually responds to doses of vitamin D of about 15 000 units per day. In 21 cases of cystinosis attending various centres (Bickel *et al.* 1973) 20 received vitamin D, and one did not need it. Thirteen patients required vitamin D constantly, usually in high doses. A beneficial effect was claimed in only one patient, was denied in eight, and considered equivocal in the remainder.

Hypokalaemia and acidosis can usually be corrected by potassium and bicarbonate supplements, although in individual patients it may be difficult to restore normal serum bicarbonate concentration.

Renal transplantation. Of 24 cystinotic children who received 27 kidney transplants (Renal Transplant Registry 1975), nine patients were well three years after transplant, four were well but had been followed up for short periods and five had died. A similar prognosis was reported by Goodman (1973*a*). Further experience of renal transplantation was reported by Malekzadeh *et al.* (1977), West *et al.* (1977) and Broyer *et al.* (1981).

Renal transplant would seem to offer the best hope of survival for patients with cystinosis and severe renal failure.

Visual function however continues to deteriorate after renal transplantation (Broyer *et al.* 1981). Transplanted kidneys tend to accumulate cystine especially if the donors are heterozygotes (West *et al.* 1977).

DISORDERS OF AROMATIC AMINO ACIDS

Phenylketonuria and the hyperphenylalaninaemias

The commonest inborn error of metabolism known to cause mental handicap, phenylketonuria (PKU), was first identified by the Norwegian physician and biochemist Fölling (1934*a,b*). He followed up his observation that urine of two retarded siblings turned green on mixing with an aqueous solution of ferric chloride, by identifying phenylpyruvic acid, and by detecting the condition in eight more retarded individuals out of 430 whom he tested. He also noted a tendency for these patients to have a mousy odour, a fair complexion, and dermatitis. Moreover, he showed that overloading rabbits with phenylalanine caused phenylpyruvic acid to appear in their urine, and suggested that the disorder was a genetic defect of phenylalanine metabolism. This was confirmed by Jervis (1947), who showed that patients were unable to convert phenylalanine into tyrosine. The disorder was named phenylketonuria by Penrose (1935).

Hydroxylation is the main metabolic pathway for phenylalanine (Fig. 5.2) at normal low blood phenylalanine concentrations (Kaufman 1976). The demonstration by Udenfriend and Cooper (1952) that phenylalanine hydroxylation is catalysed by an enzyme system was followed by a report that the activity of this system is reduced in PKU (Jervis 1953). The work of Kaufman (1957, 1959, 1971, 1976), however, demonstrated the complexity of this multicomponent system, and has led to the identification of several types of PKU and hyperphenylalaninaemia.

Treatment of PKU by low-phenylalanine diet was shown to be effective in the prevention of mental handicap, but in order to achieve the best results, treatment had to start in the first few weeks (Bickel *et al.* 1954; Hudson *et al.* 1970). This was possible only by establishing the diagnosis before the onset of symptoms and therefore by screening of newborn (Guthrie and Susi 1963; Bickel *et al.* 1980; Veale 1980). Still controversial is the optimal time to terminate treatment, and the management of pregnancies in affected mothers which result in a high incidence of abortion and foetal damage.

The classification of the hyperphenylalaninaemias is important for practical management because some do not require treatment (transient or benign hyperphenylalaninaemias), the majority will develop severe mental handicap which is preventable by low-phenylalanine diet (classical PKU), while some cases – about one to three per cent of the number of classical PKU – will develop severe brain damage in the first year, in spite of treatment by phenylalanine deprivation. These forms have been called 'malignant hyperphenylalaninaemias' and are due to deficiency of tetrahydrobiopterin, a cofactor of the phenylalanine hydroxylase reaction. Two main types have been identified (Fig. 5.2): one due to deficiency in the activity of dihydropteridine reductase, the enzyme which catalyses regeneration of tetrahydrobiopterin (BH_4), and the other to a deficiency in the biosynthesis of dihydrobiopterin and BH_4 from guanosine triphosphate (GTP). Several enzyme-catalysed steps are involved in the metabolic path of the latter (Gal *et al.* 1978; Tanaka *et al.* 1981) including

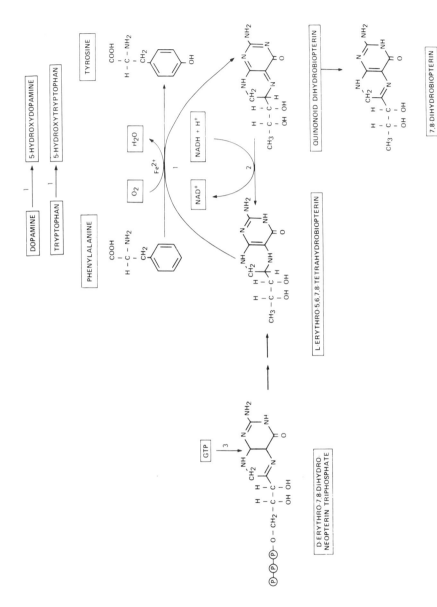

Fig. 5.2. The phenylalanine hydroxylation system: (1) Phenylalanine 4-hydroxylase (2) Dihydropteridine reductase (3) GTP cyclohydrolase I.

GTP cyclohydrolase, which is rate-limiting in human brain, and D-erythro-7,8-dihydroneopterin triphosphate synthetase. BH_4 is also a cofactor in the enzyme-catalysed hydroxylation of tryptophan to 5-hydroxytryptophan, on the pathway leading to the synthesis of 5-hydroxytryptamine (serotonin), and the hydroxylation of tyrosine to dopamine. BH_4 deficiency therefore causes deficiency of the neurotransmitters serotonin and dopamine, which may be involved in the pathogenesis of the severe neurological damage which occurs in BH_4 deficiency.

Since treatment of the BH_4-deficient hyperphenylalaninaemias with BH_4 (and empirically also with supplements of L-dopa, carbidopa and 5-hydroxytryptophan) can prevent brain damage if started early, all children with PKU should be screened when first detected by a BH_4 administration test. A fall in serum phenylalanine concentration within 4–8h is characteristic although not always marked. Other tests include assay of dihydropteridine reductase in cultured fibroblasts; and determination of biopterin excretion in urine.

Clinical features, classification, and diagnosis

Most infants with hyperphenylalaninaemia are identified through screening programmes and are usually clinically normal. When a positive screening test is reported, serum phenylalanine and tyrosine concentrations should be determined. Hyperphenylalaninaemia is defined as a serum concentration of phenylalanine of over 150 μmol/l (2.5 mg/dl) at five to seven days (Guttler and Wamberg 1977), although the majority of cases requiring dietary treatment will have values of 1212 μmol/l (20 mg/dl), or higher.

'Benign' hyperphenylalaninaemia

(a) Transient hyperphenylalaninaemia. Most infants with benign hyperphenylalaninaemia have only a modest increase of serum phenylalanine ranging from 150–606 μmol/l (2.5 to 10 mg/dl). If serum tyrosine is also raised (from normal values of 50–150 μmol/l to 150–300 μmol/l), hyperphenylalaninaemia is probably benign, and serum tyrosine and phenylalanine values will return to normal by one to six months of age (Partington *et al.* 1968; Light *et al.* 1966, 1973). The cause is unknown (Kaufman 1976) but may be developmental immaturity of enzyme systems. The incidence of tyrosinaemia is increased among pre-term infants. Tyrosinaemia may be reduced by administration of ascorbic acid (Levine *et al.* 1941). It has been suggested that hyperphenylalaninaemia associated with prematurity is due to immaturity of *p*-hydroxyphenylpyruvate oxidase (Levine *et al.* 1941; Light *et al.* 1966).

Hyperphenylalaninaemia can also occur in the different forms of hereditary tyrosinaemia (q.v.).

(b) Persisent hyperphenylalaninaemia. In some patients hyperphenylalanin-aemia remains, and is associated with some intolerance (relaxed tolerance) to phenylalanine loads but no phenylketonuria. Phenylalanine hydroxylase activity in liver biopsies from 17 such patients was shown to be reduced to 1.8–34.5 per cent of normal (Bartholomé *et al.* 1975). Three of these 17 patients, however, were seriously retarded.

Classical PKU

Children with serum phenylalanine concentrations above 606 μmol/l (10 mg/dl) should be treated by dietary restriction of phenylalanine, but tyrosinaemia and BH_4 deficiency should be excluded.

(a) Infants and children. Infants with classical PKU who are placed on well-controlled phenylalanine-low diets during the first few weeks after birth usually develop normally. Untreated infants are rare. Most have a musty odour attributed to increased phenylacetic acid, and about a quarter have facial eczema. During the first year there is gradual dilution of colour so that the skin and hair become noticeably less pigmented than those of their unaffected sibs at comparable age. This has been attributed to reduced melanin synthesis because of inhibition of tyrosinase activity by phenyl-alanine accumulation.

It has been estimated that untreated patients lose approximately 50 points off their IQ during their first 12 months (Koch *et al.* 1974). This causes delay or failure in walking and speaking. In the majority, the final IQ is less than 20. In their second and subsequent years, patients develop convulsions, hyperactivity, enlarged prominent maxillae with widened spaces between the upper teeth, microcephaly, and retarded linear growth.

(b) Adults with PKU. Four women with PKU and normal intelligence were reported by Perry *et al.* (1973). Screening of adults has revealed an incidence of PKU of about 1 in 80 000. These patients were mentally handicapped (Levy *et al.* 1970*b*).

(c) Maternal PKU. It is now well recognized that the majority of infants of mothers with PKU are damaged (Scriver and Rosenberg 1973; Komrower *et al.* 1979; Buist *et al.* 1979). In addition to mental retardation, which occurs in over 90 per cent and which often is not prevented by postconception treatment, these infants have had cardiac anomalies, microcephaly and growth retardation. Moreover, the abortion rate is increased. In one pregnancy, the combination of preconception and postconception low-phenylalanine diet resulted in the birth of a normal infant (Nielsen *et al.* 1979). Perry *et al.* (1973) found a correlation between

maternal levels of plasma phenylalanine and mental retardation in the offspring. Thus when fasting plasma concentrations were less than 606 μmol/l (10 mg/dl) none of the offspring was mentally damaged. When fasting concentrations were 606–1212 μmol/l (10–20 mg/dl), 91 per cent of children were mentally damaged; when fasting levels were above 1212 μmol/l (20 mg/dl), 99 per cent of their children were brain damaged.

Evidence from an international survey (Lenke and Levy 1980), indicated a relation between the degree of hyperphenylalaninaemia during the pregnancy and the defects in the foetus. Thus if the blood concentration was greater than 970 μmol/l (16 mg/dl), 25 of 93 women (27 per cent) miscarried in over 30 per cent of their pregnancies, while there were no spontaneous abortions in women with concentrations of 660–909 μmol/l (11 to 15 mg/dl). Mental retardation in 260 offspring of 112 mothers occurred in 92 per cent when maternal phenylalanine concentrations were greater than 1212 μmol/l (20 mg/dl), 21 per cent when 186–606 μmol/l (3–10 mg/dl). Congenital heart disease in offspring of 123 mothers occurred in 17 per cent of offspring when maternal phenylalanine concentrations were greater than 1212 μmol/l (20 mg/dl), in 19 per cent when 970–1151 μmol/l (16 to 19 mg/dl) and in 8 per cent when 667–909 μmol/l (11 to 15 mg/dl), and in none when below 10 mg/dl. However, the authors found no consistent relation between dietary treatment and prevention of abnormality in 34 pregnancies in which the mothers were on low-phenylalanine diets; but when treatment was begun early there was a trend towards higher IQ and larger head circumference.

Breast-fed normal infants of phenylketonuric mothers may have hyperphenylalaninaemia during the first week (Rampini 1973).

Tetrahydrobiopterin deficiency; malignant hyperphenylalaninaemia

Hyperphenylalaninaemia resulting from BH_4 deficiency may be due to deficiency of dihydropteridine reductase (Fig. 5.2) (Danks *et al.* 1975*a*; Kaufman *et al.* 1975; Brewster *et al.* 1979; Rey *et al.* 1977), or to a defect in dihydrobiopterin synthesis (Leeming *et al.* 1976; Kaufman *et al.* 1978; Schaub *et al.* 1978; Curtius *et al.* 1979), which in one case was probably due to GTP cyclohydrolase deficiency (Niederwieser *et al.* 1982*a*). Clinically, after a few months, in spite of a phenylalanine-low diet, there is progressive mental retardation and neurological regression, muscular hypotonia or hypertonia, hyperpyrexia, and death, usually in the first year. Convulsions are typical of dihydropteridine reductase deficiency, and have not been reported in defects of biopterin synthesis. A dramatic fall of serum phenylalanine occurs four to eight hours after administration of synthetic tetrahydrobiopterin, and although only few data are available, there is an encouraging clinical improvement on treatment.

Pathology of classical PKU

There is a decrease in brain lipids of myelin, especially galactolipids (Agrawal and Davison 1973; Shah *et al.* 1972) but also of proteolipids (Menkes 1968).

Myelin in eight patients was reduced and sometimes spongy (Malamud 1966).

Molecular defects in classical PKU

Phenylalanine hydroxylase (Fig. 5.2) was undetectable in aspiration liver biopsies from 13 patients with classical PKU (Bartholomé *et al.* 1975) and only 0.27 per cent of normal in another (Friedman *et al.* 1973). The deficiency of this enzyme interrupts the major pathway of phenylalanine metabolism, causing an accumulation of phenylalanine in the blood and urine and an overflow of metabolites into minor pathways in patients on normal dietary intake of phenylalanine (Kaufman 1976).

The study of Bartholomé *et al.* (1975) suggests the existence of a spectrum of severity for biochemical and clinical phenotypes of hyper-phenylalaninaemia depending on the degree of residual hepatic phenyl-alanine hydroxylase activity. They were able to correlate the requirements for phenylalanine-low diet to enzyme activities and suggest a 'cut off point' at about 8 per cent of normal hepatic hydroxylase activity. Thus four patients with hepatic phenylalanine activities of 8.7 per cent to 34.5 per cent, aged two to four and a half years, who were not treated, were developmentally normal, while three of 13 patients with activities of 1.8 per cent to 6.2 per cent were severely retarded.

A correlation between the clinical phenotype and residual enzyme activity was also found by Trefz *et al.* (1978). They measured the extent of hydroxylation of phenylalanine to tyrosine *in vivo* after intravenous injection of heptadeutero 1-phenylalanine. Residual activities correlated closely with those measured by direct enzyme assay of liver biopsies. Six patients with classical PKU had phenylalanine hydroxylase activities 0.3 per cent to 1.5 per cent of normal, while nine patients with hyperphenyl-alaninaemia had activities 2.2 per cent to 20.2 per cent of normal.

The balance of evidence suggests that in humans phenylalanine hydroxylase activity is limited to liver. Reports that activity occurs in cultured human fibroblasts (Hoffbauer and Schrempf 1976) and serum (Hoffbauer *et al.* 1976) have not been confirmed.

Although the majority of patients with classical PKU have immuno-logically cross-reacting material to phenylalanine hydroxylase antiserum in the liver (Bartholomé and Ertel 1978), one patient did not (Friedman 1973).

The phenylalanine hydroxylation system

The hydroxylation reaction (Fig. 5.2) involves conversion of phenylalanine to tyrosine by hydroxylation in the *para* position and stoichiometric utilization of oxygen and the cofactor BH_4. The reaction products are tyrosine, quinonoid dihydrobiopterin and water. Intracellular tetrahydrobiopterin is regenerated from quinonoid dihydrobiopterin by two routes. Firstly, by an important pathway tightly coupled with the hydroxylation reaction, catalysed by dihydropteridine reductase and utilizing either NADH or NADPH as hydrogen donor; and secondly by tautomerization of quinonoid dihydrobiopterin, which is unstable, to the more stable 7,8-dihydrobiopterin, followed by reduction of the last-named compound to tetrahydrobiopterin catalysed by dihydrofolate reductase with NADPH acting as hydrogen donor. The relatively small contribution of this pathway to regeneration of tetrahydrobiopterin, and conversely the importance of dihydrobiopterin reduction catalysed by dihydrobiopterin reductase, is suggested by the drastic consequences of dihydropteridine reductase deficiency.

Two other factors stimulate the hydroxylase reaction *in vitro*, but their activity *in vivo* has not been demonstrated; these are phenylalanine hydroxylase stimulator and lysolecithin.

The importance of *de novo* synthesis of dihydrobiopterin to maintain an adequate pool of tetrahydrobiopterin is illustrated by the occurrence of the second type of tetrahydrobiopterin deficiency and malignant hyperphenylalaninaemia when the synthetic pathway from guanosine triphosphate (Fig. 5.2) is interrupted.

Accumulating metabolites and diagnosis of classical PKU and BH_4 deficiency

In newborn children with classical PKU, blood phenylalanine concentration is usually over 1212 μmol/l (20 mg/dl) but can be as low as 272 μmol/l (4.5 mg/dl) if the protein intake is low (Holtzman *et al.* 1974; Güttler and Hansen 1977*a*). Other metabolites which accumulate in PKU (Fig. 5.3) include phenyllactic acid (Zeller 1943), *o*-hydroxyphenylacetic acid (Armstrong *et al.* 1955), phenylethylamine (Oates *et al.* 1963), and phenylacetic acid which becomes conjugated to glutamine to yield phenylacetyl glutamine (Woolf 1951).

Elevated phenylalanine concentrations inhibit absorption of tryptophan from the intestine (Scriver and Rosenberg 1973). In PKU there is diminished urinary excretion of 5-hydroxyindole acetic acid and reduced concentration of blood 5-hydroxytryptamine (Pare *et al.* 1957). However, there is an increase of indole pyruvic acid (Schreier and Flaig 1956) and of indole lactic acid (Armstrong and Robinson 1954) which are probably formed by bacteria from intestinal tryptophan.

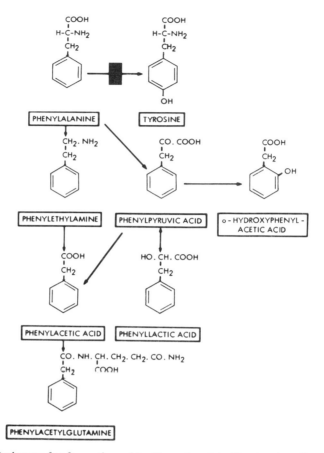

Fig. 5.3. Pathways for formation of 'spillover' metabolites in phenylketonuria.

Since 1–3 per cent of patients with persistent hyperphenylalaninaemia have a form of BH_4 deficiency rather than classical PKU, it is essential to test all newly detected patients both by BH_4 challenge, and for urine pterin excretion. In both types of BH_4 deficiency (dihydropteridine reductase deficiency and deficiency of biopterin synthesis) there is a fall in plasma phenylalanine concentration after BH_4 administration, though this is not always marked. Curtius *et al.* (1979) recommend oral administration of BH_4 hydrochloride, 8 μmol/kg, 1h before a meal and measurement of blood phenylalanine 4 or 6h after. Niederwieser *et al.* (1982*b*) recommend a single oral test dose of BH_4 hydrochloride, 7.5 mg/kg. Curtius *et al.* (1979) showed that in two patients with deficiency of biopterin synthesis, blood phenylalanine concentrations were as sharply reduced after oral administration of L-sepiapterin as by BH_4.

Patients with deficiency of dihydropteridine reductase can be distinguished from those with deficiency of biopterin synthesis by demonstration of reduced dihydropteridine reductase activity in either liver biopsies or cultured fibroblasts in the former (Kaufman *et al.* 1975; Danks *et al.* 1975*a*; Brewster *et al.* 1979) and by the pattern of biopterins in urine and blood. Biopterins can be measured as growth requirements of *Crithida fasciculata* (Dewy and Kidder 1971). This assay gives an aggregate of 7,8-dihydro-biopterin and BH_4. Alternatively, biopterins can be separated by high voltage electrophoresis and estimated individually (Curtius *et al.* 1979). More recently, high performance liquid chromatography (Niederwieser *et al.* 1980) has been automated (Niederwieser *et al.* 1982*b*) allowing rapid measurement of biopterin and neopterin. Variants of BH_4 deficiency can be distinguished by using a two-dimensional plot of percentage of biopterin and neopterin in urine, against urine biopterin/creatinine ratios. In dihydropteridine reductase deficiency there is an increase in *Crithida* factors in blood and urine (Danks *et al.* 1978); and an increase in urinary dihydrobiopterin and dihydroxanthopterin (Curtius *et al.* 1979; Schlesinger *et al.* 1979). In contrast, patients with deficiency of biopterin synthesis have decreased urinary *Crithida* factors (Leeming *et al.* 1976; Kaufman *et al.* 1978) and a considerable excretion of neopterin in the urine, but no trace of biopterins (Curtius *et al.* 1979). In the single four-year old girl reported by Niederwieser *et al.* (1982*a*) with severe mental retardation, very low concentrations of neopterin were found in the urine, indicating a block in the first reaction in the conversion of GTP to BH_4 (Gal *et al.* 1978), due probably to deficiency of GTP cyclohydrolase. Neurotransmitters or their metabolites 5-hydroxyindole acetic acid, homovanillic acid, valyl-mandelic acid, noradrenaline, and adrenaline are considerably decreased in plasma and urine in both classical PKU and in BH_4 deficiency (Kaufman *et al.* 1978; Leeming *et al.* 1976). In classical PKU the values return to normal after blood phenylalanine has been reduced by dietary restriction of phenylalanine. Failure of neurotransmitters to rise after blood phenylalanine has returned to normal is said to be diagnostic of tetrahydro-biopterin deficiency (Danks *et al.* 1978).

The diagnosis of classical PKU usually depends on the demonstration of persistent elevation of plasma phenylalanine concentrations of 606 μmol/l (10 mg/100 ml) or more for two or three days, the absence of response to BH_4 and the absence of tyrosinaemia. For practical purposes, it is not necessary to measure hepatic phenylalanine hydroxylase, although this has been done for research, either directly on aspiration liver biopsies (Friedman *et al.* 1973; Bartholomé *et al.* 1975) or indirectly by assay of tritium- or deuterium-labelled phenylalanine derivatives *in vivo* (Trefz *et al.* 1978; Fell *et al.* 1978; Curtius *et al.* 1977). When the diagnosis of classical PKU is established a few days or weeks after birth, the persistence of phenylalanine intolerance can be confirmed at about three months of age by a 'milk challenge'. In this test the infant is again hospitalized and given 700 ml of milk diluted 1:1 daily for three days. This is equivalent to approximately 180 mg of phenylalanine/kg body weight. In classical PKU there is a steady rise in serum phenylalanine concentration and appearance in the urine of phenylalanine, phenylpyruvic acid and other phenylalanine derivatives (see 'Molecular defects', p. 182). The test should be terminated prematurely if there is a sharp rise of serum phenylalanine to above 1816 μmol/l (30 mg/100 ml). A second test can be made at 12–15 months of age.

When neonatal hyperphenylalaninaemia is found to coexist with tyrosinaemia, it is usually transient and benign. The innocent nature of the neonatal tyrosinaemia can be demonstrated by its response to ascorbic acid (Levine *et al.* 1941). However, these children should be followed up since some turn out to have a form of hereditary tyrosinaemia (q.v.).

Genetic counselling and prenatal diagnosis

Heterozygotes

Classical PKU (and probably hyperphenylalaninaemia with reduced tolerance to phenylalanine) has an autosomal recessive mode of inheritance. One may speculate that classical PKU with cross-reacting material to phenylalanine hydroxylase (PH) represents homozygosity of the PH^- allele (deficient PH activity) or double heterozygosity of the PH^- allele and a PH^0 (PH activity absent – no cross-reacting material) allele. The range of hepatic residual activities of 1.8 per cent to 34.5 per cent (Bartholomé *et al.* 1975) could represent the existence of several types of PH^- allele and possibly heterozygotes for classical PKU, or other double heterozygotes at the PH locus.

Methods used for attempts at detection of heterozygotes have included determination of molar ratios of plasma phenylalanine to tyrosine, and measurement of phenylalanine following phenylalanine loading (Hsia *et al.* 1956). However, the quantitative contribution of *para*-hydroxylation, transamination, and incorporation into proteins as mechanisms of removal of phenylalanine from the extracellular pool varies according to the phenylalanine concentration (Kaufman 1976). Thus at blood phenylalanine concentrations of less than 200 μmol/l, more than 95 per cent of phenylalanine is metabolized by hydroxylation. At concentrations above 1000 μmol/l, that is, after phenylalanine loading, it is metabolized equally by transamination and hydroxylation. Others have used statistical methods for analysis of loading tests (Güttler and Hansen 1977*b*), or have analysed metabolites of phenylalanine (Blau *et al.* 1973; Koepp and Hoffman 1975).

The molar ratio of phenylalanine to tyrosine in a single blood sample at noon before lunch (to standardize circadian variation), especially when plotted against the serum phenylalanine concentration, allowed discrimination of 42 of 43 heterozygotes. The one misclassified heterozygote was reclassified correctly on a repeat test (Rosenblatt and Scriver 1968).

Using the same data, Gold *et al*. (1974) calculated that by plotting the values graphically it is possible to construct an elliptical area dividing the bivariate plane into two regions, so that subjects whose values lie in the ellipse have a probability of 0.1 of being heterozygotes, the rest of the plane being assigned to bearers of a mutant allele which affects phenylalanine oxidation.

Griffin and Elsas (1975) were able to separate 18 obligate heterozygotes for PKU from 59 controls using semi-fasting midday plasma phenylalanine to tyrosine concentrations and multivariate linear discriminant analysis. Paul *et al*. (1978) have found that the best separation between obligate heterozygotes and controls was obtained by a linear discriminant function involving the logarithms of the serum concentrations of phenylalanine, tyrosine and tryptophan. The theoretical overlap was 3.75 per cent, but for women who were pregnant, or on oral contraceptives, the overlap increased to 8.23 per cent.

The high incidence of carriers, which is in the region of 1 in 50, or more, in Celtic populations raises the possibility of some heterozygote advantage (Woolf 1978). Increased fertility of heterozyote mothers has also been proposed but not confirmed (Saugstad 1973, 1977). Similarly, increased survival of unaffected offspring of heterozygous mothers was suggested (Saugstad 1977) but again not confirmed (Paul *et al*. 1979*a*). Indeed a tendency to low IQ (Thallhammer *et al*. 1977; Ford and Berman 1977; Bessman *et al*. 1978) has been suggested for PKU heterozygotes.

Prenatal diagnosis

Prenatal diagnosis of classical PKU has been performed in two pregnancies in different families with an affected child. A cloned human PH gene probe was used to analyse DNA isolated from cultured amniotic cells. The foetuses were predicted accurately to be homozygous and heterozygous for the PKU gene, respectively. The authors (Lidsky *et al*. 1985) estimated that prenatal diagnosis of PKU by DNA analysis is possible in 90 per cent of at-risk families using ten polymorphic restriction sites at the PH locus. Prenatal diagnosis cannot be established by amniotic cell PH assay since there is no measurable phenylalanine hydroxylase activity in normal amniotic cells, and no abnormality has been reported in foetal blood or amniotic fluid of affected pregnancies. Since hepatic phenylalanine activity reaches adult levels at seven weeks' gestation (Woo *et al*. 1974) prenatal diagnosis by foetal liver biopsy is theoretically feasible. Two abnormalities have been reported in cultured fibroblasts from patients with PKU, which suggested that the defect might be expressed in cultured amniotic cells. The first was a deficiency in the activity of phenylalanine hydroxylase (Hoffbauer and Schrempf 1976) and the second that dihydropteridine reductase activity is inhibited in the presence of 2 mmol/l phenylalanine (Schlesinger *et al*. 1976; Cotton 1977). However, we and others have been unable to demonstrate any phenylalanine hydroxylase activity in normal cultured fibroblasts and Güttler *et al.* (1977) were unable to confirm the phenylalanine effect on dihydropteridine reductase.

The report that there is phenylalanine hydroxylase activity in human placenta (Matalon *et al.* 1977*a*) suggested the possibility of placental biopsy for prenatal diagnosis.

Since dihydropteridine reductase deficiency can be demonstrated in cultured fibroblasts, prenatal diagnosis of this form of hyperphenyl-alaninaemia by enzyme assay of cultured amniotic cells is probably possible. A pregnancy at risk for dihydropteridine reductase deficiency has been monitored. Normal enzyme activity was found in cultured amniotic cells, and an unaffected foetus was accurately predicted (Firgaira *et al.* 1983).

The gene for phenylalanine hydroxylase has been assigned provisionally to chromosome 1 by Berg and Saugstad (1974) because of positive lod scores at low recombination frequencies ($z=1.25$ at $0=0.05$) with phospho-glucomutase-1. A close linkage between PKU and the amylase loci (which are sited on chromosome 1) was demonstrated by Kamarýt *et al.* (1978) by combining the lod scores of Amy 1 (salivary amylase) and Amy 2 (pancreatic amylase). Rao *et al.* (1979) assigned the PKU gene to the phosphoglucomutase-1-amylase segment of chromosome 1. The gene for dihydropteridine reductase has been assigned to chromosome 4 (Kuhl *et al.* 1979).

Paul *et al.* (1979*b*), however, were unable to demonstrate linkage between the PKU locus and 15 markers including two assigned to chromosome 1, namely phosphoglucomutase 1 and Amy 2. These authors suggest that the contradictory data on gene location may be due to genetic heterogeneity in PKU.

Neonatal screening and incidence

Screening all newborn for PKU is an established national policy in the United Kingdom, it is enforced by law in the majority of American states and has become a voluntary service in many other countries. Besides allowing detection and early treatment, and therefore prevention of brain damage, newborn screening has established important differences in ethnic populations, and is highly cost-effective. Recommendations for screening for inborn errors of metabolism include the following criteria:

(a) The method should be simple, inexpensive, and sufficiently sensitive to detect the condition, while avoiding more than a low proportion of false positive results.

(b) It should be monitored by quality control.

(c) The condition being screened should be treatable and sufficiently common to justify screening.

(d) Collection of samples and reporting of results to personnel who will initiate action must be effectively organized.

(e) Treatment and long-term management should be carried out at specialized medical centres.

(f) Genetic and dietary counselling should be available.

By 1977, over 34 million newborn from 27 countries had been screened for PKU by the Guthrie test (Veale 1980). This test (Guthrie and Susi 1963) depends on the ability of phenylalanine in a heel-prick blood sample collected on filter paper to overcome the inhibitory effect which ß-2-thienylalanine has on the growth of *Bacillus subtilis*. The area of growth under standard conditions in agar medium is a measure of blood phenylalanine concentration. This Guthrie test screens only for PKU but can be modified to test for excess in the blood of galactose, tyrosine, histidine, leucine, valine, and proline. Chromatography of heparinised blood samples (Scriver *et al.* 1964) has the advantage that moderate elevations in the concentration of any of the amino acids can be detected in a single test, but requires greater technical skill than the Guthrie test.

Analysis of screening of 34 355 708 newborn from 27 centres (Veale 1980) reveals marked geographical variations in the frequency of PKU and non-PKU hyperphenylalaninaemia. Classical PKU has an overall incidence of 1 in 11 298 newborn.

In England, screening of 2 331 832 newborn revealed an incidence of 1 in 16 538. The Celtic races had high incidences, as follows: Belfast, 1 in 4504; Dublin 1 in 6315; Scotland, 1 in 8222; while the lowest incidences were found in Denmark (1 in 64 339), Japan (1 in 61 841) and Mexico (1 in 40 972). Marked variations were also evident in the ratio of PKU to non-PKU hyperphenylalaninaemia which varied from 1.04 in France to 19.00 in Italy and 6.71 in England.

Treatment

As discussed under 'Diagnosis' (p. 179), infants with positive screening tests should be admitted to hospital for monitoring of blood phenylalanine, blood tyrosine, and urinary phenylpyruvic acid and testing for tetrahydrobiopterin deficiency. In practice, patients with blood phenylalanine concentrations of over 606 μmol/l (10 mg/dl) should be treated at any rate during the first three months of life. If phenylalanine challenge shows normal tolerance, treatment can be discontinued.

Classical PKU

The aim is to control the blood concentration of phenylalanine while maintaining normal nutrition. This is achieved by partial replacement of dietary proteins by low-phenylalanine substitutes. In practice, two types of preparation are used, either mixtures or amino acids from which phenylalanine has been excluded (Aminogran, Allen and Hanbury; PKU

Aid, Scientific Hospital Supplies) or protein hydrolysates from which most of the phenylalanine has been removed (Minafen, Cow and Gate; Cymogran, Allen and Hanbury; Lofenalac, Bristol Laboratories, UK; Albumaid XP, Scientific Hospital Supplies). Natural protein is added so as to maintain blood phenylalanine concentrations that are not too high to cause brain damage or too low to cause phenylalanine deficiency. The features of phenylalanine deficiency caused by excessive restriction have been well documented (Medical Research Council 1963; Pitt 1967; Hanley *et al.* 1970; Smith *et al.* 1973; Mann *et al.* 1965; Clayton 1975). They include low intelligence, rashes, hair loss, poor growth, diarrhoea, vomiting, and frequent infections. Clearly deficiency of other nutrients may have contributed. On the other hand, children with PKU whose blood phenylalanine concentrations were above 606 μmol/l (10 mg/dl) for a total time of more than 0.02 years during their first two years had significantly lower IQ than those with better control (Smith *et al.* 1973). Preferred ranges of blood phenylalanine have varied from preprandial concentrations of 182–485 μmol/l (3–8 mg/dl) (Smith *et al.* 1975) to fasting concentrations of up to 303–909 μmol/l (5–15 mg/dl) (Hanley *et al.* 1970).

Blood phenylalanine levels are monitored every one or two weeks during the first year and evey two or four weeks later. Details of the diet and comparison between various regimens are given by Smith *et al.* (1973) and Clayton (1975).

Termination of phenylalanine-low diet at six years of age in 138 patients with PKU in an American collaborative study was followed by significant decreases in IQ during the following two years (Williamson *et al.* 1979). A fall of 11 points in IQ scores during four years following termination of treatment was reported by Cabalaska *et al.* (1977) in 22 patients with PKU who had been treated for 56 months. Significant falls in IQ were also reported in 47 patients with PKU who terminated their diets between five and 15 years of age, but smaller non-significant falls in IQ occurred in 22 patients who were changed from a low phenylalanine diet to a relaxed low phenylalanine diet. No fall in IQ scores had occurred in patients in either group whilst on strict low-phenylalanine diets (Smith *et al.* 1978).

In contrast, no fall in IQ was reported by Holtzman *et al.* (1974) or Koff *et al.* (1979) after strict diet was discontinued at four and five years, respectively. The balance of evidence suggests that where possible, treatment should be continued at least during the first 10 years. Waisbren *et al.* (1980) reviewed 19 studies that had analysed psychological assessments on children with PKU who had discontinued dietary treatment. They conclude that research should be directed into differentiating children who should remain on diet from those who may safely terminate.

The international survey of infants born to mothers with PKU discussed earlier (Lenke and Levy 1980) indicated a relation between the degree of hyperphenylalaninaemia and the incidence of abortion or foetal damage.

The results of this study suggest that during pregnancy the maternal blood phenylalanine concentration should be kept below 909 μmol/l (15 mg/dl) and preferably nearer 606 μmol/l (10 mg/dl). Otherwise as many children with mental handicap may be born as were prevented by treating their mothers.

Tetrahydrobiopterin (BH$_4$) deficiency

Patients with malignant hyperphenylalaninaemia do not respond to dietary restriction of phenylalanine alone. Oral or intravenous administration of BH$_4$ reverses the biochemical abnormalities (Bartholomé *et al.* 1977; Schaub *et al.* 1978; Curtius *et al.* 1979). However, since there is evidence that BH$_4$ does not readily enter the brain (Gal *et al.* 1976*b*), it is advisable to combine phenylalanine restriction and BH$_4$ administration with neuro-transmitter replacement until more is known about long-term results of treatment. Curtius *et al.* (1979) found that two patients with deficiency of biopterin synthesis needed 4 μmol (1.25 mg) of BH$_4$/kg body weight per day, given orally in a mixture with ascorbic acid (10 mg/kg/day), and a patient with dihydropteridine reductase deficiency needed 8 μmol (2.5 mg) of BH$_4$/kg/day. More recently Niederwieser *et al.* (1982*b*) recommended BH$_4$-dihydrochloride, 22 mg/kg. Danks *et al.* (1978) recommend L-dopa, 7.5–10 mg/kg/day, carbidopa, 1 mg/kg/day and 5-hydroxytryptophan, 4–8 mg/kg/day – these amounts divided into three equal doses during the day. Initial amounts should be lower, working up to full dosage over two weeks. Since possible toxic effects of carbidopa in children are unknown, they suggest checking blood cell counts, renal function and metaphyseal bone growth at six-monthly intervals. Available evidence indicates that if therapy is started early, prognosis is good (Danks *et al.* 1978; Bickel 1980; Niederwieser *et al.* 1982*b*).

Disorders of tyrosine metabolism

Elevated blood tyrosine concentrations are commonly found as a benign, transient, condition in newborn infants (especially if born preterm) which may respond to ascorbic acid. In screening programmes this has been found in about 0.5–1.5 per cent of infants.

A considerable impact in our thinking on disorders of tyrosine metabolism in the past was made by the careful studies on a man described by Grace Medes (1932). No similar case has been reported since. For this reason it has been suggested that the term 'tyrosinosis' be restricted to the disorder in this single patient. The term 'tyrosinaemia type 1' has been suggested (Buist *et al.* 1974*a*) for the combination of hepatic cirrhosis and renal tubular damage with hypermethioninaemia and hypophosphataemic rickets starting during infancy. This disorder is probably not a primary defect of tyrosine metabolism but is possibly due to deficiency of fumaryl acetoacetate (FAA) hydrolase and maleylacetoacetate (MAA) hydrolase. Not everyone has accepted this classification. For example, tyrosinaemia type 1 is also known simply as hereditary tyrosinaemia by Bergeron *et al.* (1974), and Lindblad *et al.* (1977); or as tyrosinosis by Goldsmith (1983) who retains tyrosinaemia II, leaving type 1 vacant. In some cases of tyrosinaemia type I hepatic *p*-hydroxyphenylpyruvate oxidase (hydroxylase) activity and immunologically cross-reacting material to this enzyme is less than two per cent of normal (Lindblad *et al.* 1977). In other cases residual activity of *p*-hydroxyphenylpyruvate oxidase has been as high as 10–30 per cent of normal (La Du and Gjessing 1978; Gentz and Lindblad 1972). Some cases improve on restriction of tyrosine. In type I tyrosinaemia, deficiencies have been reported in the activity of several other hepatic enzymes including methionine activating enzyme, cystathionine synthase and phenylalanine hydroxylase (Gaull *et al.* 1970). It is possible that hepatic and renal damage, together with deficiency of several enzymes, may arise because of increased concentration of toxic tyrosine metabolites.

Tyrosinaemia type II is much rarer. The characteristic clinical features are chronic keratitis and conjunctivitis with hyperkeratosis of the palms and soles (Richner–Hanhart syndrome) in the majority, starting in the first few years, and mental handicap in some patients. Deficiency of either soluble tyrosine transaminase (Kennaway and Buist 1971) or *p*-hydroxyphenlypyruvate oxidase (Faull *et al.* 1977) has been reported.

Transient neonatal hypertyrosinaemia

Temporary increases in blood tyrosine concentration is a common finding in normal newborn, especially if birth weight is low and gestation short. One per cent of 135 439 infants in a screening programme had blood tyrosine levels of 4 mg/dl or higher. In these, levels of above 10 mg/dl were found in 60 per cent of infants with birth weight below 2500 g, but only in 8 per cent of the others. Blood tyrosine concentrations above 8 mg/dl were found in 1.5 per cent of 3,222 infants with birth weights above 2500 g by Wong *et al.* (1967*b*). The higher incidence in premature infants has also been reported by Levy *et al.* (1969*b*) and Light *et al.* (1973). There may be associated hyperphenylalaninaemia (q.v.). Most authors consider that transitory tyrosinaemia is due to immaturity of liver hydroxyphenylpyruvate oxidase and tyrosine aminotransferase and a relatively high protein intake (Halvorsen 1980). In most, but not all, instances ascorbic acid administration corrects the transitory tyrosinaemia (Levine *et al.* 1941; Light *et al.* 1966; Scriver *et al.* 1977*a*). This may be related to the high requirement of *p*-hydroxyphenylpyruvate oxidase for this vitamin. Elevated tyrosine levels usually return to normal in a few weeks to six months but often infants are treated with ascorbic acid, 100 mg/day, and by restriction of dietary protein to less than 3 g/kg/day.

Whilst transient tyrosinaemia is usually benign (Partington *et al.* 1968; Light *et al.* 1973), anorexia and lethargy (Light *et al.* 1973), prolonged jaundice, and increased blood concentrations of phenylalanine, galactose and cholesterol (Halvorsen 1980) have been reported in the short term, while in the longer term some intellectual impairment (Bakker *et al.* 1975), and acidosis (Danks *et al.* 1975*b*) have been reported. Clearly further study is required to resolve the controversy.

Tyrosinosis

The single patient in this category was a 49-year-old Russian Jew with myasthenia gravis (Medes 1932). Careful metabolic studies showed a constant urinary excretion of *p*-hydroxyphenylpyruvic acid even during fasting and the appearance in the urine of tyrosine, *p*-hydroxyphenyllactic acid, and 3,4-dihydroxyphenylalanine (dopa) on progressively increasing tyrosine intake. Medes (1932) proposed a metabolic block at the level of *p*-hydroxyphenylpyruvic acid oxidase. This case has been discussed at length by La Du and Gjessing (1978), who point out that this suggestion is difficult to correlate with the predominant metabolite being *p*-hydroxyphenylpyruvic acid with very little *p*-hydroxyphenyllactic acid, since the latter is found in relatively large amounts in conditions where the activity of phenylpyruvic acid oxidase is decreased.

Hereditary tyrosinaemia type I; tyrosinaemia with hepatorenal dysfunction

Clinical features

First recognized as a distinct entity by Baber (1956), the disorder has been the subject of numerous reports. In the majority of patients, commencing in infancy, there is an insidious onset (Halvorsen *et al.* 1966) of failure to thrive and abdominal enlargement. Characteristically infants have the Fanconi syndrome with hypophosphataemic vitamin D resistant rickets, liver damage leading to cirrhosis, haemorrhages, oedema, and dyspnoea. More rarely the onset is acute within a few weeks after birth with liver failure causing death in the first eight months (Larochelle *et al.* 1967). Different clinical presentations occur in the same family (Halvorsen *et al.* 1966). Hepatoma can complicate cirrhosis (Weinberg *et al.* 1970). Some patients have been mentally handicapped, but this is rare. There is usually leucopenia, thrombocytopenia, hypoproteinaemia, reduced blood clotting factors derived from the liver, and hyperbilirubinaemia. Porphyria-like crises with abdominal pain and neuropathy may accompany porphyrin accumulation (Strife *et al.* 1977).

Molecular defects

There is usually, but not always, an elevation of blood tyrosine and methionine (Scriver and Rosenberg 1973). Features of the Fanconi syndrome are characteristic, including hypophosphataemia, hyperphosphaturia, glucosuria, proteinuria, and aminoaciduria. The pattern of aminoaciduria is different to that of the Fanconi syndrome associated with cystinosis. The amino acids excreted excessively in the urine in decreasing amounts are tyrosine, proline, threonine, alanine, glycine, phenylalanine, α-aminobutyric acid, isoleucine, methionine, and leucine. Aminoaciduria is more pronounced and extensive in the acute than in the chronic form (Scriver and Rosenberg 1973; La Du and Gjessing 1978). A large proportion of patients have increased serum α_1-fetoprotein (Bélanger *et al.* 1972). There is an increased excretion in the urine of *p*-hydroxyphenyllactic acid, *p*-hydroxypyruvic acid, and glyceraldehyde (Tomer *et al.* 1977).

δ-aminolaevulinic acid is also increased in urine. Succinylacetoacetate and succinylacetone (Fig. 5.4) are present in serum and urine (Lindblad *et al.* 1977; Fällström *et al.* 1979).

The deficiency of several hepatic enzymes has been reported. Earlier reports indicated reduced activity of *p*-hydroxyphenylpyruvate oxidase (La Du 1967*a*). In two patients Lindblad *et al.* (1977) found activities of 1 and 2 per cent of control values. Moreover, they found no immunoprecipitating material to enzyme antiserum in liver of two patients. However, residual hepatic activity of *p*-hydroxyphenylpyruvate oxidase has been reported to be as high as 10–30 per cent of normal (La Du and Gjessing 1972; Gentz and Lindblad 1972). Moreover, deficiency has been reported of methionine-activating enzyme (Gaull *et al.* 1970), cystathionine synthase (which would explain hypermethioninaemia), tyrosine aminotransferase, and phenylalanine hydroxylase, as well as of *p*-hydroxyphenylalanine oxidase in liver of two patients.

More recently, Lindblad *et al.* (1977) and Fällström *et al.* (1979), in attempting to link a defect in porphyrin (evidenced by the increased δ-aminolevulinate excretion) and tyrosine metabolism, showed that the activity of liver and erythrocyte porphobilinogen synthase was decreased. Further, they showed that the addition of serum or urine from patients with hereditary tyrosinaemia inhibits δ-aminolaevulinic acid dehydratase (porphobilinogen synthase) activity *in vitro*. The inhibitor was isolated and identified as succinylacetone, which they showed to be a metabolite of tyrosine formed by decarboxylation of succinylacetoacetate (which was also present in excess in urine from their patient). The authors demonstrated a deficiency of fumarylacetoacetase (fumarylacetoacetate (FAA) hydrolase) in the liver of nine patients, which would explain accumulation of fumarylacetoacetate, its reduction product succinylacetoacetate, and succinylacetone (Fig. 5.4).

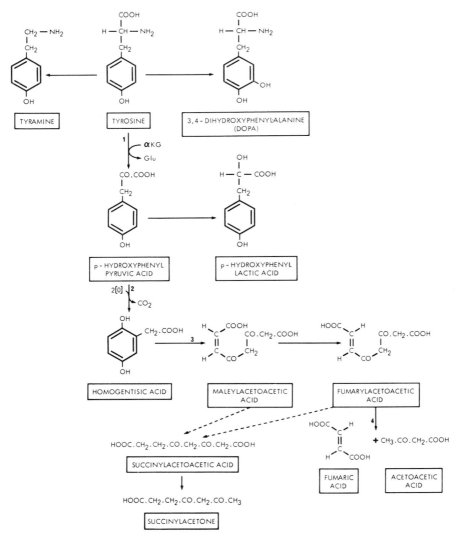

Fig. 5.4. Metabolism of tyrosine: (1) Tyrosine transaminase (2) *p*-hydroxyphenyl-pyruvic acid oxidase (3) Homogentisic acid oxidase (4) Fumarylacetoacetase.

The authors suggest that decreased FAA hydrolase activity may be the primary defect, and more recently this was supported by the finding of decreased activities in fibroblasts and lymphocytes of patients and intermediate activities in parents (Kvittingen *et al.* 1983).

The persistence of liver α_1-fetoprotein production in hereditary tyrosinaemia prompted Guguen-Guillouzo *et al.* (1979) to investigate whether it was an isolated foetal liver function. They found persistence of foetal isoenzyme patterns of liver pyruvate kinase and aldolase in five tyrosinaemic livers, but attributed this to liver regeneration processes rather than to a biochemical alteration of the disease.

Pathology

Cirrhosis of the liver with extensive fibrosis and hepatocyte regeneration are typical. Tubular degeneration and islet-cell hyperplasia (Perry *et al.* 1965) have been reported.

Diagnosis

The clinical picture of cirrhosis and rickets is distinctive but the acute form may simulate fructose 1-phosphate aldolase deficiency, fructose 1,6-diphosphatase deficiency, neonatal hepatitis, septicaemia, or galactosaemia. The diagnosis is confirmed by the demonstration of tyrosinaemia, methioninaemia, tyrosyluria, and especially of *p*-hydroxyphenyllactic acid in the urine, hypophosphataemia, hyperphosphaturia, the characteristic aminoaciduria (see 'Molecular defects', p. 194), glucosuria, proteinuria, and δ-aminolevulinicaciduria, and FAA hydrolase deficiency in fibroblasts or lymphocytes.

Newborn screening and genetic counselling

The slow rise in serum tyrosine concentration after birth in patients with hereditary tyrosinaemia and the relative frequency of the transient neonatal form makes screening for hereditary tyrosinaemia by blood tyrosine assay ineffective in the first week. Thus Alm and Larsson (1979) had 540 false positives, detected only one, and missed at least four among 640 259 screened in Sweden. Other methods for screening include assay of urinary amino acids or phenolic acids (Buist and Jhaveri 1973) or of serum α_1-fetoprotein (Belanger *et al.* 1973). A more satisfactory procedure would be to repeat the blood assay by thin-layer or paper chromatography at about four weeks. This would also assist in diagnosis of any other inborn errors which were missed at one week.

The incidence varies widely. An incidence of 1:120 000 was found in Sweden and of 1 in 685 in Quebec where the prevalence of heterozygotes is 1 in 14 (Bergeron *et al.* 1974).

The acute and chronic disorders have been described in siblings of both sexes and have an autosomal recessive mode of inheritance. Heterozygote detection may be possible by assay of FAA hydrolase activity in fibroblasts or lymphocytes (Kvittingen *et al.* 1983). Prenatal diagnosis of three affected foetuses was reported by Gagné *et al.* (1982) by demonstrating succinylacetone (15–30 μg/l) in amniotic fluid between 15 and 21 weeks' gestation. In two of the foetuses the livers of the abortuses were assayed for enzymes of the tyrosine degradation pathway. At 18 weeks' gestation FAA hydrolase was undetectable in one and, at 21 weeks' gestation was 31.8 per cent of two control 18-week foetuses, or 6 per cent of normal adults. No increase in succinylacetone was found in maternal urines. In 10 other pregnancies at risk no succinylacetone could be demonstrated in amniotic fluids (sensitivity limit 5 μg/l), and all 10 foetuses were confirmed as unaffected postnatally.

Since FAA hydrolase is present in normal amniotic cells (Kvittingen *et al.* 1983), enzyme assay of these cells could now be used to substantiate measurement of succinylacetone concentration in the amniotic fluid.

Treatment

Low tyrosine, low methionine diet (Michals *et al.* 1978) cause clinical improvement in some patients, but not in others (Bodegaard *et al.* 1969). Large doses of vitamin D (20 000–50 000 IU/day) expedite healing of rickets. Hepatic homotransplantation may be considered for hepatoma (Fisch *et al.* 1978). Slördahl *et al.* (1979) reported successful treatment of a patient by cysteine supplementation combined with a low protein diet with strict low tyrosine and low phenylalanine intake.

Hereditary tyrosinaemia type II; persistent hypertyrosinaemia; tyrosinaemia without hepatorenal dysfunction; Richner–Hanhart's syndrome

Only about 16 cases of this rare syndrome have been published. In the majority there has been bilateral keratitis with corneal ulceration, usually in the first few years. Mental handicap occurred in about half the patients, some of whom had microcephaly and convulsions. In the majority of cases there has been hyperkeratosis of the palms and soles. It has been pointed out (Goldsmith *et al.* 1973) that these lesions resemble those in patients with the Richner–Hanhart syndrome. These patients usually have a higher blood tyrosine concentration than patients with the hepatorenal type of tyrosinaemia (q.v.) and tyrosyluria characterized by the excretion of *p*-hydroxyphenyllactic acid and *p*-hydroxyphenlypyruvic acid (Fig. 5.4). Deficiency of soluble hepatic tyrosine transaminase has been reported by Faull *et al.* (1977), Goldsmith *et al.* (1979), and Lemonnier *et al.* (1979). Activity of liver *p*-hydroxyphenylpyruvate oxygenase was deficient in one patient (Faull *et al.* 1977). The latter, but not the former defect would explain the occurrence of tyrosyluria. However, it is possible that this syndrome may be caused by deficiency of either enzyme.

Treatment by low-tyrosine, low-phenylalanine diet appears to be effective in curing the keratitis (Goldsmith and Reed 1976; Hunziker 1980). Serum tyrosine should be maintained below 10 mg/dl. The tendency to tyrosinaemia on discontinuation of treatment has been found to be persistent (Buist *et al.* 1974*a*). Inheritance is probably autosomal recessive.

Parental consanguinity has been reported in about one third of families (Goldsmith and Reed 1976; Garibaldi *et al.* 1977; Goldsmith *et al.* 1979).

Animal model

Tyrosine overfeeding in rats causes a condition characterized by keratitis and blisters on the paws (Schweizer 1947; Yamamoto *et al.* 1979).

A disorder which appears to be analagous to tyrosinaemia type II occurs in the mink. It is associated with reduced hepatic tyrosine aminotransferase activity and quantity. Inheritance is autosomal recessive (Schwartz and Schackelford 1973; Goldsmith *et al.* 1981).

Alcaptonuria

The darkening of urine on standing was sufficiently unusual to lead to the early recognition of this disease (Boedeker 1859), to the early identification of the pigment as homogentisic acid (Walkow and Baumann 1891) and to the classical suggestion by Garrod (1909) that '. . . the splitting of the benzene ring in normal metabolism is the work of a special enzyme, that in congenital alcaptonuria the enzyme is wanting . . .'. This was confirmed 49 years later by the demonstration of homogentisic acid oxidase deficiency (La Du *et al.* 1958). The disease is a rare cause of arthritis, which is relatively well-known because of its historical associations, and unusual features of pigmentation of urine and cartilage. The name 'alcaptonuria' was derived from the designation by Boedeker (1859) of the substance 'alcapton' in the urine which took up oxygen in an alkaline medium, while the term 'ochronosis' was used by Virchow (1866) in describing the pigment in the connective tissues of a patient which appeared grey or black to the naked eye, but ochre under the microscope. The connection between alcaptonuria and ochronosis was not recognized until much later (Albrecht 1902; Osler 1904).

Clinical features

The main clinical features are darkening of the urine on standing due to an increased concentration of homogentisic acid, pigmentation of the cornea, of the ear cartilages and of some other sites, and severe arthritis in adults. Darkening of the diapers was noted to be slight at 38h after birth but marked at 52h by Garrod (1901). This may be due to postnatal maturation of enzymes required for tyrosine oxidation (Kretchmer *et al.* 1956). Although darkening of the urine is a characteristic feature, some patients do not notice this symptom (Cooper 1951; Minno and Rogers 1957). The reason may be that the urine only darkens when alkaline. Acid urines, or urines with high ascorbic acid concentrations may not darken for several hours (Sealock *et al.* 1940). Ochronosis is often first detected during the third or fourth decade as a greyish discolouration of the cornea, usually between the cornea and canthi, or of the pinna, where it is progressive, usually starting in the concha and antihelix and spreading to the tragus. Pigment also occurs in sweat, and skin in the axillae and genitalia. Arthritis affects mainly the large joints, causing painful limitation of movement of the shoulders, knees and hips. Radiological changes are characteristic (Pomeranz *et al.* 1941). There is narrowing of the intervertebral spaces with degeneration and calcification of the discs leading to some fusion of the vertebrae. The hips and shoulders show osteoarthritic changes. Arthritis is progressive and can cause severe disability. In men, arthritis appears to have an earlier onset and a more severe course than in women.

Molecular defect

The disorder is due to absent activity of homogentisic acid oxidase, the enzyme which catalyses the conversion of homogentisic acid (2,5-dihydroxyphenylacetic acid) to maleylacetoacetic acid (Fig. 5.4). This has been demonstrated both for liver (La Du *et al.* 1958) and kidney (Zannoni *et al.* 1962). The reaction is in the main pathway for tyrosine metabolism. This is shown by the quantitative accumulation in human liver of homogentisic acid from tyrosine when homogentisic acid oxidase is inhibited by α, α'-dipyridyl (La Du *et al.* 1958) although normally no homogentisic acid can be demonstrated in urine or plasma. The severity of the block was also shown by demonstrating that whereas normal individuals excrete only 3.2 per cent of intravenously administered ^{14}C-homogentisic acid in the urine, in two alcaptonuric patients over 90 per cent was excreted unchanged (Lustberg *et al.* 1969). Homogentisic acid oxidase utilises atmospheric oxygen (Crandall *et al.* 1954). The enzyme requires ferrous iron, maintained in the reduced form by ascorbic acid (Suda and Takeda 1950*a,b*).

Plasma homogentisic acid concentration is normal in alcaptonuric patients, and even after intravenous injection of 1.0 g of homogentisic acid it returns to normal after 30 min (Leaf and Neuberger 1948). Homogentisic acid administered orally to a patient with alcaptonuria was cleared at a rate of 400–500 ml of blood/min. Since this is about the renal blood flow rate there must be almost complete clearance by both glomerular filtration and tubular excretion (Neuberger *et al.* 1947).

About 4–8 g of homogentisic acid is excreted daily by alcaptonurics. This gives an estimate of the amount of tyrosine which is usually metabolized to maleylacetoacetic acid, which is split into fumaric acid and acetoacetic acid. In normal individuals, over 95 per cent of the carboxyl group of homogentisic acid is converted to carbon dioxide within 12h (Lustberg *et al.* 1969).

The pigment which forms in urine is the result of non-enzymic oxidation and starts on the surface of a sample, spreading downwards. It reduces

Benedict's reagent and silver in photographic paper (Neuberger *et al.* 1947). Pure alkaline homogentisic acid solutions darken when exposed to oxygen (Milch 1957). The pigment causing ochronosis is probably also an oxidation product of homogentisic acid, and may be polymeric. In theory, the oxidation may be non-enzymic as in urine, but it may be catalyzed by homogentisic acid polyphenol oxidase (Zannoni *et al.* 1969).

The relation between ochronosis and arthritis is also unknown. Intra-articular injection of homogentisic acid in rabbits causes lesions similar to those seen in alcaptonuria (Moran and Yunis 1962). Homogentisic acid has been shown to inhibit hyaluronidase (Greiling 1957) and lysyl hydroxylase (Murray *et al.* 1976).

Pathology

Pigmentation of cartilage is striking, and may be black in places. Costal, laryngeal, and tracheal cartilages are affected (Galdston *et al.* 1952; Lichtenstein and Kaplan 1954; Cooper and Moran 1957). Pigmentation also occurs in ligaments and tendons, and to a lesser extent in the viscera, endocardium and intima. The pigment is deposited both intra- and extra-cellularly.

Diagnosis

The demonstration of excessive quantities of homogentisic acid in the urine differentiates patients with alcaptonuria from those with other types of pigmentation and arthritis. A semi-qualitative test for homogentisic acid in the urine is the reduction of molybdates (Briggs 1922). More specific is the identification by paper chromatography after extraction in ether (Knox and Le May-Knox 1951) or an enzymatic assay (Seegmiller *et al.* 1961).

Genetic counselling

The disorder has an autosomal recessive mode of inheritance in the great majority of cases. Parent to offspring inheritance has been reported but in view of the frequency of consanguinity (Hogben *et al.* 1932), this may be due to the mating of homozygous and heterozygous individuals (Khachadurian and Abu Feisal 1958). Dominant inheritance with incomplete penetrance has been suggested (Milch 1957). In adults, but not in infants, males have been reported to be affected more frequently than females (Hogben *et al.* 1932). Attempts at heterozygote detection by measurement of homogentisic acid excretion after oral loading with tyrosine have been unsuccessful (Roth and Felgenhauer 1968). In Nothern Ireland the incidence has been estimated to be about 1 in 250 000 individuals (Knox 1958).

Treatment

Neither reduction of dietary phenylalanine and tyrosine, nor ascorbic acid administration (Sealock *et al.* 1941; Galdston *et al.* 1952) has been of benefit.

Oculocutaneous albinism

Oculocutaneous albinism is characterized by a genetic absence or reduction of melanin pigment in the skin, hair and eyes. Six distinct types are recognized. All have an autosomal recessive mode of inheritance and share to varying degrees the characteristic features of nystagmus, reduced visual acuity, UV photosensitive dermatosis and photophobia. In schools for the partially sighted in the United States, 9 per cent of students were suffering from oculocutaneous albinism (Witkop *et al.* 1970, 1973). Squamous cell carcinoma in sun-exposed skin occurs especially in Africa (Okoro 1975; King *et al.* 1980), but melanoma has also been reported (Alpert and Damjanov 1978). The types are distinguished by specific biochemical and clinical phenotypes. The two commonest are tyrosinase-negative (ty-neg), with tyrosinase deficiency, and tyrosinase-positive (ty-pos) types. In a third form, the Chediak–Higashi syndrome, there is incomplete pigment loss, susceptibility to infections, neutropenia and the presence of giant intracellular organelles which are probably distended lysosomes. In a fourth type, the Hermansky–Pudlak syndrome, there is a haemorrhagic diathesis due to abnormal platelets and accumulation of ceroid-like material in certain cells. In a fifth variety, the yellow type albino or yellow mutant, patients have yellow, yellow-red, or reddish hair. The sixth and rarest form, the Cross syndrome, is characterized by a severe neurological disorder, including mental retardation, gingival fibromatosis, and microphthalmia. Not discussed in this review are a Swiss family with a dominant form of oculocutaneous albinism (Frenk and Calme 1977), brown albinism (King *et al.* 1980), Rufous albinism (Walsh 1971), the autosomal dominant oculocutaneous albinoidism, piebaldism (Jimbow *et al.* 1975), the Waardenburg syndrome (Bard 1978), or the ocular type of albinism, which exists in both autosomal recessive and X-linked forms.

In addition to enzymic and clinical features, the different types of albinism are also distinguished by characteristic failure of maturation of melanosomes or melanin-containing bodies (see below).

Development of melanosomes and melanocytes

In normal skin, melanin is synthesized by specialized cells, the dendritic melanocytes (melanodendrocytes), as pigment granules known as melanosomes. Skin melanocytes originate during early embryogenesis from the neural crest, and migrate to their final destination while retinal melanocytes originate from the optic cup. Epidermal melanin units consist of melanocytes with dendritic processes which distribute melanin to adjacent skin cells, or keratinocytes (Cruickshank and Harcourt 1964) which are themselves unable to synthesize melanin. Melanosomes are membrane-bound lysosome-like organelles with high acid phosphatase activity (Seiji and Kikuchi 1969). They are synthesized as small vesicles about 0.5μm in diameter, with a membrane but no other distinct morphology designated as stage I melanosomes (Seiji *et al*. 1963). Stage II contain crosslinks and an internal lattice-like structure, and filaments which are parallel to the long axis of the elongated melanosomes. In ty-neg albinism there is a maturation arrest at the stage when melanosomes are still unpigmented. In stage III early melanin formation can be detected by electron microscopy. The organelles in stages I–III are sometimes called premelanosomes. Stage IV are fully mature, densely pigmented melanosomes.

Melanin synthesis

Melanocytes synthesize two main types of melanin, eumelanin which is black-brown and pheomelanin which is yellow-red. Both are derived from tyrosine, by a series of reactions, only some of which have been elucidated (Fig. 5.5). Both eumelanin and pheomelanin can be produced by the same melanocyte. The first two steps in melanin synthesis are the hydroxylation of tyrosine to 3,4-dihydroxyphenylalanine (dopa) and the oxidation of the hydroxyl groups to ketone groups to form dopa quinone. Both reactions are thought to be catalysed by the copper-containing enzyme tyrosinase, which in mammalian tissues occurs in four forms, T_1 to T_4. T_4 is relatively insoluble (Holstein *et al*. 1967; Burnett *et al*. 1969; Hearing *et al*. 1981). All the isoenyzmes are immunologically indistinguishable and have a molecular weight of 65 000 (Ohtaki and Miyazaki 1973; Miyazaki and Ohtaki 1975). Dopa quinone can then undergo non-enzymic cyclization and oxidation to melanin. Pheomclanin synthesis involves polymerization with sulphydryl-containing compounds, such as cysteine or other compounds, while synthesis of eumelanin involves conversion of dopa quinone to indole-5,6-quinone as an intermediate (Prota and Thomson 1976; Prota 1980).

Genetic control has been studied in mice where it is postulated that there is only one tyrosinase locus, the c locus for T_1, and that other tyrosinase forms are produced by post-translational modifications. In the mouse, about 62 loci can influence pigmentation (Silvers 1979).

Tyrosinase-negative oculocutaneous albinism

In this form no melanin is detectable from birth in skin and hair, which appear white, nor in the irides, which have a greyish-blue appearance. On transillumination of the eyeballs, no pigment opacities are seen in the irides. Brown pigmented naevi do not occur. Nystagmus is marked, and strabismus common. The pupillary reflex to direct light appears red. There is marked and often progressive loss of visual acuity (Jay *et al*. 1976).

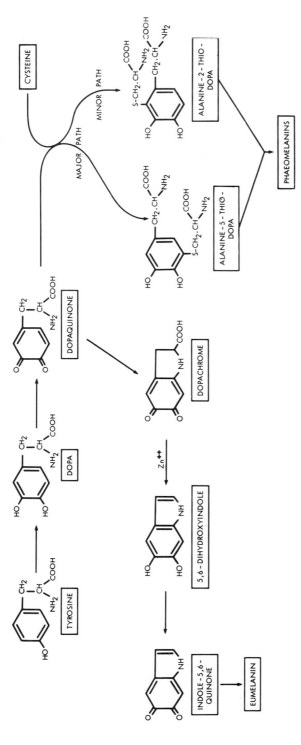

Fig. 5.5. Biosynthesis of melanin.

When freshly epilated hairbulbs are incubated with either tyrosine or dopa at 37°C for 12h, no pigment is formed, but pigment does form in hairbulbs of ty-pos patients (Witkop 1971). Electron microscopy of ty-neg hairbulbs shows a melanocyte maturation arrest, with numerous stage I and II melanosomes but no stage III or IV melanosomes.

The concentrations of serum tyrosine, copper and ß-melanocyte-stimulating hormones are within normal limits (Witkop *et al.* 1970). The disorder occurs in 1 per 35 000 US persons, but is more common in American Negroes (1:28 000) than in whites (1:39 000) (Witkop *et al.* 1974). In Caucasian heterozygotes, but rarely in blacks, the iris is abnormally translucent, but this sign is not sufficiently reliable for carrier detection. Heterozygotes may be detected by demonstrating deficient conversion of tritiated tyrosine to tritiated water by anagen hairbulbs (King and Witkop 1976, 1977).

Tyrosinase-positive oculocutaneous albinism

In patients with ty-pos albinism, unlike those with ty-neg albinism, there is some visible pigment, which increases with age and the extent of which depends on racial origin in older individuals. Thus negro ty-pos albinos may have appreciable skin, hair and eye colour, resembling pigmentation of fair-skinned white individuals. Caucasian newborn are unpigmented but with increasing age there is some pigmentation of skin and hair, with irides becoming blue or even brown. A red eye reflex is present in caucasians but may disappear in negroes. On transillumination of the eyeballs the irides have a cartwheel appearance. Freckles and pigmented naevi occur normally. Strabismus is common. Nystagmus is less marked than in ty-neg patients.

Freshly epilated hairbulbs turn black when incubated with tyrosine or dopa (Witkop *et al.* 1971). Electron microscopy of the hairbulbs reveals numerous melanosomes in stages I–III but only occasionally in stage IV.

Serum copper and ß-melanocyte-stimulating hormone concentrations are normal, but serum tyrosine concentration is at or below the lower range of normal (Witkop *et al.* 1970). Salivary tyrosine concentration is normal in heterozygotes, but about half normal in affected homozygotes (Zipkin *et al.* 1964).

An area of skin pigmentation can be induced by incubating a damaged area of skin with a solution of either tyrosine or dopa for one week followed by ultraviolet radiation (Witkop *et al.* 1971). However, oral dopa (5.5 g for 100 days) failed to induce pigmentation.

The frequency of ty-pos albinism has been estimated as 1 in 33 000 in the US with blacks (1:15 000) being more frequently affected than whites (1:40 000). Very high frequencies have been found in some ethnic communities. For example, 1 in 85 in the Brandywine isolate in Maryland, USA (Witkop *et al.* 1966) and 1 in 143 in Panama (Keeller 1953).

Chediak–Higashi syndrome

In this form of incomplete oculocutaneous albinism, the colour of the skin is creamy or grey, the hair a shiny blonde or brunette and the fundus paler than normal. Nystagmus, squint and photophobia are common. Recurrent infections in childhood are typical (Wolff *et al.* 1972). Later there is anaemia, neutropenia, and thrombocytopenia. After the first few years there is neurological regression characterized by convulsions and neuropathy of peripheral and cranial nerves, usually in the first or second decade. Death usually occurs in childhood and is due to infection, haemorrhage, or a lymphoma-like malignancy (Dent *et al.* 1966). There is often hepatosplenomegaly, gingivostomatitis, and lymph-node enlargement.

Characteristic of the disorder is the presence of giant granules identified as lysosomes in leucocytes where they stain for acid phosphatase (White 1966). These organelles have been found in Schwann cells, pancreas, liver, kidney, skin and other tissues (Myers *et al.* 1963; Lockman *et al.* 1967). Neutrophils have a reduced rate of migration and defective chemotaxis (Wolff *et al.* 1972). Melanocytes produce melanosomes, but these are packed into giant organelles (White 1966; Lockman *et al.* 1967; Windhorst *et al.* 1968).

Hermansky–Pudlak syndrome

Characteristic features of oculocutaneous albinism of variable severity are associated with a haemorrhagic diathesis presenting as episodes of severe bruising, bleeding after dental extraction, haemoptysis, and subdural haematoma. The degree of albinism depends to a large extent on the ethnic background and is severe in Irish patients (White *et al.* 1971) and mild in Puerto Rican patients (Witkop *et al.* 1974). The haemorrhagic tendency is probably due to platelet abnormalities. The platelets have reduced serotonin content – to about 10 per cent of normal – and a marked reduction in serotonin uptake *in vitro* (Maurer *et al.* 1972; Hardisty and Mills 1972; Hardisty *et al.* 1972). In the majority of patients the addition of ADP, collagen or thrombin to platelets causes a normal first wave of aggregation, but absent or markedly decreased second wave (White *et al.* 1971). Electron microscopy of platelets shows considerable diminution of electron-dense bodies, the storage organelles for serotonin. There is marked reduction of platelet ADP and of the ADP/ATP ratio (Hardisty and Mills 1972; Rao *et al.* 1974).

The mechanisms of platelet aggregation are complex and are now thought to depend on the synthesis of prostaglandin-aggregating substances from arachidonic acid (Cohen 1980). It has been suggested that platelets in the Hermansky–Pudlak syndrome are deficient in the second wave of aggregation because they lack a 'storage pool' of nucleotides, serotonin, and calcium (White and Witkop 1972). However the aggregation defect can be corrected by low concentrations of epinephrine (Rao *et al.* 1981) independently of ADP release, or of prostaglandin synthesis. In summary: in spite of considerable research and speculation the basic defect is not understood.

Some heterozygotes have a low platelet serotonin content (Gerritsen *et al.* 1977).

A striking abnormality in Hermansky–Pudlak syndrome is accumulation of a ceroid-like material in reticuloendothelial cells, buccal mucosa, leucocytes, and in urinary sediment. The material probably comes from phagocytosed erythrocytes. Inclusions occur in circulating monocytes (White *et al.* 1973).

Electron microscopy of hairbulbs shows relatively few melanosomes at stage IV but numerous ones at stages I–III. There are many small premelanosomes lacking cross-lattice matrix, similar to those seen in normal individuals with red or yellow hair (Witkop *et al.* 1974).

The yellow-type albinism

These patients resemble ty-pos albinos and are born with white hair, but develop yellow or yellow-red hair during the first six months (Nance *et al.* 1970; Witkop 1971). The disorder is further distinguished from the ty-pos type of albinism because incubation of hairbulbs with tyrosine or dopa fails to darken them until cysteine is added, when a yellow-red colour of pheomelanin develops.

Electron microscopy reveals almost complete absence of stage IV melanosomes. The fundamental defect is unknown.

Cross syndrome

This exceptionally rare disorder has been reported in a consanguineous Amish family in Ohio (Cross *et al.* 1967). It is characterized by hypopigmentation, small eyes and cloudy corneas at birth. At three months the patients developed choreoathetosis. They were physically and mentally retarded, did not sit up, were below the third centile for height and weight, and had gingival fibromatosis.

The hairbulb incubation test is positive (Witkop *et al.* 1974). Electron microscopy of hairbulbs showed considerable reduction in the numbers of melanosomes but presence of all states of maturation.

Prenatal diagnosis of oculocutaneous albinism

Oculocutaneous albinism has been diagnosed prenatally by electron microscopic examination of foetal skin samples collected during foetoscopy at 20 weeks' gestation (Eady *et al.* 1983). Melanosome development in hairbulb melanocytes had arrested at stage II. The authors found numerous stage IV melanosomes in four age-matched control foetuses. The diagnosis was confirmed in the abortus at 22 weeks' gestation.

Electron microscopic studies by Haynes and Robertson (1981) indicated that normally pigmented melanosomes can be visualized in hair follicles of the scalp, but not in forearm skin, by 16 weeks' gestation.

Animal models

Ty-neg albinism in humans is probably homologous to c/c albinism of mice and rats, where the c-locus is linked to that of ß-haemoglobin (Popp 1962). Segregation of a type of albinism with sickle cell anaemia has been reported in humans (Massie and Hartman 1957). Albinism occurs in numerous species including lepidoptera, salamanders, rodents, carnivores, fish, birds, and marsupials, but many animal forms of albinism are probably not analogous to the human forms (Creel 1980).

Some features of the Cross syndrome, namely albinism because of a deficiency of melanocytes, and microphthalmos, occur in *mi-mi* mice, homozygous for genes of the microphthalmia locus (Searle 1968). Disorders analogous to the Chediak–Higashi syndrome occur in Aleutian mink and cattle (Padgett *et al.* 1965; Padgett 1968). Pale colour and giant lysosomes also occur in beige and slate grey mice (Lutzner *et al.* 1965), and have been reported in cats (Kramer *et al.* 1977), and in a killer whale (Taylor and Farrell 1973).

Treatment of albinism

Avoidance of sunlight reduces the tendency to photosensitive dermatosis, and probably to skin cancers. Preparations containing *p*-aminobenzoic acid are useful as barriers to u.v. light-induced skin damage. Photophobia can be diminished by wearing tinted glasses or contact lenses (Taylor 1978).

The haemorrhagic diathesis of patients with the Hermansky–Pudlak syndrome can be temporarily alleviated, for example prior to surgery, by platelet transfusions.

DISORDERS OF THE UREA CYCLE

Ammonia is produced continuously from the amino groups of amino acids and purines and from the amide groups of glutamine and asparagine. Some is re-utilized for synthesis of non-essential amino acids and of pyrimidines, but the majority is converted to urea and excreted in the urine.

It may be seen from Fig. 5.6 that the two nitrogen atoms of the single urea molecule synthesized at each turn of the cycle enter at different points. One is derived from ammonia and is converted to carbamyl phosphate which combines with ornithine to form citrulline; the other is derived from aspartate which condenses with citrulline to form argininosuccinate. The metabolic pathways for urea synthesis are catalysed by the five enzymes of the Krebs–Henseleit urea cycle (Fig. 5.6). Genetic deficiencies occur for each of the five enzymes. The main clinical features

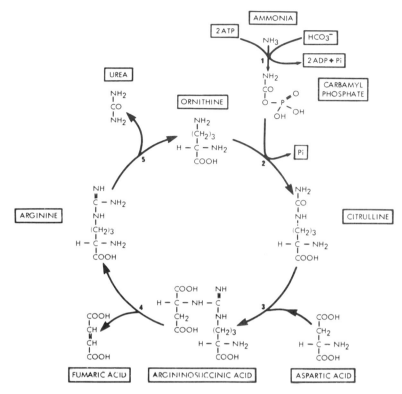

Fig. 5.6. The urea cycle: (1) Carbamyl phosphate synthase (CPS) (2) Ornithine carbamyl transferase (OCT) (3) Argininosuccinate synthase (4) Argininosuccinase (5) Arginase.

are hyperammonaemia, intolerance to dietary protein and mental handicap. Episodic drowsiness and coma due to ammonia intoxication are also characteristic and resemble features of hyperammonaemia which occur in hepatic failure. Indeed, the liver is the major site for urea synthesis (Ratner 1973), although a small contribution possibly comes from other tissues and cells.

The evidence for such contributions however is usually the demonstration that some of the urea cycle enzymes are active in non-hepatic cells, rather than demonstration of urea production itself. For example, in kidney there appears to be no evidence for urea synthesis, but there is activity of argininosuccinic acid synthetase (ASAS), argininosuccinic acid lyase (ASAL), and arginase (Ratner 1976). Although the kidney converts citrulline into arginine (Tizianello *et al.* 1980), the very low activities of renal carbamyl phosphate synthetase (CPS) and ornithine carbamyl transferase (OCT) suggest that any urea synthesis must be very limited.

Similarly in brain, OCT activity is undetectable and CPS activity is very low (Jones *et al.* 1961; Buniatin and Davtian 1966). In rat intestine ornithine can be synthesized from glutamate (Ross *et al.* 1978; Hensle and Jones 1981) but the absence of CPS activity indicates that the urea cycle is not fully operational.

The direct relation between the degree of hyperammonaemia and severity of the neurological disturbance suggests that the cerebral dysfunction may be the direct consequence of ammonia intoxication. Several explanations have been offered including a decrease of energy production, as a consequence of reduced ATP synthesis by the citric acid cycle, which is depleted of 2-oxoglutarate by conversion first to glutamate and then to glutamine because of the excess ammonia. Some evidence for this hypothesis comes from the experiments of Cooper *et al.* (1979) who found that ^{13}N-ammonia administered to rats by carotid infusion was preferentially converted to the amide group of glutamine. An abrupt rise in blood glutamine, and a fall of glutamate was observed following ammonia administration to rats by Hindfelt (1975). However, about one quarter of ammonia perfused to dog brain is converted to alanine, the remainder to glutamine (Benzi *et al.* 1977).

A toxic effect on brain function by ammonia has been demonstrated directly as reversal of postsynaptic inhibition (Raabe 1981).

Citrulline may also have toxic effects, since it has been shown to inhibit brain glucose utilization and lactate production *in vitro* (Okken *et al.* 1973).

Patients with urea cycle enzyme deficiency usually do not have decreased blood or urine urea concentrations. This is probably because surviving patients have sufficient residual enzyme activity or because the enzyme can have low activity in some viscera or cells but be normal in others (Glick *et al.* 1976; Hambraeus *et al.* 1974).

The first two enzymes, carbamyl phosphate synthetase I (CPS I) and OCT are intramitochondrial, while the other three are located in the cytosol. Ornithine therefore is formed in the cytosol, but in order to participate in the OCT reaction has to be transported into the mitochondria by a specific energy-dependent transport system. Citrulline formed by the OCT reaction apparently diffuses passively out of the mitochondria (Gamble and Lehninger 1973).

Carbamyl phosphate synthetase deficiency

Clinical features

There has been considerable clinical heterogeneity in patients with deficiency of carbamyl phosphate synthetase (CPS). Symptomatic onset is usually in the first week, but has been as late as six weeks (Odievre *et al.* 1973). The commonest symptoms are drowsiness, vomiting, and hypotonia, accentuated by protein-containing feeds. Death has occurred from two to fifteen months (Oberholzer and Palmer 1976; Van Gennip *et al.* 1980). Survivors may have severe mental retardation and neurological complications (Batshaw *et al.* 1975) but treated survivors can develop normally (Arashima and Matsuda 1972; Odievre *et al.* 1973). Individual patients have had hyperglycinaemia, cyclic neutropenia, and acidosis, which are more characteristic of the organic acidurias (Freeman *et al.* 1970), hypothermia and opisthotonus (Gelehrter and Snodgrass 1974), and spastic quadriparesis and akinetic seizures (Batshaw *et al.* 1975).

Molecular defects

There is deficiency of CPS I which catalyses the intramitochondrial synthesis of carbamyl phosphate from ammonia, bicarbonate and ATP, utilizing acetylglutamate as activator (Fahien *et al.* 1964). There is also a cytosolic carbamyl phosphate synthetase (CPS II) which is involved in pyrimidine synthesis, does not require acetylglutamate as activator, and derives nitrogen for carbamyl phosphate synthesis preferentially from glutamine rather than from ammonia. It is one thousand times less active than CPS I in rat liver (Jones *et al.* 1961; Jones 1980). The cytosolic localization, substrate specificity and the low activity explain lack of compensation by CPS II in patients with CPS I deficiency.

CPS I deficiency has been demonstrated in liver (Mantagos *et al.* 1978) but possible deficiency in leucocytes (Wolfe and Gatfield 1975) has been disputed (Rabier *et al.* 1979). In liver, residual activity has varied from zero (Gelehrter and Snodgrass 1974; Wilson and Masters 1977; Mantagos *et al.* 1978) to 22 per cent of normal (Freeman *et al.* 1970).

Hyperammonaemia has been present in all patients and may be responsible for a modest increase in blood glutamine and alanine in some patients. Blood urea and orotic acid concentrations are usually normal but have been reported to be low (Arashima and Matsuda 1972).

Diagnosis

The disorder should be suspected in infants with hyperammonaemia. The urine should also be investigated for organic acids, since methylmalonic acidaemia and other organic acidurias are known to cause hyperammonaemia. The diagnosis can be confirmed by demonstrating reduced activity of CPS I in rectal (Matsushima and Orri 1981) or duodenal (Hoogenraad *et al.* 1980) biopsy. Liver biopsy is best avoided in the neonatal period.

Genetic counselling

Since the disorder affects infants of both sexes with normal parents, it probably has an autosomal recessive mode of inheritance. This is supported by the reports that sibs of affected infants have had similar clinical phenotypes (Freeman *et al.* 1970; Gelehrter and Snodgrass 1974). Since the CPS activity is not present in normal amniotic cells amniocentesis is unlikely to be useful for prenatal diagnosis.

Ornithine carbamyl transferase (ornithine transcarbamylase) deficiency

Ornithine carbamyl transferase (OCT) deficiency is a sex-linked dominant disorder with a fatal neonatal clinical course in the great majority of hemizygous males, and a variable course in heterozygous females.

Clinical features

Hemizygous males – severe type; classical OCT deficiency in males. About three-quarters of affected males have died in the first four weeks after birth (Campbell *et al.* 1971, 1973; Kang *et al.* 1973; Snyderman *et al.* 1975; Schuchmann *et al.* 1980). The majority of the survivors have had severe mental retardation and neurological impairment. Within a few hours or the first two days after birth there is lethargy, poor feeding, and tachypnoea with grunting respiration. Convulsions, hypothermia, unresponsiveness, and death usually occur during the first week. These patients have hepatic OCT activities of less than 2 per cent of normal (Campbell *et al.* 1973).

Hemizygous males – mild variants. A few patients have been reported with later symptomatic onset and milder course or even normal development. Where measured, these patients have been found to have higher residual OCT activities than patients with the 'classical' severe type. Thus Yudkoff *et al.* (1980) reported a boy who presented with irritability and drowsiness at 11 months following an 'antecedent infection'. He had hyperammonaemia which responded to restriction of dietary protein to 1.5 g/kg body weight. Liver biopsy OCT activity was 14 per cent of control values. At 18 months he had no neurological abnormality. Levin *et al.* (1969*a*) described a boy who presented at six months of age with vomiting, failure to thrive, and episodic drowsiness. His hepatic OCT activity was 25 per cent of normal. He improved on restriction of dietary protein and had normal growth and development at two years.

A different course occurred in two boys who had been clinically normal and died of acute hyperammonaemia at six and eight years (Cathelineau *et al.* 1973; Aylsworth *et al.* 1975).

Heterozygous females. There is a very wide spectrum of clinical severity in heterozygous females. However, even those with the most severe fatal course have not died in the neonatal period, as is typical for classical OCT deficiency in hemizygous males. Common presenting features are vomiting, and feeding difficulties, often with screaming and headaches, especially after protein-containing meals. Onset has been during the first decade and sometimes from infancy. The more severely affected have had hepato-splenomegaly, drowsiness and coma leading to death (Hopkins *et al.* 1969*b*; Matsuda *et al.* 1971). Milder cases have had headache, nausea or tiredness after protein-containing meals (Russell *et al.* 1962; Short *et al.* 1973; Palmer *et al.* 1974).

Batshaw *et al.* (1980) reported seven asymptomatic heterozygotes from the same family who had mild mental handicap.

Pathology

The brain of a boy with classical OCT deficiency who died at six years showed almost total atrophy of cerebral cortex and considerable loss of underlying white matter. The brain of a heterozygous female whose symptomatic onset was at three years with death at seven showed only slight swelling (Bruton *et al.* 1970). Both these brains, and those of other children with OCT deficiency (Hopkins *et al.* 1969*b*; Levin *et al.* 1969*a*; Campbell *et al.* 1973) histologically have had a glial reaction characterized by the appearance of numerous Alzheimer type II astrocytes. These cells also occur in patients with argininosuccinic aciduria. They can be produced experimentally in rat brain where they are related to the plasma ammonia concentration (Cavanagh and Kyu 1969) and are therefore probably produced in response to hyperammonaemia.

Liver histology has been found to be normal in male hemizygous infants, but females and adult males have had steatosis, focal necrosis and stellate portal scarring.

Molecular defects

Hyperammonaemia is a constant finding in severely affected males and females. In mildly affected cases elevation of blood ammonia may occur only after protein-containing meals. As in CPS deficiency, there may be a secondary increase in plasma glutamine and alanine concentration. In contrast to CPS deficiency, however, there is an increase in the concentration of blood and urine orotic acid, and other pyrimidine derivatives including uracil and uridine (Bachman and Colombo 1980; Van Gennip *et al.* 1980; Webster *et al.* 1981). Crystals of orotic acid in the urine have been reported and suggested the diagnosis (Levin *et al.* 1969*a,b*).

It has been shown (Natale and Tremblay 1969) that intramitochondrial carbamyl phosphate is available for extramitochondrial pyrimidine synthesis. It seems likely therefore that in OCT deficiency carbamyl phosphate accumulates and is converted to pyrimidines including orotic acid. It has been suggested that this process is accentuated by inhibition of orotidine 5-phosphate:orotate phosphoribosyltransferase by accumulated

UMP (Van Gennip *et al.* 1980). Orotic aciduria does not occur in patients with CPS deficiency, presumably because the excess ammonia cannot be converted into carbamyl phosphate.

Diagnosis

The disorder should be suspected in patients with hyperammonaemia, especially if there is orotic aciduria, and confirmed by demonstrating OCT deficiency in duodenal (Hoogenraad *et al.* 1980) or possibly rectal (Matsushima and Orii 1981) biopsies. Liver biopsy is probably better avoided in the newborn.

Genetic counselling

X-linked dominant inheritance was identified by Short *et al.* (1973) by analysis of four family pedigrees. Lyonization was demonstrated cyto-chemically in liver of a female carrier by showing two populations of cells, one with and one without OCT activity (Ricciuti *et al.* 1976). X-linked dominant inheritance explains why protein intolerance is frequently present in mothers but not fathers of patients, and the severity of the disease in males and the usually milder and more variable course in affected females (Palmer *et al.* 1974).

Genetic heterogeneity in males is probably allelic and accounts for the few cases with the milder variants and appreciable hepatic OCT activity. Female heterozygotes can be detected by demonstrating elevated excretion of urinary orotic acid following protein administration (Hokanson *et al.* 1978; Batshaw *et al.* 1980).

Prenatal diagnosis has been undertaken in two pregnancies by assay of OCT in foetal liver biopsies aspirated through a foetoscope. One foetus was affected, and was aborted, while the other had normal hepatic OCT activity and continued to term. The prenatal predictions were both confirmed (Rodeck *et al.* 1982).

The report that OCT activity is present in uncultured amniotic cells (Nadler and Gerbie 1969) has not been confirmed.

Citrullinaemia; argininosuccinate synthetase deficiency

The majority of patients with citrullinaemia have had either a severe neonatal illness or a subacute type with symptomatic onset later during infancy. A few patients have been atypical.

Clinical features

The neonatal type. The clinical features resemble those of classical OCT deficiency but either sex is affected. After the first day or two following birth there is poor feeding, lethargy and irritability. Convulsions are common. After a few days drowsiness is followed by coma and death, usually in the first week (Van der Zee *et al.* 1971; Wick *et al.* 1973), but with treatment infants can survive up to eight months (Danks *et al.* 1974*b*; Thoene *et al.* 1976).

The subacute (or infantile) type. This form may present during the first year with neurological symptoms such as convulsions, ataxia, and tremors, with recurrent vomiting and failure to thrive, or with delayed development. Hepatomegaly has been reported (Scott-Emuakpor *et al.* 1972; Buist *et al.* 1974*b*).

Variants. A late-presenting variant of citrullinaemia has been reported in about 25 Japanese patients. Symptomatic onset is in childhood or adult life with non-specific complaints including tiredness and confusion especially after meals, vomiting, diarrhoea, and hallucinations. The disorder usually has a slowly progressive course leading to neuroses, psychoses, disorder of motor function, myoclonic and generalized convulsions. Some patients have had hepatomegaly. The oldest survivor is mentally handicapped at 28 years (Tsujii *et al.* 1976). Death has been reported between 22 (Takamizawa *et al.* 1973; Saheki *et al.* 1981) and 46 (Yamauchi *et al.* 1980; Saheki *et al.* 1980, 1981).

An unusual patient with citrullinaemia and hyperammonaemia, was a 21-year-old male who developed spastic paraparesis due to myelopathy; no enzyme studies were made (Miyazaki *et al.* 1971).

A few asymptomatic non-Japanese patients have been reported (Whelan *et al.* 1976). A boy with normal growth and development was found to have citrullinaemia on routine screening (Wick *et al.* 1973).

Pathology

In the neonatal type, the liver may be infiltrated with fat and show cellular necrosis (Van der Zee *et al.* 1971; Wick *et al.* 1973; Matsuda *et al.* 1976). The brain is oedematous and shows degenerative changes of neurones and myelin, with delayed myelination (Wick *et al.* 1973).

Molecular defects

Plasma and urinary citrulline concentrations are considerably elevated. Intermittent hyperammonaemia is sometimes associated with hypo-argininaemia. There is deficiency of argininosuccinate synthetase (ASAS) (Fig. 5.6). In liver, activity may be undetectable (Van der Zee *et al.* 1971; Roerdink *et al.* 1973) or as high as 14 and 20 per cent of normal even in the neonatal type (Wick *et al.* 1973). Activity of ASAS in brain has been either present (Roerdink *et al.* 1973) or absent (Van der Zee *et al.* 1971). In cultured fibroblasts, activity is either absent or considerably reduced. The residual enzyme in the patient of Kennaway *et al.* (1975) had a high K_m for citrulline and aspartate. ASAS deficiency can also be demonstrated in cultured lymphocytes (Spector *et al.* 1975*a*). Six citrullinaemia cell lines investigated by Tsung-Sheng *et al.* (1983) lacked immunologically detectable enzyme antigen.

Genetic counselling and prenatal diagnosis

The neonatal type has an autosomal recessive mode of inheritance as evidenced by the occurrence in both sexes and in sibs of clinically normal parents and by the presence of raised plasma citrulline in some parents (Wick *et al.* 1970). The evidence for autosomal recessive inheritance is even stronger for the subacute form since parents have reduced fibroblast ASAS activities (Kennaway *et al.* 1975).

Unlike patients with deficiency in the activity of CPS or OCT, those with citrullinaemia show abnormal plasma and urinary amino acid concentrations, namely elevated citrulline concentration.

The diagnosis is confirmed by demonstrating reduced activity of ASAS in cultured fibroblasts or cultured lymphocytes (see 'Molecular defects', p. 214). Prenatal diagnosis of the infantile type has been undertaken in at least 10 pregnancies by determination of the rate of incorporation of ^{14}C-ureido-citrulline into TCA-precipitable material of cultured amniotic cells (Tedesco and Mellman 1967).

In two cases the foetus was affected (Cathelineau *et al.* 1981*a*; Fleisher *et al.* 1983), and in eight unaffected (Roerdink *et al.* 1973; Christensen *et al.* 1980; Fensom and Benson 1981, unpublished observation; Jacoby *et al.* 1981). In the pregnancy reported by Fleisher *et al.* (1983) where the foetus was affected there was an elevation in amniotic fluid citrulline concentration. It will be interesting to see whether this latter assay will prove to be reliable for prenatal diagnosis in the same way as an increased level of amniotic fluid argininosuccinic acid seems to be useful in the prenatal diagnosis of another urea cycle enzymopathy – argininosuccinic aciduria (Goodman *et al.* 1973*b*; Hartlage *et al.* 1974; Fleisher *et al.* 1979).

The gene for argininosuccinate synthetase has provisionally been assigned to chromosome 9 (Carrit 1977).

Somatic cell hybridization studies between cultured fibroblasts from 10 citrullinaemia strains failed to produce any complementation. Unfortunately, clinical data are only given for six of these, all of which had the neonatal type with symptomatic onset between one and eight days (Cathelineau *et al.* 1981*b*). It would have been more interesting to hybridize cells of the neonatal with those of the subacute and variant types. Evidence for interallellic but not intergenic complementation has been presented by Kennaway and Curtis (1981) by studying heterokaryons formed from eight patients with clinically different forms of citrullinaemia.

To study the genetic defect at the molecular level a complementary

DNA (cDNA) clone containing the entire length of argininosuccinate synthetase messenger RNA (mRNA) except for about 50 bases was hybridized to total RNA from cultured fibroblasts (Tsung-Sheng *et al.* 1983). After digestion with single stranded nuclease the cDNA fragments were fractionated by electrophoresis. Their data indicate that a minimum of three out of five non-consanguineous patients represented compound heterozygotes; while one patient whose parents were first cousins with a heterozygous pattern was homozygous for an allele which produced an RNA with a single stranded-nuclease detectable defect.

Argininosuccinic aciduria; argininosuccinase deficiency

Deficiency of argininosuccinase (argininosuccinate lyase) (Fig. 5.6) causes accumulation of argininosuccinic acid (ASA) in blood, urine, and cerebrospinal fluid and sometimes hyerammonaemia and arginine deficiency.

Clinically patients can be classified into neonatal, subacute (or infantile) and late-onset types.

Clinical features

Neonatal type. At least 21 patients with this form have been reported. The majority have died in the first three weeks (Glick *et al.* 1976; Francois *et al.* 1976; Collins *et al.* 1980; John *et al.* 1980; Perry *et al.* 1980; Batshaw *et al.* 1981). During the first few days infants feed poorly; there is persistent vomiting, and lethargy leading to coma and death unless the patients are treated. Convulsions and hepatomegaly have been noted in some patients. Therapy by restriction of dietary protein and administration of arginine has caused considerable clinical improvement and survival to 18 months with a developmental level of 10 months. In this patient, abnormally fragile hair was noted at two weeks (Collins *et al.* 1980).

Subacute (infantile) and late-onset types. Patients who present during the first few months have failure to thrive and feeding difficulties, later complicated by psychomotor retardation and convulsions.

In the commonest form, patients present with delayed development often in the second year. Other common symptoms are convulsions, feeding difficulties, and irritability. As with the other types lethargy, intermittent ataxia and convulsions due to hyperammonaemia are provoked by a high protein intake, especially during infections. About half the reported patients have had abnormally fragile hair. The majority of patients have had IQ of 30–60, but some patients have low-normal or even normal intelligence. Treatment by protein restriction and arginine supplementation after detection by routine screening in infancy can lead to normal growth and development (Hambraeus *et al.* 1974; Qureshi *et al.* 1978; Schutgens *et al.* 1979*a*; Andria *et al.* 1980).

Pathology

In the neonatal form where death is in the first few weeks, the brain has shown spongy degeneration (Baumgartner *et al.* 1968) and anoxic changes (Glick *et al.* 1976). Older individuals have had Alzheimer type II cells (see OCT deficiency) in the cerebral cortex, basal ganglia, and other regions of the brain (Solitaire *et al.* 1969). The liver is often enlarged but shows no specific changes apart from some fat deposition.

The mechanism for fragile hair (trichorrhexis nodosa) in argininosuccinic aciduria has not been elucidated.

Molecular defects

Patients have considerable elevation of ASA in blood and urine, and usually an even higher elevation in the CSF. There is often hyper-ammonaemia, especially after protein-containing meals, and sometimes orotic aciduria (see OCT deficiency). As in other hyperammonaemic states, blood glutamine and alanine may be increased.

Argininosuccinase (ASA lyase) deficiency has been demonstrated in liver (Pollitt 1973; Billmeier *et al.* 1974; Glick *et al.* 1976; Van der Heiden 1976) and red blood cells (Tomlinson and Westall 1964; Glick *et al.* 1976); and in some patients in brain (Kint and Carton 1968) and kidney (Carton *et al.* 1969). However, argininosuccinase deficiency was present in liver but not in brain or kidney in an infant with the neonatal form (Glick *et al.* 1976). Argininosuccinase deficiency has also been demonstrated in cultured fibroblasts (Goodman *et al.* 1973*b*; Pollitt 1973; Van der Heiden *et al.* 1976; Fensom *et al.* 1980).

Diagnosis

The diagnosis can be made by showing an excess of ASA in urine. Since the free acid is unstable, conversion to its anhydrides is advisable before assay using an aminoacid analyser. This is achieved conveniently by acidifying the urine to pH2 and allowing it to stand for 2.5 hours (Coryell *et al.* 1964).

The diagnosis can be confirmed by demonstrating reduced argininosuccinase activity in red cells or cultured fibroblasts (see 'Molecular defects', above). Care should be taken to exlude mycoplasma contamination as we have shown that this can lead to the appearance of argininosuccinase in mutant cells (Fensom *et al.* 1980).

Genetic counselling

Deficiency of argininosuccinase activity has an autosomal recessive mode of inheritance. Heterozygotes have reduced argininosuccinase activity in cultured fibroblasts (Billmeier *et al.* 1974) and in red cells (Tomlinson and Westall 1964; Shih and Efron 1972).

Somatic cell hybridization studies between five fibroblast strains with ASA lyase deficiency revealed two complementation groups, one strain complementing with four others (Cathelineau *et al.* 1981*b*). Unfortunately, clinical data are provided for only two of the patients, both of whom are of the neonatal type; one of these (symptomatic onset three days) complemented with another (symptomatic onset two days).

The gene for ASA lyase has been provisionally assigned to chromosome 7 (Naylor *et al.* 1978).

The incidence has been estimated as 1:70 000 births in Massachusetts by screening of 633 331 newborns (Levy *et al.* 1980), but only about 1:330 000 in New South Wales, Australia, in about one million newborn tested.

Prenatal diagnosis of an affected foetus was reported by Goodman *et al.* (1973*b*), by demonstrating both an elevated argininosuccinic acid concentration in the amniotic fluid, and deficiency of ASA lyase in amniotic cells, after six weeks of culture. Argininosuccinic acid was also elevated in maternal urine. The pregnancy was terminated at 22 weeks' gestation and the prediction confirmed by demonstrating severe reduction of foetal liver ASA lyase activity. These authors measured amniotic cell activity by incubating cells in arginine-free medium containing ^{14}C-ureido-citrulline, followed by chromatography of cell lysates for separation and radioactive assay of arginine and ASA. A more sensitive assay was described by Jacoby *et al.* (1972) who, following the incubation of cultured amniotic cells in arginine-free medium containing ^{14}C-ureido-citrulline, measured the radioactivity incorporated into trichloroacetic acid-precipitable cell protein.

Two pregnancies at risk for argininosuccinic aciduria were reported by Fleisher *et al.* (1979). In one pregnancy amniotic fluid argininosuccinic acid concentration was elevated, and incorporation of ^{14}C-citrulline by cultured amniotic cells was reduced. The pregnancy was terminated at 18 weeks and the diagnosis confirmed by enzyme assay of foetal tissues. In the other pregnancy both parameters were within normal limits. The pregnancy was allowed to continue with the birth of an unaffected infant.

Amniotic fluid and maternal urine argininosuccinic acid was elevated at 39 weeks' gestation in a pregnancy where the foetus was affected (Hartlage *et al.* 1974). Further experience is needed to assess whether elevated argininosuccinic acid in amniotic fluid supernatant or maternal urine will prove to be reliable for prenatal diagnosis.

N-acetylglutamate synthetase deficiency

A single patient has been reported with deficiency of hepatic *N*-acetylglutamate synthetase (AGAS) (Bachman *et al.* 1981). This mitochondrial enzyme catalyses the reaction: Acetyl-CoA + glutamate → *N*-

acetylglutamate ꟷ CoA (Lusty 1978; Powers 1981). As noted earlier the reaction product *N*-acetylglutamate, is an activator for CPS I.

This patient presented with hyperammonaemia on the third day. A liver biopsy CPS activity was within normal limits, but AGAS activity was undetectable.

Combined deficiency of hepatic CPS I and OCT

Combined deficiency of hepatic CPS I and OCT activities has been reported in three patients (Matsushima *et al.* 1979; Raijman 1979). Surprisingly one of these patients had normal intelligence at seven years of age (Matsushima *et al.* 1979).

Treatment of urea cycle enzyme defects other than hyperargininaemia

The aim of treatment is to eliminate accumulated ammonia, and precursors, reduce ureagenesis by low protein or low amino acid diet, and to correct hypoargininaemia. When there is severe hyperammonaemia causing life-threatening symptoms, haemodialysis (Donn *et al.* 1979) charcoal haemoperfusion (Chavers *et al.* 1980) or peritoneal dialysis (Batshaw and Brusilow 1980) may be used to reduce blood ammonia concentration. Batshaw and Brusilow (1980) found a 77 per cent fall in blood ammonia in 10 infants following peritoneal dialysis, but only a nonsignificant fall following exchange transfusion.

Administration of arginine (4 mmol/kg/day) corrects hypoargininaemia, causes a fall in blood ammonia, and promotes excretion of waste nitrogen by stimulating synthesis and subsequent excretion of argininosuccinic acid in argininosuccinic aciduria, or of citrulline in citrullinaemia. Furthermore, in these diseases there is an increased requirement of arginine because of loss of arginine 'carbon skeletons' as argininosuccinic acid, or citrulline, respectively (Brusilow and Batshaw 1979).

A recent addition to treatment of hyperammonaemia has been the administration of sodium benzoate intravenously or orally (250 mg/kg/ day). This caused a significant fall of plasma ammonia in six of eight episodes of hyperammonaemia (Brusilow *et al.* 1980; Batshaw and Brusilow 1980). The rationale is that hippuric acid nitrogen can partially substitute for urea nitrogen since the former is derived from glycine nitrogen, which is synthesized *de novo* from ammonia nitrogen either directly via the reverse glycine cleavage pathway or via transamination from glutamate. Phenylacetate administration also accelerates nitrogen excretion in the form of phenylacetylglutamine (Brusilow *et al.* 1980).

Attempts should be made to minimize gluconeogenesis to reduce nitrogen production by providing a high calorie intake in the form of carbohydrate or lipids.

Protein restriction needs to be severe in neonatal types, limiting intake to 0.5 to 1.0 gm/kg/day, especially in OCT and CPS deficiency, but can be more moderate (1.5 mg/kg/day) and monitored by blood ammonia assay in milder variants and older individuals, especially with citrullinaemia and argininosuccinic aciduria. Substitution of amino acids in the diet by α-keto or α-hydroxy analogues will also reduce hyperammonaemia (Mitch and Walser 1977; McReynolds *et al*. 1978; Brusilow *et al*. 1979).

Hyperargininaemia; argininaemia; arginase deficiency

Hyperargininaemia is due to arginase deficiency (Fig. 5.6). It is probably the rarest of the urea cycle enzyme deficiencies.

Clinical features

All untreated patients had similar clinical features of vomiting, irritability, poor feeding, drowsiness, convulsions, and coma with symptomatic onset usually in infancy. Survivors have had spastic diplegia and have been severely mentally retarded (Terheggen *et al*. 1975a; Snyderman *et al*. 1977; Cederbaum *et al*. 1977; Michels and Beaudet 1978).

Molecular defects

There is deficiency of arginase activity in erythrocytes (Terheggen *et al*. 1969; Cederbaum *et al*. 1977; Snyderman *et al*. 1977), liver (Beaudet and Michels 1977; Michels and Beaudet 1978; Cederbaum *et al*. 1979a) and leucocytes (Cederbaum *et al*. 1979a; Marescau *et al*. 1979a). Several workers have reported that they were able to detect arginase activity in cultured fibroblasts both from controls and from patients with proven arginase deficiency in erythrocytes (Van Elsen and Leroy 1975, 1978; Beaudet and Michels 1977). However, Cederbaum and Spector (1978) have not been able to demonstrate arginase activity in normal cultured fibroblasts. Plasma arginine concentration is considerably elevated, while plasma ornithine concentration is sometimes reduced. Elevated blood ammonia concentration may occur (Terheggen *et al*. 1969; Qureshi *et al*. 1981a). Orotic aciduria has been observed with normal blood ammonia concentration (Qureshi *et al*. 1981a; Naylor and Cederbaum 1981) and may be associated with increased excretion of uridine and uracil (Naylor and Cederbaum 1981). Increased excretion of N-acetylarginine, γ-guanidino-butyric acid, argininic acid and α-ketoguanidinovaleric acid has been reported (Terheggen *et al*. 1975a; Marescau *et al*. 1979b). Following loading with arginine (Terheggen *et al*. 1975a; Snyderman *et al*. 1977), or protein (Cederbaum *et al*. 1977) there is an excessive sharp rise in blood arginine concentration followed by a slow fall.

Diagnosis

The diagnosis is established by showing an elevated plasma arginine concentration, and severe deficiency of arginase activity in erythrocytes (see 'Molecular defects', p. 220).

Genetic counselling

The occurrence of the disease in siblings of both sexes with normal parents suggests an autosomal recessive mode of inheritance. This is supported by the report that some parents have reduced erythrocyte and leucocyte arginase activities (Cederbaum *et al.* 1977; Snyderman *et al.* 1977). Prenatal diagnosis has not yet been reported, but since cultured fibroblasts from affected patients have been reported to have normal arginase activities, arginase assay on foetal blood samples collected at foetoscopy may be the method of choice. This is supported by the report that foetal and adult erythrocyte arginase appears to have the same properties (Spector *et al.* 1975*b*, 1980).

Treatment of hyperargininaemia

When the diagnosis is made at birth, it is possible to prevent the brain damage which has been a consistent feature of untreated or late-treated patients (Snyderman *et al.* 1977). The principle of treatment is to exclude arginine from the diet by feeding on a mixture of essential amino acids. Snyderman *et al.* (1977) gave enough of the mixture equivalent to 2 g of protein/kg body weight and supplemented the diet with fat and carbohydrate to supply a total of 125 calories/kg, vitamins and minerals. After four months, fruits, low-protein vegetables and cereals were introduced, and later other low-protein foodstuffs, but never exceeding 5 g of protein (300 mg arginine) per day. Blood arginine and blood ammonia concentrations should be monitored regularly. Arginine intake should be reduced during infections.

In a 16-year-old girl, Qureshi *et al.* (1981*a,b*) were able to reduce fasting plasma ammonia nitrogen and glutamine and urine orotate by a combination of a low protein intake (0.25–0.37 g/kg/24h) and administration of sodium benzoate for acylation of glycine (see treatment of urea cycle enzyme defects other than hyperargininaemia).

Attempts at enzyme replacement by intravenous injection of purified Shope papilloma virus was attempted in three sisters. No fall in serum arginine was observed (Terheggen *et al.* 1975*b*).

THE HYPERORNITHINAEMIAS

Hyperornithinaemia occurs in two types of genetic disorder. One type is associated with progressive blindness due to gyrate atrophy of the choroid and retina, and deficency of ornithine ketoacid transaminase (ornithine- δ - aminotransferase), without hyperammonaemia. In the second type there is intolerance to protein-containing food, hyperammonaemia, and homo-citrullinaemia. It has been proposed that the defect in this type is in the transport of ornithine into the mitochondria, which are morphologically abnormal. There may be heterogeneity since in different patients deficiencies have been reported in the activities of either fibroblast ornithine decarboxylase or liver and leucocyte carbamyl phosphate synthetase I. In addition hyperornithinaemia associated with renal tubular dysfunction, prolonged jaundice, and mild or moderate mental retardation, but not hyperammonaemia, has been reported in a brother and sister. These patients had a deficiency of hepatic ornithine ketoacid transaminase – that is the same enzyme which is deficient in gyrate chorioretinal atrophy, but it is possible that this may have been secondary to liver damage.

Ornithine concentration rises when blood is allowed to stand at room temperature because of conversion of arginine catalysed by red cell arginase. Hyperornithinaemia also occurs during isoniazide therapy (Perry and Hansen 1978).

Gyrate atrophy of the choroid and retina; ornithine ketoacid transaminase (ornithine- δ -aminotransferase) deficiency

Clinical features

Night blindness and restriction of the peripheral visual fields usually with onset in the third decade is the characteristic presentation. Initially, there is constriction of the visual field, chorioretinal atrophy at the periphery of the fundus, and myopia. Adaptation to dark is poor and electroretinographic responses arc small. As the disease progresses, there is further peripheral constriction of fields of vision, and areas of chorioretinal atrophy become extended and lead to blindness usually in the fifth decade. Cataract formation is common (Francois 1979; Takki and Milton 1981).

Electroretinograms show progressive diminution in amplitude until there is complete lack of activity.

Morphological abnormalities have been demonstrated in liver mito-chondria which become elongated and segmented (McCulloch *et al.* 1978), and in voluntary muscle. In the latter, tubular aggregates occur in type 2 fibres visualized after staining with NADH-tetrazolium reductase (Sipila *et al.* 1979) which are not however specific for this disorder (Engel *et al.* 1970).

Molecular defects

Hyperornithinaemia in this disorder was first reported in nine patients from Finland (Simmel and Takki 1973). Plasma ornithine is increased 10–

20 fold above normal (Takki 1974; McCulloch and Marliss 1975; Berson *et al.* 1976).

The disorder is due to deficiency in the activity of ornithine ketoacid transaminase (ornithine- δ -transaminase) (Fig. 5.7) which has been demonstrated in cultured fibroblasts (Trijbels *et al.* 1977; Shih *et al.* 1978) and phytohaemagglutinin (PHA)-stimulated lymphocytes (Valle *et al.* 1977). In cultured fibroblasts (Shih *et al.* 1978) from five patients, cells from one patient only showed a progressive increase in ornithine ketoacid transaminase activity, reaching heterozygous values when the concentration of the reaction cofactor, pyridoxal phosphate, was increased in the assay medium from 0.04 to 0.4 μmol/ml. Similar observations were reported by Kennaway *et al.* (1980). The enzyme is bound to the mitochondrial matrix and catalyses the pyridoxal phosphate-dependent transamination of ornithine and 2-oxoglutarate to $\triangle$'-pyrroline 5-carboxylic acid and glutamic acid. The reaction is reversible but the equilibrium constant is 70-fold in favour of the forward direction (Strecker 1965). Other pathways for ornithine metabolism include entry into the urea cycle catalysed by ornithine transcarbamylase or into polyamine synthesis by decarboxylation catalysed by ornithine decarboxylase. Presumably these are rate-limiting for ornithine elimination and cannot dispose of the increased ornithine load when there is deficiency in transamination.

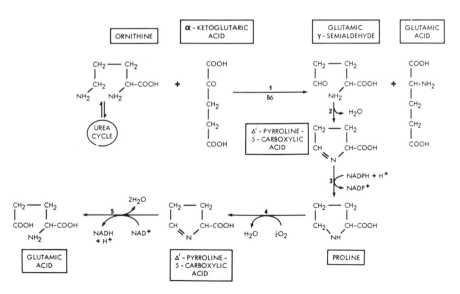

Fig. 5.7. Pathways of ornithine and proline metabolism: (1) Ornithine ketoacid transaminase (2) Spontaneous reaction (3) Pyrroline 5-carboxylic acid reductase (4) Proline oxidase (5) Pyrroline 5-carboxylic acid dehydrogenase.

There is no hyperammonaemia or homocitrullinuria. Hypolysinaemia may occur, probably because of increased renal clearance of lysine (Valle *et al.* 1980*b*).

Plasma glutamic acid and glutamine concentrations are sometimes reduced (McCullogh *et al.* 1978; Yatziv *et al.* 1979; Kennaway *et al.* 1980). The cyclic δ-lactam of ornithine or 3-aminopeperid-2-one has been identified in the urine (Oberholzer and Briddon 1978), and occurs also in other forms of hyperornithinaemias.

Diagnosis

The disorder is diagnosed by demonstrating hyperornithinaemia, without hyperammonaemia, and a marked deficiency in the activity of ornithine ketoacid transaminase in cultured fibroblasts or PHA-stimulated lymphocytes (see 'Molecular defects', p. 222).

Genetic counselling

The disorder has an autosomal recessive mode of inheritance. This was established by pedigree analysis of 14 families (Takki and Simmell 1976) and by demonstrating a partial enzyme defect in cultured fibroblasts (Shih *et al.* 1978), or PHA-stimulated lymphocytes (Valle *et al.* 1977) from heterozygotes.

Prenatal diagnosis should be possible by assay of ornithine ketoacid transaminase in cultured amniotic cells using either a colorimetric assay (Shih *et al.* 1978) or radioisotopic methods (Trijbels *et al.* 1977).

The variability in clinical severity (Kaiser-Kupfer *et al.* 1980*a*) and in residual activity and response of fibroblast ornithine ketoacid transaminase to pyridoxine suggest the existence of genetic heterogeneity. However, complementation did not occur after hybridization of cells from ten patients, suggesting that the mutations affect the same allele (Valle *et al.* 1979*a*; Shih *et al.* 1981). About half the reported patients have been Finnish.

Treatment

Pyridoxine in pharmacological doses of 500–1000 mg/day has reduced plasma ornithine, and increased plasma lysine concentrations in four patients (Berson *et al.* 1981; Weleber and Kennaway 1981). Clinically there was no further visual deterioration while on treatment. One patient had an improved electroretinogram (Weleber and Kennaway 1981).

Lowering of plasma ornithine has also been achieved by dietary restriction of arginine by lowering the protein intake to 0.2 g/kg per day. This has been accompanied by improved visual acuity in two patients (Kaiser-Kupfer *et al.* 1980*b*, 1981; McInnes *et al.* 1981).

Promotion of renal excretion of ornithine by administration of lysine or α-aminoisobutyric acid (Valle *et al.* 1980*b*) has also produced some lowering of plasma ornithine.

Creatine supplementation has also been tried for therapy (Sipila *et al.* 1981) because of the hypothesis that gyrate atrophy may be due to deficiency of creatine and creatine phosphate resulting from reduced biosynthesis because of inhibition of glycine transaminase by ornithine accumulation (Sipila 1980). Of the seven patients treated for one year, four had progression of fundal abnormalities, even though histological abnormalities in muscle improved in all patients.

Hyperornithinaemia with hyperammonaemia, homocitrullinaemia and protein intolerance

Patients with recurrent stuporose episodes probably due to hyperammonaemia following protein intake have been reported by Shih *et al.* (1969), Wright and Pollitt (1973), Fell *et al.* (1974), and Gatfield *et al.* (1975). All have had mild to severe mental handicap. Homocitrullinaemia was present and decreased on supplementation with ornithine or arginine (Fell *et al.* 1974). 3-aminopiperid-2-one has been detected in the urine (Fell and Pollitt 1978). No consistent enzyme deficiency has been established. Gatfield *et al.* (1975) reported that electron microscopy of a liver biopsy showed large mitochondria with bizarre shapes and unusual periodicity of the membranes of about 40 nm. They reported deficiency of carbamyl phosphate synthetase I in leucocytes of five patients and in liver of one of them. On the other hand, Shih and Mandell (1974) reported a deficiency of ornithine decarboxylase in the skin fibroblasts of the patient reported by Shih *et al.* (1969). Because of improvement following supplementation with arginine or ornithine and because of the abnormal mitochondria, it was proposed that the disorder might be due to a defect of transport of ornithine into the mitochondria (Fell *et al.* 1974; Gatfield *et al.* 1975). An approximate equal sex ratio, and a high rate of parental consanguinity (Gatfield *et al.* 1975) suggests autosomal recessive inheritance.

Hyperornithinaemia with liver damage and tubular dysfunction; ornithine ketoacid transaminase deficiency

Bickel *et al.* (1968) reported a brother and sister with mild and moderate mental retardation respectively, prolonged jaundice and abnormal EEG, in whom blood ornithine concentration was about three times higher than normal. In addition, there was generalized aminoaciduria and glucosuria, indicating defective renal tubular reabsorption, but no hyperammonaemia or homocitrullinuria. A deficiency of hepatic ornithine ketoacid transaminase was demonstrated. The relation between this disorder and gyrate chorioretinal atrophy in which deficiency of the same enzyme has been demonstrated remains to be established. Allelic mutations at the same locus, and mutation of different molecular forms of the enzyme or of common subunits remain theoretical possibilities. However a deficiency due to severe liver damage is also a possible explanation.

DISORDERS OF LYSINE AND HYDROXYLYSINE METABOLISM

Persistent hyperlysinaemia; saccharopinuria; deficiency of lysine-keto-glutarate reductase; deficiency of saccharopine dehydrogenase

The main pathway for lysine catabolism to acetyl-CoA and carbon dioxide is through the formation of saccharopine (Fig. 5.8). Inherited deficiency of the first two enzymes in this pathway, namely lysine-ketoglutarate reductase, which catalyses saccharopine synthesis, and saccharopine dehydrogenase, which catalyses epsilon transamination, results in persistent hyperlysinaemia. In four families, a combined deficiency of both enzymes

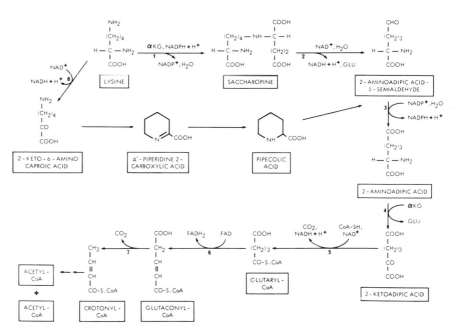

Fig. 5.8 Metabolic pathways of lysine: (1) Lysine: α-ketoglutarate reductase (2) Saccharopine dehydrogenase (3) 2-aminoadipic acid semialdehyde de-hydrogenase (4) 2-aminoadipic acid transaminase (5) 2-ketoadipic acid de-hydrogenase (6) Glutaryl-CoA dehydrogenase (7) Glutaconyl-CoA decarboxylase (8) Lysine: NAD oxidoreductase.

has been reported in cultured fibroblasts (Dancis *et al.* 1969; Dancis *et al.* 1976; Cederbaum *et al.* 1979*b*). In two other patients, accumulation of saccharopine as well as lysine has been demonstrated, and the patients have been reported as having saccharopinuria. One had severe reduction in the activity of saccharopine dehydrogenase and a moderate reduction in lysine-ketoglutarate reductase (Fellows and Carson 1974), while the other had moderate reduction in the activity of saccharopine dehydrogenase (Simell *et al.* 1973). It appears, therefore, that patients with hyperlysinaemia and lysine-ketoglutarate reductase deficiency also have deficiency of saccharopine dehydrogenase. It is possible, but not proven, that patients with saccharopinuria and deficiency in saccharopine dehydrogenase activity will also have deficiency of lysine-ketoglutarate reductase.

Clinical features

Clinically, there has been considerable variation. Moreover the existence of identical biochemical abnormalities in apparently healthy siblings and in a cousin raises the possibility that the disorder is harmless, and the clinical associations fortuitous (Woody 1964; Woody *et al.* 1966; Van Gelderen and Teijema 1973).

Severe mental retardation and petit mal have been reported (Armstrong and Robinow 1967; Ghadimi *et al.* 1965).

Molecular defects

Deficiency of lysine-ketoglutarate reductase and saccharopine dehydrogenase has been present in the majority of patients, (see above) but not all (Dancis 1974). The concentration of lysine in the plasma and cerebrospinal fluid is increased and may be up to five times normal. The urinary excretion of lysine is increased about three to 20 times normal.

Five of seven patients with the combined enzyme defects had saccharopinuria (Cederbaum *et al.* 1979*b*). Elevated levels have been reported of other metabolites in the pathways of lysine catabolism, including pipecolic acid, 2-aminoadipic acid (Woody 1964; Ghadimi *et al.* 1965) and 6-*N*-acetyllysine (Armstrong and Robinow 1967; Woody *et al.* 1967).

A seven-year-old boy with hyperlysinaemia and the combined enzyme defects was found to have intermittent cystinuria with increased urinary excretion of cystine and arginine, in addition to lysine. It was postulated that lysinaemia produced dibasic aminoaciduria as a result of competition for renal tubular reabsorption (Cederbaum *et al.* 1979*b*).

Genetic counselling and treatment

The disorder has an autosomal recessive mode of inheritance as evidenced by the equal sex ratio, and the frequency of parental consanguinity

(Ghadimi *et al.* 1964; Woody 1964; Simell *et al.* 1973). In families where the enzyme defects are expressed in cultured fibroblasts, prenatal diagnosis by enzyme assay of cultured amniotic cells is probably possible. However, as discussed above, the existence of the disorder in normal individuals suggests that it may prove to be harmless. This will also influence the decision as to whether or not to treat individual patients. Cederbaum *et al.* (1979*b*) limited protein and lysine intake. In a mentally retarded male patient, restriction of daily dietary lysine to 5.5 mg/kg body weight for 2.5 years was thought to have improved the patient's social behaviour, but not his mental development.

Hyperlysinaemia with hyperammonaemia; deficiency of L-lysine NAD oxidoreductase

Patients have been described with periodic modest hyperlysinaemia, with episodic ammonia intoxication and hyperargininaemia (Colombo *et al.* 1964, 1967; Oyanagi *et al.* 1970). These patients developed convulsions and coma provoked by high protein or lysine intake. Activities of the urea cycle enzymes were normal, but the activity of L-lysine NAD oxidoreductase (Fig. 5.8), which catalyses the deamination of lysine to 2-oxo-6-amino-caproic acid (Rothstein and Miller 1954), was reduced in a liver biopsy to 22 per cent of normal value. Colombo *et al.* (1964, 1967) suggested that increased plasma concentrations of ammonia were due to inhibition of arginase with consequent accumulation of ammonia because of reduced elimination via the urea cycle. They demonstrated that loading with lysine was followed by coma one hour after ingestion, a rise in blood ammonia and a fall in arginase activity of the red blood cells. It is not known why hyperlysinaemia is not accompanied by hyperammonaemia in patients with deficiency of lysine-ketoglutarate reductase or saccharopine dehydrogenase.

2-aminoadipic aciduria and 2-ketoadipic aciduria

In the intermediary metabolic pathway of lysine to glutaryl-CoA, 2-aminoadipic acid is converted to 2-ketoadipic acid. (Fig. 5.8).

Considerable accumulation of 2-aminoadipic acid with varying amounts of accumulation of 2-ketoadipic acid has been described in a mildly mentally handicapped 10.5-year-old boy (IQ 86), his normal nine-year-old brother (Fischer *et al.* 1974; Fischer and Brown 1980 – though the keto-acid was missed in the earlier publication), a mentally retarded 10-year-old girl (2-ketoadipic acid was present in her urine in trace quantities only) (Casey *et al.* 1978), a mentally retarded 14-year-old male and his sister who had normal intelligence (Wilson *et al.* 1975), and in a 14-month-old girl

(Przyrembel *et al.* 1975). A six-year-old mentally handicapped boy was reported to have isolated 2-aminoadipic aciduria, but since only ninhydrin-positive substances in the urine were assayed, any 2-ketoadipic acid would have been undetected. Studies on cultured fibroblasts (Przyrembel *et al.* 1975) have shown diminished conversion of ^{14}C-2-ketoadipic acid and ^{14}C-aminoadipic acid (Wendel *et al.* 1975) to $^{14}CO_2$, suggesting that the metabolic block is at the level of oxidative decarboxylation, catalysed by 2-ketoadipic acid dehydrogenase, rather than in the previous step catalysed by 2-aminoadipic acid aminotransferase.

Loading with tryptophan in one of these patients increased the concentration of 2-aminoadipic acid. Since the metabolism of tryptophan joins that of lysine at 2-ketoadipic acid, this suggests integrity of 2-aminoadipic acid transaminase, and further, supports the hypothesis that there is deficiency of 2-ketoadipic acid dehydrogenase (Fig. 5.8). 2-aminoadipic acid aminotransferase appears to be inseparable from kynurenine aminotransferase (Tobes and Mason 1975) which catalyses the transamination of kynurenine and 3-hydroxykynurenine to kynurenic acid and 3-hydroxykynurenic acid, respectively. Normal increases of the two acids in the urine of two brothers with 2-aminoadipic aciduria, following tryptophan loading, therefore is further evidence of the integrity of 2-aminoadipic acid aminotransferase. Failure to demonstrate more than trace quantities of 2-ketoadipic acid in the patient reported by Casey *et al.* (1978) is unexplained. Neither administration of pyridoxine nor of thiamine corrected the biochemical abnormality in that patient. However, since the condition can exist in individuals with normal intelligence, it is probably benign and treatment therefore unnecessary.

2-aminoadipic aciduria was present consistently in two brothers (Gregerson *et al.* 1977), and terminally in another patient (Goodman *et al.* 1977) with glutaric aciduria.

The glutaric acidurias

Glutaric aciduria type I is characterized clinically by a progressive disorder of voluntary movement including choreoathetosis presenting in early childhood. There is accumulation of glutaric, 2-hydroxyglutaric, and glutaconic acids because of deficiency in the activity of glutaryl-CoA dehydrogenase.

In glutaric aciduria type II, there is a severe progressive neonatal illness presenting within a few hours after birth, with rapid deterioration leading to death usually within the first four days. There is accumulation of numerous organic acids, including glutaric acid, possibly because of deficient activity of several acyl-CoA dehydrogenases.

Glutaric aciduria type I

Clinical features

Typically there is progressive choreoathetosis, hyperkinesia and dysarthria with or without spasticity and mental retardation presenting in the first or second year (Goodman *et al.* 1975; Brandt *et al.* 1978; Kyllerman and Steen 1977; Brandt *et al.* 1979). By the second half of the first decade the motor disorder may be so severe that the patients are completely helpless even though intelligence is normal (Brandt *et al.* 1979). A fatal course has been reported which resembled Reye's syndrome (Goodman *et al.* 1977).

Pathology

Neuronal loss has been reported localized to the putamen and caudate nuclei and to a lesser extent in the cerebral cortex. Fatty changes were present in the liver, kidney, and myocardium (Goodman *et al.* 1977). Histological examination of an affected foetus at 22.5 weeks' gestation revealed slight alteration in the size distribution of nuclei in the stratum which contained many prominent and small darkly staining nuclei (Goodman *et al.* 1980*a*).

Molecular defects

Glutaric acid is the free acid derived from glutaryl-CoA which is an intermediary metabolite in the catabolic pathways of tryptophan, lysine and hydroxylysine. In glutaric aciduria type I there is a deficiency of the activity of the mitochondrial FAD-dependent enzyme glutaryl-CoA dehydrogenase which catalyses the conversion of glutaryl-CoA to gluta-conyl-CoA (Fig. 5.8) (Goodman *et al.* 1975). It has not been possible to separate the activity of this enzyme from the enzyme catalysing the next catabolic step, namely the decarboxylation of glutaconyl-CoA to crotonyl-CoA (Goodman *et al.* 1975). The enzyme can be considered, therefore, to catalyse the conversion of glutaryl-CoA to crotonyl-CoA and CO_2. The identity of these enzymes is further supported by the observation that in glutaric aciduria type I the activity of both enzymes is deficient (Gregersen *et al.* 1977). Deficiency of glutaryl-CoA dehydrogenase activity has been demonstrated in leucocytes, fibroblasts, and liver (Goodman and Kohloff 1975; Christensen and Brandt 1978). There is accumulation of glutaric, 2-hydroxyglutaric and glutaconic acids (Goodman *et al.* 1977; Gregersen *et al.* 1977).

Diagnosis

The diagnosis is suggested by the demonstration of accumulated meta-bolites in the urine (see 'Molecular defects', above) by gas–liquid chromatography and mass spectrometry of the peaks collected by liquid partition chromatography after conversion to trimethylsilyl derivatives, and confirmed by demonstration of a severe deficiency in the activities of leucocyte or fibroblast glutaryl-CoA dehydrogenase.

Treatment

Some temporary improvement has been reported by treatment with a low protein diet, riboflavin, and Lioresol (4-amino-3-(4-chlorophenyl) butyric acid). The last-named is an analogue of GABA (4-aminobutyric acid). The rationale is that symptoms of glutaric aciduria type I are attributable to damage of the basal ganglia which have the highest concentration of GABA and GABA synthetase, the latter being inhibited by glutaric acid (Stokke *et al.* 1976; Brandt *et al.* 1979). Response to treatment however has been variable (Goodman *et al.* 1975; Kyllerman and Steen 1977; Whelan *et al.* 1979; Brandt *et al.* 1979).

Genetic counselling

The disorder has an autosomal recessive mode of inheritance, as evidenced by the demonstration of reduced activity of glutaryl-CoA dehydrogenase in leucocytes and fibroblasts of heterozygotes (Goodman and Kohloff 1975; Christensen and Brandt 1978). Parental consanguinity has been reported (Brandt *et al.* 1979).

Two pregnancies at risk for glutaric aciduria type I have been monitored by amniocentesis (Goodman *et al.* 1980*a*). In both, amniocentesis was carried out at 15 weeks' gestation. In the first pregnancy an affected foetus was predicted because of an elevated concentration of amniotic fluid glutaric acid (4.35 μg/ml; controls, mean 0.28 $\pm$ S.D. 0.11) and because cultured amniotic cells had undetectable glutaryl-CoA dehydrogenase activity.

The pregnancy was terminated at 22.5 weeks and the diagnosis confirmed in the abortus by the demonstration of markedly increased glutaric acid concentration in the liver, kidney and brain, and of considerable reduction in hepatic and renal glutaryl-CoA dehydrogenase activities.

In the second pregnancy, a normal foetus was accurately predicted since amniotic fluid glutaric acid was undetectable and cultured amniotic cell glutaryl-CoA dehydrogenase activity was within normal limits.

Glutaric aciduria type II and ethylmalonic–adipic aciduria

This rare disorder has presented with respiratory distress and tachypnoea within a few hours of birth. Hypoglycaemia, acidosis, hyperammonaemia, and convulsions have preceded death, usually within four days, and have occurred in spite of aggressive treatment (Przyrembel *et al.* 1976; Gregersen *et al.* 1980; Sweetman *et al.* 1980). However, a single case which may have had the same disorder survived until five months (Goodman *et al.* 1979). The skin, urine, and blood have been reported to have a disagreeable odour likened to sweaty feet.

In addition to glutaric acid, a complex pattern of other organic acids has been reported in the urine. In the patients described by Pryzrembel *et al.* (1976) and Gregersen *et al.* (1980) and in two patients described by Sweetman *et al.* (1980), the predominant organic acids were glutaric and lactic acids. There was also considerable elevation in the urinary excretion of other dicarboxylic acids, including ethylmalonic and suberic acids, and of other hydroxy acids including 2-hydroxybutyric, 2-hydroxyisovaleric, 2-hydroxyglutaric, and *p*-hydroxyphenyllactic acids. Urinary volatile acids were also increased. These included isobutyric, butyric, 2-methylbutyric, and isovaleric acids. Sarcosine was increased in the urine in three patients (Goodman *et al.* 1979; Sweetman *et al.* 1980; Gregersen *et al.* 1980).

In the plasma, the concentrations of glutaric acid and other dicarboxylic acids were increased in two of the three patients (Przyrembel *et al.* 1976; Sweetman *et al.* 1980). Elevated also were the plasma concentrations of the hydroxy acids and 2-hydroxybutyric acids, of long-chain fatty acids and of volatile acids. In two patients there were elevations in the plasma and urine amino acids citrulline, lysine, taurine, ornithine, and proline, while several other amino acids were increased in urine, but not plasma, suggesting renal tubular damage.

Przyrembel *et al.* (1976) reported that cultured fibroblasts from their patient had defective conversion of ^{14}C-labelled glutaric acid, valine, leucine, isoleucine, and 2-ketoisocaproic acid to ^{14}CO$_2$, but that conversion of ^{14}C-pyruvic acid to ^{14}CO$_2$ was normal.

Activity of glutaryl-CoA dehydrogenase has been found to be normal in fibroblasts (Goodman *et al.* 1979; Gregersen *et al.* 1980) but deficient in liver and kidney (Goodman *et al.* 1979).

The disorder could result from a defect in a common factor of numerous dehydrogenases, for example in an electron transferring flavoprotein required for transferring electrons from fatty acyl-CoA dehydrogenase and from sarcosine oxidase to the electron transport chain and to molecular oxygen in mitochondria. Alternatively, there may be a defect in synthesis or transport of a cofactor or enzyme subunit common to the enzymes. The enzyme blocks and metabolic consequences resulting from deficiency in the dehydrogenases implicated in glutaric aciduria type II are illustrated in the paper by Sweetman *et al.* (1980).

All but one (Gregersen *et al.* 1980) proven cases have been males, and there has been no parental consanguinity. Inheritance therefore could be either sex-linked or autosomal recessive. In the patients reported by Mantagos *et al.* (1979), and Goodman *et al.* (1980*b*) there were only small excesses of glutaric acid in the urine, and while the latter group labelled the disorder 'glutaric aciduria type II', the former group emphasized the increased excretion of ethylmalonic and adipic acids by labelling the condition 'ethylmalonic–adipic aciduria'.

Hydroxylysinuria

This disorder was first reported in a brother and sister. The patients were born to non-consanguineous parents and had very similar clinical features (Benson *et al.* 1969*b*, 1970; Swift *et al.* 1973). Progress was normal until 12–19 months after birth when neurological signs were noted. These included trembling, hyperkinesis, and myoclonic spasms. Developmental progress ceased and there was rapid neurological regression. They developed epilepsy (major, minor and myoclonic) and antisocial behaviour such as destructiveness. After a few years, regression ceased. The boy's IQ at 18 years and the sister's IQ at 15 years were 40 and 30, respectively, but they had no other neurological signs. They had persistent non-peptide hydroxylysinuria (10–12.3 mg/24h; normal, undetectable) even when on a non-meat diet. Since no hydroxylysine could be demonstrated in the blood, a renal origin for the hydroxylysinuria was considered possible. The brother also had excessive excretion of urinary citrulline.

A tolerance test was carried out on one of the patients and on a control by the oral administration of D-hydroxylysine (25 mg/kg body weight). The patient excreted more than twice as much free hydroxylysine in the first hour (17.9 per cent of load) as the control (7.5 per cent of load). However, during the first 24h both patient and control excreted similar proportions (40.8 per cent and 49.6 per cent). The plasma levels were measured for the first 2.5h after the oral load and were found to be similar in patient and control. The authors (Swift *et al.* 1973) concluded that these data do not suggest that there is a block in further metabolism of hydroxylysine.

Parker *et al.* (1970*a*) reported a four-year-old girl with hydroxylysinuria who had similar clinical features (hyperactivity, myoclonic and major convulsions, and mental handicap). Her blood hydroxylysine concentration was 0.13 mg/100 ml. (normally undetectable).

Two further unrelated male patients aged 17 and 18, both of non-consanguineous parents, with severe mental handicap, major convulsions in childhood, but not later, with hydroxylysinuria, were reported by Hoefnagel and Pomeroy (1970). Only a trace of hydroxylysine was detected in fasting plasma.

Hydroxylysinuria with raised serum-free hydroxylysine ranging from 0.3–0.6 mg/100 ml was reported in a five-year-old girl with trisomy-21 by Goodman and Browder (1970) and by Goodman *et al.* (1972), who suggested that hydroxylysinuria is due to a mutation of a protein responsible for the degradation of free hydroxylysine.

Hyperpipecolic aciduria

Only two patients have been reported with this disorder. However, the clinical and biochemical features were sufficiently similar to suggest that they suffered from the same entity (Gatfield *et al.* 1968; Thomas *et al.* 1975). The two male patients appeared normal initially, but regressed neurologically from about six months of age until death at two and two and a half years. In addition to severe mental retardation, both patients had generalized hypotonia, hepatomegaly, and abnormal optic discs (pale; dysplastic) and eye movements (spontaneous horizontal nystagmus, conjugate jerking, and pendular motions). Serum pipecolic acid concentration was increased approximately ten times above normal mean values. The concentration of other amino acids including lysine were normal in the serum. One patient only (Gatfield *et al.* 1968) had mild generalized amino aciduria. Pipecolic acid was present in the urine and in cerebrospinal fluid in low concentration.

Oral loading with DL-pipecolic acid produced higher serum concentrations and slower rates of fall in patients than controls, but loading with L-lysine caused normal increases in serum lysine without increasing pipecolate levels. The last observation suggests that pipecolic acid is not a major metabolite of lysine (Fig. 5.8).

Necropsy was performed in only one patient (Gatfield *et al.* 1968) and revealed demyelination at all levels of the nervous system, with some lipid accumulation in glia, and mild neuronal degeneration. Histological examination of the enlarged liver showed fibrosis with areas of focal necrosis. The brain contained four types of lysine dipeptides (Gatfield and Taller 1971). No beneficial therapy was identified. The mode of inheritance is unknown.

A lower elevation of serum pipecolic acid than that occurring in hyperpipecolic acidaemia was reported in two patients with Zellweger's syndrome (Danks *et al.* 1975c).

Disorders of collagen

Collagen is largely responsible for the structure and strength of connective tissues including tendons, cartilage, fascia, and bone. It is also found in glomerular filtration membrane, heart valves, and cornea. It is the most abundant protein in the body. Recently defects in its biosynthesis or structure have been identified in several types of Ehlers–Danlos syndrome (EDS) and osteogenesis imperfecta (Prockop *et al.* 1979; Bornstein and Byers 1980; Minor 1980).

Structure of collagen

The basic structure of collagen molecules is the triple helix consisting of three polypeptide subunits or α-chains (Fig. 5.9a). Each α-chain is twisted individually in a left-handed helix with three amino acids per turn and the three chains are twisted round each other in a right-handed super-helix about 300 nm long and 1.5 nm in diameter. The α-chains contain about 1000 amino acid residues and are helical except in small sequences at each end (the telopeptides). Every third amino acid is glycine. Thus the formula of an α-chain excluding the telopeptides can be represented as $(X-Y-Gly)_{333}$. About 100 of the X residues are proline and about 100 of the Y residues are hydroxyproline. The presence of glycine in the third space is essential since it is the only amino acid small enough to be accommodated in each chain where the three chains come together in the centre of the

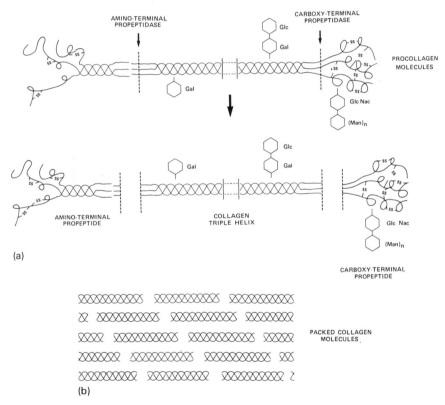

(a)

(b)

Fig 5.9. (a) Triple helical structure of collagen: conversion of procollagen to collagen by amino-terminal and carboxy-terminal propeptidases. (b) Mechanism of 'D spacing'. Each collagen molecule has an axial stagger of 1, 2, 3, or 4 D periods where each collagen molecule is 4.4 D units long. On electron microscopy, the spaces between the ends of the molecules are stained, and appear dark (0.6D), while the areas of molecular overlap (0.4D) appear light, giving alternate light and dark zones, or periodicity.

superhelix. The cylic residues of proline and 4-hydroxyproline give strength and stability to the molecule and are important for maintenance of the rigid helical structure. About two-thirds of the X and Y positions are occupied by other amino acids, some of which have hydrophobic or charged residues with side chains pointing away from the centre of the superhelix. These are important in cross-linking both of the intramolecular α-chains and in linking triple helices to each other. The cross links are from aldehyde residues synthesized extracellularly from ϵ-amino groups of lysyl or hydroxylysyl residues, a reaction which is catalysed by lysyl oxidase. The cross links are of two major kinds: firstly, intramolecular cross-link between two α-chains by aldol condensation of two aldehydes; and secondly, by formation of a Schiff base between the aldehyde group (of lysine, hydroxylysine or glycosylated hydroxylysine) and an ϵ-amino group of a second lysine, hydroxylysine or glycosylated hydroxylysine. Further reactions have been described which have the effect of stabilizing the Schiff base. These include a shift in the double bond to form a ketone, or biosynthesis of a new peptide bond by hydration and oxidation, or by reaction of histidine with both an aldol and a hydroxylysine residue. The latter cross links have been attributed to artefacts during borohydride reduction but they have also been isolated from collagen using other methods.

Types of collagens, subunit composition and distribution

The nomenclature for subunit composition of collagen is unsatisfactory. Type I collagen is the major connective tissue protein of skin, tendons, bone ligaments, dentine and arteries and represents up to 90 per cent of body collagen. It is made up of two identical chains, $\alpha_1(I)$ and one different chain, $\alpha_2(I)$. Its formula is therefore $[\alpha_1(I)]_2 \alpha_2(I)$. Type II collagen is largely found in cartilage but has been detected in the cornea, vitreous body and retina. It consists of three identical $\alpha_1(II)$ chains written as $[\alpha_1(II)]_3$. Type III collagen represents 10 per cent of adult skin collagen and 50 per cent of foetal skin collagen. It is also found in large arteries, intestine, muscle, lung and liver; its structure is $[\alpha_1(III)]_3$. Basement membranes of renal glomeruli and Descemet's membrane of the cornea contain Type IV collagen which contains $\alpha_1(IV)$ and $\alpha_2(IV)$ subunits, but has an undetermined structure (Tryggvason *et al.* 1980; Alitalo *et al.* 1980). Morphologically, type IV collagen is laminar and not fibrillar. The subunits contain globular regions that are cross-linked by disulphide bonds and lack triple helical structure.

Two other collagen subunits designated αA and αB have been discovered in small quantities in placenta (Burgeson *et al.* 1976) and other tissues (Kumamoto and Fessler 1980; Welsh *et al.* 1980). Whether these minor subunits aggregate into collagen (sometimes designated type V)

αA $(\alpha B)_2$, $(\alpha A)_3$, $(\alpha B)_3$ or $(\alpha A)_2$ αB is not yet established. Burgeson *et al.* (1976) favour the $(\alpha A)_2$ αB combination. The subunit structures of the principal collagen molecules and features of primary structure of subunits are summarized in Table 5.2

Structure and arrangement of fibrils

The triple helices in collagen fibrils are arranged parallel to each other and are staggered by slightly less than one quarter of their lengths (Fig. 5.9b). More precisely, each molecule for type I collagen has an axial stagger of 1, 2, 3 or 4 D periods where D = 1/4.4 of the molecular length (i.e. each collagen molecule is 4.4D units long). On electron microscopy, this staggering gives rise to cross-striation often referred to as 'D spacing'. The collagen fibrils are packed in parallel bundles in tendons and ligaments but form a lattice-like pattern in cartilage and intervertebral discs where they are lubricated by glycoprotein. In the cornea, the fibrils cross each other at right angles – a structure which is compatible with its lucid clarity. In bone the fibrils are arranged along the lines of stress like girders of a bridge and are reinforced with mineral salts; in skin and fascia they are interwoven.

Biosynthesis and secretion of procollagen

Initially, collagen is synthesized as a larger collagen precursor known as procollagen with extensions known as propeptides at each end (Fig. 5.9a). The procollagen chain $\alpha_1(I)$ (pro $\alpha_1(I)$) has an amino terminal extension with a molecular weight of about 20 000 and a carboxyl terminal extension with

Table 5.2 *Subunit structure of the principal collagen molecules, and features of primary structure of subunits*

Collagen type	Subunit structure	Features of primary structure
I	$[\alpha_1(I)]_2\alpha_2(I)$	Low hydroxylysine content; low glycosylation
II	$[\alpha_1(II)]_3$	High hydroxylysine content; high glycosylation
III	$[\alpha_1(III)]_3$	Low hydroxylysine, high hydroxyproline content; low glycosylation; disulphide bonds near carboxyl terminal
IV	Contains $\alpha_1(IV)$ and $\alpha_2(IV)$ chains	High hydroxylysine content; highly glycosylated; contains 3-hydroxyproline residues and some collagen extensions as globular regions
V	Contains αA and αB chains; possibly $A(\alpha B)_2$	High hydroxylysine content; highly glycosylated

a molecular weight of about 35 000 (Fessler and Fessler 1978; Bornstein and Traub 1979; Bailey and Etherington 1980).

Both the amino terminals and carboxy terminals of procollagen type I–III contain cysteine residues, which are absent from collagens of the same types. At the carboxy-terminals of the propeptides these residues make both inter- and intra-chain disulphide bonds, but in the amino-terminals of the propeptides, only intra-chain bonds. The carboxy-terminal propeptide is glycosylated with *N*-acetylglucosamine, mannose, and other sugars (Fig. 5.9a) (Kivirikko and Myllyla 1979). The function of propeptide terminals includes direction of helix formation and feedback inhibition of collagen synthesis (Wiestner *et al*. 1979).

The messenger RNA for type I collagen has been isolated and parts of the message have been cloned (Lehrach *et al*. 1978; Sobel *et al*. 1978).

Intracellular modification of procollagen

Initially, the propeptides themselves are synthesized with an additional 'signal' sequence at the amino-terminal which apparently directs movement of the polypeptides through the rough endoplasmic reticulum. These sequences are removed by proteases in the cisternae of the endoplasmic reticulum. The procollagen is then hydroxylated by three enzymes: two convert prolyl residues to either 4-hydroxyproline or 3-hydroxyproline, and the third converts lysyl residues to hydroxylysine. The chains then become glycosylated, firstly by the addition of galactose to hydroxylysyl residues catalysed by galactosyl transferase, and then by addition of glucose to some of the newly attached galactose residues, catalysed by glucosyl transferase.

A further intracellular modification is the formation of inter-chain disulphide bonds of the propeptides. It appears that hydroxylation of prolyl residues and formation of disulphide bonds are essential for helical formation.

Secretion to the extracellular space through the Golgi apparatus (Olsen and Prockop 1974) depends on the procollagens adopting a triple-helical formation. Non-helical molecules accumulate in the cisternae of the endocytoplasmic reticulum and are secreted only slowly.

Extracellular modification

The conversion of procollagen to collagen by removal of propeptide terminals is catalysed by procollagen aminoprotease (aminopeptidase) which catalyses excision of the aminopropeptide terminal, and procollagen carboxyprotease (carboxypeptidase) which catalyses excision of the carboxypropeptide terminal (Fig. 5.9a) (Leung *et al*. 1979).

Following cleavage of the propeptide terminals the collagen molecules spontaneously adopt a fibrillar structure which resembles that of mature collagen fibrils but has poor tensile strength until cross-linking has taken place. This requires several steps, the first of which is oxidative deamination of ε-amino groups of some lysyl and hydroxylysyl residues catalysed by lysyl oxidase to yield aldehyde groups. The formation of cross-links between either two aldehyde groups or between an aldehyde and amino group, and further reactions were described above under 'Structure of collagen' (p. 234).

Clinical disorders of collagen

In Ehlers–Danlos syndrome (EDS) types I–III (McKusick 1972), which have dominant inheritance, no biochemical abnormality has yet been demonstrated. EDSI or the *gravis* type is characterized by hyperextensile skin and joints, fragile and easily bruised skin, and friable tissues, including foetal membranes (causing frequent preterm birth).

In EDSII or the *mitis* type, hyperextensibility of the skin and joints is less marked. Tissues are not extensively friable.

In EDSIII there is marked hyperextensibility of the joints but only slight or absent skin changes.

(a) Ehlers–Danlos syndrome type IV; ecchymotic form; arterial type

EDS type IV differs from other types of EDS since connective tissues are fragile but only minimally hyperextensible (McKusick 1972). There is a tendency to sudden rupture of arteries or of the intestinal tract causing sudden death. The skin bruises easily and becomes thin, pale, and shiny with a prominent venous pattern.

In EDSIV a deficiency has been reported of type III collagen from zero to a few per cent of normal (Pope *et al.* 1975; Clark *et al.* 1980).

The pathogenesis is consistent with this deficiency since type III collagen is found predominantly in skin, large arteries and intestine. Cyanogen bromide (CNBr) cleaves collagen at accessible methionine residues. Analysis of CNBr-induced peptides from skin collagen of five patients with EDSIV by polyacrylamide gel electrophoresis in sodium dodecyl sulphate showed absence or reduction of a peptide consisting of 113 amino acids which is characteristic of collagen type III. Moreover, ^{14}C-labelled collagen of fibroblasts from five patients with EDS type IV produced deficient amounts of collagen type III. Amino acid analysis of aortic tissues showed ratios of hydroxylysine to other residues of 3.3 in controls, but only 0.5 in a patient with EDSIV – indicating considerable reduction of collagen, most of which is normally type III. Reduced synthesis of type III collagen may be demonstrated in cultured fibroblasts (Aumailley *et al.* 1980).

It is interesting that although the pattern of inheritance of EDS IV is usually dominant, Pope *et al.* (1975) reported partial deficiency of type III collagen in fibroblast cultures from the parents, suggesting that in these patients inheritance was autosomal recessive. Genetic heterogeneity of EDSIV therefore seems likely.

Prenatal diagnosis by demonstration of lack of collagen type III synthesis by cultured amniotic cells is theoretically possible.

(b) Ehlers–Danlos syndrome type V or X-linked form

The outstanding feature is usually hyperextensible skin, though some patients have had mild problems with fragile skin, easy bruising and poor wound healing (Beighton 1968). In most cases any joint hypermobility has been limited to the digits.

In patients with X-linked EDSV, who in addition to hyperextensible skin had kyphosis, pes planus and 'floppy mitral valve syndrome', a deficiency of cultured fibroblast lysyl oxidase activity was reported by Di Ferrante *et al.* (1975). The deficiency was both in fibroblasts and in dialysed and lyophilized culture medium. However, not all patients with EDSV have had lysyl oxidase deficiency (Siegel *et al.* 1979). Moreover, lysyl oxidase deficiency has also been reported in patients with X-linked cutis laxa without other features of EDS (Byers *et al.* 1980).

X-linkage of EDSV has been established by pedigree analysis.

Prenatal diagnosis by demonstration of deficient cultured amniotic cell lysyl oxidase activity is theoretically possible, when the index patient has been shown to have the enzyme deficiency.

(c) Hydroxylysine-deficient collagen disease; lysylprotocollagen hydroxylase deficiency; Ehlers–Danlos syndrome type VI; ocular form

Clinical features

Five patients in whom the diagnosis has been confirmed have had hyperextensible skin and joints, recurrent dislocations, kyphoscoliosis of variable severity, hypotonia, delayed motor development, microcornea, bluish sclerae, excessive bruising and atrophic scars (Krane *et al.* 1972; Pinnell *et al.* 1972; Sussman *et al.* 1974; Steinmann *et al.* 1975*a*; Elsas *et al.* 1978).

The diagnosis has been confirmed by demonstrating a reduction of skin hydroxylysine content to less than 10 per cent of normal. In the two patients reported by Pinnell *et al.* (1972), hydroxylysine content of bone and fascia was diminished, but that of cartilage was normal. Cultured fibroblasts from patients had marked decrease in lysyl hydroxylase activity; that is 10–17 per cent of control values (Elsas *et al.* 1978) or 8–9 per cent (Steinman *et al.* 1975*a*), while fibroblasts from parents had a partial deficiency (66 per cent and 39 per cent) (Elsas *et al.* 1978).

In patients of Pinnell *et al.* (1972) urinary total hydroxylysine excretion was reduced.

The disorder has an autosomal recessive mode of inheritance and theoretically could be diagnosed prenatally by demonstrating deficient cultured amniotic cell lysyl hydroxylase activity.

(d) Ehlers–Danlos syndrome type VII

Three patients with hypermobile joints, multiple dislocation of joints – especially bilateral dislocation of the hips – small stature and stretchable, velvety skin were reported by Lichtenstein *et al.* (1973).

These patients were shown to have increased amounts of procollagen in extracts of skin and tendons. This was demonstrated by electrophoretic separation of the acid-extractable components on SDS electrophoresis, followed by densitometric scanning of Coomassie blue-stained zones. Initially the defect was thought to be due to a deficiency of procollagen aminopeptidase as occurs in dermatosparaxic cattle (see 'Animal models', p. 242). On reinvestigation of new biopsy samples, a structural mutation in the pro α_2(I) chain was found, which prevented normal enzymatic removal of the aminopropeptide terminal from pro α_2 chains (Steinmann *et al.* 1980).

(e) Osteogenesis imperfecta

This is a heterogenous group of disorders characterized by osteoporosis leading to fractures, sometimes with additional features such as blue sclerae, hearing loss, and defective teeth.

Classification in most cases is based on clinical and genetic criteria, with only a few patients having been characterized at molecular level. The largest group, which has been designated osteogenesis imperfecta type I (OI I), has dominant inheritance of osteoporosis, multiple fractures, kyphoscoliosis, blue sclerae, and presenile conductive or sensorineural hearing loss (Quisling *et al.* 1979; Sillence *et al.* 1979a,b). A second group, osteogenesis imperfecta type II (OI II), presents in the newborn period with neonatal fractures. The patients die before, or soon after, birth. They have broad, crumpled femora. Most are thought to have autosomal recessive inheritance. A third group, osteogenesis imperfecta type III (OI III), with fractures at birth, develop severe progressive limb and spinal deformity. Blueness of the sclerae is less marked than the first group and tends to diminish with age. Inheritance of this group is unknown; there may be dominant and recessive genotypes. A fourth group of Sillence *et al.* (1979a,b) has an autosomal dominant mode of inheritance of osteoporosis leading to fractures with normal sclerae.

Molecular defects

In a patient with the congenital form (OI II), Penttinen *et al.* (1975) and Steinmann *et al.* (1979) showed that cultured skin fibroblasts had reduced type I/type III collagen ratios. Reduced synthesis of type I collagen, which is the predominant form in bone, would be consistent with these data.

Reduced type I/type III collagen ratios in various forms of OI have also been reported by Turakainen *et al.* (1980) and Krieg *et al.* (1981). In another patient with the congenital form (OI II), Trelstad *et al.* (1977) found increased contents of hydroxylysine but normal ratios of type I/type III collagen synthesized by cultured fibroblasts.

In cultured fibroblasts from a 21-month-old boy with moderately severe OI born from consanguineous parents, Nicholls *et al.* (1979) found total deficiency of type I collagen α_2 chain, with the parents having a partial deficiency.

Excess mannose in the carboxy propeptide terminal in skin fibroblasts from a patient with multiple fractures at birth was reported by Peltonen *et al.* (1980).

Genetic counselling and prenatal diagnosis

It is clear that logical classification of OI will only be possible when the precise molecular defects have been identified for the heterogenous clinical types. At present, genetic counselling must depend on whether there is clinical or biochemical evidence of dominant or recessive inheritance by study of the family pedigree or by demonstrating biochemical abnormalities in the index patient, which exists partially in both parents.

Where abnormalities of collagen synthesis can be demonstrated in cultured fibroblasts from the index patients, prenatal diagnosis is theoretically possible using the same techniques on cultured amniotic cells. However, prenatal diagnosis of lethal perinatal osteogenesis imperfecta has been reported (Byers *et al.* 1981; Shapiro *et al.* 1982) using a combination of ultrasonographic studies and foetal radiographs. The ultrasonographs revealed fractures and short limbs at 17 and a half weeks' gestation and progressive hydrocephalus at 19 and 21 weeks' gestation. A foetal radiograph at 21 weeks revealed that the lower limb bones were deformed. The pregnancy was terminated after collection of amniotic fluid, and amniotic cells, foetal skin fibroblasts and foetal bone cells were cultured. All three cell types were found to synthesize collagen, but, in contrast to controls, the cells from the affected foetus retained a considerable proportion of the type I collagen synthesized. Furthermore, the constituent pro α-chains of the type I procollagen secreted into the culture medium had altered electrophoretic mobilities.

Elejalde and Elejalde (1983) also reported prenatal diagnosis of the lethal perinatal form by ultrasonographic examination of the foetus at 17 weeks' gestation.

Animal models

Mottled mice

Certain genotypes at the *mottled* locus on the X-chromosome of mice are associated with fragile skin, emphysema, bony abnormalities, and aortic aneurysms and rupture.

These mice have defective collagen cross-linking due to deficiency of reactive-aldehyde-derived linkages. Affected males have been shown to have 5–30 per cent of normal lysyl oxidase activity (Rowe *et al.* 1974). These mice also have a disturbance in the transport of copper (which is essential for lysyl oxidase activity), leading to diminished tissue copper concentration in brain and liver (Hunt 1974). However, in other tissues there is accumulation of copper but this is unavailable because it is bound to specific proteins (Port and Hunt 1979). These mice are more similar to humans with Menkes' disease (q.v.) than EDSV.

Dermatosparaxis

This connective tissue disorder of cows and sheep with an autosomal recessive mode of inheritance presents with skin which is easily torn. It has also been reported in a Himalayan cat (Counts *et al.* 1980). A defect has been demonstrated in the removal of the amino terminal propeptide from procollagen. This produces collagen fibrils with irregular cross-sections and with decreased tensile strength. The tissues of dermatosparaxic cattle have severe deficiency in the activity of procollagen aminopeptidase (Lapiere *et al.* 1971). As stated above, defective conversion of procollagen to collagen also occurs in human EDSVII where the defect appears to be a structural mutation of the pro α_2 chain.

Lathyrism

Lathyrism is a disorder of connective tissue, characterized by kypho-scoliosis, diminished tensile strength of collagen and the formation of aortic aneurysms, caused by feeding ß-aminopropionitrile (BAPN) to growing animals. BAPN and other nitriles *in vitro* and *in vivo* inhibit lysyl oxidase activity (Narayanan *et al.* 1972). The resulting deficiency in reactive aldehyde groups causes defective cross-linking, and decreased strength of collagen (Levene and Gross 1959) and elastin (Foster *et al.* 1979).

DISORDERS OF PROLINE AND HYDROXYPROLINE METABOLISM

Two forms of hyperprolinameia have been characterized: Type I due to deficiency in the activity of proline oxidase (dehydrogenase); and type II due to deficiency in the activity of Δ'-pyrroline 5-carboxylic acid dehydrogenase.

Hyperprolinaemia type I; proline oxidase deficiency

In the early pedigrees described with raised plasma proline without an increased excretion of Δ'-pyrroline 5-carboxylic acid (Fig. 5.7), there was an association with renal dysplasia (Schafer *et al.* 1962; Kopelman *et al.* 1964; Efron 1965) and other disorders including aniridia and ocular dystrophy (Fusco *et al.* 1976). Other reported defects include epilepsy and mental retardation (Scriver *et al.* 1961; Piesowicz 1968). However, since the condition can occur in families without any clinical abnormalities (Fontaine *et al.* 1970; Mollica *et al.* 1971), the above-named associations are probably coincidental.

Iminoglycinuria (raised excretion of glycine, proline and hydroxyproline) occurs, as in normal people when plasma proline concentation is increased to between 0.51 mmol/l (Kopelman *et al.* 1964) and 2.68 mmol/l (Pavone *et al.* 1975). The normal range of plasma proline is 0.1–0.45 mmol/l (Scriver and Rosenberg 1973).

Incubation of liver extract from an affected individual with ^{14}C-proline showed decreased production of Δ'-pyrroline 5-carboxylic acid, but normal degradation of hydroxyproline (Efron 1965) indicating deficiency of proline oxidase (dehydrogenase) activity. The condition has an autosomal recessive mode of inheritance. Heterozygotes cannot be detected by oral loading tests with L-proline (Fontaine *et al.* 1970; Harries *et al.* 1971).

Treatment by low proline diet has been described (Harries *et al.* 1971) but since the condition is thought to be benign this is probably unnecessary.

Hyperprolinaemia type II; Δ'-pyrroline 5-carboxylic acid dehydrogenase deficiency

In the first cases reported there were notable associations with mild mental handicap, convulsions, and abnormal EEGs (Efron 1967; Emery *et al.* 1968; Selkoe 1969). The condition, however, may be benign since it occurs in normal healthy individuals (Goodman *et al.* 1974; Pavone *et al.* 1975), but the existence of variant forms has not been excluded.

The normal range of plasma proline is 0.10–0.45 mmol/l (Scriver and Rosenberg 1973) while individuals with hyperprolinaemia type II have had values usually higher than those with type I, from 1.0 to 3.5 mmol/l. Imminoglycinuria is constant (Goodman *et al.* 1974). Other compounds excreted include Δ'-pyrroline-3-hydroxy-5-carboxylic acid, Δ'-pyrroline 5-carboxylic acid (Applegarth *et al.* 1974; Goodman *et al.* 1974) and N-pyrrole 2-carboxylic acid (Applegarth *et al.* 1977).

Following oral loading of 4-hydroxy-L-proline (100 mg/kg) there is considerable delay in clearance of hydroxyproline from blood, accompanied by great excess in excretion of Δ'-pyrroline-3-hydroxy-5-carboxylate (Goodman *et al.* 1974; Similä 1979).

The condition is due to deficiency in the activity of Δ'-pyrroline 5-

carboxylic acid dehydrogenase (Fig. 5.7) which has been demonstrated in cultured fibroblasts and leucocytes (Valle *et al*. 1974, 1976, 1979*b*).

The condition has an autosomal recessive mode of inheritance. Heterozygotes have normal plasma proline concentrations but show a partial enzyme deficiency in leucocytes (Valle *et al*. 1976).

Prolidase deficiency; iminodipeptiduria; hyperimidodipeptiduria

Eight patients have been described with a variable combination of unusual facies, ptosis, prominent cranial sutures, and ulcerative dermatitis with scars. The majority also had splenomegaly, thrombocytopenia with purpura, and demineralization of the long bones. Some patients had hepatomegaly, one had corneal clouding with cataracts, while another had severe oedema. Most had some mental handicap, while one was severely retarded.

They all had in common increased excretion in the urine of numerous di- or tripeptides in which proline or hydroxyproline residues occupied the C-terminal position; for example, glutamylproline, glutamylhydroxyproline and tyrosylprolylproline (Goodman *et al*. 1968; Buist *et al*. 1972; Powell *et al*. 1974; Powell and Maniscalco 1976; Jackson *et al*. 1975*a*; Kodama *et al*. 1976*a*; Sheffield *et al*. 1977). The excretion of bound hydroxyproline in the urine was also considerably increased.

A six-year-old girl (Umemura 1978) and a 26-year-old man (Isemura *et al*. 1979) with similar biochemical abnormalities and with reduced erythrocyte and leucocyte prolidase activity had, however, normal clinical phenotypes. Goodman *et al*. (1968) found that collagen had normal amino acid content, but there was a marked increase in the aldehyde content.

After oral ingestion of gelatin, patients excreted more bound hydroxy-proline in the urine than controls. Heterozygotes had intermediate values (Powell and Maniscalco 1976).

Prolidase activity (measured using glycylproline as substrate) was absent or very low in leucocyte and erythrocyte lysates (Powell *et al*. 1974; Jackson *et al*. 1975*a*; Powell *et al*. 1977) or cultured fibroblasts (Sheffield *et al*. 1977) from patients, while prolinase activity (measured using prolyl-glycine as substrate) was normal.

The disorder has an autosomal recessive mode of inheritance, confirmed by the observations that parents had about half normal prolidase activities in erythrocytes and leucocytes (Powell *et al*. 1974; Umemura 1978; Isemura *et al*. 1979).

DISORDERS OF ß-AMINO ACID METABOLISM

Under this heading we discuss disorders of ß-alanine, carnosine and ß-aminoisobutyric acid metabolism.

The great majority of ß-alanine in the human body is either in the form of the dipeptide carnosine (ß-alanyl-L-histidine) which occurs mainly in skeletal muscle, or anserine (ß-alanyl-1-methyl-L-histidine). It also forms part of the coenzyme A molecule in the pantothenic acid component. The latter is probably derived entirely from dietary sources. Very little ß-alanine is free. Indeed, in the free state it is a neuroinhibitor acting on the cortex, brain stem, and spinal cord (Krnjević 1965).

Carnosinaemia; carnosinase deficiency

Five male patients have been reported with carnosinaemia. They appeared normal at birth but developed myoclonic jerks followed by generalized convulsions during the first six months. All had slowly progressive neurological regression leading to spasticity and in some cases blindness and deafness (Perry *et al.* 1967*c*; van Heeswijk *et al.* 1969; Murphey *et al.* 1973). A healthy sister of two affected brothers showed similar biochemical abnormalities (van Heeswijk *et al.* 1969; Murphey *et al.* 1973). Plasma carnosine concentration was increased in all subjects. Carnosinuria was constant and was influenced by dietary intake. Serum carnosinase activity was reduced to 0–12.5 per cent of normal in all patients studied. The normal sister of two affected brothers had serum carnosinase activity of 1 per cent of normal. Carnosinase activity was decreased in livers and spleens in two patients (Perry *et al.* 1967*c*; Murphey *et al.* 1973).

Two forms of carnosinase activity can be demonstrated on starch gel electrophoresis of normal tissues. In tissues of patients there is absence of the rapidly migrating component, but normal activity of the slower form (Murphey *et al.* 1973).

The demonstration of decreased carnosinase activity in parents of three patients suggests an autosomal recessive inheritance. Lack of clinical involvement in the girl with carnosinase deficiency is unexplained.

Normally carnosine is not detectable in plasma or urine in fasting individuals, but can appear after eating carnosine-containing foods (Perry *et al.* 1967*c*).

On a meat-free diet, plasma carnosinaemia persisted in affected patients, but the urinary excretion of carnosine decreased from 117–295 μmol per 24h to 57–92 μmol/24h.

Hyper ß-alaninaemia

The only child with this disorder was a male infant who presented with somnolence from birth and *grand mal* seizures from the age of six weeks. Tendon reflexes were sluggish. There were feeding difficulties and he died at five months of age (Scriver *et al.* 1966*b*).

The concentration of ß-alanine in plasma was consistently elevated (0.020–0.051 μmol/ml), while normally it is undetectable. There was an elevated excretion in the urine of ß-alanine, taurine, ß-aminoisobutyric acid, and GABA. ß-alanine was present in the cerebrospinal fluid at a concentration of 0.045 μmol/ml.

ß-alanine accumulation could, in theory, have been due to deficiency of ß-alanine:2-oxoglutarate aminotransferase activity, whilst ß-aminoiso-butyric acid and taurine (which is also a ß-aminoacid, but a sulphonic rather than carboxylic acid) may have been increased in the urine because of competition for tubular reabsorption with ß-alanine at ß-amino-acid-specific sites.

Administration of pyridoxine decreased excretion of ß-alanine, ß-amino-isobutyric acid, and taurine.

ß-alanine, ß-aminoisobutyric acid, taurine, GABA, and homocysteine have been found in the urine of patients treated with 6-azauridine triacetate (Slavik *et al.* 1973).

ß-aminoisobutyric aciduria

About 4–10 per cent of individuals with European ancestry and about 25–46 per cent of Japanese or Chinese ancestry persistently excrete more D-ß-aminoisobutyric acid (BAIB) than others in their populations. BAIB excretors are symptomless. Klujber *et al.* (1969) found a higher incidence of excretors amongst patients with Down's syndrome.

The plasma concentration of BAIB is about two to four times higher in excretors than non-excretors (Armstrong *et al.* 1963; Solem *et al.* 1975).

In excretors there is a marked increase in renal clearance of BAIB. Oral loading increases urinary excretion of BAIB in non-excretors but not in excretors. Following loading of BAIB (20 mg/kg) 90–96 per cent was recovered in urine of excretors within 8h, but only 55–72 per cent from urine of non-excretors.

In normal persons the ratio of L-BAIB:D-BAIB is 4:1, whilst in excretors it is 1:99 (Solem 1974; Solem *et al.* 1974).

Kakimoto *et al.* (1969) found that the activity of D-BAIB:pyruvate aminotransferase was low or absent in livers of seven excretors and was present but variable in eight controls (0.06–0.73 μmol/h/g). The cut-off point of urine BAIB was taken as 0.4 mmol/g creatinine.

DISORDERS OF FOLIC ACID METABOLISM

Derivatives of pteroylglutamic acid or folic acid serve as cofactors in several one-carbon transfer reactions. The remethylation of homocysteine to methionine has already been discussed. Other reactions include the biosynthesis of deoxythymidylic acid from deoxyuridylic acid, the introduc tion of carbon atoms 2 and 8 during the biosynthesis of purines, the glycine cleavage reaction (discussed under nonketotic hyperglycinaemia), and the salvage of a one-carbon moiety during histidine catabolism. The numerous derivatives of folic acid are produced by a three-stage reduction of its pyrazine ring, by six possible substitutions at either or both N^5 and N^{10}, or by additions of glutamic acid residues linked to the gamma carboxyl group of the glutamic acid residues forming varying lengths of a gamma glutamyl peptide chain, or polyglutamate.

Of five main groups of disturbances of folate metabolism, deficiency of 5, 10-methylenetetrahydrofolate reductase and of 5-methyltetrahydro-folate:homocysteine methyltransferase have been discussed under homocystinaemia/homocystinuria. In a rare third group there is a congenital defect in absorption of folate from the small intestine suggesting the existence of a specific carrier mechanism for folate. This defect presents with failure to thrive and megaloblastic anaemia in the first year (Luhby *et al.* 1961; Lanzkowsky 1970; Santiago-Borrero *et al.* 1973). In this section we discuss formiminotransferase-cyclodeaminase deficiency, and dihydrofolate reductase deficiency.

Formiminoglutamic aciduria; formiminotransferase deficiency; (?) cyclo-deaminase deficiency

There has been considerable clinical heterogeneity among the patients reported.

Five patients with mental and physical retardation were reported by Arakawa (1970, 1974) and Arakawa *et al.* (1963). They had atrophy of the cerebral cortex with dilation of the cerebral ventricles, abnormal EEGs, increased urinary excretion of formiminoglutamic acid (FIGlu) after histidine loading (Fig. 5.11), and abnormally high serum cobalamin concentrations. In four, neurological deterioration had been noted by 18 months of age. The fifth patient was an adult in a psychiatric institution. One child had folate-responsive megaloblastic anaemia. All but the adult had high serum folate concentration.

Liver biopsy formiminotransferase activity was 14–50 per cent of control normal values.

Niederwieser *et al.* (1974, 1976) reported two sisters with massive excretion of FIGlu and hydantoin-5-propionic acid in the urine which was not influenced by treatment with high doses of folic acid. One had normal intelligence and one was mentally retarded. Serum folate concentration and intestinal absorption of folates were normal.

Perry *et al.* (1975a) reported a brother and sister: the boy at 3.25 years had normal intelligence but was clumsy, walked with a broad base and had delayed speech development, but normal EEG. His eight-year-old sister had the same clinical features but her EEG was abnormal. Serum folate concentrations were normal, but high doses of folic acid produced a 75 per cent decrease in FIGlu excretion.

A 42-year-old female with intolerance to dietary carbohydrate, weight loss and normal intelligence and deficiency of erythrocyte and jejunal formiminotransferase was reported by Herman *et al.* (1969). Her urinary FIGlu excretion was about 10 per cent of that of patients reported by Niederwieser *et al.* (1974, 1976) and Perry *et al.* (1975*a*) and responded to folate therapy. Russell *et al.* (1977) reported a seven-year-old boy with hyperkinesis, speech retardation, and EEG abnormalities. He had normal serum B_{12} and folate concentrations. The increased FIGlu excretion was reduced when he was treated with methionine.

The enzymes formiminotransferase and cyclodeaminase in mammals are inseparable and consist of a single octameric bifunctional protein with two catalytic sites (MacKenzie 1979). They catalyse the last two steps in the salvage of the one-carbon unit in histidine catabolism, which may be written as:

Tetrahydrofolate (THF) + FIGlu → 5-formimino THF + glutamic acid and: 5-formimino TFH → 5,10-methenyl THF + NH_3

Since the assay carried out by Arakawa's group would not distinguish between these two activities, a possible explanation for the defect in the Japanese patients is a mutation causing slight loss of activity in the transferase, but considerable loss in the cyclodeaminase. This would lead to accumulation of 5-formimino THF or derivatives which would explain the rise in serum folate concentrations and moderate increased excretion of FIGlu. A block at the transferase site would cause a much greater accumulation of FIGlu and could explain a considerable FIGlu excretion in the patients described by Niederwieser *et al.* (1974, 1976) and Perry *et al.* (1975*a*).

Dihydrofolate reductase deficiency

Dihydrofolate reductase (DR) catalyses two sequential reactions, the reduction of folic acid to dihydrofolic acid, and the reduction of dihydrofolic acid to yield tetrahydrofolic acid. In both reactions, NADPH is the hydrogen donor.

The first patient reported with DR deficiency presented at six weeks of age with congenital megaloblastic anaemia (Walters 1967) which responded to 5-formyltetrahydrofolic acid (5-formyl THF) but not folic acid therapy. Liver DR activity was reduced to 35 per cent of normal. When followed up to the age of 19 years, he had mild mental handicap and was kept in custody because of repeated delinquency. Cultured fibroblast DR activity was within the normal range (Erbe 1979).

Two unrelated patients with congenital megaloblastic anaemia were reported by Tauro *et al.* (1976). In one, liver DR activity was absent when assayed conventionally, but normal when the reaction was carried out in the presence of 0.6M KCl.

The other patient had a partial deficiency in hepatic DR activity which rose from 0.27 to 0.49 nmol dihydrofolate reduced/min/mg protein (normal 1.0–1.7) in the presence of 0.6M KCl. Both children responded to intramuscular administration of 5-formyl THF.

DISORDERS OF TRYPTOPHAN METABOLISM

Xanthurenic aciduria; kynureninase deficiency

An increased incidence of asthma, urticaria, anaemia, and diabetes mellitus was reported in three members in each of two families and in six members in three generations in a third family by Knapp (1960).

A different clinical presentation was reported in two mentally retarded girls by O'Brien and Jensen (1963) and in two siblings by Tada *et al.* (1967*a*; 1968). Plasma amino acid concentrations were normal when fasting except for taurine, which was elevated in one patient but decreased on pyridoxine treatment (O'Brien and Jensen 1963).

Following oral tryptophan loading, there was a several-fold increase above normal in the urinary excretion of kynurenine, 3-hydroxykynurenine (Knapp 1960), kynurenic acid and acetylkynurenine (O'Brien and Jensen 1963; Tada *et al.* 1967*a*). The best discriminants were 3-hydroxykynurenine: 3-hydroxyanthranilic acid ratios.

In all patients the administration of pharmacological doses of pyridoxine or pyridoxal phosphate resulted in normal excretion patterns following tryptophan loading (Knapp 1960; O'Brien and Jensen 1963; Tada *et al.* 1967*a*).

Kynureninase catalyses the conversion of 3-hydroxykynurenine to 3-hydroxyanthranilic acid, and also of kynurenine to anthranilic acid utilizing pyridoxine as cofactor (Fig. 5.10). It may be speculated, therefore, that xanthurenic aciduria is due to reduced affinity of kynureninase apoenzyme for its coenzyme pyridoxal phosphate.

Hydroxykynureninuria

Only four patients from two families have been reported. The first patient had diarrhoea due to enterocolitis and stomatitis from 10 days of age. Fever and diarrhoea persisted for five months. Later she developed

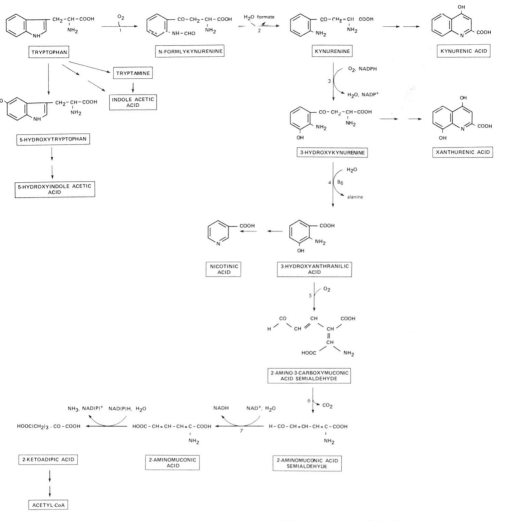

Fig. 5.10. Pathways of tryptophan degradation: (1) Tryptophan 2,3-dioxygenase (2) Formamidase (3) Kynurenic 3-monooxygenase (4) Kynureninase (5) 3-hydroxyanthranilic acid 3,4-dioxygenase (6) Aminocarboxymuconic acid semialdehyde decarboxylase (7) 2-aminomuconic acid semialdehyde dehydrogenase.
    For details of further metabolism of 2-ketoadipic acid see Fig. 5.8.

haemolytic anaemia and hepatosplenomegaly, but this disappeared by four months (Komrower *et al.* 1964; Komrower and Westall 1967). Growth was stunted, and she had skeletal deformities including hemivertebrae, fusion of several vertebrae and ribs, and fragmentation of the epiphysis of one femoral head. The IQ was estimated as 76. The patient developed sensitivity to sunlight unless treated with a vitamin mixture. This may be attributed to deficiency of nicotinic acid biosynthesis (see below).

A further three patients from one family had chronic gingivo-stomatitis and mental retardation (Reddi *et al.* 1978).

All four patients had considerably increased excretion in the urine of kynurenine, xanthurenic acid and 3-hydroxykynurenine while on a normal diet.

Oral loading with L-tryptophan (0.1 g/kg) caused a rise in the same substances which was more sustained than in controls. Anthranilic acid and 3-hydroxyanthranilic acid, which are normally derived from kynurenine and 3-hydroxykynurenine, were not detected in the urine.

No abnormality was reported in plasma. The biochemical features and nicotinic acid deficiency are consistent with deficiency of kynureninase. Pyridoxine, the cofactor of kynureninase did not correct the biochemical abnormalities when given either orally or by injection. Inheritance is probably autosomal recessive.

DISORDERS OF HISTIDINE METABOLISM

Histidinaemia

Until recently, histidinaemia was considered a rare metabolic disorder (La Du 1978). Screening programmes for the newborn, however, have revealed incidences similar to those of phenylketonuria, that is, ranging from 1 in 10 000 to 1 in 17 000 (see below). The disorder was first described by Ghadimi *et al.* (1961) who found accumulation of histidine in blood and urine of two sisters, one of whom had a speech defect. A deficiency of histidase activity in skin was reported by La Du *et al.* (1962, 1963).

Clinical features

Thalhammer (1973) reviewed 56 cases in 43 families that had not been discovered by neonatal screening. Only about half these patients showed evidence of brain damage, although they were biochemically similar to those who were considered to be normal clinically. In 11 of the 43 families there was more than one child with histidinaemia and in six of these families one of the histidinaemic children was brain damaged but the other was not. Behaviour disturbances and speech impairment have frequently been reported. Ghadimi (1981) reported 'flighty attitude' and 'bad temper', in addition to a mild degree of mental retardation and scholastic failure. Routine medical examination and conventional IQ testing may fail to differentiate between histidinaemic patients and controls, while more sensitive assessments may reveal behaviour disturbances and underachievement. Thus Gatfield *et al.* (1969) were able to elicit brain damage in apparently normal but affected siblings of an index case, only by a computer-analysed assessment of the 'Neuropsychologic Deficit Index' which evaluates multiple performance tests. Other commonly reported features include a tendency to childhood infections, short stature, abnormal EEG findings or convulsions. Tada *et al.* (1982) found no evidence of speech disturbance in 77 untreated cases of histidinaemia. Their IQ ranged from 82 to 143, the average being 105. The authors conclude that histidinaemia does not cause mental retardation.

In contrast to the hazards of children born to untreated mothers with phenylketonuria (q.v.), it is interesting that a boy born to a histidinaemic mother on a normal protein intake during the pregnancy was mentally and physically normal up to 4.5 years (Neville *et al.* 1971). This is in keeping with the experience of Tada *et al.* (1982) who reported normal postnatal development of 12 children born to histidinaemic mothers.

Molecular defects

There is severe deficiency in the activity of histidase in stratum corneum of the skin (La Du *et al.* 1962) and activity is absent in liver (Auerbach *et al.* 1967). Histidase catalyses the deamination of histidine, to yield urocanic acid, which is the first step in the major metabolic pathway of histidine (Fig. 5.11). Kuroda *et al.* (1980) found higher fasting serum concentrations, and higher increases in response to histidine loads in five patients with skin histidase activities less than 10 per cent of normal, than in four with activities about 20 per cent of normal.

Fasting blood histidine levels have varied from four to ten times normal, but considerably higher concentrations, up to 40 times normal, are reached after oral histidine loads (Auerbach *et al.* 1962).

Blood serotonin concentration has been reported to be low, but was restored to normal after treatment with low methionine diet (Corner *et al.* 1968).

Histidine concentration in the urine is considerably elevated to approximately six to tenfold of normal (Ghadimi *et al.* 1962).

In histidinaemia, the raised blood histidine concentration leads to 'spill over' to normally minor metabolic pathways leading to accumulation in the urine of imidazolepyruvic acid (Auerbach *et al.* 1962; La Du *et al.* 1963), imidazolelactic acid and imidazoleacetic acid.

Following oral loading with histidine affected individuals have increased blood histidine concentration persisting for several hours (Woody *et al.* 1965; Clarance and Bowman 1966; Cain and Holton 1968). Following histidine loading in normal individuals there is an increased excretion of formiminoglutamic acid (FIGlu) in the urine. In patients with histidinaemia characteristically there is no such rise (Auerbach *et al.* 1962; La Du *et al.*

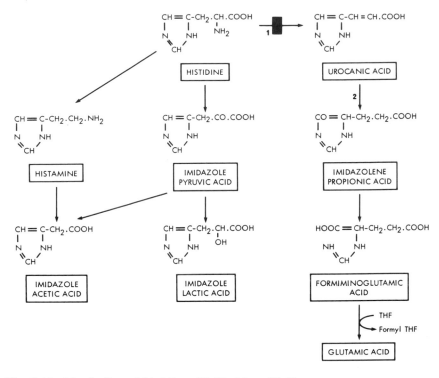

Fig. 5.11. Metabolism of histidine: (1) Histidase (2) Urocanase.

1963; Cain and Holton 1968). In histidinaemic patients following loading with urocanic acid, the product of the histidase-catalysed reaction, predictably there is a normal increase in the metabolites imidazolepyruvic acid, imidazolelactic acid, and imidazoleacetic acid (Auerbach *et al.* 1962; La Du *et al.* 1963; Holton *et al.* 1964).

Metabolites distal to the block, which are found in low concentrations in normal urine, have been absent in histidinaemic urine. These include urocanic acid, FIGlu and imidazole propionic acid (Auerbach *et al.* 1962).

Diagnosis

The ferric chloride test is a useful screening procedure for imidazole-pyruvic acid. A green colour is produced on mixing the urine with a 10 per cent aqueous solution of ferric chloride. A similar colour change occurs within 30s of dipping a Phenistix (Ames) into the urine. Since occasionally in affected patients during fasting these tests can be weakly positive or negative, they should be carried out 1–2h after oral loading with 3–5 g of L-histidine. Loading tests have been reported by Woody *et al.* (1965) and Clarance and Bowman (1966). The diagnosis is confirmed by demonstrating an increased concentration of blood and urine histidine and a reduced histidase activity in the stratum corneum layer of the skin. A sensitive method for assay of histidase has been described by Kuroda *et al.* (1980).

Genetic counselling and treatment

The disorder has an autosomal recessive mode of inheritance. Heterozygotes have reduced histidase activity (La Du 1967*b*). Attempts at detecting carriers by histidine loading have had limited success. Thus enhanced elevation of blood histidine especially in females (Ghadimi *et al.* 1962; Holton *et al.* 1964), enhanced urinary excretion of histidine (Ghadimi *et al.* 1962), and diminished excretion of FIGlu (Auerbach *et al.* 1962; La Du *et al.* 1963; Cain and Holton 1968) have been reported in parents following loading.

Newborn screening programmes have revealed an unexpectedly high incidence (Thalhammer 1980). Estimates have included 1 in 8000 (Japan; Tada *et al.* 1982), 1 in 11 000 (New Zealand; Veale *et al.* 1972), 1 in 20 000 (USA; Levy *et al.* 1974*a*), and 1 in 16 000 (Austria; Thalhammer 1980).

The question of phenotypic expression is still unresolved, so that the necessity for treatment by dietary deprivation of histidine is uncertain. Some authors (Thalhammer 1980; Ghadimi 1981) consider that it has been established that mild or moderate brain damage, speech defects and convulsions are caused in some histidinaemic patients. The apparent absence of symptoms in others may be due to the existence of some residual hepatic histidase activity, analogous to hyperphenylalaninaemia with diminished tolerance to phenylalanine and residual hepatic phenylalanine hydroxylase activity. Other authors (Levy *et al.* 1974) have not carried out treatment in the past because they consider that the causal relation between histidinaemia and brain damage has not yet been proved (Neville *et al.* 1972; Lancet 1974).

Until the controversy is resolved it would seem safer to limit histidine intake so as to reduce blood histidine levels to below 4 mg/dl (normal range 0.3 to 2.6 mg/dl; Stein and Moore 1954) for at least the first two years. Care should be taken to avoid excessive histidine reduction since infants on a diet lacking in histidine have been reported to have reduced growth and to develop rashes (Snyderman *et al.* 1963; Snyderman 1965).

The desirability for prenatal diagnosis must be considered against the background that the disorder may be benign. Theoretically, prenatal diagnosis may be possible by demonstrating a deficiency in histidase activity in cultured epithelial-like amniotic cells (Melançon *et al.* 1971). Normal fibroblast-like amniotic cells lack histidase activity.

Homocarnosinosis

A mother and three children had an isolated approximately 20-fold increased concentration of homocarnosine, the brain specific dipeptide of γ-aminobutyric acid and histidine, in the cerebrospinal fluid. Both parents and a fourth child were normal but the two brothers and one sister with homocarnosinosis developed spastic paraplegia, mental retardation, and retinal pigmentation with onset at five to 29 years of age.

Inheritance is unknown, but may be autosomal dominant, or sex-linked (Sjaastad *et al.* 1976).

DISORDERS OF GLYCINE METABOLISM

Non-ketotic hyperglycinaemia

Clinical features

During the first few days after birth there is hypotonia, lethargy, often followed by myoclonic convulsions, apnoeic attacks, and sometimes death. The majority of survivors are severely mentally handicapped with little motor or social development. There is spastic hypertonicity of voluntary muscles. Major or myoclonic convulsions are common. EEGs are grossly abnormal, often showing features of hypsarrhythmia (Baumgartner *et al.* 1969; Baumgartner *et al.* 1975; Nyhan 1983). A milder course has been reported (Frazier *et al.* 1978).

Molecular defects

There is elevation of glycine concentrations in blood, urine, cerebrospinal fluid and brain (Perry *et al.* 1975*b*). Activity of the glycine cleavage system is undetectable in brain (Perry *et al.* 1977), and low in liver (Perry *et al.* 1977; Kølvraa 1979). This complex mitochondrial multienzyme system catalyses the reaction:

Glycine + tetrahydrofolate (THF) + NAD$^+$ = 5,10-CH$_2$THF + NADH + H$^+$ + CO$_2$ + NH$_4^+$.

The system consists of four proteins designated P-, H-, T-, and L-proteins; and four cofactors, pyridoxal 5'-phosphate bound to the P-protein which is glycine decarboxylase; THF, which interacts with the T-protein; flavine adenine dinucleotide (FAD), bound to the L-protein, which has lipoamide dehydrogenase activity; and NAD. The H-protein contains lipoic acid and is the aminomethyl carrier. This reaction is the major catabolic pathway for glycine, and therefore for serine, which is largely converted to glycine in a reaction catalysed by serine hydroxy-methyltransferase. Recombination experiments with bacterial glycine-cleavage protein fractions and glycine-cleavage systems from brain of affected patients indicate inactivity of the H-protein component (Perry *et al.* 1977).

Diagnosis

The diagnosis is established by demonstration of the characteristic metabolic abnormalities (see 'Molecular defects', p. 256). Non-ketotic hyperglycinaemia is differentiated from ketotic hyperglycinaemia associated with either propionic acidaemia or methylmalonic acidaemia by the absence of ketoacidosis and hyperammonaemia, and of either propionic acidaemia or methylmalonic acidaemia.

Injection of 1-^{14}C-glycine into patients showed decreased conversion to $^{14}CO_2$, but normal conversion to ^{14}C-serine. Separate injection of 2-^{14}C-glycine showed considerable conversion to labelled serine (Ando *et al.* 1968; Baumgartner *et al.* 1969). It is not known however whether this test is sufficiently reliable to assist diagnosis.

Treatment

Although sodium benzoate administration may lower blood glycine concentration, even when started from birth, it does not prevent brain damage (Krieger and Hart 1974).

Strychnine has a high affinity for central nervous system glycine receptors and therefore is a glycine antagonist. Some benefit following strychnine administration has been reported by Gitzelman *et al.* (1977), and Arneson *et al.* (1979) but not confirmed by McDermott *et al.* (1980).

Genetics

The disorder has an autosomal recessive mode of inheritance. The defect has not been demonstrated in either leucocytes or fibroblasts and has not been diagnosed prenatally. Foetal liver biopsy and assay of glycine-cleavage system activity remains a theoretical possibility for prenatal diagnosis.

Sarcosinaemia; hypersarcosinaemia; sarcosine dehydrogenase deficiency

Sarcosine or *N*-methylglycine is formed by demethylation of dimethylglycine (DMG) and is itself demethylated to glycine in the one-carbon cycle. The latter reaction is catalysed by sarcosine dehydrogenase. These two reactions produce the majority of active methyl groups.

Clinical features

Since an intelligent 10-year-old boy with small stature had sarcosinaemia (Gloriex *et al.* 1971), it is possible that the abnormality is benign. However, the first two reported patients, a boy who died at 18 months of age with pneumonitis and spongy degeneration of the brain, and his sister, were both mentally retarded (Gerritsen and Waisman 1965, 1966). The brother and sister reported by Willems *et al.* (1971) were also mentally handicapped and in addition had multiple anomalies and small stature.

A 17-year-old boy reported by Tippett and Danks (1974) was severely mentally retarded, but two affected unrelated sibs were only mildly mentally slow.

Molecular defects and diagnosis

The disorder is due to deficiency of sarcosine dehydrogenase (S.D.), which has been demonstrated in liver (Gerritsen 1972).

Leucocytes and fibroblasts from controls have no demonstrable activity (Gloriex *et al.* 1971; Scott *et al.* 1970*a*).

The diagnosis can be established by demonstrating excessive concentration of sarcosine in the blood and urine using an amino acid analyser.

Genetic counselling

The disorder has an autosomal recessive mode of inheritance.

Loading with sarcosine produced abnormal elevation of serum sarcosine in heterozygotes (Hagge *et al.* 1967; Willems *et al.* 1971).

DISORDERS OF AMINO ACID TRANSPORT

Cystinuria

Cystinuria is a disorder of transport of the four dibasic amino acids, cystine, lysine, ornithine, and arginine. This leads to an excessive loss of these amino acids in the urine. Clinically the disorder presents with complications of renal calculi caused by crystalluria due to the poor solubility of cystine in urine.

In a careful study of 25 families, Harris *et al.* (1955*a*,*b*) differentiated two distinct abnormal phenotypes: phenotype 1, with greatly increased excretion of cystine, lysine, arginine, and ornithine, in which cystine stone formation was frequent, and phenotype 2, in which there was a moderately increased excretion of cystine and lysine, whilst excretion of arginine and ornithine was normal or only slightly increased. In the second phenotype, stone formation occurred very rarely. Pedigree analysis of these families was consistent with the hypothesis that phenotype 1, the most common form, represents the homozygote for a rare recessive gene. In this type, designated recessive cystinuria, the heterozygotes excreted normal amounts of cystine and dibasic amino acids. Phenotype 2, designated incompletely recessive cystinuria, represents the heterozygote of a different rare abnormal gene.

A different classification was proposed by Rosenberg *et al.* (1966) who reported that three types of homozygous forms could be differentiated according to the ability for intestinal absorption by homozygotes, and excretion in the urine of cystine and the other dibasic amino acids by heterozygotes (Table 5.3).

In type I there is no active transport of cystine, lysine, and arginine in intestinal mucosa by homozygotes. This can be demonstrated by failure of plasma cystine concentration to rise after oral loading. Heterozygotes have normal urinary excretion of dibasic acids. In type II homozygotes there was absent intestinal transport for lysine, but normal transport for cystine, while heterozygotes excreted increased quantities of dibasic amino acids. In type III intestinal transport of cystine, lysine, and arginine was present but reduced – there was a near normal rise in plasma cystine concentration after oral cystine loading. Heterozygotes for type III have a mild increase of dibasic aciduria. Morin *et al.* (1971) could not find any cases of type III cystinuria but described double heterozygotes of type I/type II with features resembling those of type III.

Crawhall *et al.* (1969*b*) state that type I cases of Rosenberg *et al.* (1966) correspond to recessive cystinuria of Harris *et al.* (1955*a*), while type II and III are subgroups of incompletely recessive cystinuria.

Clinical features

The outstanding clinical presentation is that of cystine crystalluria and calculus formation. Although calculi can present from one to 80 years of age (Scriver and Rosenberg 1973), the commonest presenting age is the third and fourth decades (Boström and Hambraeus 1964), with males and females being equally susceptible.

About 50 per cent of the stones are almost pure cystine, 40 per cent also containing calcium oxalate, calcium phosphate, or magnesium ammonium phosphate. Paradoxically, about 10 per cent of the calculi are cystine-free (Boström and Hambraeus 1964). The majority of stones are radiopaque because of their calcium content.

Complications of recurrent nephrolithiasis include renal colic, haematuria, hydroureter, hydronephrosis, recurrent urinary infections, and eventually, in some cases, death through renal failure.

Several workers have indicated an increased association between mental handicap and cystinuria. Thus in patients with mental handicap the incidence of cystinuria is about 1:1000, that is about 10 times greater than in the general population (Berry 1962; Scriver *et al.* 1970).

The mean height of 44 patients with cystinuria has been found to be significantly below average. This has been attributed to decreased intestinal absorption and increased renal loss of the essential amino acid lysine (Collis *et al.* 1963).

Table 5.3 *Characteristics of three types of homozygous cystinuria**

| Types of homozygous cystinuria | Exretion of cystine, lysine arginine, ornithine | | Active intestinal transport in homozygotes | | |
	Homozygotes	Heterozygotes	cystine	lysine	arginine
I	increased +++	normal	absent	absent	absent
II	increased ++	increased ++	present	absent	not studied
III	increased ++	increased +	present but reduced	present but reduced	present but reduced

* After Rosenberg *et al.* (1966)

Molecular defects

In urine of pH 4.5–7.0, the solubility of cystine is about 300 mg/l. Since cystinuria homozygotes may excrete 400–850 mg of cystine/l, crystallization of cystine and formation of calculi may occur at any time.

The defects in dibasic amino acid transport in gut and kidney have been studied intensively (Scriver and Rosenberg 1973). It appears that in the intestine, the four dibasic amino acids are transported by a common simple system which is sodium- and energy-dependent (Thier *et al.* 1965). This concept is supported by studies on intestinal absorption of cystine by cystinuric patients (Milne *et al.* 1961), and by *in vitro* uptake experiments (Thier *et al.* 1964, 1965).

In kidney, however, the situation is not so simple. The existence of a common site for renal tubular reabsorption of all four dibasic amino acids is shown by the fact that in cystinuria, which is a monogenic disorder, there are abnormalities in tubular reabsorption of all four dibasic amino acids. In addition, however, there is evidence for the existence of sites specific for the reabsorption of cystine only, and of other sites specific for the reabsorption of lysine, arginine, and ornithine only. Finally, there is evidence for the renal tubular secretion (excretion) of cystine only.

A detailed review is beyond the scope of this book, but the following points are of interest. The existence of reabsorption sites for cystine only is suggested by the report of a family with defective tubular reabsorption for cystine only (Brodehl *et al.* 1966), while the existence of reabsorption sites for lysine, arginine, and ornithine only is suggested by the report of four families with defective tubular reabsorption of lysine, arginine, and ornithine, but normal reabsorption of cystine (Perheentupa and Visakorpi 1965; Whelan and Scriver 1968*b*; Oyanagi *et al.* 1970). Evidence for the existence of a mechanism of tubular excretion of cystine but not for the other dibasic amino acids includes the observation that in patients with cystinuria, the renal clearance of cystine, in contrast to the other dibasic amino acids, may equal or exceed the glomerular filtration rate.

The concept that in cystinuria there are abnormalities in tubular reabsorption of all four dibasic amino acids was first proposed by Dent and Rose (1951). From this model it may be predicted that increasing the filtered load of one of these amino acids should cause competiton for reabsorption sites and lead to a decrease in the renal tubular reabsorption of the other dibasic amino acids. Such competition was indeed demonstrated by Robson and Rose (1957), who showed that intravenous administration of lysine increased excretion of the other dibasic amino acids in normal individuals, but failed to do so in cystinurics.

In contrast, Lester and Cusworth (1973) demonstrated increased urinary excretion of cystine, arginine, and ornithine in three of four cystinurics on continuous infusion of lysine. Kato (1977) showed that after infusion of small intravenous loads of lysine and arginine in normal controls, almost 100 per cent were absorbed in renal tubules, but reabsorption was low in cystinurics. After high loads, however, percentage reabsorption decreased in controls but was increased to almost normal in cystinurics. The percentage tubular reabsorption of cystine and other dibasic amino acids following high lysine and arginine loads decreased in both controls and cystinurics. Kato (1977) concluded that the data indicated the existence of two transport systems, one of high and the other of low capacity, and that the low capacity system was deficient in cystinuria.

Absorption defects for intestine and kidney have been demonstrated by *in vitro* techniques. Intestinal uptake of cystine, lysine and arginine is impaired and the three processes show mutual inhibition (Thier *et al.* 1964, 1965; Rosenberg *et al.* 1966, 1967; Morin *et al.* 1971).

In vivo, oral loading with cystine causes no plasma cystine rise in homozygote cystinurics types I and II; and slow to normal rise in homozygotes of type III. Heterozygotes of type I and double heterozygotes types I/III have smaller increases than homozygotes of type III (Rosenberg *et al.* 1965, 1966; Morin *et al.* 1971).

Cysteine (Rosenberg *et al.* 1965) and dipeptides containing cysteine and the other dibasic amino acids are absorbed by cystinurics as in normal controls (Hellier *et al.* 1970, 1973).

Fasting plasma concentration of cystine, cysteine, lysine, ornithine, and arginine (Morin *et al.* 1971; Rosenberg *et al.* 1965) and homoarginine (Cox and Cameron 1974) may be reduced.

There is an increased excretion in the urine of the mixed disulphide of cysteine and homocysteine (Frimpter 1961), homoarginine (Cox and Cameron 1974), and citrulline (Milne *et al.* 1962).

Renal clearances for cystine and other dibasic amino acids are increased in patients with cystinuria and dibasic amino aciduria (see above) (Frimpter *et al.* 1962; Crawhall *et al.* 1967, 1968b; Lester and Cusworth 1973), while the renal tubular absorption for these amino acids is decreased (Frimpter *et al.* 1962; Morin *et al.* 1971; Silk *et al.* 1975).

Treatment

The introduction of D-penicillamine (Crawhall *et al.* 1967; McDonald and Henneman 1965) has revolutionized the treatment of cystinuric stone-formers who do not react adversely to the drug.

Crawhall and Thompson (1965) showed that penicillamine reacted with cystine so that the concentration of cystine in plasma and therefore in urine was reduced.

However, several reactions have occurred which make penicillamine therapy unsuitable for a proportion of cystinurics. These include hypersensitivity to the drug, presenting as rashes, fever, and lymphodenopathy about 10 days after the start of therapy (Strickland 1972). Antihistamines and steroids may be required in view of the slow excretion of penicillamine. Proteinuria, or even the nephrotic syndrome, has also occurred but has proved reversible on withdrawing the drug (Felts *et al*. 1968).

Although mothers have continued penicillamine during pregnancy (Crawhall *et al*. 1967; Laver and Fairley 1971) and given birth to normal infants, a teratogenic effect cannot be ruled out, since an exposed foetus was born with a generalized connective tissue defect (Mjølnerod *et al*. 1971). A causal relation may be suspected since it is known that penicillamine chelates copper, the coenzyme of lysyl oxidase required for the synthesis of reactive aldehyde groups which are necessary for the cross-linking of collagen (q.v.).

Other methods of treatment have included attempts at keeping the urinary cystine in solution by making the urine alkaline by taking large doses of bicarbonate and by diluting the urine by a high fluid intake. Dent *et al*. (1965) treated 18 patients with a water intake high enough to produce 3 l of urine per day, but one third of the patients showed evidence of progressive urolithiasis, or failure of stones to dissolve.

Genetic counselling

The disorder is heterogenous. The homozygotes of the three types designated by Rosenberg *et al*. (1966) cannot be differentiated by assay of urinary cystine or other dibasic amino acids. Discrimination between heterozygotes and normal individuals is improved by plotting logarithmically the urinary excretion of lysine versus cystine, and of ornithine versus arginine or by applying canonical variate analysis (Crawhall *et al*. 1969*b*).

Methods for prenatal diagnosis have not yet been established. It is known, however, that the urinary concentration of cystine of affected neonates is increased at birth. The disorder can therefore be detected by screening of newborn. The incidence of cystinuria has been cited as 1 in 100 000 in Sweden, (Böstrum and Hambraeus 1964), but is probably higher. In has been estimated as 1 in 4000 in Australia (Turner and Brown 1972), 1 in 2500 in Libyan Sephardic Jews (Weinberger *et al*. 1974), and 1 in 2000 in England (Woolf 1967).

Hyperdibasicaminoaciduria

In hyperdibasicaminoaciduria there is normal cystine concentration in the urine but an increase in the excretion of lysine, ornithine, and arginine. Clinically the disorder is heterogenous. Whelan and Scriver (1968*b*) reported a pedigree of 33 persons of whom 13 members in two generations had dominantly inherited hyperdibasicaminoaciduria. The affected individuals had no clinical abnormality and were presumably heterozygotes. However, severe mental handicap with dysarthria and athetosis occurred in other families in presumptive homozygotes (Oyanagi *et al.* 1970; Kihara *et al.* 1973*b*).

A different phenotype was reported in Finland with similar hyperdibasicaminoaciduria but with hepatomegaly, diarrhoea, vomiting, and failure to thrive (Kekomäki *et al.* 1967). In the Finnish patients there was clinical improvement when dietary protein was reduced, and arginine supplements were administered. As discussed in the section dealing with cystinuria, this disorder is probably due to a mutation of a gene controlling renal tubular reabsorption of the three dibasic amino acids, but not of cystine. Inheritance is probably autosomal recessive, heterozygotes having hyperdibasicaminoaciduria but no clinical manifestations.

Hypercystinuria

In this disorder there is increased excretion of cystine but not of other dibasic amino acids, in the urine. Cystine calculi have not been reported in this apparently benign condition (Brodehl *et al.* 1966). Parents of affected individuals do not have hypercystinuria. The method of inheritance is unknown. In spite of its rarity and apparent innocence, the disorder is of interest since it supports the concept, discussed in the section dealing with cystinuria, of the existence of separate renal tubular sites specifically for cystine.

Iminoglycinuria; familial or renal iminoglycinuria

Iminoglycinuria is an inherited disorder of renal tubular transport of proline, hydroxyproline, and glycine causing increased excretion in the urine of these amino acids, while their concentration in the plasma is normal. The condition has been found in numerous healthy individuals (Scriver 1978). Reported associations with mental handicap (Goodman *et al.* 1967), blindness, deafness (Fraser *et al.* 1968; Rosenberg *et al.* 1968; Tancredi *et al.* 1970), or other abnormalities are probably coincidental.

The renal clearance for glycine, proline, and hydroxyproline are increased (Scriver 1968; Rosenberg *et al.* 1968; Tancredi *et al.* 1970).

In some, but not all, families there is also a defect in intestinal transport of these amino acids (Scriver 1978).

Accumulation of proline into leucocytes, and incorporation of proline into collagen of cultured fibroblasts from a homozygote affected individual were similar to those of controls (Tada *et al.* 1966). However Scriver (1983) has pointed out that low concentrations of substrate (proline, hydroxyproline, and glycine) are largely mediated by membrane carriers of a low-

K_m system which functions normally in iminoglycinuria; and on a high K_m system which is deficient in iminoglycinuria. Thus in the experiments carried out by Tanaka *et al.* (1976) the low concentration of proline used both for leucocytes (0.006mM) and for fibroblasts (0.05mM) does not rule out the possibility that transport by the high-K_m system was defective.

The disorder has an autosomal mode of inheritance with most heterozygotes having hyperglycinuria (excretion more than 2000 μmol/24h) but some with normal excretion of proline and hydroxyproline (Scriver 1978), suggesting that the condition is heterogenous. The incidence is about 1:15 000 live births.

A transient form occurs in normal newborn and may last for up to six months (O'Brien and Butterfield 1963; Brodehl and Gellissen 1968). The ontogeny of amino acid transport in rat kidney has been studied by Baerlocher *et al.* (1971a).

Lowe's syndrome; oculocerebrorenal syndrome

This sex-linked disorder is characterized in hemizygote males by congenital cataracts, muscular hypotonia, and severe mental handicap. Tendon reflexes are weak or absent. Later about half the patients develop clinical hypophosphataemic rickets and progressive renal tubular damage. Death usually occurs from infection or renal insufficiency in the first or second decade (Lowe *et al.* 1952). Seventy cases were reviewed by Abbassi *et al.* (1968).

There is considerable evidence of renal tubular dysfunction. The most marked abnormality is a generalized non-specific aminoaciduria which begins and increases between approximately the fourth and eighth months (Dent and Smellie 1961; Hooft *et al.* 1964). A fivefold increase in amino acid excretion was reported between the 12th and 67th day following birth (Abbassi *et al.* 1968). Aminoaciduria often decreases after about the fifth year.

Other frequent features are rickets, phosphaturia, and hypophosphataemia due to a high renal clearance for phosphate, and renal tubular acidosis associated with defective renal ammonia production, bicarbonate conservation, and defective renal hydrogen ion secretion.

Bartsocas *et al.* (1969) reported reduced uptake of lysine and arginine, but normal glycine uptake by intestinal mucosa *in vitro*, whilst Robillard *et al.* (1973) found normal lysine but reduced phenylalanine uptake.

Heterozygotes do not have hyperaminoaciduria either without or with ornithine loading, and appear not to have an increased frequency of lens opacities (Holmes *et al.* 1969).

Several post mortem examinations have been reported. There is extensive demyelination (Habib *et al.* 1962) and atrophy of the brain.

Testicular atrophy and azoospermia were reported by Matin and Sylvester (1980).

Akasaki *et al.* (1978) reported that urine from two patients had a sialic acid content four to five times higher than normal and a remarkable undersulphation of chondroitin sulphate. More recently, Fukui *et al.* (1981) reported that after incubation of cultured fibroblasts from patients with radiosulphate there was a marked undersulphation of glycosamino-glycans, mainly chondroitin sulphates and dermatan sulphate. If confirmed, this observation may prove to be the basis of prenatal diagnosis by sulphation studies of cultured amniotic cells.

Hartnup disease

In this disorder there is a defect in the intestinal and renal tubular absorption of α-monoamino-monocarboxylic acids. The eponym is derived from a 12-year-old boy Edward Hartnup, the first patient in whom the clinical and biochemical features were recognized (Baron *et al.* 1956; Jepson 1971).

Clinical features

The majority of patients develop a photosensitive rash before the age of 10 (Jepson 1971) but some reported patients show the biochemical but not the clinical phenotype (Pomeroy *et al.* 1968). The rash is pellagra-like, red and scaly, chiefly on the parts exposed to the sun. Intermittent, reversible, and sometimes progressive neurological signs occur. Tahmoush *et al.* (1976) have reported progressive neurological regression including ataxia, optic atrophy, and mental deterioration in three siblings. Typical neurological features include cerebellar ataxia, with jerky unsteady gait, intention tremors, sudden fainting, and diplopia (Baron *et al.* 1956). Psychiatric abnormalities have included emotional instability, hallucinations and delirium. Some of the patients have been mentally retarded. It is not certain whether there is a causal relationship, but this seemed possible in three siblings with progressive mental deterioration and neurological regression. Necropsy in one of these revealed diffuse cerebral atrophy with neuronal loss (Tahmoush *et al.* 1976).

Molecular defects

There is considerable elevation in the urinary excretion of α-amino acids excepting the iminoglycine (glycine, proline, and hydroxyproline) and the dibasic (cystine, lysine, ornithine, and arginine) groups.

The renal clearances are increased for the amino acids which are excreted in excessive quantities, i.e. histidine, alanine, valine, tyrosine, leucine, isoleucine, methionine, phenylalanine, serine, and asparagine

(Jepson 1971; Tarlow *et al*. 1972). The reabsorption of these amino acids in the renal tubules is reduced from the normal 90 per cent to 50–80 per cent (Scriver 1969).

Amino acids which are excreted in increased amounts in the urine have normal or slightly lowered plasma concentrations (Halvorsen *et al*. 1969; Tarlow *et al*. 1972).

Most authors have found an increased excretion in the faeces of those amino acids which have increased excretion in the urine, either under basal conditions (Scriver 1965) or after loading (Wong and Pillai 1966; Seakins and Ersser 1967). Decreased intestinal absorption of affected amino acids has been demonstrated by lower increases in their plasma concentrations following oral loads. On the other hand, absorption of the iminoglycine and dibasic amino acids was normal in a patient studied by Tahmoush *et al*. (1976).

The increased faecal amino acid concentration leads to increased faecal concentration of bacterial tryptophan metabolites and to their absorption and subsequent excretion. Thus increased urinary excretion has been demonstrated of indoxylsulphate (indican), indoxyl-ß-glucuronide, indoleacetic acid, indoleacetyl-L-glutamine, indoleacetylglucuronide and indoleacroylglycine (Weyers and Bickel 1958). The bacterial origin of these metabolites is consistent with the observation that excessive excretion of indole derivatives can be abolished by administration of neomycin (Milne *et al*. 1960; Wong and Pillai 1966).

The intestinal transport defect may be demonstrated *in vitro* as reduced uptake of tryptophan and other involved amino acids (Tarlow *et al*. 1972; Tahmoush *et al*. 1976).

A method whereby patients with Hartnup's disease are able to maintain near normal plasma concentrations of those amino acids for which they exhibit severely deficient absorption from the gut was demonstrated by Asatoor *et al*. (1970). These workers showed that there was almost undetectable absorption of tryptophan when given orally mixed with glycine, but that there was good absorption of tryptophan when administered in the form of the dipeptide glycyl-tryptophan. This suggests that the human intestine has different sites for the absorption of dipeptides to those for the absorption of free amino acids.

Diagnosis

The diagnosis is easily confirmed by demonstration of the characteristic aminoaciduria (see 'Molecular defects', p. 266).

Treatment

Treatment is by high protein diet and nicotinamide supplements.

Genetics

In the great majority of families, the disorder has an autosomal recessive mode of inheritance, without clinical or biochemical abnormalities in heterozygotes. A possible exception is the pedigree reported by Oyanagi *et al.* (1967) in which a brother and sister of four siblings from a second-cousin marriage had severe typical clinical and biochemical features, their brother and sister had biochemical but no clinical phenotype, their parents had neither, but the two symptomless grandfathers had the biochemical without the clinical phenotype. Since the obligate heterozygotes, that is the parents of the four children, had no biochemical or clinical abnormalities, it is possible that the grandparents with aminoaciduria were homozygotes for the Hartnup gene.

The incidence has been estimated as 1 in 18 000 live births in Massachusetts (Levy *et al.* 1980).

Glucoglycinuria

The association of isolated glucosuria and hyperglycinuria was reported in 14 members of a single Swiss family of 45 members (Käser *et al.* 1962). The index case had features of cystic fibrosis but the other affected individuals were normal, indicating that the condition is probably benign.

Glucosuria and glycinuria occurred with normal plasma glucose and glycine concentrations. The inheritance of the biochemical phenotype or trait was autosomal dominant with a recurrence risk of 1:2.

Disorders of the γ-glutamyl cycle

Amino acids are believed to cross membranes such as those of the renal tubular epithelium by energy-dependent, and often specific, processes. These processes, as in the case of transport of several types of molecules, are thought to begin by binding of molecules to an active site of the cell membranes, which may be specific for a single molecule or for a group of molecules. Amino acids may be bound in a complex to a carrier and moved inside the cell, where they are released. Energy is required for this transport and for release from the carrier.

Meister (1973, 1981) suggested that the γ-glutamyl cycle is responsible for amino-acid transport in mammalian kidney, and possibly brain and other tissues. In this cycle, glutathione (γ-glutamyl-cysteinyl-glycine) donates its γ-glutamyl residue which acts as a carrier in amino acid transport. The reactions in this cycle (Fig. 5.12) are catalysed by six enzymes. The first, γ-glutamyl transpeptidase, is membrane-bound and catalyses the conversion of the amino acid on the cell surface to γ-glutamyl-amino acid, by interaction with glutathione and release of

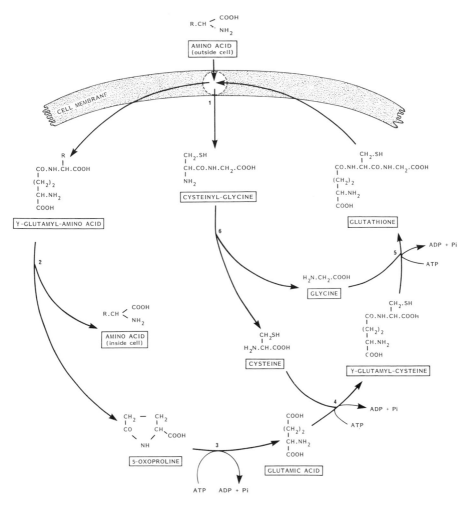

Fig. 5.12. The γ-glutamyl cycle: (1) γ-glutamyl transpeptidase (2) γ-glutamyl cyclotransferase (3) 5-oxoprolinase (4) γ-glutamyl synthase (5) Glutathione synthase (6) Peptidase.

cysteinyl glycine. The release of the amino acid (and of 5-oxoproline) inside the cell is catalyzed by the second enzyme, γ-glutamyl cyclotransferase. The conversion of 5-oxoproline to glutamic acid is catalysed by 5-oxoprolinase, which utilizes ATP yielding ADP and inorganic phosphate. A further molecule of ATP is utilized at each of the next two sequential reactions; first the γ-glutamyl cysteine synthase reaction, in which glutamic acid combines with cysteine, released under the influence of peptidase from the glutathione-derived cysteinyl-glycine to yield γ-

glutamyl-cysteine; and in the second reaction, γ-glutamyl cysteine, the product of the first reaction, binds glycine, the second residue released from cysteinyl glycine, to form glutathione. Thus, glutathione is re-synthesized, and an amino acid has been transported into the cell with the energy expenditure of three molecules of ATP; i.e.: $3ATP \rightarrow 3ADP + 3P_i$ per turn of the cycle.

Enzyme defects have been described for γ-glutamyl transpeptidase, glutathione synthase and for glutamyl cysteine synthase (Fig. 5.12).

Gamma-glutamyl transpeptidase deficiency; glutathionuria

Only three patients have been described with this condition. The first, who had a psychiatric disorder, was reported only briefly (O'Daly 1968). The second was a 33-year-old male with moderate mental retardation (IQ 62) but no other abnormality (Goodman *et al.* 1971; Schulman *et al.* 1975). The third was a moderately mentally retarded young woman (IQ 62) with severe behaviour problems including destructiveness, aggression, depression, and suicidal tendencies (Wright *et al.* 1979). The glutathione concentration in the urine was increased to over 1000 times normal.

The activity of the enzyme γ-glutamyl transpeptidase (γ-GTP) (Fig. 5.12) is considerably reduced in urine, plasma, leucocytes, and cultured fibroblasts (Schulman *et al.* 1975; Wright *et al.* 1979).

Glutathione can be detected by the cyanide–nitroprusside test. On hydrolysis with hydrochloric acid (6 mol/l for 4h at 105°C) it yields equimolar amounts of glutamic acid, glycine and half cystine which can be identified on an amino acid analyser.

γ-GTP catalyses a step in the γ-glutamyl cycle (Fig. 5.12) of Meister (1981) which was postulated as a mechanism for the transport of amino acids across membranes. However, since renal amino acid clearance appears to be normal, it cannot be an important mechanism for renal tubular reabsorption of amino acids (Schulman *et al.* 1975; Wright *et al.* 1979). Moreover, Pellefigure *et al.* (1976) found that the initial uptake kinetics for cysteine, glutamine, methionine, and alanine, which are all good γ-glutamyl acceptors, were normal for γ-GTP deficient fibroblasts. It may be that the transport role for γ-GTP is mainly for cerebral peptides, which as pointed out by Wright *et al.* (1979) are of psychiatric and behavioural significance. Since parents have reduced γ-GTP activities in cultured fibroblasts, the disorder probably has an autosomal recessive mode of inheritance.

5-oxoprolinuria; pyroglutamic aciduria; glutathione synthase deficiency

The most common clinical features have been a chronic metabolic acidosis due to accumulation of 5-oxoproline (pyroglutamic acid), with onset usually shortly after birth, an increased rate of haemolysis, and, in some cases, eventually neurological disorder. The disorder is caused by a generalized deficiency of glutathione synthase (Fig. 5.12). Jellum *et al.* (1970) reported a 19-year-old acidotic male from Norway with cerebral and cerebellar damage, and moderate mental retardation. He had a reduced urea excretion. Two sisters, aged four and seven years were developmentally normal (Hagenfeldt *et al.* 1974, 1978). Both sisters had increased concentrations of 5-oxoproline in urine and plasma, markedly decreased erythrocyte glutathione concentration, and increased rates of haemolysis. They had metabolic acidosis which required treatment with bicarbonate.

In these sisters a severe deficiency was demonstrated in the activity of glutathione synthase in skin fibroblasts, erythrocytes and, in the second sister, in placental tissue (Wellner *et al.* 1974). Parental erythrocyte glutathione synthase activities were both lower than normal controls, an observation which was confirmed by Larsson *et al.* (1976) both in the same parents and in parents of a five-year-old affected German boy with mild mental retardation. His sister had died undiagnosed at six months of age with severe acidosis and anaemia requiring several blood transfusions. A deficiency in leucocyte, erythrocyte, and fibroblast glutathione synthase was reported in an affected boy by Spielberg *et al.* (1977). Both parents had a partial deficiency.

Marstein *et al.* (1976) reported an adult patient with 5-oxoprolinuria with undetectable glutathione or, surprisingly, γ-glutamyl-cysteine. Moreover, the erythrocyte amino acid concentration was considerably increased, and the authors considered this to indicate 'strong evidence of a role for γ-glutamyl transpeptidase in the transport of amino acids into erythrocytes'. However, on restudying the two sisters described above, Hagenfeldt *et al.* (1978) found no derangement of amino-acid transport and no accumulation of γ-glutamyl-cysteine which might have acted as substitute for glutathione as glutamate donor. It appears that under normal conditions glutathione acts as a feedback inhibitor to the initial step in its own biosynthesis, catalysed by γ-glutamylcysteine synthase (Fig. 5.12). In patients with 5-oxoprolinuria, lack of this feedback causes excessive formation of γ-glutamylcysteine and its metabolites 5-oxoproline and cysteine (Larsson and Mattsson 1976; Richman and Meister 1975). It is not clear why γ-glutamyl-cysteine, the substrate of glutathione synthesis, is not excreted in excessive quantities. Wellner *et al.* (1974) suggest that its conversion to 5-oxoproline (catalysed by γ-glutamyl transpeptidase and γ-glutamylcyclo-transferase) is more rapid than the hydrolysis of 5-oxoproline to glutamic acid (catalysed by 5-oxoprolinase).

The disorder has an autosomal recessive mode of inheritance. Prenatal diagnosis should be possible by enzyme assay of cultured amniotic cells.

Gamma-glutamyl-cysteine synthase deficiency

A 38-year-old man and his 36-year-old sister had both suffered from anaemia with intermittent jaundice since birth and had both undergone cholecystectomies for cholelithiasis. Their parents were not consanguineous. At 42 years, the brother had an ataxic gait, speech impairment and myoclonic spasms of the right calf. Progressive spinocerebellar degeneration was diagnosed. At 44 years of age, the sister had some terminal dysmetria and mild impairment of co-ordination. They tended to have low haemoglobin concentration, and reticulocytosis. Both patients had severe deficiency of erythrocyte glutathione concentration and of erythrocyte γ-glutamyl-cysteine synthase activity (Fig 5.12). The mother, a son of the brother, and a daughter of the sister had intermediate reduction of glutathione concentration and γ-glutamyl-cysteine synthetase activity in their red cells (Konrad *et al*. 1972; Richards *et al*. 1974). Inheritance is therefore probably autosomal recessive. According to Meister (1973), these patients had reduced glutathione content of leucocytes and muscles, as well as erythrocytes, and exhibited a significant generalized neutral and dibasic aminoaciduria. He considers that these findings support the role of the γ-glutamyl cycle in renal transport of amino acids.

Generalized aminoaciduria

In addition to specific defects in renal tubular reabsorption of amino acids discussed above a generalized non-specific aminoaciduria can be caused by decreased renal tubular reabsorption. In these disorders the plasma concentration of amino acids is usually normal, but tyrosinaemia (q.v.) is an exception.

Generalized aminoaciduria may accompany specific metabolic disorders. Those discussed in this book are listed in Table 5.4. Other causes include nutritional rickets (Chisolm and Harrison 1962) or, exceptionally, X-linked hypophosphataemic rickets (Dent and Harris 1956), though typically in the

Table 5.4 *Causes of generalised aminoaciduria discussed elsewhere in this book*

Lowe's syndrome
The Fanconi syndrome
Cystinosis
γ-glutamyl transpeptidase deficiency
Galactosaemia (transferase deficiency)
Fructose intolerance
Fructose 1,6-diphosphatase deficiency
Tyrosinaemia
Wilson's disease

latter disorder amino acid excretion is normal. Generalized aminoaciduria has also been described in xeroderma pigmentosum (Hatano *et al.* 1968), ichthyosis vulgaris (Schnyder *et al.* 1968), and in a type of glycogen storage disease with the Fanconi syndrome, but of unknown cause (Fanconi and Bickel 1949); it may be a transient phenomenon following intravenous hyperalimentation (Kim *et al.* 1978), or in premature infants receiving high protein diet (Przyrembel *et al.* 1973). It is commonly found in patients with mental retardation and it is often transient. The cause is unknown (Carson and Neill 1962; van Gelderen and Dooren 1963).

Several toxic agents can cause generalized aminoaciduria, including lead (Chisholm 1962; Goyer *et al.* 1972), cadmium (Goyer *et al.* 1972), uranium (Clarkson and Kench 1956), lysol (a mixture of cresols) (Spencer and Franglen 1952), the drug methyl-3-chromone (Otten and Vis 1968) and outdated tetracycline, containing anhydro-4-epi-tetracycline (Lowe and Tapp 1966; Brodehl *et al.* 1968).

6 The genetic organic acidaemias and disorders of branched-chain amino acid and fatty acid metabolism

The genetic organic acidaemias are disorders characterized by the accumulation in the blood of non-amino organic acids. The majority of organic acids are efficiently excreted in the urine and may be identified by gas chromatography–mass spectrometry (GCMS). Since only a few centres have facilities and expertise for such analysis, it is important to identify indications for referring urine samples for screening.

The clinical criteria which would lead to suspicion of an organic acidaemia are:

(a) Metabolic acidosis of unknown origin;

(b) Vomiting of unknown cause;

(c) An unusual odour of the urine or patient;

(d) An acute illness in infancy, especially if aggravated by protein-containing feeds or infection

(e) An undiagnosed progressive neurological disorder, especially if the manifestations include mental handicap, convulsions, hypotonia or extra-pyramidal signs.

The history of a sibling with a similar clinical course would strengthen the significance of the above criteria and increase the urgency for investigation.

Disorders of the branched-chain amino acids leucine, isoleucine and valine are considered in this chapter for convenience, since the defects in their catabolism (with the exceptions of hypervalinaemia and hyperleucine–isoleucinaemia) cause organic acidaemias.

The 2-ketoadipic acidurias, glutaric acidurias, and ethylmalonic–adipic aciduria are considered under 'Disorders of lysine and hydroxylysine metabolism' (p. 226).

DISORDERS OF BRANCHED-CHAIN AMINO ACID METABOLISM

Catabolism of branched-chain amino acids

The catabolism of the three branched-chain amino acids valine, leucine, and isoleucine is similar for the first three reactions only. The first reaction is the transamination with 2-oxoglutarate utilizing pyridoxine as cofactor.

The existence of separate genetic defect for valine, and a distinct defect for leucine and isoleucine suggests that transamination of the 5-carbon atom amino acid is catalysed by a different enzyme to that catalysing transamination of the two 6-carbon amino acids.

The second reaction is oxidative decarboxylation by a sequence of enzyme-catalysed reactions which is believed to be similar to that of the pyruvate dehydrogenase complex (PDHC) and to that of the 2-ketoglutarate complex (Petit *et al.* 1978). The PDHC (Fig. 8.6) is probably the most complicated enzyme yet characterized and is described in Chapter 8 (p. 372). Defects of branched-chain ketoacid decarboxylation cause classical maple syrup urine disease. The identity of PDHC with branched-chain ketoacid decarboxylase is suggested by the reports of patients with deficient activity of a component of the PDHC, namely dihydrolipoyl dehydrogenase (Enz 3) who had both congenital lactic acidosis and a defect in branched-chain amino acid metabolism (Robinson *et al.* 1977; Taylor *et al.* 1978).

The third reaction common to the catabolism of the three branched-chain amino acids is the dehydrogenation of the branched-chain acyl CoAs, isobutyryl-CoA from valine, isovaleryl-CoA from leucine, and 2-methylbutyryl-CoA from isoleucine. These reactions are blocked, together with several other dehydrogenation reactions, in glutaric aciduria type II (q.v.), and ethylmalonic–adipic aciduria. This may be due to a block in the mitochondrial electron transport from an electron-transferring flavoprotein. The dehydrogenation of isovaleryl-CoA is blocked selectively in patients with isovaleric acidaemia because of a deficiency of a specific isovaleryl-CoA dehydrogenase. This enzyme was isolated and characterized from rat liver mitochondria by Ikeda *et al.* (1981). The activity of butyryl-CoA dehydrogenase is normal in cultured fibroblasts from patients with isovaleric acidaemia (Rhead and Tanaka 1980).

Further reactions in the catabolism of the three branched chain amino acids proceed independently (Fig. 6.1(b)),and are subject to separate genetic disorders.

Hypervalinaemia

The first child with hypervalinaemia was reported by Wada *et al.* (1963). He was a Japanese neonate who presented with vomiting and poor sucking when a few days of age. At two months of age he had delayed mental and physical development, nystagmus, and hyperkinesis. He was three years old when last reported and had severe physical and mental retardation.

On investigation, he was found to have isolated elevation of blood valine, but not ketoaciduria. This is consistent with a defect in the first step in the degradation of valine, that is transamination to the 2-ketoacid,

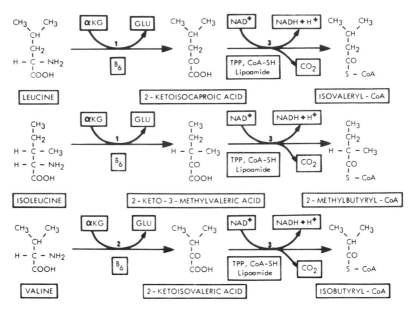

Fig. 6.1(a). Initial steps in the metabolism of branched-chain amino acids: (1) Leucine-isoleucine: α-ketoglutarate transaminase (2) Valine: α-ketoglutarate transaminase (3) Branched-chain keto acid decarboxylase.

namely 2-ketoisovaleric acid (Fig. 6.1(a)). This was confirmed by demonstrating that the leucocytes of the patient were unable to transaminate valine, but were able to transaminate leucine and isoleucine normally (Dancis *et al.* 1967). The parents, who were not consanguineous, had normal rises of blood valine and blood 2-ketoacids after oral doses of valine (Wada *et al.* 1963). There was no evidence of deficiency of pyridoxine and no improvement following the administration of high doses of pyridoxine, the coenzyme for transamination reactions (Tada *et al.* 1967*b*).

Two other children with hypervalinaemia without elevation of blood or urine leucine or isoleucine are siblings with consanguineous parents who presented with vomiting, failure to thrive, and mental retardation at three and four years of age (Reddi *et al.* 1977). Their blood and urine valine concentrations were more than five times normal. Both parents also showed elevated blood and urine valine. Enzyme studies were not reported.

This rare disorder is of biochemical interest since it suggests that in humans the transaminase for valine is distinct from the transaminase(s) for leucine and isoleucine (Fig. 6.1(a)) in contrast to porcine myocardium in which a single transaminase has been characterized for all three branched-

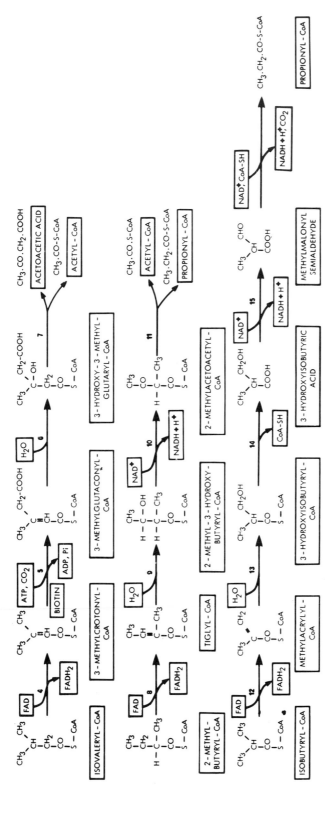

Fig. 6.1(b). Further steps in the metabolism of branched-chain amino acids: (4) Isovaleryl-CoA dehydrogenase (5) 3-methylcrotonyl-CoA carboxylase (6) 3-methylglutaconyl-CoA hydratase (7) 3-hydroxy-3-methylglutaryl-CoA lyase (8) and (12) Acyl-CoA dehydrogenase (9) and (13) Enoyl-CoA hydratase (crotonase) (10) 3-hydroxyacyl-CoA dehydrogenase (11) 2-methylacetoacetyl-CoA thiolase (14) 3-hydroxyisobutyryl-CoA hydrolase (deacylase) (15) 3-hydroxyisobutyrate dehydrogenase.

chain amino acids (Taylor and Jenkins 1966; Aki *et al.* 1967). This explanation has received support from the description of two siblings with hyperleucine–isoleucinaemia (q.v.) and a partial defect in the transamination of leucine and isoleucine, but normal transamination of valine.

Hyperleucine–isoleucinaemia

Two siblings with raised plasma and urine leucine and isoleucine concentrations, but normal valine concentrations were reported by Jeune *et al.* (1970). They presented with delayed physical and mental development, convulsions, retinal degeneration, and nerve deafness. One died at three and a half years, the other was alive at two and a half years. Leucocyte leucine and isoleucine transaminase activities were reduced to about half normal levels.

This report provides further evidence that in humans there is a separate transaminase for valine on the one hand and leucine and isoleucine on the other (Fig 6.1(a)) (see 'Hypervalinaemia' p. 275).

Maple syrup urine disease

In 'classical' maple syrup urine disease (MSUD) there is a severe neonatal illness characterized by anorexia and apathy by the end of the first week, soon followed by hypertonicity, opisthotonus, and death if untreated. In the first four infants described, the urine had the characteristic odour of maple syrup (Menkes *et al.* 1954).

Episodic variant forms have been described (Morris *et al.* 1961; Kiil and Rokkones 1964; Valman *et al.* 1973) with later onset and milder course, which have been referred to as 'intermittent branched-chain ketoaciduria'.

A third 'intermediate' type has been described with a non-intermittent course (Schulman *et al.* 1970c; Fischer and Gerritsen 1971) but it is not certain whether this variant is genetically distinct from the 'intermittent' variant or whether the differences can be explained on dietary protein intake.

A fourth variant responds to thiamine adminstration (Scriver *et al.* 1971; Pueschel *et al.* 1979).

In all these variants there is excessive excretion in the urine of amino acids which have a methyl group replacing a hydrogen atom in the main carbon atom chain (valine, leucine, and isoleucine) and a deficiency of the enzyme branched-chain ketoacid decarboxylase.

A fifth type of disorder characterized by accumulation of branched-chain ketoacids and amino acids is due to deficiency of dihydrolipoyl dehydrogenase (Enz 3) (Fig. 8.6) and, as mentioned above, is associated with a concomitant lactic acidosis (Robinson *et al.* 1977; Taylor *et al.* 1978).

Clinical features

Classical maple syrup urine disease. By the end of the first week following birth there is anorexia, apathy, vomiting, and often a high-pitched cry, followed within a few days by transient hypertonicity with opisthotonic posture alternating with flaccidity. Ocular nystagmus and convulsions are common. The urine has a distinctive odour of maple syrup which appears to be pathognomonic. If untreated, infants usually lose their tendon and Moro reflexes and die within a few weeks, but occasionally survive with severe mental retardation for one or two years. With treatment instituted in the first week of life, affected children can grow into fertile adults, but in practice (except when physicians are alerted because of previously affected siblings or by a positive urine screening test) the diagnosis is often delayed. Survivors who have started diet after a few weeks have usually had severe psychomotor retardation.

Variant forms of maple syrup urine disease; intermittent and intermediate branched-chain ketoaciduria. A variant form of MSUD was described by Morris *et al.* (1961) in a girl who had been free from symptoms until 15 months of age when she developed transient episodes of ataxia and convulsions. During these attacks she developed maple syrup odour and branch-chain ketonuria. An intermittent course was described in two siblings (Kiil and Rokkones 1964), one of whom was largely without symptoms until eight years of age, when he developed clinical features of MSUD and died. Several other patients with the intermittent variant have been reported (Valman *et al.* 1973; Zaleski *et al.* 1973). In other cases the course has been mild but not intermittent (Schulman *et al.* 1970c). One patient was a physically healthy 19-year-old female with an IQ of 76 who continuously excreted excessive amounts of branched-chain ketoacids in the urine (Fischer and Gerritsen 1971). Since then, numerous cases with a mild and intermediate course have been described (Kodama *et al.* 1976b; Duran *et al.* 1978b).

An attempt has been made by Dancis *et al.* (1972) to correlate the severity of the disease, as defined by dietary tolerance, with the level of residual activity of branched-chain ketoacid decarboxylase in cultured skin fibroblasts. These authors concluded that the level of enzyme activity in fibroblasts reflects the ability of the individual to degrade branched-chain ketoacids. They proposed a classification into three grades: Grade 1, where enzyme levels are from 0 to 2 per cent of normal — this is the classical MSUD where dietary amino acids must be limited to the minimal amounts necessary to maintain growth; Grade 2, with enzyme levels 2 to 8 per cent of normal where patients can tolerate a protein intake of 1.5 to 2.0 g/kg of body weight; Grade 3, with enzyme levels greater than 8 per cent of normal (the highest is about 15 per cent) where dietary restriction is unnecessary.

Molecular defects

The disease is due to a deficiency in oxidative decarboxylation of the branched chain keto acids 2-ketoisovaleric acid, 2-ketoisocaproic acid, and 2-keto-3-methyl-n-valeric acid, derived from transamination of valine, leucine and isoleucine, respectively (Fig. 6.1(a)) (Dancis *et al.* 1960*a*; Rudiger *et al.* 1972). As far as is known, decarboxylation of the branched-chain keto acids proceeds by a mechanism similar to that for pyruvate and 2-oxoglutarate by a multienzyme complex. This is discussed in the section dealing with lactic acidosis.

It has not been possible to attribute with confidence the oxidative decarboxylation defect to any of the components of this multienzyme system but Rudiger *et al.* (1972) who studied kidney and liver from a patient with the classical form, and Elsas *et al.* (1974*a*) who studied cultured fibroblasts, concluded that the mutant enzyme was the decarboxylase. Singh *et al.* (1977) investigated decarboxylation of the branched-chain keto acids for cofactor requirements in a specialized lysed system described by Elsas *et al.* (1974*a*). They reported heterogeneity in cofactor requirements of five classical MSUD strains and one intermittent MSUD strain. This was in accord with their complementation analysis using heterokaryons from fibroblasts of four patients with classical MSUD and one with the intermittent variety. The results indicated amelioration of the defect in two combinations, indicating that the defect in these strains is located at different functional subunits of the multienzyme complex. They also reported different cofactor requirements for the degradation of each of the three branched-chain keto acids, suggesting the presence of three decarboxylating enzymes for each of the three 2-keto acids in agreement with conclusions of Lyons *et al.* (1973) and Elsas *et al.* (1974*a*). The results are consistent with the explanation that mutations of different functional subunits each controlled by single genes are responsible for the different variants of MSUD.

Children with deficiency of dihydrolipoyl dehydrogenase (Enz 3) have lactic acidosis in addition to branched-chain ketoaciduria and aminoaciduria (Robinson *et al.* 1977, 1980, 1981; Taylor *et al.* 1978).

The concentrations of branched-chain amino acids (Westal *et al.* 1957) and keto acids (Menkes 1959; Mackenzie and Wolf 1959; Dancis *et al.* 1960*a*) are considerably elevated in plasma and urine. The direct correlation between the plasma concentrations of branched chain keto acids and amino acids suggests that the transamination reaction is reversible (Langenbeck *et al.* 1978).

Other metabolites which accumulate include L-alloisoleucine, which is produced from isoleucine by keto–enol tautomerization and transamination of 2-keto-3-methylvaleric acid (Norton *et al.* 1962; Snyderman *et al.* 1964), and 2-hydroxyisovaleric acid, the hydroxy analogue of 2-keto isovaleric acid (Lancaster *et al.* 1974; Tanaka *et al.* 1980*a*,*b*).

Hypoglycaemia of varying severity has been reported both in the classical and variant forms of the disease, and appears to be due to reduced gluconeogenesis from amino acids, and not to hyperinsulinism (Haymond *et al.* 1973, 1978).

Pathology

Reduced synthesis of total lipids and proteolipids has been reported (Menkes *et al.* 1965; Prensky and Moser 1966) usually when patients have not received early therapy, since treated patients have had normal lipids (Menkes and Solcher 1967). However, cerebral lipid composition was normal in a patient who died at three years of age who had not received dietary therapy until 35 days of age.

Diagnosis

After about the first week, an accumulation of branched-chain 2-keto acids can be demonstrated in urine using gas chromatography alone (Tanaka *et al.* 1980*a*,*b*) or combined with mass spectrometry (Sternowsky *et al.* 1973; Lancaster *et al.* 1974). Raised concentrations of branched-chain amino acids occur in blood, urine and other secretions (Snyderman 1974; Dancis and Levitz 1978).

The final diagnosis rests on demonstration of a deficiency of branched-chain ketoacid decarboxylase activity in leucocytes (Dancis *et al.* 1960*b*) or cultured fibroblasts. The reaction can be carried out in the wells of a microtitre plate (Wendel *et al.* 1973; Fensom *et al.* 1978).

Genetic counselling and prenatal diagnosis

Classical MSUD and variants have an autosomal recessive mode of inheritance. Screening of nearly 12.5 million newborn has revealed an incidence of one case of either classic or intermediate variant forms per 226 760 births. Thiamine-responsive and intermittent forms are not usually detected (Naylor 1980). Heterozygotes have on average lower enzyme activities in cultured fibroblasts, leucocytes and lymphocytes than controls, but there is considerable overlap, which makes carrier detection unreliable (Dancis *et al.* 1965; Langenbeck *et al.* 1971; Shih *et al.* 1974).

In three pregnancies at risk for MSUD monitored by Dancis (1972), branched-chain 2-ketoacid decarboxylase activities of cultured amniotic cells were normal. Unaffected foetuses were accurately predicted. In a fourth pregnancy there was low activity in epithelial-like amniotic cells. The pregnancy was allowed to continue with the birth of an affected child.

Wendel *et al.* (1973) described a microassay in which amniotic cells growing in a microwell are incubated with ^{14}C-labelled 2-ketoisocaproic or 2-ketoisovaleric acid. Decarboxylation is assayed as the amount of radioactivity trapped as $^{14}CO_2$ on glass fibre discs, previously soaked in sodium hydroxide, and placed covering the microwell. Using this method these authors were able to diagnose an affected foetus after nine days of amniotic cell culture following amniocentesis at 17 weeks' gestation.

Prenatal monitoring for MSUD by enzyme assay of cultured amniotic cells has been reported by Elsas *et al.* (1974*b*), Cox *et al.* (1978), and Fensom *et al.* (1978), and by Wendel and Clausen (1979). Prenatal diagnosis by assay of amniotic fluid branched-chain keto acids or amino acids is not feasible (Wendel *et al.* 1980*a*).

Treatment

Treatment for classical MSUD must be considered to be an emergency in view of the very rapid deterioration that follows symptomatic onset during the first few days after birth. The earliest treated patients are usually those where the paediatrician has been alerted because of a previously affected child. When detection is by a mass screening procedure for the newborn, success depends on speed of implementing and maintaining dietary therapy.

Acute branched-chain ketoacidosis can be treated successfully by either peritoneal dialysis (Gaull 1969; Wendel *et al.* 1980*b*) or by exchange transfusions (Hammersen *et al.* 1978).

In 16 of 23 classic cases, the infants were successfully started on diet mainly during the first two weeks. On follow-up, however, two had died of an acute metabolic crisis following infection (one at 27 months and the other at about 13 years of age). Of the 14 other treated survivors, one has an IQ of 95 at 10 years, and another (a Samoan child detected in New Zealand) an IQ of 42 at three years of age. The fate of the other 12 treated patients is not reported. Of the seven untreated infants, six died before treatment could be started and the seventh at six months of age (Naylor 1980). Screening for MSUD should be as early as possible, but if the test is to be combined with screening for phenylketonuria, it should not be earlier than about five days, since with progressively earlier testing an increasing proportion of phenylketonurics will be missed.

Dietary therapy involves limiting the intake of the three branched-chain amino acids, but allowing the small requirements for growth. At the start of therapy branched-chain amino acids are totally omitted for a few days until the plasma concentration returns to normal. A suitable food formula is MSUD Aid (Scientific Hospital Supplies Ltd., Liverpool, UK), together with added vitamins and minerals (see treatment of phenylketonuria). It has been observed that plasma isoleucine becomes normal first, valine next, and almost invariably leucine last, sometimes taking up to 10 days (Snyderman 1975). Dietary therapy in seven patients is described by Snyderman *et al.* (1964), and in a child by Westal (1963).

Treatment by thiamine in pharmacological doses of 150 to 1000 mg/day should also be tried (Duran *et al.* 1978*b*) since it may cause biochemical improvement in mild variants after about three weeks (Elsas *et al.* 1981). Thiamine-responsive patients have also been reported by Scriver *et al.* (1971), Fischer and Gerritsen (1971), Kodama *et al.* (1976*b*), Danner *et al.* (1978) and Pueschel *et al.* (1979).

Isovaleric acidaemia

Clinically, infants present with features similar either to those of classical MSUD with acidosis and vomiting, usually in the first week, or with a chronic intermittent variant. The first two described sibs (Tanaka *et al.* 1966; Budd *et al.* 1967) had an odour of 'sweaty feet' or 'cheese'. Without treatment in the majority there is rapid neurological regression with death occurring within a few weeks in over half the cases. The survivors tend to have a cyclical neurological illness with ataxia, tremor, lethargy, and coma. Episodes are often accompanied by vomiting and acidosis and may occur spontaneously or be precipitated by infections and high protein intake (Tanaka 1975). Moderate mental retardation, leucopenia, anaemia, and thrombocytopenia are common (Levy *et al.* 1973) but some untreated cases have had near normal or normal intelligence (Blaskovics *et al.* 1978; Cohn *et al.* 1978).

There is a metabolic defect in the pathway for degradation of leucine with defective conversion of isovaleric acid to 3-methylcrotonic acid because of reduced activity of the enzyme isovaleryl-CoA dehydrogenase (Fig. 6.1(b)). This was suspected because of the metabolic abnormality, and supported by the demonstrations that the production of $^{14}CO_2$ from 2-^{14}C-leucine by cultured fibroblasts from three patients was 0.2–1.5 per cent of controls (Tanaka *et al.* 1976); and that oxidation of 1–^{14}C-isovaleric acid by leucocytes was reduced (Tanaka *et al.* 1966; Budd *et al.* 1967). However, direct demonstration of isovaleryl-CoA dehydrogenase deficiency in isolated fibroblast mitochondria (Rhead and Tanaka 1980) was only possible after a specific tritium-release assay had been developed (Rhead *et al.* 1981). The deficiency results in the accumulation of blood isovaleric acid which during symptomatic episodes can be more than 1000 times above normal. There is also considerable increase in the urinary excretion of isovaleric acid, both free and as isovaleryl glycine, and of 3-hydroxyisovaleric acid (Ando *et al.* 1973).

About three quarters of patients with the chronic intermittent form have developed normally in spite of recurrent attacks of vomiting and drowsiness, accompanied by the 'sweaty feet' odour (Cohn *et al.* 1978; Yudkoff *et al.* 1978; Duran *et al.* 1979).

Hybridization of cultured fibroblasts from four patients with the severe form, and six patients with the mild intermittent form failed to achieve complementation (Dubiel *et al.* 1980), consistent with the demonstration that isovaleryl-CoA dehydrogenase consists of four identical subunits (Ikida *et al.* 1981). This model would predict that in any patient each subunit would have an identical mutation.

The disorder has an autosomal recessive mode of inheritance. Heterozygotes have a reduced rate of $2\text{-}^{14}\text{C}$-leucine oxidation in fibroblasts (Tanaka *et al.* 1976; Blaskovics *et al.* 1978). The evidence on whether heterozygotes excrete excess isovalerylglyine after leucine loading (Guibaud *et al.* 1973) is conflicting (Malan *et al.* 1977).

Prenatal diagnosis by assay of cultured amniotic cells should be possible (Blaskovics *et al.* 1978), but has not yet been reported.

Treatment

Low protein diet to approximately 1.5 g/kg per day, or low leucine diet to approximately 115 mg/kg per day appears to reduce the incidence and severity of ketotic attacks (Lott *et al.* 1972; Levy *et al.* 1973).

Since plasma glycine tends to be lower during an acidotic crisis than during a remission, Krieger and Tanaka (1976) considered the possibility that glycine was depleted following conjugation with isovaleryl-CoA to yield isovaleryl-glycine. They demonstrated that after leucine loading in a patient, administration of glycine caused a doubling in the rate of excretion of isovaleryl-glycine, and a lower serum isovaleric acid concentration. Similar findings were reported by Yudkoff *et al.* (1978). Glycine administration (250 mg/kg per day) produced marked biochemical improvement and normal development in two children reported by Cohn *et al.* (1978).

3-methylcrotonic aciduria; 3-methylcrotonylglycinuria; disorders of 3-methylcrotonyl-CoA carboxylation; biotin-responsive multiple carboxylase deficiency; holocarboxylase synthetase deficiency; biotinidase deficiency

Dehydrogenation of isovaleryl-CoA yields 3-methylcrotonyl-CoA which is converted to glutaconyl-CoA in a biotin-dependent carboxylase reaction (Fig. 6.1(b)).

Patients with 3-methylcrotonic aciduria have had varied clinical features. About half have presented with a neurological disorder and the remainder with metabolic acidosis, usually (but not always) during the first nine months of age. The majority responded with clinical and biochemical improvement to biotin administration. Some biotin-responsive patients have 'combined' deficiency of three mitochondrial carboxylases in cultured fibroblasts, namely 3-methylcrotonyl-CoA carboxylase, propionyl-CoA carboxylase, and pyruvate carboxylase.

Burri *et al.* (1981) suggest that the clinical features of 'combined' deficiency patients may be classified as an infantile form with onset of vomiting and ketoacidosis, followed by coma and death, in the newborn period when untreated; and a juvenile form presenting at 3–6 months with features of acrodermatitis enteropathica, including alopecia, kerato-conjunctivitis, and ataxia. Both types respond to biotin in dosage of 10 mg or more per day. Cultured fibroblasts from patients with the infantile form have low activities of the three carboxylases, which can be increased by addition of biotin to the culture medium; while cultured fibroblasts from patients with the juvenile form have normal carboxylase activities even with low concentrations of biotin in the culture medium. Patients in the juvenile group have low plasma and urine biotin, originally suggested to be due to a defect in absorption or transport of biotin (Thoene *et al.* 1981) but more recently shown to be due to an abnormality in biotin metabolism, namely a deficiency in biotinidase (Wolf *et al.* 1983).

Clinical features

Children with neurological features have had generalized convulsions at seven weeks (Lehnert *et al.* 1979), infantile spasms at 4–12 weeks (Finnie *et al.* 1976; Leonard *et al.* 1981), developmental regression and hypotonia at six–10 months (Eldjarn *et al.* 1970; Cowan *et al.* 1979; Keeton and Moosa 1976), optic atrophy at 19 months, and ataxia at 21 months (Cowan *et al.* 1979). Symptoms attributable to metabolic acidosis, such as tachypnoea, drowsiness, and coma, have been prominent in infants presenting spontaneously at 51 h or after a mild febrile illness at 14 months (Leonard *et al.* 1981). Other features have included erythematous rash starting in the buttocks (Gompertz *et al.* 1971) or face (Leonard *et al.* 1981), eczema around the mouth and nose (Leonard *et al.* 1981), periorificial candida dermatitis (Cowan *et al.* 1979), sparse hair (Leonard *et al.* 1981), alopecia (Cowan *et al.* 1979), and keratoconjunctivitis.

Response to biotin

The first patient to be treated with biotin (Eldjarn *et al.* 1970; Stokke *et al.* 1972) failed to respond. This may have been due to the small dose of 0.25 mg/day rather than 5–10 mg/day given to responsive patients. The first biotin response was in a five-month-old child with vomiting, acidosis, and rash (Gompertz *et al.* 1971) who showed a dramatic improvement on biotin, 10 mg/day, with disappearance of clinical abnormalities and of excess 3-methylcrotonylglycine and tiglylglycine in the urine. At 10 years of age, the patient is still on biotin and is mentally and physically normal

(Leonard *et al.* 1981). Such unequivocal responses to biotin have been reported in several patients.

Good responses with apparently normal subsequent progress were reported by Lehnert *et al.* (1979) and in cases 1 and 3 by Leonard *et al.* (1981). Some patients with good biochemical response to biotin, however, have had neurological sequelae. Thus a nine-year-old boy of first-cousin Pakistani parents presented with respiratory distress and delayed motor development. There was a striking biochemical response to biotin, but he had pale optic discs and persistence of retardation (Gompertz *et al.* 1973*a*). Case 2 of Leonard *et al.* (1981), also of consanguineous Pakistani parents, but female, presented at one month of age with salaam spasm, followed a few days later by generalized convulsions. She had a metabolic acidosis with raised blood pyruvate and lactate. She responded well to biotin but at five and three-quarter years had a spastic diplegia and no recognizable speech.

Considerable 'catching up' was reported in case 4 of Leonard *et al.* (1981) who at 14 months had a development level appropriate to six months, and after treatment with biotin at 26 months of age had developmental abilities within the range appropriate for 17 and 14 months.

Molecular defects

There is a block in the catabolism of leucine in the carboxylation of 3-methylcrotonyl-CoA to 3-methylglutaconyl-CoA due to a deficiency of the mitochondrial biotin-dependent enzyme 3-methylcrotonyl-CoA carboxylase (Fig. 6.1(b)). This was suggested by Stokke *et al.* (1972) and proved by Gompertz *et al.* (1973*b*). Characteristically there is excretion in the urine of 3-hydroxyisovaleric acid, 3-methylcrotonic acid, and 3-methylcrotonyl-glycine.

In the patient previously reported by Gompertz *et al.* (1971), Sweetman and Nyhan (1974) also identified methylcitrate and 3-hydroxypropionate in the urine which, together with the report that the same patient also excreted elevated quantities of tiglylglycine (2-methylcrotonyl glycine) (Gompertz *et al.* 1971), suggested a defect in the catabolism of propionate as well as of leucine. This was confirmed by the demonstration that cultured fibroblasts also had a defect in the activity of another mito-chondrial carboxylase – propionyl-CoA carboxylase (Sweetman and Nyhan 1974; Bartlett and Gompertz 1976). Activities of both enzymes can be restored to normal by the addition of biotin to the culture medium (Bartlett and Gompertz 1976; Bartlett *et al.* 1980). Kinetic studies have shown that the stimulated enzymes have normal Michaelis constants, suggesting that the holoenzymes are normal (Weyler *et al.* 1977). These last authors suggested that the primary defect is in the enzyme holo-carboxylase synthetase, which catalyzes the activation of biotin and its transfer to the apocarboxylase proteins to which it becomes covalently bound. Burri *et al.* (1981) reported the holocarboxylase synthetase activity in cultured fibroblasts from a patient with the infantile form to be 30–40 per cent of normal. Moreover the K_m for biotin was about 60 times higher than normal.

In the late-onset form of the disease Wolf *et al.* (1983) found a profound deficiency of biotinidase in the serum of three patients in two families, with intermediate activities in the parents. Biotinidase is the enzyme responsible for cleavage of biocytin (biotin-ε-lysine), a normal product of the carboxylase-catalysed degradation, resulting in regeneration of free biotin. It was suggested that biotin supplementation in these patients is required because of their inability to regenerate biotin from endogenous biocytin.

It appears from the published data that biotin-responsive multiple carboxylase deficiency is the commonest cause of 3-methylcrotonyl-glycinuria.

A deficiency in the activity of three mitochondrial carboxylases, that is, of pyruvate carboxylase, 3-methylcrotonyl-CoA carboxylase, and propionyl-CoA carboxylase, was demonstrated in cultured fibroblasts from a patient with 3-methylcrotonylglycinuria and 3-hydroxyisovaleric aciduria. The patient was a girl aged four years who had had two episodes of metabolic acidosis precipitated by infections. The abnormal metabolites disappeared from the urine after biotin administration. The specific activities of the three carboxylases returned to normal when biotin was added to the culture medium. It may be speculated that deficiency of pyruvate carboxylase may be related to elevation of blood pyruvate and lactate which has been reported in some patients (Roth *et al.* 1976; Leonard *et al.* 1981).

Other metabolites reported in the urine of some patients include 2-hydroxybutyrate (Roth *et al.* 1976), 3-hydroxybutyrate (Roth *et al.* 1976; Leonard *et al.* 1981), and 2-oxoglutarate (Finnie *et al.* 1976).

Propionyl-CoA carboxylase activity was normal in liver in one patient (Finnie *et al.* 1976) and in cultured fibroblasts from another (Case 4, Leonard *et al.* 1981).

Prenatal diagnosis of 3-methylcrotonic aciduria should be possible in the infantile form by detection of 3-methylcrotonyl-CoA carboxylase deficiency in amniotic cells cultured in biotin-free medium, and possibly by measurement of amniotic fluid metabolites.

3-methylglutaconic aciduria

At the time of writing, seven children with 3-methylglutaconic aciduria have been reported, but only three, including two sibs, have been described in any detail. The enzyme defect has not been proved, but the pattern of urinary metabolites is consistent with a deficiency of 3-methylglutaconyl-CoA hydratase which catalyses a step in the metabolism of leucine (Fig. 6.1(b)), namely the hydration of the unsaturated acid of 3-methylglutaconyl-CoA, thus converting it to 3-hydroxy-3-methylglutaryl CoA.

The first child was reported briefly by Robinson *et al.* (1976). She was normal until nine months of age, then ceased to grow and had progressive neurological and mental deterioration. At three years, she had hypotonia, bizarre protrusion of the right arm, and self-mutilation. She had an abnormal EEG, but no metabolic acidosis or hypoglycaemia. There was considerable elevation of 3-methylglutaconic acid in the urine, which was reduced when on a diet containing less than 1.5 g protein/kg body weight. Plasma concentration was not elevated. Muscle 3-methylglutaconyl-CoA hydratase activity was probably normal.

The brother and sister reported by Greter *et al.* (1978) had progressive regression from five and six months of age, respectively. The boy developed hyperkinetic movements. At 11 months his motor development was at the six months level. At six years he had moderate dementia, dyskinesia, slight spastic paraparesis, pale discs, and neurogenic hearing impairment. The girl at 10 months had lost the ability to creep which she had acquired by five months. She had choreathetosis. At three and a half years she had optic atrophy and suspected spasticity. At four and a half years she had slight dementia, ankle clonus, and extensor plantar responses, and by nine years had spastic paraparesis and dyskinetic neurological syndrome.

Both siblings had elevated urinary concentrations of 3-methylglutaconic acid (as two isomers, possibly produced by *cis–trans* isomerism) and its hydrogenation product 3-methylglutaric acid. Excretion of these metabolites increased when leucine was given by mouth. Neither parent excreted the metabolites. The authors report that they have found another two cases of 3-methylglutaconic aciduria among 300 retarded children screened.

The authors estimated that the amount of 3-methylglutaconic acid found in the urine of their patients corresponds to no more than two per cent of the daily intake of leucine.

Two brothers aged seven and five years with considerably increased excretion in the urine of 3-methylglutaconic acid and 3-hydroxyisovaleric acid were reported by Duran *et al.* (1982). The only clinical abnormality was speech retardation. The authors comment that they attempted to assay 3-methylglutaconyl-CoA hydratase in leucocytes and cultured fibroblasts, but this was unsuccessful since even normal cells did not show sufficient activity.

If these patients have deficiency of 3-methylglutaconyl-CoA hydratase, it is surprising that they do not exhibit symptoms of either acidosis or hypoglycaemia, which are characteristic in patients with deficiency of 3-hydroxy-3-methylglutaryl-CoA lyase (see p. 290).

Duran *et al*. (1978c) warned of possible errors in confusing metabolites found in 3-methylglutaconic aciduria with those found in patients with 3-hydroxy-3-methylglutaryl-CoA lyase deficiency.

3-hydroxyisobutyryl-CoA hydrolase (deacylase) deficiency

The only patient reported with this disorder was a male infant whose parents were Egyptian first cousins (Brown *et al*. 1982). He had dysmorphic facies, multiple vertebral abnormalities and Fallot's tetralogy. He was hypotonic and failed to thrive or to develop motor skills, and died at three months of age. Necropsy confirmed the cardiac disorder and revealed agenesis of the cingulate gyrus and of the corpus callosum. Metabolic screening established the presence in the urine of methacrylic acid conjugates of cysteine and cysteamine, namely *S*-(2-carboxypropyl) cysteine and *S*-(2-carboxypropyl) cysteamine. Since the coenzyme A (CoA) ester of methacrylic acid is a metabolite of valine (Fig. 6.1(b)) the authors assayed in liver and cultured fibroblasts the activity of enoyl-CoA hydratase (crotonase) (which catalyses the conversion of methacrylyl-CoA to 3-hydroxyisobutyryl-CoA), and the activity of the enzyme catalysing the next metabolic step, namely 3-hydroxyisobutyryl-CoA hydrolase. The latter enzyme is a deacylase specific to valine metabolism. The reaction involves hydrolysis of the CoA ester with release of CoA, which does not occur in the metabolism of leucine or isoleucine. The activity of the latter enzyme was markedly deficient in the patient, while that of enoyl-CoA hydratase was normal. Such deficiency would be expected to cause accumulation of both 3-hydroxyisobutyryl-CoA and methacrylyl-CoA as these compounds are in equilibrium in aqueous solvents (Robinson *et al*. 1957). Several possible explanations are considered for the formation of the conjugates of methacrylyl-CoA including spontaneous reaction between the sulphydryl bonds of cysteine and the double bond of methacrylyl-CoA, which is a highly reactive substance with thiol compounds.

The authors speculate that accumulation of methacrylyl-CoA in the cells could inhibit enzymes with sulphydryl groups, bind to cofactors such as lipoic acid, and deplete pools of cysteine and glutathione. The teratogenic effect of methacrylate esters has been demonstrated in rats, which develop a high frequency of foetal death, skeletal and other malformations similar to those observed in the patient (Singh *et al*. 1972).

Since both parents of the patient had reduced fibroblast 3-hydroxyiso-butyryl-CoA hydrolase activities, and were first cousins, autosomal recessive inheritance is probable.

Amniocentesis was carried out when the mother became pregnant again. Cultured amniotic cell 3-hydroxyisobutyryl-CoA hydrolase activity was approximately half that of control mean values, and similar to that of the parents' fibroblasts. A heterozygous foetus was predicted, and the pregnancy was completed with the birth of a normal male infant.

3-hydroxy-3-methylglutaryl-CoA lyase deficiency

In this form of organic aciduria there is deficiency of 3-hydroxy-3-methylglutaryl-CoA lyase (HMG-CoA lyase) which catalyses the final step in the degradation of leucine, namely the intra-mitochondrial cleavage of 3-hydroxy-3-methylglutaryl-CoA (HMG-CoA) to acetoacetic acid and acetyl-CoA (Fig. 6.1(b)). The main features are metabolic acidosis and severe (Faull *et al.* 1976*a,b*), or even fatal (Schutgens *et al.* 1979*b*) hypoglycaemia presenting in infancy, as early as 48h after birth (Schutgens *et al.* 1979*b*; Leonard *et al.* 1979*a,b*). Two infants have had biochemical evidence of liver damage consistent with Reye's syndrome (Faull *et al.* 1976*a,b*; Leonard *et al.* 1979*a,b*). One patient died before the disorder was characterized (Schutgens *et al.* 1979*b*), one has thrived physically but following a mild upper-respiratory tract infection at nine months, while on a diet of fresh cow's milk and solids (introduced at six months) developed an acute illness with metabolic acidosis, hypoglycaemia, and convulsions. She sustained severe neurological sequelae, including a hemiplegia and choreoathetosis. Two have developed normally (Faull *et al.* 1976*a,b*; Duran *et al.* 1979).

HMG-CoA lyase deficiency has been demonstrated in leucocytes (Wysocki and Hähnel 1976*a*; Duran *et al.* 1979), cultured fibroblasts (Wysocki and Hähnel 1976*b*) and placenta (Duran *et al.* 1979). The metabolic consequences are absence of acetoacetate formation and therefore of ketonuria, even after 24h vomiting, and accumulation of products of leucine metabolism proximal to the block (Fig. 6.1(b)). These include 3-methylcrotonic acid, 3-methylglutaconic acid, and 3-hydroxy-3-methylglutaric acid (HMG) in the urine, and other products of side reactions, including 3-hydroxy-3-methylbutyric and 3-methylglutaric acids (Wysocki *et al.* 1976). Duran *et al.* (1979) suggest that inability to form glucose-sparing ketone bodies is the cause of hypoglycaemia.

The disorder has an autosomal recessive mode of inheritance, as evidenced by reduced enzyme activity in parents of affected children (Duran *et al.* 1979). Since there is enzyme activity in cultured fibroblasts, the disorder can probably be diagnosed prenatally by direct assay of cultured amniotic cells. However, Duran *et al.* (1979) established prenatal diagnosis by demonstrating a considerable increase of HMG and 3-methylglutaconic acid in maternal urine at 23 weeks' gestation. The pregnancy continued and resulted in the birth of a mature neonate (birthweight 3500 g). The first voided urine had characteristic organic aciduria. The diagnosis was confirmed by demonstrating severe deficiency of leucocyte HMG-CoA lyase activity.

3-ketothiolase deficiency; 2-methylacetoacetyl-CoA thiolase (3-keto-acyl-CoA thiolase) deficiency

In this form of organic aciduria there is increased excretion in the urine of 2-methylacetoacetate, tiglylglycine and 2-methyl-3-hydroxybutyrate, which are metabolites in the catalytic pathway of isoleucine (Fig. 6.1(b)). The accumulation of these metabolites is consistent with deficiency in the activity of the thiolase which catalyses cleavage of 2-methylacetoacetyl-CoA to acetyl-CoA and propionyl-CoA. This has recently been confirmed in fibroblasts (Schutgens *et al.* 1982).

Clinical features

The clinical presentations have differed widely but appear to conform to three main types. The first is represented by the single child reported by Keating *et al.* (1972) who had persistent vomiting a few days after birth and later developed cardiomyopathy (Henry *et al.* 1981). In the second type, symptoms appear after the first year. Thus four children from two families reported by Daum *et al.* (1973) and two sisters reported by Robinson *et al.* (1979) had no symptoms attributable to metabolic acidosis during the first year in spite of having had several infections, but developed acidosis, vomiting and, in some cases, lethargy and coma following infections in the second and third years. Two of these children, who were unrelated, died at 12 and 30 months. An affected brother of one of the fatal cases has 'psychomotor damage' at eight years, while his second affected sister was physically and mentally normal at five years (Daum *et al.* 1973).

In the third type of presentation there are no symptoms until after the fourth year. Gompertz *et al.* (1974a) reported a boy with episodes of headache, vomiting, and abdominal pain starting from seven years.

The boy reported by Schutgens *et al.* (1982) can also be included in the last category since he had no symptoms during the first four years, after which he developed acute ataxia and diplegia attributed to mumps encephalitis, but had frequent vomiting, hypotonia, and hyperventilation from four and a half years. However, his clinically normal 36-year-old father also excreted excess urinary 2-methyl-3-hydroxybutyrate and tiglylglycine. The father recalled that he suffered more than others during mild infections. Since both father and son had undetectable 2-methylacetoacetyl-CoA thiolase activities in their cultured fibroblast the authors suggested that both were homozygous affected.

Molecular defects

The disorder involves interruption of the pathway in isoleucine catabolism (Fig. 6.1(b)). The pattern of organic acids which accumulate is consistent with a deficiency in the catalysis of thiolytic cleavage of 2-methylaceto-acetyl-CoA. Direct proof of this defect by demonstration of undetectable activity of 2-methylacetoacetyl-CoA thiolase in cultured fibroblasts was possible only after the synthesis of the substrate 2-methylacetoacetyl-CoA (Schutgens *et al.* 1982). Impaired oxidation of isoleucine by cultured fibroblasts (Daum *et al.* 1973; Hillman and Keating 1974; Gompertz *et al.* 1974*a*) and defective conversion of tiglyl-CoA to either propionyl-CoA (Gompertz *et al.* 1974*a*) or to citrate are consistent with this enzyme defect.

Three ketoacyl-CoA thiolases have been identified in mammalian liver, one in the cytosol specific for acetoacetyl-CoA, and two in the mitochondria, one of which is also specific for acetoacetyl-CoA (A-isomer) while the other is active against several 3-ketoacyl-CoA substrates. Schutgens *et al.* (1982) have identified an extrahepatic mitochondrial acetoacetyl-CoA thiolase (B-isomer) which they believe is the deficient form in this disorder.

Other metabolites which have been identified in excessive quantities in the urine include butanone, which is formed by spontaneous decarboxylation of 2-methylacetoacetate (Gompertz *et al.* 1974*a*), tiglylglycine and tiglic acid (Daum *et al.* 1973).

Deficiency of cytosolic acetoacetyl-CoA thiolase: this was reported in a single patient with severe neurological symptoms by De Groot *et al.* (1977).

DISORDERS OF PROPIONIC ACID AND METHYLMALONIC ACID METABOLISM

The last reaction which is specific to the metabolism of isoleucine is splitting of 2-methylacetoacetyl-CoA to acetyl-CoA and propionyl-CoA. The latter is carboxylated to yield D-methylmalonyl-CoA, in a reaction which is catalysed by the biotin-dependent enzyme propionyl-CoA carboxylase (Fig. 6.2(a)), which may become deficient as a result of mutations of either this enzyme, or of holocarboxylase synthetase, which leads to failure of attachment of biotin to the carboxylase. As previously described, this also causes deficiency in the activity of 3-methylcrotonyl-CoA carboxylase and of pyruvate carboxylase. D-methylmalonyl-CoA then becomes converted to L-methylmalonyl-CoA in the methylmalonyl-CoA racemase reaction. L-methylmalonyl-CoA is next converted to succinyl-CoA which enters the tricarboxylic acid cycle. The conversion to succinyl-CoA is catalysed by methylmalonyl-CoA mutase, utilizing 5-

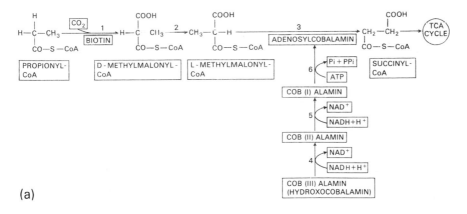

(a)

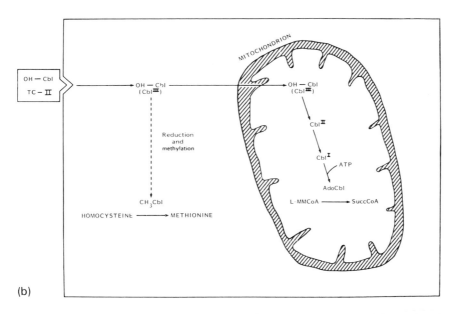

(b)

Fig. 6.2. (a) Pathways of propionate and cobalamin metabolism: (1) Propionyl-CoA carboxylase (2) Methylmalonyl-CoA racemase (3) Methylmalonyl-CoA mutase (4) Cob(III)alamin reductase (5) Cob(II)alamin reductase (6) Cobalamin adenosyl transferase.

(b) Conversion of hydroxycobalamin to methylcobalamin or adenosylcobalamin. (Modified from Rosenberg 1978): OHCbl (CblIII), Hydroxycobalamin (cobalamin III); TC-II, Transcobalamin II; CblII, Cobalamin II; CblI, Cobalamin I; CH₃Cbl, Methylcobalamin; AdoCbl, Adenosylcobalamin; L-MMCoA, L-methylmalonyl-CoA; SuccCoA, Succinyl-CoA.

deoxyadenosylcobalamin as cofactor, and is therefore deficient both in mutations affecting the mutase, and in disorders in the synthesis of the cofactor. Other substances which are metabolized to propionyl-CoA include methionine, threonine, odd-numbered fatty acids, and cholesterol. Valine is a precursor of methylmalonyl-CoA.

Propionic acidaemia

The first clinical reports and metabolic studies of a child with propionic acidaemia were by Childs *et al*. (1961), Nyhan *et al*. (1961) and Childs and Nyhan (1964), who named the disorder idiopathic hyperglycinaemia, but did not detect the accumulation of propionic acid in the blood or urine. Clinically there was a cyclical illness characterized by lethargy, dehydration, and coma, induced by protein meals. Propionic acidaemia was first reported by Hommes *et al*. (1968) in a patient without hyperglycinaemia who died at four days of age, and by Hsia *et al*. (1969) in the sister of the first patient (Childs *et al*. 1961). Hsia *et al*. (1969) showed that her leucocytes could not oxidise propionate to CO_2 but could oxidise methylmalonate. More recently, a child with biotin-responsive propionic acidaemia has been reported by Barnes *et al*. (1970). Combined defects have been reported of three mitochondrial carboxylases, including propionyl-CoA carboxylase, which also respond to biotin (Bartlett *et al*. 1980). Five different mutant classes of fibroblasts have been identified using complementation techniques (Wolf *et al*. 1980). The mechanism of hyperglycinaemia in propionic acidaemia and in methylmalonic acidaemia remains to be elucidated.

Clinical features

Clinical features of 65 patients have been reviewed by Wolf *et al*. (1981). Within a few days of birth, the infants develop deep, rapid respiration, refuse feeds, and become drowsy, dehydrated, and then comatose. Vomiting is frequent and has led to the incorrect diagnosis of pyloric stenosis. Convulsions and hepatomegaly are common. In children who survive, acute ketotic crises (as in maple syrup urine disease) are provoked by infections. There is neutropenia, which has been related to hyper-glycinaemia, and thrombocytopenia with purpura during the first six months and, in older children, osteoporosis (Childs and Nyhan 1964). Convulsions and electroencephalographic abnormalities are common. Mental retardation and early death occur in the majority.

Molecular defects

Hsia *et al*. (1969) first demonstrated a deficiency of oxidation of 3-^{14}C-propionate to [14]CO_2 in leucocytes, but normal oxidation of methylmalonate and succinate, indicating deficiency in conversion of propionyl-CoA to D-methylmalonyl-CoA because of a deficiency in the activity of the mitochondrial, biotin-dependent enzyme propionyl-CoA carboxylase (PCC) (Fig. 6.2(a)).

The enzyme defect has also been demonstrated in preparations of liver mitochondria (Gompertz *et al.* 1970), and cultured skin fibroblasts (Hsia *et al.* 1971; Gompertz *et al.* 1975). A biotin-responsive patient was described by Barnes *et al.* (1970); and the combined biotin-responsive deficiency of three mitochondrial carboxylases, namely PCC, 3-methylcrotonyl-CoA carboxylase, and pyruvate carboxylase by Bartlett and Gompertz (1976), Weyler *et al.* (1977), and Bartlett *et al.* (1980).

Genetic heterogeneity amongst patients with propionic acidaemia has been demonstrated by the characterization of five mutant classes designated *bio* (biotin responsive), *pccA*, *pccB*, *pccC* and *pccBC*, using cell hybridization techniques for production of heterokaryons (Gravel *et al.* 1977; Wolf *et al.* 1980). Three distinct patterns of complementation kinetics were reported by Wolf *et al.* (1980) who suggested that *bio* and *pcc* mutations affect different genes; that complementation between *pccA* and either *pccC* or *pccBC* lines is intergenic and involves subunit exchange and synthesis of new pcc molecules; and that complementation between *pccB* and *pccC* mutants is interallelic.

Serum propionic acid concentration may be 5.4 mM (Hommes *et al.* 1968), which is more than 1000 times that found in normal infants. An excess of odd-numbered fatty acids with 15 and 17 carbon atoms has been reported in liver (Hommes *et al.* 1968) and red cells (Gompertz *et al.* 1970).

The most striking feature is massive ketoacidosis, which is significantly greater than that seen in diabetes mellitus, occurring in neonatal infants who are normally resistant to ketogenesis. The ketoacidosis is due mainly to accumulation of propionic acid, acetoacetic acid, and 3-hydroxybutyric acid.

In addition, there is accumulation of methylcitrate, which can be formed from intramitochondrial condensation of propionyl-CoA with oxaloacetate (Ando *et al.* 1972); of propionylglycine, which can be formed from propionic acid and glycine (Rasmussen *et al.* 1972); of 3-hydroxypropionate, an intermediate in the ß-oxidation of propionate (Kaziro and Ochoa 1964); of tiglic acid, a derivative of isoleucine proximal to the block in propionic acidaemia; and of tiglylglycine (Rasmussen *et al.* 1972).

There is also a large excess of medium and longer chain ketones in the urine (Menkes 1966). These include butanone (which can be derived from the metabolism of isoleucine), pentanone, and hexanone, which are not usually found in metabolic acidosis due to diabetes mellitus or starvation.

There is frequently hyperglycinaemia and hyperglycinuria (Harris *et al.* 1981) which may be due to inhibition of the glycine cleavage pathway (Shafai *et al.* 1978; Kølvraa 1979).

Hyperammonaemia is common and may contribute to cerebral damage (Wolf *et al*. 1978; Harris *et al*. 1980). It may be due to inhibition of carbamyl phosphate synthetase by propionyl-CoA and tiglyl-CoA (Coude *et al*. 1979).

Hypoglycaemia has occurred in a few patients. Protein intolerance was investigated by Childs and Nyhan (1964), who found that administration of leucine, isoleucine, valine, threonine, and methionine each precipitated ketoacidosis. Deficiency of propionyl-CoA carboxylase together with deficiencies of 3-methylcrotonyl-CoA carboxylase and pyruvate carboxylase resulting from deficiency of holocarboxylase synthetase which catalyses the binding of biotin to the mitochondrial carboxylase has already been discussed under 3-methylcrotonicaciduria. When propionyl-CoA carboxylase deficiency is part of the multiple carboxylase deficiency syndrome, response to biotin is prompt, and can be demonstrated in cultured fibroblasts (Saunders *et al*. 1979; Wolf 1979), leucocytes (Wolf and Hsia 1978) or in patients (Barnes *et al*. 1970; Bartlett *et al*. 1980). However, isolated propionyl-CoA carboxylase deficiency is either unresponsive or only slightly responsive to biotin supplementation (Wolf 1980).

Diagnosis

The clinical features of severe neonatal acidosis are an indication for urgent assessment by gas–liquid chromatography. When a massive excretion of propionic acid is established, definitive diagnosis can be made by assay of PCC activity in leucocytes or cultured fibroblasts by demonstrating absence of propionyl-CoA-dependent fixation of $^{14}CO_2$ after incubation with ^{14}C-NaHCO$_3$ and propionyl-CoA and removing protein by precipitation (Gompertz *et al*. 1975). There is also a deficiency of incorporation of label from 1-^{14}C-propionate into cold trichloroacetic acid-precipitable protein in cultured fibroblasts (Morrow *et al*. 1976; Willard *et al*. 1976), an effect which is also demonstrable in fibroblasts from patients with methylmalonic acidaemia.

Genetic counselling and prenatal diagnosis

Propionic acidaemia has an autosomal recessive mode of inheritance. About half normal activity was found in parents and a sister of a patient reported by Hsia *et al*. (1971), but normal values were found in parents and other relatives by Gompertz *et al*. (1975).

The first prenatal diagnosis of an affected foetus was by amniocentesis in a pregnancy at 26 weeks' gestation because the diagnosis of the index case had not been made until then. Cultured amniotic cells showed deficient PCC activity but the pregnancy was allowed to continue and ended in the birth of an affected child who died at one year of age (Gompertz *et al*. 1973c; Gompertz 1975). Prenatal diagnosis has also been reported by gas chromatographic and mass spectrometric analysis of amniotic fluid which revealed excess of methylcitrate (see 'Molecular defects', p. 294) (Sweetman *et al.* 1979). This was confirmed by the demonstration of reduced PCC activity in cultured amniotic cells.

In another pregnancy at risk for propionic acidaemia, however, with an elevated amniotic fluid methylcitrate, the cultured amniotic cells showed normal PCC activity. An affected female infant was born at 37 weeks' gestation. These discordant results can be explained by assuming contamination of the amniotic fluid by maternal cells (Buchanan *et al.* 1980). The authors comment that at San Diego they have performed prenatal diagnosis for propionic acidaemia in six pregnancies. In all three affected pregnancies there was a raised concentration of methylcitrate. Currently, therefore, it would seem that the demonstration of elevated amniotic fluid methylcitrate may turn out to be useful for prenatal diagnosis of propionic acidaemia, but until further experience is gained, tests on the amniotic fluid supernatant should be confirmed by direct assay of cultured amniotic cells (Buchanan *et al.* 1980; Fensom *et al.* 1984*a*). Only one newborn with hyperglycinaemia and propionic acidaemia was detected in screening of 350 000 infants in Massachusetts (Levy 1973).

Treatment

During severe ketoacidotic crises, often precipitated by infection, sodium bicarbonate should be administered intravenously, and dietary protein discontinued. Abnormal metabolites should be eliminated rapidly by haemodialysis, peritoneal dialysis or exchange transfusions (Robert *et al.* 1972; Russell *et al.* 1974). Biotin, 5 mg daily orally, causes clinical and biochemical improvement in patients with the combined carboxylase defects (Barnes *et al.* 1970; Bartlett *et al.* 1980). Prognosis in the biotin-unresponsive cases is poor even if treatment is started immediately after birth and carefully controlled (Gompertz 1975; Wolf *et al.* 1981). Ketoacidosis is reduced, but not abolished, when the patient is treated by a low protein diet (0.5 to 1.5 g/kg per day).

The methylmalonic acidaemias

The methylmalonic acidaemias are a heterogenous group of disorders with differences in clinical phenotypes and in the causal metabolic defects. Methylmalonic acidaemia in humans was first detected in patients with pernicious anaemia (Cox and White 1962) and is a useful test for demonstrating vitamin B_{12} (cobalamin) deficiency in establishing the diagnosis of pernicious anaemia in adults. Methylmalonic acidaemia in infants was first reported by Oberholzer *et al.* (1967) and Stokke *et al.* (1967). There is defective conversion of L-methylmalonyl-CoA to succinyl-

CoA (Fig. 6.2(a)) with accumulation of methylmalonyl-CoA. In about half the cases the disorder is due to a mutation of the methylmalonyl-CoA mutase apoenzyme. In the remainder, there is clinical and biochemical response to the administration of cobalamin, the cofactor to the mutase reaction, and considerably better prognosis with treatment. The cobalamin-responsive cases tend to present later in the first year with a less severe illness. Unlike most other types of vitamin-responsive enzymopathies, in cobalamin-responsive methylmalonic acidaemia there is a defect in the biosynthesis or availability of the active form of the vitamin or cofactor, rather than a mutation of the apoenzyme creating increased requirements of the vitamin.

By enzymatic assay of cultured fibroblasts and by somatic cell hybridization techniques, six mutants have been identified to date. Two are mutations of the mutase apoenzyme and have been designated mut⁻ and mut⁰; two (cblA, cblB) involve a defective step in the conversion of hydroxycobalamin (OH Cbl) to adenosylcobalamin (AdoCbl) inside the mitochondria (Fig. 6.2(b)); and in two others (cblC, cblD) there is deficient synthesis of both AdoCbl and methylcobalamin (MeCbl). The latter is the cofactor of the enzyme 5-methyltetrahydrofolate:homocysteine methyltransferase (5-methyltranferase), which catalyses one of the reactions for remethylation of homocysteine to methionine. In these last two disorders (cblC, cblD deficiency), there is both methylmalonic acidaemia and homocystinaemia. These disorders are distinct from combined methylmalonic acidaemia and homocystinaemia arising from defective absorption of cobalamin from the gut (q.v.).

In the literature there is also a report (Kang *et al.* 1972) of a single patient, who died of methylmalonic acidaemia at 11 days of age, in whom the defective enzyme was thought to be methylmalonyl-CoA racemase (Fig. 6.2(a)); subsequent studies, however, established a mutase defect (Willard and Rosenberg 1979*b*).

Clinical features

Forms not associated with abnormalities of sulphur amino acid metabolism. In patients with the cobalamin-unresponsive form there is a severe illness in the neonatal period with tachypnoea and grunting respiration, usually starting within a few days of birth. There is vomiting, dehydration, and drowsiness with metabolic acidosis and ketosis, leading to coma and death usually with pulmonary infection (Morrow *et al.* 1969). Onset later in the first year tends to occur in some patients with the cobalamin-responsive types who present with failure to thrive, lethargy, vomiting, and delayed development. Affected infants of either type are liable to recurrent infections to which they react unfavourably with ketosis, drowsiness, and sometimes coma. Hyperglycinaemia and hyperglycinuria (Oberholzer *et al.*

1967; Rosenberg 1978) occur in the majority of patients and hypoglycaemia in about half. There may be thrombocytopenia and neutropenia but no overt cobalamin deficiency, as evidenced by normal serum cobalamin concentration, and absence of megaloblastic anaemia. In cobalamin-responsive patients, administration of either adenosylcobalamin (Rosenberg *et al.* 1968; Lindblad *et al.* 1969), or cyanocobalamin causes clinical improvement with marked reduction of methylmalonic aciduria, correction of acidosis, and disappearance of thrombocytopenia and neutropenia, and in the longer term correction of growth retardation and developmental delay.

The clinical presentation, response to cobalamin therapy, and long-term outcome in 45 patients with methylmalonic acidaemia (15 mut^0, 5 mut$^-$, 14 cblA and 11 cblB) has been reviewed retrospectively by Matsui *et al.* (1983). The long-term outcome appeared to be most favourable for the cblA patients, 69 per cent of whom were alive and well at the time of review, 23 per cent alive with neurological sequelae, and 8 per cent dead. Prognosis was worst for the mut^0 patients, 40 per cent of whom were alive but brain damaged, and 60 per cent dead. The cblB and mut$^-$ patients fared less well than the cblA group, but better than the mut^0.

Forms associated with abnormalities of sulphur amino acids. These have been described under: Combined deficiency of 5-methyltetrahydrofolate: homocysteine methyltransferase and of methylmalonyl-CoA mutase (p. 160).

Molecular defects

Mutase deficiency. Deficiency of liver methylmalonyl-CoA mutase activity in patients with methylmalonic acidaemia was shown to be either severe (less than 1 per cent of normal) and unresponsive to AdoCbl, or not so marked, and restored to normal levels by addition of AdoCbl (Morrow *et al.* 1969). It was argued that in the former there was a defect in methylmalonyl-CoA mutase apoenzyme and in the latter a defect in the biosynthesis of AdoCbl, or in its availability. Deficiency of the mutase apoenzyme has been subdivided into one type designated mut^0, with absent activity (Morrow *et al.* 1975; Willard and Rosenberg 1977, 1980) and in another type, designated mut$^-$, where the mutase apoenzyme has reduced affinity for AdoCbl (Willard and Rosenberg 1977, 1980; Morrow *et al.* 1978) with activities two–75 per cent of control values.

Defects in cobalamin biosynthesis. Using somatic cell hybridization techniques, four complementation subgroups have been identified in cultured fibroblasts from cobalamin-responsive patients. In two (designated cblA and cblB), the defect involves the conversion of hydroxycobalamin (OH-

Cbl, cobalamin III, the Roman numeral indicating the valency of the cobalt atom, i.e. Co^{3+}) to adenosylcobalamin within the mitochondria (Fig. 6.2(b)). This involves sequential reduction of cobalamin III to cobalamin II, and then to cobalamin I by one, or possibly two, Cbl reductases using NADH as hydrogen donor, followed by adenosylation catalysed by the enzyme ATP:cob(I)alamin adenosyltransferase. Deficiency in the activity of the last-named enzyme has been demonstrated in the cblB mutant. In the other two types of mutants (cblC, cblD) with defects of cobalamin biosynthesis (Gravel *et al.* 1975; Willard and Rosenberg 1980), there is deficient synthesis of both AdoCbl and MeCbl. The latter is cofactor to the cytosolic enzyme 5-methyltetrahydrofolate:homocysteine methyltransferase, which catalyses remethylation of homocysteine to methionine, and concomitant demethylation of the cosubstrate 5-methyltetrahydrofolate to tetrahydrofolate (see Fig. 5.1). The metabolic consequences are modest methylmalonic aciduria without ketoacidosis, homocystinuria, and hypomethioninaemia.

The defects in cblC and cblD mutations have not been specifically localized. It has been shown that in these mutants there is normal uptake of hydroxycobalamin (OH Cbl) into the cell, bound to the serum protein transcobalamin II (TC II) and mediated by specific binding to cell surface receptors; and normal subsequent release of OH Cbl by lysosomal hydrolysis of the TCII-Cbl complex (Youngdahl-Turner *et al.* 1978, 1979; Willard *et al.* 1978). However, it has been shown that intracellular binding of Cbl to protein, which in normal cells is usually MeCbl to methyltransferase (Mellman *et al.* 1979*a,b*), is markedly deficient in cells belonging to cblC or cblD complementation groups. For further discussion see 'Molecular defects and diagnosis of combined deficiency of 5-methyltetrahydrofolate:homocysteine methyltransferase, and of methylmalonyl-CoA mutase' (p. 160).

An additional cause of methylmalonic aciduria combined with homocystinuria is deficient absorption of vitamin B_{12} in spite of normal gastric intrinsic factor glycoprotein and normal serum B_{12} binding capacity (Gräsbeck *et al.* 1960; Imerslund and Bjornsted 1963; Mohamed *et al.* 1965). The disorder responds to parenteral administration of vitamin B_{12} (Hollowell *et al.* 1968, 1969). This is also discussed under defects of homocysteine remethylation.

Diagnosis

The methylmalonic acidaemias should be suspected in infants with metabolic acidosis, and confirmed by demonstrating excess methylmalonic acid in urine or plasma by gas–liquid chromatography either alone or combined with mass spectrometry. A micromethod using gas–liquid chromatography is described by Nakamura *et al.* (1976).

A useful screening test using intact cultured fibroblasts is measurement of incorporation of ^{14}C-propionate into protein (precipitated by cold trichloroacetic acid) (Morrow *et al.* 1976; Willard *et al.* 1976). The method tests the integrity of the whole metabolic pathway from propionate to the tricarboxylic acid cycle and subsequent conversion of its components via amino acids to protein.

The more specific assay for methylmalonyl-CoA mutase for cultured fibroblasts is described by Morrow *et al.* (1969) and Willard and Rosenberg (1977); for assay of ATP:cob(I)alamin adenosyl transferase activity, by Fenton and Rosenberg (1978) and for conversion of OH Cbl to Ado Cbl by Mahoney *et al.* (1975a). Methods for analysis of cultured fibroblasts by complementation groups are reviewed by Willard and Rosenberg (1979a,b, 1980).

Genetic counselling and prenatal diagnosis

The mutase defects mut^0 and mut$^-$ and the cblA, cblB and cblC complementation group defects have an autosomal recessive mode of inheritance. This is evidenced by the existence of similar numbers of affected males and females, by absence of the disease in parents, and for the cblB defect (Fenton and Rosenberg 1978) and mut^0 and mut$^-$ defects (Willard and Rosenberg 1979a,b) by the reports of reduced activities of the adenosyltransferase or mutase enzymes in the parents. The only two children in the cblD complementation group are males (Gravel *et al.* 1975; Youngdahl-Turner *et al.* 1978) so that X-linked inheritance currently remains a possibility. Cultured fibroblasts from four of five heterozygous for the mut$^-$ defect had evidence of codominant expression *in vitro*; that is, they exhibited complex mutase kinetics consistent with the presence of both the normal mutase and mut$^-$ mutase. For the latter an apparent K_m for AdoCbl has been estimated as 1000 times that of normal fibroblast mutase (Willard and Rosenberg 1979a).

The gene for 5-methyltransferase has been assigned to chromosome 1 (Mellman *et al.* 1979a).

Prenatal diagnosis by demonstrating accumulation of methylmalonic acid was first reported by Morrow *et al.* (1970). They noted a trace of amniotic fluid methylmalonic acid at 17 weeks' and 0.4 mg/100 ml at 25 weeks' gestation. Maternal urine also had elevated methylmalonic acid (approximately 7 mg/day) at 25 weeks' gestation.

Subsequently it has been reported that maternal urine may show normal levels of methylmalonic acid in affected pregnancies (Mahoney *et al.* 1975b) but accumulation has been demonstrated in amniotic fluid at 17 weeks' (Gompertz *et al.* 1974b) and at 19 weeks' (Ampola *et al.* 1975). Although at the time of writing demonstration of accumulated methyl-malonic acid appears reliable for prenatal diagnosis, as in all other disorders discussed it is desirable to confirm abnormalities in the amniotic fluid supernatant by direct assay of cultured amniotic cells. This was first reported by Mahoney *et al.* (1975*b*) in two pregnancies. In one at risk for defective AdoCbl synthesis the disorder was excluded in cultured amniotic cells collected by amniocentesis at 12 weeks' gestation, by demonstration of normal production of $^{14}CO_2$ from ^{14}C-propionate, and ^{14}C-methylmalonate, as well as by a normal intracellular B_{12} content; in the other, amniocentesis was at 16 weeks, and an affected foetus was correctly predicted by demonstrating that cultured amniotic fluid cells had less than 2 per cent of mutase activity and undetectable production of $^{14}CO_2$ from ^{14}C-propionate or ^{14}C-methylmalonate.

Ampola *et al.* (1975) accurately diagnosed a foetus with methylmalonic acidaemia due to defective AdoCbl synthesis, and treated during the last two months of pregnancy by administration of cyanocobalamin to the mother, first 10 mg orally, then 5 mg intramuscularly per day. The second treatment was followed by a progressive decrease of methylmalonate in maternal urine, and the birth of an affected child whose growth and development were normal until three and a half months of age.

Treatment

Treatment of acute metabolic acidosis is described below (see treatment of organic acidaemias). After collection of blood samples for assay of vitamin B_{12} and transcobalamin concentrations, and of urine for assay of organic acids and homocystine, treatment should be started by daily intramuscular injection of one mg of hydroxycobalamin to assess whether the disorder is responsive. Urinary excretion of methylmalonic acid is measured daily. Responsive patients may have a reduction from pretreatment levels of 10–50 mmol (1–5g) per day to about 1–3 mmol (100–300 mg) per day as compared to normal daily excretion of less than 0.05 mmol (5mg). In patients who are not in an acute acidosis, hydroxycobalamin injections can be reduced from daily to twice weekly in some cases (Mahoney and Rosenberg 1970).

Treatment of organic acidaemias

Acute acidosis

The aim is to restore acid-base balance by intravenous bicarbonate solution, to eliminate dietary protein which is a source of organic acids and branched-chain amino acids, and to maintain sufficient supply of calories by administration of glucose or other carbohydrate. The rise of plasma bicarbonate concentration, which is initially reduced, and of blood pH, which can drop to under 7.0 are used to monitor alkali therapy. In life-threatening acidosis, peritoneal dialysis (Russell *et al.* 1974) or haemodialysis help to reduce acidosis. Infections are a common provoking cause of acute acidotic episodes and should therefore be treated promptly.

Maintenance therapy

When the acute episode has been controlled, protein can be gradually reintroduced, initially no more than 0.5 mg/kg body weight, slowly increased to 1.5 g/kg body weight in milder forms.

Specific treatments will depend on the type of disorder; for example, in 3-hydroxy-3-methyglutaryl-CoA lyase deficiency, lethal hypoglycaemia (Schutgens *et al.* 1979*b*) was prevented by regular supplementation of glucose (10 per cent w/v, 30 ml, seven times a day) (Duran *et al.* 1979). The parents should be warned that even trivial infections can precipitate acidosis, and should be advised to seek early antibiotic therapy where appropriate.

Vitamin supplementation

Therapy with thiamin for MSUD, with vitamin B_{12} for methylmalonic acidaemia, and with biotin for 3-methylcrotonylglycinuria and the combined carboxylase deficiency type of propionic acidaemia has already been described.

DICARBOXYLIC ACIDURIA

Suberylglycinuria; systemic carnitine deficiency

Specific enzyme defects have not yet been confirmed in patients with dicarboxylic acidurias, except for the glutaric acidurias which are considered under 'Disorders of lysine and hydroxylysine metabolism' (p. 226).

A male infant of first cousins died with severe metabolic acidosis and lactic aciduria at 15h of age. An earlier brother had died at 26h with respiratory distress, hypoglycaemia and metabolic acidosis (Borg *et al.* 1972; Linstedt *et al.* 1976). The urine contained excess of the dicarboxylic acids adipic (C-6), suberic (C-8), sebacic (C-10), dodecanedioic (C-12) and tetradecanoic (C-14). The authors suggested that the disorder could be due to a defect in medium-chain acyl-CoA dehydrogenase.

A child with proximal myopathy, vomiting, hepatomegaly, hypoglycaemia, encephalopathy, carnitine deficiency and an increased excretion in the urine of dicarboxylic acids including adipic, suberic and octenedioic acids was studied by Karpati *et al.* (1975), and Engel and Angelini (1973). The decreased serum carnitine was implicated with a decrease in transport of fatty acids between cytoplasm and mitochondria. It was suggested that ω -oxidation may have compensated for impaired ß-oxidation.

A boy who died at two and a half years after an illness characterized by attacks of hypoglycaemia, hypotonia and encephalopathy was reported by

Gregersen *et al.* (1976). During attacks he excreted excessive amounts of adipic, suberic and sebacic acids, but between attacks urinary organic acids were normal, apart from isolated excessive excretion of suberylglycine. A similar pattern was reported in one of three affected siblings (Truscott *et al.* 1979*a*).

DISORDERS OF FATTY ACID METABOLISM

Muscle carnitine palmitoyltransferase deficiency

Although glycogen is the chief source of energy during short periods of strenuous muscular exercise, fatty acids become important during moderate, sustained exercise, or at rest. It is not surprising, therefore, that disturbances of voluntary muscle should arise in disorders of long-chain fatty acid metabolism. Hitherto, only defects of carnitine palmitoyl-transferase (CPT) activity have been characterized. CPT catalyses the transfer of fatty acids into the mitochondria. It has been suggested that the enzyme exists in two forms, CPT I and CPT II, the first catalysing the synthesis of palmitoyl carnitine on the outer surface of the inner mitochondrial membrane and the second the conversion of palmitoyl carnitine to palmitoyl-CoA and carnitine on the inside of the inner mitochondrial membrane (Hoppel and Tomec 1972).

Several clinical syndromes have been reported in association with reduced CPT activity. Di Mauro and Di Mauro (1973) reported a 29-year-old man who from the age of 16 years had had episodic muscle cramps and 'pigmentosuria' which often, but not always, followed physical exercise by some hours. The authors comment that prolonged, rather than strenuous, exercise appeared to cause symptoms. These circumstances would coincide with the switch of the preferential utilization to fatty acids as substrates, from glycogen as stated above. The patient's only brother had identical symptoms. There was a normal rise in venous blood lactate following ischaemic exercise. Muscle biopsy CPT activity was measured by three different methods which yielded values of 0–20 per cent of controls. Production of $^{14}CO_2$ from ^{14}C-palmitate was undetectable at intervals up to 30 min. Normal values are not cited, but *in vitro* normal mitochondria oxidised $^{14}CO_2$-palmitate at a mean rate of 1.3 nmol of $^{14}CO_2$/min/mg of mitochondrial protein. Since the utilization of palmitate by isolated mitochondria from the patient was more impaired than utilization of palmitoyl carnitine, the authors suggest that the defect affected CPT I more severely than CPT II. Bank *et al.* (1975) reported two brothers aged 29 and 33 years with myoglobinuria and renal failure, but no muscle cramps. There was marked decrease in CPT activity. The brothers were asymptomatic when on a high carbohydrate diet.

A similar disorder may have afflicted the 18-year-old twin sisters with cramps and myoglobinuria reported by Engel *et al.* (1970). Symptoms were usually, but not always, induced by exercise, but could also be prevented by fasting or a high-fat diet. The authors suggested that the disorder was caused by a defect in the utilization of long-chain fatty acids by skeletal muscle. They demonstrated a normal rise in blood lactate during ischaemic exercise, but did not assay muscle CPT.

Adrenoleucodystrophy; sudanophilic leucodystrophy; Schilder's disease; Addison disease and cerebral sclerosis; 'bronzed Schilder's' disease; adrenomyeloneuropathy

Clinical features

First described by Siemerling and Creutzfeld (1923), adrenoleucodystrophy (ALD) is an X-linked disorder characterized by progressive demyelination of cerebral white matter and reduced adrenal response to ACTH. Adrenomyeloneuropathy (AMN) is a related condition with a later onset and involvement of spinal cord or peripheral nerves, rather than brain (Griffin *et al.* 1977).

Most commonly, the clinical onset of ALD is between four and eight years of age. In a review of 17 cases, Schaumberg *et al.* (1975) reported that behavioural changes were the initial manifestation in 10 patients. These consisted of poor school performance, gradually failing memory, dementia or emotional outbursts. Other early features were spastic gait, hemiparesis, hyperreflexia, dysarthria, or dysphasia. Visual disturbances were common and included homonymous hemianopia, visual agnosia and loss of visual acuity. Hearing loss was the initial manifestation in two patients. Duration of the disease from onset of neurological symptoms to death varied between nine months and nine years.

Blaw (1970) reported that over half the patients with ALD had a history of adrenal insufficiency for as long as four years before the onset of neurological disease, but Schaumberg *et al.* (1975) and Moser *et al.* (1980*b*) found serious neurological disability in the absence of clinically evident adrenal insufficiency. Nevertheless, even though adrenal failure may not be clinically evident, almost all patients show a diminished adrenocortical response to ACTH stimulation.

ALD with onset in the neonatal period has been described by Ulrich *et al.* (1978) and Manz *et al.* (1980). Two further cases, a brother and sister who died at one and a half and two years, respectively, are mentioned by Moser *et al.* (1980*b*). These children showed very small adrenals with lipid inclusions, moderate leucodystrophy, and severe involvement of the thymus.

Some female carriers of ALD show a progressive paraparesis and sphincter disturbances, but no signs of adrenal insufficiency have been found so far (Marmion *et al.* 1979; O'Neill *et al.* 1980; Moser *et al.* 1980*b*).

In contrast to ALD, AMN presents in the third decade as a progressive paraparesis combined with sensory changes and a peripheral neuropathy (Griffin *et al.* 1977). There is adrenal insufficiency and there may also be impaired gonadal function. Progression of the disease occurs over 20 years or more, cerebellar ataxia and intellectual deterioration being late manifestations. Males are predominantly affected, and the mode of inheritance is probably X-linked (see 'Genetic counselling', p. 307). A reported occurrence of both ALD and AMN in the same kindred (Davis *et al.* 1979) indicates a close relationship between the two diseases.

Pathology

Severe demyelination, particularly in the occipital and posterior parietal areas with a spread of the destruction in a caudal-rostral direction, is the main pathological finding in ALD. In the adrenal gland ballooned cortical cells are apparent, many having a striated cytoplasm, with macrovacuoles, in the zona reticularis and fasciculata (Schaumberg *et al.* 1975).

Similar ultrastructural changes to those occurring in ALD occur in AMN (Schaumberg *et al.* 1977).

Biochemical features

Although at the time of writing the fundamental biochemical defects in ALD and AMN have not been elucidated, work carried out by Moser and his collaborators has led to the development of diagnostic procedures for confirming the diseases in suspected hemizygotes, identifying carriers and carrying out prenatal diagnosis, as well as providing insight into the metabolic processes in which the basic defects may reside.

The work orginated from the observation that a specific excess of very long-chain fatty acids (longer than C_{22}) occurs in the brain and adrenal gland cholesterol esters (Igarashi *et al.* 1976*a*) and brain gangliosides (Igarashi *et al.* 1976*b*) of ALD patients. The accumulated fatty acids are unbranched and of carbon chain length 23 to 32, but with acids of chain length 25 and 26 most prominent. Significant excesses of these fatty acids were subsequently reported in cultured fibroblasts (Kawamura *et al.* 1978; Moser *et al.* 1980*c*) and cultured muscle cells (Askanas *et al.* 1979; McLaughlin *et al.* 1980).

In cultured fibroblasts the increase was most apparent when expressed as the ratio of $C_{26:0}$ (hexacosanoic acid) to $C_{22:0}$ (docosenoic acid), the content of the latter acid being normal in ALD and AMN. Thus, the mean ratio for 26 ALD patients was 0.705 ± 0.192 S.D.; for 10 AMN patients, 0.728 ± 0.161 S.D.; for 2 neonatal ALD patients, 0.504; for 24 healthy controls, 0.0766 ± 0.032 S.D.; and for 39 controls with diseases other than ALD, 0.0803 ± 0.050 S.D. (Moser *et al.* 1980*b*). No overlap occurred between the disease and control groups. In fibroblasts from 22 obligate and six possible heterozygotes the mean ratio was 0.415 ± 0.223 S.D. but three of these values overlapped the control range. It was thus concluded that the assay would probably identify nearly 90 per cent of ALD carriers.

More recently, Moser *et al.* (1981) have shown that measurement of very long-chain fatty acids in plasma using capillary gas–liquid chromatography can also be used to diagnose ALD and AMN and to assist in identification of carriers.

Since the most striking biochemical abnormality in ALD, and the one to be demonstrated first (Igarashi *et al.* 1976*a*) was the presence of excess of very long-chain fatty acids in the cholesterol esters of brain and adrenal cortex, it was postulated that the disease was due to a defect in cholesterol ester metabolism. Studies in two laboratories, however, failed to confirm this (Ogino *et al.* 1978; Michels and Beaudet 1980; Ogino and Suzuki 1981). Moser *et al.* (1980*b*) reasoned that since the abnormal fatty acid accumulation occurs in several classes of lipids, including gangliosides (Igarashi *et al.* 1976*b*), sphingomyelin (Kawamura *et al.* 1978), and free fatty acids (Ramsey *et al.* 1979), in addition to cholesterol esters, the basic defect is more likely to involve the metabolism of the acids themselves. Evidence for this hypothesis came from studies of $^{14}CO_2$ production from $(1-^{14}C)$-labelled fatty acids by cultured fibroblasts and adrenal cortex (Singh *et al.* 1981). It was found that the CO_2 released by fibroblast homogenates from ALD patients was on average 8.2 per cent of control values when $(1-^{14}C)$-lignoceric acid $(C_{24:0})$ was used as substrate and 14 per cent of control when $(1-^{14}C)$-hexacosanoic acid was the substrate. Decreased oxidation of $(1-^{14}C)$-oleic acid $(C_{18:1})$ was less pronounced in the ALD fibroblasts (two strains in fact showing an increase over controls). The CO_2 production by adrenal glands from an ALD patient was 5 per cent of simultaneous controls with $(1-^{14}C)$-oleic acid as substrate.

While these findings suggest a deficiency of very long chain fatty acid oxidation in ALD, further studies are required to elucidate the specific enzymatic reaction involved.

Genetic counselling and prenatal diagnosis

For the identification of heterozygotes, Moser *et al.* (1981) recommend that the $C_{26:0}$ to $C_{22:0}$ ratio be measured in both cultured fibroblasts and plasma in order to minimize the chance of a false negative result.

Evidence for X-linkage of ALD, and for inactivation of the locus involved, has been provided by Migeon *et al.* (1981) who cloned skin fibroblasts from heterozygotes of three families. Two types of clone were obtained in which the cells had either a normal ratio of $C_{26:0}$ to $C_{22:0}$ fatty acids, or an excess of $C_{26:0}$ to $C_{22:0}$, similar to that seen in cells from affected males. Moreover, it was found that in most of the heterozygotes studied there were significantly more clones of abnormal than normal type, indicating an *in vitro* selective advantage for the mutant cells. Migeon *et al.* (1981) also demonstrated a close linkage between the ALD and glucose 6-phosphate dehydrogenase (G6PD) loci because in two families there were no recombinants among a total of 18 informative offspring of doubly heterozygous mothers. This linkage provides a means for prenatal diagnosis of ALD in families with cosegregating variants at the ALD and G6PD loci. More directly, prenatal diagnosis of an affected foetus is also possible by demonstrating abnormal $C_{26:0}$ fatty acids in cultured amniotic cells of male karyotype (Moser *et al.* 1982).

One of the obligatory ALD heterozygotes studied by Migeon *et al.* (1981) (II-6 in family 3) showed clinical signs which could be interpreted as diagnostic of AMN. Since cloning of her fibroblasts revealed the presence of both normal and mutant cells, female cases of AMN may well be clinically affected ALD heterozygotes. However, the possibility of another form of AMN showing a mode of inheritance other than X-linked cannot be excluded.

It is interesting that if the selective advantage of mutant cells *in vitro* in ALD heterozygotes also occurs *in vivo*, which seems likely from the studies of Migeon *et al.* (1981), an explanation is provided for the fact that many heterozygotes show some clinical features of the disease. The situation contrasts with that in some other X-linked disorders (e.g the Lesch–Nyhan syndrome) in which *in vivo* selection appears to favour normal cells (Nyhan *et al.* 1970).

Treatment

No successful treatment of ALD has been reported but dietary restriction of very long chain fatty acids is a possible course of action. After oral administration of deuterium-labelled hexacosanoic acid to a terminally ill ALD patient, Kishimoto *et al.* (1980) showed at necropsy that label was introduced into cholesterol esters of white matter, thus providing evidence for dietary origin of at least some of the accumulating fatty acids. However, since it is known that endogenous synthesis of these long chain acids also occurs (Cassagne *et al.* 1978; Murad and Kishimoto 1978), the effectiveness of their exogenous restriction – given the design of a nutritious diet – will depend on the endogenous synthesis being a relatively unimportant cause of long chain fatty acid accumulation in ALD.

7. Disorders of purine and pyrimidine metabolism

Hyperuricaemia (gout)

In the great majority of patients with gout, the cause of purine overproduction is not known. Possible mechanisms of hyperuricaemia causing clinical gout are discussed in relation to the Lesch–Nyhan syndrome and glycogen storage disease type I, and increased activity of phosphoribosylpyrophosphate synthetase (PRPPS). Other causes of 'secondary' gout include disorders associated with myeloproliferation and lymphoproliferation, such as polycythaemia, either primary (vera) or secondary, multiple myeloma and myeloid leukaemia (Yü 1965), or drugs such as thiazide diuretics (Healey and Hall 1970).

Clinical features

Acute gouty arthritis has a sudden onset and involves most commonly the metatarso-phalangeal joint of the great toe, but also ankle, wrist, knee, or other joints. Often tissues around the metatarso-phalangeal joint are swollen and the overlying skin has a dusky red colour. There is sometimes a history of intensive exercise involving the affected joint the day before the attack. For example, running or prolonged walking may precipitate an attack affecting the metatarso-phalangeal joints, while playing squash or gardening may provoke an attack at the wrist. The arthritis usually remits spontaneously within a few days or weeks, or more quickly with treatment (q.v.).

Without effective treatment, a variable number of years after the onset of acute gouty arthritis about two-thirds of patients develop chronic gouty arthritis; in some cases this follows an acute attack. Chronic gout may also be complicated by further acute attacks. In chronic gouty arthritis, there is usually persistent pain, especially after use of affected joints. In severe cases there is considerable painful limitation of joint movement. With persistent hyperuricaemia, urate deposits (tophi) develop in tissues. These tend to form palpable nodules in periarticular tissues, such as cartilage and bone near epiphyses, and may grow to several centimetres in diameter. Tophi in the helical and antihelical cartilages of the ears are also characteristic. Local necrosis surrounding tophi may cause destruction of subcutaneous tissues and skin with sinus formation. Extensive bone destruction is often evident radiologically, giving a 'punched-out' appearance, especially in the bone adjacent to the articular cartilage of the inter-phalangeal joints of the hands and feet.

A particularly serious complication is the deposit of urates in the kidneys, leading to progressive renal failure. Yü and Gutman (1967) reported that 22 per cent of 1258 patients with primary gout, and 42 per cent of 59 patients with secondary gout developed urolithiasis.

Pathology

The primary pathological lesions in gout are local inflammatory and foreign-body reactions to urate deposits which are surrounded by lymphocytes and giant cells. Local necrosis of periarticular bone and cartilage follows. In the kidneys, urate crystals and surrounding reactions are found in the interstitial tissue of the cortex and medula and in the pyramids.

Treatment and genetic counselling

See under the Lesch–Nyhan syndrome, glycogen storage disease type I, and PRPPS overactivity.

The Lesch–Nyhan syndrome

The main features of this X-linked disease include a motor disorder, compulsive aggressiveness including self-mutilation, mental retardation and hyperuricaemia. It is due to a deficiency in the activity of hypoxanthine–guanine phosphoribosyl transferase (HGPRT).

Clinical features

In hemizygote males the clinical presentation is usually delayed motor development in the first six months, and choreoathetosis in the second six months. This is followed by clinical features of spasticity with exaggerated tendon reflexes and adductor spasm of the lower limbs. The disorder was first described by Lesch and Nyhan (1964).

A characteristic feature is an irresistible urge to inflict painful self-mutilation manifested after the second year, often as biting off pieces or tips of fingers. An interesting feature is that affected children appear to be aware of their inability to prevent causing damage to themselves and may show a desire to be restrained by becoming agitated when protective devices are removed. Sometimes the craving may be for unilateral self-destruction, so that only one elbow need be splinted. Though common, this feature is not specific and has been absent in a few cases (Partington and Hennen 1967) and occurs in other disorders for example, in children with indifference to pain (Gillespie and Perucca 1960). Indiscriminate aggressiveness towards others is also common.

About half the reported cases have had fits. Intellectual retardation is usually severe to moderate but may be absent (Scherzer and Ilson 1969).

Other features described in patients with Lesch–Nyhan syndrome include nephrolithiasis (Nyhan *et al.* 1965), megaloblastic anaemia (Van der Zee *et al.* 1969) and immunological dysfunction, associated with a decrease in B-cell lymphocytes and IgG concentration (Allison *et al.* 1975). Most patients exhibit growth retardation and are under the third centile in weight (Partington and Hennen 1967). Typical gouty arthritis has been reported (Riley 1960) but is rare.

Pathology

Sodium urate deposition in shrunken kidneys has been observed in patients at necropsy (Partington and Hennen 1967; Crussi *et al.* 1969). No consistent change has been reported in the central nervous system, either to the naked eye or histologically. In one case there was loss of cells in the granular layer of the cerebellar cortex (Seegmiller 1968).

Molecular defects

Increased synthesis of uric acid is a constant finding. This has been demonstrated as 200-fold increase in the rate of *de novo* synthesis measured as the incorporation of radiolabelled glycine into uric acid over a period of several days (Lesch and Nyhan 1964; Nyhan *et al.* 1965). The deficient enzyme HGPRT (Seegmiller *et al.* 1967) normally catalyses the transfer of the phosphoribosyl moiety of 5-phosphoribosyl 1-pyrophosphate (PRPP) to the 9 position of hypoxanthine to yield inosine monophosphate (IMP) or of guanine to yield guanosine monophosphate (GMP) (Fig. 7.1(a)). These reactions are part of a recycling or 'salvage pathway', which is much more efficient in terms of energy expenditure for maintenance of the purine pool than *de novo* synthesis of purines from ATP, ribose 5-phosphate, glutamine, glycine, aspartate, formate, and bicarbonate. Activity of HGPRT is absent or very low in fibroblasts (Fujimoto and Seegmiller 1970; Kelley and Meade 1971), leucocytes (Kelley *et al.* 1969; Henderson *et al.* 1974) and erythrocytes (Arnold *et al.* 1972).

Several variants have been reported from different families which are characterized by altered properties of the residual enzyme. These include altered thermostability (Kelley *et al.* 1967), electrophoretic mobility (Bakay and Nyhan 1972), substrate specificity (Kelley *et al.* 1967), and enzyme kinetics (McDonald and Kelley 1971). Normally, the excretion of xanthine in the urine is three to 20 times higher than that of hypoxanthine but in the Lesch–Nyhan syndrome, where conversion of hypoxanthine to IMP is absent, the ratio of hypoxanthine to xanthine is reversed to values approximating 2:1 to 3:1 (Balis *et al.* 1967). Cross-reacting material (CRM)

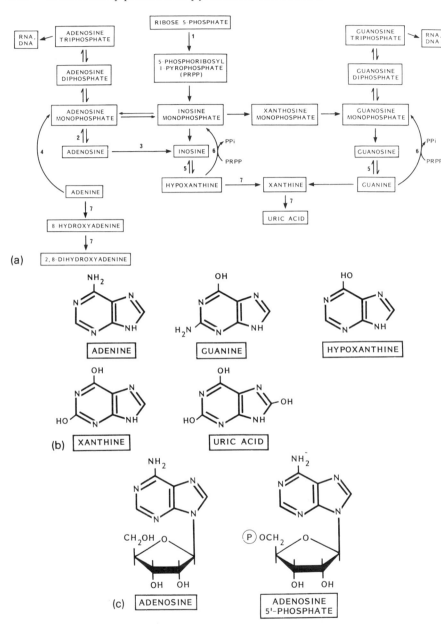

Fig. 7.1. (a) Purine metabolism; *de novo* biosynthesis and salvage pathways: (1) 5-phosphoribosyl 1-pyrophosphate synthetase (2) Adenosine kinase (3) Adenosine deaminase (4) Adenine phosphoribosyl transferase (5) Purine nucleoside phosphorylase (6) Hypoxanthine-guanine phosphoribosyl transferase (7) Xanthine oxidase.

(b) The structures of the purine bases. (c) The structures of adenosine and adenosine 5'-phosphate.

to antibodies against normal HGPRT has been found in only one of 14 cases studied (Upchurch *et al.* 1975).

In erythrocytes (Greene *et al.* 1970) and cultured fibroblasts (Rosenbloom *et al.* 1968) from affected patients, the concentration of PRPP is increased. In the case of fibroblasts, this increase appears to result from decreased utilization, presumably because it is a cosubstrate in the HGPRT reaction. Elevation of PRPP causes an increase in *de novo* purine synthesis (Kelley *et al.* 1970) and may be the principal cause of such enhancement in the Lesch–Nyhan syndrome. The mechanism of stimulation of *de novo* purine biosynthesis seems to be activation of a rate-limiting key enzyme amidophosphoribosyltransferase by conversion of a catalytically inactive high molecular form to an active lower molecular form (Holmes *et al.* 1973). Another theoretical mechanism is release of inhibition of amido-phosphoribosyltransferase as a result of possible decreased intracellular concentrations of a product of the HGPRT reaction, GMP, which normally encourages conversion of the active to the inactive form of the enzyme (Holmes *et al.* 1973).

Hyperuricaemia is present in the majority of patients, but may not be constant (Partington and Hennen 1967). The pyrimidine nucleotides UTP and CTP have been reported to be increased up to six-fold in lymphoblasts from a patient with Lesch–Nyhan syndrome (Nuki *et al.* 1973). This may be due to stimulation of the first step in *de novo* synthesis of pyrimidines catalysed by glutamine-dependent carbamyl phosphate synthetase II.

Several erythrocyte enzymes have been reported to have increased activities in patients with Lesch–Nyhan syndrome. These include adenine phosphoribosyl transferase (Fig. 7.1(a)) (Seegmiller *et al.* 1967), IMP dehydrogenase, which catalyses conversion of IMP to xanthosine mono-phosphate (XMP) (Pehlke *et al.* 1972), orotate phosphoribosyltransferase and orotidine 5'-phosphate decarboxylase (Beardmore *et al.* 1973).

Increased activity of PRPP synthetase (Fig. 7.1(a)) has been reported in cultured fibroblasts and lymphoblasts (Reem 1975) but not in erythrocytes. It is noteworthy that in some families with gout and purine overproduction (Zoref *et al.* 1977; Takeuchi *et al.* 1981), PRPP synthetase activity is increased and exhibits resistance to feedback inhibition by GDP and ADP. As a result, there is an increased concentration of PRPP and resulting stimulation of *de novo* purine synthesis causing hyperuricaemia and gout.

A less striking deficiency in the activity of HGPRT than in the Lesch–Nyhan syndrome has been reported in patients with gout, but without neurological involvement (Kelley *et al.* 1969).

Increased activity of dopamine ß-hydroxylase has been reported in patients with self-mutilation (Rockson *et al.* 1974).

The relation between the biochemical and clinical phenotypes is poorly understood. In rats, the administration of the purine caffeine has been shown to cause self-mutilation (Boyd *et al.* 1965) but the mechanism is unknown. The possibility that the deficiency of GMP or IMP or the accumulation of neurotoxic substances may damage the CNS is speculative.

Several somatic cell hybridization studies have yielded interesting information on HGPRT. For example, Sekiguchi and Sekiguchi (1973) have demonstrated interallelic complementation as the appearance of HGPRT activity in hybrids derived from diploid clones of HGPRT-deficient Chinese hamster cells, indicating the existence of at least two alleles which can affect HGPRT activity. Other workers have demonstrated the appearance of HGPRT activity in mouse A9 cells which are HGPRT deficient (HGPRT⁻) after transformation with Chinese hamster cells which are HGPRT⁺. Transformed cells, unlike untransformed A9 cells, can be recovered because of their capacity to survive in HAT medium (containing hypoxanthine, aminopterine, and thymidine) (McBride and Ozer 1973; Goss and Harris 1975).

Particularly interesting are experiments which appear to indicate derepression of the murine HGPRT locus in HGPRT⁻ murine cells after fusion with either human HGPRT⁺ (Watson *et al.* 1972) or HGRPT⁻ cells (Bakay *et al.* 1975).

In contrast to permanent transformation by cell hybridisation, temporary correction of HGPRT⁻ cells can be produced by contact with HGPRT⁺ cells (Cox *et al.* 1970) or extracts from HGPRT⁺ cells (Ashkenazi and Gartler 1971). Several mechanisms have been suggested for such metabolic co-operation including transfer of IMP (Cox *et al.* 1972) or of a pronase-sensitive corrective factor, possibly a protein (Ashkenazi and Gartler 1971).

Diagnosis

The demonstration of hyperuricaemia often leads to the diagnosis, but normal blood uric acid concentration does not rule out the disorder (Partington and Hennen 1967). Excretion of uric acid over a 24-hr period needs to be corrected for body weight and was normal in one of five patients (Mitchener 1967). A screening method which is probably as reliable is the ratio of uric acid to creatinine in a random sample of urine compared to control values of the same age (Kaufman, *et al.* 1968).

Definitive diagnosis is established by demonstrating a deficiency of HGPRT (see 'Molecular defects', p. 311). HGPRT activity is usually measured by incubating cell extracts with ³H-hypoxanthine and PRPP and separating the product ³H-inosinic acid for radioassay by paper (De Bruyn *et al.* 1976), thin-layer (Crabtree and Henderson 1977), or column (Nyhan *et al.* 1967), chromatography, or by electrophoresis on cellulose acetate (Kizaki and Sakurada 1976). A spectrophotometric assay in which

IMP production is measured as NADH by linking with IMP dehydrogenase was described by Newcombe and Willard (1971).

Autoradiography can also be used to demonstrate HGPRT activity in intact fibroblasts or leucocytes as incorporation of ³H-hypoxanthine into nucleic acids (Rosenbloom *et al.* 1967).

Genetic counselling

An X-linked mode of inheritance has been established by pedigree analysis, lack of male to male inheritance, absence of reported female patients (Shapiro *et al.* 1966; Nyhan *et al.* 1967), and the demonstration by cloning of two populations of cells in heterozygous females, one with and one without HGPRT activity (Migeon *et al.* 1968). The gene has been localized to the long arm of the X-chromosome (Shows and Brown 1975).

Carrier females have been identified using autoradiography (Rosenbloom *et al.* 1967) but this method is subject to error because of metabolic co-operation, that is by correction of cells lacking the enzyme by adjacent cells with normal HGPRT activity. Attempts at overcoming this difficulty have included the use of selective media containing 8-azaguanine or 6-thioguanine, which cannot be incorporated into nucleic acids of HGPRT⁻ cells, but can be incorporated into nucleic acids of normal cells which then cannot grow and replicate (Fujimoto *et al.* 1971).

Two populations of cells in carriers have also been demonstrated by hair root analysis (Gartler *et al.* 1971).

The incidence of Lesch–Nyhan syndrome has been estimated as 1:300 000 (Seegmiller 1976).

Female heterozygotes usually have no symptoms but may excrete excessive uric acid in their urine (Hoefnagel *et al.* 1965) and have accelerated *de novo* purine synthesis as measured by incorporation of ¹⁴C-glycine into uric acid (Emmerson and Wyngaarden 1969).

Prenatal diagnosis was first reported by Fujimoto *et al.* (1968), who demonstrated two populations of amniotic cells, one of which could, and the other could not, incorporate ³H-hypoxanthine into nucleic acids, and accurately predicted a heterozygote. DeMars *et al.* (1969) diagnosed an affected foetus at 28 weeks' gestation. The pregnancy continued and ended with the birth of affected undiagnosed twins. A correct prenatal diagnosis of an affected foetus was reported by van Heeswijk *et al.* (1972) two weeks after amniocentesis at 18 weeks' gestation, using a quantitative microassay.

Treatment

All attempts at preventing the progressive neurological disorder have proved unsuccessful.

Allopurinol, by inhibiting xanthine oxidase activity, will lower urine and blood uric acid concentrations and thus prevent urolithiasis. This treatment is sometimes complicated by xanthine stone formation (Greene *et al.* 1969*b*), but this can usually be controlled by monitoring uric acid levels so that the dose of allopurinol is kept to the minimum which will prevent hyperuricaemia; and by pressing fluids to increase the urinary flow.

The administration of adenine, GMP, AMP, orotic acid, α-methyldopa, or dietary purine restriction have produced no benefit.

Exchange blood transfusion produced a small temporary fall in serum urate concentration but no improvement in neurological function (Watts *et al.* 1974).

Patients should be restrained appropriately to diminish self-mutilation. This includes use of splints for elbows, extraction of teeth and protection by 'crash-helmets'. Choreoathetosis is sometimes improved by diazepam, haloperidol, or barbiturates.

Gout with increased phosphoribosyl pyrophosphate synthetase activity

Gout with hyperuricaemia due to accelerated *de novo* purine biosynthesis, resulting probably from increased activity of phosphoribosylpyrophosphate synthetase (PRPPS) (Fig. 7.1(a)) has been described in five families (Henderson *et al.* 1968; Sperling *et al.* 1972; Becker *et al.* 1973, 1980; Takeuchi *et al.* 1981). This is associated with an increased intracellular concentration of 5-phosphoribosyl 1-pyrophosphate (PRPP) (Becker *et al.* 1973), which is a high-energy sugar, and an important intermediate in the synthesis of both purines and pyrimidines. The enzyme catalyses the synthesis of PRPP from ATP and ribose 5-phosphate.

Increased activity of PRPPS has been demonstrated in erythrocytes (Takcuchi *et al.* 1981) and in cultured fibroblasts (Zoref *et al.* 1975). In the latter, activity was resistant to feedback inhibition by guanosine 5'-diphosphate and adenosine 5'-diphosphate (Zoref, *et al.* 1975; Becker *et al.* 1980).

PRPPS from affected male and female subjects is inactivated more rapidly at 55°C than the enzyme from normal individuals. The rate of inactivation in affected females is intermediate between that in affected males and normal individuals (Takeuchi *et al.* 1981).

Mothers of affected males have been reported to have increased PRPPS activity and to have two populations of cells, one with normal and one with elevated PRPPS activity consistent with X-linkage. This has been shown in lymphocytes by autoradiography (Takeuchi *et al.* 1981), and in cultured fibroblasts by demonstrating survival and enrichment in the number of cells with high PRPPS activity in selective media containing 6-methyl-mercaptopurine-riboside, which does not allow survival of cells with normal PRPPS activity (Zoref *et al.* 1977).

Uric acid lithiasis occurred in the family described by Sperling *et al.* (1972).

Treatment of hyperuricaemia empirically is by administration of the xanthine oxidase inhibitor allopurinol (Rundles *et al.* 1966), or, in patients who are intolerant to this drug, by uricosuric agents such as probenecid (Gutman and Yü 1957) or sulphinpyrazone (Kuzell *et al.* 1964).

Adenine phosphoribosyltransferase deficiency

Clinical features

Complete deficiency of adenine phosphoribosyltransferase (APRT) activity has been reported in five children, four of whom presented with renal calculi, without arthritis, or neurological or haematological abnormalities (Cartier and Hamet 1975; Van Acker *et al.* 1977; Barratt *et al.* 1979; Hirsch-Kauffman and Doppler 1981). The fifth child was the asymptomatic brother of an index patient (Van Acker *et al.* 1977).

Since the parents and other members of the families who were probable heterozygotes for APRT deficiency had no evidence of gout, the association between partial APRT deficiency, gout, and hyperuricaemia in a female (Emmerson *et al.* 1975) may have been coincidental.

Symptomatic onset was at six, nine, twelve and eighteen months, with passage of gravel, abdominal colic, dysuria, pain in the urethra, and urinary tract infections.

The calculi were radiotranslucent.

Molecular defects

The disorder is due to complete deficiency of APRT (Fig. 7.1(a)) which has been demonstrated in erythrocytes. Hirsh-Kauffmann and Doppler (1981) reported 0.02 per cent of normal activity in cultured fibroblasts, but were unable to demonstrate any immunologically cross-reacting material in the patient's erythrocytes.

In the absence of APRT activity adenine cannot be re-utilized by the salvage pathway and is oxidized first to 8-hydroxyadenine and then to 2,8-dihydroxyadenine, both reactions being catalysed by xanthine oxidase.

In the urine, adenine and its oxidation metabolites, which are not detectable in control urines, account for 25–34 per cent of total purine metabolites. About 50 per cent of the adenine metabolites is 2,8-dihydroxyadenine, the others being adenine and 8-hydroxyadenine.

The total excretion of purines including uric acid is within normal limits (Barratt *et al.* 1979). This is in sharp contrast to the situation in the Lesch–Nyhan syndrome (q.v.) in which purine excretion is considerably increased, and suggests that adenine nucleotides (e.g. the products of the APRT reaction) are not important as inhibitory regulators in feedback control of, *de novo* purine production.

Urolithiasis results from crystallisation of 2,8-dihydroxyadenine which has a solubility of only 1–3 mg/l H_2O.

Diagnosis

The identification of 2,8-dihydroxyadenine has been described by Simmonds *et al.* (1976). Calculi consist of approximately 90 per cent 2,8-dihydroxyadenine, the remainder being mainly uric acid. Adenine and its oxidation products in urine may be identified by u.v. scanning after isotachophoretic separation.

Assay of APRT activity in red cell lysates is described by Cartier and Hamet (1968).

Genetic counselling

The disorder has an autosomal recessive mode of inheritance. This is shown by the presence of a partial enzyme deficiency in parents, who, however, have normal adenine metabolism. One of the five patients had consanguineous parents (Barratt *et al.* 1979). The incidence of hetero-zygosis has been estimated as 0.5 per cent to 1.0 per cent (Emmerson *et al.* 1977; Fox 1977).

Since APRT deficiency is expressed in cultured fibroblasts, the disorder could probably be diagnosed prenatally by assay of cultured amniotic cells.

The gene for APRT is located on chromosome 16 (Tischfield *et al.* 1974*a*). Nineteen strains of cultured human trisomy 16 cells had 69 per cent greater mean APRT activity than control cells. This value does not differ significantly from the predicted 50 per cent increase (Marimo and Giannelli 1975).

Treatment

Administration of the xanthine oxidase inhibitor allopurinol will decrease oxidation of adenine. At a dose of 20 mg/kg/24 h, 2,8-dihydroxyadenine could no longer be detected in the urine (Barratt *et al.* 1979). With this treatment, no further stone formation has occurred.

Myoadenylate deaminase deficiency

Over 25 patients have been reported with deficiency of muscle AMP (myoadenylate) deaminase. Typical presentation is excessive fatigue of voluntary muscle, with cramps and muscle pains, after strenuous exercise. The first five patients were described by Fishbein *et al.* (1978). Other patients were reported by DiMauro *et al.* (1980), Schumate *et al.* (1980), and Sabina *et al.* (1980). Onset of symptoms has ranged from the second to the seventieth year.

There is absent activity of muscle AMP deaminase (isoenzyme A). Any residual activity is probably due to contamination by blood (Fishbein *et al.* 1981) since immunological studies reveal absence of cross-reacting material (Fishbein *et al.* 1980, 1981).

The activity of AMP deaminase in fibroblasts, leucocytes and erythrocytes in these patients is normal (Fishbein *et al.* 1978, 1980; DiMauro *et al.* 1980).

The enzyme deficiency leads to a block in the purine nucleotide cycle, AMP → IMP → AMP, one turn of which generates one molecule of fumarate and ammonia, and utilizes one molecule of aspartate, the net effect being the deamination of aspartate (Tornheim and Lowenstein 1972).

The mechanism of exercise-induced muscle dysfunction is not understood, but failure to synthesize increased amounts of either IMP or ammonia (Tornhein and Lowenstein 1974, 1975) may be involved as a mechanism for delayed regeneration of ATP.

Other causes of exercise-related muscle disturbances are discussed under 'Muscle cramps and myoglobinuria' (Chapter 8, p. 353).

Since the sex incidence is approximately equal, and parents are unaffected, autosomal recessive inheritance is probable. The mother of an affected child had reduced activity of muscle AMP deaminase (DiMauro *et al.* 1980).

Adenosine deaminase deficiency; combined immunodeficiency disease

Combined immunodeficiency disease (CID) is characterized clinically by recurrent severe bacterial, viral, and monilial infections with onset from about three to six months of age, leading to death usually by two to five years. There is deficiency of both B and T cell-mediated immunity. Chronic diarrhoea, pulmonary infections, and perioral candidiasis are almost invariable. Death has been caused by pulmonary abscesses, pneumonia due to *Pneumocystis carinii*, varicella, measles, and generalized vaccinia or B.C.G. infection following vaccination against smallpox or with B.C.G., respectively. Affected children are often below the third centile for weight.

In some cases, inheritance of CID is autosomal recessive with consanguinity in as many as a third of the patients (Hitzig and Willi 1961), while in other families it is sex-linked (Gitlin and Craig 1963); in others the disease is sporadic.

Deficiency in the activity of red cell adenosine deaminase (ADA) (Fig. 7.1(a)) was reported in two females with CID by Giblett *et al.* (1972) and in one female with CID by Dissing and Knudsen (1972). The deficiency was discovered on routine investigation of genetic markers prior to bone marrow transplant. The latter authors did not report the association until stimulated to do so by the report of Giblett *et al.* (1972). The enzyme catalyses the deamination of adenosine to inosine, and of deoxyadenosine to deoxyinosine. ADA deficiency is present in about half the patients with CID and was reported in 12 of 22 patients by Meuwissen *et al.* (1975). All reported cases have had an autosomal recessive mode of inheritance. The enzyme deficiency has also been demonstrated in serum (Meuwissen *et al.* 1975), cultured fibroblasts (Hirschhorn *et al.* 1976), lymphocytes, and liver (cited by Meuwissen *et al.* 1975), and at 22 weeks' gestation in foetal kidney, liver, and spleen (Ziegler *et al.* 1981).

No definite difference in the clinical features has been reported in patients with CID who are either ADA positive or negative. Radiologically, only one of 13 patients had a characteristic thymus shadow. This was present at seven months, but had disappeared by two years and eight months (Meuwissen *et al.* 1975). Multiple skeletal abnormalities are present in about three-quarters of the patients and are found only in those with severe CID. These include pelves which are broad with squared, flared ilia, long horizontal acetabular roofs, broad ischia, concave at their lower margins, and small sacrosciatic notches. The vertebrae are flat or slightly convex and in the thoracic and lumbar spine exhibit platyspondily. In half the patients the anterior rib ends are concave and flared (Meuwissen *et al.* 1975).

Meuwissen *et al.* (1975) reported malfunction or deficiency of T-cells in four of five ADA-negative children; none of 12 patients had normal PHA-responsive lymphocytes and only one of nine exhibited a normal mixed leucocyte reaction. The same authors reported malfunction or deficiency of B-cells in four of five patients, with reduced IgG in four, reduced IgM in ten and reduced IgA in nine of twelve ADA-negative children. Other immunological abnormalities include failure to develop anti-A or anti-B, or antibody responses to diphtheria, pertussis, tetanus, or typhoid vaccines; persistence of a positive Schick test after immunization against diphtheria; absence of delayed hypersensitivity response to candida, streptokinase–streptodornase, histoplasmin, and purified protein derivative; and failure to become sensitised to dinitrochlorobenzene (Giblett *et al.* 1972).

Genetic polymorphism of erythrocyte ADA was discovered by Spencer *et al.* (1968), who demonstrated three phenotypes representing homozygous or heterozygous expression, designated ADA1, ADA2 or ADA2–1, of the alleles ADA^1 or ADA^2. Their data indicated that ADA^2 has a frequency of 0.06 in Europeans, 0.04 in Negroes and 0.11 in Asiatic Indians. Rarer alleles were described by Hopkinson *et al.* (1969), Dissing and Knudsen (1969), and Detter *et al.* (1970). Segregation of a 'silent' allele which in the homozygous state was associated with CID was described by Chen *et al.* (1974).

The gene for ADA has been assigned to chromosome 20 by mouse/human somatic-cell hybrids (Tischfield *et al.* 1974*b*). Regulatory control of the human ADA locus was demonstrated by Siciliano *et al.* (1978) who showed that fusion of ADA–deficient human choriocarcinoma cells with mouse fibroblasts produced hybrids with human ADA^1 and ADA^2 activity.

Histological studies have revealed striking differences between the thymus of patients who are ADA positive or negative. Thus although the thymus is considerably reduced in size in both groups, characteristically the thymus in ADA-positive patients lacks Hassall's corpuscles and differentiation between cortical and medullary zones. Such changes have been attributed to arrested development at an embryonic stage. In contrast, in ADA-negative patients, Hassall's corpuscles and differentiated medullary epithelial cells are present. These changes have been interpreted as being consistent with normal early development and later, possibly even postnatal, thymic involution. This explanation is consistent with observations that in the early stages of the disease there may be evidence of normal B-cell function, for example the presence of some antibodies and normal levels of immunoglobulins at nine months of age that later decreased to very low levels (Giblett *et al.* 1972). Another child had at seven months of age a normal thymus radiological shadow which later disappeared (Meuwissen *et al.* 1975). However, in one patient lymphopenia has been reported at birth (Hirschhorn and Martin, 1978).

The relation between ADA deficiency and the immunological dysfunction is not known. It has been suggested that immunodeficiency results from accumulation and toxicity of adenosine or adenine nucleotides, including ATP and cyclic AMP which inhibit purine or pyrimidine biosynthesis (Green and Chan 1973; Kredich and Herschfield 1983). More convincing is the possibility that deoxyadenosine rather than adenosine is toxic to proliferating lymphoid cells. Thus urinary purines from a child with ADA deficiency have been shown to be deoxyadenosine, rather than adenosine (Simmonds *et al.* 1979). Moreover, deoxyadenosine is inhibitory *in vitro* at the much lower concentration of 1 μmol/l, compared with adenosine, which becomes inhibitory at 100 μmol/l.

Prenatal diagnosis of ADA deficiency has been reported after amniocentesis at 28 weeks' gestation (Hirschhorn *et al.* 1975) by assay of cultured amniotic cell ADA activity. Amniocentesis at 14 and 16 weeks' gestation was also reported in successive pregnancies at risk for ADA deficiency. Assay of cultured amniotic cell ADA activity showed the pregnancies to be normal, and homozygous deficient, respectively (Ziegler *et al.* 1980).

Heterozygotes have had reduced ADA activities in erythrocytes (Giblett *et al.* 1972; Chen *et al.* 1974) and leucocytes (Ziegler *et al.* 1980).

Treatment

Two types of treatment have proved beneficial; firstly, bone marrow transplantation (BMT), which, when an HLA-compatible, MLR-negative sibling donor is available, is the method of choice; and secondly, erythrocyte exchange transfusion. The former aims at permanent engraftment of stem cells which contain ADA. The latter treatment requires repeated transfusions since erythrocytes have a finite survival.

BMT has restored complete immunological function (Parkman *et al.* 1975; Pahwa *et al.* 1978). In addition, evidence of ADA activity following BMT is the subsequent reduction of erythrocyte deoxy ATP (Chen *et al.* 1978). However, although in view of the immunodeficiency the graft is usually not rejected, graft-versus-host reaction remains a dangerous possibility (Hirschhorn 1980).

In most cases, however, an HLA-compatible, MLR-negative sibling donor is not available. In these, repeated exchange transfusion may restore some immunological function, but not as completely as following BMT (Hirschhorn 1980). Successful correction of the immunodeficiency by frozen irradiated red blood cells from normal donors has been reported by Polmar *et al.* (1976), but failure has been reported by others (Ziegler *et al.* 1980). Ziegler *et al.* (1980) suggested that response to enzyme replacement, whether by erythrocyte or plasma infusion, may be more likely when there is appreciable residual ADA activity. Moreover, in responsive patients, lymphocytic mitogenic responses may be restored by exogenous ADA *in vitro* (Polmar *et al.* 1975), in contrast to unresponsive patients (Ziegler *et al.* 1980). According to Ziegler *et al.* (1980), other features predictive of unresponsiveness are early clinical presentation with associated severe skeletal abnormalities. It is interesting that treatment with infusions of plasma and frozen irradiated erythrocytes which was begun at seven months of age, was followed by resolution of skeletal abnormalities, but not of the immunological malfunction.

Purine nucleoside phosphorylase deficiency; T-cell immunodeficiency

The clinical features have been variable in the first four families described, and not as severe as in ADA deficiency (Stoop *et al.* 1976, 1977; Giblett *et al.* 1975; Cohen *et al.* 1976; Griscelli *et al.* 1976; Biggar 1977). In three families there was a history of recurrent pyogenic, monilial, and viral infections, including otitis media, pneumonia, candidiasis and, in two families, progressive generalized vaccinia. In the first family reported (Giblett *et al.* 1975), there was also anaemia and lymphopenia, and in another family radiological thymus shadow was absent (Stoop *et al.* 1976). In all four families there was evidence of defective T-cell immunity, including absent or decreased stimulation of peripheral lymphocytes by mitogens, such as pokeweed antigen or phytohaemagglutinin, or by allogenic cells. As in the case of ADA deficiency, some patients show a progressive decrease in T-cell immunity (Stoop *et al.* 1977). Abnormalities of B-cell function have also been reported to a lesser extent. These include a decreased number of B-cells, and antibody deficiency.

All affected individuals tested have had severe deficiency or absence of red cell and lymphocyte purine nucleoside phosphorylase (PNP) (Fig. 7.1(a)), the enzyme which catalyses cleavage of purine nucleosides to free bases, with concomitant phosphorylation of ribose by inorganic phosphate to yield ribose 1-phosphate.

The disorder probably has an autosomal recessive mode of inheritance, since it has been reported in sibling daughters whose parents were healthy and had approximately 50 per cent PNP activity (Giblett *et al.* 1975; Stoop *et al.* 1976).

The enzyme defect leads to an accumulation of the nucleosides inosine and guanosine in serum and urine (Siegenbeek van Heukelom *et al.* 1976, 1977). It has been suggested that such nucleoside accumulation might be the cause of the immunodeficiency (Stoop *et al.* 1977). It is noteworthy that, since the defect in PNP is one metabolic step from that in the Lesch–Nyhan syndrome, one of the affected children had retardation in motor development and a spastic tetraparesis (Stoop *et al.* 1977).

Xanthine oxidase deficiency; hereditary xanthinura

The commonest presentation has been nephrolithiasis with its attendant complications, which occurs in about a third of patients. The diagnosis is usually suspected because of a low serum uric acid concentration, which is the direct effect of the enzyme deficiency. Xanthine oxidase catalyses successive oxidation of hypoxanthine to xanthine, and of xanthine to uric acid (Fig. 7.1(a)). The disorder can present in childhood, as in the girl aged four and a half years reported by Dent and Philpot (1954), and followed up by Dickinson and Smellie (1959), but about half the patients present after the age of 20 years.

A few patients have had pains in voluntary muscles due to deposits of xanthine and hypoxanthine crystals (Chalmers *et al.* 1969*a, b*; Parker *et al.* 1970*b*).

The demonstration by Dent and Philpot (1954) that in affected patients the chief urinary purine was xanthine rather than uric acid suggested a deficiency of xanthine oxidase activity. This was confirmed by Watts *et al.* (1964), who demonstrated severe deficiency in jejunal mucosa and liver biopsies. The activity could not be restored by addition of molybdenum, ferric iron, or flavine adenine dinucleotide, all of which are cofactors in the electron flow during purine oxidation (Olson *et al.* 1974). Watts *et al.* (1965) were unable to demonstrate xanthine oxidase activity in normal leucocytes. This observation is at variance with the data of Castro–Mendoza *et al.* (1972) who found activity in leucocytes from normal controls, but not in those from a patient.

The total concentration of plasma oxypurines has been either normal or variably elevated, but fractionation in a patient revealed that xanthine comprised 90 per cent (Sorensen *et al.* 1972).

Urinary excretion of purines has also been variable but typically there is a large excess of urinary xanthine and hypoxanthine, with xanthine predominating, and a decrease of uric acid.

It has been determined that the solubility of xanthine ranges from 67 mg/l of urine at pH 5.8 to 165 mg/l at pH 8.1. Since the xanthinuria in affected patients is in the order of 100 to 500 mg/l, crystallization with calculus formation is a constant threat.

Inheritance is probably autosomal recessive since the disorder has been described in a brother and sister, while most potential heterozygotes are without symptoms but may excrete excessive quantities of urinary oxypurines (Wilson and Tapia 1974).

Treatment is unsatisfactory and includes a high fluid intake, alkalization of the urine by oral alkalis and in some cases the administration of allopurinol to inhibit residual xanthine oxidase activity and decrease conversion of hypoxanthine to xanthine (Engleman *et al.* 1964; Holmes *et al.* 1974). The former is more soluble than the latter, and is re-utilized by the salvage pathway catalysed by HGPRT, about 99 times faster than the latter.

Combined deficiency of sulphite oxidase and of xanthine oxidase was discussed under disorder of sulphur amino acid metabolism. Patients with this disorder may respond to molybdenum, 300 to 500μg per day (Abumrad *et al.* 1979).

Hereditary orotic aciduria

This disorder is a rare cause of vitamin B_{12}-unresponsive megaloblastic anaemia in young children. About half the patients have presented in the first year, the remainder between one and five years of age. The most common presenting features are lethargy and failure to thrive. In addition, there is usually moderate developmental delay with IQ scores of 70–80 (Becroft *et al.* 1969) but one seven and a half-year-old affected girl had normal growth and an IQ of 133 (Tubergen *et al.* 1969). Heavy crystalluria occurs almost invariably and can cause ureteral and urethral obstruction (Huguley *et al.* 1959), and haematuria (Tubergen *et al.* 1969).

Although immunological function has not been studied in detail, isolated observations suggest that there may be undue susceptibility to infection; thus, one patient died of varicella (Huguley *et al.* 1959) and another had absence of IgA and IgD (Haggard and Lockhart 1967).

Typically, there is severe hypochromic anaemia with haemoglobin concentrations as low as 4.6 g/dl (Becroft and Phillips 1965). Anisocytosis and poikilocytosis are marked. There is almost always leucopenia but no thrombocytopenia. The bone marrow shows megaloblastic changes.

A unique feature of the disorder is that in the great majority of patients there is reduced activity of two enzymes acting sequentially in the pathway of *de novo* pyrimidine synthesis. These are orotate phosphoribosyltransferase (OPRT) which catalyses the synthesis of orotidine 5′-phosphate from 5-phosphoribosyl 1-pyrophosphate (PRPP) and orotate (Fig. 7.2); and orotidine 5′-phosphate decarboxylase (ODC) which catalyses the decarboxylation of orotidine 5′-phosphate to uridine 5′-phosphate. Patients with the double enzyme deficiency are designated as having hereditary orotic aciduria type 1. In a single patient, with typical clinical features, there was an isolated deficiency of ODC (Fox *et al.* 1969, 1973) designated as hereditary orotic aciduria type II; however, following treatment with uridine, OPRT activity fell to 2 per cent of normal.

In the initial report (Huguley *et al.* 1959), erythrocytes showed 20–25 per cent residual activity of both OPRT and ODC. Rogers *et al.* (1968) measured the conversion of orotic acid to uridylic acid as $^{14}CO_2$ released from $6-^{14}C$-orotic acid which requires the activity of both OPRT and ODC. There was wide variation in activity of 19 controls who had activities ranging from 2.4 to 14.9 nmol $CO_2/10^9$ erythrocytes/h (units), but a clear deficiency in the index case, who had a value of 0.2 units. The same authors found biopsied liver OPRT and ODC activities of 1.5 per cent and 22 per cent of control mean values, respectively. In cultured fibroblasts Krooth (1965) found OPRT and ODC activities in one patient to be 9.1 per cent and 2.5 per cent of control mean values, respectively.

Excretion of orotic acid in the urine is in the region of 0.5 to 1.5 g/day, which represents an approximately 1000–2000-fold increase over normal. Urinary excretion of the ribonucleoside orotidine is variably increased up to 40-fold or more over normal (Fox *et al.* 1973).

The mechanism whereby the enzyme deficiency causes such a dramatic increase in orotic acid excretion is not known. It seems probable that there is an overflow from raised plasma concentrations but this does not appear to have been demonstrated directly. It may be calculated readily that such output of pyrimidines must be associated with overproduction, but the cause of enhanced *de novo* synthesis has not been elucidated. Stimulation

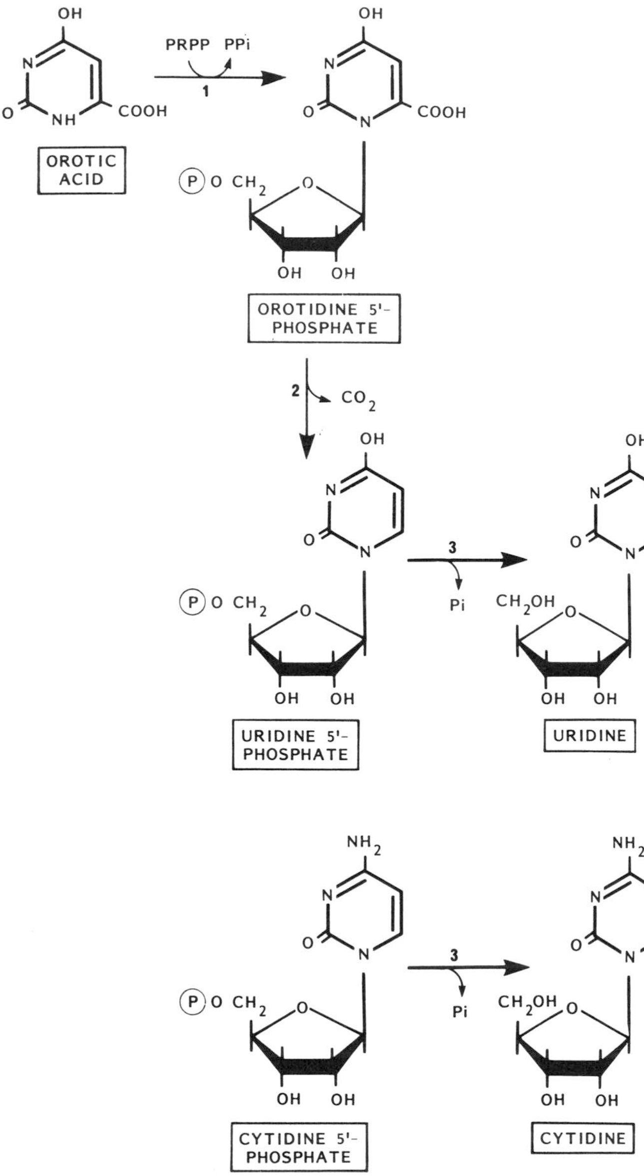

Fig. 7.2. Enzymes catalysing *de novo* pyrimidine synthesis: (1) Orotate phosphoribosyltransferase (2) Orotidine 5'-phosphate decarboxylase (3) Pyrimidine 5'-nucleotidase.

by PRPP of carbamyl phosphate synthetase II, which catalyses the first reaction of pyrimidine *de novo* synthesis, is an attractive possibility and is analogous to the suggested explanation for accelerated purine biosynthesis in the Lesch–Nyhan syndrome (q.v.). However, no increase in PRPP concentration was demonstrated in erythrocytes from one patient (Fox and Kelley 1971).

Orotic aciduria types I and II have an autosomal recessive mode of inheritance. Heterozygotes of both types have a partial deficiency in the activities of both OPRT and ODC, and an elevated excretion of orotic acid (Fox *et al.* 1973; Tubergen *et al.* 1969). We are not aware of attempts at prenatal diagnosis, but this is probably possible by demonstrating deficiency of OPRT and ODC in cultured amniotic cells.

In theory, reduction in the activity of two enzyme reactions resulting from a single gene defect could occur if a single protein could catalyse both reactions, but in calf thymus the two enzymes have been separated. However, after separation OPRT becomes unstable (Kasbekar *et al.* 1964). Thus if ODC is responsible for the integrity of OPRT, a mutation of the former might lead to inactivity of both enzymes. The two enzymes are present in the cytosol in a complex together with four other enzymes involved in the *de novo* synthesis of pyrimidines and it is possible that a mutation leading to disruption of this complex might lead to loss of activity of more than one enzyme. Other theoretical explanations for the double enzyme defect include a mutation affecting a common subunit, as occurs in ß-hexosaminidase A and B in Sandhoff's disease. Alternatively, if the loci of the two enzymes are on the same chromosome and closely linked, it is conceivable that at meiosis the OPRT locus on one of the chromosome pairs will pair aberrantly with the ODC locus of the other. By crossing over at the incorrectly placed synapsis, this could give rise to two new alleles specifying two types of polypeptide chains, one in each chromosome, which could lead to deficiency of both enzymes. This would be analagous to the situation found in patients with the Lepore haemoglobins in which aberrant pairing of ß- and δ-haemoglobin loci and crossing over give rise to a new gene δ-ß in one chromosome and ß-δ in the other, resulting in deficiency of the normal gene products HbA ($\alpha_2\beta_2$) and HbA$_2$ ($\alpha_2\delta_2$).

A further possibility is the existence of a mutation affecting regulation of enzyme activity.

There is evidence both for and against a regulatory defect. In favour is the demonstration that, in cells from patients with hereditary orotic aciduria, inhibitors of pyrimidine synthesis such as 6-azauridine can lead to an increase in the activities of OPRT and ODC to normal levels (Pinsky and Krooth 1967*a*,*b*). Evidence against a regulatory defect is the demonstration that in hereditary orotic aciduria both OPRT and ODC have abnormal electrophoretic mobilities and heat stabilities, suggesting that there is an alteration in enzyme structure (Worthy *et al.* 1974).

Orotic aciduria occurs in several disorders with hyperammonaemia, most notably in deficiency of ornithine transcarbamylase (Levin *et al.* 1969*a*,*b*) but also in argininosuccinic aciduria and citrullinaemia. Orotic aciduria may also be induced by several drugs, including 6-azauridine (Buttoo *et al.* 1965) and allopurinol.

Treatment

The anaemia does not respond to iron, vitamin B_{12}, folic or folinic acid. Administration of uridine, however, in doses of 150 mg/kg body weight/ day cures megaloblastosis and causes a brisk reticulocytosis, a steady increase in the concentrations of haemoglobin and of leucocytes and accelerates growth, but does not improve intelligence. It would be interesting to know whether intellectual impairment could be prevented by uridine supplementation from birth.

Xeroderma pigmentosum

Xeroderma pigmentosum (XP) is characterized clinically by skin which is abnormally sensitive to light, and by multiple neoplasms, and neurological and ocular complications. Several distinct genetic types are recognized, all of which exhibit defects in repair of ultra-violet radiation (UVR)-induced or chemically-induced DNA lesions.

Clinical features

Exposed areas of the skin appear normal at birth but soon exhibit excessive erythema, usually after solar exposure. There is delayed skin response to UVR at the short UVR wavelength (290–320 nm) of the solar spectrum (Ramsay and Giannelli 1975). Pigmented macules and dryness appear progressively during the first decade, when the skin becomes atrophied and develops telangiectasia (Figs 7.3 and 7.4). Usually in the second and third decade there is multiple neoplasia of the skin (Fig. 7.4), most commonly basal and squamous cell carcinomata, but also keratocanthoma, adenocarcinoma, melanomata, sarcomata, neuromata and angiosarcomata, causing early death. About 14 per cent of patients (Elsässer *et al.* 1950) develop progressive neurological complications. These are designated eponymously as the de Sanctis–Cacchione (1932) syndrome. There is choreoathetosis, ataxia, spasticity, and sometimes sensorineural deafness, microcephaly, and mental retardation. EEG abnormalities are common (Larmande and Tinsit 1955). Ocular lesions include recurrent conjunctivitis, keratitis, photophobia, and scarring of the eyelids with entropion.

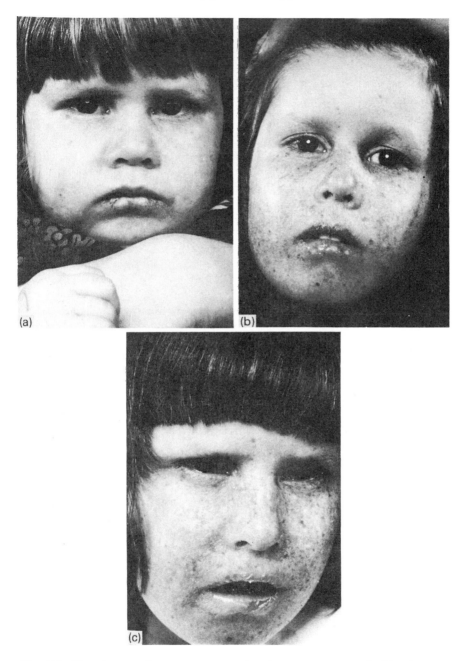

Fig. 7.3. Xeroderma pigmentosum; progressive appearance of skin dryness, atrophy and telangiectasia in a girl: (a) two and a half years (b) five years (c) six years.

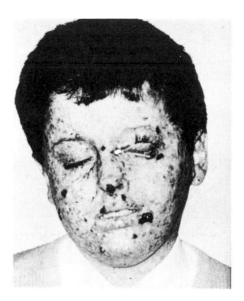

Fig. 7.4. Xeroderma pigmentosum in an eight-year-old boy; multiple neoplasia of the skin and right ectropion.

Pathology

The non-malignant areas of the skin show acantholysis and irregular pigmentation with increased numbers of melanocytes (Lynch *et al.* 1967). In patients with neurological complications there is cerebral atrophy, especially of the temporal and frontal lobes, with loss of neurones (Robbins *et al.* 1974).

Molecular defects

UVR wavelengths of 290–320 nm cause several types of damage to DNA, but the formation of dimers between adjacent pyrimidine bases on the same strand (Fig. 7.5) are considered the most critical (Setlow 1968). In normal cells these pyrimidine dimers are removed by the excision repair process (Fig. 7.5). The dimers which distort the adjacent DNA are identified and the damaged strand is incised and then removed with the dimer by excision catalysed by endonucleases. The resulting gap in the strand is filled by synthesis of new DNA (unscheduled DNA synthesis) catalysed by a DNA polymerase and a ligase.

In 80–90 percent of XP patients there is a block in the incision step so that they do not excise pyrimidine dimers (Setlow *et al.* 1969; Cleaver and Trosko 1970). In the remainder of patients, designated XP-variants, excision repair is normal (Cleaver 1972). These patients have a defect of a poorly understood type of DNA repair mechanism known as 'post-replication' or 'daughter strand' repair where UVR-induced dimers are not excised but bypassed during DNA replication (Lehmann *et al.* 1975).

When fibroblasts from different XP patients are hybridized, the defect in DNA repair may be corrected or not. If it is corrected, the patients presumably have a different genetic defect and are said to belong to a different complementation group. If the defect is not corrected, the patients are said to belong to the same complementation group and are considered to have the same genetic defect. By international agreement, five complementation groups of XP have been designated A–E. However, two patients do not belong to any of these groups and complement each

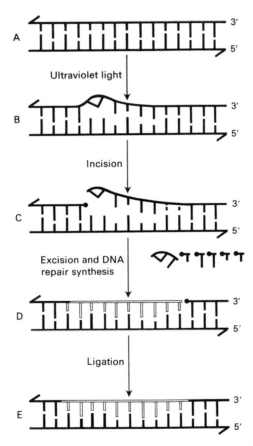

Fig. 7.5. Scheme of excision repair of UVL-induced pyrimidine dimers (unscheduled DNA synthesis) catalysed by sequential actions of endonucleases, polymerase and ligase.

other. There are therefore at least seven complementation groups (Takebe 1978; Bootsma 1978). When 16 patients investigated in this laboratory (Pawsey *et al.* 1979) were analysed with 16 other unrelated patients (De Weerd-Kastelein 1974; Kraemer *et al.* 1975), the majority were allocated to groups A (11 patients), C (13 patients) and D (six patients). The only patient in group B had the phenotype of Cockane's syndrome, including neurological complications (Robbins *et al.* 1974). Other patients with neurological complications were in complementation groups A (nine of 11 patients) or D (two of six patients). There were no neurological complications in patients in group C or in the single patient in group E.

Differences in the kinetics of complementation have been demonstrated for cells in groups A, C, D, and E (Giannelli and Pawsey 1974, 1976; Pawsey *et al.* 1979), supporting the hypothesis that these groups are genetically distinct.

In addition to the defect in excision repair of pyrimidine dimers, XP cells exhibit defects in certain chemical-induced DNA lesions produced by agents such as polycyclic hydrocarbons and aromatic amides. Abnormal responses include reduced survival or reduced repair synthesis (Pawsey *et al.* 1979).

The precise defects in the various complementation groups have not been characterised. Thus Cook *et al.* (1978) reported that XP fibroblasts and lymphocytes of complementation groups A, C or D could not excise UVR-damaged DNA, while Mortelmans *et al.* (1976) showed that extracts of cells from XP complementation groups A, C, or D could excise pyrimidine dimers from purified u.v.-irradiated DNA, but extracts of group A cells could not excise dimers of unfractionated cell-free preparations.

Little is known about possible heterogeneity amongst XP variants. These disorders are important in that they demonstrate that even patients with normal excision-repair function develop XP, and that therefore both excision repair and post-replication repair mechanisms are required to prevent the clinical XP phenotypes.

Diagnosis

The diagnosis of XP can be proved by assessment of unscheduled DNA synthesis in cultured fibroblasts after damaging DNA by exposure to a 30 W germicidal Hg vapour arc (95 per cent 254 nm u.v.radiation) and incubating with ^{3}H-thymidine. DNA repair is measured in non-replicating cells by counting the silver grains over the nuclei, visualized by autoradiography (Giannelli and Pawsey 1976). A considerable reduction of nuclear grains is seen in XP cell nuclei when compared with normal control cell nuclei.

In the 10–20 per cent of patients with normal unscheduled DNA synthesis who are XP variants, the diagnosis can be established by demonstrating an abnormal peak of low molecular weight DNA chains synthesized after u.v. radiation by centrifugation in an alkaline sucrose density gradient (Lehmann *et al.* 1975). This slow maturation of DNA chains is considerably inhibited by caffeine, when compared to similarly treated control cells.

It has been claimed that some cases have normal or almost normal excision and post-replication repair mechanisms, but the clinical phenotypes were not described (Lehmann *et al.* 1977; Takebe *et al.* 1977).

Genetic counselling

So far as is known, XP and XP variants have an autosomal recessive mode of inheritance. XP can be diagnosed prenatally by demonstrating defective DNA incision-repair in cultured amniotic cells (Ramsay *et al.* 1974; Halley *et al.* 1979).

No clinical evidence of skin abnormality was found in nearly 100 heterozygotes in Egypt (Hashem *et al.* 1980; Cleaver *et al.* 1981).

Carrier detection by measurement of unscheduled DNA synthesis is not reliable. Thus of 14 parents studied by Pawsey *et al.* (1979), four had values above 100 per cent of controls, and all 79 per cent or higher. Values of 98–116 per cent were reported in three parents by Stitch (1975) and of 42–48 per cent in two parents by Cleaver (1970).

As far as we know, prenatal diagnosis of XP variant has not been reported, but this should be possible using the technique of Lehmann *et al.* (1975).

Treatment

Prevention of skin lesions is possible only by protection from u.v. radiation. Strict avoidance of sunlight should be mandatory as soon as the diagnosis is established. This can be aided by the wearing of sun-barrier substances such as *p*-aminobenzoic acid or its derivatives, which absorb u.v. radiation.

8. Disorders of carbohydrate metabolism

THE GLYCOGEN STORAGE DISEASES

Glycogen storage diseases (GSD) are a group of hereditary disorders characterized by abnormal concentration and/or structure of glycogen in the tissues. The nomenclature is clear for types I–V where the enzyme defects, eponymous designations, and clinical and biochemical features are well established. In the other types, however, there is considerable diversity in classification.

Preliminary screening procedures

The characteristic clinical features of GSD types II, IV, V, and VII (as defined below) are sufficiently distinctive to suggest the diagnosis which should then be confirmed by appropriate enzyme studies. There is no general agreement, however, on screening procedures for patients who present often with hepatomegaly, and with metabolic disturbances including hypoglycaemia, ketosis, acidosis, hyperlipidaemia, hyperuricaemia, and elevated blood lactate concentration. Often the clinician will request liver biopsy and enzyme assays to distinguish between GSD types IA, IB, and III, and GSD due to a defect of the hepatic phosphorylase system. Since these assays are time-consuming, expensive, and carried out only by a few specialized enzyme laboratories, it is highly desirable to establish the most likely diagnosis by screening procedures before requesting enzyme assays.

Several schemes have been suggested. Fernandes *et al.* (1974) and Fernandes (1980) suggest starting by measuring blood lactate concentration during a glucose tolerance test. A fall of an excessively elevated blood lactate level is indicative of GSD I, and no further preliminary test is necessary. An excessive rise of an initially normal lactate level indicates GSD III or a defect of the phosphorylase system. Further differentiation may be possible by glucagon challenge after an overnight fast. A normal rise of blood glucose indicates a phosphorylase-b-kinase deficiency, while a flat curve suggests either GSD III or a defect of liver phosphorylase. The last two disorders may be differentiated by repeating the glucagon test after feeding. Blood glucose levels remain low in phosphorylase deficiency, but rise to normal levels in GSD III.

In patients where the initial response of blood lactate to glucose loading was equivocal further information may be obtained by a galactose tolerance test. Dunger and Leonard (1982) propose a simpler scheme based on a study of the response to glucagon of blood glucose and lactate concentration in 40 patients with hepatic glycogen storage disease. An analysis of their data indicated that following glucagon challenge all their patients with GSD types IA, IB, and III had either a rise in a blood glucose concentration of less than 2 mmol/l, or a blood lactate level greater than 2.4 mmol/l either fasting or during the test. These criteria would have detected 85 per cent of all their 40 patients (13 with type IA; 5 with IB; 12 with III; 10 with phosphorylase kinase deficiency). However, they consider that patients who have fasting hypoglycaemia (blood glucose less than 2.5 mmol/l) and lactic acidosis (blood lactate more than 2.4 mmol/l) do not need further screening and should be investigated for GSD IA or IB by liver biopsy.

The response to glucagon by children with hepatic GSD has also been studied by Perkoff *et al.* (1962), Sokal *et al.* (1961), Hug (1962), and Lowe *et al.* (1962).

Glycogen storage disease type IA; Von Gierke disease; hepatorenal glycogen storage disease; glucose 6-phosphatase deficiency

Clinical features

This disorder usually presents in the first year, or subsequently, with convulsions due to hypoglycaemia or with abdominal enlargement due to massive hepatomegaly. Recurrent hypoglycaemia is a major problem. There is stunting of growth, with the height usually being below the third centile, but with careful treatment a growth spurt to the 10th centile has been reported (Stacey *et al.* 1980). Moderate obesity is common and possibly induced by frequent glucose supplementation. The spleen is not enlarged, but there is marked enlargement of the kidneys, usually not detectable by palpation because of hepatomegaly. The fundi often show numerous yellow areas in the region of the macula (Fine *et al.* 1968). Hyperlipidaemia predisposes to xanthomata on the extensor surfaces of the limbs, and hyperuricaemia to clinical gout. A bleeding diathesis due to platelet dysfunction has been reported (Corby *et al.* 1974).

Radiological evidence of osteoporosis is common. Rickets has also been reported (Stacey *et al.* 1980). Both these skeletal abnormalities may be due to acidosis which is due to a high blood lactate concentration and is a constant finding after fasting. However, blood lactate concentration decreases rapidly after glucose administration.

The effect of this disorder on life span is not well documented, but individual patients have survived to 34 and 40 (Van Creveld 1961), and 35 and 35 (Alepa *et al.* 1967).

Pathology

Liver biopsies in three patients showed adenomata. Using a radioisotope scanning technique, evidence of adenomata was found in seven of eight patients, one of whom developed hepatic carcinoma (Howell *et al.* 1976).

Molecular defects

The disorder is due to a severe deficiency of hepatic glucose 6-phosphatase (Fig. 8.1) (Cori and Cori 1952).

Liver glucose 6-phosphate concentration is only slightly increased (Öckerman 1965). Liver glycogen concentration is variably raised by at least 50 per cent above normal values of 2.5–6.0g/100g wet weight (Hug 1980). Although earlier reports had indicated that serum ketones were considerably elevated (Howell *et al.* 1962) and that ketonuria was often present (Van Creveld 1961) it has since been shown that ketogenesis is in fact decreased and that fasting blood ketone concentration is normal (Fernandes and Pikaar 1972; Binkiewicz and Senior 1973).

Fasting hypoglycaemia which does not respond to administration of glucagon, epinephrine or galactose is characteristic. Glucose tolerance tests on the other hand give a 'diabetic response'. The reason for this is unknown. Blood insulin concentration is decreased to approximately half normal levels and there is a reduced insulin response to glucose or arginine administration (Lockwood *et al.* 1969).

Hyperuricaemia is a common complication (Kelley *et al.* 1968). There is also a marked increase in the blood concentration of pyruvic acid, fatty acids, triglycerides, phospholipids and cholesterol (Howell *et al.* 1962). Hepatic and extrahepatic lipoprotein lipase activities are reduced (Fernandes 1980).

The mechanisms of these biochemical changes are not well understood. It may be argued that the enzyme defect is directly responsible for hypoglycaemia and would lead to an accumulation of glucose 6-phosphate. The chief metabolic pathways for glucose 6-phosphate would then be either to lactate and pyruvate via glycolysis with subsequent entry from pyruvate to acetyl-CoA and the Krebs cycle, or to ribose via the hexose monophosphate shunt. These reactions would generate an abundance of NADH, NADPH and acetyl-CoA which could cause an accelerated synthesis of fatty acids, cholesterol, triglycerides, and phospholipids, thus causing an increase in their blood concentration. The increase in blood free fatty acids could be accentuated by hypoglycaemia-induced mobilization from adipose tissue.

The cause of hyperuricaemia is also not clear. The early suggestion that hyperlacticacidaemia led to competitive inhibition between the renal tubular secretion of lactate and urate (Jeandet and Lestradet 1961) cannot be the only explanation since some hyperuricaemic patients do not have decreased uric acid excretion (Jeune *et al.* 1959). Moreover, glucose infusions increase the uric acid clearance before any marked fall in the blood lactate concentration (Fine *et al.* 1966).

The demonstration that incorporation of $1\text{-}^{14}C$-glycine into uric acid is accelerated indicates that the rate of *de novo* purine synthesis is increased (Kelley *et al.* 1968). It has been suggested that this may be due to an increased synthesis of phosphoribosylpyrophosphate (PRPP) because of the accelerated production of ribose from the hexose monophosphate shunt. The increase of PRPP would then accelerate the first reaction in the *de novo* purine synthesis pathway, namely the coupling of PRPP with glutamine to form 5-phosphoribosyl-1-amine, catalysed by 5-PRPP amido-transferase (Howell 1968).

Some alternative theoretical explanations for hyperuricaemia in GSD were suggested by Hers (1980) as follows: he cites the increase in blood uric acid concentration caused by fructose in normal individuals (Van den Berghe 1978) and argues that this effect would be enhanced in GSD I since fructose 1-phosphate cannot be metabolized through to glucose in the liver; and he states that uric acid formation in the liver would be increased by stimulation of inosine 5-phosphate deaminase, because of decreased inhibition resulting from reduced free phosphate concentration (in the absence of dephosphorylation of glucose 6-phosphate).

Diagnosis

The demonstration of the characteristic biochemical abnormalities described above would strongly suggest the disorder. Several other tests assist in establishing the diagnosis. The most useful is the demonstration of a sharp fall of fasting hyperlacticacidaemia following glucose administration, which is characteristic of glycogenosis due to glucose 6-phosphatase deficiency, but does not occur in glycogenosis due to deficiency of either amylo-1,6-glucosidase (debrancher, type III), liver phosphorylase (type VI) or phosphorylase kinase (type VIII) where blood lactate concentration is normal initially, and rises normally or excessively following glucose administration (Fernandes 1980).

However, not all patients with GSD IA have elevated blood lactate concentration. Dunger and Leonard (1982) found that all 13 patients they investigated following glucagon challenge had either a reduced increase in blood glucose concentraiton (less than 2 mmol/l), or a raised blood lactate level (over 2.4 mmol/l) either fasting or during the glucagon test.

If correction of hyperlacticacidaemia is unequivocal, other screening tests are unnecessary. Definitive diagnosis can be established by demon-

338

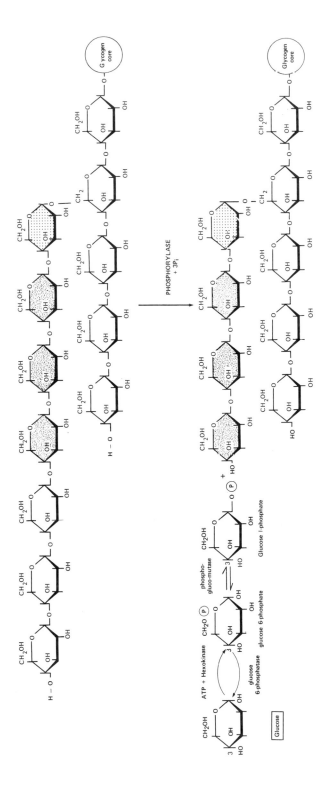

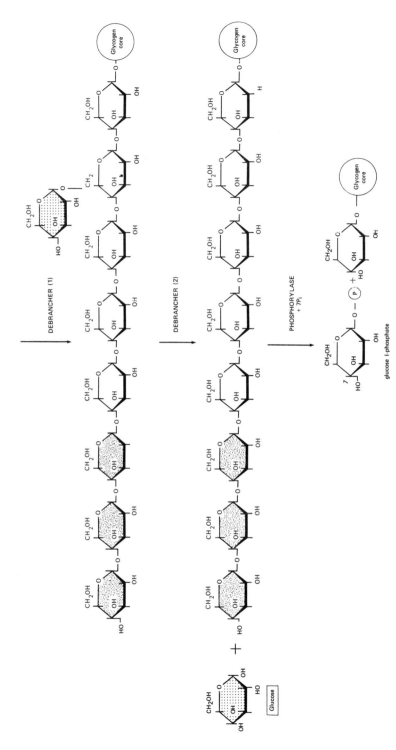

Fig. 8.1. Glycogenolysis: debrancher (1) oligo-(1,4→1,4)-glucantransferase; Debrancher (2) amylo-1,6-glucosidase.

strating severe deficiency of glucose 6-phosphatase activity in a liver biopsy. The sample can also be used for assay of liver glycogen (see 'Molecular defects', p. 336).

Other screening tests which are used in some centres include demonstration of failure of the normal hyperglycaemic response following administration of adrenalin, glucagon (Howell *et al.* 1962), or galactose (Schwartz *et al.* 1957).

An original test for glucose 6-phosphatase activity *in vivo* has been reported by Van Hoof *et al.* (1972). The test depends on the observation that after intravenous injection the half-life of universally-labelled ^{14}C-glucose is 1.6 times longer than that of 2-^{3}H-glucose. The more rapid loss of 2-^{3}H-glucose may be explained on molecular readjustment during the futile cycle in the liver: Glucose $\rightarrow$ glucose 6-phosphate $\rightarrow$ fructose 6-phosphate $\rightarrow$ glucose 6-phosphate $\rightarrow$ glucose. This is because the tritium in position 2 of the glucose is lost in the reversible conversion of glucose 6-phosphate to fructose 6-phosphate catalysed by phosphohexose-isomerase. Since in patients with glucose 6-phosphatase deficiency the final reaction is blocked, there is a slower fall of ^{3}H/^{14}C ratio in blood glucose. The mean ratio for eight normal subjects was 59.6 per cent after 60 min (100 per cent at 0 min) and for seven patients with GSD IA, 93.2 per cent after 60 min (Van Hoof *et al.* 1972).

Genetic counselling

The disorder has an autosomal recessive mode of inheritance. Parents of affected children may have reduced glucose 6-phosphatase activity in the intestine (Field and Drash 1967) and platelets (Negishi *et al.* 1974).

As far as we are aware, prenatal diagnosis has not been reported. Two methods have been suggested which might in theory be successful. These are assay of glucose 6-phosphatase activity in a placental biopsy (Matalon *et al.* 1977*b*) or in cultured amniotic epithelial cells after stimulation by dibutyryl cyclic AMP (Negishi and Benke 1977).

The reported incidence of all types of glycogen storage diseases varies widely from 1:50 000 to 1:246 000.

Treatment

Frequent feeds during the day and nasogastric glucose drips at night prevent hypoglycaemia, and reduce the elevated blood lactate, triglyceride, and cholesterol concentrations (Fernandes 1980). Several methods for glucose administration have been tried. Stacey *et al.* (1980) found accelerated growth, reduction of liver size, and correction of the blood coagulation defect and biochemical abnormalities, including a fall in serum uric acid, most marked during 24-hr intravenous glucose supplementation.

However, this was less practical than 24-hr nasogastric infusion. By either route the flow was regulated by a peristaltic pump powered by a battery harnessed on the chest so that the patient was fully ambulant. Diurnal hourly glucose supplementation and nocturnal continuous nasogastric infusion were also satisfactory. By either the intravenous or the oral route, the dose of glucose was 0.3–0.44g/kg/h.

Fernandes (1980) showed that the intervals between feeds can be increased by food substances such as starch or glucose in curd which release glucose slowly. He suggests that the diet should consist of starch and glucose to yield 50–70 per cent of energy with restriction of lactose and sucrose; fat, 15–35 per cent of energy as polyunsaturated fat such as corn oil; and protein, 15 per cent of energy. When acidosis and hyperuricaemia are resistant to glucose supplementation, the former can be corrected by oral administration of sodium bicarbonate, 1–2g daily; and the latter by allopurinol (see 'Treatment of Lesch–Nyhan syndrome', p. 315).

Portacaval shunting preceded by intravenous alimentation has been used in the past (Folkman *et al.* 1972; Starzl *et al.* 1973) since it has been shown to reduce liver glycogen in experimental animals (Starzl *et al.* 1965). However, this has now been largely abandoned in view of the high mortality, and the demonstration that results are similar to nocturnal intragastric feeding (Burr *et al.* 1974; Fernandes 1980).

Glycogen storage disease type IB

In a minority of patients with the clinical and biochemical features of glycogen storage disease (GSD) type I, conventional *in vitro* assays of liver glucose 6-phosphatase have shown normal activity. This disorder has been designated GSD type IB.

An important observation was reported by Narisawa *et al.* (1978) who demonstrated that if liver glucose 6-phosphatase from a patient with GSD type IB was assayed on intact microsomes, deficient activity (13.8 per cent of control mean) could in fact be demonstrated. However, after disruption of microsomes with deoxycholate (0.2 per cent), the activity was in the high range of normal, suggesting that the catalytic activity of glucose 6-phosphatase is present but masked by a defective microsomal component. Further evidence of glucose 6-phosphatase deficiency in GSD type IB comes from the studies of Sann *et al.* (1980) who demonstrated the deficiency *in vivo* in one patient by the method of Van Hoof *et al.* (1972) (see GSD type IA, 'Diagnosis', p. 340). In this case the ^{3}H/^{14}C ratio was similar to that obtained for seven patients with GSD IA.

These reports suggest therefore that in GSD IB there is a defect in the transport of glucose 6-phosphate from the cytoplasm to the active site of the enzyme glucose 6-phosphatase.

Narisawa *et al.* (1978) have suggested that in GSD type IB there is a defect of the transport component of the microsomal membrane. Lange *et al.* (1980) consider that glucose 6-phosphatase of liver and kidney requires three integral microsomal membrane components: (1) a glucose 6-phosphate specific translocase, designated T_1, that mediates penetration of glucose 6-phosphate into the cisternae; (2) a relatively non-specific phosphohydrolase located on the luminal surface of the membrane; and (3) a phosphate translocase, T_2, that mediates both the efflux of inorganic phosphate, and penetration of pyrophosphate into the microsomes. Since the latter function was normal in intact liver microsomes from a patient with GSD type IB (as evidenced by normal pyrophosphorylase activity), the authors localize the defect in GSD type IB to the T_1 component of the glucose 6-phosphatase system. Thus patients with GSD type I could be classified as either type IA or IB according to the method of assay. It is not possible to establish from the literature what proportion of patients diagnosed as having glucose 6-phosphatase deficiency are of the GSD type IB variety. Indeed in one family (Beaudet *et al.* 1980) one sibling was diagnosed as having GSD type IA and another as having type IB. It is highly likely that both had the type IB variant.

The only additional features which have been suggested in GSD type IB are neutropenia (McCabe *et al.* 1980) and impaired neutrophil migration (Beaudet *et al.* 1980), but insufficient cases of leucocyte function have been reported to assert that leucocyte numbers and function differ in GSD types IA and IB.

The reports that affected siblings have been of both sexes and have had normal parents, and that parents of one of the few type IB families in the literature were first cousins (Sann *et al.* 1980) is consistent with autosomal recessive inheritance.

Treatment is as for GSD type IA.

Glycogen storage disease (glycogenosis) type II; α-1,4-glucosidase deficiency; acid maltase deficiency

Deficiency of acid maltase or α-1,4-glucosidase (α-glucosidase) is found in patients with two distinct clinical phenotypes. Firstly, the infantile type known as Pompe's disease or generalized glycogenosis, in which the cardinal features are severe hypotonia and cardiomegaly presenting in the first year, with death usually in the first or second year; and secondly the muscular type in which there is a slowly progressive proximal myopathy. The age of onset in the second type is variable. When the disorder presents in early childhood, survival is rarely beyond the second decade, but usually there is a slower course in patients with later symptomatic onset.

Pompe's disease

Clinical features

Hypotonia is usually first noticed at two or three months of age and is associated with weakness of voluntary muscles and delayed motor development, with persistence of 'head lag' in the supine position.

There is considerable cardiac enlargement.

The ECG shows characteristic giant QRS complexes and shortened PR interval.

Death from cardiac failure usually occurs in the first year (Ehlers and Engle 1963) but may be in the second.

Pathology

There is considerable increase in the glycogen concentration in voluntary and cardiac muscles and in most other tissues. In the central nervous system, glycogen storage occurs mainly in the motor cells (Smith *et al.* 1967).

Massive cardiomegaly is probably due to glycogen infiltration of the myocardium which is evident histologically. Endocardial fibroelastosis is common.

Electronmicrographs show that unlike the case in other types of glycogen storage diseases, glycogen in liver and kidney is stored not only outside the lysosomes but also in membrane-bound vesicles which are probably distended lysosomes (Baudhuin *et al.* 1965; Garancis 1968).

Molecular defects

The disorder is due to a generalized deficiency of lysosomal α-glucosidase (Hers 1963). This deficiency has been demonstrated in skeletal muscle (Angelini and Engel 1972) and cultured fibroblasts (Nitowsky and Grunfield 1967; Reuser *et al.* 1977). In the kidney, deficiency has been demonstrated using glycogen as substrate (Brown *et al.* 1970), but not using the artificial substrate 4-methylumbelliferyl-α-D-glucoside (Salafsky and Nadler 1973).

The general increase in tissue glycogen concentration is discussed above under 'Pathology'.

In contrast to patients with GSD type IA, phosphorolytic glycogenolysis in the cytosol is intact. There is therefore a normal hyperglycaemic response to glucagon and epinephrine, and no hypoglycaemia.

Diagnosis

Since we have investigated two patients with normal activity of α-glucosidase in mixed leucocytes, and isolated lymphocytes and granulocytes using 4-methylumbelliferyl-α-D-glucoside as substrate, we routinely assay cultured fibroblasts using the same substrate and have consistently demonstrated severe deficiency in α-glucosidase activity, measured at pH 4, in affected patients. Reuser *et al.* (1977) reported absent activities of α-glucosidase in fibroblasts from patients with Pompe's disease. Using the natural substrate, glycogen, Koster *et al.* (1974) have reported the leucocyte assay to be reliable.

Genetic counselling

Pompe's disease has an autosomal recessive mode of inheritance. Studies for detection of carriers by assay of α-glucosidase activity in urine, fibroblasts, leucocytes, and skeletal muscle are summarized by Loonen *et al.* (1981*a*). The authors conclude that there is some overlap in values between heterozygotes and controls for all methods used. However, mean values for heterozygotes are intermediate between mean control values and values for Pompe's patients. We have found that the best discriminant is the ratio of activities of α-glucosidase/ß-galactosidase in cultured fibroblasts, (unpublished observations).

The structural gene for acid α-glucosidase has been assigned to chromosome 17 using somatic cell hybrids (Solomon *et al.* 1979; D'Ancona *et al.* 1979).

Loonen *et al.* (1981*b*) have reported a family in which some members had the generalized form of Pompe's disease, and others the late onset, muscular type (q.v.).

Prenatal diagnosis by assay of α-glucosidase activity of cultured amniotic cells (Fensom *et al.* 1976) has now been reported in numerous pregnancies.

α-glucosidase in the amniotic fluid supernatant has properties of the kidney enzyme which is present in Pompe's disease, and therefore its measurement is unreliable for prenatal diagnosis (Salafsky and Nadler 1971).

Treatment

Attempts at enzyme replacement have been unsuccessful. De Barsy and Van Hoof (1974) found that purified human placental α-glucosidase was not taken up by muscle of two patients with Pompe's disease. Since low density lipoprotein (LDL) has specific high affinity receptors on many types of cells, Williams and Murray (1980) linked purified human liver α-glucosidase to LDL and administered it by intravenous infusion to a 13-month-old female with Pompe's disease on two occasions without producing clinical improvement.

Muscular or late onset form of glycogen storage disease (glycogenosis) type II

There is slowly progressive weakness of the proximal muscles of the limbs. When clinical onset is in early childhood, survival beyond the second decade is unusual. However, onset after childhood is associated with a slower course and survival for many years (Engel *et al.* 1973). Whereas cultured fibroblast α-glucosidase activity is almost undetectable in generalized glycogenosis, residual activity of 7–22 per cent of normal has been reported in patients with the muscular form of the disease in whom the degree of enzyme deficiency correlates with the clinical severity (Reuser *et al.* 1977). In the family with members who had either generalized glycogenosis or the muscular form (Loonen *et al.* 1981*b*) of the disease, one of the patients with the latter had fibroblast α-glucosidase activities of 15–20 per cent of control values. The authors speculate that he may have been a double heterozygote, carrying alleles for both the generalized and muscular forms.

As for the generalized form, the muscular disorder has an autosomal recessive form of inheritance. Prenatal diagnosis by assay of cultured amniotic cell α-glucosidase should be possible.

Glycogen storage disease type III; amylo-1,6-glucosidase (debrancher) deficiency; oligo-(1,4→1,4)-glucantransferase deficiency; limit dextrinosis; Cori's disease; Forbes' disease

Clinical features

Hepatomegaly, stunted growth and sometimes splenomegaly (Brandt and de Luca 1966) are the main clinical features. Hypoglycaemia is unusual (Hug 1980), but may occur (Fernandes and Van de Kamer 1968). A unique feature is spontaneous disappearance of hepatomegaly in the second decade in some patients (Van Creveld and Huijing 1965; Brown and Brown 1968). In contrast to patients with GSD I, there is no renal enlargement. The disorder usually has a benign course, but cirrhosis of the liver (Brandt and de Luca 1966) and myopathy (Brunberg *et al.* 1971) have occurred in some patients.

Molecular defects

The disorder is due to enzymatic deficiency of the glycogen 'debranching' system. The debranching process occurs in the cytosol and is distinct from the random hydrolysis of glycogen by the lysosomal α-glucosidase, the enzyme deficient in GSD II. Two enzymes are known to be involved (Fig. 8.1). The first operates when phosphorylase has catalysed the sequential removal of glucose oligo- α-1,4→ α-1,4-linked residues from the non-reducing end of the glycogen branches, up to approximately four glucose residues from the glucose $1 \rightarrow$ 6-glucose branch points. In the liver, approximately one-third of glucose residues can be removed by phosphorylase activity, leaving phosphorylase limit-dextrin. For further glycogen degradation, approximately three of the four glucosyl residues from the non-reducing end of one branch are transferred to another branch to which they become covalently bound in $1 \rightarrow 4$ linkage. This reaction is catalysed by oligo-$(1,4\rightarrow1,4)$-glucantransferase. The removal of the remaining single glucose residues attached to the elongated chains, representing approximately 8 per cent of the glycogen molecule, is catalysed in the second reaction of the debranching system by amylo-1,6-glucosidase. Further degradation is then catalysed by phosphorylase from the non-reducing ends of the branches, until the debranching system is required again at the next branch points. Deficiency of the glycogen debranching system has been demonstrated in liver (Brown and Brown 1968; Hug 1980), cultured fibroblasts (Justice *et al.* 1970; Di Mauro and Mellman 1973; Di Mauro *et al.* 1973), and leucocytes (Williams *et al.* 1963; Williams and Field 1968; Huijing 1964, 1975). Erythrocyte glycogen concentration is usually (Sidbury *et al.* 1961; Van Hoof and Hers 1967) but not always (Van Hoof and Hers, 1967; Williams and Field 1968) increased.

In a meticulous study of GSD III, Van Hoof and Hers (1967) classified 45 patients into six biochemical subgroups, designated IIIA to IIIF according to the degree of debrancher deficiency in both liver and muscle as demonstrated by five methods. The methods were measurement of:

(i) release of glucose from phosphorylase limit-dextrin, that is from glycogen which had previously been treated with phosphorylase to liberate α-1,4-linked glucose residues, and which required debrancher activity for release of further glucose molecules (Fig. 8.1);

(ii) incorporation of ^{14}C-glucose into glycogen, that is by the reverse debrancher reaction, since brancher enzyme activity requires UDP-glucose rather than free glucose;

(iii) the oligo-$(1,4\rightarrow1,4)$-glucantransferase activity;

(iv) release of glucose from 1,6-linked glucose residues of the glucosyl Schardinger dextrin;

(v) release of glucose from 6^3-α-glucosyl maltotetraose.

The activities of these reactions which characterize the subgroups are shown in Table 8.1.

Immunologically cross-reacting material to debrancher enzymes has been demonstrated in liver and muscle (Dreyfus *et al.* 1974).

Characteristically, after meals the outer branches consisting of α-1,4-linked residues elongate normally, but on fasting these are hydrolysed up to a few glucose residues from the α-1,6-linked branch points (Fig. 8.1). Sidbury *et al.* (1961) and Van Hoof and Hers (1967) demonstrated

Table 8.1 *Biochemical subgroups of GSD III (Van Hoof and Hers, 1967)*

Method of debrancher activity assay	A		B		C		D		E		F	
	L	M	L	M	L	M	L	M	L	M	L	M
(i) Glucose release from phosphorylase limit dextrin	2^-	2^-	2^-	0	0	0	2^-	0	2^-	N	2^-	2^-
(ii) Incorporation of ^{14}C-glucose	2^-	2^-	0	N	2^-	2^-	N±	N±	2^-	–	0	0
(iii) Oligo-(1,4→1,4)-glucantransferase activity	2^-	2^-	0	N	–	–	0	0	2^-	N	2^-	–
(iv) Glucose release from Schardinger dextrin	0	0	2^-	N±	N	2^-	–	N	2^-	N	0	–
(v) Glucose release from 6^3-α-glucosyl maltotetraose	2^-	0	2^-	N	±	nd	–	nd	nd	nd	nd	nd

Key: 0 = absent
2^- = very low
– = reduced or moderately reduced
± = slightly reduced
N = normal
nd = not done
L = liver
M = muscle

increased concentration of erythrocyte glycogen with short outer branches in some patients with GSD III. However, in muscle biopsies glycogen concentration and structure were normal in seven of 29 patients (Brown and Brown 1968).

Liver glycogen concentration is usually increased, for example to 9.3 g/ 100 g wet weight from normal values of 2.5–6.0 g/100 g (Hug 1980) to as high as 17.3 g/100 g (Brown and Brown 1968). Van Hoof and Hers (1967) reported that muscle glycogen concentration was elevated in GSD IIIA, but not in IIIB, IIIE or IIIF, while Hug (1980) quotes a value of 6.0 g/100 g (normal 0.1–1.5). Glycogen accumulation has also been reported in the myocardium (Levin *et al.* 1967).

In contrast to GSD I, severe lactic acidosis is not characteristic, but fasting ketosis is marked, especially in young children. It decreases during the first decade, being rare after seven years of age (Fernandes and Pikaar 1972; Fernandes 1980).

The outer branches of glycogen molecules consist of glucose residues, linked by α-1,4-glycosidic bonds (Fig. 8.1), which elongate rapidly after feeding with glucose; or with galactose and fructose which are converted to glucose. Thus in patients with GSD III mobilization provoked by epinephrine and glucagon is normal 2 h after feeding, but absent after a 14-h-fast, when glycogen molecules will have short outer branches and cannot be degraded further without cleavage of the 1,6-glycosidic branch points (Brown and Brown 1968).

Diagnosis

The diagnosis is confirmed by demonstration of reduced amylo-1,6-glucosidase activity in cultured fibroblasts or leucocytes (see 'Molecular defects', above). Serum transaminases have increased activities in childhood but return to normal in survivors. Hyperlipidaemia occurs in some patients, but is not as marked as in patients with GSD I. Hyperuricaemia and gout do not occur (van Creveld and Huijing 1965).

Liver glucose 6-phosphatase activity has consistently shown a moderate reduction in activity (Hug 1980).

Genetic counselling

The disorder has an autosomal recessive mode of inheritance. Some (Williams *et al.* 1963; Williams and Field 1968; Chayoth *et al.* 1967) but not all (Huijing *et al.* 1968) studies have demonstrated reduced activity of leucocyte debrancher activity in heterozygotes.

GSD III is the commonest form of GSD in North African Jews, in whom an incidence of 1:5 420 has been estimated (Levin *et al.* 1967).

In view of the benign nature of the disorder, prenatal diagnosis may not be justified. However, this should be possible since debrancher activity is present in normal amniotic cells (Justice *et al.* 1970).

Treatment

The disorder has a good prognosis and tends to remit spontaneously during the second decade (see 'Clinical features', p. 345). Treatment during childhood is aimed at preventing hypoglycaemia by a high protein diet.

Glycogen storage disease type IV; α-1,4-glucan: α-1,4-glucan 6-glucosyl transferase (brancher) deficiency; amylopectinosis; Andersen's disease

Infants with this rare form of GSD fail to thrive shortly after birth. There is progressive hepatosplenomegaly with onset a few months after birth. Liver failure is accompanied by cirrhosis, ascites, and portal hypertension, leading to death usually before the age of two years (Andersen 1956; Fernandes and Huijing 1968). One child survived to four years of age (Sidbury *et al.* 1962). Hypotonia and muscular atrophy are usually present, and can be severe (Zellweger *et al.* 1972; McMaster *et al.* 1979).

Molecular defects

The disorder is due to deficiency of the brancher enzyme α-1,4-glucan: α-1,4-glucan 6-glucosyl transferase (Fig. 8.2) which has been demonstrated in liver (Brown and Brown 1966; Fernandes and Huijing 1968; Reed *et al.* 1968), cultured fibroblasts (Howell *et al.* 1971) and leucocytes (Brown and Brown 1966; Fernandes and Huijing 1968).

The disorder is classified as a GSD because glycogen has an abnormal structure, in spite of the fact that its concentration in the liver is normal (Brown and Brown 1966), or reduced (Sidbury *et al.* 1962). However, Fernandes and Huijing (1968) reported a single patient with increased liver glycogen concentration.

In GSD type IV, deficiency of brancher enzyme leads to formation of liver glycogen with long outer chains, and relatively few branch points (6 per cent rather than 8 per cent), resembling amylopectin (Illingworth and Cori 1952), but having somewhat different chain sizes (Mercier and Whelan 1973). The occurrence of even this reduced proportion of branch points is unexplained, but may be catalysed by other, as yet uncharacterized, enzyme(s), or by reversal of the brancher reaction (Mercier and Whelan 1973). The presence of abnormal liver glycogen was confirmed by Brown and Brown (1966) who, however, found glycogen with normal structure in muscles. In rabbit muscle, the purified branching enzyme acts directly on glycogen enlarged by sequential addition of α-1,4-linked glucose residues (transferred from UDP-glucose) by transferring a length

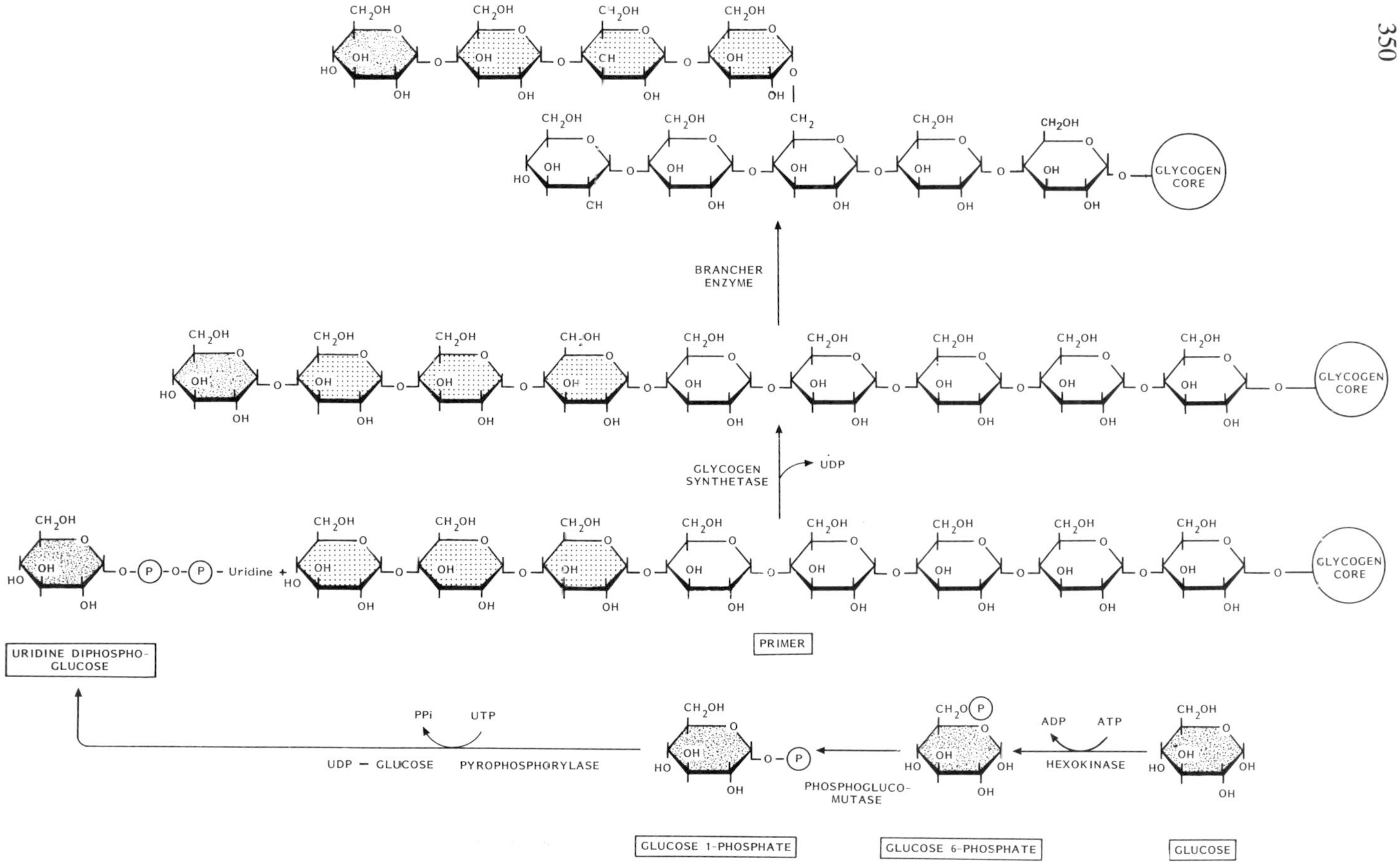

Fig. 8.2. Glycogen synthesis.

of terminal oligosaccharides from the non-reducing end of the chain to elsewhere in α-1,6 glucosidic linkage (Fig. 8.2). The preferred length of transferred oligosaccharide was seven glucosyl units. It is interesting that in the reversed debrancher reaction, that is, when glycogen branches are enlarged by addition of glucose residues in α-1,4 glucosidic linkage from glucose 6-phosphate, catalysed by phosphorylase in liver, the oligosaccharide transferred during branching also consists preferentially of seven glucosyl units (Verhue and Hers 1966). Branching in the latter system occurs when the average donor chains have elongated to 11–21 glucosyl units.

Diagnosis

The diagnosis can be established by demonstrating severe deficiency of brancher enzyme activity in leucocytes or cultured fibroblasts (see 'Molecular defects', above).

The abnormal structure of hepatic glycogen can be demonstrated as increased chromogenicity of an iodine–potassium iodide complex with an absorption maximum of 525 nm, rather than 440 nm, or by demonstration of long outer chains which are degraded to nearly 50 per cent by purified phosphorylase rather than to the normal value of about 36 per cent.

Genetic counselling

The disorder has an autosomal recessive mode of inheritance. Reduced brancher activity has been demonstrated in leucocytes of parents from affected children by Sidbury *et al.* (1962), but normal values were reported in parents of two other patients (Howell *et al.* 1971). Both parents from one patient had reduced brancher activity in cultured fibroblasts (Howell *et al.* 1971).

Pregnancies in mothers from two families at risk for GSD IV have been monitored for brancher activity in cultured amniotic cells, and both foetuses were found to be affected (Barbara I. Brown, Washington University School of Medicine, St. Louis, 1980; personal communication).

Glycogen storage disease type V; muscle phosphorylase deficiency; McArdle's disease

Characteristically, the natural evolution of the symptomatology starts with excessive tiredness in the second decade. Muscular cramps and myoglobinuria occur in about half the patients in the second to fourth decades, and are followed by progressive myopathy (McArdle 1951, 1969; Schmid and Mahler 1959; Schmid *et al.* 1959; Rowland *et al.* 1966). Although ECG changes occur (Salter 1968), cardiac failure has not been reported.

An interesting observation is the decline of muscle cramps if strenuous muscular exercise is continued. This has been attributed to increasing sources of energy derived from increasing concentrations of fatty acids and increased blood flow during prolonged exercise (Pernow *et al.* 1967).

McArdle demonstrated that in these patients venous blood lactate concentration fails to rise, as in normal individuals, following exercise of voluntary muscle made temporarily ischaemic by arterial occlusion. He pointed out that this observation was consistent with a metabolic block in the conversion of glycogen to lactate. The enzyme defect was identified as a deficiency of muscle phosphorylase by Mommaerts *et al.* (1959), Schmid *et al.* (1959), and Schmid and Mahler (1959), who also demonstrated an increased muscle glycogen concentration and an intact phosphorylase kinase activity. The latter observation was confirmed by Hug (1980) who reported that muscle phosphorylase kinase activity may be increased up to seven times normal.

Liver glycogen concentration and phosphorylase activity are within normal limits. Predictably, therefore, these patients do not have a tendency to hypoglycaemia and exhibit normal hyperglycaemic responses to glucagon and epinephrine.

This disorder therefore has provided further evidence of independent genetic control of muscle and hepatic phosphorylases.

The diagnosis is suspected when venous blood lactate fails to rise during ischaemic exercise, and confirmed by demonstrating severe deficiency of phosphorylase activity in a muscle biopsy. This latter test is necessary to distinguish the disorder from muscle phosphofructokinase deficiency (GSD VII) and other causes of exercise-induced muscle cramps (see below).

The phosphorylase activity which reappears in cultured muscle cells from patients (Roelofs *et al.* 1972; Mitsumoto 1979) has been shown to have electrophoretic and immunochemical properties of the foetal iso-enzyme, and not of the adult skeletal muscle enzyme (Sato *et al.* 1977; Di Mauro *et al.* 1978).

The disorder has an autosomal mode of inheritance, which is usually recessive but sometimes dominant (Dawson *et al.* 1959). Immunologically cross-reactive material to phosphorylase is absent in some patients, but present in others (Dreyfus *et al.* 1974).

The only proven beneficial management is avoidance of strenuous exercise, to prevent painful cramps and possible renal failure from myoglobinuria (Bank *et al.* 1972).

A variant of muscle phosphorylase deficiency presents at a few weeks of age with progressive weakness of voluntary muscles. In three patients death by respiratory muscle failure occurred in the first few months (Di Mauro and Hartlage 1978; Miranda *et al.* 1979; De La Marza *et al.* 1980).

Muscle cramps and myoglobinuria

Muscle cramps during exercise, excessive fatiguability, and sometimes myoglobinuria occur in McArdle's disease (glycogen storage disease type V) due to muscle phosphorylase deficiency, and in muscle phosphofructo-kinase deficiency (glycogen storage disease type VII). They are attributable to impairment of glycogenolytic and glycolytic energy pathways, respectively. Deficiencies in muscle carnitine palmitoyltransferase (q.v.) or in the muscle (M) subunits of phosphoglycerate mutase, or of lactic dehydrogenase, have also been associated with some of these symptoms and may also be attributed to the reduced ability of voluntary muscle to generate energy during muscular exercise. The pathogenesis of these symptoms in myoadenylate deaminase deficiency (q.v.) is less well understood.

Muscle phosphoglycerate mutase deficiency

Di Mauro *et al.* (1981) reported a 52-year-old man with muscle pains, 'stiffness', and weakness after 15 to 30 min of intense exercise. These episodes were often followed by pigmenturia but the urine was not tested for myoglobin. He had gouty tophi at the wrists and elbows.

An abnormally reduced rise of venous lactate concentration followed ischaemic exercise of the forearm. From histochemical and electron-microscopical studies, it was concluded that there was a mild increase in muscle glycogen. Muscle phosphoglycerate mutase, the glycolytic enzyme which catalyses the conversion of 3-phosphoglycerate to 2-phosphoglycerate (Fig. 9.1), was reduced to 5.7 per cent of the lowest control value. Electrophoretic, heat-stability and mercury inhibition studies indicated that the small residual activity of the enzyme could be accounted for by the brain BB isoenzyme of phosphoglycerate mutase, suggesting that the genetic defect involved the M subunit which normally predominates in muscle, and which can be demonstrated as a slow anodal-migrating MM band in normal muscle, which has only faint BB and MB bands (Omenn and Cheung 1974). The authors suggested that the residual BB activity could account for the reduced, but not absent, rise of venous lactate after ischaemic forearm exercise.

Muscle lactic dehydrogenase deficiency

A single patient has been reported with complete deficiency of lactate dehydrogenase (LDH) M subunit (Kanno *et al.* 1980). He was an 18-year-old male who presented with exertional pigmenturia and excessive fatigue, but not muscle cramps. The pigment was identified as myoglobin. LDH is a tetramer consisting of H and M subunits aggregating into five isoenzymes as follows: LDH I (H_4), LDH II (H_3M_1), LDH III (H_2M_2), LDH IV (H_1M_3), and LDH V (M_4). Electrophoretic study of serum revealed a single band of LDH I and total absence of LDH II–V. The LDH M-subunit was also absent from erythrocytes, leucocytes and the vastus intermedialis.

LDH M-subunit was also absent in three of five siblings (two male and one female). The parents and other two siblings (one male and one female) had reduced erythrocyte M-subunit content as revealed by an increase in the LDH H/M ratio.

As discussed earlier, during exercise, glycolysis is greatly accelerated in skeletal muscle. This is accompanied by an efflux of lactate formed from pyruvate, catalysed by LDH in skeletal muscle.

In the propositus, during ischaemic forearm muscle exercise (hand grasping with sphygmomanometer set 20 mm Hg above systolic blood pressure), there was a considerably reduced rise of serum lactate, and markedly exaggerated rises of pyruvate concentrations in venous blood. During the period 5–12 h following the exercise there was a rise in serum creatine phosphokinase (CPK) to a maximum approximately 200-fold higher than in three normal controls whose serum CPK showed no detectable increase. The rise in CPK was accompanied by simultaneous myoglobinuria.

The partial deficiency of M-subunit in both parents and in some siblings of both sexes suggests an autosomal recessive mode of inheritance.

The natural history of the disorder is not known at present, but prenatal diagnosis should be possible by demonstration of absent LDH I isoenzyme in cultured amniotic cells, or foetal leucocytes or erythrocytes.

Glycogen storage disease due to deficiency of liver phosphorylase or its activating systems

Several conflicting classifications of defects of liver phosphorylase and its activating systems have been proposed. At the time of writing the ensuing confusion is unresolved, and has arisen largely because several cases in the literature have been investigated incompletely. While most authorities including McKusick (1978) designate liver phosphorylase deficiency as GSD VI, Illingworth and Brown (1964) and Hers and van Hoof (1968) suggest that this category should include all hepatomegalic glycogenoses which are not types I, III, or IV. Ryman (1980) designated sex-linked liver and muscle phosphorylase kinase deficiency as GSD VIA, and autosomal liver phosphorylase or phosphorylase kinase deficiency as GSD VIB. However, Hug (1980) states that muscle tissue has been normal in all his patients with phosphorylase kinase deficiency, whether sex-linked or autosomal recessive. He argues powerfully that classification should adhere to the rule that a Roman numeral should be assigned to each enzyme defect or, if this is not known, to a different clinical syndrome. In his classification, GSD VI and GSD VII remain as hepatic phosphorylase and muscle phosphofructokinase deficiencies, respectively. GSD VIII is restricted to a disorder which has been proven in only one child (although suspected in two others) with hepatomegaly, truncal ataxia, progressive degeneration of the brain, and increased hepatic and cerebral glycogen concentration, with cerebral glycogen in the form of giant α-particles within axon cylinders and synapses. Total liver phosphorylase was normal, although most was in the inactive form, despite the fact that the phosphorylase activating system was normal. No specific enzyme defect was demonstrated (Hug *et al.* 1965; 1967). In the classification by Hug (1980) hepatic phosphorylase kinase deficiency is designated as IXA when inheritance is autosomal recessive, and as IXB when sex-linked. GSD X refers to a single female with marked hepatosplenomegaly and later minimal muscle cramps, and increased glycogen concentration of liver and muscle, whose total phosphorylase content of both tissues was normal but present completely in the inactive *b* forms. Phosphorylase kinase activity of muscle and liver could be demonstrated only if activated cAMP-dependent kinase was added. The latter was prepared from I-strain (phosphorylase kinase deficient) mouse muscle and cAMP. Since in this patient cAMP-dependent phosphorylations were intact, the defect was attributed to a deficiency of cAMP-dependent kinase activity. GSD XI (Hug 1980) is characterized by stunting of growth, vitamin D-resistant rickets, marked hepatomegaly, and increased glycogen concentration in liver and kidneys. No enzyme deficiency has yet been elucidated. The classifications of glycogen storage diseases types VI–XI are summarized in Table 8.2.

There is general agreement that muscle phosphorylase occurs in a totally inactive *b* form which is fully activated by AMP. In contrast, liver phosphorylase *b* was thought to be inactive, even in the presence of AMP (Wosilait and Sutherland 1956). However, it has since been demonstrated that it may be activated by AMP in the presence of sulphate and fluoride ions, and that it is almost completely inhibited by caffeine (Lederer and Stalmans 1976). By exploiting these properties, it is possible to measure total phosphorylase activity (*a* and *b*), and phosphorylase *a* activity. Other workers prefer to measure total liver phosphorylase using the hepatic physiological system, namely in the presence of ATP and Mg^{2+} (Schimke *et al.* 1973; Hug 1980). Hug (1980) stressed the importance of correct handling of the liver biopsy which should be frozen on dry ice or in liquid nitrogen immediately after its removal from the body. Necropsy specimens are not suitable for liver phosphorylase assay since the enzyme is unstable. To prove the existence of liver phosphorylase deficiency, he recommends that phosphorylase assay be carried out:

(a) in the presence of purified phosphorylase kinase, since under these

Table 8.2 *Classifications of glycogen storage diseases types VI–XI*

Enzyme defect	McKusick (1978)	Ryman (1980)	Hug (1980)
Liver phosphorylase	VI (23270)	—	VI
Liver/muscle phosphorylase kinase (sex-linked)	—	VIA	—
Liver phosphorylase or phosphorylase kinase (autosomal)	—	VIB	—
Liver phosphorylase kinase (sex-linked)	VIII (30600)	—	IXB
Liver phosphorylase kinase (autosomal recessive)	—	—	IXA
Muscle phosphofructokinase	VII (23280)	VII	VII
Unknown	—	—	VIII
cAMP-dependent kinase	—	—	X
Unknown	—	—	XI

circumstances absent activity would indicate hepatic phosphorylase deficiency; and

(b) in the presence of purified crystalline rabbit muscle phosphorylase *b*, since the appearance of previously deficient phosphorylase activity would demonstrate the integrity of the human hepatic activating system, and provide further evidence of true hepatic phosphorylase deficiency.

Phosphorylase activation results from a series of sequential reactions (Fig. 8.3) which commence by stimulation of adenylate cyclase by interaction of a membrane receptor with glucagon in the liver (or epinephrine in muscle). Adenylate cyclase catalyses the synthesis of cAMP, which activates cAMP-dependent protein kinase which in turn activates phosphorylase kinase. The last-named catalyses the conversion of phosphorylase *b* to the active *a* form which in the presence of inorganic phosphate catalyses the sequential removal of 1,4-bound glucosyl residues from the outer branches of the glycogen molecule.

Hers (1959) reported two boys and one girl with hepatic glycogenosis with normal liver glucose 6-phosphatase and amylo-1,6-glucosidase activities, but low liver phosphorylase activities which could not be stimulated by ATP and Mg^{2+}. These patients were assigned to GSD VI by Stetten and Stetten (1960) and labelled as having hepatic phosphorylase deficiency by Lamy *et al.* (1960). In retrospect, in view of the complexity of the phosphorylase activating system, which could at the time not be analysed, the latter diagnosis must be considered unproven.

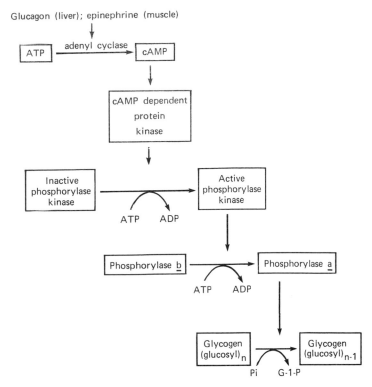

Fig. 8.3. Glycogenolytic cascade.

Glycogen storage disease type VI; liver phosphorylase deficiency; Hers' disease (McKusick: 23270)

There is marked hepatomegaly, a benign course and usually no hypogly-caemia, acidosis, or hyperlipidaemia. There is a tendency to ketosis on fasting (Fernandes and Pikaar 1972). In contrast to patients with hepatic phosphorylase kinase deficiency (GSD VIII), there is no hyperglycaemic response to glucagon, although there is an increase in the excretion of cAMP in the urine. As discussed above, the diagnosis can be established only after the phosphorylase activating systems have been shown to be intact (Koster *et al.* 1973; Fernandes *et al.* 1974).

Leucocyte phosphorylase activity has been reported to be reduced in some patients (Williams and Field 1961; Hülsmann *et al.* 1961) but this is not reliable for diagnosis (Hug 1980). Koster *et al.* (1973) reported considerably reduced phosphorylase activity in leucocytes but normal activity in cultured fibroblasts. There are insufficient reports to evaluate the reliability of measurement of phosphorylase activities in haemolysates by the method of Lederer *et al.* (1975), but reduced activity was reported in a patient with partial phosphorylase deficiency by Lederer and Stalmans (1976). Other patients with partial phosphorylase deficiency and normal liver phosphorylase kinase activity were reported by Drummond *et al.* (1970) and Hug *et al.* (1974).

Histologically, the hepatocytes are enlarged by glycogen and osmiophilic lipid droplets. Fibrosis of the liver and widening of the portal tracts may occur. Electron microscopy reveals large areas of glycogen α-particles which displace intracellular organelles (Lamy *et al.* 1960; Hug *et al.* 1974; De Bruijn *et al.* 1975).

The disorder has an autosomal recessive mode of inheritance.

Glycogen storage disease type VIII; liver phosphorylase kinase deficiency (McKusick: 30600; type IXB of Hug, 1980)

In the majority of patients, the disorder has an X-linked mode of inheritance, and is probably the commonest form of GSD (de Barsy and Lederer 1980). Of 32 patients studied by Huijing and Fernandes (1969), only three were girls and considered to be heterozygotes. Deficiency of phosphorylase kinase was demonstrated in leucocytes and erythrocytes. Muscle phosphorylase kinase activity is in the normal range (Hug 1980). There is a normal hyperglycaemic response to glucagon.

An extensive pedigree reported by Huijing and Fernandes (1969) is consistent with X-linked inheritance.

Lederer *et al.* (1975) were able to distinguish sex-linked and autosomal recessive forms of GSD VIII by assay of phosphorylase kinase in haemolysates. Hug (1980) and de Barsy and Lederer (1980) also recognize a form with autosomal recessive inheritance.

Cultured fibroblast phosphorylase kinase was reduced to 20–56 per cent of control values in a patient with the sex-linked form. Activity in uncloned fibroblasts of the mother of this patient was normal, but seven of the 46 single-cell clones had activity within the hemizygous range, favouring an X-linked mode of inheritance (Migeon and Huijing 1974).

Clinically, both forms of GSD VIII present with marked hepatomegaly without splenomegaly. There is no acidosis or hyperuricaemia but fasting hypoglycaemia may occur rarely. The course is benign. The liver size may return to normal in adults.

Phosphorylase kinase deficiency in the mouse is also X-linked (Gross and Mayer 1974; Gross 1975).

Glycogen storage disease type VII; muscle phosphofructokinase deficiency (McKusick 23280)

The symptoms are similar to those of GSD V (McArdle's disease) and were first reported by Tarui *et al.* (1965) in three sibs, one of whom had had myoglobinuria. Muscle pains on exercise may commence in childhood (Tobin *et al.* 1973). A mild benign haemolytic anaemia has been reported (Tarui *et al.* 1969). As in GSD V, ischaemic muscle exercise fails to elevate venous blood lactic acid concentration.

Muscle phosphofructokinase (PFK) activities were 3 per cent or less of normal. Deficiency involves the M subunit (Meienhofer *et al.* 1979; Vora *et al.* 1980a; see red cell PFK deficiency, p. 389). Other muscle changes included decreased fructose 1,6-diphosphate, and elevated glucose 6-phosphate and fructose 6-phosphate, concentrations. Erythrocytes contained approximately 50 per cent of normal PFK activity, consistent with the observation that about 50 per cent of erythrocyte PFK is of the muscle type (Tarui *et al.* 1969). Elevations of serum glutamic oxaloacetate transaminase, lactic dehydrogenase, and creatine phosphokinase activities have been reported (Tobin *et al.* 1973).

In a patient with GSD VII no immunologically cross-reacting material to human muscle PFK antibody could be demonstrated (Layzer *et al.* 1967).

The disorder probably has an autosomal recessive form of inheritance in patients described by Tarui (1965) and Layzer *et al.* (1967), but sex-linked inheritance seems possible in the patient reported by Serratrice *et al.* (1969).

Liver glycogen synthase deficiency; UDPG: glycogen-glucosyl transferase deficiency

Fasting hypoglycaemia and a tendency to ketosis were reported by Lewis *et al.* (1963), and Dykes and Spencer-Peet (1972) in identical twins and in an unrelated nine-year-old girl with a similar disorder by Aynsley-Green *et al.* (1977). The twins had a diabetic-type glucose tolerance test, and had a normal hyperglycaemic response to glucagon after feeding, but not during fasting. Hepatic glycogen concentration was reduced.

The authors attributed the disorders to the deficiency of hepatic UDPG: glycogen-glucosyl transferase. However, because of the proven ability of these children to synthesize glycogen after feeds, the causal relation of the enzyme deficiency and clinical phenotype requires further study.

Treatment of 'ketotic hypoglycaemia' aims at preventing hypoglycaemic convulsions and brain damage by frequent feeds and a high protein diet.

DISORDERS OF GALACTOSE METABOLISM

In this section we discuss galactokinase deficiency, classical galactosaemia and uridine diphosphate galactose 4'-epimerase deficiency. Clinical disorders associated with deficiencies of lysosomal ß-galactosidases are described under GM$_1$-gangliosidosis, Krabbe's disease and MPS IVB, and of α-galactosidase under Fabry's disease. Lactase deficiency and congenital glucose-galactose malabsorption are discussed under 'Disorders of carbohydrate absorption', p. 377.

Galactokinase deficiency

The only consistent clinical feature of galactokinase deficiency (Fig. 8.4) has been the occurrence of nuclear cataracts either before or shortly after birth. Hyperbilirubinaemia (Cook *et al.* 1971) and pseudotumour cerebri (Litman *et al.* 1975) have occurred in individual cases and are of interest because the former is typical of, and the latter has been observed in, transferase defect galactosaemia. However, these findings are probably coincidental since they have not occurred in other patients (Gitzelmann 1967; Monteleone *et al.* 1971; Vecchio *et al.* 1976).

The pathogenesis of the cataracts has been attributed to accumulation of the sugar alcohol galactitol formed from accumulated galactose, under the influence of aldehyde (aldose) reductase. This concept is supported by the demonstration of the protective effect of an aldose reductase inhibitor on

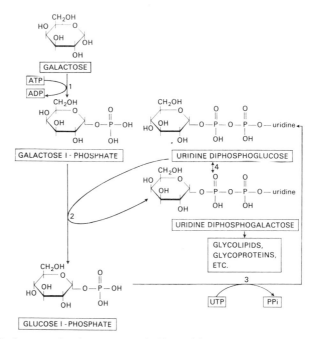

Fig. 8.4. Pathways of galactose metabolism: (1) Galactokinase (2) Galactose 1-phosphate uridyl transferase (3) Uridine diphosphoglucose pyrophosphorylase (4) Uridine diphosphogalactose 4-epimerase.

cataract formation in galactose-exposed rabbit lenses (Kinoshita *et al.* 1968). The process may be compared to the formation of cataracts in diabetics by the glucose alcohol sorbitol. The damage to the lens appears to result from a series of events involving swelling of the lens and consequent fibre disruption (Kinoshita 1965).

Galactokinase deficiency has an autosomal recessive mode of inheritance. Heterozygotes have reduced activities of galactokinase in erythrocytes (Mayes and Guthrie 1968) and cultured fibroblasts (Zacchello *et al.* 1972; Benson *et al.* 1976*b*) and an increased tendency to cataracts (Monteleone *et al.* 1971; Beutler *et al.* 1973*a*; Kaloud *et al.* 1973).

There is an increased excretion of both galactose and galactitol in the urine. Following galactose loading, there is an excessive and prolonged rise in the blood galactose concentration.

The first three patients reported with galactokinase deficiency came from gipsy families (Gitzelmann 1967; Thalhammer *et al.* 1968; Linneweh *et al.* 1970), but the majority of patients have had non-gipsy ethnic origins (Cook *et al.* 1971; Pickering and Howell 1972). There is a report of a patient where galactokinase deficiency was transient (Vigneron *et al.* 1970). Screening of 6 million newborn from North America, Europe and Asia has revealed an incidence of 1:1 000 000 (Levy 1980*a*). A galactokinase polymorphism has been detected among North American blacks who have lower erythrocyte activities than whites (Tedesco *et al.* 1972, 1977).

In culture, fibroblasts derived from patients with galactokinase deficiency grow normally in media containing glucose, but fail to grow when the only hexose is galactose (Benson *et al.* 1976*b*).

Classical galactosaemia; deficiency of galactose 1-phosphate uridyl transferase; transferase deficiency

Clinical features

Infants appear normal at birth but become lethargic and thrive poorly during the second week. There is anorexia with loss of weight, jaundice, hepatomegaly, vomiting, and sometimes oedema and ascites. Unless milk feeding is discontinued, the condition is usually progressive and rapidly fatal. Survival is prolonged when lactose-free intravenous feeding is substituted for milk, but the patients relapse on attempted regrading to a milk diet. After a few weeks, nuclear cataracts develop. These may clear if galactose-free diet is commenced, but otherwise become irreversible.

Some untreated or partially treated patients survive with cataracts, cirrhosis of the liver, mental handicap, and more subtle changes in visual perception (Donnell *et al.* 1980; Komrower *et al.* 1956; Komrower and Lee 1970; Nadler *et al.* 1969; Fishler *et al.* 1972). Even in treated patients, the weight lags behind that of normal controls during the first 18 years. Height is also retarded but becomes normal by 18 years (Donnell *et al.* 1980).

Although treatment, by dietary restriction, causes rapid reversal of the acute illness, intellectual development is not always satisfactory. A more favourable outcome has been noted when treatment is started early. Thus in 16 patients where treatment was started in the first seven days after birth, the mean IQ was 99. Patients in whom treatment was started at longer intervals after birth show a progressive fall in mean IQ, which in five patients, in whom treatment was delayed between four and eleven months, was 62.

In spite of the great importance of early diagnosis and treatment of babies with galactosaemia paediatricians often consider screening of newborn unnecessary. This is because of the assumptions that galactosaemia always presents in the neonatal period with at least jaundice and vomiting, and that these clinical features will lead to recognition of the disorder. These assumptions have been shown to be incorrect by Levy (1980*a*).

Molecular defects

The disorder is due to a deficiency in the activity of galactose 1-phosphate uridyl transferase (Fig. 8.4) that can be demonstrated in haemolysates (Kalckar *et al.* 1956; Schwarz *et al.* 1956) and cultured fibroblasts (Benson *et al.* 1979*b*).

In contrast to galactokinase deficiency, where there is a block in the synthesis of galactose 1-phosphate from galactose, in transferase defect galactosaemia there is accumulation of galactose 1-phosphate which has been incriminated for toxic effects on brain, liver, and renal tubules (see below). In both disorders there is accumulation of galactose and galactitol in the blood and urine, and cataract formation, probably due to galactitol, as discussed under galactokinase deficiency.

Increased erythrocyte galactose 1-phosphate was reported by Donnell *et al.* (1980), in all 12 specimens of cord blood from homozygous transferase-deficient newborn, even though their mothers had had dietary restriction of galactose during pregnancy. The toxic effect of elevated foetal blood galactose 1-phosphate has been suggested as a possible cause of prenatal brain damage. However, the latest IQs of the 12 galactosaemic children with elevated cord blood galactose 1-phosphate concentrations had a mean score of 102, only two scoring below 97.

Gitzelmann and Hansen (1974) point out that accumulation of galactose and galactose 1-phosphate can occur in galactosaemic subjects even when the only dietary hexose is glucose. They incriminate galactose biosynthesis by the pyrophosphorylase pathway and suggest that this uncontrolled process can act as a mechanism of self-intoxication. Moderate or severe hypoglycaemia has occured in some patients.

In addition to galactosuria, there is glycosuria and generalized amino-aciduria when the patient is on a galactose-containing diet, probably as a result of damage to the renal tubules by galactose 1-phosphate.

In culture, fibroblasts derived from patients with galactosaemia grow poorly when the only hexose in the medium is galactose (Krooth and Weinberg 1961).

Diagnosis

The diagnosis can be suspected if galactosuria is demonstrated, and is confirmed by assay of transferase activities in haemolysates (see 'Molecular defects', above.) which reveals values less than four per cent of control mean. When values of five to ten per cent are encountered further studies should be undertaken to establish whether the patient is homozygous for a variant transferase allele associated with decreased activity, or a double heterozygote for either two types of these alleles or of one of these alleles and of the classical galactosaemia allele (see below).

Genetic counselling

Classical transferase deficiency galactosaemia has an autosomal recessive mode of inheritance. Carriers have decreased transferase activities in haemolysates and cultured fibroblasts (Benson *et al.* 1979*b*; Fensom *et al.* 1979*a*) with little or no overlap with normal controls and none with galactosaemic patients.

As discussed above, although the acute clinical features are readily reversible by placing the patient on a galactose-free diet, intellectual development is not always optimal. We have found that in families where previously affected children are handicapped, or have died, parents usually request prenatal diagnosis in subsequent pregnancies (Benson *et al.* 1979*b*).

Prenatal diagnosis is possible by demonstrating deficient transferase activity in cultured amniotic cells (Fensom *et al.* 1974; Fensom and Benson 1975; Benson *et al.* 1979*b*). Methods for assay of transferase involving optical measurement of NADP reduction are unreliable because of a high rate of endogeneous NADP reduction in amniotic cells (Fensom and Benson 1975).

Increased concentration of amniotic fluid galactitol appears to be a useful method of confirming the diagnosis (Allen *et al.* 1980) but further experience is required to establish whether this method is reliable.

In pregnancies presenting in advanced gestation, foetal transferase activity can be measured directly on foetal blood collected at foetoscopy (Fensom *et al.* 1979*a*).

The incidence of transferase deficiency galactosaemia has been estimated as 1:62 000 by screening of six million newborn in North America, Europe, and Asia (Levy 1980a).

Several allelic variants for the transferase locus have been reported. All are associated with decreased transferase activity except for the Los Angeles variant which in homozygotes may lead to activities of up to 140 per cent of control normal values (Ng *et al.* 1973). The commonest polymorphism, designated the Duarte variant after a suburb of Los Angeles, was the first described and is associated with an approximate activity of 44 per cent of control values in Duarte homozygotes, and of 75 per cent in Duarte/normal heterozygotes (Ng *et al.* 1973). Estimates of the gene frequency of the Duarte variant range from 0.032, giving the frequency of heterozygotes as 6.4 per cent (Beutler 1973), to 0.080, corresponding to a frequency of heterozygotes of 16.0 per cent (Pflugshaupt *et al.* 1976). Since double heterozygotes are associated with transferase activities intermediate between those of their constituent alleles, for accurate prenatal diagnosis it is sometimes necessary to establish the genotype of the parents by a combination of electrophoresis and activity measurements of haemolysates, since anodal electrophoretic mobilities are similar for the Los Angeles and Duarte variants but faster than for the normal transferase enzyme (Benson *et al.* 1979b).

Other transferase variants which may be associated with clinical symptoms of galactosaemia in homozygotes include the Rennes variant (Hammersen *et al.* 1975) and the Münster variant reported by Matz *et al.* (1975). Variant alleles which may produce some clinical and biochemical features of galactosaemia in double heterozygotes when the coexistent allele is that of classical galactosaemia include the Negro (Segal *et al.* 1965; Baker *et al.* 1966), Duarte (Benson *et al.* 1979b), Indiana (Chacko *et al.* 1971) and Chicago variants (Chacko *et al.* 1977). The Indiana variant is unstable. No symptoms have been associated with the Berne variant, which has a gene frequency of 0.0009 (Scherz *et al.* 1976).

Treatment of disorders of galactose metabolism

Transferase deficiency

Without exclusion of milk and other galactose-containing feeds, infants with transferase defect have a severe, sometimes fulminating, illness with death in the first few weeks (see 'Clinical features', p. 361). This rapidly fatal course can be prevented if diagnosis is prompt and treatment initiated early. Long-term prognosis is discussed above.

Diagnosis can be confirmed within a few hours of birth if the cord blood of infants with affected siblings is tested for transferase activity. Where laboratory facilities are not immediately available, the neonate should be kept on a galactose-free diet until the results of enzyme assay are available.

Where there is no previously affected sibling, early diagnosis depends either on a positive screening test, or on the vigilance of paediatricians who should consider the diagnosis in any sick neonate, especially, but not only, when there is jaundice or hepatomegaly.

As for other inborn errors of metabolism requiring special diets which exclude a harmful component, it is essential that the patient should receive sufficient vitamins, minerals, and other nutrients. Suitable 'milk-like' preparations containing soya bean protein, vegetable oils, vitamins, and minerals are available. Proprietary brands include Nutramigen (Mead Johnson) and Wysoy (Wyeth). It is important that lactose, as well as galactose, is excluded as far as possible. The efficacy of the diet should be monitored by regular assays of blood galactose 1-phosphate (Donnell and Bergren 1975).

With older children it becomes progressively more difficult to exclude some lactose- (or galactose-) containing foods such as cakes, biscuits, chocolate, ice-cream, etc., but the importance of strict dietary control must be explained to the parents and reinforced.

The principle that after a few years of age, galactose restriction can be relaxed (Komrower 1973; Donnell and Bergren 1975) is vigorously opposed by Brandt (1980), who points out that transferase deficiency persists throughout life, that the ability of patients to metabolize galactose does not improve with age, and that galactose 1-phosphate is very toxic. Brandt concludes that the lactose- (and galactose-) free diet should be lifelong. More recently, Donnell *et al.* (1980) stated that discontinuation of the restricted diet was never recommended because adaptation to galactose was never documented. They point out that concern about the development of cataracts was an important factor in the decision against discontinuation.

Pregnancy

When a mother has given birth to a child with transferase deficiency galactosaemia, and wishes to have further children with the same biological father, it has become accepted that the mother should be on a restricted lactose intake during the pregnancy. However, the data of Donnell *et al.* (1980) showed that even when maternal lactose intake is restricted, the cord blood galactose 1-phosphate concentration was elevated in all 12 homozygous affected newborn. In contrast, obligatory heterozygous foetuses of homozygous affected mothers who have been on lactose-restricted diets during pregnancy have normal cord blood galactose 1-phosphate concentration (Sardharwalla *et al.* 1980).

Duarte galactosaemia transferase deficiency

Some Duarte galactosaemia double heterozygotes have neonatal hyper-bilirubinaemia (Benson *et al.* 1979*b*) and reduced galactose tolerance. Lactose restriction for at least two years is advisable (Brandt 1980).

Galactokinase deficiency

The principle of dietary exclusion is as for transferase deficiency galactosaemia, but the object is prevention or reversal of cataract formation, rather than prevention of hepatic, cerebral and renal damage. Biochemical monitoring is by assay of plasma galactose concentrations.

Uridine diphosphate galactose 4′-epimerase deficiency

Deficiency of UDP-galactose 4′-epimerase (epimerase) restricted to blood cells (Fig. 8.4) was discovered in a mass newborn screening programme (Gitzelmann 1972; Gitzelmann *et al.* 1973). Nine cases from four families were reviewed by Gitzelmann *et al.* (1976) and Gitzelmann and Hansen (1980); these were five newborn, a five-year-old and three sisters in their sixties who were all well without treatment. The condition can therefore be considered to be benign. In Switzerland, the five newborn were detected during screening of 230 000 infants, giving an estimated frequency of 1:46 000, or about the same as that for classical galactosaemia. Epimerase activity was undetectable in red cells of all nine subjects, and in leucocytes of all seven subjects tested, except in a 14-month-old girl, who was found to have developed epimerase activity during an upper respiratory tract infection (Gitzelmann *et al.* 1976).

Residual lymphocyte epimerase activity was 6.5 per cent of control mean value in one of three subjects tested; the remainder had undetectable lymphocyte activities. Epimerase activities were within the normal range in liver biopsies from both subjects tested, and in fibroblasts of all four subjects tested. Surprisingly, normal epimerase activity appeared in a long-term lymphoblast line derived from epimerase-deficient lymphocytes, and in phytohaemagglutinin-stimulated epimerase-deficient lymphocytes (Mitchell *et al.* 1975). Partially purified mutant epimerase from either lymphoblast lines or cultured fibroblasts had reduced heat stability and exhibited a higher NAD requirement for stability at 40°C (Gitzelmann and Haigis 1978). These observations are consistent with the explanation that in the mutant epimerase there is a structural defect in all tissues, but deficiency can be demonstrated only in cells with a low rate of epimerase synthesis, or where NAD concentration is too low to stabilize the enzyme. The deficiency has an autosomal recessive mode of inheritance. Heterozygotes have reduced erythrocyte epimerase activities. In two of four families there was parental consanguinity (Gitzelmann and Steinmann 1973; Gitzelmann *et al.* 1976).

DISORDERS OF FRUCTOSE METABOLISM

In this section we discuss essential fructosuria, hereditary fructose intolerance, and fructose 1,6-diphosphatase deficiency. Muscle phosphofructokinase deficiency is considered under the heading of glycogen storage disease type VII, while sucrase–isomaltase deficiency is considered under disorders of carbohydrate malabsorption.

Fructose, in the form of the disaccharide sucrose, is an important source of energy. Its initial metabolism is by one of two major pathways which vary in importance in different tissues. In the liver, and to a lesser extent in the intestine and kidney, fructose is phosphorylated to fructose 1-phosphate by ATP. The reaction is catalysed by fructokinase (ketohexokinase), the enzyme which is deficient in essential fructosuria. Fructose 1-phosphate (Fig. 8.5) is then split into glyceraldehyde and dihydroxyacetone phosphate in the aldolase reaction. About 90 per cent of glyceraldehyde is phosphorylated by ATP to D-glyceraldehyde 3-phosphate, catalysed by glyceraldehyde kinase (Sillero *et al.* 1969), which can then be further metabolized to pyruvate. The last pathway accounts for about 20 per cent of fructose absorbed from the intestine in man (Bergström *et al.* 1972).

The second major initial pathway for fructose metabolism is the hexokinase-catalysed phosphorylation by ATP which is most active in adipose or muscle tissues. The product, fructose 6-phosphate, can be phosphorylated by ATP to fructose 1,6-diphosphate catalysed by phosphofructokinase. The reverse reaction, however, is catalysed by fructose 1,6-diphosphatase. The importance of this step is evident from the severe effects resulting from fructose 1,6-diphosphatase deficiency. The reaction is important since it forms part of the gluconeogenetic pathway by which glucose (and glycogen) are synthesized from pyruvate, alanine, and glycerol. Fructaldolase B is deficient in hereditary fructose intolerance.

Essential fructosuria

Essential fructosuria is a benign condition and is usually detected by testing the urine for reducing substances. Fructosuria can readily be distinguished from glucosuria since it gives no reaction to tests which rely on glucose oxidase for detection of glucose. It is due to deficiency of fructokinase activity in liver, kidney, and intestinal mucosa (Schapira *et al.* 1961–62).

After fructose loading (1g/kg body weight, up to 50 g), there is an

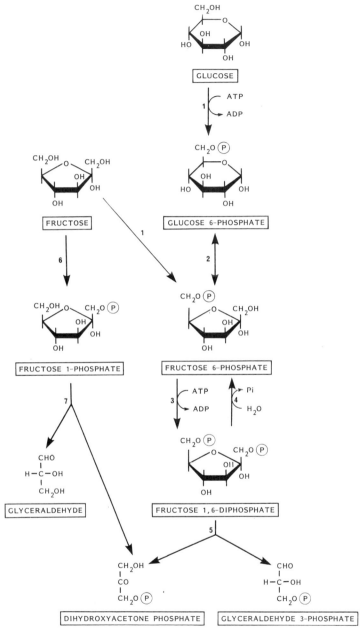

Fig. 8.5. Pathways of fructose metabolism: (1) Hexokinase (2) Glucose 6-phosphate isomerase (3) Phosphofructokinase (4) Fructose 1,6-diphosphatase (5) Aldolase (6) Fructokinase (ketohexokinase) (7) Fructose 1-phosphate aldolase.

excessive rise of blood fructose, but blood glucose, lactic acid, and uric acid fail to show their normal rise (Steinmann *et al.* 1975*b*).

Essential fructosuria has an autosomal recessive mode of inheritance. Heterozygotes do not show abnormal fructose loading levels (Leonidas 1965), but fructokinase activities in carriers do not appear to have been reported. The incidence has been estimated as 1:130 000 (Lasker 1941).

Hereditary fructose intolerance

Hereditary fructose intolerance (HFI) was recognized as a severe clinical disorder of fructose metabolism (Chambers and Pratt 1956; Froesch *et al.* 1957) several years before the demonstration that it was caused by deficiency of fructaldolase B. Perheentupa *et al.* (1972) found 100 cases in the literature, while Odievre *et al.* (1978) analyzed diagnosis and management of 55 cases.

Clinical features

Clinical and laboratory features of 20 children from 17 families, with HFI have been reported by Baerlocher *et al.* (1978). Acute symptoms followed ingestion of fructose either from sucrose added to bottle milk feeds in infants or from fruit or vegetables at weaning, or from sweets in childhood. The symptoms were acute after exposure to fructose feeds, and chronic after repeated exposure.

Acute symptoms were more severe with younger than older children and consisted of sweating, trembling, dizziness, nausea, vomiting, lethargy, and convulsions.

Chronic symptoms included anorexia, vomiting, failure to thrive, jaundice, hepatomegaly, oedema, ascites, and haemorrhages. Older children had developed an aversion to sweets.

Molecular defects

The disorder is due to deficiency of fructaldolase B which has been demonstrated in liver (Hers and Joassin 1961; Steinmann *et al.* 1975*b*), kidney (Kranhold *et al.* 1969), and intestine (Baerlocher *et al.* 1978). The deficiency in the liver is greater when fructose 1-phosphate rather than fructose 1,6-diphosphate (Fig. 8.5) is the substrate (Steinmann *et al.* 1975*b*).

The majority of children investigated had fructosaemia and fructosuria, and all had some biochemical evidence of hepatic malfunction, including elevated serum transaminase, bilirubin, alkaline phosphatase and prolonged prothrombin time.

Defective renal tubular reabsorption of amino acids produced a generalized hyperaminoaciduria. Other abnormalities included hypoglycaemia, hyperuricaemia, hypophosphataemia, hypokalaemia, metabolic acidosis, and, in some, increased concentrations of serum tyrosine and methionine.

Diagnosis

The diagnosis can be suspected when acute clinical and biochemical features are reversed by restriction of dietary fructose. Further evidence can be demonstrated by an intravenous fructose (0.2–3 g/kg) (or sorbitol) tolerance test which is followed immediately by a fall in blood glucose concentration to a minimum value after 40 min. The diagnosis can be confirmed by demonstrating severe reduction in the activity of fructaldolase B in a liver or intestinal biopsy (see 'Molecular defects', p. 369).

Genetic counselling

The disorder has an autosomal recessive mode of inheritance but aldolase activities in the liver of parents have been within normal limits (Raivio *et al.* 1967). The report by Beyreiss *et al.* (1968) that continuous intravenous fructose infusion causes a higher blood fructose concentration in parents than in controls could not be confirmed by Baerlocher *et al.* (1980). Reports consistent with dominant inheritance in some families (Köhlin and Mehlin 1968; Schulte and Lenz 1977) do not rule out the possibility that the affected parent was a homozygote for the mutant HFI gene.

The incidence in Switzerland has been estimated as 1:20 000 (Gitzelmann and Baerlocher 1973).

Treatment

Treatment by placing the patient on a fructose-free diet is a matter of urgency since haemorrhage and liver failure can be fatal (Baerlocher *et al.* 1978).

There is rapid improvement of the acute clinical features, but hepatomegaly resolves over several months (Odievre 1976).

Fructose 1,6-diphosphatase deficiency

As discussed below, fructose-1,6-diphosphatase (FDPase) is an important enzyme in gluconeogenesis, which is an important source of glucose, especially in the newborn (Pagliara *et al.* 1973). When this pathway is blocked, failure in conversion of pyruvate to glucose during fasting results in lactic acidaemia and hypoglycaemia.

Since the initial description by Baker and Winegrad (1970) several other patients have been reported (Greene *et al.* 1972; Eagle *et al.* 1974; Hopwood *et al.* 1977). Clinical features and laboratory findings of 16 patients were analysed by Baerlocher *et al.* (1980).

Clinical features

Half the patients have presented in the first four days after birth, and a quarter before and a quarter after six months of age. The commonest presenting symptom has been hyperventilation due to the metabolic acidosis. Hepatomegaly was present in a third of the patients when first seen but later developed in nearly all. Other common features included hypotonia, lethargy, vomiting, failure to thrive, and hypoglycaemic convulsions.

Molecular defects and diagnosis

The diagnosis should be suspected in infants when there is fasting hypoglycaemia and hyperlacticacidaemia.

Intravenous fructose loading causes hypoglycaemia and hyperlacticacidaemia, maximal after 45 minutes, and a fall in plasma bicarbonate concentration (Steinmann and Gitzelmann 1976).

The disorder is confirmed by demonstration of reduced FDPase (Fig. 8.5) activity in liver biopsies, where values have ranged from 0 to 30 per cent of normal values, or in intestines (Greene *et al.* 1972) or kidney (Hülsmann and Fernandes 1971). Values for cultured fibroblasts do not appear to have been reported. Deficiency of FDPase activity in leucocytes of patients was reported by Melançon *et al.* (1973) and by Schrijvers and Hommes (1975), using a spectrophotometric assay, but Cahill and Kirtley (1975) were unable to demonstrate activity in normal leucocytes. More recently, deficient activity was found in leucocytes of a Lebanese patient (Dr D. Alexander, American University of Beirut, personal communication) using both the spectrophotometric assay and a new radiochemical method (Janssen and Trijbels 1982).

Genetic counselling

The disorder has an autosomal recessive mode of inheritance. Reduced FDPase activities have been demonstrated in livers from three parents (Saudubray *et al.* 1973; Gitzelmann *et al.* 1973). Parents in three of the reported families were consanguineous (Hülsmann and Fernandes 1971; Baerlocher *et al.* 1971b; Greene *et al.* 1972). Dominant inheritance was suggested in a single case where a mother and daughter were both affected (Taunton *et al.* 1978). However, the possibility could not be ruled out that the mother was homozygous for autosomal recessive mutant genes. By assaying leucocytes using both the spectrophotometric and radiochemical methods, Dr D. Alexander (personal communication) noted some overlap between heterozygotes and controls but was able to identify two heterozygotes unequivocally. The radiochemical assay gave the least overlap between heterozygotes and controls.

Treatment

The principle of dietary management should be to administer frequent meals and restrict fructose, sorbitol, and, if necessary, triglycerides. Folic acid appeared to produce an improvement in two patients (Greene *et al.* 1972), but did not help the patient reported by Saudubray *et al.* (1973).

DISORDERS OF PYRUVATE METABOLISM AND GLUCONEOGENESIS; LACTIC ACIDOSIS

Deficiency of pyruvate dehydrogenase

Defining deficiencies of the pyruvate dehydrogenase complex (PDHC) (Fig. 8.6) is complicated by the fact that the complex consists of three separate catalytic enzymes and two regulatory enzymes. The chief reaction is the irreversible decarboxylation of pyruvate to yield acetyl-CoA (Randle *et al.* 1979; Reed *et al.* 1980).

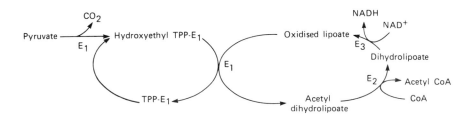

E_1 Pyruvate decarboxylase

E_2 Dihydrolipoate acetyl transferase

E_3 Dihydrolipoate dehydrogenase (lipoamide dehydrogenase)

Fig. 8.6. Reactions of the pyruvate dehydrogenase complex.

The complex, which is intramitochondrial, consists of about 150 polypeptide chains and has a molecular weight of about 7–10 million. The catalytic enzymes consist of pyruvate decarboxylase (Enz 1) which is thought to be the site of regulation, by end product inhibition, by acetyl-CoA and NADH (Garland and Randle 1964; Tsai *et al.* 1973), and by phosphorylation by a kinase (to inactivate) and dephosphorylation by a phosphatase to activate (Linn *et al.* 1969*a*,*b*) the complex. The kinase catalyses the transfer of a phosphate group from ATP to serine groups in the α-subunits of Enz 1 (Yeaman *et al.* 1978) which is a tetramer consisting of two different subunits $(\alpha_2\beta_2)$. Enz 1 is thiamin-pyrophosphate-dependent and catalyses the transfer of a hydroxyethyl group from pyruvate (which is decarboxylated) to oxidised lipoate, to yield acetyl dihydrolipoate.

Enz 2 is an acetyl transferase which catalyses the transfer of the acetyl residue from acetyl dihydrolipoate to CoA to yield acetyl-CoA and dihydrolipoate. Enz 3 is a dihydrolipoyl dehydrogenase (lipoamide dehydrogenase) required to oxidise dihydrolipoate utilising NAD^+ as hydrogen acceptor. The cycle then recommences by the Enz 1-catalysed reaction.

Clinical features

The first patient in whom deficiency of PDHC (PDHC⁻) was demonstrated in a muscle biopsy had intermittent ataxia, mild acidosis, and hyperpyruvic-acidaemia (Blass *et al.* 1970). Other patients have presented with two very different syndromes, namely either severe lactic acidosis at birth (Farrell *et al.* 1975) or with ataxia in the second decade (Blass *et al.* 1976*b*; Robinson *et al.* 1980). In the former type residual activity was almost undetectable, but in the latter group, a high level of residual activity of 40–50 per cent of normal was found (Blass *et al.* 1976*a*; Kark and Rodriguez-Budelli 1979). Features of subacute necrotizing encephalomyelopathy (Leigh 1951) have been reported (Blass *et al.* 1976*a*).

Molecular defects and diagnosis

The diagnosis of PDHC⁻ has been established on four types of evidence (Blass 1980):

Deficiency of PDHC activity. This has usually been demonstrated as deficient release of $^{14}CO_2$ from 1-^{14}C-pyruvate (Blass *et al.* 1970; Farrell *et al.* 1975); however, specific deficiency of dihydrolipoyl dehydrogenase (Enz 3) was demonstrated in patients with lactic acidosis (Robinson *et al.* 1977, 1980, 1981) and in a patient with ataxia (Kark and Rodriguez-Budelli 1977). Reduced activation of liver PDHC from an infant with lactic acidaemia was interpreted to indicate a deficiency of activating phosphatase (Robinson and Sherwood 1975).

The deficiency has been demonstrated in cultured fibroblasts (Blass *et al.* 1976*b*) liver, brain, and muscle (Farrell *et al.* 1975; Robinson and Sherwood 1975; Robinson *et al.* 1977). Deficiency of PDHC activity has been demonstrated in all tissues studied except in the patient reported by Willems *et al.* (1974) in whom there was PDHC deficiency in liver but not cultured fibroblasts.

Demonstration of deficiency of pyruvate oxidation by intact cells. This method depends on the demonstration of diminished release of $^{14}CO_2$ from 1-^{14}C-pyruvate or from 2-^{14}C-pyruvate by intact cultured fibroblasts, leucocytes, or muscle biopsy material. This method, by itself, is not specific for PDHC$^-$ because reduced conversion of ^{14}C-pyruvate to $^{14}CO_2$ could also be caused by decreased transport of ^{14}C-pyruvate into the mitochondria, or by dilution of ^{14}C-pyruvate by an enlarged intracellular pyruvate pool. However, the demonstration of both deficiency of PDHC activity and of deficient pyruvate oxidation by intact cells in several patients (Blass *et al.* 1970, 1976*b*; Kark *et al.* 1974; Strömme *et al.* 1976) indicates that the latter observation may indicate PDHC$^-$ under physiological conditions.

Accumulation of metabolites in vivo. Accumulation of pyruvate and its metabolites is consistent with but not conclusive evidence of PDHC$^-$. Furthermore, the degree of accumulation depends on the diet (Falk *et al.* 1976). Blood pyruvate concentration is usually only moderately increased to 150–300 μmol/l (normal less than 100 μmol/l). CSF pyruvate concentration may also be elevated (Blass *et al.* 1970). Blood lactate (Blass *et al.* 1970) and alanine (Lonsdale *et al.* 1969) concentrations may also be increased.

Beneficial effects of a high-fat, low-carbohydrate diet. The beneficial effects of a ketogenic high-fat, low-carbohydrate diet have been documented by Falk *et al* (1976). The deleterious effect of a high carbohydrate diet has also been reported (Cederbaum *et al.* 1976).

Genetic counselling

Disorders producing PDHC$^-$ probably have an autosomal recessive mode of inheritance. Parents have had values of PDHC activity which have been either at or below the low range of normal (Blass *et al.* 1970; Cederbaum *et al.* 1976; Falk *et al.* 1976). Both sexes have been affected and in one family the parents were consanguineous (Blass *et al.* 1972). In families where deficiency of PDHC or of one of its components can be demonstrated in cultured fibroblasts, prenatal diagnosis by direct enzyme assay of cultured amniotic cells should be possible.

Treatment

The beneficial effect of a ketogenic high-fat, low-carbohydrate diet in some, but not all patients (Robinson *et al.* 1977) has already been mentioned. This should be started in hospital with careful biochemical monitoring.

Thiamin pyrophosphate is a cofactor of the Enz 1 component of PDHC and has been used for treatment of PDHC$^-$. One patient has shown unequivocal improvement on high doses (Wick *et al.* 1977). Improvement in a few patients has been reported following treatment by citrate (Oka *et al.* 1976), cholinergic agonists (Rodriguez-Budelli *et al.* 1978), and steroids (Blass *et al.* 1971).

Defects of gluconeogenesis

Gluconeogenesis is the conversion of lactate and amino acids to glucose. Although it is a continuous process, it is especially important for maintaining normoglycaemia during the first few days after birth (Pagliara *et al.* 1973) and is accelerated during fasting and after glucagon administration (Exton and Perk 1972).

Most of the reactions of glycolysis are reversible and utilized during gluconeogenesis but there are three which are not:

1. During glycolysis, the ATP-generating conversion of phosphoenol pyruvate to pyruvate is catalysed by pyruvate kinase. Two other enzymes are required for the reconversion in gluconeogenesis. Firstly, the intra-mitochondrial biotin-dependent pyruvate carboxylase-catalysed conversion of pyruvate to oxaloacetate involving CO_2-fixation and ATP, and secondly the GTP-dependent conversion of oxaloacetate to phosphoenolpyruvate, involving decarboxylation catalysed by phosphoenolypyruvate carboxy-kinase which in man exists both in mitochondria and in the cytosol.

2. During glycolysis, the conversion of fructose 1-phosphate to fructose 1,6-diphosphate (by ATP) is catalysed by phosphofructokinase. In gluconeogenesis, the reverse reaction is catalysed by fructose 1,6-diphos-phatase.

3. During glycolysis, the conversion of glucose to glucose 6-phosphate (by ATP) is catalysed by glucokinase. In gluconeogenesis, the reverse reaction is catalysed by glucose 6-phosphatase.

Deficiency of fructose 1,6-diphosphatase has been considered under disorders of fructose metabolism and of glucose 6-phosphatase under glycogen storage disease type I. In this section we discuss disorders of the other two enzymes catalysing reactions which reverse glycolysis, namely deficiency of pyruvate carboxylase and of phosphoenolpyruvate carboxy-kinase.

Multiple biotin-dependent carboxylase deficiency, involving the car-boxylases of propionyl-CoA and 3-methylcrotonyl-CoA as well as pyruvate, is discussed elsewhere. In this section we review patients with heterogenous neurological disorders presenting during the first two and a half years after birth with reasonable evidence of pyruvate carboxylase deficiency. However, since the possibility of multiple carboxylase deficiency (q.v.) was not always considered, it is not possible to rule out this diagnosis in some of these patients.

Pyruvate carboxylase deficiency

Clinical features

Convulsions and retarded development (Brunette *et al.* 1972; Tada *et al.* 1978), hypotonia (van Biervliet *et al.* 1977; Tada *et al.* 1973, 1978) and acidosis (van Biervliet *et al.* 1977; De Vivo *et al.* 1977; Atkin *et al.* 1979) have been the commonest clinical features.

Features of subacute necrotizing encephalomyelopathy (Leigh 1951) have been reported (Farmer *et al.* 1973).

Molecular defects and diagnosis

All patients have had elevated blood lactate concentrations (900–40 000 μmol/l; normal: less than 100 μmol/l), but definitive diagnosis depends on the demonstration of pyruvate carboxylase deficiency in leucocytes or cultured fibroblasts (Atkin *et al.* 1979; Marsac *et al.* 1982). Surprisingly, hypoglycaemia, which is severe in other deficiencies of enzymes implicated in gluconeogenesis, has occurred in only a few patients (Brunette *et al.* 1972; van Biervliet *et al.* 1977). The patient reported by Atkin *et al.* (1979) had no detectable pyruvate carboxylase activity in his leucocytes (normal 0.070–0.208 nmol HCO_3^- fixed/min/mg protein) and markedly reduced activity in his fibroblasts.

Genetic counselling

The reports that affected siblings of both sexes have had unaffected parents are consistent with an autosomal recessive mode of inheritance. This is confirmed by the demonstration that parents had reduced pyruvate carboxylase activities in cultured fibroblasts (De Vivo *et al.* 1977; Atkin 1979) and lymphocytes (Atkin 1979; Marsac *et al.* 1982).

Prenatal diagnosis of pyruvate carboxylase deficiency by demonstration of reduced cultured amniotic cell activity was reported by Marsac *et al.* (1982). Following amniocentesis at 17 weeks' gestation cultured amniotic cell pyruvate carboxylase activity was 0–10 per cent of control values in all four laboratories in which the assays were carried out. Since propionyl-CoA carboxylase and 3-methylcrotonyl-CoA carboxylase activities were normal, the diagnosis of multiple carboxylase deficiency could be ruled out.

Treatment

Administration of glutamate (Tang *et al.* 1972) and aspartate (De Groot and Hommes 1973), possibly by correcting deficiency of oxaloacetate, has produced lowering of blood lactate concentration in some, but not all, patients.

High doses of thiamin have also lowered blood lactate concentration in some patients (Tada *et al.* 1973; Moosa and Hughes 1974; Lonsdale *et al.* 1969).

Phosphoenolpyruvate carboxykinase deficiency

Deficiency of hepatic phosphoenolypyruvate carboxykinase was reported in a patient by Fiser *et al.* (1974) and in two others by Hommes *et al.* (1976). The outstanding feature was recurrent severe hypoglycaemia starting between three and 19 months of age. Two were males and one was female. In the two patients reported by Hommes *et al.* (1976), there was ECG evidence of left ventricular hypertrophy and cardiomegaly. In spite of vigorous treatment by glucose administration, the patients died a few days or months after presentation. Necropsy revealed massive fatty deposition in liver and kidneys.

MISCELLANEOUS CARBOHYDRATE DISORDERS

Essential pentosuria; xylitol dehydrogenase deficiency

Individuals with pentosuria are asymptomatic. Life expectancy is normal (Lasker 1955). The condition is usually discovered when the urine is tested for reducing substances. Reported cases have almost exclusively been Ashkenazi Jews. The cause is a deficiency in the conversion of L-xylulose to xylitol catalysed by xylitol dehydrogenase, utilizing NADP as hydrogen acceptor, which has been demonstrated in erythrocytes (Wang and Van Eys 1970).

Irrespective of dietary intake, the daily excretion of urine pentose varied from 1.1 to 3.7 g/24 h in five affected adults (Enklewitz and Lasker 1935). Normal excretion is up to 0.060g/24 h. The pentose was identified as L-xylulose by Levene and La Forge (1914). Patients previously considered to have arabinosuria probably excreted L-xylulose, since no further cases of arabinosuria have been reported since 1928 (Lasker 1950). L-xylulose is an intermediate in the glucuronic acid oxidation pathway, which in humans appears to serve no useful function (Touster *et al.* 1957).

The incidence has been estimated as 1:2 000–2 500 in American Jews and 1:5 000 among Jews in Israel (Goodman 1979).

DISORDERS OF INTESTINAL CARBOHYDRATE ABSORPTION

The symptoms of carbohydrate malabsorption are particularly severe in young infants, who characteristically have watery diarrhoea and excoriated perianal skin following ingestion of the offending sugar. Loss of fluids and electrolytes may lead to life-threatening dehydration. Abdominal distension and vomiting are common.

In general, symptoms are milder in older children and adults who tend to pass frequent loose motions, sometimes intermittently, and suffer from abdominal colic or irritable bowel syndrome.

Although some of the unabsorbed sugars are excreted in the stools and are detectable as 'reducing substances', most are degraded by intestinal bacteria to hydrogen, carbon dioxide (Calloway *et al.* 1966) and organic acids such as acetic, butyric, propionic, and lactic. An osmotic gradient from unabsorbed sugars and their breakdown products causes transfer of water and electrolytes from plasma into the intestinal lumen, thus contributing to the intestinal loss.

Defects in carbohydrate absorption can be primary or secondary.

Primary absorptive defects

Congenital lactase deficiency

This disorder usually presents in infants, with severe watery diarrhoea following milk feeds, which can cause marasmus and death unless lactose is removed from the diet. The disorder was described by Holzel *et al.* (1959).

Normal intestinal mucosa contains three types of ß-galactosidase (lactase, lysosomal ß-galactosidase and hetero ß-galactosidase). Three of four children studied by Asp and Dahlqvist (1974) had absent lactase activity and the fourth had severe deficiency. The other ß-galactosidases had normal activity. Considerably reduced or absent activities were also found in four patients by Freiburghaus *et al.* (1976).

Ontogenic late onset (acquired) lactase deficiency

As in several mammals (Rubino *et al.* 1964), in the majority of non-Caucasian ethnic populations, intestinal lactase activity falls after weaning to about 5–10 per cent of control levels so that lactose malabsorption is normal in adults. This has been observed in Uganda (Cook and Kajubi 1966), America (Bayless and Rozensweig 1966) and Nigeria (Ransome-Kuti *et al.* 1975). The majority of Caucasians, however, retain high lactase activities. The cause of the fall of lactase activity is not related to lactose intake (Knudson *et al.* 1968; Johnson *et al.* 1974), and is probably genetic. Family studies suggest that in adults the ability to digest lactose has an autosomal dominant type of inheritance while 'inability' has an autosomal recessive mode of inheritance.

Variable but relatively mild symptoms occur, after ingestion of small quantities of lactose in individuals with low lactase activity, often starting in early childhood. It has been suggested that there is a relation between the pattern of milk intake and ethnic distribution of late-onset lactase deficiency (Simoons 1970).

Sucrase–isomaltase deficiency

Sucrase–isomaltase activity is concentrated in complexes of the intestinal brush-border (Gray 1975). Separation of the two enzymes leads to lability of the isomaltase component. It has been suggested therefore that sucrase deficiency is the primary defect (Conklin *et al.* 1975). The disorder has an autosomal recessive mode of inheritance and has been found in 0.2 per cent of North Americans (Peterson and Herber 1967) and 10 per cent of Greenland Eskimos (McNair *et al.* 1972).

Watery diarrhoea occurs in infants after ingestion of sucrose and is therefore absent in exclusively breast-fed infants. Diagnosis can be confirmed by enzyme assay of intestinal mucosal biopsies. Gray *et al.* (1976) found that neither enzyme activity nor immunologically cross-reacting material was present in seven patients, but they were present in considerably reduced quantities in two others.

It is not understood why some patients present in adult life (Neale *et al.* 1965).

Heterozygotes have reduced intestinal sucrase/lactase ratios (Kerry and Townley 1965).

Congenital glucose/lactose malabsorption

Newborn with this rare disorder develop severe diarrhoea on the first two days after birth following feeding with glucose or lactose. Absorption of fructose, xylose, leucine, and alanine is normal (Schneider *et al.* 1966; Phillips and McGill 1973; Hughes and Senior 1975).

The activities of disaccharidases of the intestinal mucosa are within normal limits (Phillips and McGill 1973; Wimberley *et al.* 1974; Meeuwisse and Dahlqvist 1968).

The molecular basis of this disorder has not yet been characterized.

Trehalase deficiency

Trehalose (1-O-α-D-glucopyranosyl-α-D-glucopyranoside) occurs in plants including mushrooms. Deficiency of intestinal brush-border trehalase has been described in two members of the same family who presented with intolerance to mushrooms (Madzarovová-Nohejlova 1973). Inheritance may have been autosomal dominant.

Diagnosis of primary disorders of carbohydrate absorption

A careful history often reveals the onset of symptoms with introduction of the relevant sugars to the diet. Excluding these sugars results in rapid improvement, thus indicating the diagnosis.

Examination of the watery component of the stools immediately after passage, collected on non-absorbent material, may reveal reducing substances, but these may occur when there is diarrhoea due to other causes (Davidson and Mullinger 1970; Soeparto *et al.* 1972).

Faecal pH is usually low, but this finding can also be non-specific (Soeparto *et al.* 1972; McMichael *et al.* 1965).

Sugar and loading tests may show 'flat' blood sugar concentration curves, and precipitate symptoms, but the former are often found in normal individuals (Newcomer and McGill 1967).

Barium–lactose meals give an abnormal radiological appearance but also frequent false positive results (Morrison, *et al.* 1974).

The only completely diagnostic test is direct enzyme assay of jejunal mucosa (Townley *et al.* 1965), but this may be unnecessary if sugar intolerance can be clearly demonstrated by the effects of feeding and withdrawals.

Treatment of primary disorders of carbohydrate absorption

Treatment is by excluding the relevant sugars from the diet.

Dehydrated seriously ill infants will require intravenous feeding until rehydrated.

Secondary disorders of carbohydrate absorption

These occur much more frequently than primary disorders of carbohydrate absorption. They are due to damage of the intestinal mucosa by one of several factors, the commonest of which is bacterial or viral gastroenteritis (Barnes *et al.* 1974). Other causes include giardiasis (Hoskins *et al.* 1967), intestinal resection in newborn (Burke and Anderson 1966) and sprue (Plotkin and Isselbacher 1964).

PRIMARY HYPEROXALURIA

Amongst patients with calcium oxalate nephrolithiasis, two types of disorder have been characterized by demonstrating enzyme defects. It is likely, however, that other types exist.

In type I, or 'glycolic aciduria', there is deficiency in the activity of 2-oxoglutarate:glyoxylate carboligase. In type II, or 'L-glyceric aciduria', there is deficiency of D-glyceric dehydrogenase. Both types have autosomal recessive inheritance.

Oxalate metabolism in primary hyperoxaluria

Oxalate catabolism

In humans, oxalate is not metabolized and serves no known useful function. When ^{14}C-oxalate is administered intravenously to normal men, 89–99 per cent appears in the urine (Akcay and Rose, 1980). Expiratory CO_2 is not labelled. In rats also it is excreted unchanged (Weinhouse and Friedmann 1951).

Since oxalate is not catabolized, primary hyperoxaluria must be due to persistent overproduction of oxalate, since affected patients do not have increased intestinal absorption.

Oxalate synthesis

Oxalate is synthesized from either ascorbate or glyoxylate. Approximately 35–50 per cent of urinary oxalate is derived from ascorbic acid. Increasing ascorbic acid intake, however, does not increase urinary oxalate excretion unless over 4 g of ascorbic acid are ingested per day (Briggs *et al.* 1973).

The oxidation of glyoxylate to oxalate can be catalysed by lactic dehydrogenase in the presence of NAD (Sawaki *et al.* 1967) or by either of the flavoproteins, glycolic acid oxidase (Richardson and Tolbert 1961) or xanthine oxidase (Booth 1938). The last-named enzyme is probably relatively unimportant in oxalate synthesis, since administration of the inhibitor allopurinol does not diminish urinary oxalate excretion (Gibbs and Watts 1966). Glyoxylate is synthesized from three precursors: glycine, glycolic acid, and 2-oxo-4-hydroxyglutarate. Isotope studies show that glycine is a major precursor of urinary oxalate. However, there is no evidence that synthesis of oxalate from glycine is increased in hyperoxaluria. Thus in a control subject 40 per cent of oxalate was derived from glycine and in two patients with primary hyperoxaluria, 50 and 32 per cent. It appears, therefore that the glycine → glyoxylate → oxalate pathway was not enhanced in these patients.

In patients with primary hyperoxaluria, intravenously administered 1-^{14}C-glyoxylate (Frederick *et al.* 1963) and 1-^{14}C-glycolate (Hockaday *et al.* 1964) were poorly converted into [14]CO_2, but there was increased excretion as urinary ^{14}C-oxalate and ^{14}C-glycolate.

Glyoxylate metabolism

Glyoxylate accumulation is a possible cause of enhanced oxalate synthesis. Reactions in which glyoxylate is involved are therefore relevant to the possible causes of primary hyperoxaluria. Eight such reactions observed in man and nine others in microorganisms and plants are reviewed by Williams and Smith (1968*a*). Of particular relevance is the conjunction of glyoxylate with 2-oxoglutarate, which is decarboxylated in the presence of thiamin pyrophosphate, producing 2-hydroxy-3-keto adipate. The reaction is catalysed by the carboligase which has been shown to be deficient in primary hyperoxaluria type I.

Primary hyperoxaluria type I; glycolic aciduria

This is by far the commoner form. Patients present usually in the first five years with renal colic, haematuria, and passage of small calculi due to urolithiasis. A few patients first present with stunted growth and renal failure. The majority of patients die of renal failure in the first two decades.

Rarer manifestations include a gout-like acute arthritis (McLaurin *et al.* 1961) and although there is often hyperuricaemia, there have been crystalline deposits in synovial membranes of calcium oxalate, not sodium urate. Atrioventricular block is another complication, possibly due to deposition of calcium oxalate in the atriovenricular cardiac conducting tissues (Coltart and Hudson 1971; Pikula *et al.* 1973). A few patients survive into adult life, while others die in the first year.

Before the onset of renal failure, the 24 h excretion of oxalate is rarely below 100 mg/24 h (normal: 10–50mg/1.8 m^2 surface area), but has been as high as 400 mg/24 h (Hockaday *et al.* 1965). With progressive renal-failure, oxalate excretion decreases and may reach normal values.

Oxalate crystals can often be seen in the urine.

Williams and Smith (1968*a*) demonstrated considerably reduced activity of the soluble 2-oxoglutarate:glyoxylate carboligase (OGC), the enzyme which catalyses the conjugation of 2-oxoglutarate and glyoxylate to form 2-hydroxy-3-keto-adipic acid in liver, spleen, and kidney (Fig. 8.7).

The activity of mitochondial OGC was found to be normal in liver (Crawhall and Watts 1962), liver, kidney, and spleen (Koch *et al.* 1967) and skeletal muscle (Bourke *et al.* 1972) from type I patients.

However, Koch *et al.* (1967) found deficiency of the soluble OGC in liver, kidney and spleen in five patients with type I hyperoxaluria, while Bourke *et al.* (1972) found no deficiency in soluble OGC of skeletal muscle in one patient. The last-named authors concluded that these findings suggested heterogeneity among type I primary hyperoxaluria.

Frederick *et al.* (1963) demonstrated a reduced rate of conversion of 1-^{14}C-glyoxylate to respiratory $^{14}CO_2$ in patients with type I hyperoxaluria and in three of their four parents, suggesting a partial defect in heterozygotes. The patients also exhibited an increased conversion of 1-

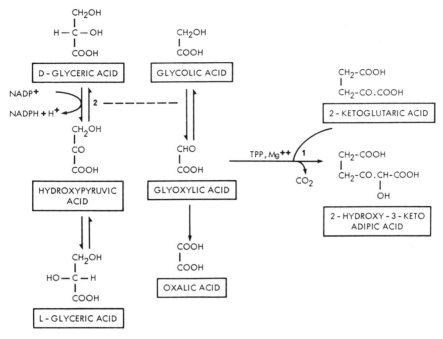

Fig. 8.7. Enzyme steps involved in primary hyperoxaluria: (1) 2-ketoglutarate: glyoxylate carboligase (2) D-glycerate dehydrogenase.

^{14}C-glyoxylate to ^{14}C-glycine and ^{14}C-glycolate. Type I patients also excrete excessive quantities of glycolic and glyoxylic acids.

Primary hyperoxaluria type II; L-glyceric acidura

In this rare form of primary hyperoxaluria there is also excessive excretion of glyceric acid in the urine: 225–638 mg/24h/1.73 m^2 surface area (Gibbs and Watts 1969) (normal urine: undetectable).

Williams and Smith (1968*b*) demonstrated considerably reduced activity of D-glyceric acid dehydrogenase (Fig. 8.7) in leucocytes of three affected siblings. This enzyme catalyses the reversible oxidation of D-glyceric acid to hydroxypyruvic acid, utilising either NADP or NAD as hydrogen acceptor.

The cause of the hyperoxaluria, however, is not clear. A theoretical possibility is that the enzyme deficiency may block the reduction of glyoxylate to glycolate, in addition to the reduction of hydroxypyruvate to D-glycerate. Such blocks would cause accumulation of glyoxylate and consequently encourage oxalate synthesis. Accelerated conversion of ^{14}C-

glyoxylate to ^{14}C-oxalate has in fact been demonstrated in a patient with
D-glyceric dehydrogenase deficiency (Williams and Smith 1968*b*).

The further possibility that accumulated hydroxypyruvate is converted
directly to oxalate, causing hyperoxaluria, seems unlikely in view of the
low excretion of glycolate in these patients.

Pathology of primary hyperoxaluria

The kidneys show characteristic changes of chronic obstruction and
infection in addition to nephrocalcinosis. They are often small and are
gritty because of calcium oxalate deposits. Histologically, crystals are
visible and there is intestinal fibrosis. Crystals are also present in the walls
of arterioles (Scowen *et al*. 1959; Daniels *et al*. 1960).

Oxalosis, or deposit of calcium oxalate, also occurs in bone, cartilage,
bone marrow, testes, myocardium, blood vessels, thyroid, thymus,
parathyroids, pituitary, and adrenals (Burke *et al*. 1955; Neustein *et al*.
1955; Daniels *et al*. 1960) and may occur in the nervous system (Scowen *et
al*. 1959; Moorhead *et al*. 1975).

Calcium oxalate crystals can be identified from their chemical properties
(Johnson 1956; Edwards 1957) or by X-ray diffraction (Dunn 1955;
Godwin *et al*. 1958).

Treatment

Numerous attempts have been made to inhibit oxalate synthesis, but the
only consistently successful method of reducing hyperoxaluria has been
high doses of pyridoxine (Smith and Williams 1967; Gibbs and Watts 1970;
Watts *et al*. 1979). Administration of aldehyde dehydrogenase inhibitors
has been reported to reduce oxalate excretion (Solomons *et al*. 1967*a*,*b*)
but this could not be confirmed by Zarembski *et al*. (1967) or Smith *et al*.
(1969).

Magnesium oxide has reduced oxalate excretion in patients with oxalate
nephrolithiasis (Gershoff and Prien 1967; Dent and Stamp 1970) and in
pyridoxine-deficient rats (Borden and Lyon 1969).

As with other causes of recurrent nephrolithiasis, drinking large
quantities of water is important to increase the urine volume.

Haemodialysis and peritoneal dialysis are not effective in reducing
hyperoxaluria permanently (Zarembski *et al*. 1969; Saxon 1973).

Renal transplanation is of temporary benefit but calcium oxalate crystals
deposit in the donor kidneys and in other tissues (Halverstadt and Wenzl
1974; Morgan *et al*. 1974).

9. Erythrocyte enzyme deficiencies and haemolytic anaemias

Mature red blood cells have lost their DNA and cannot synthesize enzymes or other proteins; they have no mitochondria and therefore cannot generate energy by synthesis of ATP by oxidative phosphorylation. Erythrocytes survive about 120 days, and during this lifespan there is gradual enzyme denaturation. There is evidence that the integrity of erythrocyte structure and function depends on the continued generation of energy by the reactions catalysed by the cells' ageing enzymes (Seaman *et al.* 1980).

The increased permeability to potassium of pyruvate kinase-deficiency erythrocytes (Nathan *et al.* 1965) is in accord with the view that the decreased concentration of ATP cannot maintain the energy-dependent gradient of electrolytes between the erythrocytes and plasma (Bowdler and Prankerd 1964; Mentzer *et al.* 1971).

In some inherited erythrocyte enzyme defects, even these limited systems for ATP generation become attenuated, and the ensuing energy crisis leads to shortening of the erythrocyte lives, premature disintegration and haemolytic anaemia. The known erythrocyte enzyme deficiencies associated with haemolytic anaemias involve the glycolytic pathway, hexose monophosphate (HMP) shunt, and glutathione or nucleotide metabolism.

In some of these disorders the enzyme defect is confined to the erythropoietic tissues. In others, the deficiency is more generalized.

The HMP shunt docs not directly generate energy. It is thought to protect erythrocytes against oxidative degeneration of its haemoglobin by its oxidative functions of reducing a molecule of NADP from the metabolites of every molecule of glucose metabolized by glucose 6-phosphate dehydrogenase and a further molecule metabolized by 6-phosphogluconate dehydrogenase (Fig. 9.1). The reduced coenzymes are re-oxidized during the simultaneous reduction of oxidized glutathione (GSSG) to reduced glutathione (GSH) catalysed by glutathione reductase (Fig. 9.1). Deficiencies of the two enzymes involved in the synthesis of glutathione have also been implicated in haemolytic anaemia. These are γ-glutamylcysteine synthetase which catalyses the synthesis of γ-glutamylcysteine from glutamic acid and cysteine; and glutathione synthetase which catalyses the conjugation of γ-glutamylcysteine with glycine to form glutathione.

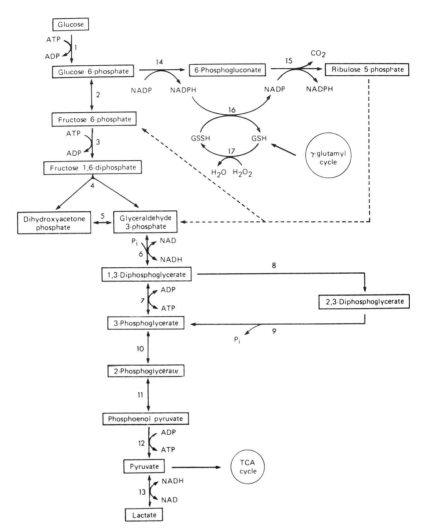

Fig. 9.1. The glycolytic pathway, hexosemonophosphate shunt and glutathione metabolism: (1) Hexokinase (2) Glucose 6-phosphate isomerase (3) Phosphofructokinase (4) Aldolase (5) Triosephosphate isomerase (6) Glyceraldehyde 3-phosphate dehydrogenase (7) Phosphoglycerate kinase (8) Diphosphoglyceromutase (9) Diphosphoglycerate phosphatase (10) Phosphoglyceromutase (11) Phosphopyruvate hydratase (12) Pyruvate kinase (13) Lactate dehydrogenase (14) Glucose 6-phosphate dehydrogenase (15) Phosphogluconate dehydrogenase (16) Glutathione reductase (17) Glutathione peroxidase.

Deficiency of adenylate kinase is also associated with haemolytic anaemia and a decrease of ATP-synthesis since it catalyses the reaction: 2 ADP → ATP + AMP.

By far the commonest enzymopoenic anaemias are associated with deficiency of G6PD and pyruvate kinase.

DEFICIENCIES OF ERYTHROCYTE ENZYMES OF GLYCOLYSIS

Hexokinase deficiency

Hexokinase catalyses the phosphorylation of glucose to glucose 6-phosphate by ATP. Two to four isoenzymes have been reported in red cells (Rogers *et al.* 1975) with different electrophoretic mobilities and substrate specificities. Deficiency of hexokinase has been reported in association with haemolytic anaemia (Valentine *et al.* 1967; Beutler *et al.* 1978; Board *et al.* 1978).

Genetic polymorphism has been described (Necheles *et al.* 1970; Keitt 1969; Paglia *et al.* 1981).

Inheritance was probably autosomal recessive in the patient reported by Valentine *et al.* (1967), but father to son transmission was reported by Necheles *et al.* (1970). Although the father may have been homozygous for autosomal recessive hexokinase mutant genes, the pattern of inheritance in another family was also consistent with dominant inheritance (Siimes *et al.* 1979).

Pyruvate kinase deficiency

There is a wide range in severity of the clinical phenotype. In most cases the anaemia and jaundice are first noticed in the first few years (Tanaka and Paglia 1971; Van Eys and Garms 1971) but the spectrum can range from kernicterus and death in the newborn (Bowman *et al.* 1965) to a symptom-free course until adult life (Nixon and Buchanan 1967; Van Eys and Garms 1971). Early symptomatic onset tends to be associated with greater clinical severity. Thus, anaemia in early childhood often requires regular blood transfusions but usually becomes less severe later, so that transfusions are needed less frequently or not at all. As in most chronic haemolytic anaemias there is an increased tendency to gallstone formation (Keitt 1966), splenomegaly (Bowman and Procopio 1963), haemochromatosis (Reeves *et al.* 1963) and rarely, leg ulcers (Vives-Corrons *et al.* 1980).

Haematology

Moderate or severe anaemia and reticulocytosis and macrocytosis are usually present, being least marked after splenectomy. Macrocytosis can be explained by the greater mean cell volume of the circulating reticulocytes (Oski and Diamond 1963; Grimes *et al.* 1964). Slight anisocytosis and poikolocytosis are common and acanthosis can occur (Oski *et al.* 1964; Baughan *et al.* 1968*a*).

Following splenectomy, haemoglobin concentration and reticulocyte concentrations rise (Grimes *et al.* 1964; Keitt *et al.* 1966). The indirect serum bilirubin concentration is usually elevated (Bowman and Procopio 1963).

Erythrocytes from most, but not all, affected individuals (Grimes *et al.* 1964) exhibit increased haemolysis (autohaemolysis) when incubated *in vitro*. This phenomenon is corrected by the addition of ATP but not of glucose.

Molecular defects and diagnosis

The diagnosis is confirmed by demonstrating a moderate to severe reduction in the activity of red cell pyruvate kinase, which catalyses the conversion of phosphoenol pyruvate to pyruvate and the regeneration of ATP from ADP. A spectrophotometric assay is described by Tanaka (1969), using NADH oxidation to measure conversion of phosphoenol-pyruvate to pyruvate and then to lactate in the presence of excess linking lactic dehydrogenase, which utilizes NADH as hydrogen donor. In view of the shortened erythrocyte lifespan, there is a relatively young mean cell age and therefore many red cell enzymes are less degraded and more active than normal (Loder and de Gruchy 1965). As expected in erythrocytes there is usually a deficiency of ATP, one of the pyruvate kinase reaction products. However, reticulocytes have a high ATP generation rate from mitochondrial oxidative phosphorylation, so that whole haemolysate ATP concentration is variable, and may be low, normal or high (Oski and Diamond 1963; Grimes *et al.* 1964; Tanaka and Paglia 1971).

The concentration of glycolytic pathway metabolites between glucose 6-phosphate and phosphoenol pyruvate, including 2,3-diphosphoglycerate, are increased (Grimes *et al.* 1964; Oski and Bowman 1969; Van Eys and Garms 1971).

Genetic counselling

The disorder has an autosomal recessive mode of inheritance. Hetero-zygotes have reduced pyruvate kinase activity in red cells (Tanaka 1969) but no increased tendency to haemolysis or anaemia. However, they may have moderately low erythrocyte ATP and high 2,3-diphosphoglycerate and phosphoenol pyruvate concentrations (Busch *et al.* 1966; Keitt 1966; Miwa and Nishima 1974).

Pyruvate kinase deficiency has now been detected in most races, and is particularly common in central and northern Europe. In Japan it occurs more frequently than G6PD deficiency (Miwa and Nishina 1974). A particularly severe form of the disorder, which can cause kernicterus and neonatal death, occurs in the Amish community of Mifflin County, Pennsylvania (Bowman *et al.* 1965).

Three isoenzymes of pyruvate kinase (PK) have been characterized in humans (Bigley *et al.* 1968). PK I occurs in erythrocytes and liver and is the form which is deficient in pyruvate kinase deficiency haemolytic anaemia. Patients also have a partial deficiency of PK I in liver (Bigley and Koler 1968). PK II is found in kidney, and PK III in liver, leucocytes, platelets, kidney, muscle, and brain. Residual hepatic activity in liver and normal activity elsewhere is due to normal activity of PK II and PK III.

Numerous allelic variants of erythrocyte pyruvate kinase have been identified. However, data from different laboratories are sometimes difficult to compare in view of different electrophoretic conditions (Bigley *et al.* 1968; Paglia and Valentine 1970; Staal *et al.* 1971), especially since electrophoretic mobility alters during erythrocyte ageing (Paglia and Valentine 1970). However, unequivocal electrophoretic variants have been reported (Löhr *et al.* 1968; Nakashima 1973). Other properties exploited for the identification of allelic variants include enzyme activation by the allosteric effector fructose 1,6-diphosphate (Munro and Miller 1970; Paglia and Valentine 1971) and the affinity of the enzyme for the substrate phosphoenol pyruvate. These variants can have K_m values which are high (Löhr *et al.* 1968; Paglia *et al.* 1968; Paglia and Valentine 1971; Munro and Miller 1970), low (Brandt and Hanel 1971; Staal *et al.* 1972), or normal (Staal *et al.* 1972).

Double heterozygotes for the mutant genes have been reported (Paglia *et al.* 1968, 1972; Yamada *et al.* 1974).

Glucose phosphate isomerase deficiency

Glucose phosphate isomerase (GPI) catalyses the interconversion of glucose 6-phosphate and fructose 6-phosphate (Fig. 9.1). Deficiency of GPI causes haemolytic anaemia (Baughan *et al.* 1968*b*; Kahn *et al.* 1978; Zanella *et al.* 1980). Splenectomy results in some improvement (Matsumoto *et al.* 1973).

The disorder has an autosomal recessive mode of inheritance. Haemolytic anaemia has been reported in GPI deficiency associated with numerous allelic variants (Blume *et al.* 1972; Schröter *et al.* 1974; Paglia *et al.* 1975; Detter *et al.* 1968).

Red cell phosphofructokinase deficiency

Phosphofructokinase in its smallest active form is a tetramer. In muscle each subunit is of the M variety. Erythrocytes contain a mixture of the five possible tetramers consisting of the muscle (M) and liver (L) subunits as follows: L_4, L_3M, L_2M_2, LM_3 and M_4.

In glycogen storage disease type VII there is deficiency of the M subunit (Meienhofer *et al.* 1979; Vora *et al.* 1980*a*) so that erythrocytes contain only the L_4 tetramer (Vora *et al.* 1980*b*). Clinically there is a severe myopathy and mild haemolysis. In patients with haemolysis, but without myopathy (Miwa *et al.* 1972; Waterbury and Frankel 1972) one may speculate that there is deficiency of the L subunit, though this remains to be established. A mild haemolytic anaemia has been reported in muscle phosphofructokinase deficiency (Tarui *et al.* 1969).

Aldolase deficiency

Hereditary intolerance to fructose due to deficiency of fructaldolase B in liver, kidneys, and intestine is described under disorders of fructose metabolism (p. 369).

Red cell aldolase deficiency has been reported as a further cause of haemolytic anaemia (Chapman 1969; Beutler *et al.* 1973*b*). The mode of inheritance is uncertain.

Triose phosphate isomerase deficiency

Triose phosphate isomerase (TPI) catalyses the interconversion of dihydroxyacetone phosphate with glyceraldehyde 3-phosphate. In addition to haemolytic anaemia, deficiency of TPI causes severe progressive neurological deterioration starting shortly after birth, and leading to death usually in the first or second decade. Sudden death due to cardiac involvement has been reported (Schneider *et al.* 1965; Valentine *et al.* 1966; Schneider *et al.* 1968*c*). When red cells are incubated with glucose there is accumulation of dihydroxyacetone phosphate. TPI deficiency has been demonstrated in erythrocytes, leucocytes, cultured fibroblasts, muscle, plasma, and CSF (Schneider *et al.* 1968*c*). The structural gene for one subunit of the enzyme is located on chromosome 12 (Jongsma *et al.* 1974).

Phosphoglycerate kinase deficiency

Phosphoglycerate kinase (PGK) catalyses the energy-generating reaction: 1,3-diphosphoglycerate + ADP → 3-phosphoglycerate + ATP. PGK deficiency is a sex-linked condition. The structural gene is situated on the long arm of the X chromosome (McKusick 1980). Hemizygote males usually have a severe neurological disorder and haemolytic anaemia, and female heterozygotes a variable but often milder haemolytic anaemia. In males, there is usually mental handicap and sometimes death in childhood (Valentine *et al.* 1969; Chen *et al.* 1971; Konrad *et al.* 1973).

Deficiency of PGK has been demonstrated in erythrocytes and leucocytes (Valentine *et al.* 1968).

Splenectomy may improve the haemolytic anaemia (Valentine *et al.* 1969).

The amino acid sequence of phosphoglycerate kinase has now been determined. Three variants have been shown to be due to single amino acid substitutions as follows: PGK-II, asparagine replaces threonine at position 352; PGK-München, asparagine replaces aspartic acid at position 268; PGK-Uppsala, proline replaces arginine at position 206 (Huang *et al.* 1980).

Other disorders of glycolysis

Haemolytic anaemia has been associated with deficiency of 2,3-diphosphoglycerate mutase (Schröter 1970). No haemolytic anaemia occurred in a Japanese with diabetes in his sixth decade with deficiency of erythrocyte lactic dehydrogenase due to inability to synthesize the H subunit (Miwa *et al.* 1971; Kitamura *et al.* 1971). However, there was accumulation of triose phosphates in the red cells.

DEFICIENCIES OF ERYTHROCYTE ENZYMES OF THE HEXOSE MONOPHOSPHATE SHUNT AND GLUTATHIONE METABOLISM

The relation between oxidation–reduction reactions of NADP/NADPH in the HMP shunt and of GSSG/GSH in glutathione metabolism was discussed at the beginning of this chapter.

Glucose 6-phosphate dehydrogenase deficiency

The clinical effects of glucose 6-phosphate dehydrogenase (G6PD) (Fig. 9.1) deficiency are almost completely linked to abnormalities of the red cells and complications of haemolysis. Characteristically, haemolytic crises are induced by drugs, broad beans, infections, birth, or other factors. Numerous genetic variants of G6PD are known, some associated with severe and others with mild or no clinical features. These variants show marked ethnic distributions. The World Health Organization Scientific Group published recommendations for the standardization of procedures for characterization of variants (Betke *et al.* 1967). Adoption of these techniques has allowed numerous laboratories around the world to identify G6PD variants.

Clinical features

Clinical features fall into three groups. Firstly, there are those without any symptoms or with very mild symptoms; secondly, those with haemolytic crises when exposed to drugs or other exogenous factors; and thirdly, those who have congenital non-spherocytic haemolytic anaemia (CNSHA). Females of this X-linked trait may be affected severely or partially (Gross *et al.* 1958). Parameters required to identify the variants include the activity and electrophoretic mobility of erythrocyte G6PD; the Michaelis constant for glucose 6-phosphate and NADP; the activity of the enzyme towards other substrates such as 2-deoxyglucose 6-phosphate, galactose 6-phosphate, and NAD; the thermal stability, pH optimum, and susceptibility to sulphydryl reagents; and the rate of elution on ion-exchange columns.

Using these criteria, Beutler and Yoshida (1973) and Yoshida and Beutler (1978) tabulated the properties of 142 G6PD variants which were fairly completely characterized according to the techniques standardized by the WHO Scientific Group (Betke *et al.* 1967), and a further 43 variants for which insufficient information was available, or their uniqueness was questionable.

G6PD variants associated with mild or no symptoms. This category includes G6PD Baltimore–Austin and Ibadan–Austin (Long *et al.* 1965) which have normal activity but slow electrophoretic mobilities, and G6PD Seattle which has moderate enzyme deficiency and slow electrophoretic mobility (Kirkman *et al.* 1965).

G6PD variants causing haemolytic crises when exposed to some exogenous factors. This category includes the common mild deficiency found in individuals of African ancestry including American negroes (G6PDA⁻) (Yoshida *et al.* 1967), the severe deficiency of the Mediterranean type (Kattamis *et al.* 1969) and the Chinese moderate to severe deficiency, or Canton type (McCurdy *et al.* 1966).

The haemolytic effect of primaquine, 30 mg daily, has been studied under controlled conditions in subjects with the A⁻ type (Dern *et al.* 1954). After two or three days, the urine darkened and in mild cases this was the only abnormality observed. In more severely affected cases, jaundice and pains in the back and abdomen occurred. Anaemia and reticulocytosis developed and Heinz bodies appeared. Heinz bodies are haemoglobin denaturation products attached to the red cell membranes (Fertman and Fertman 1955). After about one week, even when primaquine administration was continued, the symptoms and haematological abnormalities remitted.

The improvement that occurs even when exposure to primaquine is continued has been attributed to the development of a younger generation of erythrocytes with higher G6PD activities as the older cells are destroyed and erythropoiesis is accelerated (Beutler *et al.* 1954).

Some compounds which have caused haemolysis are listed in Table 9.1.

Besides these compounds, infections (Tugwell 1973), and diabetic coma (Gellady and Greenwood 1972) can provoke haemolytic crises in subjects with G6PDA⁻. Kernicterus has been reported in G6PDA⁻ infants (Eshaghpour *et al.* 1967), but this is much more common in the severe Mediterranean form (Weatherall 1960) and in Chinese (Jim and Chu

Table 9.1 *Some compounds which have caused haemolysis in G6PD-deficient subjects (Beutler, 1978)*

Antimalarials	Primaquine
	Pamaquine
	Pentaquine
	Quinocide
	Quinacrine (Atabrine)
Analgesics	Acetanilid
	Acetylsalicylic acid (4–12 g)
	Acetophenetidin (Phenacetin)
Sulphonamides and Sulphones	Sulphanilamide
	Sulphapyridine
	Diphenylsulphone
	N-acetylsulphonilamide
	Sulphacetamide
	Thiazolesulphone
	Salicylazosulphapyridine
	Sulphamethopyridazine (Kynex)
Other Antibacterials	Chloramphenicol (in G6PD Mediterranean; not in other variants)
	Furazolidone
	Furmenthol
	Nitrofurantoin
	Nitrofurazone
Others	Fava Beans (in G6PD Mediterranean)
	Naphthalene
	Methylene blue
	Naldixic acid
	Phenylhydrazine
	Ascorbic acid (very large doses)
	Quinine (in G6PD Mediterranean; not in G6PDA⁻)
	Quinidine (in G6PD Mediterranean; not in G6PDA⁻)
	Niridazole
	Nitrite

1963). Infectious hepatitis can be particularly severe in patients with G6PD deficiency (Choremis *et al.* 1966; Fried *et al.* 1977).

In patients with the Mediterranean form, contact with the fava bean (Kattamis *et al.* 1969) or compounds shown in Table 9.1 usually causes severe jaundice which, unlike that in G6PDA⁻ is not self-limiting. Severe haemolysis can cause renal failure.

G6PD variants causing congenital nonspherocytic haemolytic anaemia. Patients with CNSHA and G6PD deficiency usually have a persistent haemolytic anaemia and mild splenomegaly with reticulocytosis. Exposure to drugs listed in Table 9.1 and infection can aggravate the anaemia. Neonatal jaundice often occurs. Many G6PD variants causing CNSHA have unstable erythrocytes.

Molecular defects and diagnosis

The symptomatic variants are characterized by decreased G6PD activity in mature erythrocytes. Following primaquine administration to an individual with G6PDA⁻, the older red cells with lower than average G6PD activity are haemolysed first (Beutler *et al.* 1954), and the average erythrocyte G6PD activity rises (Herz *et al.* 1970). It is important, therefore, to be aware of the effect of haemolysis when activity is measured.

Assay of erythrocyte G6PD has been standardized by the WHO Scientific Group (Betke *et al.* 1967).

In erythrocytes with complete G6PD deficiency, there is no synthesis of NADPH from glucose metabolism, which causes a deficiency in reduced glutathione (GSH) concentration. Following exposure of subjects with G6PDA⁻ to primaquine, haemolysis is preceeded by a fall in GSH concentration, accountable by a decrease in the aged erythrocyte GSH. When the aged erythrocytes are destroyed, GSH concentration rises to at least normal levels (Kosower 1968).

The mechanism of haemolysis is not understood. *In vitro* primaquine and other drugs have no increased lytic effect (George *et al.* 1966). One may speculate therefore that primaquine-induced lysis depends on some damage to the G6PD-deficient erythrocyte which shortens the life-span *in vivo*, presumably by encouraging destruction of the erythrocytes by the reticulo-endothelial cells.

The mechanisms of haemolysis provoked by infection, diabetic ketosis, and birth are equally obscure.

Amino acid composition and sequencing studies have revealed single amino acid substitutions in G6PD between the normal (B⁺) and the common negro variant (A⁺) where aspartic acid replaces asparagine (Yoshida 1967), and between G6PDB⁺ and G6PD Hektoen, where tyrosine replaces histidine (Yoshida 1970).

Genetic counselling

G6PD deficiency has an X-linked mode of inheritance. Linkage, with a five per cent recombination frequency, has been shown for deuteranopic colour blindness (Porter *et al*. 1962). The structural gene for G6PD has recently been cloned (Persico *et al*. 1981).

According to the Lyon hypothesis, two populations of cells, each with a different G6PD variant, have been demonstrated by cloning fibroblasts from heterozygotes (Davidson *et al*. 1963), and by separating erythrocytes (Beutler and Baluda 1964). Moreover, tumours in heterozygotes arising from single cells, and which can therefore be considered as clones, contain only one of the two G6PD variants (Linder and Gartler 1965; Fialkow 1976). When chronic myeloid leukaemia develops in heterozygotes, both leucocytes and erythrocytes originate from the same clone (Fialkow 1974, 1976).

Detection of heterozygotes by demonstrating Lyonization (by visualization of G6PD activity in single cells) has been useful. Methods employed have included methaemoglobin elution (Gall *et al*. 1965) and tetrazolium-linkage (Fairbanks and Lampe 1968).

Prenatal diagnosis should be possible by demonstration of G6PD deficiency in cultured amniotic cells, since enzyme variants are expressed in cultured fibroblasts (Gartler *et al*. 1962; Chan *et al*. 1965).

G6PD deficiency has been found in nearly all ethnic populations. High frequencies have been reported in Angola (17–27 per cent), Nigeria (10–27 per cent), Tanzania (2–28 per cent), Kurdistan (58 per cent), Caucasus (28 per cent), Saudi Arabia (0–65 per cent), Thailand (7–33 per cent), Greece (1–32 per cent), Italy (0–35 per cent), and Papua New Guinea (0–33 per cent) (Betke *et al*. 1967).

It has been suggested that the high gene frequencies of G6PD deficiency might be maintained as an X-linked polymorphism by a balance of selective forces in a population endemic for falciparum malaria. Resistance to malaria was confirmed by the demonstration that in erythrocytes from heterozygotes for G6PDA$^-$, deficient cells were less parasitized than normal cells (Luzzatto *et al*. 1969). Resistance to falciparum infections appears to be limited to heterozygotes (Bienzle *et al*. 1972).

6-phosphogluconate dehydrogenase (6-PGD) deficiency

6-PGD catalyses the reaction: 6-phosphogluconate + NADP → ribulose 5-phosphate + NADPH + CO_2 (Fig. 9.1). Severe deficiency of erythrocyte 6-PGD has been reported in association with haemolytic anaemia (Idelsen *et al*. 1973). Partial deficiency has been associated either with mild haemolysis (Lausecher *et al*. 1965; Scialom *et al*. 1966) or no haemolysis (Parr and Fitch 1967; Brewer and Dern 1964).

Gluthathione reductase (GR) deficiency

Deficiency of erythrocyte GR has been reported in association with diverse clinical syndromes, including haemolysis and neurological disease (Carson *et al.* 1961; Waller 1968). It is unclear whether these syndromes represent specific genetic entities or not, since Beutler (1969*a,b*) has shown that glutathione reductase deficiency can be due to deficiency of the cofactor flavine adenine dinucleotide, probably resulting from dietary deficiency of riboflavin. The structural gene for glutathione reductase is located on the short arm of chromosome 8 (McKusick 1980).

Glutathione peroxidase deficiency

Erythrocyte glutathione peroxidase catalyses the oxidation of GSH, utilizing peroxides as hydrogen donors. Deficiency of the enzyme has been associated with haemolytic anaemia (Boivin *et al.* 1969; Hopkins and Tudhope 1973).

Deficiency of glutathione synthase is discussed under disorders of the γ-glutamyl cycle.

DEFICIENCIES OF ERYTHROCYTE ENZYMES OF PYRIMIDINE METABOLISM

Generalized deficiency of enzymes of purine and pyrimidine metabolism are discussed in chapter 7. Only deficiency of erythrocyte pyrimidine 5′-nucleotidase is considered in this section since it is associated with haemolysis.

Hereditary erythrocyte pyrimidine 5′-nucleotidase deficiency

Deficiency of erythrocyte pyrimidine 5′-nucleotidase activity (Valentine *et al.* 1976*a*) is associated with a hereditary nonspherocytic haemolytic anaemia, basophilic stippling in the erythrocytes (Valentine *et al.* 1973; Ramot 1976), and increased concentration of pyrimidine nucleotides. It is interesting that in haemolytic anaemia due to lead poisoning there is also deficiency of erythrocyte pyrimidine 5′-nucleotidase (Valentine *et al.* 1976*b*).

It has been suggested that lack of dephosphorylation of pyrimidine nucleotides derived from degradation of RNA causes them to diffuse slowly and therefore accumulate in red cells (Valentine *et al.* 1976*a*). Harley (University of Cape Town, personal communication 1978), however, found that erythrocytes from patients (and controls) could synthesize uridine nucleotides from uridine or orotic acid, and argues that in affected patients accumulated pyrimidine nucleotides could, to some extent, be synthesized *de novo*. He found that administration of allopurinol produced an increase, rather than a decrease in erythrocyte pyrimidine nucleotides. Parents have partial enzyme deficiency (Valentine *et al.* 1976*a*; Ramot 1976), and the disorder has an autosomal mode of inheritance.

10. Disorders of porphyrin and haem

HEREDITARY METHAEMOGLOBINAEMIA

NADH dehydrogenase (diaphorase) deficiency; NADH cytochrome b_5 reductase deficiency

Normally about 0.4 per cent of the total haemoglobin is present in the methaemoglobin form (Van Slyke *et al*. 1946), suggesting that the capacity of the cell to convert methaemoglobin (MetHb) to haemoglobin is about 250 times greater than the reverse capacity (Scott 1968). MetHb is an oxidation product of haemoglobin in which the sixth coordination position of the ferric haem is bound to a water molecule in the acid form which predominates physiologically (absorption peaks 631 nm and 500 nm) or to a hydroxyl ion in the alkaline form (absorption peaks 575 nm and 540 nm).

In vitro, however, in the presence of oxygen, haemoglobin is gradually converted to methaemoglobin since the standard oxidation–reduction potential of Hb_2/MetHb (25°C, pH 7) is 0.165 V, or strongly in favour of MetHb production.

In vivo, MetHb reduction normally prevents accumulation of MetHb. The most important pathway for MetHb reduction is from the hydrogen donor NADH catalysed by NADH cytochrome b_5 reductase (NADH dehydrogenase). However, *in vitro* the reduction of MetHb in the presence of NADH and NADH dehydrogenase occurs very slowly, but the reaction is accelerated more than 70-fold in the presence of cytochrome b_5 (0.8 μM) (Hultquist and Passon 1971). The reduction of cytochrome b_5 by NADH is catalysed by cytochrome b_5 reductase. However, the coelution of the latter enzyme with NADH-reductase from Sephadex gel columns suggests that a single protein is responsible for both catalytic properties (Passon and Hultquist 1972; Schwartz and Reiss 1974). This is supported by the observation that both activities are considerably reduced in red cells from patients with NADH dehydrogenase deficiency (Kitao *et al*. 1974).

The reaction may be written: NADH $\xrightarrow{e^-}$ cytochrome b_5 reductase $\xrightarrow{e^-}$ cytochrome b_5 $\xrightarrow{e^-}$ methaemoglobin, where e^- is an electron. Substances other than methaemoglobin can act as terminal electron acceptors. Thus the dye dichlorophenolindophenol is reduced by NADH 9000 times faster than by MetHb and is used to measure erythrocyte NADH reductase activity (Scott 1960).

In vitro potassium ferrocyanide greatly accelerates MetHb reduction by

NADH. Hegesh and Avron (1967) found that maximal stimulation occurred when the ferrocyanide/MetHb ratio was 4/1 and suggested that this indicated that ferrocyanide interacted with MetHb in some way which facilitated MetHb reduction, rather than acting as an electron carrier from NADH to MetHb.

Two forms of NADH reductases I and II were isolated by Scott *et al.* (1965), who calculated that NADH reductase I was responsible for 95 per cent of erythrocyte methaemoglobin reduction *in vivo*.

Other mechanisms which can reduce MetHb probably have very small physiological significance. Thus *in vitro* ascorbic acid and reduced glutathione reduce MetHb non-enzymically, but patients with scurvy, or with glutathione deficiency due to glutathione synthetase deficiency do not have methaemoglobinuria.

The ability of the various mechanisms for reduction of MetHb can be studied in preparations of erythrocytes in which all haemoglobin has been converted to MetHb by incubation with nitrite. Binding of glutathione by *N*-ethylmaleimide does not impair the rate of MetHb reduction, thus providing further evidence that glutathione is not essential for MetHb reduction. The availability of NAD is clearly essential for the function of NADH production from NAD dehydrogenase reduction of MetHb. The major source of NADH production from NAD is the glyceraldehyde 3-phosphate dehydrogenase (GAPD) reaction of the glycolytic pathway. In conditions when pyruvate is decarboxylated there is a stoichiometric production of two molecules of NADH from one molecule of glucose. In conditions favouring conversion of pyruvate to lactate, catalysed by lactic dehydrogenase utilizing NADH as hydrogen donor, however, glycolysis produces no net change in the NADH/NAD ratio.

Evidence for these processes may be demonstrated *in vitro* in nitrite-treated erythrocytes. Thus, addition of iodoacetate, which inhibits GAPD activity and therefore generation of NADH, blocks reduction of MetHb (Gibson 1948). In contrast, addition of fluoride, which inhibits enolase activity and therefore pyruvate synthesis, stimulates reduction of MetHb.

Erythrocytes also contain NADPH-dehydrogenase. However, since NADPH does not reduce MetHb and since there appears to be no physiological intermediary enzyme acceptor to transfer electrons from NADPH to MetHb, it is unlikely that NADPH acts in reducing MetHb under physiological conditions. This is supported by the observation that patients with G6PD deficiency and therefore NADPH deficiency do not develop methaemoglobinaemia. Moreover, the single patient reported with NADPH dehydrogenase deficiency did not have methaemoglobin-aemia (Sass *et al.* 1967).

The situation is changed dramatically in the presence of methylene blue, which acts as an artificial electron transferring agent thus:

$$\text{NADPH} \xrightarrow{c} \text{methylene blue} \xrightarrow{c} \text{MetHb}$$

The kinetics of MetHb reduction have been studied by Sannes and Hultquist (1978) and Abe and Sugita (1979). Intravenous injection of methylene blue activates the NADPH-dependent MetHb-reducing pathway and causes MetHb reduction (Gibson 1948; Keitt *et al.* 1966). This reaction is limited in patients with NADPH deficiency due to G6PD deficiency (Jaffé 1966) or NADPH reductase deficiency (Sass *et al.* 1967; Sass 1968).

Deficiency of both soluble and microsomal cytochrome b_5 reductase in patients with methaemoglobinaemia and mental handicap (Leroux *et al.* 1975; Kaplan *et al.* 1979) suggests that both these enzymes are coded by the same structural gene. This is supported by the immunological identity of the two enzymes (Goto-Tomura *et al.* 1976).

Clinical features

The most characteristic symptom of NADH dehydrogenase deficiency is congenital grey-coloured cyanosis. There are usually no other symptoms until methaemoglobin is in the region of 30–40 per cent, when the patients may experience tiredness or exertional dyspnoea.

The association of methaemoglobinaemia with mental handicap and neurological deficit has been documented, but in view of the fact that the large majority of individuals with methaemoglobinaemia are asymptomatic this may be fortuitous (Fialkow *et al.* 1965; Jaffé 1966), but a causal relation is supported by the observation that in a high proportion of subjects with mental handicap NADH dehydrogenase activity is undetectable rather than reduced (see below).

Molecular defects and diagnosis

Blood NADH is usually determined by the MetHb ferrocyanide reductase method of Hegesh *et al.* (1968). Activity has been found to be absent in three-quarters of patients with methaemoglobinaemia and mental handicap (Fialkow *et al.* 1965; Beauvais and Kaplan 1978; Kaplan *et al.* 1979). Severe deficiency of microsomal cytochrome b_5 reductase was demonstrated in fibroblasts, leucocytes, muscle, brain, and liver in a patient (Leroux *et al.* 1975) and in leucocytes or fibroblasts in others (Beauvais and Kaplan 1978; Kaplan *et al.* 1979). Reduced NADH dehydrogenase activity and methaemoglobinaemia have been reported where the anodal electrophoretic mobility of the residual enzyme was normal (Kaplan and Beutler 1967; Bloom and Zarkowsky 1969), fast (Schwartz *et al.* 1972), or slow (Hsieh and Jaffé 1971).

Several mutant types of NADH-reductase are abnormally labile. These include the La Tronche, Boston slow, and Puerto Rico variants (Schwartz *et al.* 1972; Hsieh and Jaffé 1971).

Genetic variants of NADH dehydrogenase with abnormal electrophoretic mobilities but normal activities were reported by Hopkinson *et al.* (1970) in a study of 2783 healthy individuals and were designated Dia 2, Dia 3, Dia 4, and Dia 5.

Young infants have NADH-dehydrogenase activities considerably lower than adults.

Methaemoglobinaemia can be diagnosed by demonstrating the characteristic absorbance peaks (pH 6.5) at 500 and 631 nm. The decrease in optical density after addition of potassium cyanide can be used for determination (Evelyn and Malloy 1938).

Genetic counselling

The disorder has an autosomal recessive mode of inheritance. Heterozygotes have a level of NADH dehydrogenase activity intermediate between homozygotes and normal individuals (Jaffé 1966).

Erythrocytes from heterozygotes show an intermediate rate of reduction of nitrite-induced methaemoglobin *in vitro* (Jaffé 1966).

Prenatal diagnosis of two foetuses at risk for NADH-cytochrome b_5 reductase deficiency was undertaken by Junien *et al.* (1981). In the first case enzyme activity of cultured amniotic cells was in the heterozygous to normal range, and an unaffected foetus was accurately predicted. In the second case cultured amniotic cells had undetectable enzyme activity. The pregnancy was terminated and the diagnosis of NADH-cytochrome b_5 reductase deficiency was confirmed in several tissues and cultured fibroblasts from the abortus.

The structural gene for soluble NADH cytochrome b_5 reductase has been assigned to chromosome 22 (Junien *et al.* 1978).

Double heterozygotes with methaemogobinaemia have been reported (Bloom and Zarkowsky 1969; Hsieh and Jaffé 1971).

Treatment

The majority of individuals with hereditary methaemoglobinaemia do not require treatment. If the patients request reduction of the blood MetHb concentration for cosmetic reasons, this can usually be achieved by administration of oral ascorbic acid (500 mg daily) or i.v. methylene blue (1 mg/kg/2 weeks).

Acute toxic methaemoglobinaemia (methaemoglobinuria)

Normal infants, heterozygotes, and especially homozygotes for NADH-dehydrogenase deficiency, who all have lower than normal adult activities of the enzyme, are particularly liable to acute toxic methaemoglobinaemia when exposed to drugs or chemicals which stimulate formation of MetHb. Artificial milks made up with water rich in nitrate, which is converted to nitrite by intestinal bacteria, have caused epidemics of methaemoglobinuria in infants, and an appreciable mortality. Nappies washed in aniline-containing disinfectants or labelled with ink have also caused infantile toxic methaemoglobinuria. Malarial chemophrophylaxis in Vietnam also produced methaemoglobinaemia in heterozygotes for NADH dehydrogenase deficiency (Cohen *et al.* 1968).

In view of the possible fatal outcome of acute toxic methaemoglobinaemia, treatment should be commenced urgently. Intravenous injection of methylene blue (see above) causes a rapid reduction in blood MetHb concentration, provided NADPH production is not impaired by G6PD deficiency. In patients with G6PD deficiency, exchange blood transfusion may be life-saving, by removing MetHb and the chemical or drug which induced the crisis.

M haemoglobin defects

Methaemoglobinaemia can also result from defects in the globin component of haemoglobin which renders it resistant to reduction. The structure of several of these defective haemoglobins has been determined (Gerald and Efron 1961; Hayashi *et al.* 1968; Udem *et al.* 1970). All M haemoglobins have amino acid substitutions affecting the haem–globin bond, so that there is a change in the amino acid involved, resulting in new, but stable, bonds producing structures which inhibit haem iron reduction. In the nomenclature of Lehman and Huntsman (1972), helical regions are designated A–H. The number following indicates the position of the amino acid residue in the helical region, while the number in brackets represents the position of the amino acid in the α - or ß-chain. In four M haemoglobins there is replacement by tyrosine of the following amino acids; the proximal histidine of HbM_{Iwate} F8 (87) (α-chain); or of $HbM_{Hyde\ Park}$ F8 (92) (ß-chain); or of the distal histidine of HbM_{Boston} E7 (58) (α-chain); or of $HbM_{Saskatoon}$ E7 (63) (ß-chain). In each case the phenolic groups of tyrosine form ferric complexes with the haem iron. In a fifth mutant $HbM_{Milwaukee}$ E11 (67), the valine in the ß-chain is replaced by glutamic acid, which is bonded to the ferric iron through its additional carboxylic group (Hayashi *et al.* 1968; Udem *et al.* 1970).

DEFECTS OF CATALASE AND PEROXIDASE

Acatalasaemia

The majority of individuals with moderate or severe hereditary reduction of catalase activity in erythrocytes and other cells and tissue are asymptomatic. Gangrenous ulceration around the tooth sockets, combined with tooth loss, occurs in the minority, with considerable variation in severity.

Catalase contains four haematin groups, one for each of four subunits. It has two main catalytic actions, both involving the splitting of hydrogen peroxide (H_2O_2) to water. In the catalase reaction, a second molecule of hydrogen peroxide is the hydrogen donor, the products being water and molecular oxygen, thus:

$$H_2O_2 + H_2O_2 = 2H_2O + O_2$$

Catalase, however, can also act as a peroxidase when other molecules are hydrogen donors, thus:

$$\text{Hydrogen donor} + H_2O_2 = \text{oxidized donor} + 2H_2O$$

Hydrogen donors include formate, ethanol, and methanol. The reaction with formate is:

$$HCOOH + H_2O_2 = CO_2 + 2H_2O$$

Erythrocyte catalase is one of the enzymes that prevent 'oxidative stress' and the oxidation of haemoglobin to methaemoglobin by peroxide. Peroxide is generated from a number of reactions, including the superoxide dismutase-catalysed conversion of free oxygen radicals to peroxide and oxygen, thus:

$$2O\cdot_2^- + 2H^+ \rightarrow H_2O_2 + O_2 \text{ (McCord and Fridovich, 1969).}$$

Thus damage from oxygen radicals is normally prevented first by the superoxide dismutase reaction and second by the catalase reaction, in competition with the glutathione peroxidase reaction, which plays a major role. It is probable that under physiological conditions, splitting of peroxide is catalysed mainly by glutathione peroxidase and that catalase activity is important only when peroxide formation is enhanced. Peroxide formation is also catalysed by uricase, D-amino acid oxidase and glycolate oxidase. Ionizing radiation and some drugs are other sources of peroxide formation.

Clinical features

Takahara (1968) wrote that after excising a foul-smelling gangrenous area extending from the molars of the right upper jaw into the maxillary sinus and the nasal cavity from an 11-year-old girl in 1947, he poured hydrogen peroxide on the wound for cleansing. To his great surprise, '. . . the blood coming in contact with the hydrogen peroxide immediately turned a brownish-black colour and the usual bubbles did not appear. I thought that by mistake the nurse might have handed me a bottle of silver nitrate . . .'. The explanation turned out to be that her blood lacked catalase and therefore could not release oxygen bubbles from the peroxide which converted haemoglobin to methaemoglobin. The patient was the second of seven siblings. Four of them had mouth gangrene and lacked blood catalase activity.

Oral gangrene characteristic of the disease, or Takahara's disease, starts as small painful ulcers around the neck of a tooth or in the tonsillar lacunae during the first 10 years, even in infancy. The progressive nature of the ulceration varies considerably in severity. In severe cases, gangrene and inflammation spread into the maxilla and cause loss of teeth. The lesions are thought to spread because bacteria in the crevices around the teeth or in the tonsillar lacunae produce hydrogen peroxide which oxidizes haemoglobin to methaemoglobin, leading to ulceration locally.

The disorder responds well to locally applied antibiotics but healing is accompanied by marked recession of the gums (Takahara 1968).

Initially, the disorder was found only in individuals from Japan but subsequently patients have been reported from Switzerland (Aebi *et al.* 1968), and elsewhere.

Molecular defects and diagnosis

Acatalasaemia is heterogenous and has an autosomal recessive mode of inheritance. In the Japanese form, only about half the homozygotes have oral gangrene. Catalase activity in the blood of homozygotes is 0.1–0.2 per cent of normal (Takahara *et al.* 1960; Takahara 1968) and in the majority of families heterozygotes have approximately 40–60 per cent activity. In some Japanese families, however, heterozygotes have had values over-lapping with those of normal homozygote controls (Hamilton and Neel 1963).

In Switzerland, homozygotes are asymptomatic and do not have oral gangrene (Takahara's disease). The deficiency is present in some red cells, but not in others producing red cell mosaicism. In this form of the disorder, catalase is unstable. Activity in homozygotes is 50–70 per cent of normal (Aebi *et al.* 1968).

In a healthy family of mixed Scandinavian and British extraction, catalase activity and stability were normal, but heterozygotes had a catalase with increased electrophoretic mobility (Baur 1963).

Initially, Takahara named the disorder acatalasaemia, but subsequently it was demonstrated that deficiency in the activity of catalase was also expressed in cultured fibroblasts (Krooth *et al.* 1962), liver and spleen (Takahara 1968) and the designation 'acatalasia' has been suggested.

However, residual liver catalase activity of nearly 40 per cent was reported by Takahara *et al.* (1960).

The activity of catalase from homozygotes is higher in cultured fibroblasts than in red cells. Sadamoto (1966) and Krooth (1967) found activities of about 2–4 per cent of normal in the Japanese type, and about 15 per cent of normal in the Swiss type. Acatalasaemic red cells contain an immunologically reacting catalase-like protein (Aebi *et al.* 1964; Shibata *et al.* 1967).

The concentration of methaemoglobin in acatalasaemic blood is normal. As discussed earlier, catalase probably does not exert a protective role in preventing transformation of haemoglobin to methaemoglobin, until the peroxidative activities of glutathione peroxidase and the linking enzymes to the pentose phosphate shunt are exceeded (Aebi *et al.* 1968).

Genetic counselling

The disorder has an autosomal recessive mode of inheritance. The Swiss form is asymptomatic, while only 50 per cent of individuals with the Japanese form develop symptoms.

In the Japanese form, heterozygotes may be detected by blood catalase assay, but in the Swiss type, although mean catalase activity is reduced, there is some overlap between heterozygotes and individuals who are homozygous for the wild gene (see 'Molecular defects', above).

It is probable that the disorder could be diagnosed prenatally by assaying catalase activity in cultured amniotic cells.

Takahara (1967) estimated that the incidence of hypoacatalasaemia among Asians was 0.25 per cent in Japanese and 1.29 per cent in Koreans. The incidence of acatalasaemia was 0.004 per cent in 73 661 individuals screened in Switzerland.

Treatment

Extraction of all teeth and removal of the tonsils, together with application of antibiotics are effective in the treatment of Takahara's disease. Granulating tumours and gangrenous areas should be excised. Healing occurs at a normal rate (Takahara 1968).

Animal models

Five strains of mice with mutant alleles for catalase were identified by screening (Feinstein 1970). One of these appeared to have a mutation which resembled the human form.

Polymorphism of blood and liver catalase in mice has been reported by Heston *et al.* (1965).

Normal ducks have blood catalase activities in the range of that found in Japanese homozygotes for acatalasia, i.e. about 0.2 per cent of human mean normal levels.

Myeloperoxidase deficiency

Neutrophils have several antimicrobial mechanisms (Babior 1978). Those most studied involve the reduction products of oxygen superoxide ($O\cdot_2^-$) (Babior *et al*. 1973) and hydrogen peroxide (H_2O_2) (Root and Metcalf 1977). The antibacterial and antifungal properties of $O\cdot_2^-$ and H_2O_2 are considerably enhanced in the presence of the azurophilic granule enzyme myeloperoxidase (MPO) and halide cofactors (Klebanoff 1968; Rosen and Klebanoff 1979). Patients who have absent or reduced ability to generate $O\cdot_2^-$ have chronic granulomatous disease (Babior 1978) and are susceptible to infections with fungi and catalase-positive bacteria.

Of 15 individuals from 12 families with myeloperoxidase deficiency, only three had an increased susceptibility to infections (Lehrer and Cline 1969; Moosman and Bojanowsky 1975; Cech *et al*. 1979). In a further 26 subjects identified by screening about 60 000 individuals, half had partial and half complete MPO deficiency (Parry *et al*. 1981). Only four had infections and only two of these had major infections, one totally MPO-deficient with candidiasis, the other partially MPO-deficient with bacteraemia.

These authors demonstrated that *in vitro* killing of *Candida albicans* by MPO-deficient cells was markedly deficient, but killing of *Staphylococcus aureus* was only slightly defective.

The disorder has an autosomal recessive mode of inheritance.

INDIRECT HYPERBILIRUBINAEMIA· THE CRIGLER–NAJJAR SYNDROMES

Bilirubin is a waste product derived from haem, the prosthetic group of haemoglobin (Berk *et al*. 1969). The conversion of haem to the bile pigments biliverdin and bilirubin in man occurs in cells of the reticuloendothelial system such as macrophages (Gemsa *et al*. 1973). Bilirubin passes into the plasma, where it is bound to serum albumen in two or three sites; tightly to the primary site with an association constant of $10^8 M^{-1}$, and loosely to the others (Odell 1959; Brodersen 1980). Clinically it has been useful to distinguish the 'direct' and 'indirect' types of serum bilirubin, according to whether the pigment couples with diazotized sulphanilic acid immediately, or after treatment with alcohol, respectively (Van den Bergh and Mueller 1916). Indirect bilirubin is produced from haem by a series of reactions involving fissure of the reduced haem ring in the presence of molecular oxygen to yield biliverdin, followed by the formation of bilirubin by reduction. The latter is indirect bilirubin. Its concentration in the serum is raised in haemolytic anaemia. The formation of direct bilirubin from indirect bilirubin requires its absorption into hepatocytes, followed by its conjugation, that is, esterification of one or two of its carboxyl groups to

glucuronic acid, catalysed by bilirubin:UDP-glucuronate glucuronyl transferase in the smooth endoplasmic reticulum to yield bilirubin mono- and
diglucuronides. Failure in these steps also leads to an elevation of serum
indirect (i.e. unconjugated) bilirubin. The role of bilirubin-binding
proteins within the hepatocyte is not understood. Ligandin, or Y protein,
which binds to bilirubin and several other organic compounds, has been
purified and partially characterized (Fleischner *et al.* 1972; Kamisaka *et al.*
1975) and has been identified as glutathione *S*-transferase B (Habig *et al.*
1974). The next process in the production of biliary bilirubin is the
secretion of direct (i.e. glucuronate-conjugated) bilirubin into the bile.
Defects in the latter process cause regurgitation of bilirubin into the blood,
leading to an elevation of serum direct bilirubin, commonly caused by
hepatitis or extra hepatic biliary duct obstruction.

Gilbert's disease

This disease (Gilbert and Lereboullet, 1901; Gilbert *et al.* 1907), is
characterized by mild chronic fluctuating elevations of unconjugated serum
bilirubin presenting as jaundice in the neonatal period, or more commonly
after puberty (Foulk *et al.* 1959). Symptoms are non-specific and probably
coincidental (Berk *et al.* 1981). The syndrome probably represents a
heterogenous group of disorders resulting from diverse causes including a
defect in pigment clearance (Okolicsanyi *et al.* 1981) and reduction of
hepatic bilirubin:UDP-glucuronyltransferase activity (Arias *et al.* 1969;
Black and Billing 1969).

Administration of phenobarbitone to patients with Gilbert's disease
reduces the serum bilirubin concentration and increases the plasma
clearance of bilirubin without increasing the activity of liver glucuronyltransferase (Black *et al.* 1974). The occurrence of the disorder in more than
one generation is consistent with an autosomal dominant mode of
inheritance (Damashek and Singer 1941; Powell, *et al.* 1967).

Crigler–Najjar syndrome type I

In this severe disorder, jaundice, associated with a raised serum indirect
bilirubin concentration, occurs in the first to third days after birth. Five of
six infants from three intermarried families developed kernicterus and died
in the first 15 months. The sixth survived, with jaundice, but without
neurological damage, until the occurrence of kernicterus, followed by
death during the sixteenth year (Crigler and Najjar 1952).

The majority of other affected patients have also died of kernicterus,
although a few have survived with neurological sequelae (Blaschke *et al.*
1974; Wolkoff *et al.* 1979). There is no anaemia or hepatosplenomegaly.

Pathology

There is yellow staining of the basal ganglia characteristic of kernicterus (Huang *et al.* 1970). The liver is not enlarged. Only minor changes are reported in the Kupffer cells and hepatocytes (Blaschke *et al.* 1974; Wolkoff *et al.* 1979), with prominent smooth endocytoplasmic reticulum (De Brito *et al.* 1966; Minio-Paluello *et al.* 1968).

Molecular defects and diagnosis

The disorder is due to absent activity of hepatic bilirubin:UDP-glucuronyl-transferase (Arias *et al.* 1969; Blaschke *et al.* 1974). Reduced formation of glucuronides from chloral hydrate, trichloroethanol, and salicylates (Childs *et al.* 1959) has been demonstrated.

There is no haemolysis or reticulocytosis. Most liver function tests are normal, but serum indirect bilirubin concentrations are permanently elevated, varying between 15 and 48 mg/dl. Serum direct bilirubin is undetectable.

The faeces have low faecal urobilinogen concentration but are normal in colour.

Genetic counselling

The disorder has an autosomal recessive mode of inheritance. There is a high rate of parental consanguinity (Childs *et al.* 1959; Arias *et al.* 1969). Some parents have reduced ability to glucuronate salicylate (Childs *et al.* 1959) and menthol (Arias *et al.* 1969).

In view of the low activity of hepatic glucuronyltransferase activity in normal human foetuses (Brown and Zuelzer 1958), prenatal diagnosis by foetal liver biopsy may not be possible.

Treatment

Reduction of serum indirect bilirubin has been achieved by exchange transfusion, plasmaphoresis, and peritoneal dialysis (Blaschke *et al.* 1974; Berk *et al.* 1975*a*; Wolkoff *et al.* 1979). The administration of oral cholestyramine to bind bile pigments has not proved successful (Blaschke *et al.* 1974). Phototherapy has also been attempted without persistent benefit (Wolkoff *et al.* 1979).

Crigler–Najjar syndrome type II

In this disorder, first reported by Arias (1962), there is a milder degree of familial jaundice due to raised serum indirect bilirubin than occurs in Crigler–Najjar syndrome type I. Onset of jaundice has been at one to 30 years of age. Serum indirect bilirubin concentrations are usually below 20 mg/dl (Berk *et al.* 1975*b*). Neurological damage due to kernicterus has occurred in a few patients, (Arias 1962; Gollan *et al.* 1975) but not in the great majority of patients (Gollan *et al.* 1975; Gordon *et al.* 1976). In bile, in contrast to Crigler–Najjar syndrome type I, some bilirubin monoglucuronide is present (Gordon *et al.* 1976; Fevery *et al.* 1977). It appears therefore that the primary defect is an inability to catalyse the esterification of a second carboxyl group of bilirubin with glucuronide. Reduced glucuronide formation has been demonstrated *in vivo* in some, but not all patients (Arias *et al.* 1969; Hunter *et al.* 1971). Reduced or absent conjugation has also been demonstrated *in vitro* (Gollan *et al.* 1975).

In contrast to the type I disorder, administration of high doses of phenobarbital (60–180 mg daily) reduces serum indirect bilirubin concentration, possibly by stimulation of hepatic microsomal glucuronyltransferase activity (Yaffe *et al.* 1966; Gollan *et al.* 1975).

In some families both parents and siblings have had elevated serum indirect bilirubin concentrations (Hunter *et al.* 1971; Gollan *et al.* 1975). Inheritance is therefore either autosomal recessive (Hunter *et al.* 1971) or dominant with incomplete penetrance (Arias *et al.* 1969). However, in some families detection of non-icteric carriers has proved unsuccessful by menthol-conjugation tests (Arias *et al.* 1969).

Animal model – the Gunn rat

Livers of the mutant strain of Wistar rats described by Gunn (1938) have considerably reduced ability to form bilirubin glucuronides (Schmid *et al.* 1958). There is a reduced affinity of UDP-glucuronyltransferase for UDP-glucuronic acid causing defective formation of microsomal UDP-glucuronyltransferase.

Affected animals frequently develop neurological (Blanc and Johnson 1959) and pathological (Schutta and Johnson 1969) features of kernicterus. The disorder therefore has several features in common with Crigler–Najjar disease type I.

THE HEREDITARY PORPHYRIAS

The porphyrias are a group of rare or very rare disorders which may be genetic or acquired. They are characterized by an excessive production of porphyrins (or their precursors) implicated in the biosynthesis of haem. They are divided into hepatic and erythropoietic porphyrias, according to whether excessive production of porphyrins is predominantly in liver or the erythropoietic system. Apart from very few disorders, for example, familial hypercholesterolaemia, and possibly Crigler–Najjar syndrome type II (see Chapter 1), they are the only known inborn errors of metabolism with an autosomal dominant mode of inheritance.

Haem synthesis

Haem is the prosthetic group of several proteins involved with molecular oxygen. These are concerned with oxygen transport (e.g. haemoglobin) and electron transfer (e.g. mitochondrial cytochromes). Other haemo-proteins include myoglobin, microsomal cytochromes, catalase, and tryptophan oxygenase.

δ-aminolevulinic acid synthetase

The rate-limiting reaction in haem synthesis is the biosynthesis of δ-aminolevulinic acid (δ-ALA) from succinyl-CoA (derived mainly from the tricarboxylic acid cycle and partly from propionate metabolism) and glycine. α-amino-ß-keto adipic acid is a reaction intermediate which becomes decarboxylated to yield δ-ALA. The reaction is catalysed in the mitochondria by δ-ALA synthetase in the presence of pyridoxal phosphate. The enzyme is subject to end-product inhibition by haem (Kaplan 1971; Whiting and Elliot 1972) and is induced in several of the porphyrias, presumably as a secondary event of decreased haem synthesis. The very short half-life of δ-ALA synthetase, of one or two hours, may be due to degradation by a specific protease (Aoki 1978).

δ-ALA dehydratase (porphobilinogen synthetase)

The next step in the biosynthesis of haem is the condensation of two molecules of δ-ALA to form a monopyrrole, porphobilinogen (PBG). This reaction is catalysed by the cytosolic enzyme δ-ALA dehydratase in the cytoplasm, in the presence of a sulphydryl donor (Tsukamoto *et al.* 1979).

Uroporphyrinogen (UROgen) I synthetase; (PBG deaminase); UROgen III cosynthase

The condensation of four molecules of PBG to the tetrapyrrolic porphyrins to yield UROgen III in the pathway to haem synthesis requires the cytoplasmic enzymes UROgen I synthetase and UROgen III cosynthase. It was previously believed that the next step in the pathway was oxidation of UROgen III to uroporphyrin (URO) III. However, it is now established that the URO III is a by-product which cannot be converted to haem.

The enzyme UROgen I synthetase alone will catalyse the condensation of four pyrrole molecules to UROgen I. However, this pathway does not lead to haem synthesis.

Condensation of the four pyrrole rings starts from ring A (Fig. 10.1) with sequential addition of rings B, C, and D. In the biosynthesis of the type III tetrapyrrole, ring D is reversed and undergoes intramolecular rearrangement. These reactions are catalysed by the combined action of UROgen I synthetase and UROgen III cosynthase (Battersby *et al.* 1979).

Recently, UROgen I synthetase from erythrocytes has been separated into five types, designated A–E, by DEAE cellulose chromatography, polyacrylamide gel electrophoresis, and isoelectric focusing (Andersen and Desnick 1980).

UROgen decarboxylase

The conversion of UROgen III to coproporphyrinogen (COPROgen) III by decarboxylation of all four acetic acid side chains is catalysed by the cytosolic enzyme UROgen decarboxylase. During the stepwise decarboxylation intermediate porphyrins are hepta–, hexa- and pentacarboxylated. It has been shown (Jackson *et al.* 1975*b*) that decarboxylation starts with the acetic acid in the group D ring of UROgen III and is followed by sequential decarboxylation of rings A, B, and C.

COPROgen oxidase

COPROgen III can be converted by the by-product coproporphyrin (COPRO)III or continue on the synthetic pathway to haem inside the mitochondria. The conversion to protoporphyrin (PROTO) IX by oxidative decarboxylation of two propionyl groups of rings A and B to vinyl groups utilising molecular oxygen is catalysed by COPROgen oxidase (Sano and Granick 1961; Yoshinaga and Sano 1980).

PROTOgen oxidase

Although the intramitochondrial oxidation by molecular oxygen of PROTOgen IX to PROTO IX can proceed spontaneously, the reaction is catalysed by PROTOgen oxidase (Poulson 1976).

Ferrochelatase

Ferrochelatase, which is situated in the inner mitochondrial membrane, catalyses the final reaction in haem biosynthesis, namely the conversion of PROTO IX to haem by the incorporation of iron (Bugany *et al.* 1971).

Hepatic porphyrias

The clinical phenotypes of the different forms of hepatic porphyria show many common features. Neurovisceral, neurological, and mental abnormalities are limited to those hepatic porphyrias with over-production of porphyrin precursors δ-aminolevulinic acid (δ-ALA) and porphobilinogen (PBG). On the other hand, cutaneous photosensitivity occurs in porphyrias characterized by over-production of porphyrins.

Three types of hepatic porphyria are universally considered to be genetically determined. These are acute intermittent porphyria (AIP), hereditary coproporphyria (HCP) and variegate porphyria (VP). A fourth

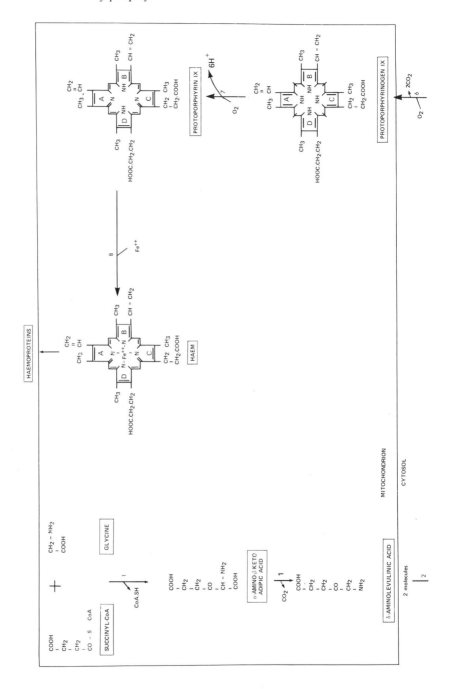

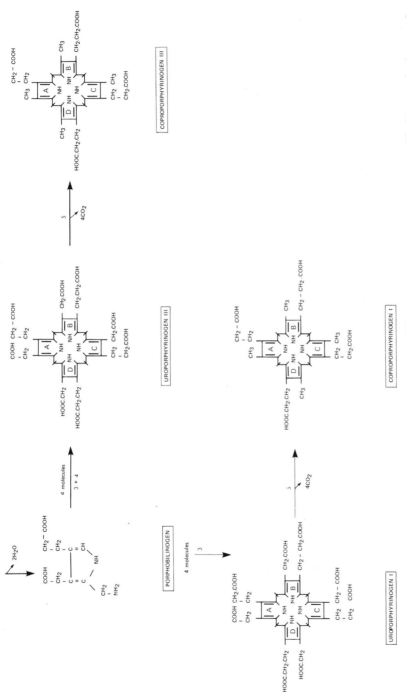

Fig. 10.1 Haem synthesis: (1) δ-aminolevulinic acid synthetase (2) δ-amino-levulinic dehydratase (3) Uroporphyrinogen I synthetase (4) Uroporphyrinogen III cosynthetase (5) Uroporphyrinogen decarboxylase (6) Coproporphyrinogen oxidase (7) Protoporphyrinogen oxidase (8) Ferrochelatase.

type, porphyria cutanea tarda (PCT), has sometimes been referred to as acquired hepatic porphyria owing to lack of convincing evidence that there is a genetic basis. However, the demonstration of reduced uroporphyrin (URO) decarboxylase activity in liver and erythrocytes in patients, and in erythrocytes in relatives, together with reports of familial occurrence, now indicates that PCT is also genetically determined.

Acute intermittent porphyria; Swedish type; uroporphyrinogen I synthetase (PBG deaminase) deficiency

Clinical features

The clinical features of acute intermittent porphyria have been reviewed by Waldenstrom (1957), Mustajoki and Koskelo (1976), and Goldberg (1959).

Episodes of neurovisceral disturbances typically begin in early adult life and are exceptional in childhood. Attacks of abdominal pain, vomiting, and constipation occur in the majority of patients. Abdominal pain is usually colicky and of variable severity and localization. The attacks vary in duration from days to weeks and may be intermittent. Intestinal distension may occur and mimic acute obstruction. The vomiting may lead to malnutrition, dehydration, and renal failure.

The gastrointestinal symptoms may be attributed to disturbances of the autonomic nervous system. Other neurological manifestations include abnormalities of peripheral or cranial nerve function and mental disturbances. The latter occur in about half the patients and manifest as emotional disturbances, confusion, or visual hallucinations (Goldberg and Rimington 1962).

There is a wide spectrum in the severity of clinical manifestations. For example, it is well established that some carriers are without symptoms (Waldenstrom 1957; Meyer *et al.* 1972; Sassa *et al.* 1974*a*); others are well between attacks apart from minor symptoms, often described as 'indigestion'.

A characteristic feature of AIP is sensitivity of patients to certain drugs, steroids, or other agents which can precipitate acute episodes.

Drugs which can provoke acute attacks in patients with acute porphyria (classically phenobarbital) are listed by Wetterberg (1975) and Rifkind (1976). The majority have been found to induce δ-ALA synthetase in chick embryo liver cells either *in vitro* or *in ovo*. Steroids are also well recognized inducers of acute attacks in patients with AIP.

It is relevant, therefore, that several links have been established between steroids and AIP. Direct evidence includes the reports that patients with AIP have increased concentration of several 17-oxosteroids in their blood and urine (Goldberg *et al.* 1969; Paxton *et al.* 1974), and deficient conversion of testosterone to its 5 α-metabolites, presumably due to deficiency of testosterone 5 α-reductase. However, the significance of this information is obscure since patients with male pseudohermaphroditism and 5 α-reductase deficiency (p. 441) do not have porphyria. It has been established, however, that steroid metabolites are potent inducers of δ-ALA synthetase, and can stimulate porphyrin biosynthesis in embryonic avian liver cells (Sassa *et al.* 1979*a*).

Less direct evidence linking steroid metabolism with AIP includes the rarity of symptoms before puberty, the greater frequency of acute attacks in females than males, the association in some females of acute attacks with a phase of the menstrual cycle and with either administration of oestrogens, including oral contraceptives, or by withholding contraceptives. Acute exacerbation may also be precipitated by fasting (Knudson *et al.* 1977). AIP is the only form of porphyria in which cutaneous photosensitivity does not occur.

Molecular defects and diagnosis

There is a deficiency in the activity of UROgen I synthetase in liver (Strand *et al.* 1970), erythrocytes and cultured fibroblasts (Meyer 1973; Strand *et al.* 1972; Sassa *et al.* 1974*a,b*). UROgen I synthetase catalyses the condensation of four molecules of PBG to the cyclic tetrapyrrole uroporphyrinogen III.

Deficiency of UROgen I synthetase causes an accumulation and an increased excretion in the urine of the porphyrin precursors PBG and δ-ALA, which are accentuated by a secondary enhancement in the activity of δ-ALA synthetase. The latter enzyme catalyses the condensation of glycine with succinyl CoA to form δ-ALA utilizing pyridoxal phosphate as cofactor; this is the only rate-limiting reaction in the biosynthetic pathway to haem formation. Hepatic δ-ALA synthetase is sensitive to feedback inhibition by haem (Scholnick *et al.* 1972). Derepression in haem synthesis therefore can explain the increased activity in AIP.

Some individuals with decreased erythrocyte UROgen I synthetase activity do not have clinical features of the disease (Meyer *et al.* 1972).

Characteristically, the urine contains a considerable excess of PBG and δ-ALA which usually rise and fall approximately in parallel, and increase considerably during acute attacks (Bonkowsky *et al.* 1971). Between attacks urine PBG excretion is much lower and may even be normal (Waldenstrom 1957; Meyer *et al.* 1972). Faecal porphyrin excretion may be mildly elevated (Watson 1960).

The relation between the neurological symptoms and deficiency of UROgen I synthetase is not understood.

The diagnosis is best established by assay of erythrocyte UROgen I synthetase activity since asymptomatic carriers may have normal excretion in the urine of PBG and δ-ALA (Meyer *et al.* 1972; Taddeini and Watson 1968).

Symptoms of AIP also occur in patients with tyrosinaemia type I (p. 193) (Gentz *et al.* 1965, 1969*a,b*). This may be attributed to inhibition of δ-ALA dehydratase (porphobilinogen synthetase) by succinylacetone which accumulates because of fumarylacetoacetase deficiency. δ-ALA dehydratase activity has been less than 5 per cent and 1 per cent of controls in erythrocytes and liver of patients with tyrosinaemia, respectively (Lindblad *et al.* 1977).

Genetic deficiency of δ-ALA dehydratase with (Doss *et al.* 1977) or without (Bird *et al.* 1979) symptoms of AIP has also been reported.

Genetic counselling
AIP has an autosomal dominant mode of inheritance. Homozygotes probably do not survive. The female to male ratio is approximately 3:2 (Waldenstrom 1957).

Prepubertal or asymptomatic carriers can be detected by demonstrating deficiency of erythrocyte UROgen I synthetase activity (Meyer *et al.* 1972). As discussed above, assay of urinary PBG or δ-ALA is not reliable.

The incidence in Lapland was estimated as 1:1000 (Waldenstrom 1957) but this population contained a single family with 137 affected individuals. Other studies by measurement of urine PBG which will not have detected some carriers give incidences of 1.5 to 20:100 000. Higher incidences have been found amongst mentally ill patients (Kaelbling *et al.* 1961).

Prenatal diagnosis has been reported by demonstrating deficiency of UROgen I synthetase in cultured amniotic cells (Sassa *et al.* 1975).

Hereditary coproporphyria

Clinical features
The neurovisceral, neurological, and mental manifestations of hereditary coproporphyria (HC) are usually indistinguishable from those of AIP (p. 414) (Goldberg *et al.* 1967; Jaeger *et al.* 1975). However, about one third of patients also have photosensitivity. Many heterozygotes for this disorder with autosomal dominant inheritance are free from symptoms (Connon and Turkington 1968).

Molecular defects and diagnosis
The most constant chemical abnormality is the considerable excess of faecal COPRO III (Goldberg *et al.* 1967; Lomholt and With 1969). There is also an increased excretion in the urine of COPRO III, δ-ALA and PBG, but chiefly during acute attacks. During the quiescent stage, urinary excretion may be normal. An increased excretion of 17-oxosteroids has been reported (Paxton *et al.* 1974).

As in patients with AIP, the activity of hepatic δ-ALA synthetase is increased, consistent with defective inhibition by the end-product, haem. The increase in excretion of COPRO III and of the porphyrin precursors δ-ALA and PBG is consistent with a block in conversion by oxidative decarboxylation of COPROgen III to PROTO IX. This reaction is catalysed by the mitochondrial enzyme coproporhyrinogen oxidase. An approximate 50 per cent reduction in the activity of this enzyme has been demonstrated in liver, cultured fibroblasts (Elder *et al.* 1976) and leucocytes (Grandchamp and Nordmann 1977) and is probably the basic defect in HC.

Genetic counselling

The disorder has an autosomal dominant mode of inheritance. It seems likely that about half of heterozygotes are asymptomatic (Goldberg *et al.* 1967; Haeger-Aronson *et al.* 1968; Jaeger *et al.* 1975). As far as we are aware, prenatal diagnosis has not been attempted. However, since deficiency can be demonstrated in the activity of cultured fibroblast coproporphyrinogen oxidase in heterozygotes, it seems probable that in affected pregnancies a deficiency could be demonstrated in cultured amniotic cells.

Variegate porphyria; South African genetic porphyria

The name 'variegate' refers to the varied clinical features which may be exclusively or predominantly cutaneous, neurovisceral, or absent. The 'South African' designation refers to the studies of Dean (1971) who traced 236 cases of variegate porphyria (VP) in 13 families which had descended from a Dutch couple who had settled in South Africa in 1688. However, patients have since been identified from most parts of the world.

Clinical features

These involve the skin and the nervous system. Over 80 per cent of patients have skin involvement with onset usually in the third decade (Eales *et al.* 1975). This is characterized by excessive fragility on mild trauma, mainly of skin exposed to light. Erosions and abrasions often become vesicular. Without infection these lesions normally heal leaving pigmented moderate scars. The skin changes, including pigmentation, are usually progressive. Infection exacerbates the skin damage and scarring. Photosensitivity is usually mild but can be severe when there is concurrent liver damage. Hypertrichosis of the face occurs in women.

The neurological features are similar to those described for AIP (q.v.) and may be precipitated by similar agents, including drugs and steroids.

Molecular defects and diagnosis

There is excessive faecal excretion of protoporphyrin and coproporphyrin

(Dean and Barnes 1959; Eales *et al.* 1975). This is continuous and occurs even when symptoms are mild (Grosser and Eales 1973) or absent, but is most marked during neurovisceral attacks. Faecal porphyrins are ether-insoluble, probably peptide-bound compounds referred to as the X-porphyrin fraction. Peptide porphyrin conjugates are also found in patients with porphyria cutanea tarda, but these are probably of different structure (Elder *et al.* 1974).

During acute neurovisceral attacks there is considerable excretion in the urine of the porphyrin precursors δ-ALA and PBG, and of porphyrins, some of which are peptide-bound. Affected patients without symptoms or with only cutaneous involvement have normal or only slightly elevated urinary excretion of porphyrins or their precursors (Dean 1971; Eales *et al.* 1975).

During acute neurovisceral attacks, the plasma may have a red fluorescence because of the high liver-derived coproporphyrin concentration.

No enzyme defect has yet been demonstrated but the pattern of urinary porphyrins and precursors is consistent with a block in the conversion of PROTOgen to haem. This is catalysed by two mitochondrial enzymes acting sequentially, PROTOgen oxidase and ferrochelatase. The former has not yet been assayed, while the activity of the latter is normal in muscle of affected patients.

Genetic counselling

VP has an autosomal dominant mode of inheritance. Extensive studies have shown that when one parent is affected, the disorder is inherited by approximately half the daughters and sons (Dean 1971). Amongst South African whites the incidence is about 1:333.

Porphyria cutanea tarda

Clinical features

Porphyria cutanea tarda (PCT) usually presents with excessive fragility of the skin in areas exposed to the sun. Abrasions, ulcers, or vesicles, which can be infected, occur in the fifth or sixth decades. The majority of patients are chronic alcoholics with liver injury (Elder *et al.* 1972; Pimstone 1975). Exceptionally, PCT occurs in childhood (Uys and Eales 1963). There is often a history of dark faeces and of reddish-brown urine. Sensory neuropathy has been reported. Hypertrichosis sometimes occurs, frequently on the face and forearms.

Two types of hepatic tumours are associated with PCT: firstly, carcinomata associated with cirrhosis (see 'Pathology', below) and secondly, either benign or malignant tumours which fluoresce in u.v. light, in contrast to the non-tumour liver tissue which does not (Tio *et al.* 1957; Thompson *et al.* 1970; Waddington 1972). This observation and the report that surgical removal of these tumours cures PCT (Tio *et al.* 1957) indicate that in these patients, abnormal porphyrin metabolism is restricted to the tumour.

Pathology

The characteristic pathological findings are alcoholic cirrhosis and excessive hepatic deposition of iron (siderosis) (Timme 1971; Pimstone 1975). At necropsy, carcinoma of cirrhotic liver is found in about half the patients. There is a large excess of porphyrins in hepatocytes, causing fluorescence under u.v. light.

Molecular defects and diagnosis

The urine contains an excess of uroporphyrin, mainly of the isomer type I and of smaller quantities of hepta- and hexa-carboxylic acid porphyrins of the isomer type III and of coproporphyrin and pentacarboxylic porphyrins, isomers of both type I and type III (Dowdle *et al.* 1970; Elder 1977). Rarely there is an excess of the porphyrin precursors δ-ALA and PBG (Dowdle *et al.* 1970). Acidified urine exhibits a pink fluorescence under u.v. light (Schmid *et al.* 1954). Faecal porphyrin concentration may be normal or considerably increased and consists of ether-soluble coproporphyrins and derivatives (Elder *et al.* 1972; Elder 1975) and ether-insoluble peptide-bound conjugates of heptacarboxylic porphyrin and uroporphyrin ('x-porphyrin') (Elder *et al.* 1974).

Liver porphyrins are considerably increased and consist mainly of uroporphyrin and heptacarboxylic acid porphyrins (Dowdle *et al.* 1970). Erythrocyte porphyrins concentration is normal (Schmid *et al.* 1954).

Deficiency in the activity of hepatic UROgen decarboxylase to about 25 per cent of normal was reported by Kushner and Barbuto (1975) and Kushner *et al.* (1976) in six patients with PCT. These patients and six of nine asymptomatic relatives also had deficiency of UROgen decarboxylase in their erythrocytes. The enzyme catalyses the decarboxylation of the four acetic acid side chains of UROgen III to yield COPROgen III.

Deficiency in the activity of UROgen decarboxylase would not therefore be expected to cause an increased concentration of urinary and faecal COPROs, which are distal to the block. However, secondary enzyme deficiencies may arise from hepatic siderosis. This has been demonstrated for UROgen III cosynthase and for UROgen I synthetase. Such secondary inhibition could explain accumulation of substances not directly affected by the primary genetic defect.

Deficiency of UROgen decarboxylase in patients with PCT was also reported in liver by de Verneuil *et al.* (1978).

The relation between alcoholism, cirrhosis, siderosis, and clinical expression of PCT is not understood. Only a small minority of chronic alcoholics with cirrhosis or siderosis of the liver develop PCT (Hällén and Krook 1963). A possible explanation is that individuals with deficiency of UROgen decarboxylase are more likely to develop PCT when there is alcoholic liver damage or siderosis. This suggestion is in accord with the late presentation and with the higher incidence of males who tend to be alcoholic more frequently than females, and with the rarity of familial occurrence (Dahlin *et al.* 1973; de Verneuil and Nordmann 1981). The beneficial effect of iron depletion on the clinical phenotype (without affecting deficiency of UROgen decarboxylase activity) illustrates the role of iron overloading in the pathogenesis. PCT is rarely induced by polychlorinated phenol, oestrogens, and oral contraceptives.

Genetic counselling

Initially considered to be an acquired disease secondary to hepatic damage (Uys and Eales 1963; Waldenstrom 1957), the reports that PCT has occurred in some sibs of both sexes (Taddeini and Watson 1968; Eales *et al.* 1975), and in three generations of a single family (Benedetto *et al.* 1978) suggest autosomal dominant inheritance.

PCT is probably the commonest type of porphyria and is particularly prevalent among the Bantu population of South Africa (Lamont *et al.* 1961; Eales *et al.* 1975; Pimstone 1975).

Hepatoerythropoietic porphyria

It has been suggested that patients with the severe form of cutaneous porphyria – hepatoerythropoietic porphyria – may be homozygotes for the PCT gene, since uroporphyrinogen decarboxylase activities in three patients were 7 per cent of normal in erythrocytes and 8 per cent of normal in cultured fibroblasts. In the father of one of these patients, the activity was reduced to 62 per cent of normal, a value consistent with his being a heterozygote for PCT.

Treatment

Depletion of iron by repeated removal of blood results in reduction of urine porphyrin concentration to normal levels and disappearance of excessive skin fragility for some years, but the remission is shortened if iron is administered. Administration of chelating agents also encourages iron loss (Turnbull *et al.* 1973).

Erythropoietic porphyrias

Congenital erythropoietic porphyria

Clinical features

The main clinical features of congenital erythropoietic porphyria (CEP) are a severe photosensitive dermatosis and haemolytic anaemia. In the majority of reported patients, photosensitive skin lesions have occurred in the first few years of age. Exceptionally, they are first noticed in adults (Kramer *et al*. 1965; Pain *et al*. 1975). Blisters occur, especially in the non-covered areas of skin after exposure to sunlight. Infections may cause extensive scarring and tissue loss, including loss of the eyelids and parts of the pinnae, and fingers.

Haemolytic anaemia of variable severity is present in most but not all patients (Chatterjea 1964). The anaemia may be fatal (Simard *et al*. 1972). Splenomegaly is present in the majority of patients.

Other clinical features include hypertrichosis, a brownish stain of the teeth (which emit red fluorescence under u.v. light), and pink or red urine, which may occur from birth, or even prenatally, leading to a staining of the amniotic fluid (Nitowsky *et al*. 1978).

Molecular defects and diagnosis

The most characteristic feature is a considerable increase in the excretion of type I isomers of uroporphyrin (URO) and coproporphyrin (COPRO) in the urine (Aldrich *et al*. 1951; Taddeini and Watson 1968; Meyer and Schmid 1978) with smaller excess of type III isomers (Watson *et al*. 1964; Taddeini and Watson 1968) but not of the porphyrin precursors δ-aminolevulinic acid (δ-ALA) or porphobilinogen (PBG) (Watson and Schwartz 1941). Plasma, circulating erythrocytes, and faeces also have excess content of URO and COPRO, mainly of the type I isomers. Erythrocyte protoporphyrin content is usually normal but exceptionally increased (Hofstad *et al*. 1973). The urobilinogen content of the faeces is usually increased (Schmid *et al*. 1955).

The enzymatic basis for CEP is unknown. The pattern of abnormal porphyrin distribution and production is consistent with an increased synthesis of porphyrins, predominantly of the type I isomers, and to a lesser extent of type III isomers. Moore *et al*. (1978) found that normal erythrocytes produced such small quantities of series I COPRO isomers that the ratio of COPRO I to COPRO III could not be accurately determined. In contrast, two patients with CEP produced increased, measurable amounts of COPRO and had series I to series III ratios of 3.6 and 4.2. In the same patients, these authors found a slight to marked increase in the activities of leucocyte δ-ALA synthetase, coproporphyrogen (COPROgen) oxidase, and ferrochelatase; and of erythrocyte δ-ALA dehydratase, uroporphyrinogen (UROgen) I synthetase and UROgen decarboxylase.

Romeo and Levin (1969) found that in patients with CEP, the activity of UROgen III cosynthase was decreased to about one-third to one-tenth of normal. The explanation that the overactivity of the haem pathway with a relatively greater activity of type I than type III isomer production might result from constitutive regulator (Watson *et al.* 1964) or structural gene mutation, lacks experimental proof.

Moreover, Levin (1968, 1971) and Romeo and Levin (1969) showed that the activity of UROgen III cosynthase decreased while the UROgen synthetase reaction proceeded *in vitro* and therefore suggested that in CEP the basic defect was a defect of UROgen III cosynthase. These observations would explain the imbalance in the rate of synthesis between type I and type III isomers, but not the cause of hyperactivity of enzymes of haem biosynthesis.

CEP has an autosomal recessive mode of inheritance (Meyer and Schmid 1978).

Haematology

In addition to the variable anaemia, there is usually a reticulocytosis and the appearance of normoblasts in peripheral blood. The bone marrow shows erythroid hyperplasia (Schmid *et al.* 1955). Bone marrow erythroblasts exhibit fluorescence, mainly in the nuclei (Varadi 1958; Gross 1964).

Gray *et al.* (1950*a,b*) reported that about half the erythrocytes have a normal lifespan, while the remainder have a considerably reduced survival to about 20 days.

It has been suggested that anaemia in CEP is due both to an autoimmune haemolysis (Chatterjea 1964) and to decreased erythropoiesis (Gray *et al.* 1950*b*; Kramer *et al.* 1965).

Variants of CEP

A boy with red urine from birth, a haemolytic anaemia and splenomegaly at seven weeks, and photosensitive dermatosis at two years (Hofstad *et al.* 1973) had a distribution of porphyrins quite atypical for CEP. Detailed analysis in this patient was reported by Eriksen and Seip (1973), Rimington and With (1973), and Eriksen and Eriksen (1974). The disorder has an autosomal recessive mode of inheritance (Meyer and Schmid 1978) and can be diagnosed *in utero* by demonstrating increased amniotic fluid porphyrin concentration (Nitowsky *et al.* 1978).

In contrast to the predominance of type I isomers found in typical patients with CEP, in this boy more than 50 per cent of urinary porphyrins belonged to the type III series and were mainly hepta-, hexa-, and pentacarboxylic porphyrins. Moreover, in erythrocytes, the predominant porphyrin was protoporphyrin (PROTO).

Two other patients with similar clinical features, increased urinary excretion of penta-carboxylic acid porphyrins, and increased erythrocyte PROTO content were reported by Piñol Aguadé *et al.* (1969, 1975).

A girl with dermal photosensitivity from 10 years of age was found to have a 200-fold increase in red cell PROTO content and an increased faecal excretion of both PROTO and COPRO (Heilmeyer and Clotten 1964; Heilmeyer *et al.* 1966), suggesting an increased production of type III isomer porphyrins. The patient's mother, who did not have symptoms, had a similar abnormal porphyrin distribution.

Animal models for congenital erythropoietic porphyria

In bovine erythropoietic porphyria there is photosensitivity of unpigmented skin, anaemia, and splenomegaly (Watson *et al.* 1958). The chief abnormality is overproduction of uroporphyrin I in erythroid cells (Jøergenssen and With 1963). As in the human disease, there is decreased red cell UROgen III cosynthase activity. Inheritance is autosomal recessive.

An autosomal dominant form of porphyria occurs in pigs (Jøergensen and With 1955). There is staining and fluorescence of the teeth but no photosensitivity.

Porphyria in cats has similar phenotypic abnormalities to the porcine variety and also, unlike human and bovine erythropoietic porphyrias, has an autosomal dominant mode of inheritance (Glenn *et al.* 1968). Erythrocyte UROgen III cosynthase activity has been reported to be normal.

Erythropoietic porphyrias also occur in the fox squirrel where, as in the human and bovine types, there is overproduction of porphyrins of the isomer type I and considerable reduction in the activity of UROgen III cosynthase. However, affected fox squirrels are not photosensitive.

Erythropoietic protoporphyria; protoporphyria

Erythropoietic protoporphyria (EPP) is characterized clinically by dermal photosensitivity, especially to long wavelength u.v. light (Magnus *et al.* 1961), and biochemically by increased content of PROTO in erythrocytes, plasma, and faeces but not in urine. The incidence is much higher than that of CEP.

Clinical features

Symptoms usually start in the first decade. Typically within minutes or hours following exposure to sunlight there is severe pain, pricking, itching, oedema, and erythema of the skin (Magnus *et al.* 1961). Blisters, cutaneous haemorrhages, and scarring are less common (De Leo *et al.* 1976). More rarely, a chronic eczematous rash occurs (Redeker and Bronow 1964; Cripps *et al.* 1966). The wavelength of the u.v. light that causes the skin lesions is around 400 nm (Magnus *et al.* 1961) which corresponds to the maximal spectral absorption of porphyrins.

A minority of patients develop a mild anaemia (De Leo *et al.* 1976; Haeger-Aronson and Krook 1966) which can be associated with severe haemolysis and iron deposition (Scott *et al.* 1973; Porter and Lowe 1963). Cholelithiasis occurred in 12 of 34 patients with EPP (De Leo *et al.* 1976).

Cirrhosis of the liver occurs in a minority of patients and can cause death.

Molecular defects and diagnosis

The main biochemical abnormality is a considerable increase in the content of PROTO in erythrocytes and faeces and a smaller increase in plasma. In contrast to all other forms of porphyria, there is no increase in the excretion of porphyrins in urine.

Fluorescence is marked in peripheral blood reticulocytes (Clark and Nicholson 1971) but rare in erythrocytes (Kaplowitz *et al.* 1968; Cripps *et al.* 1966), and is absent or slight in bone marrow erythroblast cytoplasm (Porter and Lowe 1963; Haeger-Aronson 1963).

Piomelli *et al.* (1975) reported that PROTO diffused rapidly from erythrocytes to the plasma, approximately 40 per cent leaving the red cells per day. This loss, together with that from the bone marrow, could roughly correspond to the daily excretion of PROTO in the faeces.

There is deficiency in the activity of the mitochondrial enzyme ferrochelatase which catalyses the final step in haem synthesis (the insertion of iron into PROTO IX). This has been demonstrated in bone marrow (Bottomely *et al.* 1975) cultured skin fibroblasts (Bonkowsky *et al.* 1975; Bloomer *et al.* 1977) and in circulating nucleated red blood cells (De Goeij *et al.* 1975; Brodie *et al.* 1978). In cultured fibroblasts, the activity has been estimated as approximately 10 per cent of controls, which is surprisingly low, since the patients are heterozygotes for a dominant gene and might be expected to have activities nearer 50 per cent of normal. However, the residual functional activity has been estimated to be nearer 50 per cent of normal using more physiological conditions and intact cells (Sassa *et al.* 1979*b*). This technique demonstrates excessive accumulation of PROTO in mitogen-stimulated EPP lymphocytes incubated in the presence of δ-ALA. It may be, therefore, that the gene dosage effect mirrors the metabolic effect more closely than *in vitro* enzyme assay.

Genetic counselling

The disorder has an autosomal dominant mode of inheritance with varying penetrance. In a family studied by Sassa *et al.* (1979*b*), the mother and two daughters had some of the biochemical abnormalities, but not the clinical phenotype of EPP, while three sons had both the biochemical and clinical phenotype.

It is possible, in theory, that prenatal diagnosis could be established by foetal blood sampling and either measuring accumulation of PROTO in mitogen-stimulated foetal lymphocytes (Sassa *et al.* 1979*b*) or by assay of ferrochelatase in foetal nucleated red cells (De Goiej *et al.* 1975; Brodie *et al.* 1978).

11. Disorders of steroid metabolism

Disorders of steroid metabolism (with the exception of steroid sulphatase deficiency, see p. 126) will be considered as:

1. Adrenogenital syndromes where there are specific deficiencies in the activities of enzymes of steroid metabolism affecting both the adrenal cortex and the gonads.

2. Defective synthesis of aldosterone, where there is deficiency in the activity of enzymes of steroid metabolism affecting the adrenal cortex, but not the gonads.

3. Male pseudohermaphroditism where there is a defect in the activity or of binding of androgenic enzymes.

BIOCHEMISTRY OF STEROIDS

Classification and structure of adrenocortical steroids and sex hormones

Steroids implicated in genetic disorders of the adrenal cortex and gonads are classified according to their function into three main types: glucocorticoid (e.g cortisol), mineralocorticoid (e.g. aldosterone), and sex hormones (e.g. testosterone and oestrogens). While the glucocorticoids and mineralocorticoids are synthesized in the adrenals and not in the gonads, the sex hormones are synthesized both in the adrenals and gonads.

The main precursor of all these steroids is cholesterol, which has a basic structure consisting of four carbon rings designated A, B, C, and D, and an eight-carbon side-chain attached to ring D at C-17. It will be noted from Fig. 11.1 that C-18 and C-19 in cholesterol are methyl groups so that the first carbon of the side chain is numbered 20. In ring B, cholesterol has a double bond $\triangle 5$ and in ring A a hydroxyl group at C-3.

In glucocorticoids and mineralocorticoids, a keto-group replaces the hydroxyl group at C-3, while the double bond moves to $\triangle 4$ in ring A. Further, their side chain is shortened from eight to two carbon atoms, and there is a keto-group on C-20. Glucocorticoids but not mineralocorticoids have a hydroxyl group at C-17.

Testosterone and other virilizing steroids have lost the side chain, which is replaced by a hydroxyl group on C-17 while oestradiol and other feminizing steroids have further lost their C-19 methyl groups, due to aromatization of their A ring with reduction of their C-3 attached keto-group. Aldosterone is unique in that the aldehyde group on carbon 18

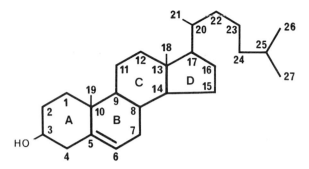

Fig. 11.1. Structure of cholesterol.

interacts with the hydroxyl group on carbon 11 to form an 11,18 cyclic hemiacetal as shown in Fig. 11.2.

Biosynthesis of adrenocortical steroids

The 17-deoxy pathway for synthesis of mineralocorticoids is similar to that of the 17-hydroxy pathway for glucocorticoids until the synthesis of pregnenolone and progesterone, which may be hydroxylated to 17-hydroxypregnenolone and 17-hydroxyprogesterone, respectively, in the 17-hydroxy pathway (Fig. 11.2).

The partial cleavage of the side chain catalysed by 20,22-desmolase (cholesterol desmolase) converts cholesterol to pregnenolone. The desmolase system is a complex consisting of several enzymes (Hochberg *et al.* 1974). Two reactions are similar to the hydroxylations involved in biosynthesis of other steroids, which also require NADPH and molecular oxygen (see below). The first reaction in the 20,22-desmolase system with the formation of 20α-cholesterol is probably rate-limiting. The second 20,22-desmolase-catalysed reaction results in the formation of 20,22-dihydroxycholesterol, and the third in the cleavage of a six-carbon fragment from the side chain utilizing further molecules of NADPH and molecular oxygen to form isocaproaldehyde and pregnenolone. Hydroxylations are the main reactions in steroidogenesis. Several of the reaction components of the various hydroxylases are similar. Electrons are transferred sequentially from NADPH to a flavoprotein dehydrogenase (adrenodoxin reductase), a non-haem iron-containing pigment, and to cytochrome P-450. The reduced P-450 cytochrome, together with molecular oxygen, complexes with the steroid substrate which becomes hydroxylated. Pregnenolone is then converted to progesterone by the combined actions of 3 ß-dehydrogenase and $\triangle$5-$\triangle$4-isomerase. Progesterone is then converted to deoxycorticosterone (by 21-hydroxylase) and then to corti-

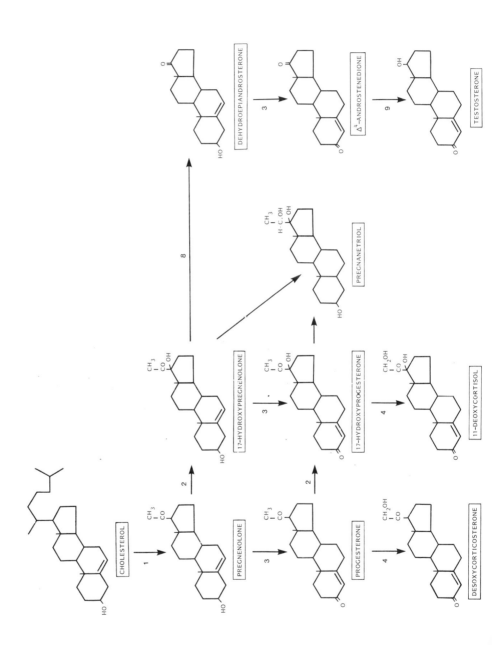

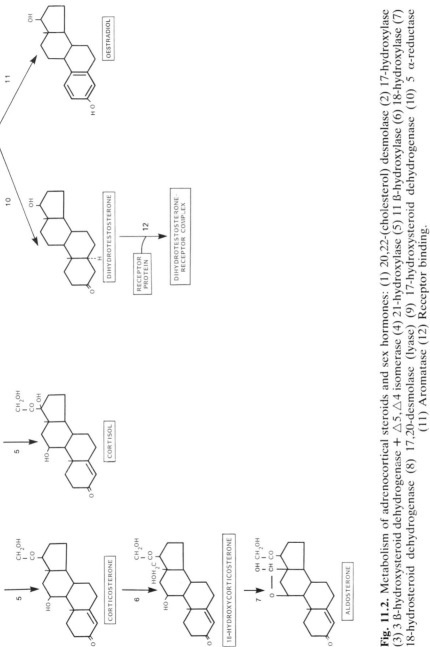

Fig. 11.2. Metabolism of adrenocortical steroids and sex hormones: (1) 20,22-(cholesterol) desmolase (2) 17-hydroxylase (3) 3 ß-hydroxysteroid dehydrogenase + △5,△4 isomerase (4) 21-hydroxylase (5) 11 ß-hydroxylase (6) 18-hydroxylase (7) 18-hydrosteroid dehydrogenase (8) 17,20-desmolase (lyase) (9) 17-hydroxysteroid dehydrogenase (10) 5 α-reductase (11) Aromatase (12) Receptor binding.

costerone (by 11-hydroxylase). Finally the sequential actions of 18-hydroxylase and 18-dehydrogenase catalyse the synthesis of aldosterone which occurs exclusively in the zona glomerulosa of the adrenal cortex (Ulick 1976).

As mentioned earlier, 17-hydroxylation at the level of pregnenolone or progesterone will divert these steroids into the 17-hydroxy (glucocorticoid) synthetic pathway. The conversion of 17-hydroxypregenolone to 17-hydroxyprogesterone is catalysed by 3 ß-dehydrogenase and $\triangle$5-$\triangle$4-isomerase.

The next two reactions in the 17-hydroxy pathway are analogous to those of the 17-deoxy pathway, namely 21-hydroxylation to 11-deoxycortisol (also known as 'substance S' or cortexolone), catalysed by 21-hydroxylase, followed by 11-hydroxylation to cortisol, catalysed by 11-hydroxylase.

It has been suggested that the zona glomerulosa and zona fasciculata of the adrenal cortex function as separate glands. This implies that ACTH stimulates secretion of cortisol, corticosterone, and androgens in the zona fasciculata; while angiotensin (in the renin–angiotensin system) stimulates secretion of aldosterone in the zona glomerulosa (New and Seaman 1970; Mason *et al.* 1979).

Biosynthesis of sex steroids

The reactions involved in the conversion of cholesterol to 17-hydroxy-pregnenolone and to 17-hydroxyprogesterone are described in the section on biosynthesis of the adrenocortical steroids. In the biosynthetic pathways of sex steroids, the 17,20-desmolase (17,20-lyase) system catalyses the replacement of the C20, C21 side chain by a 17-keto group of 17-hydroxypregnenolone to form dehydroepiandrosterone which retains the 3-hydroxy, $\triangle$5 structure of cholesterol. In the next reaction, catalysed by 3 ß-dehydrogenase and $\triangle$5-$\triangle$4-isomerase, this structure is transformed to 3-keto, $\triangle$4 with the formation of androstenedione. The latter is also synthesized from the $\triangle$4 compound 17-hydroxyprogesterone in a reaction synthesized by 17,20-desmolase. The reduction of the 17-keto group of androstenedione to form testosterone is catalysed by 17-dehydrogenase. The A ring of androstenedione and testosterone can undergo aromatization catalysed by aromatase to form the oestrogens oestrone (E_1) and oestradiol, respectively.

THE ADRENOGENITAL SYNDROMES; FEMALE PSEUDOHERMAPHRODITISM

In this section we discuss the three defects in steroid metabolism causing prenatal virilization of the female (female pseudohermaphroditism), namely: 21-hydroxylase deficiency, 11 ß-hydroxylase deficiency, and 3 ß-dehydrogenase deficiency.

Relation between symptomatology of adrenogenital syndromes and steroid abnormalities

Salt losers

Salt loss may occur in untreated patients with deficiency of 20,22-desmolase, 3 ß-dehydrogenase, 21-hydroxylase, 18-hydroxylase, and 18-dehydrogenase. All these enzymes catalyse a reaction in the biosynthetic pathway to aldosterone, and all but 18-hydroxylase and 18-dehydrogenase also to cortisol. Salt loss does not occur in patients with deficiency of 11 ß-hydroxylase which normally also catalyses a reaction in the biosynthetic pathway to aldosterone. This is due probably to an accumulation of deoxycorticosterone, since when this is reversed by administration of cortisol, salt loss may occur.

The clinical features are chronic or acute. Chronic salt loss has an insidious onset with loss of weight, anorexia, vomiting, hypotension, and sometimes Addisonian dermal pigmentation. In an acute salt-losing crisis, there is rapid deterioration and circulatory collapse which may be fatal. There is hyperkalaemia, hyponatraemia, and hyperazotaemia. Blood aldosterone and cortisol concentrations are reduced, but plasma renin activity is increased. Some survivors later develop a craving for salt.

Virilization

Virilization occurs in deficiency of 21-hydroxylase, 11 ß-hydroxylase, and 3 ß-dehydrogenase. In all these disorders there is a block in the biosynthetic pathway to cortisol which may decrease the negative feedback inhibition, leading to enhanced corticotrophin secretion, and an over-production of testosterone and other androgens which are responsible for virilization. Prenatally, in the female, there may be ambiguity of the external genitalia, varying in degree from slight enlargement of the clitoris, with or without labial fusion, to full masculinization. According to the degree of virilization, the diagnosis may be suspected at birth, because of ambiguous genitalia, or may be detected later because an apparent boy with hypospadias is abnormally tall. The internal sexual organs and, of course, sex chromosome complement, remain female.

Postnatally, in both sexes there is radiological evidence of advanced bone age and accelerated skeletal growth until epiphyseal fusion, after which height centiles fall to below the average.

In the male there is precocious development of secondary sexual characteristics, simulating puberty, but the testes remain prepubertal.

A different, milder form of virilization occurs in adrenogenital syndrome due to 3 ß-hydrogenase deficiency (q.v.) where the girls have a moderate virilization of the external genitalia at birth, probably because of accumulation of dehydroepiandrosterone (DHA), while the boys have defective masculinization because of deficiency of testosterone. In both sexes premature puberty is likely because of the raised DHA levels.

Hypertension

Hypertension may occur in deficiency of 11 ß-hydroxylase or of 17-hydroxylase. As mentioned above, the former is the only enzyme defect in the biosynthetic pathway to aldosterone in which salt loss does not occur. This may be explained by an accumulation of deoxycorticosterone, which is a mineralocorticoid and which is probably responsible for the hypertension in both disorders.

21-hydroxylase deficiency

This is the commonest form of adrenogenital syndrome, and accounts for about 90 per cent of cases in Europe.

Clinical features

Clinically, in girls there is a variable degree of congenital virilization (see above). About a quarter of affected patients of either sex are salt losers. In severe forms of the disorder, life-threatening circulatory collapse may occur after the first 10 days or so after birth.

It is not yet established whether salt-losers have a different genetic defect from non-salt-losers. Most observers have noted that severe virilization in girls is associated with salt loss. Mild and severe clinical expression can occur in the same family (Zachmann and Prader 1978, 1979). It seems probable that several mutant alleles exist, associated with different degrees of severity. There is evidence, substantiated by HLA-linkage studies (see 'Genetic counselling', p. 434), that in some families the severe form of the disorder occurs in homozygotes, and the mild form in heterozygotes.

When the diagnosis in girls is recognized early in childhood, there is little doubt that the patients should be brought up as females. Money and Ehrardt (1972), however, noted shallowness of female emotional responses to males in later life, and suggested a possible defect of virilizing steroids on the foetal brain. However, there are now numerous examples of treated females who have had successful pregnancies and normal children.

Non-salt-losing males appear normal at birth, but exhibit early and excessive development of the penis, secondary sexual characteristics and acne. Pubic and axillary hair growth, together with a deep voice and muscular development occur in the first few years. Unlike in the true precocious puberty initiated by pituitary pathology however, the testes remain infantile.

In both sexes virilization initially causes a skeletal growth spurt, but premature fusion of the epiphyses eventually leads to dwarfing of previously tall children.

Molecular defects and diagnosis

Deficiency in the activity of 21-hydroxylase has been demonstrated directly in adrenal tissue (Bongiovanni 1958; Axelrod and Goldzieher 1967). The enzyme is not expressed in erythrocytes, leucocytes, serum, or cultured fibroblasts.

There is a large excess of adrenocortical steroids proximal to the enzyme block in the pathways of steroid biosynthesis. This results from over-production, stimulated by increased ACTH secretion because of decreased feedback inhibition by cortisol. In compensated, i.e. non-salt-losing patients, the serum cortisol concentration, however, is usually normal but fails to rise normally after administration of ACTH. In salt losers, production of cortisol and its derivatives is considerably reduced (Eberlein and Bongiovanni 1958; Migeon and Kenny 1966). Salt losers usually have reduced aldosterone levels while non-salt-losers tend to have normal or raised levels (New *et al.* 1966; Bartter *et al.* 1968; Simopoulos *et al.* 1971).

Recently it has been shown that when the zona glomerulosa of the adrenal cortex is stimulated by renin in normal individuals or in non-salt-losers there is a rise in urinary and serum aldosterone concentrations. Salt-losers however show almost no such rises (Kuhnle *et al.* 1981). This is consistent with the explanation that the 21-hydroxylase deficiency in salt-losers (but not in non-salt-losers) involves the zona glomerulosa. The observation that stimulation of the zona fasciculata by ACTH causes a marked rise in serum progesterone but reduced rise of serum cortisol in both salt-losers and non-salt-losers suggests 21-hydroxylase deficiency involves the zona fasciculata in both types of adrenogenital syndrome.

Because of the defect in C-21-hydroxylation, there is a large accumulation in the adrenal cortex of C-21-methylated steroids, including the substrate of the 21-hydroxylase reaction 17-hydroxyprogesterone, its precursor progesterone, and its 11-hydroxylated derivative. In the urine, there is an accumulation of their derivatives pregnanediol, pregnanetriol and 11-oxopregnanetriol (pregnanetriolone), respectively, formed by reduction of the 3-keto groups (Hughes and Winter 1976; Pang *et al.* 1979).

The diagnosis can be established by demonstrating the abnormal patterns of plasma and urinary steroid distribution described above, and especially elevated 17-hydroxyprogesterone concentration in the plasma and increased excretion of pregnanetriol or 11-oxopregnanetriol in the urine. The latter predominates during infancy, the former in older children.

In female pseudohermaphroditism, it is wise to establish the genetic sex rapidly by the demonstration of Barr bodies, and to confirm this by chromosome analysis.

In postnatal virilization, the differential diagnosis from virilizing adrenal tumours should be considered. This can be established by the different pattern of urinary steroids and usually by the almost complete failure of suppression of the abnormal steroid excretion by adequate suppressive doses of cortisol (see 'Treatment of adrenogenital syndromes', p. 436).

Genetic counselling

The disorder has an autosomal recessive mode of inheritance. Steroid studies do not allow confident discrimination between heterozygotes and individuals who are homozygote normal. Thus, although Knorr *et al.* (1977) found significant rises of urinary 17-hydroxyprogesterone after administration of ACTH, this could not be demonstrated by others. Lee and Gareis (1975) found increased levels of plasma 17-hydroxyprogesterone following intravenous injection of ACTH. Qazi *et al.* (1971) found significant elevation of mean urinary 11-hydroxyandrosterone only in fathers of affected children, but there was considerable overlap with normal values.

It is probable that some conflicting results among heterozygotes are due to genetic heterogeneity. Thus in some families heterozygotes show not only biochemical but also clinical abnormalities (Zachmann and Prader 1979). There is close linkage between the structural genes for 21-hydroxylase and for the HLA B locus on the short arm of chromosome 6 (Pucholt *et al.* 1980; Levine *et al.* 1978; Lorenzen *et al.* 1980). When family studies allow identification of the HLA haplotype linked to the mutant 21-hydroxylase gene, heterozygote siblings can be identified with a high degree of confidence. Similar studies, by determining the foetal HLA haplotype on amniotic cells (Levine *et al.* 1979) can be used for prenatal diagnosis (Couillin *et al.* 1979; Pollack *et al.* 1979). Elevated amniotic fluid 17-ketosteroid (Jeffcoate *et al.* 1965) or pregnanetriol (Nichols and Gibson 1969; Nichols 1970) concentrations in affected pregnancies have been detected only in the terminal weeks of gestation. Serial amniotic fluid steroidal analyses from the 23rd week failed to detect an affected foetus (Merkatz *et al.* 1969). However, elevated amniotic fluid 17-hydroxy-progesterone concentration measured by radioimmunoassay after chromatographic purification has allowed prenatal diagnosis of two affected foetuses at 16 and 24 weeks' gestation (Milunsky and Tulchinsky 1977; Nagamani *et al.* 1978), of four affected female foetuses in the second trimester (Carson *et al.* 1982), and retrospectively at 17 weeks (Hughes and

Laurence 1979). Measurement of steroids in amniotic fluids by other methods is probably less reliable. However, the need for, and ethics of, prenatal diagnosis have been questioned (Sibert 1979; Steffes and Wong 1979).

The incidence of 21-hydroxylase deficiency has been reported as 1:67 000 in Maryland, USA (Childs *et al.* 1956) and 1:15 472 in Switzerland (Werder *et al.* 1980). Very high incidences have been reported among Alaskan (1:1 481) and Yupik (1:490) Eskimos, with a high incidence of salt losers (Hirschfeld and Fleshman 1969). Most surveys show a predominance of females, probably because of easier diagnosis owing to ambiguous genitalia. If one assumes that the diagnosis is missed in some non-salt-losers and in a sizeable proportion of males, the incidence is probably higher than recorded.

11 ß-hydroxylase deficiency

About 10 per cent of all patients with adrenogenital syndrome have 11 ß-hydroxylase deficiency, but in Israel and Turkey the proportion is about 50 per cent.

Eberlein and Bongiovanni (1956) reported a patient with female pseudohermaphroditism and virilization who had severe hypertension. There was an accumulation in the blood and urine of 11-deoxycortisol (substance S) and in the urine of its derivative etiocholanolone and tetrahydro S. The clinical and biochemical abnormalities were reversed by administration of cortisol.

Study of further cases has revealed that some patients have a mild defect without hypertension, and mild masculinization of females, while others have a severe defect with hypertension and marked masculinization of females. In both types both substance S and deoxycorticosterone (DOC) are increased, indicating an 11-hydroxylase block in both the 17-deoxy- and 17-hydroxysteroid pathways (Gandy *et al.* 1960; Gabrilove *et al.* 1965). It is the only form of adrenogenital syndrome in which salt loss does not occur. This lack of salt loss and hypertension are probably caused by accumulation of DOC. Treatment by cortisol (see 'Treatment of adrenogenital syndromes', below) causes a decrease in DOC levels accompanied by resolution of hypertension and sometimes by the appearance of salt loss.

Though early reports on the rate of aldosterone biosynthesis were conflicting, it has now been established that it is low in most patients. Absent aldosterone or low rates of production were reported by Bryan *et al.* (1965) and Kowarski *et al.* (1968).

Recent studies suggest that 11-hydroxylase deficiency affects the ACTH-stimulable zona fasciculata rather than the zona glomerulosa (Levine *et al.* 1980).

Prenatal diagnosis has been reported by demonstration of elevated levels of maternal urine and amniotic fluid 11-deoxycortisol and tetrahydrocortisol (Rösler *et al.* 1979; Schumert *et al.* 1980).

Inheritance is autosomal recessive.

3 ß-dehydrogenase deficiency

3 ß-dehydrogenase deficiency was first described by Bongiovanni (1962), who reported a considerable increase in $\triangle 5$ ß-hydroxy steroids in urine and later demonstrated the enzyme deficiency in adrenal cortex and gonads. The latter was confirmed by Schneider *et al.* (1975). Females have mild congenital virilization and males a hypospadias and sometimes a rudimentary phallus and vagina, which can be attributed to overproduction of dehydroepiandrosterone (DHA) and deficiency of testosterone. Marked elevation of plasma DHA, and urinary pregnenetriol are diagnostic (Bongiovanni 1980). The former is probably the cause of the precocious appearance of secondary sexual characteristics.

Salt loss occurs in the majority of patients and can lead to death during infancy, in spite of adequate treatment. Other patients have survived (Jänne *et al.* 1970; Zachmann *et al.* 1970), for example, boys to puberty (Parks *et al.* 1971; Schneider *et al.* 1975).

The disorder probably has an autosomal recessive mode of inheritance.

Treatment of adrenogenital syndromes

If untreated, virilization is progressive, so that suppressive steroid therapy should be started early. Cortisol (hydrocortisone) can be given in initial doses of 25 mg/m² surface area in divided doses. It is generally accepted that salt losers should also receive the salt retaining hormone 9 α-fludrocortisone acetate. More recently it has been suggested that control of 21-hydroxylase deficiency in non-salt losers was also improved by fludrocortisone administration which caused a fall of the blood ACTH level, and improved growth (Winter 1980). Dosage may be modified according to the clinical and biochemical responses. Biochemical monitoring is usually by demonstrating suppression of excessive excretion in the urine of steroids characteristic for the defect, e.g. 17-hydroxy-pregnenolone and pregnanetriol in 21-hydroxylase deficiency; 11-deoxy-cortisol and deoxycorticosterone in 11-hydroxylase deficiency; and $\triangle 5$ ß-hydroxy-steroids in 3 ß-dehydrogenase deficiency. Monitoring dosage by assay of blood levels of adrenal steroids – for example, by the suppression of plasma 17-hydroxyprogesterone – is also satisfactory (Korth–Schutz *et al.* 1978; Golden *et al.* 1978).

Employing the minimal dose of oral cortisol required to suppress abnormal steroid synthesis allows optimal growth.

Patients with adrenogenital syndrome usually should be reared in the roles of their genetic sex. This should be established as soon as possible. Early surgical correction of the genitalia is desirable (Jones and Scott 1971).

DEFECTIVE SYNTHESIS OF ALDOSTERONE

18-hydroxylase and 18-dehydrogenase deficiencies

Disorders associated with hypoaldosteronism and salt loss (q.v.) are caused by an enzyme deficiency involving the 17-deoxy biosynthetic pathway between cholesterol and aldosterone. Aldosterone is synthesized exclusively in the zona glomerulosa (Ulick 1976) from corticosterone by two reactions, the first catalysed by 18-hydroxylase, the second by 18-dehydrogenase. Since these affect only the synthesis of aldosterone, salt loss (q.v.) is the only manifestation. There is no virilization, abnormality of sexual differentiation, or cortisol or androgen synthesis.

Deficiency of both the 18-hydroxylation (Ulick *et al.* 1964) and the 18-dehydrogenation (Visser and Cost 1964) have been reported. As expected, the clinical features were produced by hypoaldosteronism and salt loss. Inheritance of both disorders is probably autosomal recessive.

MALE PSEUDOHERMAPHRODITISM: TESTOSTERONE DEFICIENCY: COMPLETE TESTICULAR FEMINIZATION: ANDROGEN RESISTANCE: PARTIAL TESTICULAR FEMINIZATION: 5 α-REDUCTASE DEFICIENCY

Individuals with a male chromosome complement and testes whose phenotype is feminized to varying degrees are said to have male pseudohermaphroditism. The degree of feminization can vary from having hypospadias to resembling a female anatomically in almost every way except that there are testes instead of ovaries.

Syndromes causing male pseudohermaphroditism may be considered under the following headings:

1. Deficiency of Müllerian regression factor.

2. Specific defects of testosterone and dihydrotestosterone biosynthesis, i.e. a deficiency of:

(i) 20,22-desmolase (cholesterol desmolase);

(ii) 3 ß-dehydrogenase (considered under adrenogenital syndromes);

(iii) 17-hydroxylase;

(iv) 17,20-desmolase;

(v)   17-dehydrogenase (17-reductase);
(vi) 5 α-reductase.
3. Insensitivity to testosterone:
(i)   complete testicular feminization;
(ii) familial incomplete male pseudohermaphroditism;
(iii) incomplete testicular feminization.
4. Testicular feminization with unidentified abnormality.

Normal sexual development

Embyronic derivation of the internal genitalia differs in males and females, but in both sexes they originate from the Wolffian and Müllerian ducts. The external genitalia, on the other hand, are derived from common embryonic structures in both sexes. The Wolffian ducts in both sexes operate as excretory ducts of the embryonic mesonephric kidney and are connected to the gonads. In males, the Wolffian ducts form the epididymis, the vas deferens and the seminal vesicles. The Müllerian ducts arise from the Wolffian ducts and in the male undergo regression under the influence of the Müllerian regression factor (MRF), synthesized by the embryonic seminiferous tubules (Josso 1971, 1981). In females (who do not synthesize MRF), the Müllerian ducts form the fallopian tubes, uterus, and approximately the upper one-third of the vagina.

In both sexes, the external genitalia develop from four embryonic structures: (1) the urogenital sinus which in the male forms the prostate and prostatic urethra and in the female approximately the lower two-thirds of the vagina and the urethra; (2) the genital swelling, which in the male becomes the scrotum and in the female the labia majora; (3) the genital folds, which in the male become the shaft of the penis and in the female the labia minora; (4) the genital tubercles which in the male form the glans penis and in the female the clitoris. The indifferent gonads become testes in the presence of H-Y antigen and ovaries in its absence. The synthesis of the H-Y antigen is under control of genes in the pericentric region of the Y-chromosome and in the short arm of the X-chromosome (Ohno 1978; Wachtel 1979; Watchel and Koo 1981).

Jost (1953, 1961, 1970) showed in elegant experiments that castrated male embryos developed a female phenotype. From this it may be deduced that secretions from the testis are required for male sexual development. The role of MRF in preventing the formation of fallopian tubes and uterus in males was considered above. Testosterone and dihydrotestosterone (DHT) are also required for male differentiation. Testosterone stimulates the conversion of the Wolffian duct to form the epididymis, vas deferens, and seminal vesicles, before these structures are capable of converting testosterone to the 5 α-configuration, that is to dihydrotestosterone, catalysed by 5 α-reductase (Siiteri and Wilson 1974). Dihydrotestosterone is formed subsequently in the lower urogenital tract (Wilson and Laznitzki 1971; Sulcova *et al.* 1973); it induces the conversion of the urogenital sinus to form prostate and prostatic urethra, the development of the male external genitalia from the genital tubercule and swelling (see above), and fusion of the genital folds in the shaft of the penis and of the genital swellings to form the scrotum.

Deficiency of Müllerian duct regression factor

The role of MRF in male sexual development has been described above. The factor has been partially characterized and in foetal calf testis is a glycoprotein of molecular weight 215 000 when determined by gel filtration, 124 000 when determined by density gradient sedimentation, and 123 000 by SDS-PAGE (Picard *et al.* 1978). Dissociation into a 72 000 subunit was demonstrated under reducing conditions. Inheritance is either X-linked or autosomal recessive. Predictably, in males there is persistence of the Müllerian duct derivatives – the uterus and fallopian tubes. The testes are undescended but the external and internal genitalia are otherwise normal (Sloan and Walsh 1976). The nature of the molecular defect is not understood.

Specific enzyme defects of testosterone and dihydrotestosterone biosynthesis

Deficiency of 5 α-reductase causes a block in the conversion of testosterone to DHT. All the other enzyme defects are on the biosynthetic pathway of both testosterone and DHT (Fig. 11.2). The reactions of sex steroid biosynthesis are described above. Deficiency in the activity catalysing these reactions (see below) causes varying degrees of primary hypogonadism. In genetic males, development of the external genitalia may be incomplete, causing male pseudohermaphroditism. This may vary in severity from ambiguous genitalia to complete feminization. The testes are usually undescended.

Defective synthesis of DHT may be demonstrated as failure of DHT production following administration of human gonadotrophin (Saenger *et al.* 1978; Imperato–McGinley *et al.* 1979). Plasma testosterone concentration is normal or high (Peterson *et al.* 1977).

20,22-desmolase (cholesterol desmolase) deficiency (lipoid adrenal hyperplasia)

Since 20,22-desmolase is the first enzyme-catalysed reaction in conversion of cholesterol to other steroids, deficiency impairs the synthesis of all classes of steroids (Prader and Gurtner 1955; Camacho *et al.* 1968;

Kirkland *et al.* 1973). The adrenals are considerably enlarged, up to three times the normal size, and the cortical cells are enlarged and packed with cholesterol and other lipids. Only a few patients have been reported. Males have ambiguous or female genitalia with undescended testes. Females have normal female genitalia. There is usually severe salt loss (q.v.) and death in infancy. Adrenal tissue from patients does not cleave cholesterol (Degenhart 1971). The disorder has an autosomal recessive mode of inheritance.

3 ß-dehydrogenase deficiency (see adrenogenital syndrome, p. 436)

17-hydroxylase deficiency

Patients with 17-hydroxylase deficiency synthesize testosterone, oestrogens, and cortisone at a considerably reduced rate. The compensatory increase in the 17-deoxy pathway causes accumulation of corticosterone and DOC which give rise to hypertension and hypokalaemia. Females (Biglieri *et al.* 1966) survive to adult life but do not develop secondary sexual characteristics. Males have female or ambiguous genitalia with complete hypospadias and a shallow vagina (New 1970; Madan and Schoemaker 1980). Testosterone secretion is low and fails to rise after administration of gonadotrophin. Administration of testosterone caused growth of secondary sexual hair and of the phallus (New 1970). Inheritance is probably autosomal recessive.

17,20-desmolase (17,20-lyase) deficiency

Deficiency of 17,20-desmolase in males causes male pseudohermaphroditism, i.e. primary hypogonadism without puberty, since the defect blocks production of testosterone and oestrogens. Following injection of gonadotrophin there is an absent or minimal rise of urinary testosterone and DHEA, but a rise in the already increased excretion of pregnanetriolone, a derivative of 17-hydroxyprogesterone (Zachmann *et al.* 1972; Goebelsman *et al.* 1976; Forest *et al.* 1980). No affected female has yet been reported. The mode of inheritance is not yet established.

17-dehydrogenase (17-reductase; 17 ß-hydroxysteroid dehydrogenase) deficiency

17-dehydrogenase catalyses the reduction of the 17-keto group of $\triangle^4$-androstenedione to form testosterone. The impaired conversion of $\triangle^4$-androstenedione to testosterone causes an elevated blood $\triangle^4$-androstenedione/testosterone ratio which may be used for diagnosis. In infants, however, prior stimulation by gonadotrophin may be necessary (Levine *et al.* 1980). Clinically there is male pseudohermaphroditism, normal adrenal function and, at puberty, gynaecomastia (Goebelsmann *et al.* 1973; Knorr *et al.* 1973; Givens *et al.* 1974; Levine *et al.* 1980).

Accumulation of oestrone (E₁) is probably responsible for gynaecomastia (Saez *et al.* 1971). Sexual hair growth occurs and is probably due to accumulation of androstenedione and DHA.

5 α-reductase deficiency

In this form of incomplete male pseudohermaphroditism, deficiency of 5 α-reductase impairs conversion of testosterone to DHT. In males, organs which are normally formed from the urogenital sinus and genital tubercle (which are normally under the influence of DHT) fail to develop (see 'Normal sexual development', p. 438), while those derived from the Wolffian ducts develop normally. The disorder has been called pseudo-vaginal perineoscrotal hypospadias (Opitz *et al.* 1972). The patients have female external genitalia with a vagina and usually a large clitoris, but normal male epidydimis, vas deferens, and ejaculatory ducts. The testes are intra-abdominal. In these patients, pubertal virilization occurs normally under the influence of testosterone and there is no gynaecomastia.

The enzyme defect has been demonstrated in biopsies from foreskin, labia majora, corpora cavernosa, epididymis (Walsh *et al.* 1974), and cultured foreskin fibroblasts (Wilson 1975; Pinsky *et al.* 1981). Enzyme heterogeneity has been demonstrated (Leshin *et al.* 1978; Imperato-McGinley *et al.* 1980).

The disorder has an autosomal recessive inheritance (Opitz *et al.* 1972) and is not expressed phenotypically in females. Prenatal diagnosis by enzyme assay of a foreskin biopsy collected at fetoscopy should be feasible.

Treatment

The need for orchidectomy when testicles are intra-abdominal is discussed under 'Complete testicular feminization' (below). In view of the danger of malignant change, the testicles should be removed. Since these patients are reared as females, orchidectomy is best performed before puberty to avoid virilization. Oestrogen supplementation during the time that normal puberty occurs usually promotes breast development. Vaginal reconstruction (Wabrek *et al.* 1971) may be indicated.

Androgen resistance syndromes

Complete testicular feminization

Individuals with complete testicular feminization are genetically male. The external genitalia are female. The testes are either intra-abdominal, in the inguinal canal or in the labia majora. The commonest presenting symptoms are either inguinal hernia in childhood or primary amenorrhoea. Breast development is normal for females. Secondary sexual hair is scanty or absent. The vagina is short but usually adequate. There are no fallopian tubes or uterus. The testes are prone to malignant changes (Morris 1953). Histologically, spermatogenesis is always absent, in contrast to other abdominal testes in which spermatogenesis occurs in about 50 per cent. Sertolli cell adenomata are frequent. The number of Leydig cells is increased (O'Leary 1965).

It has been estimated that 1–2 per cent of girls with inguinal hernias have testicular feminization (Jagiello and Atwell 1962; German *et al.* 1973; Pergament *et al.* 1973). Bone age is normal, but height tends to be excessive for females (Hauser 1965).

Administration of testosterone fails to stimulate growth of sexual hair, although pubic hair follicles are present (Gwinup *et al.* 1966). Lack of response to androgen treatment has been observed for nitrogen and phosphorus balance (French *et al.* 1966; Volpe *et al.* 1968; Castaneda *et al.* 1971), sebum production (Gwinup *et al.* 1966), and the response of plasma thyroxine-binding globulin (Vagenakis *et al.* 1972).

Androgen production by the testis is normal and plasma concentration of testosterone is usually above the normal male level (Jeffcoate *et al.* 1968; Tremblay *et al.* 1972). The latter probably results from elevated luteinizing hormone (LH) concentration, which may be caused by failure of feedback inhibition because of insensitivity to androgen (Faiman and Winter 1974). Aromatization of the A ring of testosterone accelerates oestradiol production, giving rise to excessive levels for males. This explains normal or generous postpubertal female-type breast development and female habitus and fat distribution.

Molecular defects and diagnosis

Skin fibroblasts from normal subjects bind with DHT because of the presence of a receptor protein (Keenan *et al.* 1974). Skin fibroblasts from patients with testicular feminization do not bind with DHT (Keenan *et al.* 1974; Griffin *et al.* 1976). These data suggest that androgen resistance must be due to defects in the androgen-binding protein. Later it was shown that in some subjects with complete testicular feminization binding of DHT to receptor protein is thermolabile. Thus at 42°C binding is about 20 per cent of that at 37°C. The effect is reversible on cooling (Griffin 1979; Pinsky *et al.* 1981). In one patient with testicular feminization, androgen receptor was missing from a testis (Tamaya *et al.* 1978). Reduced urinary excretion of DHT (Mauvais-Jarvis *et al.* 1970) is probably a secondary effect.

Genetic counselling

Testicular feminization has either an X-linked or autosomal recessive mode of inheritance (Pettersson and Bonnier 1937). X-linkage has been shown by demonstrating Lyonization in fibroblast clones from obligatory heterozygotes (Meyer *et al.* 1975).

Ascertainment artefacts give an apparent 4:1 male preponderance (Taillard and Prader 1957; Lenz 1959). Jagiello and Atwell (1962) have estimated the rate of new mutations to be 0.4–0.5×10^{-5}. The incidence has been estimated as 1:20 000–1:65 000 (Hauser 1965; Jagiello and Atwell 1962; German *et al.* 1973). Prenatal diagnosis by the demonstration of defective binding of cultured amniotic cells to DHT is a theoretical possibility. Further evidence that the foetus is affected would be a 46,XY chromosome complement in cultured amniotic cells and the visualization of female external genitalia.

Treatment

The patients have a female phenotype and disposition and in spite of having an XY sex chromosome complement, they invariably adopt female roles as sterile women with amenorrhoea. The chief hazard is malignancy in the intra-abdominal testes. Approximately one in 64 become malignant which is four times higher than in scrotal testes (MacNab 1955). Malignancy does not occur until after puberty so that orchidectomy can be delayed until immediately after puberty unless there is testicular pain from a labial or inguinal testis. If the testes are removed before puberty, oestrogen replacement should be given at the time of normal puberty to encourage breast development and female habitus. If orchidectomy is carried out after puberty, oestrogen may be necessary to diminish menopausal symptoms.

Animal models

The disorder occurs in mice and has an X-linked mode of inheritance (Lyon and Hawkes 1970). In kidney and submandibular gland there is defective incorporation of androgen into the cell nucleus and absence of DHT-binding protein (Verhoeven and Wilson 1976). Testicular feminization has also been reported in the rat (Bardin *et al.* 1970), cow (Nes 1966), dog (Schultz 1962), horse (Kieffer *et al.* 1976), and chimpanzee (Eil *et al.* 1980).

Familial incomplete male pseudohermaphroditism; Reifenstein syndrome; Rosewater syndrome; Gilbert–Dreyfus syndrome; Lubs syndrome

This disorder has an X-linked mode of inheritance. The basic defect is similar to that in complete testicular feminization, in that cultured scrotal skin fibroblasts exhibit no demonstrable binding to DHT (Griffin *et al.* 1976). The distinguishing feature from complete testicular feminization is the differing degree of feminization which occurs in genetic males in the same family. Affected subjects have a 46,XY karyotype. Thus in one family (Gardo and Papp 1974), three of four affected members were phenotypic females, while the fourth had perineoscrotal hypospadias, bifid scrotum and gynaecomastia. In another family (Wilson *et al.* 1974), eight of 11 affected members had perineoscrotal hypospadias, one had in addition a vaginal orifice and no vas deferens, while two had a small penis and a bifid scrotum. A similar range of feminization was reported by Walker *et al.* (1970). It is clear, therefore, that in this disorder there is a continuous range of phenotype from mild to complete feminization of affected males and that the earlier designations into separate syndromes as follows are no longer justified: Lubs *et al.* (1959), partial labial fusion and partial Wolffian duct development; Gilbert–Dreyfus *et al.* (1957), small phallus, hypospadias, post-pubertal gynaecomastia, and partial Wolffian duct development; Reifenstein (1947) and Bowen *et al.* (1965), perineoscrotal hypospadias, bifid scrotum, and pubertal breast development; Rosewater *et al.* (1965), post-pubertal gynaecomastia and sterility. Male pseudo-hermaphroditism due to androgen resistance is a common cause of infertility. The endocrine changes are similar to those found in complete testicular feminization, namely increased production of testosterone and oestradiol, and increased serum concentration of luteinizing hormone, probably because of hypothalamic insensitivity to androgens.

It should be possible to diagnose this disorder prenatally, as for complete testicular feminization.

Treatment

If the diagnosis is made in the newborn period, gender assignment is usually female. Management is then as described for complete testicular feminization. In addition, surgical reconstruction of the vagina (Wabrek *et al.* 1971) and removal of the phallus may be indicated.

Psychological problems have been reported when sex-orientation is changed during childhood. It is important therefore to establish gender identity as early as possible.

When the diagnosis is made later, the patient will already have established a sex orientation. If this is male, gynaecomastia may cause embarrassment at puberty and should be treated by mastectomy.

Incomplete testicular feminization

A number of patients with clinical features which differ from those of complete testicular feminization or incomplete male pseudohermaphroditism have been named 'incomplete testicular feminization' (Winterborn *et al.* 1970; Rosenfield *et al.* 1971; Madden *et al.* 1975). The patients are genetically male with a 46, XY karyotype. The external genitalia are ambiguous with partial fusion of labioscrotal folds. The Wolffian duct derivatives opening into the vagina (the epididymis, vas deferens, seminal vesicles and ejaculatory ducts) are well formed. The testes are intra-abdominal or in the inguinal canals. The vagina is short.

After puberty, patients develop breasts appropriate for normal women, and female type secondary sexual characteristics, including sexual hair and distribution of body fat. The habitus, however, may be partially virilized. Plasma testosterone and luteinizing hormone concentrations are excessive for normal men, consistent with androgen insensitivity (Madden *et al.* 1975).

In vitro formation of DHT in a skin biopsy from the labia majora (Madden *et al.* 1975) was normal. Binding of DHT to cultured fibroblasts, however, was reduced to about 50 per cent of normal, consistent with a partial deficiency of a DHT-binding protein (Griffin *et al.* 1976).

The pattern of inheritance has not been established. The treatment for the patients and the possibility for prenatal diagnosis are as described for complete testicular feminization.

Testicular feminization with unidentified abnormality

In spite of careful investigation by available techniques it has not been possible to localize the defect in several patients with androgen resistance. Clinically the phenotype is that of complete or incomplete testicular feminization. DHT-binding protein is present, and stable. Nuclear localization of testosterone and 5 α-reductase activity appear to be normal. (Amrhein *et al.* 1976; Collier *et al.* 1978; Griffin and Durrant 1981).

12. Disorders of lipoprotein metabolism

In this chapter we discuss the two types of familial hyperlipoproteinaemias caused by known abnormalities of enzyme activities, and lecithin: cholesterol acyltransferase deficiency. Salient features and key references are tabulated for the other types of familial hyperlipoproteinaemias. The sphingolipidoses are discussed in Chapter 4.

About one-fifth of patients with hypertriglyceridaemia or hyper-cholesterolaemia suffer from one of the six types of monogenic disorders comprising the familial hyperlipidaemias. In the majority the cause is unknown (Goldstein *et al.* 1973*b*).

Function of lipoproteins

Lipoproteins are complexes between cholesterol or triglycerides and proteins. The protein moiety serves to solubilize hydrophobic lipids and in certain cases to recognize specific cell surface receptors. Their function is to transport lipids from their site of synthesis or of intestinal absorption to specific target cells.

Some properties of the four main classes of lipoproteins are shown in Table 12.1

Table 12.1 *Properties of major classes of plasma lipoproteins*

Type of lipoprotein	Densities	Electrophoretic mobility	Size (nm)
Chylomicrons	1.096	Origin	80–500
Very low density lipoproteins (VLDL)	0.94–1.006	pre-ß	30–70
Low density lipoproteins (LDL)	1.019–1.063	ß	18–30
High density lipoproteins (HDL)	1.063–1.21	α	5–12

Cholesterol is a component of all cell plasma membranes and is the precursor of steroids synthesized in the adrenal cortex and gonads. It is synthesized mainly in liver and intestine and transported to other cells, chiefly in the form of low density lipoprotein (LDL) (Table 12.1). About two-thirds of cholesterol in plasma is packaged in LDL particles, but this varies with age and sex (Lipid Research Clinics 1980). The majority of cholesterol molecules in LDL particles are esterified with long chain fatty acids, mainly the diunsaturated linoleic acid. The esterification is probably catalysed by the plasma enzyme lecithin:cholesterol acyltransferase. They migrate in the ß-region. Cholesteryl esters form the core of the LDL particle which has a molecular weight of 3×10^6 and a diameter of about 18–30 nm. The cholesteryl esters are enclosed in layers of phospholipids, unesterified cholesterol and a protein, apoprotein B. About three-quarters of the particle mass is lipid.

In several mammalian species, when dietary cholesterol is restricted, the liver increases the rate of cholesterol synthesis until the liver and intestine synthesize the great majority of total cholesterol. The remainder is synthesized by cells of other body tissues.

When dietary cholesterol is increased, its rate of synthesis in liver is largely suppressed (Brown *et al.* 1981).

Suppression of cholesterol synthesis has been studied in human cultured fibroblasts. These cells each have a maximum of 20 000 to 50 000 surface receptors that bind specifically to the apoprotein B component of LDL (Goldstein *et al.* 1976). The number of receptors is regulated as discussed below. The receptors can be visualized by electron microscopy as indentations in the plasma membrane occurring in clusters. The pits are coated by a non-covalently-bound protein known as clathrin. These 'coated pits' cover about 2 per cent of the cell surface (Anderson *et al.* 1976, 1977a; Orci *et al.* 1978).

The LDL particles bind specifically to these receptors which invaginate and engulf them, forming endocytic vesicles which fuse with lysosomes (Anderson *et al.* 1977a). Inside the lysosomes the protein is hydrolysed to amino acids, while the cholesteryl esters are hydrolysed to cholesterol and fatty acids. Free cholesterol then escapes into the cytoplasm where it is used for membrane synthesis, and as a mediator of a feedback control system which regulates the intracellular cholesterol content.

The inhibitory effect of cholesterol on the cell content of cholesterol is exercised in three ways: firstly, by inhibiting the activity of microsomal 3-hydroxy-3-methylglutaryl-CoA (HMGCoA) reductase, the enzyme which catalyses the rate-limiting reaction in cholesterol synthesis, that is the oxidation of HMGCoA to mevalonate, utilizing NADPH as hydrogen donor (Brown *et al.* 1974); secondly, by activating a cholesterol esterifying enzyme, acyl-CoA: cholesterol acyl transferase, which converts free cholesterol to cholesteryl esters (Goldstein *et al.* 1974b); unlike the plasma cholesteryl esters, which are mainly esterified by the diunsaturated fatty acid linoleic acid, intracellular esters contain mainly monounsaturated oleic acid; and thirdly, by inhibiting the synthesis of LDL receptors, thus reducing the amount of entry of cholesterol-containing LDL into the cells (Brown and Goldstein 1975).

Dietary cholesterol and triglycerides are absorbed into the plasma as chylomicrons (Tytgat *et al.* 1971; Lewis *et al.* 1973) which are large lipoprotein particles (Table 12.1). In adipose tissue and lactating mammary glands about 80 per cent of the triglycerides are hydrolysed by lipoprotein lipase which is localized in the capillary endothelium (Shimada *et al.* 1981). The endothelial cells in mammary tissues have finger-like processes projecting into the capillary lumen which engulf chylomicrons. Free fatty acids are released sequentially and enter the fat cells, where they are re-esterified into triglycerides. Glycerol and some free fatty acids may escape into the plasma. The cholesterol-rich remnant of the chylomicron is taken up by liver cells. Cholesterol is then converted into bile salts, esterified for storage, or released into the circulation as very low density lipoproteins (VLDL) (Table 12.1), which are converted to LDL for transfer to cells of peripheral tissues.

Cholesterol eventually finds its way back from cell membranes to the liver, probably by transfer to high density lipoprotein (HDL) particles.

Lipoprotein analysis and abnormal patterns

Abnormal patterns of plasma lipoproteins are shown in Table 12.2. Assay of plasma total cholesterol and triglyceride concentrations, and inspection for a creamy chylomicron layer after the plasma has stood at 4°C, usually allows characterization into most patterns, but electrophoresis is required for diagnosis of type 3. Plasma types 1 and 5 both have a creamy layer after standing. In type 1, the lower part of the tube is clear; in type 5 (increased VLDL), it is turbid. In both types there is hypercholesterolaemia and hypertriglyceridaemia. Hypercholesterolaemia without hypertriglycerid-aemia or chylomicrons is typical of type 2a (increased LDL). Hyper-cholesterolaemia and hyperglyceridaemia without an increase of chylo-microns (increased LDL and VLDL) indicate 2b. Type 3 pattern is characterized by the presence of ß-migrating VLDL, the plasma usually being turbid. In pattern 4, there is hypertriglyceridaemia without hyper-cholesterolaemia (VLDL increased).

Familial lipoprotein lipase deficiency

Clinical features

The majority of patients with this rare disorder have recurrent attacks of abdominal pain due to pancreatitis, often precipitated by excessive fat intake. Pain can be severe and in undiagnosed patients often leads to exploratory laparotomy. Severe pancreatitis may be fatal. The onset of abdominal pain has varied from the first to the fifth decade.

Table 12.2 *Lipoprotein patterns[1]*

Pattern types	Chylomicrons	VLDL	LDL	HDL	Appearance of plasma	Plasma concentration Cholesterol	Triglycerides	Genetic disorder in which pattern is found
1	++	N	L	L	Creamy layer on top, clear below	+	+	Familial lipoprotein lipase deficiency
2a	N	N	+	N	Clear	+	N	Familial hypercholesterolaemia; familial combined hyperlipidaemia
2b	N	+	+	N	Clear	+	+	Familial combined hyperlipidaemia
3	N	*	**	N	Turbid	+	+	FHLP type 3
4	0	+	N	N	Usually turbid	N	+	FHLP types 3, 4 or 5; familial hypertriglyceridaemia; familial combined hyperlipidaemia
5	+	+	N	N	Creamy layer on top, turbid below	+	+	FHLP type 5

*	= ß-VLDL demonstrated by electrophoresis
**	= LDL abnormal
N	= normal
+	= increased
++	= markedly increased
O	= undetectable
L	= lower than normal
FHLP	= familial hyperlipoproteinaemia
1	= classification modified from the World Health Organisation (Beaumont *et al.* 1970)

There is frequently mild or moderate hepatosplenomegaly after excessive fat intake, but this diminishes on a low fat diet.

A characteristic feature present in about half the patients is the appearance of eruptive xanthomata which appear fairly rapidly and tend to disappear in a few weeks, sometimes leaving slight pigmentation. They are small, non-tender, yellow nodules distributed in flexures and pressure areas.

The presence of excessive chylomicrons causes pallor of the retina and sometimes an exaggerated light reflex.

There is no convincing evidence of premature atherosclerosis or other vascular disease.

Pathology

Histologically there is macrophage and chylomicron-laden foam cell infiltration of the skin, bone marrow, spleen, and liver.

Molecular defects and diagnosis

The disease is due to accumulation of chylomicron triglycerides (type 1 hyperlipoproteinaemia, Table 12.2) because of deficiency of plasma lipoprotein lipase. This enzyme was once called the 'clearing factor' because of its activation by injected heparin to reduce the slight turbidity of hypertriglyceridaemic plasma.

As discussed in 'Function of lipoproteins' (p. 446), the enzyme catalyses the hydrolysis of the ester bonds of glycerides. However, its activity is specific for the primary ester bonds, so that for complete hydrolysis of triglycerides the 2-monoglyceride must undergo isomerization to the 1- or 3-monoglyceride before hydrolysis can occur.

The plasma triglyceride concentration is considerably increased, to 2500–12 000 mg/dl (normal: less than 200 mg/dl) when patients are on average diets, but falls to normal or near normal levels with fat-free diets.

After standing for some hours at 4°C, the plasma is clear below and has a creamy layer of chylomicrons on top (Table 12.2).

When plasma triglyceride concentration becomes elevated, there is also an increase in the plasma chylomicron cholesterol concentration.

The diagnosis is strongly suggested by a decrease of raised plasma triglyceride and cholesterol concentrations when patients receive a very low fat diet. Further evidence is lack of the normal enhancement of electrophoretic mobility of α- and pre-ß-lipoproteins 10 minutes after heparin injection, produced by enrichment of fatty acids released from chylomicrons by activated lipoprotein lipase. Following heparin injection several hydrolases besides lipoprotein lipase are released into the plasma. Lipoprotein lipase is thought not to be synthesized in the liver. A hepatic lipase, also released by heparin, must be distinguished from lipoprotein lipase to establish the diagnosis. The two lipases can be assayed individually by protamine inhibition (Krauss *et al.* 1974) or by interaction with specific antisera (Huttunen *et al.* 1975). More recently, direct assay of adipose tissue lipoprotein lipase activity has been used for diagnosis. The sample may be collected by needle aspiration. This method has the advantage that adipose tissue, unlike plasma, does not contain hepatic lipase (Taskinen *et al.* 1980).

Genetic counselling

The disorder has an autosomal recessive mode of inheritance. Some parents have excessive and prolonged hypertriglyceridaemia following fatty meals, but usually normal fasting triglyceride concentration. Adipose tissue lipoprotein lipase activity is approximately half that of normal controls (Harlan *et al.* 1967). It has not yet been established whether foetal blood sampling would allow prenatal diagnosis.

Treatment

Treatment is by restriction of dietary lipids in order to reduce the hyperchylomicronaemia. Since medium chain triglycerides are absorbed into the portal circulation, and do not form chylomicrons, they may be included in the diet. Alcohol should be excluded since it enhances hyperlipidaemia (Little *et al.* 1970).

Familial hypercholesterolaemia

Familial hypercholesterolaemia has an autosomal dominant mode of inheritance; predictably, the severity is considerably greater in homozygotes than in heterozygotes.

Clinical features

Heterozygotes: Heterozygotes have hypercholesterolaemia from birth and an increasing proportion develop xanthomas up to the fourth decade. Thus in the first to fifth decades xanthomas were present in 2.6, 12.5, 69.2, 90.0 and 70.3 per cent of patients, respectively (Kwiterovich *et al.* 1974).

Tendinous xanthomas are nodules which occur often in the Achilles tendons and in the tendons of the extensor digitorum longus. Nodular xanthomas form commonly on bony prominences such as the olecranon. Subperiosteal xanthomas occur in the anterior superior tibia or on the olecranons. Palpebral xanthomas (xanthelasma) are common. Cutaneous planar xanthomas are flattened orange papules and commonly form over elbows, knees, and buttocks. Histologically, they contain numerous extra-lysosomal cholesterol-laden 'foam cells' (Bulkley *et al.* 1975).

Ischaemic heart disease occurs prematurely and is about 25 times more common than in unaffected relatives (Jensen *et al.* 1967). Slack (1969) found the onset of the clinical complications of myocardial infarction and angina pectoris at a mean age of 43 for men and 53 for women. Stone *et al.* (1974) found that coronary heart disease in males was present in 52 per cent by the age of 60 compared to 12.7 per cent in controls. The equivalent figures for females were 32.8 per cent in heterozygotes and 9.1 per cent in controls.

Recurrent attacks of arthritis and tendinitis have also been reported (Glueck *et al*. 1968).

Arcus corneae occurs in about 10 per cent before 30 years of age, and 50 per cent subsequently (Fredrickson and Levy 1972).

Homozygotes: The clinical features of homozygotes resemble those of heterozygotes but are more severe and have an earlier onset. Cardiac ischaemia, causing myocardial infarction and angina pectoris, occurs in children and adolescents, causing death in the great majority before the age of 30 (Khachadurian and Uthman 1973).

Atherosclerosis occurs in the main arteries, in the aortic (Goldstein 1972) and mitral valves, and in the endocardium, causing aortic stenosis and regurgitation, and mitral stenosis (Maher *et al*. 1958; Schettler 1969).

Arthritis and tendinitis are common complications (Khachadurian 1968). The erythrocyte sedimentation rate is persistently elevated (Khachadurian 1967).

Arcus corneae may occur before the age of 10 years (Fredrickson and Levy 1972).

Molecular defects

The lipoprotein pattern is of the 2a type (Table 12.2). As shown by ^{125}I-labelled LDL turnover studies, homozygotes have a two- to four-fold overproduction of LDL and approximately a 60 per cent reduction in the LDL fractional catabolic rate as compared with normal controls (Simons *et al*. 1975; Bilheimer *et al*. 1975, 1979). Heterozygotes have normal rates of LDL catabolism, but calculated fractional catabolic rates for LDL of about half those in normal subjects (Langer *et al*. 1972).

Mean total plasma cholesterol concentration is about four-fold higher than controls for homozygotes and about two fold higher than controls for heterozygotes (Khachadurian 1971; Khachadurian and Uthman 1973).

Patients who are homozygous for the hypercholesterolaemia gene have abnormalities of LDL receptors (see 'Function of lipoproteins' above). Three types of mutation have been reported (Goldstein and Brown 1979); the first where there is no detectable binding of LDL to cell surface receptors, the responsible allele being designated R^{bo}; the second where such binding is considerably reduced (1–10 per cent of normal) (Goldstein *et al*. 1975) with an R^{b-} allele; and the third allele ($R^{b+,io}$) specifies LDL receptors which have normal binding but lack the ability to internalize bound LDL (Goldstein *et al*. 1977).

Of 50 patients with hypercholesterolaemia, 29 were judged to be homozygotes for the R^{bo} allele, 20 who had defective receptors to be either homozygotes for the R^{b-} allele or to be R^{b-}/R^{bo} double heterozygotes; the remaining patient was considered to be an $R^{bo}/R^{b+,io}$ double heterozygote (Goldstein and Brown 1979).

In patients who are homozygotes for the R^{bo} allele, LDL cannot be engulfed and there is therefore no suppression of HMGCoA reductase activity or stimulation of cholesterol ester synthesis, with consequent cholesterol overproduction and accumulation.

In cell culture, when cholesterol is introduced into R^{bo}/R^{bo} cells by adding cholesterol dissolved in ethanol to the culture medium, HMGCoA reductase activity and cholesterol synthesis are suppressed while cholesteryl ester synthesis is stimulated (Brown and Goldstein 1976a).

Patients who are homozygotes for the R^{b-} allele have defective LDL receptor binding which varies from 2 to 25 per cent of normal. Clinically they are indistinguishable from homozygotes for the R^{bo} allele.

Fibroblasts from patients with normal LDL binding but with undetectable LDL internalization (Brown and Goldstein 1976b; Goldstein and Brown 1979) showed similar abnormalities of HMGCoA reductase, cholesterol synthesis, and cholesteryl ester synthesis to cells from R^{bo}/R^{bo} patients.

The internalization defect appears to be specific for LDL since sucrose, insulin, gamma globulin, and epidermal growth factor are internalized normally. Electronmicrographs of the fibroblasts with the internalization defect reveal that LDL receptors are not clustered in coated pits (see 'Function of lipoproteins' p. 446) but are scattered in a random manner over the cell surface (Anderson *et al.* 1977b).

Genetic counselling

The disorder has an autosomal dominant mode of inheritance. Fibroblasts from heterozygotes of the R^{bo} allele with a genotype $+/R^{bo}$ (where + represents the normal allele) express about half the normal number of LDL receptors. Both in the absence of exogenous cholesterol and in the presence of varying amounts of cholesterol added to the culture medium, heterozygous cells produce about one-half the number of functional LDL receptors as normal cells (Goldstein *et al.* 1976).

As described under 'Molecular defects' above, heterozygotes have elevated total plasma cholesterol concentrations. There is almost no overlap with values of homozygote normal or affected individuals (Khachadurian 1971; Khachadurian and Uthman 1973). Genetic heterogeneity and the existence of double heterozygotes for the LDL-receptor alleles have been suggested (Goldstein and Brown 1979).

Prenatal diagnosis of a homozygous receptor negative affected foetus was first reported by Brown *et al.* (1978). Amniocentesis was at 16 weeks' gestation. Cultured amniotic cells had less than five per cent LDL receptor activity of control cells as demonstrated by binding, uptake, and degradation of ^{125}I-labelled LDL, LDL-mediated suppression of HMGCoA reductase, and stimulation of cholesteryl ester synthesis. The pregnancy was terminated at 20 weeks' gestation. Foetal blood cholesterol level was 280 mg/100 ml, and in four control foetuses 18–47 mg/100 ml.

The frequency of heterozygotes in London has been calculated to be 1:200 from the homozygote frequency of $1:10^6$ (Carter *et al.* 1971). Goldstein *et al.* (1973*b*) have estimated the frequency of heterozygotes to be 1:500. A much higher incidence of homozygotes and heterozygotes has been reported in Afrikaaners in South Africa (1:30 000 and 1:100, respectively; Seftel *et al.* 1980).

Treatment

The only effective methods for reducing plasma cholesterol levels are dietary restriction and cholestyramine administration. These may be effective in reducing the plasma cholesterol concentration in heterozygotes but rarely succeed in homozygotes. Combined dietary restriction of cholesterol and administration of cholesterol binding resins such as cholestyramine or colestipol can reduce plasma cholesterol concentration in heterozytoes to the upper limit of normal (Hashim and Van Itallie 1965; Kuo *et al.* 1979; Levy *et al.* 1980*b*).

Other forms of familial hyperlipoproteinaemia

In four forms of familial hyperlipoproteinaemia which have been well characterized on clinical grounds and on the abnormal patterns of plasma lipoproteins, no enzyme defect has been discovered. Their main features are summarized in Table 12.3.

Familial lecithin:cholesterol acyltransferase deficiency

Clinical features

This rare disorder is characterized by corneal opacities and in some patients by anaemia, proteinuria, premature atheroma, and progressive renal failure, with hypertension.

The corneal opacities have been noted from early childhood and consist of small grey dots which are more numerous in the periphery, where they become confluent, forming an opaque ring or corneal arcus (Gjone and Berghaust 1969).

A normochromic mild haemolytic anaemia has been present in the majority of patients (Gjone 1974; Frohlich *et al.* 1978; Iwamoto *et al.* 1978) but not all (Bron *et al.* 1975). There is an increased number of target cells. Foam cells occur in the bone marrow. 'Sea-blue' histiocytes (i.e. histiocytes

Table 12.3 *Summary of main manifestations of the familial hyperlipoproteinaemias without known enzyme defects*

Disorder	Main clinical features	Plasma lipo-protein types	References
Type III (Dysbetalipoprotein-aemia)	Premature cardiac ischaemia and peri-pheral vascular occlu-sion; xanthoma; obesity; hyperglycaemia	3	Fredrickson *et al.* 1978; Uttermann *et al.* 1977; Mishkel *et al.* 1975; Zannis *et al.* 1981
Familial hypertriglycerid-aemia	Premature cardiac ischaemia; obesity hyperglycaemia	4	Fredrickson *et al.* 1978
Type V	Eruptive xanthoma; pancreatitis; obesity; hyperglycaemia; hyper-uricaemia; polyneuro-pathy	4,3 or 2b	Fallat and Glueck, 1976; Greenberg *et al.* 1977
Familial combined hyperlipidaemia	Premature cardiac ischaemia; obesity hyperglycaemia	2a, 2b or 4	Nikkila and Aro, 1973; Rose *et al.* 1973

which acquire a 'sea-blue' colour after staining with Giemsa), have been found in bone marrow and spleen (Jacobsen *et al.* 1972).

The majority (Gjone 1974) of patients with some exceptions (Bron *et al.* 1975), have proteinuria and later may develop hypertension and renal failure (Gjone 1974) which may be fatal.

Pathology

On electronmicrography, the granules of the 'sea-blue' histiocytes appear as lamellar membranes (Jacobsen *et al.* 1972). The main arteries including the renal arteries show premature atherosclerosis (Gjone 1974). Accumul-ation of lipids and foam cells occurs in vessel walls. The renal vessels are fibrotic and hyalinized.

Histological and ultrastructural studies of renal biopsies (Hovig and Gjone 1974) have revealed deposits of membranes and amorphous material in renal capillaries, which have irregular basal lamina. There is loss of endothelial cells in the glomerular tufts, and there are foam cells in the mesangium and membranous deposits in the subendothelial tissue of the glomeruli and in the interna of the renal artery and arterioles.

The liver and spleen contain lipid-laden phagocytic cells with membranous inclusions.

Molecular defects

The disorder is due to deficiency of the plasma enzyme lecithin:cholesterol acyltransferase (LCAT). This enzyme catalyses the transfer of an acyl group from lecithin (phosphatidylcholine) to cholesterol to yield cholesteryl esters and lysolecithin. Since the enzyme is activated by the major HDL apolipoprotein, A1, the reaction probably occurs mainly in HDL, with cholesteryl esters subsequently being transferred to other lipoproteins (Glomset and Norum 1973). The reaction can be assayed by following the formation of radioactive cholesteryl esters when plasma is incubated with radioactive cholesterol (Stokke and Norum 1971).

The possibility that these patients suffered from LCAT deficiency was suggested by the finding in the plasma of a high concentration of the enzyme substrates and a low concentration of the enzyme products. LCAT deficiency was then demonstrated by enzyme assay (Norum and Gjone 1967; Gjone and Norum 1968). The reduced amounts of cholesteryl esters present in the patients' plasma are probably absorbed from the intestine.

Plasma lipids have several characteristic abnormalities (Glomset *et al.* 1970, 1973, 1980*a*). As expected, all lipoprotein types, and especially the HDLP, have a low ratio of esterified to non-esterified cholesterol.

The LDL are more heterogenous than in normal individuals and contain large lamellar particles enriched with esterified cholesterol, phosphatidyl choline, vesicular particles and normal-sized particles very rich in triglycerides. However, the core triglycerides and esterified cholesteryl esters content are normal. In addition to normal-sized LDL spherical particles of 20–24 nm diameter, electronmicroscopy reveals two types of particles; very large flattened particles, approximately 100 nm in diameter, and intermediate-sized particles (40–60 nm). The HDL concentration is about one-third of normal. The HDL contain both normal-looking particles and disc-shaped particles (Forte *et al.* 1971; Torsvik 1972; Glomset *et al.* 1970). It is probable that the abnormal structure, distribution and composition of the HDL are a direct result of LCAT deficiency. Incubation of HDL from affected patients with normal LCAT converts the disc-shaped HDL into spherical HDL particles of normal appearance (Glomset and Norum 1973), designated HDL_2 and HDL_3 (Glomset *et al.* 1980*b*). These changes probably result from rearrangements due to accumulation of cholesteryl esters.

The VLDL migrate as ß-proteins on electrophoresis and their concentration is often increased. In spite of hyperlipidaemia, there is no pre-ß-lipoprotein, and only a very faint α_1-lipoprotein band (Norum and Gjone 1967; Glomset *et al.* 1973).

The majority of patients have triglyceridaemia; all have raised concentrations of unesterified cholesterol.

The relations of the pathological findings to deficiency of LCAT are not all understood. The reversal of some of these changes on incubation with normal LCAT demonstrates a causal relation. It is probable that accumulation of unesterified cholesterol and phosphatidylcholine in plasma membranes which exchange with plasma proteins also are the direct effect of LCAT deficiency.

Accumulation of the intracellular and extracellular membranous deposits may be secondary to increased plasma lipid content. Haemolytic anaemia may also be due to the low ratio of esterified to non-esterified cholesterol. Foam cells and sea-blue histiocytes, atherosclerosis and renal pathology may also be due to lipid accumulation.

Plasma of some patients contains either no immunochemically cross-reacting material to LCAT, or very little (Albers and Utermann 1981; Albers *et al*. 1981).

Diagnosis

The diagnosis is established by demonstrating deficiency of plasma LCAT activity (Norum and Gjone 1967) and the abnormal pattern of plasma lipoproteins described under 'Molecular defects' above.

Genetic counselling

Pedigrees are consistent with autosomal recessive inheritance (Teisberg and Gjone 1974; Teisberg *et al*. 1975). There is a high incidence in certain areas in Norway where the heterozygote frequency has been estimated as 4 per cent of the population.

Linkage studies indicate close association with the haptoglobin genes, with a lod score of 3.41 at a recombination fraction of 0.00 (Teisberg *et al*. 1975), localizing the gene for LCAT between the terminal point and middle of the long arm of chromosome 16 (Teisberg *et al*. 1975).

Heterozygotes have normal plasma LCAT activities and lipoprotein distribution and structure.

It is not known whether foetal blood sampling would permit prenatal diagnosis.

Treatment

Plasma transfusion has produced normalization of the plasma lipoproteins for about two weeks (Norum and Gjone 1968).

Kidney transplantation offers the best prognosis for patients with renal failure.

13. Disorders of the thyroid hormones; familial goitre

Disorders of the thyroid gland such as Graves' disease (thyrotoxicosis), nodular goitre or hypothyroidism due to autoimmune disease, are common in clinical practice. In this chapter we discuss some rare but well-defined types of familial hypothyroidism, often associated with goitre and, if untreated, with the well-known features of hypothyroidism including mental handicap and retardation of skeletal growth. In disorders where no enzyme defect has been demonstrated, we outline the salient features and cite key references, but we consider in greater detail disorders where a specific enzyme defect has been demonstrated directly or is considered probable.

Biosynthesis of thyroid hormones

Iodide (iodine ion, I^-) is absorbed from the intestine and circulates in the plasma, and is either excreted in urine or taken up (mainly by the thyroid). Small quantities are also cleared by the salivary and gastric glands. In the thyroid the quantity of inorganic iodide is about one per cent of total thyroid iodine. Its concentration however increases considerably under the influence of thyrotropic hormone (TSH).

Shortly after its arrival in the thyroid, iodide is oxidised and becomes linked to the 3 and 5 positions of the tyrosyl residues of the glycoprotein thyroglobulin. Other proteins, especially albumin, are iodinated but to a much lesser extent. The reaction is catalysed by an iodide-specific peroxidase in the presence of a peroxide-generating system which is probably rate-limiting, and is referred to as organification of iodine. Thyroglobulin is synthesized in the endoplasmic reticulum in the microvilli which project into the colloid (Benabdeljlil et al. 1967). Although the thyroglobulin molecule has a molecular weight of about 670 000 and contains about 122 tyrosyl residues, it contains only about 26 iodine atoms.

Organification results in the synthesis of monoiodotyrosine (MIT) or diiodotyrosine (DIT) in covalent linkage with thyroglobulin. The next stage in synthesis of the thyroid hormones is the coupling between molecules of iodotyrosines. The molecular mechanisms are not completely understood. The reaction is catalysed by thyroid peroxidase.

Regulation, transport, and action of thyroid hormones

Regulation of the concentrations of the thyroid hormones in the blood is mediated through the hypothalamic–pituitary–thyroid axis mainly by a negative feedback system. Thus, a fall in the concentrations of triiodo-thyronine (T_3) or tetraiodothyronine (T_4; thyroxine) in the blood causes the thyrotrophin releasing hormone (TRH) to release the glycoprotein TSH from the anterior pituitary. TRH is synthesized in the hypothalamus and passes down the hypothalamic portal system (Demeester-Mirkine and Dumont 1980). It is a tripeptide, pyroglutaminylhistidylprolineamide. TSH can be released *in vitro* by cyclic AMP and epinephrine. It is probable that stimulation of TRH is by cyclic AMP which is synthesized from ATP by a reaction catalysed by adenyl cyclase.

Interaction of TSH and the thyroid begins with binding of TSH to a receptor of the plasma membrane which stimulates synthesis of cyclic AMP within seconds, by activation of adenyl cyclase. Numerous biochemical reactions in the thyroid become stimulated by TSH and are probably mediated by cyclic AMP. These include oxidation of glucose, organification and coupling during hormone synthesis (see above), proteolysis and release of hormones, incorporation of amino acids, and activation of numerous thyroid enzymes (Stanbury 1978; De Visscher 1980).

The metabolically active hormones, T_3 and T_4, when released into the blood, are reversibly bound to carrier proteins, especially thyroxine-binding globulin (TBG). T_3 only also binds to a carrier prealbumin. TBG is a glycoprotein. One mole of TBG binds one mole of T_4.

The unbound hormones only are metabolically active, while the bound hormones are inactive until released. Only about 0.05 per cent of circulating T_4 and 0.5 per cent of T_3 are unbound.

In target cells, T_4 but especially T_3 have numerous binding sites. T_3 and, to a lesser extent, T_4 binds to nuclear protein. Nuclear-bound T_3 and possibly also T_4 stimulates the activity of DNA-dependent RNA poly-merase and therefore synthesis of messenger RNA and proteins. T_4 may be considered as a prohormone or simply a precursor of the active hormone T_3.

The effects of T_3 and T_4 on peripheral target cells are still controversial (Stanbury 1978). As described earlier, they stimulate the activity of DNA-dependent RNA polymerase, including the mitochondrial enzyme, and accelerate protein synthesis. This may be reflected clinically by the growth-promoting properties of thyroxine in hypothyroid children.

A popular explanation of the mechanism of T_3 action is that it uncouples oxidative phosphorylation; that is, it inhibits the mitochondrial electron transport-mediated synthesis of ATP from ADP and inorganic phosphate while allowing electron transport to continue.

Although uncoupling by T_3 can be demonstrated *in vitro*, there are objections to the view that it occurs *in vivo*. These include the failure of other substances which uncouple to cure hypothyroidism, and the observation that the concentration of T_4 required to uncouple *in vitro* is higher than that existing *in vivo*.

T_3 alters the activity of numerous cellular and plasma enzymes. A comprehensive review is outside the scope of this book. Examples include inhibition of NAD-linked dehydrogenases and of plasma cholinesterase, and stimulation of reductase activity catalysing reduction in the liver of the A steroid ring, cytochrome oxidase, muscle hexokinase, and carbonic anhydrase.

Iodide transport defect; genetic defect in thyroid hormogenesis (GDTH) I

This is a rare cause of familial hypothyroidism and goitre. Clinically it has presented in children as early as one month of age (Gilboa *et al.* 1963). The diagnostic criterion is demonstration of considerably reduced capacity of the thyroid to take up radioiodine from the plasma. This defect has also been shown in the salivary glands and gastric mucosa. In theory, there may be a defect in the iodine transport system, or of a carrier, or of receptors (Stanbury and Chapman 1960; Wolff *et al.* 1964). A partial defect was reported by Papadopoulos *et al.* (1970) and Medeiros-Neto *et al.* (1972).

Treatment by potassium iodide or Lugol's solution has reversed hypothyroidism and goitre within a few weeks.

Since parents of one case were consanguineous (Stanbury and Chapman 1960) and in another family two brothers were affected, inheritance is probably autosomal recessive.

GDTH IIA; organification defect I; peroxidase defect

The clinical features have been variable. Some patients (Hagen *et al.* 1971; Niepomniszcze *et al.* 1973; Pommier *et al.* 1976) presented with goitre, but had normal growth and mental development. Other patients are mentally and physically retarded (Niepomniszcze *et al.* 1973). The majority have large goitres and hypothyroidism (Valenta *et al.* 1973).

These patients all show a rapid release of labelled iodine after administration of potassium thiocyanate or chlorate. Normally very little iodine is discharged in this way. The probable explanation is that these patients have an impaired ability to iodinate tyrosyl residues of thyroglobulin, that is, they have a defect in the organification of iodine. However, several different defects have been found among patients with precipitous discharge of accumulated iodide from the thyroid after thiocyanate administration. Thus direct deficiency of thyroid peroxidase activity was demonstrated in thyroid tissue and enzyme activity was restored after addition of haematin, the prosthetic group of peroxidase, suggesting that the defect was deficient binding of haematin by the peroxidase apoenzyme (Hagen *et al.* 1971; Niepomniszcze *et al.* 1973). In other patients, demonstration of peroxidase deficiency was possible, but activity was restored by solubilization (Pommier *et al.* 1976). In some patients, peroxidase activity appears to be normal (Niepomniszcze *et al.* 1973).

Both males and females of normal parents are affected, consistent with autosomal recessive inheritance.

GDTH IIB; organification defect II; Pendred's syndrome

These patients have a milder type of organification defect, and clinically they are characterized by congenital deafness in addition to congenital goitre (Brain 1927; Nilsson *et al.* 1964). Hypothyroidism, goitre, and enhancement of discharge of radioiodine by thiocyanate or chlorate is less marked than in GDTH IIA. In fact most affected individuals are euthyroid. Normal thyroid peroxidase activity has been found by Ljunggren *et al.* (1973), Burrow *et al.* (1973) and Cave and Dunn (1975). A defect in production of peroxide was suggested by Ljunggren *et al.* (1973). Fraser *et al.* (1960) studied 113 patients in 72 families.

Inheritance is autosomal recessive (Trotter 1960).

GDTH III; defect in coupling of iodotyrosines

In this disorder there is goitre and hypothyroidism, normal synthesis of MIT and DIT, but defective coupling of iodotyrosyl residues (Morris 1964). The defect can be shown in a thyroid biopsy. Pommier *et al.* (1974) showed that peroxidase activity ($2I^- \rightarrow I_2$) was normal, iodination of thyroglobulin only slightly reduced, but that coupling activity was considerably reduced. They suggested a failure of conversion of peroxidase A to B. Thyroid MIT and DIT are increased (Stanbury *et al.* 1955).

The diagnostic test is the demonstration that T_3 and T_4 do not appear in thyroid biopsies.

GDTH IV; defect in iodotyrosine deiodinase (dehalogenase) activity

Goitrous cretinism was reported in 12 itinerant nomadic tinkers from Scotland, four of whom were sibs (McGirr and Hutchinson 1953; McGirr *et al.* 1959; Hutchinson and McGirr 1954, 1956).

In 10 of the patients there was a rapid uptake of ^{131}I by the thyroid.

Two brothers and an unrelated boy were reported from Holland

(Stanbury *et al*. 1955, 1956*a,b*). In these patients administered ^{131}I was shown to appear as MIT and DIT in the blood and urine. In contrast, control individuals excreted labelled iodine as free iodide. The thyroids of one of the brothers and of the unrelated patient were removed. Thyroid slices from both patients failed to deiodinate DIT (Querido *et al*. 1956; Choufoer *et al*. 1960; Choufoer 1961).

A partial defect has been reported (Kusakabe and Miyaki 1963, 1964).

The disorder seems to be due to iodine deficiency due to loss of MIT and DIT in the urine, because of deiodinase deficiency.

The disorder has an autosomal recessive mode of inheritance. Carriers have a defective ability of deiodinating DIT (Rochiccioli and Dutau 1974; Codaccioni *et al*. 1978).

Improvement has been reported after administration of either thyroid (Choufoer 1961), or iodide (Codaccioni *et al*. 1970).

GDTH V; plasma iodoprotein defect

A 12-year-old boy with congenital goitre and hypothyroidism whose parents were second cousins was found to have iodinated albumin-like protein in his thyroid instead of thyroglobulin (Lissitzky *et al*. 1968). His sister was euthyroid but had a goitre. Both sibs had a low blood protein-bound iodine.

This entity may be non-specific, since iodinated albumin-like proteins have been identified in the plasma of patients with heterogenous disorders of the thyroid including Graves' disease (Stanbury and Janssen 1963) and Hashimoto's disease (Owen and McConahey 1956).

Defects of thyroglobulin synthesis

Absent or diminished synthesis of thyroglobulin was reported by Riddick *et al*. (1969) in three sibs with goitre and hypothyroidism from a sibship of four. Uptake of iodine by the thyroid was normal or high. Thyroid tissue showed absence of thyroglobulin. Other patients with absent or diminished synthesis of thyroglobulin have been reported by Pittman and Pittman (1966) and Lissitzky *et al*. (1975).

Animal models

Defective incorporation of amino acids into thyroglobulin has been reported in goitrous merino sheep from New Zealand and Australia (Rac *et al*. 1968; Falconer *et al*. 1970). Afrikander cattle with congenital goitre have thyroglobulin with abnormal amino acid (Theron and Van Jaarseveld 1975) and sugar (Pammenter *et al*. 1978) composition.

Dutch hypothyroid goats have only trace amounts of thyroglobulin (Van

Voorthuizen *et al.* 1978) and considerable deficiency of mRNA for thyroglobulin (Dinsart *et al.* 1977).

Unresponsiveness to thyroid hormone

Two of six children of a consanguineous marriage reported by Refetoff *et al.* (1967, 1980) had goitre, deaf-mutism, stippled epiphyses and abnormally high blood protein-bound iodine concentrations.

The authors suggested that the disorder was due to target-cell unresponsiveness to thyroid hormones.

A 25-year-old woman reported by Lamberg (1973) had congenital goitre. The concentrations of blood thyroid hormones and TSH were about twice normal. Her response to TRH was normal. The features are consistent with unresponsiveness to thyroid hormones. Since the TSH level was increased in the presence of high blood thyroid hormone concentration, presumably the hypothalamic–pituitary system was also unresponsive.

Screening of newborn for hypothyroidism

Primary congenital hypothyroidism can be detected a few days after birth by determination of thyroid-stimulating hormone (TSH) and serum thyroxine (T_4) concentration. Follow-up studies have demonstrated that treatment of hypothyroidism before the appearance of clinical symptoms and signs protects against the mental retardation which occurs in children who are treated after a clinical diagnosis has been made (New England Congenital Hypothyroidism Collaborative 1981).

The incidence of hypothyroidism has been estimated in North America to vary between 1 in 3802 in New England to 1 in 6417 in Pittsburgh by screening of one million infants (Fisher *et al.* 1979). A higher incidence of 1 in 2222 has been found among infants with a Spanish surname in California (Frasier *et al.* 1982), while a lower incidence has been reported among black infants in Georgia (Brown *et al.* 1981).

Delayed or missed diagnosis was reported in 14 of 159 infants with hypothyroidism encountered during the first five years of the New England Regional Screening Program and the New England Congenital Hypothyroidism Collaborative (1982). Of these 14 infants, three were never tested, six were missed because of errors during processing of the blood specimens and five because the concentrations of TSH at birth were not elevated. The diagnosis of hypothyroidism is established when the confirmed serum TSH concentration is above 40 μu/ml. However, it is recommended that further assays should be made for infants with values of 20–40 μu/ml. T_4 concentrations of 6 μg/dl or less are also indicative of hypothyroidism.

Treatment

The New England Congenital Hypothyroid Collaborative (1981) suggested that L-thyroxine be administered in doses sufficient to maintain T_4 concentrations between 10 and 14 μg/dl during the first year and between 8 and 11 μg/dl thereafter. The mean dose of L-thyroxine up to 3 months was 8 $\pm$ 2.4 (SD) μg/kg/day, and for 3–12 months, 6 $\pm$ 1.3 μg/kg/day.

14. Disorders of copper metabolism

Two disorders of copper metabolism in humans are known, although in neither case has the basic biochemical defect been elucidated. Wilson's disease is inherited as an autosomal recessive condition and presents at age varying from four to 40 years. An effective treatment is available. Menkes' disease is an X-linked recessive condition which presents shortly after birth and is invariably fatal.

Wilson's disease; hepatolenticular degeneration

Clinical features

The first report of Wilson's disease was probably by von Frerichs (1861) some 50 years before the paper which led to the eponym for the disease (Wilson 1912). The patient described by von Frerichs was a young boy who suffered from severe liver disease associated with violent tremors and convulsions. He died at age 10 years, when necropsy revealed cirrhosis of the liver. Wilson's original patient suffered from a long illness characterized by muscular rigidity, tremor, and forced grimacing. Necropsy revealed cirrhosis of the liver and gross degeneration in the lenticular nucleus of the brain.

The age of onset of the disease is highly variable. Most patients present in adolescence or early childhood, but onset can be as early as four years or as late as the fifth decade. The single most diagnostic clinical feature of the disease is now regarded to be the presence of Kayser–Fleischer rings in the cornea, although these were not described in Wilson's original article. The rings appear at the margin of the cornea near the limbus and are always bilateral. They are golden brown or greenish and, although often visible to the naked eye in advanced stages of the disease, they are initially visualized only by slit lamp examination. Kayser–Fleischer rings are rarely seen in any circumstances other than Wilson's disease (Fleming *et al.* 1977; Frommer *et al.* 1977), but they may be absent in children who have only hepatic involvement (Fleischer 1912; Sternlieb 1966). The rings are only rarely seen before seven years of age (Arima and Kurumada 1962).

Although the triad of Kayser–Fleischer rings of the cornea, cirrhosis of the liver, and bizarre neurological manifestations is regarded as diagnostic of Wilson's disease, the variability of the disease is so great that not all patients develop this classic picture. Furthermore, many patients show involvement with other systems, including kidneys, blood, and bones.

The neurological features are of two forms. The most severe, and probably least common, is lenticular degeneration or the dystonic form, which occurs predominantly in young adults. It is characterized by spasticity, rigidity, drooling, dysarthria, and dysphagia. The other form, which has been called 'pseudosclerosis', can occur in patients of any age and manifests itself as flapping tremors at the wrists and shoulders. It can progress to the more severe form. In addition, psychiatric disturbances are quite common, and may be the presenting feature. They can consist of a subtle change in personality, falling off in school performance, difficulty in writing, hysteria, development of a speech defect, or schizophrenia.

The first sign of hepatic involvement is an enlarged firm liver with splenomegaly. The liver failure that ensues, if the underlying cause is not recognized and treated, is not at all distinctive, and can progress as a subacute, chronic, or acute hepatitis, or as cirrhosis. Onset of neurological symptoms may occur at any stage of the liver failure, although the patient may die before there are any neurological manifestations; Strickland *et al.* (1973) have observed that the earlier liver disease occurs, the more likely is the presentation to consist of liver disease alone.

Laboratory evidence of renal involvement in Wilson's disease is often evident, although clinically renal disease is rarely observed. The abnormalities that occur include a progressive failure of tubular reabsorption for amino acids, glucose, uric acid, calcium, and phosphate, and a defect in the capacity of the kidney to acidify the urine.

Haemolytic anaemia may occur and may be the presenting feature. It has been suggested (Cartwright *et al.* 1954) that the precipitating cause of acute episodes may be sudden release of large amounts of copper from the tissues.

Finally, a wide variety of bone lesions occurs including, rarely, vitamin D-resistant rickets.

The disease has an autosomal recessive mode of inheritance, as established by study of 30 families who had a familial pattern and high incidence (47 per cent) of parental consanguinity (Bearn 1960).

Biochemical features

The role of copper in the pathogenesis of Wilson's disease was established unequivocally by Glazebrook (1945), who reported high concentrations of this metal in the liver and brain of a patient. Later, decreased levels of the copper-containing protein ceruloplasmin were found in serum from patients (Bearn and Kunkel 1952; Scheinberg and Gitlin 1952).

Ceruloplasmin is a cupro-glycoprotein which functions as an oxidase against several substrates. Copper is essential for its catalytic function. It acts as a ferroxidase oxidizing ferrous iron to the ferric state. It also catalyses the oxidation of diamines and polyphenols by oxygen; and the transfer of copper to copper-containing enzymes, including cytochrome oxidase. Deficient activity of the last function has been reported in Wilson's disease (Shokeir and Shreffler 1969). Six genetic variants of ceruloplasmin at autosomal loci have been reported. The commonest allele, $C_p{}^B$, is universally distributed (Shreffler *et al.* 1967). A high incidence of the $C_p{}^A$ allele has been found in Africans and Greeks (Kellermann and Walter 1972).

In a comparison between serum ceruloplasmin concentrations in 230 patients and 309 normal controls, Sass-Kortsak (1974) found that 222 patients (96 per cent) had values below the 85 per cent confidence limits of the controls.

The serum copper concentration is also decreased in most patients, but not to the extent expected from the decreased ceruloplasmin levels; this is because the level of non-ceruloplasmin-bound copper is increased (Cartwright *et al.* 1954).

The urinary excretion of copper is usually increased to an extent which, in general, correlates with duration of clinical disease. Thus, young presymptomatic patients may have normal copper excretion while older patients with established disease can excrete up to 1.5 mg/24h (normal, less than 40 μg).

The copper content of liver, brain, kidney, and cornea is increased, but the most extensive information is available for the liver. Thus, Sternlieb and Scheinberg (1968) found the mean hepatic copper concentration in specimens obtained by needle biopsy to be elevated 20-fold over mean normal values in symptomatic patients, and 30-fold in presymptomatic patients. Values for heterozygotes overlapped those for controls and patients. Histochemical studies have revealed that in the early stages of the disease the copper is diffusely located in the cytoplasm of parenchymal cells, but that at later stages it may be localized in lysosomes (Goldfischer and Sternlieb 1968; Goldfischer 1965).

In the brain the basal ganglia are usually the main area for deposition of copper, but in some patients large amounts are present in the cortex.

Many studies on the fate of orally or intravenously administered radioactive copper have been carried out in relation to Wilson's disease (Cartwright *et al.* 1954; Maytum *et al.* 1961; Aspin and Sass-Kortsak 1966). Both normal subjects and patients with the disease exhibit a rapid rise in total plasma activity following an oral dose of labelled copper, followed by a fall after 1 or 2 h, although in patients the peak activity tends to be higher and the subsequent fall somewhat slower than in normals. However, in normal individuals, but not patients, a secondary rise in plasma activity occurs some 4–6h after administration of the isotope. Electrophoretic studies have shown that the radioactivity in the normal secondary rise is associated with the α-2-globulin fraction, presumably ceruloplasmin

(Bearn and Kunkel 1954), while in Wilson's disease the activity remains associated with albumin.

As radiolabel disappears from the plasma following the initial rise, it appears in the tissues, mainly in the liver. This is followed by a gradual fall in activity for normal subjects, but for patients little of the labelled copper is released from the liver. Since this cannot be attributed to a failure of incorporation of copper into ceruloplasmin, it is interpreted as indicative of a block in biliary excretion of copper.

Several hypotheses concerning the fundamental defect have been put forward. Uzman suggested (Uzman *et al.* 1956; Iber *et al.* 1957*b*) that the mutant gene causes formation of an abnormal intracellular protein that has an increased affinity for binding copper, and that is present in the liver and other tissues in which copper is increased. An alternative hypothesis (Sass-Kortsak 1965) is that a structural gene mutation causes formation of a defective carrier protein which is normally responsible for making copper available both for incorporation into ceruloplasmin and excretion into the bile. Thirdly, it has been suggested (Sternlieb *et al.* 1973) that the fundamental cause is a lysosomal defect resulting in impaired excretion of lysosomal copper into bile. Finally, an interesting and more recent suggestion is that the disease is caused by a controller gene mutation which results in continuation of the foetal and neonatal mode of copper metabolism into later life (Epstein and Sherlock 1981).

Diagnosis

Although, as pointed out above, the diagnosis can be made without equivocation in a patient with the classic symptoms of liver disease, bizarre neurological features and Kayser-Fleischer rings, the great clinical variability necessitates the use of laboratory studies for confirmatory diagnosis. Moreover, since the disease is treatable, presymptomatic testing is required for sibs of affected patients.

Demonstration of reduced serum copper and ceruloplasmin levels, and increased urinary copper are the most straightforward tests. For the urinary studies, measurement of the 24h excretion is essential (Walshe 1970). If these tests lead to equivocal results, measurement of the copper in a liver biopsy or of the rate of incorporation of radio copper into ceruloplasmin may be necessary.

In presymptomatic patients the 24h urinary copper excretion may be normal, but an increased excretion over that occurring in normals following a standard dose of penicillamine may be evidence of the disease. Low serum ceruloplasmin levels should be interpreted with caution as they may be shown by heterozygotes. Some clarification may be possible here by comparing levels in the parents, since low serum ceruloplasmin in heterozygotes tends to cluster in families (Wilson-Cox *et al.* 1972). If the non-invasive tests still prove equivocal in making the diagnosis in presymptomatic patients, a study of the liver copper content and histology is advisable (Walshe 1975).

Important studies by Chan *et al.* (1980) and Chan and Rennert (1982) have shown that cultured fibroblasts from patients with Wilson's disease contain an elevated copper content compared to control cells. The mean copper content in seven Wilson's disease strains was 276.2 ng ± 21.4 S.D. copper per mg soluble cell protein, compared to 98.5 ng ± 19.2 S.D. in 11 control strains. The cadmium content of the Wilson's disease cells, measured for reference purposes, was similar to the controls. The concentration in cells from one heterozygote was intermediate between Wilson's disease and controls. The findings could have importance for presymptomatic and possibly prenatal diagnosis, and perhaps lead to a convenient system for studying the basic defect in the disorder.

Treatment

The rationale behind treatment is to remove the excessive amounts of copper from the tissues by administration of a chelating agent. This was first proposed by Cumings (1948), but unfortunately 2,3-dimercapto-propanol (British antilewisite; BAL), the only chelating agent available at the time, was unsatisfactory (Cumings 1948; Denny-Brown and Porter 1951). However, D-penicillamine (ß,ß-dimethylcysteine), introduced by Walshe (1956), proved to be a highly effective chelating agent, resulting in striking clinical improvement in patients.

Experiences with this drug have been reviewed by Walshe (1967) and Milne *et al.* (1968). The drug is administered orally with a maximum starting dose for adults and adolescents of 1 g three times a day (Walshe 1975). For the first few weeks after starting treatment there is often an increase in neurological signs, but gradually the clinical condition improves and provided that treatment has not been left to too late a stage, the patient can expect to return eventually to normal life. If treatment is started at the presymptomatic stage, the disease can be prevented (Sternlieb and Scheinberg 1968; Levi *et al.* 1967).

The drawback to the use of penicillamine is that in about 10 per cent of patients some form of immunologically induced intolerance to the drug develops (Walshe 1982). Chemical toxicity, such as pyridoxine deficiency, can also occur. In 1969, Walshe reported that an alternative chelating agent, triethylene tetramine dihydrochloride (trien), was successful in treating a patient who had developed immune-complex nephritis six years after treatment with penicillamine was started. Recently, Walshe (1982) has summarized his experience with this drug over 13 years, with 20 patients in whom severe penicillamine intolerance had developed. Evidence was found for depletion of body copper stores coinciding with clinical improvement. In most of the patients the penicillamine-induced toxic symptoms were decreased on trien therapy, although elastosis perforans did not seem to benefit, and two patients with systemic lupus erythematosus were not helped. The author concluded that trien is a safe drug, and that it could become the drug of first choice for management of Wilson's disease if granted a product licence.

Menkes' disease; kinky hair syndrome; steely hair sydrome

Clinical features

This X-linked disease was first described by Menkes *et al.* (1962) in five male patients in a single pedigree. All subsequent patients have been males.

The patients are often born prematurely, but usually with a birth weight appropriate for gestational age. Although feeding difficulty, vomiting, and poor weight gain are common during the newborn period, normal growth and development can occur during the first six to eight weeks of life. After this time seizures, gastrointestinal and respiratory infections, hypothermia, and hypotonia occur. Severe progressive neurological deterioration becomes apparent and the seizures become more persistent as the disease progresses. The facial appearance is reported to be quite characteristic with a 'pallid skin, pudgy tissues, horizontal, tangled eyebrows and a cupid's bow upper lip' (Danks *et al.* 1974*a*). The hair is lustreless and de-pigmented and breaks off easily leaving a brittle stubble. Microscopically, pili torti are present.

Radiologically, spurs are seen on the metaphyses of the long bones and subperiosteal calcifications along the shafts. Wormian bones appear in the posterior sagittal and the lambdoidal sutures. Arteriography reveals tortuosity and elongation of the visceral, cerebral, and limb arteries and occlusion of the lumen of some of these vessels (Danks *et al.* 1972*a*, 1974*a*).

The progressive psychomotor deterioration and recurrent infections result in death of the patients at an age between six months and three years. A patient with a milder form of the disease has been reported by Procopis *et al.* (1981).

Pathology

The main pathological changes are found in the central nervous system, eyes, and arteries. In the brain, cleavages at the corticomedullary junction, a decrease of white matter – particularly in the temporal lobes – and marked generalized atrophy of the cerebellar hemispheres are seen. Light microscopy reveals severe neuronal loss in the cerebral cortex, and replacement by proliferating astrocytic glial cells and macrophages. The white matter is markedly demyelinated and there may be a glial reaction. There is a virtual absence of Purkinje cells; those remaining are altered in size and shape. Pathology of the eye and vascular system are reviewed by Wray *et al.* (1976) and Oakes *et al.* (1976), respectively.

Biochemical features

The underlying defect is a deficiency of copper in the tissues. This was first shown by Danks *et al.* (1972*a,b*), who noticed a close similarity between the hair and arterial changes in their patients and those in copper-deficient sheep and chickens. They found low concentrations of copper and ceruloplasmin (q.v.) in serum from patients, as well as low copper content in liver and brain. When two of the patients were given oral doses of copper sulphate (1 or 2.5 mg/24h), there was no significant rise in serum copper or ceruloplasmin, excluding the possibility of nutritional copper deficiency. Intravenous administration of copper, on the other hand, was later found to lead to a rise in both serum copper and ceruloplasmin to normal levels (Bucknall *et al.* 1973; Grover and Strutton 1975). The concentration of apoceruloplasmin in serum from Menkes' disease patients is normal (Matsuda *et al.* 1974).

It can be concluded from these observations that the major defect is probably in the gastrointestinal absorption of copper. Evidence for the site of the defect came from further work by Danks and his colleagues (Danks *et al.* 1973) when they showed that the copper content of intestinal mucosa is markedly increased in the disease. This suggested that copper absorption is undisturbed, but that the transport from the mucosal cell to the extracellular environment is defective. The increased copper content of cultured skin fibroblasts from patients (Goka *et al.* 1976) and increased incorporation of ^{64}Cu (Horn 1976) might also be due to a defective transport mechanism.

Several groups of workers have made more detailed studies of copper metabolism in Menkes' disease cultured fibroblasts in order to obtain further information about the fundamental defect in the disease. Beratis *et al.* (1978) and Chan *et al.* (1978) found that the efflux of ^{64}Cu was significantly reduced in mutant cells, accounting for the increased accumulation observed. Much of the ^{64}Cu incorporated was bound to a protein or proteins of molecular weight approximately 10 000 daltons, the amount bound being significantly higher in Menkes' disease cells (Beratis *et al.* 1978). The protein(s) was (were) presumed to be metallothionine (a low-molecular-weight protein, rich in sulphydryl groups, which binds zinc, cadmium, mercury, and copper; it appears to function in homeostatis of mammalian intestinal and hepatic zinc, in the detoxification of heavy metals, in storage of copper in foetal and neonatal liver, and possibly in copper metabolism). La Badie *et al.* (1981) also studied the binding of ^{64}Cu by lysates from normal and Menkes' cultured skin fibroblasts. They found that the amount of ^{64}Cu bound to the 10 000 dalton protein(s) was 2- to 3-fold higher for Menkes' lysates than for normal, although no difference in affinity for copper was detected between lysates from the two cell types. It was concluded that the increased accumulation and reduced efflux of copper in Menkes' fibroblasts results either from an increased amount of the 10 000 dalton copper-binding protein(s) or an increased capacity of the molecule(s) for copper.

Although more work is required to elucidate the fundamental defect in the disease, much of the pathogenesis can be understood in terms of a copper deficiency (for reviews see Danks *et al.* 1974*a* and Holtzman 1976). Thus, since ceruloplasmin possesses oxidase activity towards amines such as serotonin, histamine, and dopamine, a deficiency of ceruloplasmin could result in neurological disturbances. Further, deficiency of two other copper-containing enzymes, cytochrome oxidase and dopamine ß-hydroxy-lase, from neuronal cells could be a cause of nervous system dysfunction (French *et al.* 1972), while reduction in activity of superoxide dismutase might be the cause of demyelination (Prohaska and Wells 1974).

Depigmentation of hair and skin is probably due to diminished activity of tyrosinase, an enzyme in the pathway of melanin biosynthesis (q.v.). The vascular and bone lesions may result from decreased cross-linkages in collagen and elastin because of diminished oxidative deamination of ϵ-amino groups of lysine (Holtzman 1976).

Diagnosis

The laboratory diagnosis can be established by demonstrating decreased serum copper and ceruloplasmin contents, a decreased liver copper content or an increased copper content of the duodenal mucosa. Less invasive than the latter techniques, however, is demonstration of an elevated copper content in cultured skin fibroblasts. This may be established either by measuring the copper content by atomic absorption spectrophotometry (Goka *et al.* 1976) or by studying *in vitro* ^{64}Cu incorporation (Horn 1976). Using the former method, the mean copper content for eight Menkes' disease strains was 335 ng/mg protein (range 231–440) compared to a mean of 59.2 ng/mg protein for 29 control strains (range 23.0–95.4). With the latter method the median 20h incorporation for seven Menkes' cultures was 74.4 ng ^{64}Cu/mg protein (range 54.8–142), and for nine control cultures was 26.1 ng ^{64}Cu/mg protein (range 19.5–37.5).

Genetic counselling and prenatal diagnosis

It has been estimated that the incidence in Australia may be as high as 1 in

35 000 (Danks *et al.* 1972*a*). Heterozygous females do not normally show any clinical signs, but in some cases pili torti have been reported (Danks *et al.* 1972*a*; Volpintesta 1974; Collie *et al.* 1978). The last authors claim that a careful examination of hair morphology in at-risk females may be useful for identifying heterozygotes. Alternatively, examination of the hair by scanning electron microscopy may be employed (Taylor and Green 1981).

Incorporation of ^{64}Cu by fibroblasts after cloning has also been used for identifying heterozygotes (Horn *et al.* 1980) and, although time consuming, probably affords the most reliable method. This study has also provided evidence for Lyonization of the gene involved in the disease. ^{64}Cu incorporation by uncloned fibroblasts may also be of use in identifying heterozygotes (Horn 1980). Since cultures from four heterozygotes showed increasingly abnormal copper uptake after repeated freezing, suggesting selection of the mutant cell, this manipulation of the culture conditions could be useful for identifying carrier status for women displaying borderline results when first tested (Horn 1980).

The first prenatal diagnosis of an affected foetus was reported by Horn (1976) who demonstrated an abnormal incorporation of ^{64}Cu by cultured amniotic cells. Subsequently, findings in 42 at-risk pregnanciess from 22 families have been reported (Horn 1981). Ten foetuses with a male karyotype were predicted to be affected, and in all cases this was confirmed at termination by demonstration of an increased placental copper content. Prediction of 14 unaffected male foetuses was confirmed by birth of 13 normal boys and spontaneous abortion of one foetus, in which no biochemical evidence for Menkes' disease was present. Of the 18 foetuses with a female karyotype, nine were predicted to be heterozygotes on the basis of tissue culture studies or placental copper content.

Treatment

Danks *et al.* (1972*a*) suggested that treatment of Menkes' disease might be possible by parenteral administration of copper, or even by giving large oral doses. Subsequent therapy in several patients by intravenous (Bucknall *et al.* 1973; Grover and Strutton 1975; Wehinger *et al.* 1975), intramuscular (Walker-Smith *et al.* 1973), or subcutaneous (Dekaban and Steusing 1974) administration resulted in increased serum copper and ceruloplasmin levels, but with no evidence for clinical improvement. In one patient treated from age 28 days, however (Grover and Strutton 1975), an arrest of the symptomatic deterioration was claimed, raising the possibility that very early treatment might be effective. On the other hand, the fact that the disease appears to be due to a generalized defect in copper transport, which is probably already manifested *in utero* (Heydorn *et al.* 1975) makes it perhaps unlikely that parenteral copper therapy can be effective (see also Garnica *et al.* 1977).

Animal model

A useful animal model is the mottled (Mo) mutant mouse (Hunt 1974). The severity of the disease in the mouse varies according to the particular allele present at the Mo locus. Of the five known alleles, the brindled (Mo^{br}) is the most useful since affected males die around 14 days *post partum*, an approximately equivalent developmental age to that at which most Menkes' disease patients die. Copper-binding proteins from liver and kidney of normal and brindled mice have been studied by Hunt (1976) and Port and Hunt (1979).

15. Miscellaneous disorders

Pancreatic lipase deficiency

Patients with reduced pancreatic lipase have a decreased ability to digest
dietary triglycerides. Intestinal lipase activity is in the region of ten per cent
of normal. The stools are pale, bulky, and fatty (Sheldon 1964; Rey *et al.*
1966).

Deficiency of trypsinogen, trypsin, chymotrypsin and procarboxypeptidase

Trypsinogen is synthesized in the pancreas and converted to trypsin. The
latter normally activates chymotrypsinogen and procarboxypeptidase.
When there is deficiency of trypsinogen therefore there is considerable
reduction of protein digestion in the small intestine.

Affected newborn fail to thrive and develop hypoproteinaemia, oedema,
and anaemia.

Treatment by feeding protein hydrolysates rapidly reverses the clinical
features (Townes *et al.* 1967; Morris and Fisher 1967).

Intestinal enterokinase (enteropeptidase) deficiency

Enterokinase catalyses the conversion of trypsinogen to trypsin. Deficiency
of enterokinase therefore leads to deficiency of trypsin, chymotrypsin, and
carboxypeptidase. Lack of protein digestion leads to the same clinical
features and response to treatment as described for deficiency of
trypsinogen (q.v.) (Hadorn *et al.* 1969; Tarlow *et al.* 1970; Polonowsky and
Bier 1970).

Vitamin D-dependent rickets (autosomal recessive); 25-hydroxychole-calciferol 1- α -hydroxylase deficiency; and familial hypophosphataemic (Vitamin D resistant) rickets (X-linked dominant); autosomal dominant hypophosphataemia

A detailed discussion of the control of calcium and phosphate metabolism
by parathyroid hormone (PTH), vitamin D, and calcitonin, or of vitamin D
metabolites is beyond the scope of this book. The subjects are reviewed by
Rasmussen and Bordier (1974) and DeLuca *et al.* (1971). We consider only

those aspects which have direct clinical relevance to the hereditary forms of vitamin D-dependent or to resistant rickets.

In humans, vitamin D is derived mainly either from conversion of 7-dehydrocholesterol to vitamin D_3 in the skin, catalysed by u.v. light, or as a dietary supplement of vitamin D_2 (irradiated ergosterol). Vitamin D_3 may be stored unchanged in muscle or adipose tissue or can be hydroxylated in the liver to 25-hydroxy vitamin D_3 (25-(OH)D_3) or converted to the most biologically active metabolite 1,25-dihydroxyvitamin D_3 (1,25(OH)$_2D_3$), catalysed by the kidney mitochondrial enzyme 25-hydroxycholecalciferol 1-α-hydroxylase (1-hydroxylase). Two other major metabolites are 24,25-(OH)$_2D_3$ and 25,26-(OH)$_2D_3$. It is probable that analogous metabolites are formed from vitamin D_2.

There are two main types of hereditary vitamin D-dependent or resistant rickets, though variants within each group have been identified. Firstly, autosomal recessive vitamin D-dependent rickets, probably due to deficiency of 1-hydroxylase, and secondly X-linked dominant hypophosphataemic (vitamin D-resistant) rickets.

Vitamin D-dependent rickets

Two types are recognized: type I, where the plasma concentration of 1,25-(OH)$_2$D is decreased, and type II where it is increased.

Type I. Clinically, the presentation is similar to that of rickets caused by nutritional deficiency of vitamin D, usually in the first year. There is tetany, convulsions, bowing of legs, growth retardation, muscular weakness, enlargement of the wrists and ankles, frontal bossing, and enlargement of the costo-chondral junctions (rachitic rosary; Scriver 1970; Prader *et al.* 1976). Radiologically the changes are also similar to those seen in vitamin D deficiency rickets.

Characteristic biochemical abnormalities include hypocalcaemia, with normal or mildly reduced plasma phosphate concentration. There is decreased renal reabsorption of phosphate, hyperphosphaturia, and a generalized aminoaciduria. Plasma alkaline phosphatase activity is increased. Plasma PTH is elevated probably because of the hypocalcaemia (Arnaud *et al.* 1970).

Plasma concentration of 25-(OH)D is normal but that of 1,25-(OH)$_2$D is low, consistent with a deficiency of 1-hydroxylase activity which is thought to be the fundamental defect (Fraser *et al.* 1973). The disorder has an autosomal recessive mode of inheritance (Scriver 1970), but no abnormalities have been found in heterozygotes. The clinical and biochemical abnormalities are rapidly reversed by the administration of 1,25-(OH)$_2D_3$ in doses of 1–2μg/day and probably less efficiently by 25-(OH)D_3 (500–1000 μg/day) (Balsan *et al.* 1979; Prader *et al.* 1976). An animal model has been described in pigs (Harmeyer and Plonait 1967).

Type II. Clinically the presentation is similar to type I except that the majority (Rosen *et al.* 1979) of reported patients, but not all (Zerwekh *et al.* 1979), have had alopecia totalis. The most striking biochemical difference between type I and type II is the decrease in plasma 1,25-$(OH)_2D_3$ in the former and its elevation in the latter. Elevated levels in type II have been attributed to end organ (e.g. bone and intestinal mucosa) resistance to 1,25-$(OH)_2D_3$ (Sockalosky *et al.* 1980).

The disorder probably has autosomal recessive inheritance. Results of treatment with pharmacological doses of various types of vitamin D has been variable, but usually disappointing.

Hypophosphataemic (vitamin D-resistant) rickets

Clinically this disorder usually presents later than vitamin D-dependent rickets, with either reduced stature or leg deformities – usually bowing of the legs. In contrast to patients with vitamin D dependency they do not have convulsions or tetany. Radiologically rachitic changes are most marked around the knees at the lower femoral and tibial metaphyses. The disorder persists into adult life as osteomalacia. Hemizygous males tend to have more skeletal deformity than heterozygous females.

The most constant biochemical abnormality is hypophosphataemia, which in infants may be detected shortly after birth (Harrison *et al.* 1966; Stickler 1969). The plasma calcium concentration is usually normal. The plasma concentration of 25-$(OH)D_3$ is normal (Hadad *et al.* 1973), but reports on levels of other vitamin D metabolites have been conflicting. Plasma PTH concentration is normal.

The basic defect appears to be a decrease in the renal tubular reabsorption of phosphate, leading to increased excretion of phosphate in the urine, and a decreased concentration of phosphate in the blood (Dent and Harris 1956). Typically there is reduced renal tubular maximum for reabsorption of phosphate (Tm_p).

Family studies of Winters *et al.* (1958) and Burnett *et al.* (1964) have established inheritance of hypophosphataemic rickets as X-linked dominant. All subjects with rickets, male or female, have hypophosphataemia. Hypophosphataemia may be present in females (and rarely males), without rickets. Thus the skeletal abnormalities but not the hypophosphataemia can skip a generation to reappear in grandchildren.

Treatment with high doses of vitamin D usually produced incomplete healing of rickets, but often, in the long term, renal damage due to vitamin D toxicity (Stickler *et al.* 1971).

Correction of hypophosphataemia by administration of phosphate at frequent intervals, e.g. 1–4 g of phosphorus per day given in divided doses four-hourly (West 1971), appears to be a safer form of therapy, but can give rise to secondary hyperparathyroidism (Rasmussen *et al.* 1981). The most effective treatment is probably the combination of phosphate and vitamin D (Rasmussen *et al.* 1980) though this also is prone to complications.

An animal model, namely the Hyp mouse, appears to be an analogous X-linked disorder (Eicher *et al.* 1976; Tenenhouse and Scriver 1979).

Autosomal recessive hypophosphataemic osteomalacia

A much rarer form of hyposphosphataemia, with autosomal recessive inheritance has also been described. Affected children have osteomalacia, but not rickets (Scriver *et al.* 1977*b*).

Serum cholinesterase (pseudocholinesterase) deficiency

Individuals with serum cholinesterase deficiency are quite healthy unless they receive the muscle relaxant suxamethonium (succinyl dicholine) to which they are markedly sensitive. Cholinesterase is sometimes called 'Pseudocholinesterase' to stress the difference between the neurological enzyme (also found in erythrocytes) with substrate specificity mainly for acetylcholine, known as 'true cholinesterase' or acetylcholinesterase. Serum cholinesterase, in contrast, hydrolyses a variety of choline and some non-choline-esters.

Following the introduction of suxamethonium as a muscle-relaxant during surgery or electroconvulsive therapy it was reported that in contrast to the majority of individuals, in whom the action of the drug is short-lived, about 1 in 2000 individuals in Europe with serum cholinesterase deficiency experience severe prolonged muscular paralysis and apnoea (Bourne *et al.* 1952; Evans *et al.* 1952).

Family studies revealed that affected individuals were homozygous for an atypical E^a_1 allele, that is, with a genotype E^a_1/E^a_1, as compared with individuals with the normal, or usual genotype E^u_1/E^u_1. Relatives often had partial enzyme defects (Lehman and Ryan 1956) and could be allocated a genotype E^u_1/E^a_1.

The report that the affinity of the 'usual' enzyme for a variety of choline esters was far greater (lower K_m) than that of the 'atypical' enzyme (Davies *et al.* 1960) suggested that the enzyme deficiency resulted from a structural difference.

Further evidence that the atypical enzyme was structurally different from the usual enzyme was reported by Muensch *et al.* (1978). These workers labelled serum pseudocholinesterase with radioactive substrate diisopropylfluorophosphate (DFP) which binds with the active site of the enzyme at serine residues. Following tryptic digestion, the peptides derived from either the usual or atypical enzymes were separated by high voltage thin-layer electrophoresis in silica gels. The DFP-labelled peptides were identified by autoradiography. The results showed that both enzymes had two peptides with high radioactivity and some peptides with low activity. The atypical enzyme peptides had greater net positive charges (migrated cathodally) than the usual enzyme peptides (migrated anodally), consistent with amino acid substitution, probably at the active site where the substrate DFP binds. Serum cholinesterase from individuals of the intermediate phenotype contained labelled peptides which migrated both anodally and cathodally, confirming that they had both usual and atypical enzymes, and were therefore heterozygotes.

Further studies showed that the degree of inhibition of cholinesterase by dibucaine, which presumably depends on qualitative differences of the enzyme, could discriminate completely between individuals with usual, intermediate and atypical cholinesterase activities. When serum cholinesterase activities were assayed there was considerable overlap between these groups, presumably because activity depends not only on qualitative differences between the usual and atypical enzymes, but also on the amount of enzyme in the serum (Kalow and Staron 1957; Harris *et al.* 1960).

Percentage inhibition by dibucaine under standard conditions was designated the 'Dibucaine number' (DN). Individuals with a DN of about 20 are very suxamethonium-sensitive; those with a DN around 60 are intermediate and those with a value of 80 have a usual phenotype (Kalow and Staron 1957; Harris and Whittaker 1962).

A third allele for serum cholinesterase was discovered because some parents of atypical individuals had the usual enzyme only. This could be explained by postulating that the parents had a silent allele E^s_1 which could not produce functional enzyme. Thus individuals with a genotype $E^a_1 E^s_1$ could produce only the atypical enzyme. Furthermore, one would anticipate individuals with genotype $E^u_1 E^s_1$ who could produce only the usual enzyme and individuals with genotype E^s_1/E^s_1, who could produce no enzyme, or very little. These predictions have been confirmed (Liddell *et al.* 1962; Hodgkin *et al.* 1965; Goedde *et al.* 1965).

A further allele was discovered because although in the majority of families inhibition of serum cholinesterase activity by either dibucaine or fluoride segregates populations into usual, intermediate, and atypical groups, in a few families there are exceptions. This was shown to be due to the existence of a further variant resulting from another allele, designated E^f_1. The E^f_1/E^f_1 genotype is associated with a serum cholinesterase activity of about 55 per cent of that of individuals with an E^u_1/E^u_1 genotype (Harris and Whittaker 1962; Liddell *et al.* 1963; Whittaker 1967). Double heterozygotes of the four alleles occur, and can give rise to ten genotypes. Severe suxamethonium sensitivity occurs in individuals with genotypes $E^a_1/$

E^a_1, E^a_1/E^s_1, and E^s_1/E^s_1. Less severe sensitivity to suxamethonium is associated with the genotypes E^f_1/E^f_1, E^a_1/E^f_1, and E^f_1/E^s_1.

Heterogeneity among the E^s_1 allele is suggested by the observation that in some individuals it is associated with some, but reduced, catalytic activity and immunologically cross-reacting material, while in others both are absent (Altland and Goedde 1970; Scott 1973). Silent alleles are very rare. However, in Alaskan eskimos nearly two per cent are homozygous for E^s_1 and 25 per cent are heterozygous (Gutsche *et al.* 1967; Scott *et al.* 1970*b*). Moreover, among this population, Scott (1973) found segregation of two different E^s_1 alleles.

A second locus E_2 is responsible for serum cholinesterase activity in about 10 per cent of individuals. This has been shown by two-dimensional electrophoresis which demonstrates four serum cholinesterase isoenzymes C_1, C_2, C_3, and C_4 in about 90 per cent of individuals, and a fifth isoenzyme, C_5, in about 10 per cent (Harris and Robson 1963; La Motta *et al.* 1970). Most of the enzyme activity is derived from C_4, which is thought to be a tetramer of polypeptides from the E_1 locus. It has been suggested that C_5 may consist of three subunits coded by the E_1 locus and one from the E_2 locus (Scott and Powers 1974).

The presence of the C_5 isoenzyme in 10 per cent of individuals could then be explained on the assumption that they have an allele at the E_2 locus (E^+_2) which codes for an active polypeptide and that the remaining 90 per cent of individuals have an E^-_2 which codes for an inactive polypeptide.

Acetyl-CoA carboxylase deficiency

A single patient has been reported with severe deficiency of the lipogenic enzyme acetyl-CoA carboxylase. She was admitted shortly after birth with severe respiratory problems. She later had severe brain damage, myopathy, and stunted growth (Blom *et al.* 1981).

Mass spectrometric analysis of urinary organic acids revealed the presence of 2-ethyl-3-oxo-hexanoic acid, 2-ethyl-3-hydroxy-hexanoic acid and 2-ethyl-hexanedioic acid.

The authors reasoned that 2-ethyl-3-oxo-hexanoic acid was the condensation product of two butyryl-CoA molecules, just as 2-methyl-3-oxo-valeric acid results from the condensation of two propionyl-CoA molecules in propionic acidaemia (Truscott *et al.* 1979*b*); and that the accumulation of butyryl-CoA could be explained by a deficiency of malonyl-CoA which is synthesized from acetyl-CoA by the cytosolic biotin enzyme acetyl-CoA carboxylase.

In fat tissues, *de novo* fatty acid synthesis is entirely by elongation of acetyl-CoA by malonyl-CoA. In liver, however, the first elongation of acetyl-CoA into butyryl-CoA occurs chiefly by the reversal of ß-oxidation

(Lin and Kumar 1972). Further elongation of butyryl-CoA is by using malonyl-CoA as substrate. Thus deficiency of acetyl-CoA carboxylase would lead to accumulation of butyryl-CoA and arrest *de novo* fatty acid synthesis. Condensation of two molecules of butyryl-CoA would lead to the production of 2-ethyl-3-oxo-hexanoyl-CoA which, when subjected to reduction and ω-oxidation, would yield 2-ethyl-3-hydroxy-hexanoic acid and 2-ethyl-hexanedioic acid.

Deficiency of acetyl-CoA carboxylase was demonstrated in a liver biopsy (activity was 2 per cent of a rat control) and in cultured fibroblasts (activity was 10 per cent of control values).

Hyperglycerolaemia; glycerol kinase deficiency

Patients with reduced activity of ATP:glycerol 3-phosphotransferase (glycerol kinase), glyceroluria, and hyperglycerolaemia have been reported in two families. The enzyme deficiency has been demonstrated in cultured fibroblasts as well as in the tissues where phosphorylation of glycerol occurs principally *in vivo*, namely liver, kidney, and leucocytes (McCabe *et al.* 1977, 1982; Rose and Haines 1978; Guggenheim *et al.* 1980).

Clinical features differ in the two families. In the family reported by McCabe *et al.* (1977) and Guggenheim *et al.* (1980), two brothers had developmental delay from birth, without regression. There was growth failure, spasticity, and osteoporosis, with multiple fractures in the older brother. A 'non-specific' myopathy was suspected because of a marked elevation of creatine phosphokinase activity and histological appearance of a muscle biopsy which revealed fibres which were hyalinized or basophilic or opaque and hypercontracted. In the older boy, at six years of age, adrenal cortical insufficiency was detected, involving both mineralocorticoids and glucocorticoids. The younger boy died suddenly three days after an operation for correction of strabismus. At necropsy, the adrenals showed non-specific distortion of architecture.

In the family reported by Rose and Haines (1978) there was no consistent phenotype. The disorder was detected during assessment of myocardial infarction and diabetes mellitus. All affected patients were male. The pedigree was consistent with an X-linked mode of inheritance. Prenatal diagnosis should be possible by assay of glycerol kinase activity of cultured amniotic cells.

Recently it has been shown that for cultured fibroblasts the K_m for glycerol was increased 5 to 200-fold over control values but the K_m for ATP was not altered. Hybridization of cultured fibroblasts from a member of each of the two families failed to enhance glycerol kinase activities, suggesting that even though the clinical phenotypes are so different, the enzyme defects may be allelic (McCabe *et al.* 1982).

Hypophosphatasia

The outstanding clinical features depend on reduced and inadequate osseous calcification of various grades of severity.

In the foetal or congenital type, there may be prenatal fractures, deformed limbs, a soft cranium, and an unstable collapsing thoracic cage, leading to respiratory distress and rapid death.

Infantile forms have bone deformities, including sometimes craniosynostosis, and retarded physical and motor development. Teeth are lost early.

In the adult form there are similar features of skeletal deformity, especially bowing of the long bones, craniosynostosis and early loss of teeth. There may be calcification of ligaments and cartilage (Rathbun 1948; Fraser and Yendt 1955; Fraser 1957; Méhes *et al.* 1972). Radiologically there are some features resembling rickets. The metaphyses appear ragged with uncalcified areas.

There is reduced alkaline phosphatase activity in serum (to about 25 per cent of normal), liver, bone, kidney, cultured fibroblasts, and leucocytes (Kretchmer *et al.* 1958; Brydon *et al.* 1975).

Some patients with skeletal manifestations of hypophosphatasia have had normal plasma alkaline phosphatase activities (Scriver and Cameron 1969; Méhes *et al.* 1972). Scriver and Cameron (1969) noted that residual plasma alkaline phosphatase in affected patients was more heat labile than normal and had a higher K_m for phosphoethanolamine.

Alkaline phosphatase was studied in serum from 10 patients with hypophosphatasia by acrylamide gel electrophoresis (Danovitch *et al.* 1968). The concentration of the small intestinal enzyme was high, whilst those of the liver and bone enzyme were reduced or absent.

Blau *et al.* (1978) examined tissues from a very severe case who died minutes after delivery and a milder case who survived to 63 days. In the very severe patient both bone and liver alkaline phosphatase activities were absent, but some residual activity was observed in the milder case. These observations suggest the existence of genetic heterogeneity even in the congenital lethal forms of hypophosphatasia.

Phosphoethanolamine concentration is increased in the plasma by two to three times the normal level, and in the urine by three to eight times or more (Fraser *et al.* 1955; McCance *et al.* 1955; Harris and Robson 1959). Phosphoethanolamine clearance is raised in urine of patients. Inorganic pyrophosphate is about three times normal (Russell 1965; Russell *et al.* 1971). The more severe cases have had hypercalcaemia and hypercalcuria.

Millán *et al.* (1980) found total serum alkaline phosphatase activity reduced to below two standard deviations of the normal mean value in nine of 23 related adults affected with an adult form of hypophosphatasia. All nine had reduced bone isoenzyme activities, four had reduced liver isoenzyme, while smaller variable reductions were found in the intestinal isoenzyme activities. The urinary excretion of phosphoethanolamine and phosphoserine was increased in patients with low hepatic alkaline phosphatase activities, but did not correlate with activities of the bone isoenzyme. This suggests that the low liver rather than bone isoenzyme activity is related to the increased excretion of phosphoethanolamine and phosphoserine.

Histologically the defect in ossification is indistinguishable from that in rickets. At areas of bone growth there is widening of uncalcified osteoid tissue lined with osteoblasts (Sobel *et al.* 1953).

The disorder usually has an autosomal recessive mode of inheritance. Heterozygotes have a clearance of phosphoethanolamine which is intermediate to those of controls and affected patients. Intermediate values have also been reported for excretion of phosphoethanolamine, pyrophosphate, and for serum alkaline phosphatase activities. Some heterozygotes have had mild skeletal abnormalities and premature shedding of teeth. Silverman (1962) reported a family in which the father and two sons were affected. Autosomal dominant inheritance could not be ruled out.

Prenatal diagnosis has been attempted by ultrasound scanning (Rudd *et al.* 1976) and by assay of alkaline phosphatase activity in amniotic fluid or cells (Rudd *et al.* 1976; Rattenbury *et al.* 1976; Osang *et al.* 1979). No clear answer has emerged. When the foetuses were affected, alkaline phosphatase activity was absent in the fluid and normal in the cells (Hoar and Rudd 1976), normal in amniotic cells and cord blood (Rudd *et al.* 1976) or reduced in amniotic cells (Rattenbury *et al.* 1976; Blau *et al.* 1978). Moreover, there were wide variations in the same cultures (from either patients or controls) at different subcultures, with activities in normal cells sometimes being below values in amniotic cells from an affected foetus (Hoar and Rudd 1976; Blau *et al.* 1978). Mulivor *et al.* (1978) consider cultured amniotic cells alkaline phosphatase activity to be useful for prenatal diagnosis.

Treatment with vitamin D or cortisone has been unsuccessful (Fraser 1957). However, treatment of three patients with high doses of oral neutral sodium phosphate (1.25 g – 3 g/day in divided doses) produced a small rise in plasma phosphate concentration, and radiological evidence of bone calcification (Bongiovanni *et al.* 1968). More trials of phosphate therapy are clearly needed.

Salla disease

Four patients, three brothers and a female third cousin, with an uncharacterized lysosomal storage disease, designated Salla disease, have been reported by Aula *et al.* (1978, 1979). Salla is the area in Sweden where the family resided. The authors refer to 27 severely mentally retarded individuals from 10 families sharing many clinical features, living in the same area. Biochemical studies on 13 patients were described by Renlund *et al.* (1979).

The main features are retarded psychomotor development, first noted at 12 months in the three brothers and during the first year in the cousin. Deterioration was noticed in three but not in the fourth, who was the oldest and had an IQ of 34 at 31 years. At 21–27 years all four were severely mentally retarded, had a coarse facial appearance, dysarthria, and impaired motor function; three had ataxia. EEGs showed diffuse abnormalities. Skeletal changes were limited to thick calvaria. All four had vacuolated lymphocytes in peripheral blood. Electron microscopy of fresh skin biopsies showed cytoplasmic inclusions interpreted as distended lysosomes. The only biochemical abnormality discovered was an approximate 10-fold increase in the urinary excretion of free *N*-acetylneuraminic acid (sialic acid) (Renlund *et al.* 1979). No enzyme defect has been detected.

Aula *et al.* (1982*b*) reviewed a total of 20 male and 14 female patients, two to 63 years of age, from 23 families consisting of 133 sibs. They conclude that inheritance is autosomal recessive. No necropsy has yet been reported.

References

Aaron, K., Goldman, H., and Scriver, C.R. (1971). Cystinosis; new observations: 1. Adolescent (type III) form. 2. Corrections of phenotypes in vitro with dithiothreitol. In *Inherited disorders of sulphur metabolism* (ed. N.A.J. Carson and D.N. Raine), p.150. Churchill Livingstone, Edinburgh.

Abbassi, V., Lowe, C.U., and Calcagno, P.L. (1968), Oculo-cerebro-renal syndrome. *Am. J. Dis. Child.* **115**, 145.

Abderhalden, E. (1903). Familiäre Cystindiathese. *Z. Physiol. Chem.* **38**, 557.

Abe, K. and Sugita, Y. (1979). Properties of cytochrome b5 and methemoglobin reduction in human erythrocytes. *Eur. J. Biochem.* **101**, 423.

Aberg, A., Mitelman, F., Cantz, M., and Gehler, J. (1978). Cardiac puncture of fetus with Hurler's disease avoiding abortion of unaffected co-twin. *Lancet* **ii**, 990.

Abramov, A., Schorr, S., and Wolman, M. (1956). Generalized xanthomatosis with calcified adrenals. *J. Dis. Child.* **91**, 282.

Abul-Haj, S.K., Martz, D.G., Douglas, W.F., and Geppert, L.J. (1962). Farber's disease. Report of a case with observations on its histogenesis and notes on the stored material. *J. Ped.* **61**, 221.

Abumrad, N.N., Schneider, A.J., Steel, D.R., and Rogers, L.S. (1979). Acquired molybdenum deficiency. *Clin. Res.* **27**, 774A.

Achord, D., Brot, F., Gonzalez-Noriega, A., Sly, W., and Stahl, P. (1977). Human ß-glucuronidase. II. Fate of infused human placental ß-glucuronidase in the rat. *Ped. Res.* **11**, 816.

Ackman, R.G. and Hooper, S.N. (1973). Isoprenoid fatty acids in the human diet. Distinctive geographical features in butterfats and importance in margarines based on marine oils. *Can. Inst. Food. Technol. J.* **6**, 159.

Aebi, H. (1967). The investigation of inherited enzyme deficiencies with special reference to acatalasia. In *Proc. 3rd int. congress of human genetics, Chicago* (ed. J.F. Crow and J.V. Neel), p.189. Johns Hopkins University Press, Baltimore.

—— Bossi, E., Cantz, M., Matsubara, S., and Suter, H. (1968). Acatalasemia in Switzerland. In *Hereditary disorders of erythrocyte metabolism* (ed. E. Beutler), Vol 1, p.31. Grune and Stratton, New York.

—— Baggiolini, M., Dewald, B., Lauber, E., Suter, H., Micheli, A., and Frei, J., (1964). Observations in two Swiss families with acatalasia. II. *Enzymol. Biol. Clin.* **4**, 121.

Agrawal, H.C. and Davison, A.N. (1973). Myelination and amino acid imbalance in the developing brain. In *Biochemistry of developing brain* (ed. W.A. Himwich), p.143. Marcel Dekker Inc., New York.

Akasaki, M., Fukui, S., Sakano, T., Tanaka, T., Usui, T., and Yamashina, I. (1978). Urinary excretion of a large amount of bound sialic acid and of undersulfated chondroitin sulfate A by patients with the Lowe syndrome. *Clin. Chim. Acta* **89**, 119.

Akcay, T. and Rose G.A. (1980). The real and apparent plasma oxalate. *Clin. Chim. Acta* **101**, 305.

Aki, K., Ogawa, K., Shirai, A., and Ichihara, A. (1967). Transaminase of branched chain amino acids III. Purification and properties of the mitochondrial enzyme from hog heart and comparison with the supernatant enzyme. *J. biol. Chem.* **62**, 610.

Albers, J.J., Adolphson, J.L., and Chen, G. (1981). Radioimmunoassay of human plasma lecithin:cholesterol acyltransferase. *J. clin. Invest.* **67**, 141.

Albers, J.J. and Utermann, G. (1981). Genetic control of lecithin:cholesterol acyltransferase: measurement of LCAT mass in a large kindred with LCAT deficiency. *Am. J. hum. Genet.* **33**, 702.

Albrecht, H. (1902). Ueber Ochronose. *Z. Heilk.* **23**, 366.

Aldrich, R.A., Hawkinson, V., Grinstein, M., and Watson, C.J. (1951). Photosensitive or congenital porphyria with hemolytic anemia I. Clinical and fundamental studies before and after splenectomy. *Blood* **6**, 685.

Alepa, F.P., Howell, R.R., Klinenberg, J.R., and Seegmiller, J.E. (1967). Relationships between glycogen storage disease and tophaceous gout. *Am. J. Med.* **42**, 58.

Alhadeff, J.A., Andrews-Smith, G.L., and O'Brien, J.S. (1978*b*). Biochemical studies on an unusual case of fucosidosis. *Clin. Genet.* **14**, 235.

—— Cimino, G., and Janowsky, A. (1978*a*). Isoenzymes of human liver α-L-fucosidase: chemical relationship, kinetic studies, and immunochemical characterisation. *Mol. cell. Biochem.* **19**, 171.

—— Miller, A.L., Wenger, D.A., and O'Brien, J.S. (1974). Electrophoretic forms of human liver α-L-fucosidase and their relationship to fucosidosis (mucopolysaccharidosis F). *Clin. Chim. Acta* **57**, 307.

—— —— Weenas, H., Vedvick, T., and O'Brien, J.S. (1975). Human Liver α-L-fucosidase: purification, characterisation, and immunological studies. *J. biol. Chem.* **250**, 7106.

Alitalo, K., Vaheri, A., Krieg, T., and Timpl, R. (1980). Biosynthesis of two subunits of type IV procollagen and of other basement membrane proteins by a human tumor cell line. *Eur. J. Biochem.* **109**, 247.

Allen, J.T., Gillet, M., Holton, J.B., Kings, G.S., and Pettit, B.R. (1980). Evidence of galactosaemia in utero. *Lancet* i, 603.

Allen, N. and De Veyra, E. (1967). Microchemical and histochemical observations in a case of Krabbe's leukodystrophy. *J. Neuropathol. exp. Neurol.* **26**, 456.

Allen, R.H. (1975). Human vitamin B_{12}-transport proteins. *Prog. Hematol.* **9**, 57.

Allison, A.C., Hovi, T., Watts, R.W.E., and Webster, A.D.B. (1975). Immunological observations on patients with Lesch-Nyhan syndrome, and on the role of de-novo purine synthesis in lymphocyte transformation. *Lancet* ii, 1179.

Alm, J. and Larsson, A. (1979). A follow-up of a nationwide neonatal metabolic screening programme in Sweden. *Ped. Res.* **13**, 79.

Alpert, L.I. and Damjanov, I. (1978). Malignant melanoma in an albino: diagnosis supported by ultrastructure. *Mt Sinai J. Med.* **45**, 447.

Altland, K. and Goedde, H.W. (1970). Heterogeneity in the silent gene phenotype of pseudocholinesterase of human serum. *Biochem. Genet.* **4**, 321.

Alzheimer, A. (1910). Beitrage zur kenntnis der pathologischen Neurologia und ihrer Beziehung zu den Abbauvorgangen im Nervengewebe. *Nissl – Alzheimer's histol. histopathol. Arb.* **3**, 493.

Amirhakimi, G.H., Haghighi, P., Ghalambor, M.A., and Honari, S. (1976). Familial lipogranulomatosis (Farber's disease). *Clin. Genet.* **9**, 625.

Ampola, M.G., Efron, M.L., Bixby, E.M., and Meshorer, E. (1969). Mental deficiency and a new aminoaciduria. *Am. J. Dis. Child.* **117**, 66.

—— Mahoney, M.J., Nakamura, E., and Tanaka, K. (1975). Prenatal therapy of a patient with vitamin B_{12}-responsive methylmalonic acidemia. *New. Engl. J. Med.* **293**, 313.

Amrhein, J.A., Meyer, J.W. III, Jones, H.W. Jr., and Migeon, C.J. (1976). Androgen insensitivity in man: evidence for genetic heterogeneity. *Proc. natl. Acad. Sci. USA* **73**, 891.

Andersen, D.H. (1956). Familial cirrhosis of the liver with storage of abnormal glycogen. *Lab. Invest.* **5**, 11.

Anderson, P.M. and Desnick, R.J. (1980). Purification and properties of uroporphyrinogen I synthase from human erythrocytes, identification of stable enzyme-substrate intermediates. *J. biol. Chem.* **255**, 1993.

Anderson, R.G.W., Brown, M.S., and Goldstein, J.L. (1977*a*). Role of the coated endocytic vesicle in the uptake of receptor-bound low density lipoprotein in human fibroblasts. *Cell* **10**, 351.

—— Goldstein, J.L., and Brown, M.S. (1976). Localization of low density lipoprotein receptors on plasma membrane of normal human fibroblasts and their absence in cells from a familial hypercholesterolemia homozygote. *Proc. natl. Acad. Sci. USA* **73**, 2434.

—— —— —— (1977*b*). A mutant that impairs the ability of lipoprotein receptors to localise in coated pits on the cell surface of human fibroblasts. *Nature* **270**, 659.

Anderson, W. (1898). A case of angiokeratoma. *Brit. J. Dermatol.* **10**, 113.

Ando, T., Rasmussen, K., Wright, J.M., and Nyhan, W.L. (1972). Isolation and identification of methylcitrate, a major metabolic product of propionate in patients with propionic acidemia. *J. biol. Chem.* **247**, 2200.

—— Nyhan, W.L., Backmann, C., Rasmussen, K., Scott, R., and Smith, E.K. (1973). Isovaleric acidemia: identification of isovalerate, isovalerylglycine, and 3-hydroxyisovalerate in urine of a patient previously reported as having butyric and hexanoic acidemia. *J. Ped.* **82**, 243.

—— —— Gerritsen, T., Gong, L., Heiner, D.C., and Bray, P.F. (1968). Metabolism of glycine in the nonketotic form of hyperglycinemia. *Ped. Res.* **2**, 254.

Andrews, J.M. and Menkes, J.H. (1970). Ultrastructure of experimentally produced globoid cells in the rat. *Exp. Neurol.* **29**, 483.

Andria, G., Vairo, P. Strisciuglio, P., Sannolo, N., and Kosova, P. (1980). Argininosuccinico-aciduria subacuta con insolita presentazione. *Riv. Ital. Pediatria* **6**, 495.

Angelini, C. and Engel, A.G. (1972). Comparative study of acid maltase deficiency. *Arch. Neurol. Chicago* **26**, 344.

Anthony, M. and McLeay, A.C. (1976). A unique case of derangement of vitamin B_{12} metabolism. *Proc. Aust. Assoc. Neurol.* **13**, 61.

Aoki, Y. (1978). Crystallization and characterization of a new protease in mitochondria of bone marrow cells. *J. biol. Chem.* **253**, 2026.

Applegarth, D.A., Goodman, S.I., Irvine, D.G., and Jellum, E. (1977). Hyperprolinemia type II: identification of the glycine conjugate of pyrrole-2-carboxylic acid in urine. *Clin. Biochem.* **10**, 20.

—— Ingram, P., Hingston, J., and Hardwick, D.F. (1974). Hyperprolinemia Type II. *Clin. Biochem.* **7**, 14.

Arakawa, T.S. (1970). Congenital defects in folate utilisation. *Am. J. Med.* **48**, 594.

—— (1974). Congenital and acquired disturbances of histidine metabolism. *Clin. Endocrinol. Metab.* **3**, 17.

—— Ohara, K., Kudo, Z., Tada, K., Hayashi, T., and Mizuno, T. (1963). 'Hyperfolic-acidemia with formimino-glutamic-aciduria following histidine loading.' Suggested for a case of congenital deficiency in formimino transferase. *Tohoku J. exp. Med.* **80**, 370.

Arashima, S. and Matsuda, I. (1972). A case of carbamyl phosphate synthetase deficiency. *Tohoku J. exp. Med.* **107**, 143.

Arbisser, A.I., Donnelly, K.A., Scott, C.I. Jr., Di Ferrante, N.M., Singh, J., Stevenson, R.E., Aylesworth, A.S., and Howell, R.R. (1977). Morquio-like syndrome with beta-galactosidase deficiency and normal hexosamine sulfatase activity:mucopolysaccharidosis IVB. *Am. J. med. Genet.* **1**, 195.

Arias, I.M. (1962). Chronic unconjugated hyperbilirubinemia without overt signs of hemolysis in adolescents and adults. *J. clin. Invest.* **41**, 2233.

—— Gartner, L.M., Cohen, M., Ben-Ezzer, J., and Levi, A.J. (1969). Chronic non-hemolytic unconjugated hyperbilirubinemia with glycuronyl transferase deficiency: clinical, biochemical, pharmacologic, and genetic evidence for heterogeneity. *Am. J. Med.* **47**, 395.

Arima, M. and Kurumada, T. (1962). Genetical study of Wilson's disease in childhood I. Clinical and biochemical analysis of sixteen families. II. Mode of inheritance and gene frequency in Japan. *Pediatr. Univ. Tokyo* **7**, 1.

Armstrong, M.D., and Robinow, A.M. (1967). A case of hyperlysinemia. Biochemical and clinical observations. *Pediatrics* **39**, 546.

—— and Robinson, K.S. (1954). On the excretion of indole derivatives in phenylketonuria. *Arch. Biochem.* **52**, 287.

—— Shaw, K.N.F., and Robinson, K.S. (1955). Studies on phenylketonuria. II. The excretion of ß-hydroxyphenyl-acetic acid in phenylketonuria. *J. biol. Chem.* **213**, 797.

—— Yates, K., Kakimoto, Y., Taniguchi, K., and Kappe, T. (1963). Excretion of ß-aminoisobutyric acid by man. *J. biol. Chem.* **238**, 1447.

Arnaud, C., Maijer, R., Reade, T., Scriver, C.R., and Whelan, O.T. (1970). Vitamin D dependency: an inherited postnatal syndrome with secondary hyperparathyroidism. *Pediatrics* **46**, 871.

Arneson, D., Ch'ien, L.T., Chance P., and Wilroy, R.S. (1979). Strychnine therapy in nonketotic hyperglycinemia. *Pediatrics* **63**, 369.

Arnold, W.J., Meade, J.C., and Kelley, W.N. (1972). Hypoxanthine-guanine phosphoribosyltransferase. Characteristics of mutant enzyme in erythrocytes from patients with the Lesch–Nyhan syndrome. *J. clin. Invest.* **51**, 1805.

Aronson, S.M. (1964). Epidemiology. In *Tay-Sachs disease* (ed. B.W. Volk), p.118. Grune and Stratton, New York.

—— and Volk, B.W. (1962). Some epidemiologic and genetic aspects of Tay–Sachs disease. In *Cerebral sphingolipidoses: a symposium on Tay–Sachs disease and allied disorders* (ed. S.M. Aronson and B.W. Volk), p.375. Academic Press, New York.

Arstila, A.U., Palo, J., Haltia, M., Riekkinen, P., and Autio, S. (1972). Aspartylglycosaminuria I: fine structure studies on liver, kidney and brain. *Acta Neuropathol.* (Berlin) **20**, 207.

Asatoor, A.M., Cheng, B., Edwards, K.D.G., Lant, A.F., Matthews, D.M., Milne, M.D., Navab, F., and Richards, A.J. (1970). Intestinal absorption of two dipeptides in Hartnup disease. *Gut* **11**, 380.

Ashkenazi, Y.E. and Gartler, S.M. (1971). A study of metabolic cooperation utilizing human mutant fibroblasts. *Exp. Cell Res.* **64**, 9.

Askanas, V., McLaughlin, J., Engel, W.K., and Adornato, B.T. (1979). Abnormalities in cultured muscle and peripheral nerve of a patient with adrenomyeloneuropathy. *New Engl. J. Med.* **301**, 588.

Asp, N.-G. and Dahlqvist, A. (1974). Intestinal ß-galactosidases in adult low

lactase activity and in congenital lactase deficiency. *Enzyme* **18**, 84.
Aspin, N. and Sass-Kortsak, A. (1966). Radiocopper studies on a family with Wilson's disease. In *The biochemistry of copper* (ed. J. Prisach, P. Aizen, and W.E. Blumberg), p.503. Academic Press, New York and London.
Assmann, G., Fredrickson, D.S., Sloan, H.R., Fales, H.M., and Highet, R.J. (1975). Accumulation of oxygenated steryl esters in Wolman's disease. *J. Lipid Res.* **16**, 28.
Atkin, B.M. (1979). Carrier detection of pyruvate carboxylase deficiency in fibroblasts and lymphocytes. *Ped. Res.* **13**, 1101.
—— Utter, M.F., and Weinberg, M.B. (1979). Pyruvate carboxylase and phosphoenolpyruvate carboxykinase activity in leucocytes and fibroblasts from a patient with pyruvate carboxylase deficency. *Ped. Res.* **13**, 38.
Atzpodien, W., Kremer, G.L., Schnellbacher, E., Denk, R., Haferkamp, G., and Bierback, H. (1975). Angiokeratoma coporis diffusum (Morbus Fabry). Biochemische diagnostik im blutplasma. *Dtsch. Med. Wschr.* **100**, 423.
Auerbach, V.H. DiGeorge, A.M., Carpenter, G.G. and Baldrige, R.C. (1967). Histidinemia: direct demonstration of absent histidase activity in liver. *Fed. Proc.* **26**, 680.
—— —— Baldrige, R.C., Tourtelotte, C.D., and Brigham, M.P. (1962). Histidinemia. A deficiency in histidase resulting in the urinary excretion of histidine and of imidazolepyruvic acid. *J. Ped.* **60**, 487.
Aula, P., Autio, S., Raivio, K.O., and Näntö, V. (1974). Detection of heterozygotes for aspartylglucosaminuria (AGU) in cultured fibroblasts. *Hum. Genet.* **25**, 307.
—— Raivio, K.O., and Autio, S. (1976). Enzymatic diagnosis and carrier detection of aspartylglucosaminuria using blood samples. *Ped. Res.* **10**, 625.
—— Rapola, J., and Andersson, L.C. (1975b). Distribution of cytoplasmic vacuoles in blood T and B lymphocytes in two lysosomal disorders. *Virchows Arch. B. Cell. Path.* **18**, 263.
—— —— von Kuskull, H., and Ämmälä, P. (1984). Prenatal diagnosis and fetal pathology of aspartylglucosaminuria. *Am. J. med. Genet.* **19**, 359.
—— Autio, S., Raivio, K., and Rapola, J. (1982a). Aspartylglucosaminuria. In *Genetic errors of glycoprotein metabolism* (ed. P. Durand and J.S. O'Brien), p.123. Edi-Ermes, Milan; Springer Verlag, Berlin.
—— —— —— —— and Renlund, M. (1982b). Salla disease. In *Genetic errors of glycoprotein metabolism* (ed. P. Durand and J.S. O'Brien), p.185. Edi-Ermes, Milan; Springer Verlag, Berlin.
—— Rapola, J., Autio, S., Raivio, K., and Karjalainen, O. (1975a). Prenatal diagnosis and fetal pathology of I-cell disease. *J. Ped.* **87**, 221.
—— Autio, S., Raivio, K.O., Rapola, J., Thoden, C.-J., Koskela, S.-L., and Yamashina, I. (1979). 'Salla disease'. A new lysosomal storage disorder. *Arch. Neurol.* **36**, 88.
—— Raivio, K., Autio, S., Thoden, C.-J., Rapola, J., Koskela, S.-L., and Yamashina, I. (1978). Four patients with a new lysosomal storage disorder (Salla disease). *Monogr. Hum. Genet.* **10**, 16.
Aumailley, M., Krieg, T., Dessau, W., Muller, P.K., Timpl, R., and Bricaud, H. (1980). Biochemical and immunological studies of fibroblasts derived from a patient with Ehlers–Danlos syndrome type IV. Demonstrate reduced type III collagen synthesis. *Arch. Dermatol. Res.* **269**, 169.
Aurebeck, G., Osterberg, K., Blaw, M., Chou, S., and Nelson, E. (1964). Electron microscopic observations on metachromatic leukodystrophy. *Arch. Neurol.* **11**, 273.
Austin, J.H. (1957). Metachromatic form of diffuse cerebral sclerosis. 2. Diagnosis

during life by isolation of metachromatic lipids from urine. *Neurology* **7**, 716.

—— (1959). Metachromatic sulfatides in cerebral white matter and kidney. *Proc. Soc. exp. Biol. Med.* **100**, 361.

—— (1973*a*). Studies in metachromatic leukodystrophy. XI. Therapeutic considerations. In *Birth defects: Original article series IX,* **2**, 125. Williams and Wilkins, Baltimore.

—— (1973*b*). Studies in metachromatic leukodystrophy. XII. Multiple sulfatase deficiency. *Arch. Neurol.* **28**, 258.

—— and Lehfeldt, D. (1965). Studies in globoid (Krabbe) leukodystrophy. III. Significance of experimentally produced globoid-like elements in rat white matter and spleen. *J. Neuropathol. exp. Neurol.* **24**, 265.

—— Armstrong, D., and Margolis, G. (1968). Canine globoid leukodystrophy: a model demyelinating disorder. *Trans. Am. Neurol. Assoc.* **93**, 181.

—— Lehfeldt, D., and Maxwell, W. (1961). Experimental 'globoid bodies' in white matter and chemical analysis in Krabbe's disease. *J. Neuropathol. exp. Neurol.* **20**, 284.

—— Armstrong, G.D., and Shearer, L. (1965*b*). Metachromatic form of diffuse cerebral sclerosis V. The nature and significance of low sulfatase activity: a controlled study of brain, liver and kidney in four patients with metachromatic leukodstrophy (MLD). *Arch. Neurol.* **13**, 593.

—— Balasubramanian, A.S., Pattabiraman, T.N., Saraswathi, S., Basu, D.K., and Bachhawat, B.K. (1963). A controlled study of enzymatic activities in three human disorders of glycolipid metabolism. *J. Neurochem.* **10**, 805.

—— Suzuki, K., Armstrong, D., Brady, R., Bachhawat, B.K., Schlenker, J., and Stumpf, D. (1970). Studies in globoid (Krabbe) leukodystrophy (GLD) V. Controlled enzymic studies in ten human cases. *Arch. Neurol.* **23**, 502.

Autio, S. (1972). Aspartylglucosaminuria. Analysis of thirty-four patients. *J. Ment. Defic. Res. Monograph Ser.* **I**, 1.

—— Norden, N.E., Öckerman, P.A., Riekkinen, P., Rapola, J., Loukimo, T. (1973). Mannosidosis. Clinical fine structural and biochemical findings in three cases. *Acta Pediat. Scand.* **62**, 555.

Avigan, J. (1966). The presence of phytanic acid in normal human and animal plasma. *Biochim. Biophys. Acta* **116**, 391.

Avila, J.L., Convit, J., and Velazquez-Avila, G. (1973). Fabry's disease: normal α - galactosidase activity and urinary sediment glycosphingolipid levels in two obligate heterozygotes. *Br. J. Dermatol.* **89**, 149.

Axelrod, F. and Goldzieher, J.W. (1967), Steroid biosynthesis by adrenal and ovarian tissue in congenital adrenal hyperplasia. *Acta Endocr.* (Kobenhavn) **56**, 453.

Aylsworth, A.S., Swisher, C.N., and Kirkman, H.N. (1975). Lethal hyperammonemia due to partial ornithine transcarbamylase deficiency in a 6 year old male. *Am. J. Hum. Genet.* **27**, 15A.

—— Taylor, H.A., Stuart, C.E., and Thomas, G.H. (1976). Mannosidosis: phenotype of a severely affected child and characterization of α -mannosidase activity in cultured fibroblasts from the patient and his parents. *J. Ped.* **88**, 814.

Aynsley-Green, A., Williamson, D.H., and Gitzelmann, R. (1977). Hepatic glycogen synthetase deficiency. *Arch. Dis. Child.* **52**, 573.

Babarik, A., Benson, P.F., Dean, M.F., and Muir, H. (1974). Chondroitin sulphaturia with α -L-iduronidase deficiency. *Lancet* **ii**, 464.

—— —— Fensom, A.H., and Barrie, H. (1976). Corneal clouding in GM_1-generalised gangliosidosis. *Br. J. Ophth.* **60**, 565.

Baber, M.D. (1956). A case of congenital cirrhosis of the liver with renal tubular defects akin to those in the Fanconi syndrome. *Arch. Dis Child.* **31**, 335.

Babior, B.M. (1978). Oxygen-dependent microbial killing by phagocytes. *New Eng. J. Med.* **298**, 659.

—— Kipnes, R.S., and Cuanutte, J.T. (1973). Biological defence mechanisms: the production by leukocytes of superoxide, a potential bacteriocidal agent. *J. clin. Invest.* **52**, 741.

Bach, G., Bargal, R., and Cantz, M. (1979a). I-cell disease: deficiency of extracellular hydroxylase phosphorylation. *Biochem. biophys. Res. Commun.* **91**, 476.

—— Cohen, M. and Kohn, G. (1975). Abnormal ganglioside accumulation in cultured fibroblasts from patients with mucolipidoses IV. *Biochem. biophys. Res. Commun.* **66**, 1483.

—— Cantz, M., Hall, C.W., and Neufeld, E.F. (1973a). Genetic errors of mucopolysaccharide degradation. *Biochem. Soc. Trans.* **1**, 231.

—— Eisenberg, F. jun., Cantz, M., and Neufeld, E.F. (1973b). The defect in the Hunter syndrome: deficiency of sulfoiduronate sulfatase. *Proc. natl. Acad. Sci. USA* **70**, 2134.

—— Friedman, R., Weissmann, B., and Neufeld, E.F. (1972). The defect in the Hurler and Scheie syndromes: deficiency of α-L-iduronidase. *Proc. natl. Acad. Sci. USA* **69**, 2048.

—— Zeigler, M., and Kohn, G. (1980). Biochemical investigations of cultured amniotic fluid cells in mucolipidosis type IV. *Clin. Chim. Acta* **106**, 121.

—— —— Kohn, G., and Cohen, M.M. (1977). Mucopolysaccharide accumulation in cultured skin fibroblasts derived from patients with mucolipidosis IV. *Am. J. Hum. Genet.* **29**, 610.

—— —— Schaap, T., and Kohn, G. (1979b). Mucolipidosis type IV: Gangliosidase sialidase deficiency, *Biochem. biophys. Res. Commun.* **90**, 1341.

—— Kohn, G., Lasch, E.E., El Massri, M., Ornoy, A., Sekeles, E., Legum, C., and Cohen, M.M. (1978). A new variant of mannosidosis with increased residual enzymatic activity and mild clinical manifestation. *Ped. Res.* **12**, 1010.

Bachmann, C. and Colombo, J.P. (1980). Diagnostic value of orotic acid excretion in heritable disorders of the urea cycle and in hyperammonemia due to organic acidurias. *Eur. J. Pediatr.* **134**, 109.

Bachmann, C., Krähenbühl, S., Colombo, J.P., Schubiger, G., Jaggi, K.H., and Tonz, O. (1981). N-acetylglutamate sythetase deficiency: a disorder of ammonia detoxification. *New Engl. J. Med.* **304**, 543.

Baerlocher, K., Gitzelmann, R., and Steinmann, B. (1980). Clinical and genetic studies of disorders in fructose metabolism. In *Inherited disorders of carbohydrate metabolism* (ed. D. Burman, J.B. Holton, and C.A. Pennock), p.163. MTP Press Ltd., Lancaster.

—— Scriver, C.R., and Mohyuddin, F. (1971a). The ontogeny of amino acid transport in rat kidney I. Effect of distribution ratios and intracellular metabolism of proline and glycine. II. Kinetics of uptake and effect of anoxia. *Biochim. Biophys. Acta* **249**, 353, 364.

—— Gitzelmann, R., Nüssli, R., and Dumermuth, G. (1971b). Infantile lactic acidosis due to hereditary fructose 1,6-diphosphatase deficiency. *Helv. Paediatr. Acta* **26**, 489.

—— —— Steinmann, B., and Gitzelmann-Cumarasamy, N. (1978). Hereditary fructose intolerance in early childhood: a major diagnostic challenge. Survey of 20 symptomatic cases. *Helv. Paediatr. Acta* **33**, 465.

Bagdale, J.D., Parker, F., Ways, P.O., Morgan, T.E., Lagunoff, D., and Eidelman, S. (1968). Fabry's disease: a corrective clinical, morphologic, and biochemical study. *Lab. Invest.* **18**, 681.

Bailey, A.J. and Etherington, D.J. (1980). Metabolism of collagen and elastin.

In *Comprehensive biochemistry* (ed. M. Florkin, N. Neuberger, and L.L. Van Deenan), p.229. Elsevier/North-Holland Biomedical Press, Amsterdam.

Bakay, B., Nyhan, W.L. (1972). Activation of variants of hypoxanthine-guanine phosphoribosyl transferase by the normal enzyme. *Proc. natl. Acad. Sci. (Wash.)* **69**, 2523.

—— —— Croce, C.M., and Kobrowski, H. (1975). Reversion in expression of hypoxanthine-guanine phosphoribosyl transferase following cell hybridization. *J. Cell. Sci.* **17**, 567.

Baker, H.J. and Lindsey, J.R. (1974). Feline GM$_1$-gangliosidosis. *Am. J. Pathol.* **74**, 649.

—— —— McKhann, G.M., and Farrell, D.M. (1971). Neuronal GM$_1$ gangliosidosis in a Siamese cat with ß-galactosidase deficency. *Science* **174**, 838.

Baker, L., and Winegrad, A.I. (1970). Fasting hypoglycaemia and metabolic acidosis associated with deficiency of hepatic fructose-1,6-diphosphatase activity. *Lancet* **ii**, 13.

—— Mellman, W.J., Tedesco, T.A., and Segal, S. (1966). Galactosemia: symptomatic and asymptomatic homozygotes in one Negro sibship. *J. Ped.* **68**, 551.

Bakker, H.D., Wadman. S.K., Van Sprang, F.J., Van der Heiden, C., Ketting, D., and De Bree, P.K. (1975). Tyrosinemia and tyrosyluria in healthy prematures: time courses not vitamin-C-dependent. *Clin. Chim. Acta* **61**, 73.

Balis, M.E., Krakoff, I.H., Berman, P.H., and Dancis, J. (1967). Urinary metabolites in congenital hyperuricosuria. *Science* **156**, 1122.

Balsan, S., Garabedian, M., Lieberherr, M., Gueris, J., and Ulmann, A. (1979). Serum 1,25-dihydroxyvitamin D concentrations in two different types of pseudo-deficiency rickets. In *Vitamin D Basic Research and its Clinical Application* (ed. A.W. Norman, K. Schaefer, D.V. Herrath, II.-G. Grigoleit, J.W. Colburn, H.F. De Luca, E.B. Mawer, and T. Suda), p.1143. Walter De Gruyter, Berlin.

Bank, W.J., DiMauro, S., and Rowland, L.P. (1972). Renal failure in McArdle's disease. *New Engl. J. Med.* **287**, 1102.

—— —— Bonilla, E., Capuzzi, D.M., and Rowland, L.P. (1975). A disorder of muscle lipid metabolism and myoglobinuria. Absence of carnitine palmityl transferase. *New Engl. J. Med.* **292**, 443.

Barber, G.W. and Spaeth, G.L. (1967). Pyridoxine therapy in homocystinuria. *Lancet* **i**, 337.

—— and Spaeth, G.L. (1969). The successful treatment of homocystinuria with pyridoxine. *J. Ped.* **75**, 463.

Bard. L.A. (1978). Heterogeneity in Waardenburg's syndrome. *Arch. Ophthalmol.* **96**, 1193.

Bardin, C.W., Bullock, L., Schneider, G., Allison, J.E., and Stanley, A.J. (1970). Pseudohermaphrodite rat: end organ insensitivity to testosterone. *Science* **167**, 1136.

Barnes, G.L., Bishop, R.F., and Townley, R.R.W. (1974). Microbial flora and disaccharidase depression in infantile gastroenteritis. *Acta Paediat. Scand.* **63**, 423.

Barnes, N.D., Hull, D., Balgobin, L., and Gompertz, D. (1970). Biotin-responsive propionic acidaemia. *Lancet* **ii**, 244.

Baron, D.N., Dent, C.E., Harris, H., Hart, E.W., and Jepson, J.B. (1956). Hereditary pellagra-like skin rash with temporary cerebellar ataxia, constant renal aminoaciduria, and other bizarre biochemical features. *Lancet* **ii**, 421.

Barranger, J.A., Rapoport, S.I., Fredericks, W.R., Pentchev, P.G., MacDermot, K.D., Steusing, J.K, and Brady, R.O. (1979). Modification of the blood–brain barrier: increased concentration and fate of enzymes entering the brain. *Proc. natl. Acad. Sci. USA* **76**, 481.

Barratt, T.M., Simmonds, H.A., Cameron, J.S., Potter, C.F., Rose, G.A., Arkell, D.G., and Williams, D.I. (1979). Complete deficiency of adenine phosphoribosyl transferase. A third case presenting as renal stones in a young child. *Arch. Dis Child.* **54**, 25.

Barrett, A.J. and Heath, M.F. (1977). Lysosomal enzymes. In *Lysosomes* (ed. J.T. Dingle), p.19. North-Holland Publishing Co., Amsterdam.

Barriere, H. and Gillot, F. (1973). La lipogranulomatose de Farber. *Nouv. Presse Med.* **2**, 767.

Bartholomé, K. and Ertel, E. (1978). Immunological detection of phenylalanine hydroxylase in phenylketonuria. *Lancet* **i**, 454.

—— Lutz, P. and Bickel, H. (1975). Determination of phenylalanine hydroxylase activity in patients with phenylketonuria and hyperphenylalaninemia. *Ped. Res.* **9**, 899.

—— Byrd, D.J., Kaufman, S., and Milstien, S. (1977). Atypical phenylketonuria with normal phenylalanine hydroxylase and dihydropteridine reductase activity *in vitro*. *Pediatrics* **59**, 757.

Bartholomew, W.R. and Rattazzi, M.C. (1974). Immunochemical characterization of human hexosaminidase from normal individuals and patients with Tay–Sachs disease. *Int. Arch. Allergy.* **46**, 512.

Bartlett, K. and Gompertz, D. (1976). Combined carboxylase defect: biotin-responsiveness in cultured fibroblasts. *Lancet* **ii**, 804.

—— Ng, H., and Leonard, J.V. (1980). A combined defect of three mitochondrial carboxylases presenting as biotin-responsive 3-methylcrotonylglycinuria and 3-hydroxyisovaleric aciduria. *Clin. Chim. Acta* **100**, 183.

Bartsocas, C.S., Levy, H.L., Crawford, J.D., and Thier, S.O. (1969). A defect in intestinal amino acid transport in Lowe's syndrome. *Am. J. Dis. Child.* **117**, 93.

Bartter, F.C., Henkin, R.I., and Bryan, G.T. (1968). Aldosterone hypersecretion in 'non-salt-losing' congenital adrenal hyperplasia. *J. clin. Invest.* **47**, 1742.

Basner, R., Von Figura, K., Glössl, J., Klein, U, Kresse, H., and Mlekusch, W. (1979). Multiple deficiency of mucopolysaccharide sulfatases in mucosulfatidosis. *Ped. Res.* **13**, 1316.

Batshaw, M.L. and Brusilow, S.W. (1980). Treatment of hyperammonemic coma caused by inborn errors of urea synthesis. *J. Ped.* **97**, 893.

—— —— and Walser, M. (1975). Treatment of carbamyl phosphate synthetase deficiency with keto analogues of essential amino acids. *New Engl. J. Med.* **292**, 1085.

—— Roan, Y., Jung, A.I., Rosenberg, L.A., and Brusilow, S.W. (1980). Cerebral dysfunction in asymptomatic carriers of ornithine transcarbamylase deficiency. *New Engl. J. Med.* **302**, 482.

—— Painter, M.J., Sproul, G.T., Schafer, I.A., Thomas, G.H., and Brusilow, S. (1981). Therapy of urea cycle enzymopathies: three case studies. *Johns Hopkins Med. J.* **148**, 34.

Battersby, A.R., Fookes, C.J.R., Matcham, G.W., and McDonald, E. (1979). Order of assembly of the four pyrrole rings during biosynthesis of the natural porphyrins. *J.C.S. chem. Comm.* **539**.

Baudhuin, P., Hers, H.G., and Loeb, H. (1965). An electron microscopic and biochemical study of type II glycogenesis. *Lab. Invest.* **13**, 1139.

Baughan, M.A., Paglia, D.E., Schneider, A.S., and Valentine, W.N. (1968a). An unusual hematological syndrome with puruvate-kinase deficiency and thalassemia minor in the kindreds. *Acta Haematol.* **39**, 345.

—— Valentine, W.N. Paglia, D.E., Ways, P.O., Simons, E.R., and Demarsh, Q.B. (1968*b*). Hereditary hemolytic anemia associated with glucose phosphate isomerase (GPI) deficiency – a new enzyme defect of human erythrocytes. *Blood* **32**, 236.

Baumgartner, R., Ando, T., and Nyhan, W.L., (1969). Nonketotic hyperglycinemia. *J. Ped.* **75**, 1022.

—— Bachmann, C., Brechbühler, T., and Wick, H. (1975). Acute neonatal nonketotic hyperglycinemia: normal propionate and methylmalonate metabolism. *Ped. Res.* **9**, 559.

—— Scheidegger, S., Stalder, G., and Hottinger, A. (1968). Argininoernsteinsäure-Krankheit des Neugeborenen mit lethalen Verlauf. *Helv. Paediatr. Acta* **23**, 77.

—— Wick, H., Maurer, R., Egli, N., and Steinmann, B. (1979*a*). Congenital defect in intracellular cobalamin metabolism resulting in homocystinuria and methylmalonic aciduria. I. Case report and histopathology. *Helv. Paediatr. Acta* **34**, 465.

—— —— Ohnacker, H., Probst, A., and Maurer, R. (1980). Vascular lesions in two patients with congenital homocystinuria due to different defects of remethylation. *J. inher. metab. Dis.* **3**, 101.

—— —— Linnell, J.C., Gaull, G.E., Bachmann, C., and Steinmann, B. (1979*b*). Congenital defect in intracellular colbalamin metabolism resulting in homocystinuria and methylmalonic aciduria. II. Biochemical investigations. *Helv. Paediatr. Acta* **34**, 483.

Baur, E.W. (1963). Catalase abnormality in a caucasian family in the United States. *Science* **140**, 816.

Baxter, J.H. and Steinberg, D. (1967). Absorption of phytol from dietary chlorophyll in the rat. *J. Lipid. Res.* **8**, 615.

—— —— Mize, C.E., and Avigan, J. (1967). Absorption and metabolism of uniformly ^{14}C-labeled phytol and phytanic acid by the intestine of the rat studied with thoracic duct cannulation. *Biochim. Biophys. Acta* **137**, 277.

Bayless, T.M. and Rosensweig, N.S. (1966). A racial difference in the incidence of lactase deficiency. *J. Am. Med. Assoc.* **197**, 968.

Beardmore, T.D., Meade, J.C., and Kelley, W.N. (1973). Increased activity of two enzymes of pyrimidine biosynthesis 'de novo' in erythrocytes from patients with Lesch–Nyhan syndrome. *J. lab. clin. Med.* **81**, 43.

Bearn, A.G. (1960). A genetical analysis of thirty families with Wilson's disease (hepatolenticular degeneration). *Ann. Hum. Genet. (Lond.)* **24**, 33.

—— and Kunkel, H.G. (1952). Biochemical abnormalities in Wilson's disease (abstract). *J. clin. Invest.* **31**, 616.

—— —— (1954). Localisation of ^{64}Cu in serum fractions following oral administration: an alteration in Wilson's disease. *Proc. Soc. exp. biol. Med.* **85**, 44.

Beaudet, A.L. and Caskey, C.T. (1978). Detection of Fabry's disease heterozygotes by hair root analysis. *Clin. Genet.* **13**, 251.

—— and Manschreck, A.A. (1982). Metabolism of sphingomyelin by intact cultured fibroblasts: differentiation of Niemann–Pick diseases types A and B. *Biochem. biophys. Res. Commun.* **105**, 14.

—— and Michels, V. (1977). Metabolic studies in argininemia. *Hum. Hered.* **27**, 164.

—— and Nichols, B.L. (1976). Residual altered α -mannosidase in human mannosidosis. *Biochem. biophys. Res. Commun.* **68**, 292.

—— Ferry, G.D., Nichols, B.L., and Rosenberg, H.S. (1977). Cholesterol ester storage disease: clinical, biochemical and pathological studies. *J. Ped.* **90**, 910.

—— DiFerrante, N.M., Ferry, G.D., Nichols, B.L., and Mullins, C.E. (1975). Variations in the phenotypic expression of ß-glucuronidase deficiency. *J. Ped.* **86**, 388.

—— Anderson, D.C., Michels, V.V., Arion, W.J., and Lange, A.J. (1980). Neutropenia and impaired neutrophil migration in type IB glycogen storage disease. *J. Ped.* **97**, 906.

Beaumont, J.L., Carlson, L.A., Cooper, G.R., Fejfar, Z., Fredrickson, D.S, and Strasser, T. (1970). Classification of hyperlipidaemias and hyperlipoprotein-aemias. *Bull. WHO* **43**, 891.

Beauvais, P. and Kaplan, J.C. (1978). La Méthémoglobinémie congénitale récessive: étude de huit cas avec encephalopathie. Nouvelle conception nosologique. *Journ. Parisiennes Pédiatrie*, 145.

Becker, H., Aubock, L., Haidvogl, M., and Bernheimer, H. (1976). Disseminierte Lipogranulomatoe (Farber). Kasuisitischer Bericht des 16. Falles einer Cerami-dose. *Verh. Dtsch. Ges. Pathol.* **60**, 254.

Becker, M.A., Meyer, L.J., Wood, A.W., and Seegmiller, J.A. (1973). Purine overproduction in man associated with increased phosphoribosyl-pyrophosphate synthetase activity. *Science* **179**, 1123.

—— Raivio, K.O., Bakay, B., Adams, W.B., and Nyhan, W.L. (1980). Variant human phosphoribosylpyrophosphate synthetase altered in regulatory and catalytic functions. *J. clin. Invest.* **65**, 109.

Becroft, D.M. and Phillips, L.I. (1965). Hereditary orotic aciduria and megalo-blastic anaemia: a second case, with response to uridine. *Br. med. J.* **i**, 547.

Becroft, D.M.O., Phillips, L.I., and Simmonds, A. (1969). Hereditary orotic aciduria: long-term therapy with uridine and a trial of uracil. *J. Ped.* **75**, 885.

Beemer, F.A. and Delleman, J.W. (1980). Combined deficiency of xanthine oxidase and sulfite oxidase; ophthalmological findings in a 3-week-old girl. *Metab Pediatr. Ophthalmol.* **4**, 49.

Beighton, P. (1968). X-linked recessive inheritance in the Ehlers–Danlos syn-drome. *Br. med. J.* **ii**, 9.

Bélanger, L., Bélanger, M., and Larochellé, J. (1972). Existence d'alpha-1-foétoprotéine circulante chez 8 patients souffrant de tyrosinémie héréditaire. *Union med. Can.* **101**, 877.

Bell, C.E. Jr., Sly, W.S., and Brot, F.E. (1977). Human ß-glucuronidase deficiency mucopolysaccharidosis: identification of cross-reacting antigen in cultured fibroblasts of deficient patients by immunoassay. *J. clin. Invest.* **59**, 97.

Benabdeljlil, C., Michel-Bechet, M., and Lissitzky, S. (1967). Isolation and iodinating ability of apical poles of sheep thyroid epithelial cells. *Biochem. biophys. Res. Commun.* **27**, 74.

Benedetto, A.V., Kushner, J.P., and Taylor, J.S. (1978). Porphyria cutanea tarda in three generations of a single family. *New Engl. J. Med.* **298**, 358.

Benson, P.F. and Polani, P.E. (1978). Appendix A: an estimate of the number of amniocenteses for metabolic disorder required per year. In *The provision of services for the prenatal diagnosis of fetal abnormality in the United Kingdom* (*Report of the Clinical Genetics Society working party on prenatal diagnosis in relation to genetic counselling*), p.30. *Bull. Eugen. Soc. suppl.* 3.

—— Bowser-Riley, F., and Giannelli, F. (1971). ß-galactosidases in fibroblasts: Hurler and Sanfilippo syndromes. *New Engl. J. Med.* **283**, 999.

—— Dean, M.F., and Muir, H. (1972). A form of mucopolysaccharidosis with visceral storage and excessive urinary excretion of chondroitin sulphate. *Dev. Med. Child Neurol.* **14**, 69.

—— Hamerton, J.L., and Young, V. (1969*a*). Incorporation of methionine sulphur into cysteine *in vitro* by fibroblasts deficient in cystathionine synthetase. *Arch. Dis. Child.* **44**, 779.

—— Swift, P.N., and Studdy, J.D. (1970). A biochemical and clinical study of 30 families with mentally subnormal siblings. In *Errors of phenylalanine, thyroxine and testosterone metabolism* (ed. W. Hamilton and F.P. Hudson), p.42. E. and S. Livingstone, London.

—— —— and Young, V.K. (1969*b*). Hydroxylysinuria. *Arch. Dis. Child.* **44**, 134,

—— Babarik, A., Brown, S.P., and Mann, T.P. (1976*a*). GM$_1$-generalised gangliosidosis variant with cardiomegaly. *Postgrad. Med. J.* **52**, 159.

—— Brandt, N.J., Christensen, E., and Fensom, A.H. (1979*b*). Prenatal diagnosis of galactosaemia in six pregnancies — possible complications with rare alleles of the galactose 1-phosphate uridyl transferase locus. *Clin. Genet.* **16**, 311.

—— Button, L.R., Fensom, A.H., and Dean, M.F. (1979*a*). Lumbar kyphosis in Hunter's disease (MPS II). *Clin. Genet.* **16**, 317.

—— Brown, S.P., Cree, J. Fensom, A.H., Grant, A.R., and Mann, T.P. (1976*b*). Phenotypic expression of galactokinase deficiency in heterozygous and homozygous subjects: *in vivo* and *in vitro* studies. *Birth defects: Original article series XII*, **3**, 305. Alan R.Liss, New York.

—— Fensom, A.H., Crees, M.J., Lam, S.T.S., Coleman, D., Heaton, D., and Rodeck, C. (1983). Recent advances in prenatal diagnosis of inborn errors of metabolism. Presented at symposium on *Progress in perinatal medicine*, Florence, May 9th–12th, p. 165. Excepta Medica, Amsterdam.

Ben-Yoseph, Y., Burton, B.K., and Nadler, H.L. (1977). Quantification of the enzymatically deficient cross-reacting material in GM$_1$-gangliosidosis. *Am. J. hum. Genet.* **29**, 575.

Benzi, G., Arrigoni, E., Strada, P., Pastoris, O., Villa, R.F., and Agnoli, A. (1977). Metabolism and cerebral energy state: effect of acute hyperammonemia in beagle dog. *Biochem. Pharmacol.* **26**, 2397.

Beratis, N.G., Aron, A.M., and Hirschhorn, K. (1973). Metachromatic leukodystrophy: detection in serum. *J. Ped.* **83**, 824.

—— Turner, B.M., and Hirschhorn, K. (1975*b*). Fucosidosis: detection of the carrier state in peripheral blood leukocytes. *J. Ped.* **87**, 1193.

—— Price, P., LaBadie, G., and Hirschhorn, K. (1978). ^{64}Cu metabolism in Menkes and normal cultured skin fibroblasts. *Ped. Res.* **12**, 699.

—— Turner, B.M., LaBadie, G., and Hirschhorn, K. (1977). α-L-fucosidase in cultured skin fibroblasts from normal subjects and fucosidosis patients. *Ped. Res.* **11**, 862.

—— —— Weiss, R., and Hirschhorn, K. (1975*a*). Arylsulfatase B deficiency in Maroteaux–Lamy syndrome: cellular studies and carrier identification. *Ped. Res.* **9**, 475.

Berg, K. and Saugstad, L.F. (1974). A linkage study of phenylketonuria. *Clin. Genet.* **6**, 147.

Bergeron, P., Laberge, C., and Grenier, A. (1974). Hereditary tyrosinemia in the province of Quebec. Prevalence at birth and geographic distribution. *Clin. Genet.* **5**, 157.

Bergström, J., Fürst, P., Gallyas, F., Hultman, E., Sonnilsson, L.H., Roch-Norlund, A.E., and Vinnas, E. (1972). Aspects of fructose metabolism in normal man. *Acta Med. Scand.* **542**, 57.

Berk, P.D., Wolkoff, A.W., and Berlin, N.I. (1975*b*). Inborn errors of bilirubin metabolism. *Med. clin. N.A.* **59**, 803.

—— Martin, J.F., Blaschke, T.E., Scharschmidt, B.F., and Plotz, P.H. (1975*a*).

Unconjugated hyperbilirubinemia: physiological evaluation and experimental approaches to therapy. *Ann. intern. Med.* **82**, 552.

—— Howe, R.B., Bloomer, J.R., and Berlin, N.I. (1969). Studies of bilirubin kinetics in normal adults. *J. clin. Invest.* **48**, 2176.

—— Berman, M.D., Blitzer, B.L., Chretien, P., Martin, J.F., Scharschmidt, B.F., Vierling, J.M., Wolkoff, A.W., Vergalla, J., and Waggoner, J.G. (1981). Effect of splenectomy on hepatic bilirubin clearance in patients with hereditary spherocytosis: implications for the diagnosis of Gilbert's syndrome. *J. lab. clin. Med.* **98**, 37.

Berman, E.R., Kohn, G., Yatziv, S., and Stein, H. (1974*a*). Acid hydrolase deficiencies and abnormal glycoproteins in mucolipidosis III. *Clin. Chim. Acta* **52**, 115.

—— Livni, N., Shapira, E., Merin, S., and Levij, I.S. (1974*b*). Congenital corneal clouding with abnormal systemic storage bodies: a new variant of mucolipidosis. *J. Ped.* **84**, 519.

Berra, B., DiPalma, S., and Brunngraber, E.G. (1974). Altered levels of tissue gangliosides and glycoproteins in the infantile form of GM_1-gangliosidosis. *Clin. Chim. Acta* **57**, 301.

Berry, H.K. (1962). Detection of metabolic disorders among mentally retarded children by means of paper spot tests. *Am. J. Ment. Defic.* **66**, 555.

—— and Spinanger, J. (1960). Screening test for Hurler's syndrome. *J. Lab. clin. Med.* **55**, 136.

Berson, E.L., Schmidt, S.Y., and Rabin, A.R. (1976). Plasma amino acids in hereditary retinal disease. Ornithine, lysine and taurine. *Br. J. Ophthalm.* **60**, 142.

—— Shih, V.E., and Sullivan, P.L. (1981). Ocular findings in patients with gyrate atrophy on pyridoxine and low-protein, low-arginine diets. *Ophthalmology* **88**, 311.

Besançon, A.M., Belon, J.P., Castelnau, L., Dumez, Y., and Poenaru, L. (1984). Prenatal diagnosis of atypical Tay–Sachs disease by chorionic villi sampling. *Prenatal Diagnosis* **4**, 365.

Besley, G.T.N. (1977). Sphingomyelinase defect in Niemann–Pick disease, Type C, fibroblasts. *FEBS Lett.* **80**, 71.

—— (1978). Diagnosis of Niemann–Pick disease using a simple and sensitive fluorometric assay of sphingomyelinase activity. *Clin. Chim. Acta* **90**, 269.

—— and Bain, A.D. (1976). Krabbe's globoid cell leucodystrophy. *J. med. Genet.* **13**, 195.

—— —— (1978). Use of a chromogenic substrate for the diagnosis of Krabbe's disease with special reference to its application in prenatal diagnosis *Clin. Chim. Acta* **88**, 229.

—— and Gatt, S. (1981). Spectrophotometric and fluorimetric assays of galacto-cerebrosidase activity, their use in the diagnosis of Krabbe's disease. *Clin. Chim. Acta* **110**, 19.

—— Hoogeboom, A.J.M., Hoogeveen, A.M., Kleijer, W.J., and Galjaard, H. (1980). Somatic cell hybridization studies showing different gene mutations in Niemann–Pick variants. *Hum. Genet.* **54**, 409.

Bessman, S.P., Williamson, M.L., and Koch, R. (1978). Diet, genetics and mental retardation. Interaction between phenylketonuric heterozygous mother and fetus to produce nonspecific diminution of IQ: evidence in support of the justification hypothesis. *Proc. natl. Acad. Sci. USA* **75**, 1562.

Betke, K., Beutler, E., Brewer, G.J., Kirkman, H.N., Luzzatto, L., Motulsky, A.G., Ramot, B., and Siniscalco, M. (1967). Standardization of procedures for the study of glucose-6-phosphate dehydrogenase. Report of a WHO scientific group. *WHO tech. Rep. Ser. no.* 366.

Beutler, E. (1969*a*). Glutathione reductase: stimulation in normal subjects by riboflavin supplementation. *Science* **165**, 613.

—— (1969*b*). Effect of flavin compounds on glutathione reductase activity: in vivo and in vitro studies. *J. clin. Invest.* **48**, 1957.

—— (1973). Screening for galactosemia. Studies of the gene frequencies for galactosemia and the Duarte variant. *Isr. J. med. Sci.* **9**, 1323.

—— (1978). *Hemolytic anemia in disorders of the red cell metabolism.* Plenum Press, New York.

—— and Baluda, M.C. (1964). The separation of glucose-6-phosphate dehydrogenase-deficient erythrocytes from the blood of heterozygotes for glucose-6-phosphate dehydrogenase deficiency. *Lancet* **i**, 189.

—— and Kuhl, W. (1970). The diagnosis of the adult type of Gaucher's disease and its carrier state by demonstration of deficency of ß-glucosidase activity in peripheral blood leukocytes. *J. lab. clin. Med.* **76**, 747.

—— —— (1972). Purification and properties of human α -galactosidases. *J. biol. Chem.* **247**, 7195.

—— —— (1973). Absence of cross-reactive antigen in Fabry's disease. *New Engl. J. Med.* **289**, 694.

—— —— (1975). Subunit structure of human hexosaminidase verified: interconvertibility of hexosaminidase isoenzymes. *Nature* **258**, 263.

—— and Yoshida, A. (1973). Human glucose-6-phosphate dehydrogenase variants: a supplementary tabulation. *Ann. hum. Genet. (Lond.)* **37**, 151.

—— Dern, R.J., and Alving, A.S. (1954). The hemolytic effect of primaquine IV. The relationship of cell age to hemolysis. *J. lab. clin. Med.* **44**, 439.

—— Dyment, P.G., and Matsumoto, F. (1978). Hereditary nonspherocytic hemolytic anemia and hexokinase deficiency. *Blood* **51**, 935.

—— Kuhl, W., Matsumoto, F., and Panglais, G. (1976). Acid hydrolases in leukocytes and platelets of normal subjects and in patients with Gaucher's and Fabry's disease. *J. exp. Med.* **143**, 975.

—— —— Trinidad, F., Teplitz, R., and Nadler, H. (1971). ß-glucosidase activity in fibroblasts from homozygotes and heterozygotes for Gaucher's disease. *Am. J. hum. Genet.* **23**, 62.

—— Scott, S., Bishop, A., Margolis, N., Matsumoto, F., and Kuhl, W. (1973*b*). Red cell aldolase deficiency and hemolytic anemia: a new syndrome. *Trans. Assoc. Am. Physicians* **86**, 154.

—— Matsumoto, F., Kuhl, W., Krill, A., Levy, N., Sparkes, R., and Degnis, M. (1973*a*). Galactokinase deficiency as a cause of cataracts. *New Engl. J. Med.* **288**, 1203.

Beyreiss, K., Willgerodt, H., and Theile, H. (1968). Untersuchungen bei heterozygoten Merkmalsträgern für Fruktoseintoleranz. *Klin. Wochenschr.* **9**, 465.

Bickel, H. (1955). Die Entwicklung der biochemischen Läsion bei der Lignac–Franonischen Krankheit. *Helv. Paediat. Acta* **10**, 259.

—— (1978). Cited by Gitzelmann, R. and Hansen R.G. (1980), p.75.

—— (1980). Phenylketonuria: past, present, future. *J. inher. metab. Dis.* **3**, 123.

—— Gerrard, J., and Hickmans, E.M. (1954). The influence of phenylalanine intake on the chemistry and behaviour of a phenylketonuric child. *Acta Paediat. (Uppsala)* **43**, 64.

—— Guthrie, R., and Hammersen, G. (eds.) (1980). *Neonatal screening for inborn errors of metabolism.* Springer Verlag, Berlin, Heidelberg, and New York.

—— Lutz, P., and Schmidt, H. (1973). The treatment of cystinosis with diet or drug. In *Cystinosis* (ed. J.D. Schulman), p.199. *DHEW Publication no.* (NIH) **72–249**. US Government Printing Office, Washington DC.

—— Feist, D., Müller, H., and Quadbeck, G. (1968). Ornithionemie. *Dtsch. Med. Wschr.* **93**, 2247.

—— Smallwood, W.C., Smellie, J.M., and Hickmans, E.M. (1952). Cystine storage disease with aminoaciduria and dwarfism (Lignac–Fanconi disease): clinical description, factual analysis and treatment of Lignac–Fanconi disease. *Acta Paediat.* **42** (suppl 90), 27.

Bienzle, U., Lucas, A.O., Ayeni, O., and Luzzato, L. (1972). Glucose-6-phosphate dehydrogenase and malaria. Greater resistance of females heterozygous for enzyme deficiency and of males with non-deficient variant. *Lancet* **i**, 107.

Biggar, W.D. (1977). *A new form of nucleoside phosphorylase deficiency in two brothers with defective T cell function.* Presented at the national pediatric conference honoring Dr J. Anderson, 16–18 June, Minneapolis.

Bigley, R.H. and Koler, R.D. (1968). Liver pyruvate kinase (PK) isozymes in a PK-deficient patient. *Ann. hum. Genet. (Lond.)* **31**, 383.

—— Stenzel, P., Jones, R.T., Campos, J.O., and Koler, R.D. (1968). Tissue distribution of human pyruvate kinase isozymes. *Enzymol. Biol. Clin.* **9**, 10.

Biglieri, E.G., Herron, M.A., and Brust, N. (1966). 17 α-hydroxylation deficieny in man. *J. clin. Invest.* **45**, 1946.

Bilheimer, D.W., Stone, N.J., and Grundy, S.M. (1979). Metabolic studies in familial hypercholesterolemia: evidence for a gene-dosage effect *in vivo*. *J. clin. Invest.* **64**, 524.

—— Goldstein, J.L., Grundy, S.M., and Brown, M.S. (1975). Reduction in cholesterol and low density lipoprotein synthesis after portacaval shunt surgery in a patient with homozygous familial hypercholesterolemia. *J. clin. Invest.* **56**, 1420.

Billmeier, G.J., Molinary, S.V., Wilroy, R.S., Duenas, D.A., and Brannon, M.E. (1974). Argininosuccinic aciduria: investigation of an affected family. *J. Ped.* **84**, 85.

Binkiewicz, A. and Senior, B. (1973). Decreased ketogenesis in von Gierke's disease (type I glycogenosis). *J. Ped.* **83**, 973.

Bird, T.D., Hamernyik, P., Nutter, J.Y., and Labbe, R.F. (1979). Inherited deficiency of delta-aminolevulinic acid dehydratase. *Am. J. hum. Genet.* **31**, 662.

Bishop, J.O. (1974). The gene numbers game. *Cell* **2**, 81.

Bitter, M. and Muir, H. (1962). A modified uronic acid carbazole reaction. *Anal. Biochem.* **4**, 330.

Bittles, A.H. and Carson, N.A.J. (1973). Tissue culture techniques as an aid to prenatal diagnosis and genetic counselling in homocystinuria. *J. med. Genet.* **10**, 120.

—— and Carson, N.A.J. (1974). Cystathionase deficiency in fibroblast cultures from a patient with primary cystathioninuria. *J. med. Genet.* **11**, 121.

—— and Carson, N.A.J. (1981). Homocystinuria: studies of cystathionine ß-synthase, *S*-adenosylmethionine synthetase, and cystathionase activities in skin fibroblasts. *J. inher. metab. Dis.* **4**, 3.

Black, M. and Billing, B.H. (1969). Hepatic bilirubin UDP glucuronyltransferase activity in liver disease and Gilbert's syndrome. *New Engl. J. Med.* **280**, 1266.

—— Fevery, J., Parker, D., Jacobson, J., Billing, B.H., and Carson, E.R. (1974). Effect of phenobarbitone on plasma (^{14}C)-bilirubin clearance in patients with unconjugated hyperbilirubinaemia. *Clin. Sci. mol. Med.* **46**, 1.

Blakemore, W.F. (1972). GM_1-gangliosidosis in a cat. *J. comp. Pathol.* **82**, 179.

Blanc, W.A. and Johnson, L. (1959). Studies on kernicterus. *J. Neuropathol. exp. Neurol.* **18**, 165

Blaschke, T.F., Berk, P.D., Scharschmidt, B.F., Guyther, J.R., Vergalla, J., and Waggoner, J.G. (1974). Criglar–Najjar syndrome: an unusual course with development of neurologic damage at age eighteen. *Ped. Res.* **8**, 573.

Blaskovics, M.E., Ng, W.G., and Donnell, G.N. (1978). Prenatal diagnosis and a case report of isovaleric acidaemia. *J. inher. metab. Dis.* **1**, 9.

Blass, J.P. (1980). Pyruvate dehydrogenase deficiencies. In *Inherited disorders of carbohydrate metabolism* (ed. D. Burman, J.B. Holton and C.A. Pennock), p.239. MTP Press Ltd., Lancaster.

—— Avigan, J., and Uhlendorf, B.W. (1970). A defect in pyruvate decarboxylase in a child with an intermittent movement disorder. *J. clin. Invest.* **49**, 423.

—— Cederbaum, S.D., and Dunn, H.G. (1976a). Biochemical defect in Leigh's disease. *Lancet* **i**, 1237.

—— Kark, R.A.P., and Engel, W.K. (1971). Clinical studies of a patient with pyruvate decarboxylase deficiency. *Arch. Neurol.* **25**, 449.

—— —— Menon, N., and Harris, S.H. (1976b). Decreased activities of the pyruvate and ketoglutarate dehydrogenase complexes in fibroblasts from five patients with Friedreich's ataxia. *New Engl. J. Med.* **295**, 62.

—— Schulman, J.D., Young, D.S., and Hom, E. (1972). An inherited defect affecting the tricarboxylic acid cycle in a patient with congenital lactic acidosis. *J. clin. Invest.* **51**, 1845.

Blau, K., Rattenbury, J.M., Pryse-Davies, J., Clark, P., and Sandler, M. (1978). Prenatal detection of hypophosphatasia: cytological and genetic considerations. *J. inher. metab. Dis.* **1**, 37.

—— Summer, G.K., Newsome, H.C., Edwards, C.H., and Mamer, O.A. (1973). Phenylalanine loading and aromatic acid excretion in normal subjects and heterozygotes for phenylketonuria. *Clin. Chim. Acta* **45**, 197.

Blaw, M.E. (1970). Melanodermic type leucodystrophy (adrenoleucodystrophy). In *Handbook of clinical neurology* (ed. P.J. Vinken, and G.W. Bruyn, Vol. 10, p.128. American Elsevier, New York.

Blom, W., De Muinck Keizer, S.M.P.F., and Scholte, H.R. (1981). Acetyl-CoA carboxylase deficiency: an inborn error of de novo fatty acid synthesis. *New Engl. J. Med.* **305**, 465.

Bloom, G. and Zarkowsky, H. (1969). Heterogeneity of the enzyme defect in congenital methaemoglobinuria. *New Engl. J. Med.* **281**, 919.

Bloomer, J.R., Brenner, D.A., and Mahoney, M.J. (1977). Study of factors causing excess protoporphyrin accumulation in cultured skin fibroblasts from patients with protoporphyria. *J. clin. Invest.* **60**, 1354.

Blume, K.G., Hryniuk, W., Powars, D., Trinidad, F., West, C., and Beutler, E. (1972). Characterization of two new variants of glucose-phosphate-isomerase deficiency with hereditary nonspherocytic hemolytic anemia. *J. lab. clin. Med.* **79**, 942.

Board, P.G., Trueworthy, R., Smith, J.E., and Moore, K. (1978). Congenital nonspherocytic hemolytic anemia with an unstable hexokinase variant. *Blood* **51**, 111.

Bodegaard, G., Gentz, J., Lindblad, B., Lindstedt, S., and Zetterstrom, R. (1969). Hereditary tyrosinemia. III. On the differential diagnosis and the lack of effect of early dietary treatment. *Acta. Pediat. Scand.* **58**, 37.

Boedeker, C. (1859). Ueber das Alcapton: ein neuer Beitrag zur Frage: Welche Stoffe des Harms können Kupferreduction bewirken? *Rat. Med.* **7**, 130.

Bois, E., Feingold, J. Frenay, P., and Briard, M.-L. (1976). Infantile cystinosis in

France: genetics, incidence, geographic distribution. *J. med. Genet.* **13**, 434.

Boivin, P., Galand, C., Hakim, J., Rogé,J., and Gueroult, N. (1969). Anémie hémolytique avec déficit en glutathion-peroxydase chez un adulte. *Enzymol. Biol. clin.* **10**, 68.

Bongiovanni, A.M. (1958). *In vitro* hydroxylation of steroids by whole adrenal homogenates of beef, normal man, and patients with adrenogenital syndrome. *J. clin. Invest.* **37**, 1342.

—— (1962). Adrenogenital syndrome with deficiency of 3ß-hydroxysteroid dehydrogenase. *J. clin. Invest.* **41**, 2086.

—— (1980). Urinary steroidal pattern of infants with congenital adrenal hyperplasia due to 3ß-hydroxysteroid dehydrogenase deficiency. *J. Steroid Biochem.* **13**, 809.

—— Album, M.M., Root, A.W., Hope, J., Marino, J., and Spencer, D.M. (1968). Studies on hypophosphatasia and response to high phosphate intake. *Am. J. Med. Sci.* **255**,163

Bonkowsky, H.L., Tschudy, D.P., Weinbach, E.C., Ebert, P.S., and Doherty, J.M. (1975). Porphyrin synthesis and mitochondrial respiration in acute intermittent porphyria: Studies using cultured human fibroblasts. *J. lab. clin. Med.* **85**, 93.

—— —— Collins, A. Doherty, J. Bossenmaier, I., Cardinal, R., and Watson, C.J. (1971). Repression of the overproduction of porphyrin precursors in acute intermittent porphyria by intravenous infusions of hematin. *Proc. natl. Acad. Sci. USA* **68**, 2725.

Booth, V.H.. (1938). The specificity of xanthine oxidase. *Biochem. J.* **32**, 494.

Booth, C.W. and Nadler, H.L. (1974). Demonstration of heterozygous state in Hunter's syndrome. *Pediatrics* **53**, 396.

—— Chen, K.K., and Nadler, H.L. (1975). Cerebroside sulfatase activity in cultivated human skin fibroblasts and amniotic fluid cells. *J. Ped.* **86**, 560.

—— —— —— (1976). Mannosidosis: clinical and biochemical studies in a family of affected adolescents and adults. *J. Ped.* **88**, 821.

—— Gerbie, A.B. and Nadler, H.L. (1973). Intrauterine detection of GM$_1$-gangliosidosis type 2. *Pediatrics* **52**, 521.

Bootsma, D. (1978). Xeroderma pigmentosum. In *DNA repair mechanisms* (ed. P.C. Hanawalt and E.C. Friedberg), p.589. Academic Press, New York.

Borden, T.A. and Lyon, E.S. (1969). The effects of magnesium and pH on experimental calcium oxalate stone disease. *Invest. Urol.* **6**, 412.

Borg, L., Lindstedt, S., Steen, G., and Hjalmarson, O. (1972). Aliphatic C_6–C_{14} dicarboxylic acids in urine from an infant with fatal congenital lactic acidosis. *Clin. Chim. Acta* **41**, 363.

Bornstein, P. and Byers, P.H. (1980). Disorders of collagen metabolism. In *Metabolic control and disease,* 8th edn. (ed. P.K. Bondy and L.E. Rosenberg), p.1089. W.B. Saunders Co., Philadelphia.

Bornstein, P. and Traub, W. (1979). The chemistry and biology of collagen. In *The proteins*, 3rd edn. (ed. H. Neurath, R.L. Hill and C.-L. Boeder). Vol. 4, p.411. Academic Press Inc., New York.

Børresen, A.-L. and Van der Hagen, C.B. (1973). Metachromatic leukodystrophy II. Direct determination of arylsulphatase A activity in amniotic fluid. *Clin. Genet.* **4**, 442.

Borrone, C., Gatti, R., Trias, X., and Durand, P. (1974). Fucosidosis: clinical, biochemical, immunologic and genetic studies in two new cases. *J. Ped.* **84**, 727.

Bosch, E.P. and Hart, M.N. (1978). Late adult-onset metachromatic leukodystrophy: dementia and polyneuropathy in a 63 year old man. *Arch. Neurol.* **35**, 475.

Boström, H. and Hambraeus, L. (1964). Cystinuria in Sweden. VII. Clinical, histopathological and medicosocial aspects of the disease. *Acta Med. Scand. suppl.* **411**, 1.

Botstein, D., White, R.L., Skolnick, M., and Davis, R.W. (1980). Construction of a genetic linkage map in man using restriction fragment length polymorphisms. *Am. J. hum. Genet.* **32**, 314.

Bottomley, S.S., Tanaka, M., and Everett, M.A. (1975). Diminished erythroid ferrochelatase activity in protoporphyria. *J. lab. clin. Med.* **86**, 126.

Bourke, E., Frindt, G., Flynn, P., and Schreiner, G.E. (1972). Primary hyperoxaluria with normal α-ketoglutarate: glyoxylate carboligase activity. *Ann. int. Med.* **76**, 279.

Bourne, J.G., Collier, H.O.J., and Somers, G.F. (1952). Succinylcholine (succinoylcholine). Muscle relaxant of short action. *Lancet* **i**, 1225.

Bowdler, A.J. and Prankerd, T.A.J. (1964). Studies in congenital non-spherocytic haemolytic anaemias with specific enzyme defects. *Acta Haematol.* **31**, 65.

Bowen, P., Lee, C.S.N., Migeon, C.J., Kaplan, N.M., Whalley, P.J., McKusick, V.A., and Reifenstein, E.C. jun. (1965). Hereditary male pseudohermaphroditism with hypogonadism, hypospadias, and gynecomastia (Reifenstein's syndrome). *Ann. intern. Med.* **62**, 252.

Bowen, D.M. and Radin, N.S. (1969). Cerebroside galactosidase: a method for determination and a comparison with other lysosomal enzymes in developing rat brain. *J. Neurochem.* **16**, 501.

Bowman, H.S. and Procopio, F. (1963). Hereditary non-spherocytic hemolytic anemia of the pyruvate-kinase deficient type. *Ann. intern. Med.* **58**, 567.

—— McKusick, V.A., and Dronamraju, K.R. (1965). Pyruvate kinase deficiency hemolytic anemia in an Amish isolate. *Am. J. hum. Genet.* **17**, 1.

Boyd, E.M., Dolman, M., and Sheppard, E.P. (1965). The chronic oral toxicity of caffeine. *Can. J. Physiol. Pharmacol.* **43**, 995.

Brady, R.O. (1966). The sphingolipidoses. *New Engl. J. Med.* **275**, 312.

—— (1978). Glucosyl ceramide lipidosis: Gaucher's disease. In *The metabolic basis of inherited disease*, 4th edn. (ed. J.B. Stanbury, J.B. Wyngaarden, and D.S. Fredrickson), p.731. McGraw-Hill, New York.

—— Johnson, W.G., and Uhlendorf, B.W. (1971*a*). Identification of heterozygous carriers of lipid storage diseases. *Am. J. Med.* **51**, 423.

—— Kanfer, J.N., and Shapiro, D. (1965). Metabolism of glucocerebrosides. II. Evidence of an enzyme deficiency in Gaucher's disease. *Biochem. biophys. Res. Commun.* **18**, 221.

—— Uhlendorf, B.W., and Jacobson, C.B. (1971*b*). Fabry's disease: antenatal detection. *Science* **172**, 174.

—— Kanfer, J.N., Bradley, R.M., and Shapiro, D. (1966*a*). Demonstration of a deficiency of glucocerebroside-cleaving enzyme in Gaucher's disease. *J. clin. Invest.* **45**, 1112.

—— —— Mock, M.B., and Fredrickson, D.S. (1966*b*). The metabolism of sphingomyelin. II. Evidence of an enzymatic deficiency in Niemann–Pick disease. *Proc. natl. Acad. Sci. USA* **55**, 366.

—— Barranger, J.A., Furbish, F.S., Stowens, D.W., and Ginns, E.I. (1982). Prospects for enzyme replacement in Gaucher's Disease. In *Gaucher disease: a century of delineation and research* (ed. R.J. Desnick, S. Gatt, and G.A. Grabowski). p.669. Alan R Liss, New York.

—— Pentchev, P.G., Gal, A.E., Hibbert, S.R., and Dekaban, A.S. (1974). Replacement therapy for inherited enzyme deficiency. Use of purified glucocerebrosidase in Gaucher's disease. *New Engl. J. Med.* **291**, 989.

—— Gal, A.E., Bradley, R.M., Mårtensson, E., Warshaw, A.L., and Laster, L. (1967). Enzymatic deficiency in Fabry's disease: ceramide trihexosidase deficiency. *New Engl. J. Med.* **276**, 1163.

—— Tallman, J.F., Johnson, W.G., Gal, A.E., Leahy, W.R., Quirk, J.M., Dekaban, A.S. (1973). Replacement therapy for inherited enzyme deficiency: use of purified ceramidetrihexosidase in Fabry's disease. *New Engl. J. Med.* **289**, 9.

Braidman, J., Carroll, M., Dance, N., Robinson, D., Poenam, L., Weber, A., Dreyfus, J.C., Overdijk, B., and Hooghwinkel, G.J.M. (1974). Characterisation of human hexosaminidase C. *FEBS Lett.* **41**, 181.

Brailsford, J.F. (1929). Chondro-osteo-dystrophy, roentgenographic and clinical features of child with dislocation of vertebrae. *Am. J. Surg.* **7**, 404.

Brain, W.R. (1927). Heredity in simple goitre. *Q. J. Med.* **20**, 303.

Brandt, I.K. and De Luca, V.A. jun. (1966). Type III glycogenosis: a family with an unusual tissue distribution of the enzyme lesion. *Am. J. Med.* **40**, 779.

Brandt, N.J. (1980). How long should galactosaemia be treated? In *Inherited disorders of carbohydrate metabolism* (ed. D. Burman, J.B. Holton, and C.A. Pennock), p.117. MTP Press Ltd., Lancaster.

—— and Hanel, H.K. (1971). Atypical pyruvate kinase in a patient with haemolytic anaemia. *Scand, J. Haematol.* **8**, 126.

—— Brandt, S., Christensen, E., Gregersen, N., and Rasmussen, K. (1978). Glutaric aciduria in progressive choreo-athetosis. *Clin. Genet.* **13**, 77.

—— Gregersen, N., Christensen, E., Grøn, I.H., and Rasmussen, K. (1979). Treatment of glutaryl-CoA dehydrogenase deficiency (glutaric aciduria). *J. Ped.* **94**, 669.

Brante, G. (1952). Gargoylism: a mucopolysaccharidosis? *Scand. J. clin. lab. Invest.* **4**, 43.

Braunstein, G.D., Ziel, F.H., Allen, A., Van de Velde, R., and Wade, M. (1976). Prenatal diagnosis of placental steroid sulfatase deficiency. *Am. J. Obstet. Gynecol.* **126**, 716.

Brenton, D.P. and Cusworth, D.C. (1966). Homocystinuria: metabolism of [^{35}S] methionine. *Clin. Sci.* **31**, 197.

—— Cusworth, D.C., and Gaull, G.E. (1965). Homocystinuria: metabolic studies on three patients. *J. Ped.* **67**, 58.

—— —— Biddle, S.A., Garrod, P.J., and Lasley, L. (1977). Pregnancy and homocystinuria. *Ann. clin. Biochem.* **14**, 161.

—— Dow, C.J., James, J.I.P., Hay, R.L., and Wynne-Davis, R. (1972). Homocystinuria and Marfan's syndrome: a comparison. *J. Bone Joint Surg.* **54B**,277.

Brett, E.M., Ellis, R.B., Haas, L., Ikonne, J.K., Lake, B.D., Patrick, A.D., and Stevens, R. (1973). Late onset GM$_2$-gangliosidosis, clinical, pathological and biochemical studies on eight patients. *Arch. Dis. Child.* **48**, 775.

Brewer, G.J. and Dern, R.J. (1964). A new inherited enzymatic deficiency of human erythrocytes: 6-phosphogluconate dehydrogenase deficiency. *Am. J. hum. Genet.* **16**, 472.

Brewster, M.A., Whaley, S.A., and Kane, A.C. (1973). Variables in the laboratory diagnosis of Fabry's disease by measurement of methylumbelliferyl-α-galactosidase activity. *Clin. Chem.* **20**, 383.

Brewster, T.G., Moskowitz, M.A., Kaufman, S., Breslow, J.L., Milstien, S., and Abroms, I.F. (1979). Dihydropteridine reductase deficiency associated with severe neurological disease and mild hyperphenylalanimenia. *Pediatrics* **63**, 94.

Briggs, A.P. (1922). A colorimetric method for the determination of homogentisic acid in urine. *J. biol. Chem.* **51**, 453.

Briggs, M.H., Garcia-Webb, P., and Davies, P. (1973). Urinary oxalate and vitamin-C supplements. *Lancet* **ii**, 201.

Brill, P.W., Mitty, H.A., and Gaull, G.E. (1974). Homocystinuria due to cystathionine synthase deficiency: Clinical-roentgenologic correlations. *Am. J. Roentgenol. radium Ther. nucl. Med.* **121**, 45.

Brinn, L. and Glaubman, S. (1962). Gaucher's disease without splenomegaly: oldest patient on record, with review. *N. Y. J. Med.* **62**, 2346.

Brock, D.J.H., Gordon, H. Seligman, S., and Lobo, E. de H. (1971). Antenatal detection of Hurler's syndrome. *Lancet* **ii**, 1324.

Brodehl. J. and Gellissen, K. (1968). Endogenous renal transport of free amino acids in infancy and childhood. *Pediatrics* **42**, 395.

—— —— and Kowalesko, S. (1966). Isolated cystinuria (without lysine-ornithine-argininuria) in a family with hypocalcemic tetany. *Proceedings of the 3rd International congress on nephrology*. Washington.

—— —— Hagge, W., and Schumacher, H. (1968). Reversibles renales Fanconi–Syndrom durch toxisches Abbauprodukt des Tetrazyklins. *Helv. Paediatr. Acta* **23**, 373.

Brodersen, R. (1980). Binding of bilirubin to albumin. *CRC Crit. Rev. clin. lab. Sci.* **11**, 305.

Brodie, M.J., Moore, M.R., Thompson, G.G., Goldberg, A., and Holti, G. (1978). Haem biosynthesis in peripheral blood in erythropoietic protoporphyria. *Clin. exp. Dermatol.* **2**, 381.

Bron, A.F., Lloyd, J.K., Fosbrooke, A.S., Winder, A.F., and Tripathi, R.C. (1975). Primary lecithin: cholesterol acyltransferase deficiency disease. *Lancet* **i**, 928.

Brown, A.K. and Zuelzer, W.W. (1958). Studies on the neonatal development of the glucuronide conjugating system. *J. clin. Invest.* **37**, 332.

Brown, A.L., Fernhoff, P.M., Milner, J., McEwan, C., and Elsas, L.S. (1981). Racial differences in the incidence of congenital hypothyroidism. *J. Ped.* **99**, 934.

Brown, B.I. and Brown, D.H. (1966). Lack of an α-1,4-glucan: α-1,4-glucan 6-glycosyl transferase in a case of type IV glycogenosis. *Proc. natl. Acad. Sci. USA* **56**, 725.

—— —— (1968). The glycogen storage diseases. Types I, III, IV, V, VII and unclassified glycogenoses. In *Carbohydrate metabolism and its disorders*, Vol II (ed. W.J. Whelan), p.123. Academic Press, New York.

—— —— and Jeffrey, P.L. (1970). Simultaneous absence of α-1,4-glucosidase and α-1,6-glucosidase in tissues of children with type II glycogen storage disease. *Biochemistry* **9**, 1423.

Brown, G.K., Hunt, S.M., Scholem, R. Fowler, K., Grimes, A., Mercer, J.F.B., Truscott, R.M., Cotton, R.G.M., Rogers, J.G., and Danks, D.M. (1982). ß-hydroxyisobutyryl coenzyme A deacylase deficiency: a defect in valine metabolism associated with physical malformations. *Pediatrics* **70**, 532.

Brown, J.H. and Allison, J.B. (1948). Effects of excess dietary DL-methionine and/or L-arginine on rats. *Proc. Soc. exp. Biol. Med.* **69**, 196.

Brown, M.S.,and Goldstein, J.L. (1975). Regulation of the activity of the low density lipoprotein receptor in human fibroblasts. *Cell* **6**, 307.

—— —— (1976*a*). Receptor-mediated control of cholesterol metabolism. *Science* **181**, 150.

—— —— (1976*b*). Analysis of mutant strain of human fibroblasts with a defect in the internalization of receptor-bound low density lipoprotein. *Cell* **9**, 663.

—— Dana, S.E., Goldstein, J.L. (1974). Regulation of 3-hydroxy-3-methylglutaryl coenzyme A reductase activity in cultured human fibroblasts: comparison of cells from a normal subject and from a patient with homozygous familial hypercholesterolemia. *J. biol. Chem.* **249**, 789.

—— Kovanen, P.T., and Goldstein, J.L. (1981). Regulation of plasma cholesterol by lipoprotein receptors. *Science* **212**, 628.

—— —— —— Eeckels, R., Vandenberghe, K., Van den Bergh, H., Fryns, J.P., and Cassiman, J.J. (1978). Prenatal diagnosis of homozygous familial hypercholesterolaemia: expression of a genetic receptor disease in utero. *Lancet* **i**, 526.

Broyer, M., Guillot, M., Gubler, M.-C., and Habib, R. (1981) Infantile cystinosis: a reappraisal of early and late symptoms. In *Advances in nephrology* (ed. J. Hamburger, J.C. Rosnier, J.-P. Grünfeld, and M.H. Maxwell), p.137. Year Book, Chicago.

Brunberg, J.A., McCormick, W.F., and Schochet, S.S. (1971). Type III glycogenosis: an adult with diffuse weakness and muscle wasting. *Arch. Neurol.* **25**, 171.

Brunette, M.G., Delvin, E., Hazel, B., and Scriver, C.R. (1972). Thiamin responsive lactic acidosis in a patient with deficient low K_m pyruvate carboxylase activity in liver. *Pediatrics* **50**, 702.

Brusilow, S.W., Batshaw, M., and Walser, M. (1979). Use of ketoacids in inborn errors of urea synthesis. In *Nutritional management of genetic disorders* (ed. M. Winick), P.65. John Wiley and Sons, New York.

—— —— (1979). Arginine therapy of argininosuccinase deficiency. *Lancet* **i**, 124.

—— Tinker, J., and Batshaw, M.L. (1980). Amino acid acylation: a mechanism of nitrogen excretion in inborn errors of urea synthesis. *Science* **207**, 659.

Bruton, C.J., Corsellis, J.A.N., and Russel, A. (1970). Hereditary hyperammonaemia. *Brain* **93**, 423.

Bryan, G.T., Kliman, B., and Bartter, F.C. (1965). Impaired aldosterone production in 'salt-losing' congenital adrenal hyperplasia. *J. clin. Invest.* **44**, 957.

Brydon, W.G., Crofton, P.M., Smith, A.F., Barr, D.G.D., and Harkness, R.A. (1975). Hypophosphatasia: enzyme studies in cultured cells and tissues. *Biochem. Soc. Trans.* **3**, 927.

Buchanan, P.D., Kahler, S.G., Sweetman, L., and Nyhan, W.L. (1980). Pitfalls in the prenatal diagnosis of propionic acidaemia. *Clin. Genet.* **18**, 177.

Bucknall, W.E., Haslam, H.A., and Holtzman, N.A. (1973). Kinky hair syndrome: response to copper therapy. *Pediatrics* **52**, 653.

Budd, M.A., Tanaka, K., Holmes, L.B., Efron, M.L., Crawford, J.D., Isselbacker, K.J. (1967). Isovaleric acidemia: clinical features of a new genetic defect of leucine metabolism. *New Engl. J. Med.* **277**, 321.

Bugany, H., Flothe, L., and Weser, U. (1971). Kinetics of metal chelatase of rat liver mitochondria. *FEBS Lett.* **13**, 92.

Bugiani, O. (1982). Cited by Durand, P., Gatti, R., and Borrone, C. (1982). Fucosidosis. In *Genetic errors of glycoprotein metabolism* (Ed. P. Durand and J.S. O'Brien), p.81. Edi-Ernes, Milan; Springer, New York.

Buhler, F.R., Thiel, G., Dubach, V.C., Enderlin, F., Gloor, F., and Tholen, H. (1973). Kidney transplantation in Fabry's disease. *Br. med. J.* **3**, 28.

Buist, N.R.M. and Jhaveri, B.M. (1973). A guide to screening newborn infants for inborn errors of metabolism. *J. Ped.* **82**, 511.

—— Kennaway, N.G., and Fellman, J.H. (1974*a*). Disorders of tyrosine metabolism. In *Heritable disorders of amino acid metabolism* (ed. W.L. Nyhan). p.160. John Wiley and Sons, New York.

—— Lis, E.W., Tuerck, J.M., and Murphey, W.H. (1979). Maternal phenylketonuria. *Lancet* **ii**, 589.

—— Strandholm, J.J., Bellinger, J.F., and Kennaway, N.G. (1972). Further studies on a patient with iminopeptiduria: a probable case of prolidase deficiency. *Metabolism* **21**, 1113.

—— Kennaway, N.G., Hepburn, C.A., Strandholm, J.J., and Damberg, M.S. (1974*b*). Citrullinemia: investigation and treatment over a four-year period. *J. Ped.* **85**, 208.

Bulkley, B.H., Buja, L.M., Ferrans, V.J., Bulkley, G.B., and Roberts, W.C. (1975). Tuberous xanthoma in homozygous type II hyperlipoproteinemia. *Arch. Pathol.* **99**, 293.

Bundza, A., Lowden, J.A., and Charlton, K.M. (1979). Niemann–Pick disease in a poodle dog. *Vet. Pathol.* **16**, 530.

Buniatian, H.C. and Davtian, M.A. (1966). Urea synthesis in brain. *J. Neurochem.* **13**, 743.

Buonnano, F.S., Ball, M.R., Laster, D.W., Moody, D.M., and McLean, W.T. (1978). Computed tomography in late-infantile metachromatic leukodystrophy. *Ann. Neurol.* **4**, 43.

Burditt, L., Chotai, K., Halley, D., and Winchester, B.G. (1980*a*). Comparison of the residual acidic α-D-mannosidase in three cases of mannosidosis. *Clin. Chim. Acta* **104**, 201.

—— —— Hirani, S., Nugent, P.G., Winchester B.G., and Blakemore, W.F. (1980*b*). Biochemical studies on a case of feline mannosidosis. *Biochem. J.* **189**, 467.

Burgeson, R.E., Adli, F.A.E., Kaitila, I.I., and Hollister, D.W. (1976). Fetal membrane collagens: identification of two new collagen alpha chains. *Proc. natl. Acad. Sci. USA* **73**, 2579.

Burke, V. and Anderson, C.M. (1966). Sugar intolerance as a cause of protracted diarrhoea following surgery of the gastrointestinal tract in neonates. *Aust. Paediatr. J.* **2**, 219.

Burke, E.C., Baggenstoss, A.H., Owen, C.A. jun., Power, M.H., and Lohr, O.W. (1955). Oxalosis. *Pediatrics* **15**, 383.

Burkholder, P.M., Updike, S.J. Ware, R.A., Reese, O.G. (1980). Clinopathologic, enzymatic and genetic features in a case of Fabry's disease. *Arch. Pathol. lab. Med.* **104**, 17.

Burman, J.F., Mollin, D.L., Sourial, N.A., and Sladden, R.A. (1979). Inherited lack of transcobalamin II in serum and megaloblastic anaemia: a further patient. *Br. J. haematol.* **43**, 27.

Burnett, C.H., Dent, C.E., Harper, C., and Warland, B.J. (1964). Vitamin D-resistant rickets: analysis of 24 pedigrees with hereditary and sporadic cases. *Am. J. Med.* **36**, 222.

Burnett, J.B., Holstein, T.J., and Quevedo, W.C. jun. (1969). Electrophoretic variations of tyrosinase in follicular melanocytes during the hair growth cycle. *J. exp. Zool.* **171**, 369.

Burns, G.F., Cawley, J.C., Flemans, R.J., Higgy, K.E., Worman, C.P., Barker, C.R., Roberts, B.E., and Hayhoe, F.G.J. (1977). Surface marker and other characteristics of Gaucher cells. *J. clin. Pathol.* **30**, 981.

Burr, I.M., O'Neill, J.A., Karzan, D.T., Howard, L.J., and Greene, H.L. (1974). Comparison of the effects of total parenteral nutrition, continuous intragastric feeding, and portacaval shunt on a patient with Type I glycogen storage disease. *J. Ped.* **85**, 792.

Burri, B.J., Sweetman, L., and Nyhan, W.L. (1981). Mutant holocarboxylase synthetase: evidence for the enzyme defect in early infantile biotin-responsive multiple carboxylase deficiency. *J. clin. Invest.* **68**, 1491.

Burrow, G.N., Spaulding, S.W., Alexander, N.M., and Bower, B.F. (1973).

Normal peroxidase activity in Pendred's syndrome. *J. clin. Endocr. Metab.* **36**, 522.

Burton, B.K. and Nadler, H.L. (1978). Mannosidosis: separation and characterisation of two acid α -mannosidase forms in mutant fibroblasts. *Enzyme* **23**, 29.

Busch, D., Witt, I., Berger, M., and Künzer, W. (1966). Deficiency of pyruvate kinase in the erythrocytes of a child with hereditary non-spherocytic hemolytic anemia. *Acta Paediatr. Scand.* **55**, 177.

Butterworth, J. and Guy, G.J. (1977). α -L-fucosidase of human skin fibroblasts and amniotic fluid cells in tissue culture. *Clin. Genet.* **12**, 297.

—— Sutherland, G.R., Broadhead, D.M., and Bain, D. (1973). Lysosomal enzyme levels in human amniotic fluid cells in tissue culture. *Life Sci.* **13**, 713.

Buttoo, A.S., Israëls, M.C.G., and Wilkinson, J.S. (1965). Hypocholesterolaemia and orotic aciduria during treatment with 6-azauridine. *Br. med. J.* **i**, 552.

Byers, P.H., Barsh, G.S., Peterson, K.E., Phillips, J.M., Shapiro, J., Holbrook, K.A., Levin, L.S., and Rowe, D.W. (1981). Biochemical characterisation of perinatal lethal osteogensis imperfecta (OI) and its prenatal detection (Abstract no. 705). *Ped. Res.* **15**, 559.

—— Siegel, R.C., Holbrook, K.A., Narayanan, A.S., Bornstein, P., and Hall, J.C. (1980). X-linked cutis laxa. Defective cross-link formation in collagen due to decreased lysyl oxidase activity. *New Engl. J. Med.* **303**, 61.

Cabalska, B., Duczynska, N., Borzymowska, J., Zorska, K., Koslacz-Folga, A., and Bozkowa, K. (1977). Termination of dietary treatment in phenylketonuria. *Eur. J. Pediatr.* **126**, 253.

Cahill, J. and Kirtley, M.E. (1975). FDPase activity in human leucocytes. *New Engl. J. Med.* **292**, 212.

Caimi, L., Tettamanti, G., Berra, B., Sale, F.O., Borrone, C., Gatti, R., Durand, P., and Martin, J.J. (1982). Mucolipidosis IV, a sialolipidosis due to ganglioside sialidase deficiency. *J. inher. metab. Dis.* **5**, 218.

Cain, A.R.R. and Holton, J.B. (1968). Histidinaemia: a child and his family. *Arch. Dis. Child.* **43**, 62.

Cain, H., Egner, E., and Kresse, H. (1977). Mucopolysaccharidose III A. Histochemische, elektronenmikroskopische und biochemische Befunde. *Beitr. Path.* **160**, 58.

Callahan, J.W. and Khalil, M. (1975). Sphingomyelinases in human tissues. III. Expression of Niemann–Pick disease in cultured skin fibroblasts. *Ped. Res.* **9**, 914.

—— —— and Gerrie, J. (1974). Isoenzymes of sphingomyelinase and the genetic defect in Niemann–Pick disease type C. *Biochem. biophys. Res. Commun.* **58**, 384.

—— —— and Philippart, M. (1975). Sphingomyelinases in human tissues. II. Absence of a specific enzyme from liver and brain of Niemann–Pick disease type C. *Ped. Res.* **9**, 908.

Calloway, D.H., Colasito, D.J., and Mathews, R.D. (1966). Gases produced by human intestinal microflora. *Nature* **212**, 1238.

Camacho, A.M., Kowarski, A., Migeon, C.J., and Brough, A.J. (1968). Congenital adrenal hyperplasia due to a deficiency of one of the enzymes involved in the biosynthesis of pregnenolone. *J. clin. Endocr. Metab.* **28**, 153.

Cammermeyer, J. (1956). Neuropathological changes in hereditary neuropathies: manifestation of the syndrome heredopathia atactica polyneuritiformis in the presence of interstitial hypertrophic neuropathy. *J. Neuropathol. exp. Neurol.* **3**, 340.

—— Haymaker, W., and Refsum, S. (1954). Heredopathia atactica polyneuritiformis: the neuropathological changes in three adults and one child. *Am. J. Path.* **30**, 643.

Campailla, E. and Martinelli, B. (1969). Morquio's disease: modification of mucopolysacchariduria with advancing age. In *Clinical delineation of birth defects. IV. Skeletal dysplasias*, p. 172. National Foundation, March of Dimes. Alan R. Liss, New York.

Campbell, A. and Williams, E. (1967). Natural history of Refsum's syndrome in a Gloucestershire family. *Br. med. J.* **iii**, 777.

Campbell, A.G.M., Rosenberg, L.E., Snodgrass, P.J., and Nuzum, C.T. (1971). Lethal neonatal hyperammonaemia due to complete ornithine-transcarbamylase deficiency. *Lancet* **ii**, 217.

—— —— —— —— (1973). Ornithine transcarbamylase deficiency. A cause of lethal neonatal hyperammonemia in males. *New Engl. J. Med.* **288**, 1.

Campbell, S., Grundy, M., and Singer, J.D. (1976). Early antenatal diagnosis of spina bifida in a twin fetus by ultrasonic examination and alpha fetoprotein estimation. *Br. med. J.* **ii**, 676.

Cantz, M., Gehler, J., and Spranger, J. (1977). Mucolipidosis I: increased sialic acid content and deficiency of a α-N-acetylneuraminidase in cultured fibroblasts. *Biochem. biophys. Res. Commun.* **74**, 732.

—— Kresse, H., Barton, R.W., and Neufeld. E.F. (1972). Corrective factors for inborn errors of mucopolysaccharide metabolism. *Methods Enzymol.* **28**, 884.

Carey, M.C., Fennelly, J.J., and Fitzgerald, O. (1968*b*). Homocystinuria. II. Subnormal serum folate levels, increased folate clearance and effects of folic acid therapy. *Am. J. Med.* **45**, 26.

—— Donovan, D.E., Fitzgerald, O., and McAuley, F.D. (1968*a*). Homocystinuria I. A clinical and pathological study of nine subjects in six families. *Am. J. Med.* **45**, 7.

Carmel, R. and Herbert, V. (1969). Deficiency of vitamin B_{12}-binding alpha globulin in two brothers. *Blood* **33**, 1.

—— Bedros, A.A., Mace, J.W., and Goodman, S.I. (1980). Congenital methylmalonic aciduria-homocystinuria with megaloblastic anemia: observations on response to hydroxocobalamin and on the effect of homocysteine and methionine on the deoxyuridine suppression test. *Blood* **55**, 570.

Carritt, B. (1977). Somatic cell genetic evidence for the presence of a gene for citrullinemia on human chromosome 9. *Cytogenet. cell. Genet.* **19**, 44.

Carroll, M. and Robinson, D. (1973). Immunological properties of N-acetyl-ß-D-glucosaminidase of normal human liver and of GM_2-gangliosidosis liver. *Biochem. J.* **131**, 91.

—— Dance, N., Masson, P.K., Robinson, D., and Winchester, B. (1972). Human mannosidosis — the enzyme defect. *Biochem. biophys. Res. Commun.* **49**, 579.

Carson, D.J., Okuno, A., Lee, P.A., Stetten, G., Didolkar, S.M., and Migeon, C.J. (1982). Amniotic fluid steroid levels. *Am. J. Dis. Child.* **136**, 218.

Carson, N.A.J. (1975). Homocystinuria. In *The treatment of inherited metabolic disease* (ed. D.N. Raine) p.33. MTP Publishing, Lancaster.

—— and Carre, I.J. (1969). Treatment of homocystinuria with pyridoxine: a preliminary study. *Arch. Dis. Child.* **44**, 387.

—— and Neill, D.W. (1962). Metabolic abnormalities detected in a survey of mentally backward individuals in Northern Ireland. *Arch. Dis. Child.* **37**, 505.

—— Dent, C.E., Field, C.M.B., and Gaull, G.E. (1965). Homocystinuria: clinical and pathological review of ten cases. *J. Ped.* **66**, 565.

—— Cusworth, D.C., Dent, C.E., Field, C.M.B., Neil, D.W., and Westall, R.G.

(1963). Homocystinuria: a new inborn error of metabolism associated with mental deficiency. *Arch. Dis. Child.* **38**, 425.

Carson, P.E., Brewer, G.J., and Ickes, C. (1961). Decreased glutathione reductase with susceptibility to hemolysis. (abst). *J. lab. clin. Med.* **58**, 804.

Carter, C.O., Slack, J., and Myant, N.B. (1971). Genetics of hyperlipoprotein-aemias. *Lancet* **i**, 400.

Cartier, P. and Hamet, M. (1968). Les activities purine-phosphiribosyl transférasiques des globules rouges humains: technique de dosage. *Clin. Chim. Acta* **20**, 205.

—— —— (1975). Une nouvelle maladie metabolique: le déficit complet en adenine-phosphoribosyltransférase avec lithiase de 2,8-dihydroxyadénine. *C. r. Acad. Sci. Ser. D. (Paris)* **279**, 883.

—— and Perignon, J.L. (1978). Xanthinuria. *Nouv. Presse Med.* **7**, 1381.

Carton, D., De Schrijver, F., Kint, J., Van Durme, J., and Hooft, C. (1969). Case report. Argininosuccinic aciduria. Neonatal variant with rapid fatal course. *Acta Paediatr. Scand.* **58**, 528.

Cartwright, G.E., Hodges, R.E., Gubler, C.J., Mahoney, J.P., Daum, K., Wintrobe, M.M., and Bean, W.B. (1954). Studies on copper metabolism. XIII. Hepato-lenticular degeneration. *J. clin. Invest.* **33**, 1487.

Casey, R.E., Zaleski, W.A., Philip, M., Mendelson, I.S., and MacKenzie, S.L. (1978).Biochemical and clinical studies of a new case of α-aminoadipic aciduria. *J. inher. metab. Dis.* **1**, 129.

Cassagne, C., Darriet, D., and Bourne, J.M. (1978). Biosynthesis of very long chain fatty acids by the sciatic nerve of the rabbit. *FEBS Lett.* **90**, 336.

Castaneda, E., Perez, A.E., Guillen, M.A., Ramirez-Robles, S., Gual, C., and Perez-Palacios, G. (1971). Metabolic studies in a patient with testicular feminization syndrome. *Am. J. Obstet. Gynecol.* **110**, 1002.

Castro-Mendoza, H.J., Cifuentes-Delatte, L.C., and Rapado, Y.A.R. (1972). Una nueva observacion de xanthinuria familiar. *Rev. clin. Esp.* **124**, 341.

Cathelineau, L., Dinh, D.P., Briand, P., Kamoun. P. (1981*b*). Studies on complementation in argininosuccinate synthetase and argininosuccinate lyase deficiencies in human fibroblasts. *Hum. Genet.* **57**, 282.

—— Navarro, J., Polonovski, C., and Saudubray, J.-M. (1973). X-linked transmission of structural gene mutations responsible for ornithine-trans-carbamylase deficiencies. *Lancet* **i**, 261.

—— Pham Dinh, D., Boué, J., Saudubray, J.M., Farriaux, J.P., and Kamoun, P. (1981*a*). Improved method for the antenatal diagnosis of citrullinemia. *Clin. Chim. Acta* **116**, 111.

Cavanagh, J.B. and Kyu, M.H. (1969). Colchincine-line effect on astrocytes after portacaval shunt in rats. *Lancet* **ii**, 620.

Cave, W.T. jun. and Dunn, J.T. (1975). Studies on the thyroidal defect in an atypical form of Pendred's syndrome. *J. clin. Endocr. Metab.* **41**, 590.

Cech, P., Stalder, H., Widmann, J.J., Rohner, A., and Miescher, P.A. (1979). Leukocyte myeloperoxidase deficiency and diabetes mellitus associated with *Candida albicans* liver abscess. *Am. J. Med.* **66**, 149.

Cederbaum, S.D., Blass, J.P., Minkoff, N., Brown, W.J., Cotton, M.E., and Harris, S.H. (1976). Sensitivity to carbohydrate in a patient with familial intermittent lactic acidosis and pyruvate dehydrogenase deficiency. *Ped. Res.* **10**, 713.

—— and Spector, E.B. (1978). Arginase activity in fibroblasts. *Am. J. hum. Genet.* **30**, 91.

—— Shaw, K.N.F., and Valente, M. (1977). Hyperargininemia. *J. Ped.* **90**, 569.

—— —— Dancis, J. Hutzler, J., and Blaskovics, J.C. (1979b). Hyperlysinemia with saccharopinuria due to combined lysine-ketoglutarate reductase and saccharopine dehydrogenase deficiencies presenting as cystinuria. *J. Ped.* **95**, 234.

—— —— Spector, E.B., Verity, M.A., Snodgrass, P.J., and Sugarman, G.I. (1979a). Hyperargininemia wth arginase deficiency. *Ped. Res.* **13**, 827.

Chacko, C.M., Christian, J.C., and Nadler, H.L. (1971). Unstable galactose-1-phosphate uridyl transferase: a new variant of galactosemia. *J. Ped.* **78**, 454.

—— Wappner, R.S., Brandt, I.K., and Nadler, H.L. (1977). The Chicago variant of clinical galactosemia. *Hum. Genet.* **37**, 261.

Chalmers, R.A., Johnson, M., Pallis, C., and Watts, R.W.E. (1969a). Xanthinuria with myopathy (with some observations on the renal handling of oxypurines in the disease). *Q. J. Med.* **38**, 493.

—— Watts, R.W.E, Bitensky, L. and Chayen, J. (1969b). Microscopic studies on crystals in skeletal muscle from two cases of xanthinuria. *J. Pathol.* **99**, 45.

Chambers, R.A. and Pratt, R.T. (1956). Idiosyncrasy to fructose. *Lancet* **ii**, 340.

Champion, M.J., Brown, J.A., and Shows, J.A. (1978). Studies on the α-mannosidase (MAN $_B$), peptidase D (PEPD), and glucose phosphate isomerase (GPI) syntenic group on chromosome 19 in man. *Cytogenet. cell. Genet.* **22**, 186.

Chan, A.M., Lynch, M.J.G., Bailey, J.D., Ezrin, C., and Fraser, D. (1970). Hypothyroidism in cystinosis. *Am. J. Med.* **48**, 678.

Chan, T.K., Todd, D., and Wong, C.C. (1965). Tissue enzyme levels in erythrocyte glucose-6-phosphate dehydrogenase deficiency. *J. lab. clin. Med.* **66**, 937.

Chan, W.-Y. and Rennert, O.M. (1982). Prenatal and postnatal diagnosis of diseases of copper metabolism. *Ann. clin. lab. Sci.* **12**, 372.

—— Garnica, A.D., and Rennert, O.M. (1978). Cell culture studies of Menkes kinky hair disease. *Clin. Chim. Acta* **88**, 495.

—— Cushing, W., Coffman, M.A., and Rennert, O.M. (1980). Genetic expression of Wilson's disease in cell culture: a diagnostic marker. *Science* **208**, 299.

Chang, P.L. and Davidson, R.G. (1980). Complementation of arylsulfatase A in somatic hybrids of metachromatic leucodystrophy and multiple sulfatase deficiency disorder fibroblasts. *Proc. natl. Acad. Sci. USA* **77**, 6166.

Chapman, R.C. (1969). Red cell aldolase deficiency in hereditary spherocytosis. *Br. J. Haematol.* **16**, 145.

Charles, B.M., Hosking, G., Green, A., Pollitt, R., Bartlett, K., and Taitz, L.S. (1979). Biotin-responsive alopecia and developmental regression. *Lancet* **ii**, 118.

Chase, G.A. and McKusick, V.A. (1972). Founder effect in Tay–Sachs disease. *Am. J. hum. Genet.* **24**, 339.

Chatterjea, J.B. (1964). Erythropoietic porphyria. *Blood* **24**, 806.

Chavers, B.M., Kjellstrand, C.M., Wiegand, C., Ebben, J., and Mauer, S.M. (1980). Techniques for use of charcoal hemoperfusion in infants: experience in two patients. *Kidney Int.* **18**, 386.

Chayoth, R., Moses, S.W., and Steinitz, K. (1967). Debrancher enzyme activity in blood cells of families with type III glycogen storage disease. *Isr. J. med. Sci.* **3**, 422.

Chazan, S., Zitman, D., and Klibansky, C. (1978). Prenatal diagnosis of Gaucher's and Neimann–Pick diseases: assays of glucocerebrosidase and sphingomyelinase in tissue cultures using natural substrates. *Clin. Chim. Acta* **86**, 45.

Cheetham, P.S.J., Dance, N., and Robinson, D. (1975). Isoenzymes of human liver ß-galactosidase. *Biochem. Soc. Trans.* **3**, 240.

Chen, S.-H., Scott, C.R., and Giblett, E.R. (1974). Adenosine deaminase:

Demonstration of a 'silent' gene associated with combined immunodeficiency disease. *Am. J. hum. Genet.* **26**, 103.
—— Malcolm, L.A., Yoshida, A., and Giblett, E.R. (1971). Phosphoglycerate kinase: an X-linked polymorphism in man. *Am. J. hum. Genet.* **23**, 87.
—— Ochs, H.D., Scott, C.R., Giblett, E.R., and Tingle, A.J. (1978). Adenosine deaminase deficiency. Disappearance of adenine deoxynucleotides from a patient's erythrocytes after successful marrow transplantation. *J.clin. Invest.* **62**, 1386.
Chern, C.J. and Croche, C.M. (1976). Assignment of the structural gene for human ß-glucuronidase to chromosome 7 and tetrameric association of subunits in the enzyme molecule. *Am. J. hum. Genet.* **28**, 350.
Chester, M.A., Lundblad, A., and Masson, P.K. (1975). The relationship between different forms of human α-mannosidase *Biochim Biophys. Acta* **391**, 341.
—— —— Öckerman, P.-A., and Autio, S. (1982). Mannosidosis. In *Genetic errors of glycoprotein metabolism* (ed. P. Durand and J.S. O'Brien), p.90. Edi-Ermes, Milan; Springer-Verlag, New York.
Childs, B. and Nyhan, W.L. (1964). Further observations of a patient with hyperglycinemia. *Pediatrics* **33**, 403.
—— Grumbach, M.M., and Van Wyk, J.J. (1956). Virilizing adrenal hyperplasia: a genetic and hormonal study. *J. clin. Invest.* **35**, 213.
—— Sidbury, J.B., and Migeon, C.J. (1959). Glucuronic acid conjugation by patients with familial non-hemolytic jaundice and their relatives. *Pediatrics* **23**, 903.
—— Nyhan, W.L., Borden, M., Bard, L., and Cooke, R.E. (1961). Idiopathic hyperglycinemia and hyperglycinuria: a new disorder of aminoacid metabolism. *Pediatrics* **27**, 522.
Chisolm, J.J. jun. (1962). Aminoaciduria as a manifestation of renal tubular injury in lead intoxication and a comparison with patterns of aminoaciduria seen in other diseases. *J. Ped.* **60**, 1.
—— and Harrison, H.E. (1962). Aminoaciduria in vitamin D deficiency states in premature infants and older infants with rickets. *J. Ped.* **60**, 206.
Choremis, C., Kattamis, Ch.A., Kyriazakou, M., and Gavriilidou, E. (1966). Viral hepatitis in G-6-PD deficiency. *Lancet* **i**, 269.
Choufoer, J.C. (1961). Further observations on congenital hypothyroidism with defective dehalogenation of iodotyrosines. In *Advances in thyroid research* (ed. R. Pitt-Rivers), p.36. Pergamon Press, Elmsford, New York.
—— Kassenaar, A.A.H., and Querido, A. (1960). The syndrome of congenital hypothyroidism with defective dehalogenation of iodotyrosines: further observations and discussion of the pathophysiology. *J. clin. Endocr. Metab.* **20**, 983.
Choy, F.Y.M. and Davidson, R.G. (1980a). Gaucher's disease. II. Studies on the kinetics of ß-glucosidase and the effects of sodium taurocholate in normal and Gaucher tissues. *Ped. Res.* **14**, 54.
—— —— (1980b). Gaucher disease. III. Substrate specificity of glucocerebrosidase and the use of nonlabeled natural substrates for the investigation of patients. *Am. J. hum. Genet.*, **32**, 670.
Christensen, E. and Brandt, N.J. (1978). Studies on glutaryl-CoA dehydrogenase in leucocytes, fibroblasts and amniotic fluid cells. The normal enzyme and the mutant form in patients with glutaric aciduria. *Clin. Chim. Acta* **88**, 267.
—— Brandt, N.J., Philip, J., and Kennaway, N.G. (1980). Citrullinemia: the possibility of prenatal diagnosis. *J. inher. metab. Dis.* **3**, 73.
Christensen-Lou, H.O. (1966). A biochemical investigation of angiokeratoma corporis diffusum. *Acta pathol. microbiol. Scand.* **68**, 332.

Christomanou, H. (1980). Niemann–Pick disease type C: evidence for the deficiency of an activating factor stimulating sphingomyelin and glycocerebroside degradation. *Hoppe-Seylers Z. physiol. Chem.* **361**, 1489.

Clarance, G.A. and Bowman, J.K. (1966). Further case of histidinaemia. *Br. med. J.* **i**, 1019.

Clark, D. and Cuddeford, D. (1971). A study of the aminoacids in urine from dogs with cystine urolithiasis. *Vet. Res.* **88**, 414.

Clark, J.G., Kuhn, C. III, and Uitto, J. (1980). Lung collagen in type IV Ehlers-Danlos syndrome: ultrastructural and biochemical studies. *Am. Rev. respir. Dis.* **122**, 971.

Clark, K.G.A. and Nicholson, D.C. (1971). Erythrocyte protoporphyrin and iron uptake in erythropoietic protoporphyria. *Clin. Sci.* **41**, 363.

Clarke, J.T.R., Knaack, J., Crawhall, J.C., and Wolfe, L.S. (1971). Ceramide trihexosidosis (Fabry's disease) without skin lesions. *New Engl. J. Med.* **284**, 233.

Clarkson, T.W., and Kench, J.E. (1956). Urinary excretion of amino acids by men absorbing heavy metals. *Biochem. J.* **62**, 361.

Clayton, B.E. (1975). The principles of treatment by dietary restriction as illustrated by phenylketonuria. In *The treatment of inherited metabolic disease* (ed. D.N. Raine), p.1. MTP Publishing, Lancaster.

—— and Patrick, A.D. (1961). Use of dimercaprol or penicillamine in the treatment of cystinosis. *Lancet* **ii**, 909.

Cleaver, J.E. (1970). DNA repair and radiation sensitivity in human (xeroderma pigmentosum) cells. *Int. J. radiat. Biol.* **18**, 557.

—— (1972). Xeroderma pigmentosum. Variants with normal DNA repair and normal sensitivity to ultraviolet light. *J. invest. Dermatol.* **58**, 124.

—— and Trosko, J.E. (1970). Absence of excision of ultraviolet-induced cyclobutane dimers in xeroderma pigmentosum. *Photochem. Photobiol.* **11**, 547.

—— Zelle, B., Hashem, N., El-Hefnawi, M.H., and German, J. (1981). Xeroderma pigmentosum in patients from Egypt. II. Preliminary correlations of epidemiology, clinical symptoms and molecular biology. *J. invest. Dermatol.* **77**, 96.

Clinical Genetics Society (1978). *The provision of services for prenatal diagnosis of fetal abnormality in the United Kingdom.* The Eugenics Society, London.

Coates, P.M., Cortner, J.A., Mennuti, M.T., and Wheeler, J.E. (1978). Prenatal diagnosis of Wolman disease. *Am. J. med. Genet.* **2**, 397.

Codaccioni, J.L., Rinaldi, J.P., and Bismuth, J. (1978). The test of overloading of 1-diiototyrosine (DIT) in the screening of iodotyrosine dehalogenase deficiency. *Acta Endocrinol.* **87**, 95.

—— Pierron, H., Rouault, F., Aquaron, R., and Jaquet, P. (1970). Hypothyroîdie infantile par défaut D-iodotyrosine-déshalogenase. II. Résultats du traitment par l'iode de 5 cas. *Ann. Endocr. (Paris)* **31**, 1174.

Cogan, D.G., Kuwabara, T., and Moser, H. (1970). Metachromatic leuco-dystrophy. *Ophthalmologia (Basel)* **160**, 2.

—— —— —— and Hazard, G.W. (1966). Retinopathy in a case of Farber's lipogranulomatosis. *Arch. Ophthalmol.* **75**, 752.

—— —— Kinoshita, J., Sheehan, L., and Merola, L. (1957). Cystinosis in an adult. *J. Am. med. Assoc.* **164**, 394.

Cohen, A., Doyle, D., Martin, D.W. jun., and Ammann, A.J. (1976). Abnormal purine metabolism and purine overproduction in a patient deficient in purine nucleoside phosphorylase. *New Engl. J. Med.* **295**, 1449.

Cohen, H.P., Choitz, H.C., and Berg, C.P. (1958). Response of rats to diets high in methionine and related compounds. *J. Nutr.* **64**, 555.

Cohen, I. (1980). Platelet structure and function. Role of prostaglandins. *Ann. clin. lab. Sci.* **19**, 187.

Cohen, R.J., Sachs, J.R., Wicker, D.J., and Conrad, M. (1968). Methemoglobinemia provoked by malarial chemoprophylaxis in Vietnam. *New Engl. J. Med.* **279**, 1127.

Cohn, R.M., Yudkoff, M., Rothman, R., and Segal, S. (1978). Isovaleric acidemia: use of glycine therapy in neonates. *New Engl. J. Med.* **299**, 996.

Collie, W.R., Moore, C.M., Goka, T.J., and Howell, R.R. (1978). Pili torti as marker for carriers of Menkes disease. *Lancet* **i**, 607.

Collier, M.E., Griffin, J.E., and Wilson, J.D. (1978). Intranuclear binding of [^{3}H]dihydrotestosterone by cultured human fibroblasts. *Endocrinol.* **103**, 1499.

Collins, F.S., Summer, G.K., Schwartz, R.P., and Parke, J.C. (1980). Neonatal argininosuccinic aciduria — survival after early diagnosis and dietary management. *J. Ped.* **96**, 429.

Collis, J.E., Levi, A.J., and Milne, M.D. (1963). Stature and nutrition in cystinuria and Hartnup disease. *Br. med. J.* **i**, 590.

Colombi, A., Kostyal, A., Bracher, R., Gloor, F., Mazzi, R., and Tholen, H. (1967). Angiokeratoma corporis diffusum — Fabry's disease. *Helv. med. Acta* **34**, 67.

Colombo, J.P., Richterich, R., Donath, A., Spahr, A., and Rossi, E. (1964). Congenital lysine intolerance with periodic ammonia intoxication. *Lancet* **i**, 1014.

—— Burgi, W., Richterich, R., and Rossi, E. (1967). Congenital lysine intolerance with periodic ammonia intoxication: a defect in L-lysine degradation. *Metabolism* **16**, 910.

Coltart, D.J., and Hudson, R.E.B. (1971). Primary oxalosis of the heart: a cause of heart block. *Br. Heart. J.* **33**, 315.

Committee on Nutrition (1976). *Pediatrics* **57**, 783.

Conklin, K.A., Yamashiro, K.M., and Gray, G.M. (1975). Human intestinal sucrase–isomaltase. Identification of free sucrase and isomaltase and cleavage of the hybrid into active distinct subunits. *J. biol. Chem.* **250**, 5735.

Connon, J.J. and Turkington, V. (1968). Hereditary coproporphyria. *Lancet* **ii**, 263.

Constantopoulos, G., McComb, R.D., and Dekaban, A.S. (1976). Neurochemistry of the mucopolysaccharidoses. *J. Neurochem.* **26**, 901.

Conzelman, E. and Sandhoff, K. (1978). AB variant of infantile GM$_2$-gangliosidosis: deficiency of a factor necessary for stimulation of hexosaminidase A-catalyzed degradation of ganglioside GM$_2$ and glycolipid GA$_2$. *Proc. natl. Acad. Sci. USA* **75**, 3979.

Cook, G.C., and Kajubi, S.K. (1966). Tribal incidence of lactase deficiency in Uganda. *Lancet* **i**, 725.

Cook, J.G.H., Don, N.A., and Mann, T.P. (1971). Hereditary galactokinase deficiency. *Arch. Dis. Child.* **46**, 465.

Cook, P.J., Noades, J.E., Newton, M.S., and Mey, R. (1977). On the orientation of the Rh: E1-1 linkage group. *Ann. hum. Genet.* **41**, 157.

Cook, P.R., Brazell, I.A., Pawsey, S.A., and Giannelli, F. (1978). Changes induced by ultraviolet light in the superhelical DNA of lymphocytes from subjects with xeroderma pigmentosum and normal controls. *J. Cell. Sci.* **29**, 117.

Cooper, A.J.L., McDonald, J.M., Gelbard, A.D., Geldhill, R.F., and Duffy, T.E. (1979). The metabolism fate of ^{13}N labeled ammonia in rat brain. *J. biol. Chem.* **254**, 4982.

Cooper, J.A., and Moran, T.J. (1957). Studies on ochronosis. *Am. med. Assoc. Arch. Path.* **64**, 46.

Cooper, P.A. (1951). Alkaptonuria with ochronosis. *Proc. r. Soc. Med.* **44**, 917.

Corby, D.G., Putman, C.W., and Greene, H.L. (1974). Impaired platelet function in glucose-6-phosphatase deficiency. *J. Ped.* **85**, 71.

Corden, B.J., Schulman, J.D., Schneider, J.A., and Thoene, J.G. (1981). Adverse reactions to oral cysteamine use in nephropathic cystinosis. *Dev. pharm. Therap.* **3**, 25.

Cori, G.T. and Cori, C.F. (1952). Glucose-6-phosphatase of the liver in glygogen storage disease. *J. biol. Chem.* **199**, 661.

Corner, B.O., Holton, J.B., Norman, R.M., and Williams, P.M. (1968). A case of histidinemia controlled with a low histidine diet. *Pediatrics* **41**, 1074.

Coryell, M.E., Hall, W.K., Thevaos, T.G., Welter, D.A., Gatz, A.J., Horton, B.F., Sisson, B.D., Looper, J.W., and Farrow, R.T. (1964). A familial study of a human enzyme defect, argininosuccinic aciduria. *Biochem. biophys. Res. Commun.* **14**, 307.

Cotton, R.G.H. (1977). The primary molecular defects in phenylketonuria and its variants. *Int. J. Biochem.* **8**, 333.

Coude, F.X., Sweetman, L., and Nyhan. W.L. (1979). Inhibition by propionyl-coenzyme A of *N*-acetylglutamate synthetase in rat liver mitochondria: a possible explanation for hyperammonemia in propionic and methylmalonic acidemia. *J. clin. Invest.* **64**, 1544.

Couillin, P., Nicholas, H., Boué, J., and Boué, A. (1979). HLA typing of amniotic-fluid cells applied to prenatal diagnosis of congenital adrenal hyperplasia. *Lancet* **i**, 1076.

Counts, D.F., Byers, P.H., Holbrook, K.A., and Hegreberg, G.A. (1980). Dermatosparaxis in a Himalayan cat: I. Biochemical studies of dermal collagen. *J. invest. Dermatol.* **74**, 96.

Cowan, M.J., Wara, D.W., Packman, S., Ammann, A.J., Yoshiro, M., Sweetman, L., and Nyhan, W. (1979). Multiple biotin-dependent carboxylase deficiencies associated with defects in T-cell and B-cell immunity. *Lancet* **ii**, 115.

Cox, B.D. and Cameron, J.S. (1974). Homoarginine in cystinuria. *Clin. Sci. Molec. Med.* **46**, 173.

Cox, E.V. and White, A.M. (1962). Methylmalonic acid excretion: index of vitamin B_{12} deficiency. *Lancet* **ii**, 853.

Cox, R.P., Hutzler, J., and Dancis, J. (1978). Antenatal diagnosis of maple-syrup-urine disease. *Lancet* **ii**, 212.

—— Krauss, M.R., Balis, M.E., and Dancis, J. (1970). Evidence for transfer of enzyme product as the basis of metabolic cooperation between tissue culture fibroblasts of Lesch–Nyhan disease and normal cells. *Proc. natl. Acad. Sci. USA* **67**, 1573.

—— —— —— —— (1972). Communication between normal and enzyme deficient cells in tissue culture. *Exp. Cell. Res.* **74**, 251.

Crabtree, C.W. and Henderson, J.F. (1971). Rate-limiting steps in the inter-conversion of purine ribonucleotides in Ehrlich ascites tumor cells in vitro. *Cancer Res.* **31**, 985.

Crawhall, J.C. and Thompson, C.J. (1965). Cystinuria: effect of D-penicillamine on plasma and urinary cystine concentrations. *Science* **147**, 1459.

—— and Watts, R.W.E. (1962). The metabolism of $[I-^{14}C]$–glyoxylate by the liver mitochondria of patients with primary hyperoxaluria and non-hyperoxaluric subjects. *Clin. Sci.* **23**, 163.

—— Scowen, E.F., and Watts, R.W.E. (1964). Further observations on use of D-penicillamine in cystinuria. *Br. med. J.* **i**, 1411.

—— Lietmann, P.S., Schneider, J.A., and Seegmiller, J.E. (1968*b*).Cystinosis: plasma cystine and cysteine concentrations and the effect of D-penicillamine and dietary treatment. *Am. J. Med.* **44**, 330.

—— Parker, R. Sneddon, W., and Young, E.P. (1969*a*). ß-mercaptolactate-cysteine disulfide in the urine of a mentally retarded patient. *Am. J. Dis. Child.* **117**, 71.

—— Purkiss, P., Watts, R.W.E., and Young, E.P. (1969*b*). The excretion of amino acids by cystinuric patients and their relatives. *Ann. hum. Genet.* (*Lond.*) **33**, 149.

—— Scowen, E.F., Thompson, C.J., and Watts, R.W.E. (1967). Dissolution of cystine stones during D-penicillamine treatment of a pregnant patient with cystinuria. *Br. med. J.* **ii**, 216.

—— Parker, R., Sneddon, W., Young, E.P., Ampola, M.G., Efron, M.L., and Bixby, E.M. (1968*a*). Beta mercaptolactate-cysteine disulfide: analog of cystine in the urine of a mentally retarded patient. *Science* **160**, 419.

Creel, D. (1980). Review: Inappropriate use of albino animals as models in research. *Pharmacol. Biochem. Behav.* **12**, 969.

Crigler, J.F., and Najjar, V.A. (1952). Congenital familial non-hemolytic jaundice with kernicterus. *Pediatrics* **10**, 169.

Cripps, D.J., Hawgood, R.S., and Magnus, I.A. (1966). Iodine tungsten fluorescence microscopy for porphyrin fluorescence. A study on erythropoietic protoporphyria. *Arch. Dermatol.* **93**, 129.

Crocker, A.C. (1961). The cerebral defect in Tay–Sachs and Niemann–Pick disease. *J. Neurochem.* **7**, 69.

—— Vawter, G.F., Neuhauser, E.B.D., and Rosowsky, A. (1965). Wolman's disease: three new patients with a recently described lipidosis. *Pediatrics* **35**, 627.

Crome, L., Hanefeld, F., Patrick, D., and Wilson, J. (1973). Late onset of globoid cell leukodystrophy. *Brain* **96**, 84.

Cross, H.E., McKusick, V.A., and Breen, W. (1967). A new oculocerebral syndrome with hypopigmentation. *J. Ped.* **70**, 398.

Cruickshank, C.N.D. and Harcourt, S.A. (1964). Pigment donation in vitro. *J. invest. Dermatol.* **42**, 183.

Crussi, F.G., Robertson, D.M., and Hiscox, J.L. (1969). The pathological condition of the Lesch–Nyhan syndrome. Report of two cases. *Am. J. Dis. Child.* **118**, 501.

Cumings, J.N. (1948). The copper and iron content of brain and liver in the normal and in hepato-lenticular degeneration. *Brain* **71**, 410.

Curtius, H.-C., Zagalak, M.J., Baerlocher, K., Schaub, J., Leimbacher, W., and Redweik, U. (1977). *In vivo* studies of the phenylalanine-4-hydroxylase system in hyperphenylalaninemics and phenylketonurics. *Helv. paediatr. Acta* **32**, 461.

—— Niederwieser, A., Viscontini, M., Otten, A., Schaub, J., Scheibenreiter, S., and Schmidt, H. (1979). Atypical phenylketonuria due to tetrahydrobiopterin deficiency, diagnosis and treatment with tetrahydrobiopterin, dihydrobioterin and sepiapterin. *Clin. Chim. Acta* **93**, 251.

Cusworth, D.C. and Dent, C.E. (1969). Homocystinuria. *Br. med. Bull.* **25**, 42.

Cuzner, M.L. and Davison, A.N. (1968). The lipid composition of rat brain myelin and subcellular fractions during development. *Biochem. J.* **106**, 29.

Dahlin, O., Enerbäck, L., and Lundvall, O., (1973). Porphyria cutanea tarda: A genetic disease? A biochemical and fluorescence microscopical study in four families. *Acta med. Scand.* **194**, 265.

Damashek, W. and Singer, K. (1941). Familial nonhemolytic jaundice. Constitutional hepatic dysfunction with indirect van den Bergh reaction. *Am. med. Assoc. Arch. Intern. Med.* **67**, 259.

Dance, N., Price, R.G., and Robinson, D. (1970). Differential assay of human hexosaminidase A and B. *Biochim. Biophys. Acta* **222**, 662.

Dancis, J. (1972). Maple syrup urine disease. In *Antenatal diagnosis* (ed. A. Dorfman), p.317. University of Chicago Press, Chicago.

—— (1974). Abnormalities in the degradation of lysine. In: '*Heritable disorders of amino acid metabolism.*' (ed. W.L. Nyhan), p.387. John Wiley, New York.

—— and Levitz, M. (1978). Abnormalities of branched-chain amino-acid metabolism. In *The metabolic basis of inherited disease*, 4th edn. (ed. J.B. Stanbury, J.B. Wyngaarden and D.S. Fredrickson), p.397. McGraw-Hill, New York.

—— Hutzler, J. and Cox, R.P. (1979). Familial hyperlysinemia: Enzyme studies, diagnostic methods, comments on terminology. *Am. J. hum. Genet.* **31**, 290.

—— —— and Levitz, M. (1960b). Metabolism of the white blood cells in maple syrup urine disease. *Biochim. Biophys. Acta* **43**, 342.

—— —— —— (1965). Detection of the heterozygote in maple syrup urine disease. *J. Ped.* **66**, 595.

—— Levitz, M., and Westall, R.G. (1960a). Maple syrup urine disease: branched chain keto-aciduria. *Pediatrics* **24**, 72.

—— Hutzler, J., Cox, R.P., and Woody, N.C. (1969). Familial hyperlysinemia with lysine-ketoglutarate reductase insufficiency. *J. clin. Invest.* **48**, 1447.

—— —— Snyderman, S.E., and Cox, R.P. (1972). Enzyme activity in classical and variant forms of maple syrup urine disease. *J. Ped.* **81**, 312.

—— —— Woody, N.C., and Cox, R.P. (1976). Multiple enzyme defects in familial hyperlysinemia. *Ped. Res.* **10**, 686.

—— —— Tada, K., Wada, Y., Morikawa, T., and Arakawa, T. (1967). Hypervalinemia: A defect in valine transamination. *Pediatrics* **39**, 813.

D'Ancona, G.G., Wurk, J. and Croce, C.M. (1979). Genetics of type-II glycogenosis-assignment of the human gene for acid alpha-glucosidase to chromosome-17. *Proc. natl. Acad. Sci. USA* **76**, 4526.

Dancs, B.S. and Bearn, A.G. (1965). Hurler's syndrome: Demonstration of an inherited disorder of connective tissue in cell culture. *Science* **149**, 987.

—— and Degnan, M. (1974). Different clinical and biochemical phenotypes associated with ß-glucuronidase deficiency. *Birth defects; Orig. Art. Ser. X.* **12**, 251.

—— Salk, I., and Flynn, F.J. (1972). Treatment of Hurler syndrome. *Lancet* **ii**, 883.

Daniels, R.A., Michels, R., Aisen, P., and Goldstein, G. (1960). Familial hyperoxaluria. *Am. J. Med.* **29**, 820.

Danks, D.M., Cotton, R.G.H., and Schlesinger, P. (1975a). Tetrahydrobiopterin treatment of variant form of phenylketouria. *Lancet* **ii**, 1043.

—— Tippett, P., and Rogers, J. (1975b). A new form of prolonged transient tyrosinemia presenting with severe metabolic acidosis. *Acta Paediatr. Scand.* **64**, 209.

—— —— and Zentner, G. (1974b). Severe neonatal citrullinemia. *Arch. Dis. Child.* **49**, 579.

—— Cartwright, E., Stevens, B.J., and Townley, R.R.W. (1973). Menkes' kinky hair disease: Further definition of the defect in copper transport. *Science* **179**, 1140.

—— Tippett, P., Adams, C., and Campbell, P. (1975c). Cerebro-hepato-renal syndrome of Zellweger. A report of eight cases with comments upon the incidence, the liver lesion, and a fault in pipecolic acid metabolism. *J. Ped.* **86**, 382.

—— Campbell, P.E., Stevens, B.J., Mayne, V., and Cartwright, E. (1972a). Menkes' kinky hair syndrome; an inherited defect in copper absorption with widespread effects. *Pediatrics* **50**, 188.

—— Stevens, B.J., Campbell, P.E., Gillespie, J.M., Walker-Smith, J., Blomfield, J., and Turner, B. (1972b). Menkes' kinky hair syndrome. *Lancet* i, 1100.

—— Bartholomé, K., Clayton, B.E., Curtius, H., Gröbe, H., Kaufman, S., Leeming, R., Pfleiderer, W., Rembold, H., and Rey, F. (1978). Malignant hyperphenylalaninaemia — current status (June 1977). *J. inher. metab. Dis.* **1**, 49.

—— Stevens, B.J., Campbell, P.E., Cartwight, E.C., Gillespie, J.M., Townley, R.R.W., Walker-Smith, J.A., Blomfield, J., Turner, B.B. and Mayne, V. (1974a). Menkes' kinky hair syndrome. An inherited defect in the intestinal absorption of copper with widespread effects. *Birth Defects: Orig. Art. Ser.* **10**, 132.

Danner, J.D., Wheeler, F.B., Lemmon, S.K., and Elsas, L.J. (1978). *In vivo* and *in vitro* response of human branched chain α-ketoacid dehydrogenase to thiamine and thiamine pyrophosphate. *Ped. Res.* **12**, 235.

Danovitch, G.H., Baer, P.N., and Laster, L. (1968). Intestinal alkaline phosphatase activity in familial hypophosphatasia. *New Engl. J. Med.* **278**, 1253.

Daum, R.S., Lamm, P.H., Mamer, O.A., and Scriver, C.R. (1971). A 'new' disorder of isoleucine metabolism. *Lancet* ii, 1289.

—— Scriver, C.R., Mamer, O.A., Delvin, E., Lamm, P., and Goldman, H. (1973). An inherited disorder of isoleucine catabolism causing accumulation of α-methylacetoacetate and α-methyl-ß-hydroxybutyrate, and intermittent metabolic acidosis. *Ped. Res.* **7**, 149.

Davidson, A.F.G. and Mullinger, M. (1970). Reducing substances in neonatal stools detected by Clinitest. *Pediatrics* **46**, 632.

Davidson, R.G., Nitowsky, H.M., and Childs, B. (1963). Demonstration of two populations of cells in the human female heterozygous for glucose-6-phosphate dehydrogenase variants. *Proc. natl. Acad. Sci. USA* **50**, 481.

Davies, R.O., Marton A.V., and Kalow, W. (1960). The action of normal and atypical cholinesterase of human serum upon a series of esters of choline. *Can. J. biochem. Physiol.* **38**, 545.

Davis, L.E., Snyder, R.D., Orth, D.N., Nicholson, W.E., Kornfield, M., and Seelinger, D.F. (1979). Adrenoleucodystrophy and adrenomyeloneuropathy associated with partial adrenal insufficiency in three generations of a kindred. *Am. J. Med.* **66**, 342.

Dawson, D.M., Spong, F.L., and Harrington, J.F. (1959). McArdle's disease: lack of muscle phosphorylase. *Ann. intern. Med.* **69**, 229.

Dawson, G. (1972). Glycosphingolipid abnormalities in liver from patients with glycosphingolipid and mucopolysaccharide storage diseases. In *Sphingolipids, sphingolipidosis and allied disorders.* (ed. B.W. Volk and S.M. Aronson), p.395. Plenum., New York.

—— and Spranger, J.W. (1971). Fucosidosis: a glycosphingolipidosis. *New Engl. J. Med.* **285**, 122.

—— Matalon, R. and Li, Y.-T. (1973). Correction of the enzymatic defect in cultured fibroblasts from patients with Fabry's disease: Treatment with purified α-galactosidase from Ficin. *Ped. Res.* **7**, 684.

Dayan, A.D. (1967). Dichroism of cresyl violet-stained cerebroside sulfate. *J. Histochem. Cytochem.* **15**, 421.

D'Azzo, A., Hoogeveen, A., Reuser, A.J.J., Robinson, D., and Galjaard, H. (1982). Molecular defect in combined ß-galactosidase and neuraminidase deficiency in man. *Proc. natl. Acad. USA* **79**, 4535.

Dean, G. (1971). *The porphyrias: A story of inheritance and environment*, 2nd Edn. Pitman Medical, London.

—— and Barnes, H.D. (1959). Porphyria in Sweden and South Africa. *S. Afr. med. J.* **33**, 246.

Dean, K.J. and Sweeley, C.C. (1977). Fabry disease. In *Practical enzymology of the sphingolipidoses* (ed. R.H. Glew and S.P. Peters), p.173. Alan R. Liss, New York.

—— Sung, S.S., and Sweeley, C.C. (1977). The identification of α-galactosidase B from human liver as an α-*N*-acetylgalactosaminidase. *Biochem. biophys. Res. Commun.* **77**, 1411.

Dean, M.F., Benson, P.F., and Muir, H. (1973). Mobilisation of glycosaminoglycans by plasma infusion in mucopolysaccharidosis type III: two types of response. *Nature (New Biol.)* **243**, 143.

—— Muir, H., and Ewins, R.J.F. (1971). Hurler's, Hunter's, and Morquio's syndromes. *Biochem. J.* **123**, 883.

—— —— Benson, P.F., and Button, L.R. (1981). Enzyme replacement therapy by transplantation of HLA-compatible fibroblasts in Sanfilippo A syndrome. *Ped. Res.* **15**, 959.

—— —— —— —— Batchelor, J.P., and Bewick, M. (1975). Increased breakdown of glycosaminoglycans and appearance of corrective enzyme after skin transplants in Hunter syndrome. *Nature* **257**, 609.

—— Stevens, R.L., Muir, H., Benson, P.F., Button, L.R., Anderson, R.L., Boylston, A., and Mowbray, J. (1979). Enzyme replacement therapy by fibroblast transplantation: long-term biochemical study in three cases of Hunter's disease. *J. clin. Invest.* **63**, 138.

De Baecque, C.M., Suzuki, K., Rapin, I., Johnson, A.B., Withers, D.L., and Suzuki, K. (1975). GM₂-gangliosidosis, AB variant. *Acta Neuropathol.* **33**, 207.

De Barsy, T. and Lederer, B. (1980). Type VI glycogenosis: identification of subgroups. In: *Inheritied disorders of carbohydrate metabolism*. (ed. D. Burman, J.B. Holton and C.A. Pennock), p.369. MTP Press, Lancaster.

—— and Van Hoof, F. (1974). Enzyme replacement therapy with purified human acid α-glucosidase in type II glycogenosis. In: *Enzyme therapy in lysosomal storage diseases* (ed. J.M. Tager, G.J.M. Hooghwinkel and W.T. Daems), p.277. North Holland, Amsterdam.

De Berranger, P., Navel, M., Decobert, G., and Gillot, F. (1975). La forme aiguë infantile de la maladie de Gaucher. *Ann. Pédiat.* **22**, 641.

De Brito, T., Borges, M.A., and Dasilva, L.C. (1966). Electron microscopy of the liver in nonhemolytic acholuric jaundice with kernicterus (Crigler–Najjar) and in idiopathic conjugated hyperbilirubinemia (Rotor). *Gastroenterologia* **106**, 325.

De Bruijn, W.C., Fernandes, J. Hubert, J., and Koster, J.F. (1975). Liver glycogenosis. A histochemical and ultrastructural study. *Pathol. Eur.* **10**, 3.

Degenhart, H.J. (1971). A study of the cholesterol splitting enzyme system in normal adrenal and in adrenal lipoid hyperplasia. *Acta. paediatr. Scand.* **60**, 611.

De Goeij, A.F.P.M., Christianse, K., and Van Steveninck, I. (1975). Decreased haem synthetase activity in blood cells of patients with erythropoietic protoporphyria. *Eur. J. clin. Invest.* **5**, 397.

De Groot, C.J. and Hommes, F.A. (1973). Further speculation on the pathogenesis of Leigh's encephalomyelopathy. *J. Ped.* **82**, 541.

—— Luit-De Haan, G., Hulstaert, C.E., and Hommes, F.A. (1977). A patient
with severe neurologic symptoms and acetoacetyl-CoA thiolase deficiency. *Ped.
Res.* **11**, 1112.

De Groot, P.G., Westerveld, A., Meera Khan, P., and Tager, J.M. (1978).
Localization of a gene for human α -galactosidase B (= N-acetyl- α -D-
galactosaminidase) on chromosome 22. *Hum Genet.* **44**, 305.

De Groot, W.P. (1964). Genetic aspects of the thesauris-mosis lipoidica hereditaria
Ruiter–Pompen–Wyers (angiokeratoma corporis diffusum Fabry). *Dermatologica (Basel)* **129**, 281.

Dekaban, A.S. and Steusing, J.K. (1974). Menkes' kinky hair disease treated with
subcutaneous copper sulphate. *Lancet* **ii**, 1523.

—— Holden, K.R. and Constantopoulos, G. (1972). Effects of fresh plasma or
whole blood transfusions on patients with various types of mucopolysaccharidosis. *Pediatrics* **50**, 688.

—— Constantopoulos, G., Herman, M.M., and Steusing, J.K. (1976). Mucopolysaccharidosis type V (Scheie syndrome). A post mortem study by multi-
disciplinary techniques with emphasis on the brain. *Arch. Pathol. lab. Med.* **100**,
237.

De La Chapelle, A. and Miller, O.J. (1979). Report of the committee on the
genetic constitution of chromosomes 10, 11, 12, X and Y. *Cytogenet. Cell
Genet.* **25**, 47.

De La Marza, M., Patten, B.M., Williams, J.C., and Chambers, J.P. (1980).Myo-
phosphorylase deficiency: A new cause of infantile hypotonia simulating infantile
muscular atrophy. *Neurology* **30**, 402.

Deleo, V.A., Poh-Fitzpatrick, M., Matthews-Roth, M., and Harber, L.C. (1976).
Erythropoietic protoporphyria. Ten years experience. *Am. J. Med.* **60**, 8.

De Luca, H.F., Blunt, J.W., and Rikkers, H. (1971). Biogenesis of vitamin D. In:
The Vitamins (ed. W.H. Sebrell, Jun. and R.S. Harris) Vol. III, p.213.
Academic Press, New York.

Demars, R.I. and Leroy, J.G. (1967). The remarkable cells cultured from a human
with Hurler's syndrome: An approach to visual selection for *in vitro* genetic
studies. *In Vitro* **2**, 107.

—— Sarto, G., Felix, J.S., and Benke, P. (1969). Lesch–Nyhan mutation: prenatal
detection with amniotic fluid cells. *Science* **164**, 1303.

Demeester-Mirkine, N. and Dumont, J.E. (1980). The hypothalamopituitary
thyroid axis. In: *The thyroid* (ed. M. De Visscher), p.145. Raven Press, New
York.

Denny-Brown, D. and Porter, H. (1951). The effect of BAL (2,3-dimer-
captopropanol) on hepatolenticular degeneration (Wilson's disease). *New Engl.
J. Med.* **245**, 917.

Dent, C.E. and Harris, H. (1951). The genetics of cystinuria. *Ann. hum. Genet.* **16**,
60.

—— —— (1956). Hereditary forms of rickets and osteomalacia. *J. Bone Joint
Surg. (Am.)* **38-B**, 204.

—— and Philpot, G.R. (1954). Xanthinuria, an inborn error (or deviation) of
metabolism. *Lancet* **i**, 182.

—— and Rose, G.A. (1951). Amino acid metabolism in cystinuria. *Quarterly J.
Med.* **20**, 205.

—— and Smellie, J.M. (1961). Two children with the oculo-cerebral-renal
syndrome of Lowe. *Proc. R. Soc. Med.* **54**, 335.

—— and Stamp, T.C.B. (1970). Treatment of primary hyperoxaluria. *Arch. Dis. Child.* **45**, 735.

—— Friedman, M., Green, H., and Watson, L.C.A. (1965). Treatment of cystinuria. *Brit. med. J.* **i**, 403.

Dent, P.B., Fish, L.A., White, J.G., and Good, R.A. (1966). Chediak–Higashi syndrome: Observations on the nature of the associated malignancy. *Lab. Invest.* **15**, 1634.

Den Tandt, W.R. and Jaeken, J. (1979). Determination of lysosomal enzymes in saliva: Confirmation of the diagnosis of metachromatic leukodystrophy and fucosidosis by enzyme analysis. *Clin. Chim. Acta* **97**, 19.

Depape-Brigger, D., Goldman, H., Scriver, C.R., Delvin, E., and Mamer, O. (1977). The *in vivo* use of dithiothreitol in cystinosis. *Ped. Res.* **11**, 124.

Dern, R.J., Beutler, E., and Alving, A.S. (1954). The hemolytic effect of primaquine. II. The natural course of the hemolytic anemia and the mechanism of its self-limited character. *J. lab. clin. Med.* **44**, 171.

Derry, D.M., Fawcett, J.S., Andermann, F., and Wolfe, L.S. (1968). Late infantile systemic lipidosis. *Neurology* **18**, 340.

De Sanctis, C. and Cacchione, A. (1932). L'Idiozia Xerodermica. *Riv. Sper. do Freniat.* **56**, 269.

Desilva, K.L. and Pearce, J. (1973). Neuropathy of metachromatic leuko-dystrophy. *J. Neurol. Neurosurg. Psychiatr.* **36**, 30.

Desnick, R.J., Krivit, W., and Sharp, H.L. (1973a). In utero diagnosis of Sandhoff's disease. *Biochem. biophys. Res. Commun.* **51**, 20.

—— Sweeley, C.C., and Krivit, W. (1970). A method for the quantitative determination of neutral glycosphingolipids in urine sediment. *J. Lipid Res.* **11**, 31.

—— Bleiden, L.D., Sharp, H.L., and Moller, J.H. (1976b). Cardiacvalvular anomalies in Fabry's disease: Clinical, morphologic and biochemical studies. *Circulation* **54**, 818.

—— Dean, K.J., Grabowski, G.A., Bishop, D.F., and Sweeley, C.C. (1979). Enzyme therapy XII: Enzyme therapy in Fabry's disease: Differential enzyme and substrate clearance kinetics of plasma and splenic α-galactosidase iso-enzymes. *Proc. natl. Acad. Sci. USA* **76**, 5326.

—— —— —— —— —— (1980). Enzyme Therapy XVII: Metabolic and immuno-logic evaluation of α-galactosidase A replacement in Fabry's disease. In *Enzyme therapy in genetic diseases* Vol. 2 (ed. R.J. Desnick), p.393. Alan R. Liss, New York.

—— Allen, K.Y., Desnick, S.J., Raman, M.K., Bernlohr, R.W., and Krivit, W. (1973b). Enzymatic diagnosis of hemizygotes and heterozygotes. Fabry's disease. *J. lab. clin. Med.* **81**, 157.

—— —— Simmons, R.L., Woods, J.E., Anderson, C.F., Najarian, J.S., and Krivit, W. (1972). Correction of enzymatic deficiences by renal transplantation: Fabry's disease. *Surgery* **72**, 203.

—— Sharp, H.L., Grabowski, G.A., Brunning, R.D., Quie, P.G., Sung, J.H., Gorlin, R.J., and Ikonne, J.U. (1976a). Mannosidosis: clinical, morphologic, immunologic and biochemical studies. *Ped. Res.* **10**, 985.

Detter, J.C., Stamatoyannopoulas, G., Giblett, E.R., and Motulsky, A.G. (1970). Adenosine deaminase: racial distribution and report of new phenotype. *J. med. Genet.* **7**, 356.

—— Ways, P.O., Giblett, E.R., Baughan, M.A., Hopkinson, D.A., Povey, S., and Harris, H. (1968). Inherited variations in human phosphohexose isomerase. *Ann. hum. Genet.* **31**, 329.

De Verneuil, H. and Nordmann, Y. (1981). Porphyrie cutanée symptomatique. Type familial et type sporadique. *Nouv. Pr. Med.* **10**, 3541.

—— Aitken, G., and Nordmann, Y. (1978). Familial and sporadic porphyria cutanea: Two different diseases. *Hum. Genet.* **44**, 145.

De Visscher, M. (1980). *The thyroid gland*. Raven Press, New York.

De Vivo, D.C., Haymond, M.W., Leckie, M.P., Bussman, Y.L., McDougal, D.B., and Pagliara, A.S. (1977). The clinical and biochemical implications of pyruvate carboxylase deficiency. *J. clin. Endocrinol. Metab.* **45**, 1281.

De Weerd-Kastelein, E.A. (1974). In *Genetic heterogeneity in the human skin disease xeroderma pigmentosum*, Paper 3, Doctoral Thesis. Erasmus University. Bonder-Offset, B.V., Rotterdam.

Dewey, V.C. and Kidder, G.W. (1971). Assay of unconjugated pteridines. *Methods in Enzymology* **18B**, 618.

Dewhurst, N., Besley, G.T.N., Finlayson, N.D.C., and Parker, A.C. (1979). Sea blue histiocytosis in a patient with chronic non-neuropathic Niemann–Pick disease. *J. clin. Pathol.* **32**, 1121.

Dickinson, C.J. and Smellie, J.M. (1959). Xanthinuria. *Br. med. J.* **ii**, 1217.

Di Ferrante, N. (1967). The measurement of urinary mucopolysaccharides. *Analyt. Biochem.* **21**, 98.

—— (1980). *N*-acetylglucosamine-6-sulfate sulfatase deficiency reconsidered. *Science* **210**, 448.

—— and Nicholas, B.L. (1972). A case of the Hunter syndrome with progeny. *Johns Hopkins med. J.* **130**, 325.

—— Ginsberg, L.C., Donnelly, P.V., Di Ferrante, D.T., and Caskey, C.T. (1978). Deficiencies of glucosamine-6-sulfate or galactosamine-6-sulfate sulphatases are responsible for different mucopolysaccharidoses. *Science* **199**, 79.

—— Hyman, B.H., Klish, W., Donnelly, P.V., Nichols, B.L., and Dutton, R.G. (1974). Mucopolysaccharidosis VI (Maroteux–Lamy disease): Clinical and biochemical study of a mild variant case. *Johns Hopkins med. J.* **135**, 42.

—— Leachman, R.D., Angelini, P., Donnelly, P.V., Francis, G., and Almazan, A. (1975). Lysyl oxidase dificiency in Ehlers–Danlos syndrome type V. *Connect. Tissue Res.* **33**, 49.

—— Nichols, B.L., Donnelly, P.V., Neri, G., Hrgovic, R., and Berglund, R.K. (1971). Induced degradation of glycosaminoglycans in Hurler's and Hunter's syndrome by plasma infusion. *Proc. natl. Acad. Sci. USA* **68**, 303.

Dillon, M.J., England, J.M., Gompertz, D., Goodney, P.A., Grant, D.B., Hussein, H.A., Linnell, J.C., Matthews, D.M., Mudd, S.H., Newns, G.H., Seakins, J.W.T., Uhlendorf, B.W., and Wise, I.J. (1974). Mental retardation, megaloblastic anaemia, methylmalonic aciduria and abnormal homocysteine metabolism due to an error in vitamin B_{12} metabolism. *Clin. Sci. mol. Med.* **47**, 43.

Di Matteo, G., Durand, P., Gatti, R., Maresca, A., Orfeo, M., Urbano, F., and Romeo, G. (1976). Human α-fucosidase. Single residual enzymatic form in fucosidosis. *Biochim. Biophys. Acta* **429**, 538.

Di Mauro, S. and Di Mauro, P.M.M. (1973). Muscle carnitine palmityl-transferase deficiency and myoglobinuria. *Science* **182**, 929.

—— and Hartlage, P.L. (1978). Fatal infantile form of muscle phosphorylase deficiency. *Neurology* **28**, 1124.

—— and Mellman, W.J. (1973). Glycogen metabolism of human diploid fibroblast cells in culture. II. Factors influencing glycogen concentration. *Ped. Res.* **7**, 745.

—— Rowland, L.P., and Mellman, W.J. (1973). Glycogen metabolism of human diploid fibroblast cells in culture. I. Studies of cells from patients with glycogenosis types II, III and V. *Ped. Res.* **7**, 739.

—— Arnold, S., Miranda, A., and Rowland, L.P. (1978). McArdle's disease: The mystery of reappearing phosphorylase activity in muscle culture — a fetal isoenzyme. *Ann. Neurol.* **3**, 60.

—— Miranda, A.F., Khans, S., and Gitlin, K. (1981). Human muscle phosphoglycerate mutase deficiency: newly discovered metabolic myopathy. *Science* **212**, 1277.

—— —— Hays, A.P., Franck, W.A., Hoffman, G.S., Schoenfeldt, R.S., and Singh, N. (1980). Myoadenylate deaminase deficiency: Muscle biopsy and muscle culture in a patient with gout. *J. neurol. sci.* **47**, 191.

Dinsart, C., Van Voorthuizen, F., and Vassart, G. (1977). Reverse transcription of thyroglobulin 33-S mRNA. *Eur. J. Biochem.* **78**, 175.

Dissing, J. and Knudsen, J.B. (1969). A new red cell adenosine deaminase phenotype in man. *Hum. Hered.* **19**, 375.

—— and Knudsen, B. (1972). Adenosine-deaminase deficiency and combined immunodeficiency syndrome. *Lancet* **ii**, 1316.

Distler, J., Hieber, V., Sahagian, G., Schmickel, R., and Jourdian, G.W. (1979). Identification of mannose 6-phosphate in glycoproteins that inhibit the assimilation of ß-galactosidase by fibroblasts. *Proc. natl. Acad. Sci. USA* **76**, 4235.

Djaldetti, M., Fishman, P., and Bessler, H. (1979). The surface ultra-structure of Gaucher cells. *Am. J. clin. Pathol.* **71**, 146.

Dodd, M.G., Pusin, S.M., and Green, W.R. (1978). Adult cystinosis: A case report. *Arch. Ophthalmol.* **96**, 1054.

Donn, D.M., Swartz, R.D., and Thoene, J.G. (1979). Comparison of exchange transfusion, peritoneal dialysis and hemodialysis for the treatment of hyperammonemia in an anuric newborn infant. *J. Ped.* **95**, 67.

Donnell, G.N. and Bergren, W.R. (1975). The galactosaemias. In *The treatment of inherited metabolic disease* (ed. D. Raine), p.91. MTP Press, Lancaster.

—— Koch, R., Fishler, K., and Ng, W.G. (1980). Clinical aspects of galactosaemia. In *Inherited disorders of carbohydrate metabolism* (ed. D. Burman, J.B. Holton, and C.A. Pennock), p.103. MTP Press, Lancaster.

Donnelly, P.V. and Di Ferrante, N. (1975). Reliability of the Booth–Nadler technique for the detection of Hunter heterozygotes. *Pediatrics* **56**, 429.

Donnelly, W.J.C., Sheahan, B.J., and Kelly, M. (1973). Beta-galactosidase deficiency in GM_1-gangliosidosis of Friesian calves. *Res. vet. Sci.* **15**, 139.

Dorfman, A. and Lorincz, A.E. (1957). Occurrence of urinary acid mucopolysaccharides in the Hurler syndrome. *Proc. natl. Acad. Sci. USA* **43**, 443.

Doss, M., Von Tiepermann, R., Schneider, J., and Schmid, H. (1977). New type of hepatic porphyria with porphobilinogen synthase defect and intermittent acute clinical manifestation. *Klin. Wochensch.* **57**, 1123.

Dowdle, E., Goldswain, P., Spong, N., and Eales, L. (1970). The pattern of porphyrin isomer accumulation and excretion in symptomatic porphyria. *Clin. Sci.* **39**, 147.

Drayer, J.I.M., Cleophas, A.J.M., Trijbels, J.M.F., Smals, A.G.H., and Kloppenborg, P.W.C. (1980). Symptoms, diagnostic pitfalls, and treatment of homocystinuria in seven adult patients. *Neth. J. Med.* **23**, 89.

Dreborg, S., Erickson, A., and Hagberg, B. (1980). Gaucher Disease — Norrbottnian type. 1. General clinical description. *Eur. J. Pediatr.* **133**, 107.

Dreyfus, J.C., Proux, D., and Alexandre, Y. (1974). Molecular studies on glycogen storage diseases. *Enzyme* **18**, 60.

Drummond, G.I., Hardwick, D.F., and Israels, S. (1970). Liver glycogen phosphorylase deficiency. *Canad. med. Assoc. J.* **102**, 740.

Dubiel, B., Wetts, R., and Tanaka, K. (1980). Heterogeneity in diseases of leucine metabolism: Complementation studies using cultured skin fibroblasts. *Ped. Res.* **14**, 521.

Dubois, G., Harzer, K., and Baumann, N. (1977). Very low arylsulphatase A and cerebroside sulfatase activities in leucocytes of healthy members of meta-chromatic leucodystrophy family. *Am. J. hum. Genet.* **29**, 191.

—— Turpin, J.C., Georges, M.C., and Baumann, N. (1980). Arylsulfatases A and B in leukocytes: A comparative statisitical study of late infantile and juvenile forms of metachromatic leukodystrophy and controls. *Biomedicine* **33**, 2.

Ducos, J., Marty, Y., Sanger, R., and Race, R.R. (1971). Xg and X chromosome inactivation. *Lancet* **ii**, 219.

Dugal, B. (1977). Measurement of 1-aspartamido-ß-*N*-acetylglucosamine amido-hydrolase activity in human tissues. *Biochem. J.* **163**, 9.

Dulaney, J.T. and Moser, H.W. (1977*a*). Metachromatic leukodystrophy. In *Practical enzymology of the sphingolipidoses* (ed. R.H. Glew and S.P. Peters). p.137. Alan R. Liss, New York.

—— —— (1977*b*). Farber's disease (lipogranulomatosis) In *Practical enzymology of the sphingolipidoses* (ed. R.H. Glew and S.P. Peters), p.283. Alan R. Liss, New York.

—— —— (1978). Sulfatide lipidosis: Metachromatic leukodystrophy. In *The metabolic basis of inherited disease*, 4th Edn. (ed. J.B. Stanbury, J.B. Wyngaarden, and D.S. Fredrickson), p.770. McGraw-Hill, New York.

—— Milunsky, A., Sidbury, J.B., Hobolth, N., and Moser, H.W. (1976). Diagnosis of lipogranulomatosis (Farber's disease) by use of cultured fibroblasts. *J. Ped.* **89**, 59.

Dunger, D.B. and Leonard, J.V. (1982). Value of the glucagon test in screening for hepatic glycogen storage disease. *Arch. Dis. Child.* **57**, 384.

Dunn, H.G. (1955). Oxalosis: Report of a case with review of the literature. *Am. J. Dis. Child.* **90**, 58.

—— Perry, T.L., and Dolman, C.L. (1966). Homocystinuria. *Neurology* **16**, 407.

—— Lake, B.D., Dolman, C.L., and Wilson, J. (1969). The neuropathy of Krabbe's infantile cerebral sclerosis. *Brain* **92**, 329.

Dupperrat, B., Puissant, A., Saurat, J.-H., Delanoe, J., Doyard, P.-A., and Grunfeld, J.-P. (1975). Maladie de Fabry. Angiokeratomes presents à la naissance. Action de la diphenylhydantoine sur les crises douloureuses. *Ann. Dermatol. Syphil.* **102**, 392.

Duran, M., Tielens, A.G.M., Wadman, S.K., Stigter, J.C.M., and Kleijer, W.J. (1978*b*). Effect of thiamine in a patient with a variant form of branched-chain keto aciduria. *Acta Paediatr. Scand.* **67**, 367.

—— Beemer, F.A., Heiden, C.V.C., Korteland, J., De Brec, P.K., Brink, M., and Wadman, S.K. (1978*a*). Combined deficiency of xanthine oxidase and sulfite oxidase: A defect of molybdenum metabolism or transport. *J. inher. metab. Dis.* **1**, 175.

—— Shutgens, R.B.H., Ketel, A., Heymans, H., Berntssen, M.W.J., Ketting, D., and Wadman, S.K. (1979). 3-hydroxy-3-methylglutaryl coenzyme A lyase deficiency: Postnatal management following prenatal diagnosis by analysis of maternal urine. *J. Ped.* **95**, 1004.

—— Beemer, F.A., Tibosch, A.S., Bruinvis, L., Ketting, D., and Wadman, S.K. (1982). Inherited 3-methylglutaconic aciduria in two brothers — Another defect of leucine metabolism. *J. Ped.* **101**, 551.

Durand, P., Borrone, C., and Della Cella, G. (1966). A new mucopolysaccharide lipid-storage disease? *Lancet* **ii**, 1313.

—— —— —— (1969). Fucosidosis. *J. Ped.* **75**, 665.

—— Gatti, R., and Borrone, C. (1982). Fucosidosis. In *Genetic errors of glycoprotein metabolism* (ed. P. Durand and J.S. O'Brien). Edi-ermers, Milan; Springer, New York.

—— Borrone, C., Philippart, M., Della Cella, G., and Bugiani, O. (1967). Una nuova malattia da accumulo di glicolipidi. *Minerva Pediatr.* **19**, 2187.

—— Gatti, R., Borrone, C., Costantino, G., Cavalieri, S., Filocamo, M., and Romeo, G. (1979). Detection of carriers and prenatal diagnosis for fucosidosis in Calabria. *Hum. Genet.* **51**, 195.

—— —— Cavalieri, S., Borrone, C., Tondeur, M., Michalski, J.C., and Strecker, G. (1977). Sialidosis (Mucolipidosis I). *Helv. paediatr. Acta* **32**, 391.

Dykes, J.R.W. and Spencer-Peet, J. (1972). Hepatic glycogen synthetase deficiency. *Arch. Dis. Child.* **47**, 558.

Eady, R.A.J., Gunner, D.B., Garner, A., and Rodeck, C.H. (1983). Prenatal diagnosis of oculocutaneous albinism by electron microscopy of fetal skin. *J. Invest. Dermatol.* **80**, 210.

Eagle, R.B., Macnab, A.J., Ryman, B.E., and Strang, L.B. (1974). Liver biopsy data on a child with fructose-1,6-diphosphatase deficiency that closely resembled many aspects of glucose-6-phosphatase deficiency (Von Gierke's type I glycogen-storage disease). *Biochem. Soc. Trans.* **2**, 1118.

Eales, L., Grosser, T., and Sears, W.G. (1975). The clinical biochemistry of the human hepatocutaneous porphyrias in the light of recent studies of newly identified intermediates and porphyrin derivatives. *Ann. NY Acad. Sci.* **244**, 441.

Eberlein, W.R. and Bongiovanni, A.M. (1956). Plasma and urinary corticosteroids in hypertensive form of congenital adrenal hyperplasia. *J. biol. Chem.* **223**, 85.

—— —— (1958). Defective steroidal biogenesis in congenital adrenal hyperplasia. *Pediatrics* **21**, 661.

Edwards, D.L. (1957). Idiopathic familial oxalosis. *Arch. Pathol.* **64**, 546.

Efron, M.L. (1965). Familial hyperprolinemia. Report of a second case associated with congenital malformations, hereditary hematuria and mild mental retardation with demonstration of an enzyme defect. *New Engl. J. Med.* **272**, 1243.

—— (1967). Treatment of hydroxyprolinemia and hyperprolinemia. *Am. J. Dis. Child.* **113**, 166.

Ehlers, K.H. and Engle, M.A. (1963). Glycogen storage disease of myocardium. *Am. Heart J.* **65**, 145.

Eicher, E.M., Southard, J.L., Scriver, C.R., and Glorieux, F.H. (1976). Hypophosphatemia: mouse model for human familial hypophosphatemic (Vitamin D-resistant) rickets. *Proc. natl. Acad. Sci. USA* **73**, 4667.

Eil, C., Merriam, G.R., Bowen, J., Ebert, J., Tabor, E., White, B., Douglass, E.C., and Loriaux, D.L. (1980). Testicular feminization in the chimpanzee. *Clin. Res.* **28**, 624A.

Einhorn, N.H., Moore, J.R., and Rountree, L.G. (1946). Osteochondro-dystrophia deformans (Morquio's disease); observations at autopsy in one case. *Am. J. Dis. Child.* **72**, 536.

Elder, G.H. (1975). Differentiation of porphyria cutanea tarda symptomatica from other types of porphyria by measurement of isocoproporphyrin in faeces. *J. clin. Pathol.* **28**, 601.

—— (1977). Porphyrin metabolism in porphyria cutanea tarda. *Sem. Hematol.* **14**, 227.

—— Gray, C.H., and Nicholson, D.C. (1972). The porphyrias: a review. *J. clin. Pathol.* **25**, 1013.

—— Magnus, I.A., Handa, F., and Doyle, M. (1974). Faecal 'X porphyrin' in the hepatic porphyrias. *Enzyme* **17**, 29.

—— Evans, J.O., Thomas, N., Cox, R., Brodie, M.J., Moore, M.R., Goldberg, A., and Nicholson, D.C. (1976). The primary enzyme defect in hereditary coproporphyria. *Lancet* **ii**, 1217.

Eldjarn, L., Jellum, E., Stokke, O., Pande, H., and Waaler, P.E. (1970). ß-hydroxyisovaleric aciduria and ß-methylcrotonylglycinuria: a new inborn error of metabolism. *Lancet* **ii**, 521.

Elejalde, B.R. and Elejalde, M.M. (1983). Prenatal diagnosis of perinatally lethal osteogenesis imperfecta. *Am. J. med. Genet.* **14**, 353.

Elkington, A.R., Freedman, S.S., Jay, B., and Wright, P. (1973). Anterior dislocation of the lens in homocystinuria. *Br. J. Ophthalmol.* **57**, 235.

Ellenborgen, L.E. (1979). Uptake and transport of cobalamins. *Int. Rev. Biochem.* **27**, 45.

Ellis, J.E. and Patrick. D. (1976). Wolman disease in a Pakistani infant. *Am. J. Dis. Child.* **130**, 545.

Ellis, R.B., Ikonne, J.U., and Masson, P.K. (1975). DEAE cellulose microcolumn chromatography coupled with automated assay for hexosaminidase components. *Anal. Biochem.* **63**, 5.

—— Patrick, A.D., Stephens, R., and Willcox, P. (1973). Prenatal diagnosis of Tay–Sachs disease. *Lancet* **ii**, 1144.

Elsas, L.J., Miller, R.L., and Pinnell, S.R. (1978). Inherited human collagen lysyl hydroxylase deficiency: ascorbic acid response. *J. Ped.* **92**, 378.

—— Danner, D.J., Rogers, B.L., and Priest, J.H. (1974*a*). Mechanisms of thiamine stimulated branched chain aminoacid metabolism. *Ped. Res.* **8**, 431.

—— —— Lubitz, D., Fernhoff, P., and Dembure, P. (1981). Metabolic consequence in inherited defects in branched chain α-ketoacid dehydrogenase: mechanism of thiamine action. In *Metabolism and clinical implications of branched chain amino and ketoacids* (ed. M. Walser and J.R. Williamson), p.369. Elsevier/North-Holland, New York.

—— Priest, J.H., Wheeler, F.B., Danner, D.J., and Pask, B.A. (1976*b*). Maple syrup urine disease: coenzyme function and prenatal monitoring. *Metabolism* **23**, 569.

Elsässer, G., Freusberg, O., and Theml, F. (1950). Das Xeroderma pigmentosum und die Xerodermische idiotic. *Arch. Derm. syph. (Berl)* **188**, 651.

Emery, F.A., Goldie, L., and Stern, J. (1968). Hyperprolinaemia Type II. *J. ment. Defic. Res.* **12**, 187.

Emery, J.M., Green, W.R., Hulf, D.S., and Sloan, H.R. (1972). Niemann–Pick disease (type C) histopathology and ultrastructure. *Am. J. Ophthalmol.* **74**, 1144.

Emmerson, B.T. and Wyngaarden, J.B. (1969). Purine metabolism in heterozygous carriers of hypoxanthine-guanine phosphoribosyltransferase deficiency. *Science* **166**, 1533.

—— Gordon, R.B., and Thompson, L. (1975). Adenine phosphoribosyltransferase deficiency: its inheritance and occurrence in a female with gout and renal disease. *Aust. NZ. J. Med.* **5**, 440.

—— Johnson, L.J., and Gordon, R.B. (1977). Incidence of APRT deficiency. In *Proceedings of the second international symposium on purine metabolism in man, Vienna, June 1976.* (ed. M.M. Muller, E. Kaiser, and J.E. Seegmiller), p.293. Plenum Press, New York.

Engel, A.G. and Angelini, C. (1973). Carnitine deficiency of human skeletal muscle with associated lipid storage myopathy: a new syndrome. *Science* **179**, 899.

—— Gomez, M.R., Seybold, M.E., and Lambert, E.H. (1973). The spectrum and diagnosis of acid maltase deficiency. *Neurology* **23**, 95.

Engel, W.K., Bishop, D.W., and Cunningham, G.G. (1970). Tubular aggregates in type II muscle fibers: ultrastructural and histochemical correlation. *J. Ultrastruct. Res.* **31**, 507.

Engleman, K., Watts, R.W.E., Klinenberg, J.R., Sjoerdsma, A., and Seegmiller, J.E. (1964). Clinical, physiological, and biochemical studies of a patient with xanthinuria and pheochromocytoma. *Am. J. Med.* **37**, 839.

Enklewitz, M. and Lasker, M. (1935). Pentosuria in twins. *J. Am. Med. Assoc.* **105**, 958.

Epstein, C.J., Brady, R.O., Schneider, E.L., Bradley, R.M., and Shapiro, D. (1971). In utero diagnosis of Niemann–Pick disease. *Am. J. hum. Genet.* **23**, 533.

—— Yatziv, S., Neufeld, E.F., and Liebaers, J. (1976). Genetic counselling for Hunter syndrome. *Lancet* **ii**, 730.

Epstein, O. and Sherlock, S. (1981). Is Wilson's disease caused by a controller gene mutation resulting in perpetuation of the fetal mode of copper metabolism into childhood? *Lancet* **i**, 303.

Erbe, R.W. (1979). Genetic aspects of folate metabolism. In *Advances in human genetics 9* (ed. H. Harris and K. Hirschhorn), p.293. Plenum Publishing Corp., New York.

Eriksen, L. and Eriksen, N. (1974). Porphyrin distribution and porphyrin excretion in human congenital erythropoietic porphyria. *Scand. J. clin. Lab. Invest.* **33**, 323.

—— and Seip, M. (1973). Congenital erythropoietic porphyria. A family study. *Clin. Genet.* **4**, 166.

Erickson, R.P., Sandman, R., Robertson, W.V.B., and Epstein, C.J. (1972). Inefficacy of fresh frozen plasma therapy of mucopolysaccharidosis II. *Pediatrics* **50**, 693.

Eshaghpour, E., Oski, F.A., and Williams, M. (1967). The relationship of erythrocyte glucose-6-phosphate dehydrogenase deficiency to hyperbilirubinemia in Negro premature infants. *J. Ped.* **70**, 595.

Espinas, O.E. and Faris, A.A. (1969). Acute infantile Gaucher's disease in identical twins. An account of clinical and neuropathologic observations. *Neurology* **19**, 133.

Eto, Y. and Kitagawa, T. (1970). Wolman's disease with hypolipoproteinemia and acanthocytosis: clinical and biochemical observations. *J. Ped.* **77**, 862.

—— Suzuki, K., and Suzuki, Y. (1970). Globoid cell leucodystrophy: isolation of myelin with normal glycolipid composition. *J. Lipid. Res.* **11**, 473.

—— Wiesmann, U., and Herschkowitz, N.N. (1974*a*). Sulfogalactosylsphingosine sulfatase: characteristics of the enzyme and its deficiency in metachromatic leukodystrophy in human cultured skin fibroblasts. *J. biol. Chem.* **249**, 4955.

—— Rampini, S., Wiesmann, U., and Herschkowitz, N.N. (1974*b*). Enzymic studies of sulphatases in tissues of the normal human and in MLD with multiple sulphatase deficiencies: arylsulphatases A, B and C, cerebroside sulphatase, psychosine sulphatase and steroid sulphatases. *J. Neurochem.* **23**, 1161.

—— Wiesmann, U.N., Carson, J.H., and Herschkowitz, N.N. (1974*c*). Multiple sulfatase deficiencies in cultured skin fibroblasts. *Arch. Neurol.* **30**, 153.

Evans, F.T., Gray, P.W.S., Lehmann, H., and Silk, E. (1952). Sensitivity to succinyl-choline in relation to serum cholinesterase. *Lancet* **i**, 1229.

Evelyn, K. and Malloy, H. (1938). Microdetermination of oxyhemoglobin, methemoglobin and sulfhemoglobin in a single sample of blood. *J. biol. Chem.* **126**, 655.

Exton, J.H. and Park, C.R. (1972). Interaction of insulin and glucagon in the control of liver metabolism. In *Handbook of physiology 7, Endocrinology, 1* (ed. D.F. Steiner and N. Freinkel), p.437. American Physiological Soc., Washington.

Fabry, J. (1898). Ein Beitrag zur Kenntnis der Purpura haemorrhagica nodularis (Purpura papulosa haemorrhagica Hebrae). *Arch. Derm. Syph. (Berl)* **43**, 187.

Fahien, L., Schooler, J.M., Gehred, G.A., and Cohen, P.P. (1964). Studies on the mechanism of action of acetylglutamate as an activator of carbamyl phosphate synthetase. *J. biol. Chem.* **239**, 1935.

Faiman, C. and Winter, J.S.D. (1974). The control of gonadotropin secretion in complete testicular feminization. *J. clin. Endocr. Metab.* **39**, 631.

Fairbanks, V.F. and Lampe, L.T. (1968). A tetrazolium-linked cytochemical method for estimation of glucose-6-phosphate dehydrogenase activity in individual erythrocytes: application in the study of heterozygotes for glucose-6-phosphate dehydrogenase deficiency. *Blood* **31**, 589.

Falconer, I.R., Roitt, I.M., Seamark, R.F., and Torrigiani, G. (1970). Studies of the congenitally goitrous sheep: iodoproteins of the goitre. *Biochem. J.* **117**, 417.

Falk, R.E., Cederbaum, S.D., Blass, J.P., Pruss, R.J., and Carrell, R.E. (1976). Ketonic diet in the management of pyruvate dehydrogenase deficiency. *Pediatrics* **58**, 713.

Fallat, R.W. and Glueck, C.J. (1976). Familial and acquired type V hyperlipoproteinemia. *Atherosclerosis* **23**, 41.

Fällström, S.-P., Lindblad, B., Lindstedt, S., and Steen, G. (1979). Hereditary tyrosinemia — fumarylacetoacetase deficiency. *Ped. Res.* **13**, 78a.

Fanconi, G. and Bickel, H. (1949). Die chronische Aminoacidurie (Aminosäure-diabetes oder nephrotisch-glukosurischer Zwergwuchs) bei der Glykogenose und der Cystinkrankheit. *Helvet. paediatr. Acta* **4**, 359.

Farber, S., Cohen, J., and Uzman, L.L. (1957). Lipogranulomatosis. A new lipoglycoprotein 'storage' disease. *J. Mt Sinai Hosp.* **24**, 816.

Fardeau, M., Abelanet, R., Laudat, P., and Bonduelle, M. (1970). Maladie de Refsum. Étude histologique, ultrastructurale et biochimique d'une biopsie de nerf périphérique. *Rev. Neurol.* **122**, 185.

Farmer, T.W., Veath, L., Miller, A.L., O'Brien, J.S., and Rosenberg, R.M. (1973). Pyruvate decarboxylasc deficiency in a patient with subacute necrotizing encephalomyelopathy. *Neurology* **23**, 429.

Farrell, D.F., Clark, A.F., Scott, C.R., and Wennberg, R.P. (1975). Absence of pyruvate decarboxylase activity in man: a cause of congenital lactic acidosis. *Science* **187**, 1082.

—— Percy, A.K., Kaback, M.M., and McKhann, G.M. (1973). Globoid cell leucodystrophy: heterozygote detection in cultured skin fibroblasts. *Am. J. hum. Genet.* **25**, 604.

Faull, K.F., Bolton, P.D., Halpern, B., Hammond, J., and Danks, D.M. (1976*b*). The urinary organic acid profile associated with 3-hydroxy-3-methylglutaric aciduria. *Clin. Chim. Acta.* **73**, 553.

—— Gan, I., Halpern, B., Hammond, J.I.S., Cotton, R.G.H., and Danks, D.M. (1977). Metabolic studies on two patients with nonhepatic tyrosinemia using deuterated tyrosine loads. *Ped. Res.* **11**, 631.

—— Bolton, P., Halpern, B., Hammond, J., Danks, D.M., Hähnel, R., Wilkinson, S.P., Wysocki, S.J., and Masters, P.L. (1976*a*). Patient with defect in leucine metabolism. *New Engl. J. Med.* **294**, 1013.

Feinstein, R.N. (1970). Acatalasemia in the mouse and other species. *Biochem. Genet.* **4**, 135.

Fell, V., Hoskins, J.A., and Pollitt, R.J. (1978). The labelling of urinary acids after oral doses of deuterated 1-phenylalanine and 1-tyrosine in normal subjects. Quantitative studies with implications for the deuterated phenylalanine load test in PKU. *Clin. Chim. Acta* **83**, 259.

—— and Pollitt, R.J. (1978). 3-aminopiperid-2-one, an unusual metabolite in the urine of a patient with hyperammonemia, hyperornithinemia and homocitrullinuria. *Clin. Chim. Acta* **87**, 405.

—— —— Sampson, G.A., and Wright, T. (1974). Ornithinemia, hyperammonemia, and homocitrullinuria. A disease with mental retardation and possibly caused by defective mitochondrial transport. *Am. J. Dis. Child.* **127**, 752.

Fellows, F.C.I. and Carson, N.A.J. (1974). Enzyme studies in a patient with saccharopinuria: a defect of lysine metabolism. *Ped. Res.* **8**, 42.

Felts, J.H., King, J.S., and Boyce, W.H. (1968). Nephrotic syndrome after treatment with D-penicillamine. *Lancet* **i**, 53.

Fensom, A.H. and Benson, P.F. (1975). Assay of galactose-1-phosphate uridyl transferase in cultured amniotic cells for prenatal diagnosis of galactosaemia. *Clin. Chim. Acta* **62**, 189.

—— —— and Baker, J.A. (1978). A rapid method for assay of branched-chain ketoacid decarboxylation in cultured cells and its application to prenatal diagnosis of maple syrup urine disease. *Clin. Chim. Acta* **87**, 169.

—— —— and Blunt, S. (1974). Prenatal diagnosis of galactosaemia. *Br. med. J.* **iv**, 386.

—— —— Baker, J.A., and Mutton, D.E. (1980). Prenatal diagnosis of argininosuccinic aciduria: effect of mycoplasma contamination on the indirect assay for argininosuccinate lyase. *Am. J. hum. Genet.* **32**, 761.

—— —— Grant, A.R., and Jacobs, L. (1979*b*). Fibroblast α-galactosidase A activity for identification of Fabry's disease heterozygotes. *J. inher. metab. Dis.* **2**, 9.

—— —— Babarik, A.W., Grant, A.R., and Jacobs, L. (1977). Fibroblast phosphodiesterase deficiency in Niemann-Pick disease. *Biochem. biophys. Res. Commun.* **74**, 877.

—— —— Blunt, S., Brown, S.P., and Coltart, T.M. (1976). Amniotic cell 4-methylumbelliferyl-α-glucosidase activity for prenatal diagnosis of Pompe's disease. *J. med. Genet.* **13**, 148.

—— —— Rodeck, C.H., Campbell, S., and Gould, J.D.M. (1979*a*). Prenatal diagnosis of a galactosaemia heterozygote by fetal blood enzyme assay. *Br. med. J.* **i**, 21.

—— —— Crees, M.J., Ellis, M., Rodeck, C.H., and Vaughan, R.W. (1983). Prenatal exclusion of homocystinuria (cystathionine ß-synthase deficiency) by assay of phytohaemagglutinin-stimulated fetal lymphocytes. *Prenatal Diagnosis* **3**, 127.

—— —— Neville, B.R.G., Moser, H.W., Moser, A.E., and Dulaney, J.T. (1979*c*). Prenatal diagnosis of Farber's disease. *Lancet* **ii**, 990.

—— —— Chalmers, R.A., Tracey, B.M., Watson, D., King, G.S., Pettit, B.R., and Rodeck, C.H. (1984*a*). Experience with prenatal diagnosis of propionic acidaemia and methylmalonic aciduria. *J. inher. metab. Dis.* **7, Suppl. 2**, 127.

—— Chase, D., Crees, M.J., Jackson, M., McGuire, V.M., Marsh, J,. Sanguinetti, N., Vimal, C.M., and Yuen, M. (1984*b*). Biochemical studies on chorion biopsy. In *Proceedings of the international symposium on early prenatal diagnosis: present* *and future.* (ed. A. Paladini, D. Catalano, A. Di Lieto, and F. Rullo) p. 63. Naples, 12th–13th October. G.V.A. Press, Naples.

Fenton, W.A., and Rosenberg, L.E. (1978) Genetic and biochemical analysis of human cobalamin mutants in cell culture. *Ann. Rev. Genet.* **12**, 223.

Fernandes, J. (1980). Hepatic glycogenosis: diagnosis and management. In *Inherited disorders of carbohydrate metabolism* (ed. D. Burman, J.B. Holton, and C.A. Pennock), p.297. MTP Press Ltd., Lancaster.

—— and Huijing, F. (1968). Branching enzyme-deficiency glycogenosis: studies in therapy. *Arch. Dis. Child.* **43**, 347.

—— and Pikaar, N.A. (1972). Ketosis in hepatic glycogenosis. *Arch. Dis. Child.* **47**, 41.

—— and Van de Kamer, J.H. (1968). Hexose and protein tolerance tests in children with liver glycogenosis caused by a deficiency of the debranching enzyme system. *Pediatrics* **41**, 935.

—— Koster, J.F., Grose, W.F.A., and Sorgedrager, N. (1974). Hepatic phosphorylase deficiency. Its differentiation from other hepatic glycogenoses. *Arch. Dis. Child.* **49**, 189.

Fertman, M.H. and Fertman, M.D. (1955). Toxic anemias and Heinz bodies. *Medicine (Baltimore)* **34**, 131.

Fessler, J.H. and Fessler, L.I. (1978). Biosynthesis of procollagen. *Ann. Rev. Biochem.* **47**, 129.

Fevery, J., Blanckaert, N., Heirwegh, K.P.M., Preaux, A.-M., and Berthelot, P. (1977). Unconjugated bilirubin and an increased proportion of bilirubin monoconjugates in the bile of patients with Gilbert's syndrome and Crigler–Najjar syndrome. *J. clin. Invest.* **60**, 970.

Fialkow, P.J. (1970). X-chromosome inactivation and the Xg locus. *Am. J. hum. Genet.* **22**, 460.

—— (1974). The origin and development of human tumors studied with cell markers. *New Engl. J. Med.* **291**, 26.

—— (1976). Clonal origin of human tumors. *Biochim. Biophys. Acta.* **458**, 283.

—— Browder, J.A., Sparkes, R.S., and Motulsky, A.G. (1965). Mental retardation in methemoglobinemia due to diaphorase deficiency. *New Engl. J. Med.* **273**, 840.

Fiddler, M.B., Vine, D., Shapira, E., and Nadler, H.L. (1979). Is multiple sulphatase dificiency due to defective regulation of sulphohydrolase expression? *Nature, Lond.* **282**, 98.

Field, J.B. and Drash, A.L. (1967). Studies in glycogen storage disease. II. Heterogeneity in the inheritance of glycogen storage diseases. *Trans. Assoc. Am. Physicians* **80**, 284.

Fine, R.N., Strauss, J., and Donnell, G.N. (1966). Hyperuricemia in glycogen storage disease type 1. *Am. J. Dis. Child.* **112**, 572.

—— Wilson, W.A., and Donnell, G.N. (1968). Retinal changes in glycogen storage disease type 1. *Am. J. Dis. Child.* **115**, 328.

Finkelstein, J.D., Kyle W.E., and Martin, J.J. (1975). Abnormal methionine adenosyltransferase in hypermethioninemia. *Biochem. biophys. Res. Commun.* **66**, 1491.

—— Mudd, S.H., Irreverre, F., and Laster, L. (1964). Homocystinuria due to cystathionine synthetase deficiency: the mode of inheritance. *Science* **146**, 785.

—— —— —— —— (1966). Deficiencies of cystathionase and homoserine dehydratase activities in cystathioninuria. *Proc. natl. Acad. Sci. USA* **55**, 865.

Finnie, M.D.A., Cottrall, K., Seakins, J.W.T., and Snedden, W. (1976). Massive excretion of 2-oxoglutaric acid and 3-hydroxyisovaleric acid in a patient with a deficiency of 3-methylcrotonyl-CoA carboxylase. *Clin. Chim. Acta* **73**, 513.

Firgaira, F.A., Cotton, R.G.H., Danks, D.M., Fowler, K., Lipson, A., and Yu, J.S. (1983). Prenatal determination of dihydropteridine reductase in a normal fetus at risk for malignant hyperphenylalaninemia. *Prenatal Diagnosis* **3**, 7.

Fisch, R.O., McCabe, E.R.B., Doeden, D., Koep. L.J., Kohlhoff, J.G., Silverman, A., and Starzl, T.E. (1978). Homotransplantation of the liver in a patient with hepatoma and hereditary tyrosinemia. *J. Ped.* **93**, 592.

Fischer, G. and Jatzkewitz, H. (1975). The activator of cerebroside sulfatase, purification from human liver and identification as a protein. *Hoppe-Seyler's Z. Physiol. Chem.* **356**, 605.

Fischer, M.H. and Brown, R.R. (1980). Tryptophan and lysine metabolism in alpha-aminoadipic aciduria. *Am. J. med. Genet.* **5**, 35.

—— and Gerritsen, T. (1971). Biochemical studies on a variant of branched-chain ketoaciduria in a 19-year-old female. *Pediatrics* **48**, 795.

—— —— and Opitz, J.M. (1974). α-Aminoadipic aciduria, a non-deleterious inborn metabolic defect. *Hum. Genet.* **24**, 265.

Fiser, R.H., Melsher, H.L., and Fischer, D.A. (1974). Hepatic phosphoenol-pyruvate carboxykinase deficiency: a new cause of hypoglycaemia in childhood. *Ped. Res.* **8**, 432.

Fishbein, W.N., Armbrustmacher, W.M., and Griffin, J.L. (1978). Myoadenylate deaminase deficiency. A new disease of muscle. *Science* **200**, 545.

—— —— —— (1981). Myoadenylate deaminase deficiency: verification on repeat biopsy, fresh or frozen, and origin of the residual enzyme. *IRCS med. Sci. Biochem.* **9**, 103.

—— Davis, J.I., Nagarajon, K., Winkert, J.W., and Foellmer, J.W. (1980). Immunologic distinction of human muscle adenylate deaminase from the isozyme in human peripheral blood cells: implications for myoadenylate deaminase deficiency. *Arch. Biochem. Biophys.* **205**, 360.

Fisher, D.A., Dussault, J.H., Foley, T.P., Klein, A.H., La Franchi, S., Larsen, P.R., Mitchell, M.L., Murphey, W.H., and Walfish, P.G. (1979). Screening for congenital hypothyroidism: results of screening one million North American infants. *J. Ped.* **94**, 700.

Fisher, E.R. and Reidbord, H. (1962). 'Gaucher's disease'. Pathogenetic considerations based on electron microscopic and histochemical observations. *Am. J. Pathol.* **41**, 679.

Fishler, K., Donnell, G.N., Bergren, W.R., and Koch, R. (1972). Intellectual and personality development in children with galactosemia. *Pediatrics* **50**, 412.

Flament-Durand, J., Noel, P., Rutsaert, J., Toussaint, D., Malmendier, C., and Lyon, G. (1971). A case of Refsum's disease: clinical, pathological, ultra-structural and biochemical study. *Path. Europ.* **6**, 172.

Fleischer, B. (1912). Über einer der 'Pseudosklerose' nahestehende bisher unbekannte Krankheit (gekennzeichnet durch Tremor, psychische Störungen, bräunliche Pigmentierung bestimmter Gewebe, insbesondere auch der Hornhautperipherie, Lebercirrhose). *Dtsch. Z. Nervenheilkd.* **44**, 179.

Fleischer, L.D., Harris, C.J., Mitchell, D.A., and Nadler, H.L. (1983). Citrullinemia: prenatal diagnosis of an affected fetus. *Am. J. hum. Genet.* **35**, 85.

—— Tallan, H.H., Beratis, N.G., Hirschhorn, K., and Gaull, G.E. (1973). Cystathionine synthase deficiency: heterozygote detection using cultured skin fibroblasts. *Biochem. biophys. Res. Commun.* **55**, 38.

—— Longhi, R.C., Tallan, H.H., Beratis, N.G., Hirschhorn, K., and Gaull, G.E.

(1974). Homocystinuria: investigations of cystathionine synthase in cultured fetal cells and the prenatal determination of genetic status. *Pediatrics* **85**, 677.
—— Rassin, D.K., Desnick, R.J., Salwen, H.R., Rogers, P., Bean, M., and Gaull, G.E. (1979). Argininosuccinic aciduria: prenatal studies in a family at risk. *Am. J. hum. Genet.* **31**, 439.
Fleischner, G., Robbins, J., and Arias, I.M. (1972). Immunological studies of Y protein: a major cytoplasmic organic anion binding protein in rat liver. *J. clin. Invest.* **51**, 677.
Fleming, C.R., Dickson, E.R., Wahner, H.W., Hollenhorst, R.W., and McCall, J.T. (1977). Pigmented corneal rings in non-Wilsonian liver disease. *Ann. Int. Med.* **86**, 285.
Fletcher, T.F., Kurtz, H.J., and Stadlan, E.M. (1971). Experimental Wallerian degeneration in peripheral nerves of dogs with globoid cell leukodystrophy. *J. Neuropathol. exp. Neurol.* **30**, 593.
Fliegner, J.R.H., Schindler, I., and Brown, J.B. (1971). Low urinary oestriol excretion during pregnancy associated with placental sulphatase deficiency or congenital adrenal hypoplasia. *J. Obstet. Gynaecol. Br. Commonw.* **79**, 810.
Fluharty, A.L., Davis, M.L., and Kihara, H. (1974*b*). Simplified procedure for preparation of ^{35}S-labeled brain sulfatide. *Lipids* **9**, 865.
—— Stevens R.L., Sanders, D.L., and Kihara, H. (1974*a*). Arylsulfatase B deficiency in Maroteaux-Lamy syndrome cultured fibroblasts. *Biochem. biophys. Res. Commun.* **59**, 455.
—— —— Davis, L.L., Shapiro, L.J., and Kihara, H. (1978). Presence of arylsulfatase A (ARS A) in multiple sulfatase deficiency disorder fibroblasts. *Am. J. hum. Genet.* **30**, 249.
—— —— De la Flor, S.D., Shapiro, L.J., and Kihara, H. (1979). Arylsulfatase A modulation with pH in multiple sulfatase deficiency disorder fibroblasts. *Am. J. hum. Genet.* **31**, 574.
—— —— Miller, R.T., Shapiro, S.S., and Kihara, H. (1976). Ascorbic acid-2-sulphate sulfohydrolase activity in human arysulphatase A. *Biochim. Biophys. Acta* **429**, 508.
Folkman, J. Philippart, A., Tze, W.J., and Crigler, J. (1972). Portacaval shunt for glycogen storage disease: value of prolonged intravenous hyperalimentation before surgery. *Surgery* **72**, 306.
Fölling, A. (1934*a*). Utskillelse af fenylpyrodruesyre i urinen som stoffskifte-anomali i farbindelse med imbecillitet. *Nord. med. Tidskr.* **8**, 1054.
—— (1934*b*). Über Ausscheidung von Phenylbrenztraubensäure in den Harn als Stoffwechselanomalie in Verbindung mit Imbezillität. *Ztschr. Physiol. Chem.* **227**, 169.
Fontaine, G., Farriaux, J.P., and Dautrevaux, M. (1970). L'hyperprolinémie de Type 1. Etude d'une observation familiale. *Helv. Paediatr. Acta.* **25**, 165.
Ford, R.C. and Berman, J.L. (1977). Phenylalanine metabolism and intellectual functioning among carriers of phenylketonuria and hyperphenylalaninaemia. *Lancet* i, 767.
Forest, M.G., Lecornu, M., and De Peretti, E. (1980). Familial male pseudo-hermaphroditism due to 17-20-desmolase deficiency. I. In vivo endocrine studies. *J. clin. Endocr. Metab.* **50**, 826.
Forte, T., Norum, K.R., Glomset, J.A., and Nichols, A.V. (1971). Plasma lipoproteins in familial lecithin:cholesterol acyltransferase deficiency: structure of low and high density lipoproteins as revealed by electron microscopy. *J. clin. Invest.* **50**, 1141.
Fortuin, J.J.H. and Kleijer, W.J. (1980). Hybridisation studies of fibroblasts from

Hurler, Scheie, and Hurler–Scheie compound patients: support for the hypothesis of allelic mutants. *Hum. Genet.* **53**, 155.

Foster, J.A., Rich, C.B., Berglund, N., Huber, S., Mecham, R.P., and Lange, G.
(1979). The anti-proteolytic behaviour of lathyrogens. *Biochim. Biophys. Acta*
587, 477.

Foulk, W.T., Butt, H.R., Owen, C.A., and Whitcomb, F.F. (1959). Constitutional
hepatic dysfunction (Gilbert's disease): its natural history and related syndrome.
Medicine **38**, 25.

Fowler, B., Børresen, A.L., and Boman, N. (1982). Prenatal diagnosis of
homocystinuria. *Lancet* **ii**, 875.

—— Kraus, J., Packman, S., and Rosenberg, L.E. (1978). Homocystinuria:
evidence for three distinct classes of cystathionine-ß-synthase mutants in cultured
fibroblasts. *J. clin. Invest.* **61**, 645.

Fox, I.H. (1977). Purine enzyme abnormalities: a four year experience. In
Proceedings of the second international symposium on purine metabolism in man.
Vienna, June 1976 (ed. M.M. Muller, E. Kaiser, and J.E. Seegmiller), p.265.
Plenum Press, New York.

—— and Kelley, W.N. (1971). Phosphoribosyl pyrophosphate in man: biochemical
and clinical significance. *Ann. intern. Med.* **74**, 424.

Fox, R.M., O'Sullivan, W.J., and Firkin, B.G. (1969). Orotic aciduria. *Am. J.*
Med. **47**, 332.

—— Wood, M.H., Royse-Smith, D., and O'Sullivan, W.J. (1973). Hereditary
orotic aciduria. *Am. J. Med.* **55**, 791.

France, J.T. and Liggins, G.C. (1969). Placental sulfatase deficiency. *J. clin.*
Endocr. Metab. **29**, 138.

—— Seddon, R.J., and Liggins, G.C. (1973). A study of a pregnancy with low
oestrogen production due to placental sulphatase deficiency. *J. clin. Endocr.*
Metab. **36**, 1.

Francois, B., Cornu, G., and De Meyer, R. (1976). Peritoneal dialysis and
exchange transfusion in a neonate with argininosuccinic aciduria. *Arch. Dis.*
Child. **51**, 228.

Francois, J. (1972). Ocular manifestations in aminoacidopathies. *Adv.*
Ophthalmol. **25**, 28.

—— (1979). Gyrate atrophy of the choroid and retina. *Ophthalmologica* (*Basel*)
178, 311.

—— Hanssens, M. Coppieters, R., and Evens, L. (1972). Cystinosis. A clinical and
histopathologic study. *Am. J. Ophthalmol.* **73**, 643.

Francke, U. (1976). The human gene for ß-glucuronidase is on chromosome 7. *Am.*
J. hum. Genet. **28**, 357.

Frankhauser, R., Luginbühl., and Hartley, W.J. (1963). Leukodystophie vom
Typus Krabbe beim Hund. *Schweiz. Arch. Tierheilkd.* **105**, 198.

Fraser, D. (1957). Hypophosphatasia. *Am. J. Med.* **22**, 730.

—— and Yendt, E.R. (1955). Metabolic abnormalities in hypophosphatasia. *Am.*
J. Dis. Child. **90**, 552.

—— —— and Christie, F.H. (1955). Metabolic abnormalities in hypophosphatasia. *Lancet* **i**, 286.

—— Kooh, S.W., Kind, H.P., Holick, M.F., Tanaka, Y., and Deluca, H.F.
(1973). Vitamin-D dependent rickets: an inborn error of vitamin D metabolism.
New Engl. J. Med. **289**, 817.

Fraser, G.R., Morgans, M.E., and Trotter, W.R. (1960). The syndrome of
sporadic goitre and congenital deafness. *Q. J. Med.* **29**, 279.

—— Friedman, A.I., Patton, V.M., Wade, D.N., and Woolf, L.L. (1968).

Iminoglycinuria — a 'harmless' inborn error of metabolism? *Hum. Genet.* **6**, 362.

Fraser Roberts, J.A. and Pembrey, M.E. (1985). *An introduction to medical genetics*, 8th edn. Oxford Univeristy Press, Oxford.

Frasier, S.D., Penny, R., and Snyder, R. (1982). Primary congenital hypothyroidism in Spanish-surnamed infants in Southern California. *J. Ped.* **101**, 315.

Fratantoni, J.C., Hall, C.W., and Neufeld, E.F. (1968*a*). The defect in Hurler's and Hunter's syndromes: faulty degradation of mucopolysaccharides. *Proc. natl. Acad. Sci. USA* **60**, 699.

—— —— —— (1968*b*). Hurler and Hunter syndromes: mutual correction of the defect in cultured fibroblasts. *Science* **162**, 570.

—— —— —— (1969*a*). The defect in Hurler and Hunter syndromes. II. Deficiency of specific factors involved in mucopolysaccaride degradation. *Proc. natl. Acad. Sci. USA* **64**, 360.

—— Neufeld, E.F., Uhlendorf, B.W., and Jacobson, C.B. (1969*b*). Intrauterine diagnosis of the Hurler and Hunter syndromes. *New Engl. J. Med.* **280**, 686.

Frazier, D.M., Summer, G.K., and Chamberlin, H.R. (1978). Hyperglycinuria and hyperglycinemia in two siblings with mild developmental delays. *Am. J. Dis. Child.* **132**, 777.

Frederick, E.W., Rabkin, M.T., Richie, R.H. jun., and Smith, L.H. jun. (1963). Studies on primary hyperoxaluria. I. *In vivo* demonstration of a defect in glyoxylate metabolism. *New Engl. J. Med.* **269**, 821.

Fredrickson, D.S. and Levy, R.I. (1972). Familial hyperlipoproteinemia. In *The metabolic basis of inherited disease*, 3rd end. (ed. J.B. Stanbury, J.B. Wyngaarden, and D.S. Fredrickson), p.574. McGraw-Hill, New York.

—— and Sloan, H.R. (1972*a*). Glucosylceramide lipidosis: Gaucher's disease. In *The metabolic basis of inherited disease*, 3rd end. (ed. J.B. Stanbury, J.B. Wyngaarden, and D.S. Fredrickson), p.730. McGraw-Hill, New York.

—— —— (1972*b*). Sphingomyelin lipidoses: Niemann–Pick disease. In *The metabolic basis of inherited disease*, 3rd end. (ed. J.B. Stanbury, J.B. Wyngaarden, and D.S. Fredrickson), p.783. McGraw-Hill, New York.

—— Goldstein, J.L., and Brown, M.S. (1978). The familial hyperlipoproteinemias. In *The metabolic basis of inherited disease*, 4th end. (ed. J.B. Stanbury, J.B. Wyngaarden, and D.S. Fredrickson), p.604. McGraw-Hill, New York.

Freeman, J.M., Finkelstein, J.D., and Mudd, S.H. (1975). Folate-responsive homocystinuria and 'schizophrenia'. A defect in methylation due to deficient 5,10-methylenetetrahydrofolate reductase activity. *New Engl. J. Med.* **292**, 491.

—— Nicholson, J.F., Schimke, R.T., Rowland, L.P., and Carter, S. (1970). Congenital hyperammonemia: association with hyperglycinemia and decreased levels of carbamyl phosphate synthetase. *Arch. Neurol.* **23**, 430.

Freiburghaus, A.U., Schmitz, J., Schindler, M., Rotthauwe, H.W., Kuitunen, P., Launiala, K., and Hadorn, B. (1976). Protein patterns of brush border fragments in congenital lactose malabsorption and in specific hypolactasia of the adult. *New Engl. J. Med.* **294**, 1030.

French, J.H., Bootz, M., and Poser, C.M. (1969). Lipid composition of the brain in infantile Gaucher's disease. *Neurology (Minneap.)* **19**, 81.

—— Sherard, E.C., Lubell, M., Brotz, M., and Moore, C.L. (1972). Trichopoliodystrophy. I. Report of a case and biochemical studies. *Arch. Neurol.* **26**, 229.

French, F.S., Van Wyk, J.J., Baggett, B., Easterling, W.E., Talbert, L.M., Johnston, F.R., Forchielli, E., and Dey, A.C. (1966). Further evidence of a target organ defect in the syndrome of testicular feminization. *J. clin. Endocr. Metab.* **26**, 493.

534   *References*

Frenk, E. and Calme, A. (1977). Hypopigmentation oculo-cutanée familial à transmission dominante due à un trouble de la formation des mélanosomes. *Schweiz. med. Wochenschr.* **107**, 1964.

Fried, D., Gotlieb, A., and Roitman, A. (1977). Infectious hepatitis with excessive hyperbilirubinemia and a hemolytic crisis in an 8-year-old boy. *Clin. Pediatr.* **16**, 482.

Fried, K., Mathoth, Y., and Goldschmidt, E. (1963). Gaucher's disease — chronic adult type. In *The genetics of migrant and isolate populations* (ed. E. Goldschmidt), p.292. Williams and Wilkins, Baltimore.

Friedman, P.A., Fisher, D.B., Kang, E.S., and Kaufman, S. (1973). Detection of hepatic phenylalanine 4-hydroxylase in classical phenylketonuria. *Proc. natl. Acad. Sci. USA* **70**, 552.

Frimpter, G.W. (1961). The disulphide of L-cysteine and L-homocysteine in urine of patients with cystinuria. *J. biol. Chem.* **236**, pc51.

—— (1965). Cystathioninuria: Nature of the defect. *Science* **149**, 1095.

—— Haymovitz, A., and Horwith, M. (1963). Cystathioninuria. *New Engl. J. Med.* **268**, 333.

—— Horwith, M., Furth, E., Fellows, R.E., and Thompson, D.P. (1962). Innulin and endogenous amino acid renal clearances in cystinuria: evidence for tubular secretion. *J. clin. Invest.* **41**, 281.

Frisch, A. and Neufeld, E.F. (1981). Limited proteolysis of the ß-hexosaminidase precursor in a cell-free system. *J. biol. Chem.* **256**, 8242.

Froesch, E.R., Prader, A., Labhart, A., Stuber, H.W., and Wolf, H.P. (1957). Die hereditäre Fructoseintoleranz, eine bisher nicht bekannte kongenitale Stoffwechselstörung. *Schweiz. Med. Wochenschr.* **87**, 1168.

Frohlich, J., Godolphin, W.J., Reeve, C.E., and Evelyn, K. (1978). Familial LCAT deficiency. Report ot two patients from a Canadian family of Italian and Swedish descent. *Scand. J. clin. Lab. Invest.* **38**, suppl. 150. 156.

Frommer, D., Morris, J., Sherlock, S., Abrams, J., and Newman, S. (1977). Kayser–Fleischer-like rings in patients without Wilson's disease. *Gastroenterology* **72**, 1331.

Frost, P., Tanaka, Y., and Spaeth, G.I.. (1966). Fabry's disease: glycolipid lipidosis. *Am. J. Med.* **40**, 618.

Frost, R.G., Holmes, E.W., Norden, A.G.W., and O'Brien, J.S. (1978). Characterization of purified human liver acid ß-galactosidases A2 and A3. *Biochem. J.* **175**, 181.

Fryer, D.G., Winckleman, A.C., Ways, P.O., and Swanson, A.G. (1971). Refsum's disease. *Neurology* **21**, 162.

Fujimoto, W.Y. and Seegmiller, J.E. (1970). Hypoxanthine-guanine phosphoribosyltransferase deficiency: activity in normal, mutant and heterozygote-cultured human skin fibroblasts. *Proc. natl. Acad. Sci. USA* **65**, 577.

—— Subak-Sharpe, J.H., and Seegmiller, J.E. (1971). Hypoxanthine-guanine phosphoribosyltransferase deficiency: chemical agents selective for mutant or normal cultured fibroblasts in mixed and heterozygote cultures. *Proc. natl. Acad. Sci. USA* **68**, 1516.

—— Seegmiller, J.E., Uhlendorf, B.W., and Jacobson, C.B. (1968). Biochemical diagnosis of an X-linked disease in utero. *Lancet* **ii**, 511.

Fukui, S., Yoshida, H., Tanaka, T., Takashi, S., Usui, T., and Yamashina, I. (1981). Glycosaminoglycan synthesis by cultured skin fibroblasts from a patient with Lowe's syndrome. *J. biol. Chem.* **256**, 10313.

Fusco, G., Carlomagno, S., Remano, A., Rinaldi, E., Cedrola, G., Cianciaruso, L., Curto, A., Rosolia, S., and Auricchio, G. (1976). Type-1 hyperprolinemia in a family suffering from aniridia and severe dystrophia of ocular tissues. *Ophthalmologica* **173**, 1.

Gabrilove, J.L., Sharma, E.C., and Dorfman, R.I. (1965). Adrenocortical 11-beta-hydrolase deficiency and virilism first manifest in the adult woman. *New Engl. J. Med.* **272**, 1189.

Gagné, R., Lescault, A., Grenier, A., Laberge, C., Melancon, S.B., and Dallaire, L. (1982). Prenatal diagnosis of hereditary tyrosinaemia: measurement of succinylacetone in amniotic fluid. *Prenatal Diagnosis* **2**, 185.

Gahl, W.A., Bashan, N., Tietze, F., Bernardini, I., and Schulman, J.D. (1982). Cystine transport is defective in isolated leukocyte lysosomes from patients with cystinosis. *Science* **217**, 1263.

Gal, A.E., Pentchev, P.G., and Fash, F.J. (1976*a*). A novel chromogenic substrate for assaying glucocerebrosidase activity. *Proc. Soc. exp. Biol. Med.* **153**, 363.

—— Brady, R.O., Hibbert, S.R., and Pentchev, P.G. (1975). A practical chromogenic procedure for the detection of homozygotes and heterozygous carriers of Niemann–Pick disease. *New Engl. J. Med.* **293**, 632.

—— —— Pentchev, P.G., Furbish, F.S., Suzuki, K., Tanaka, H., and Schneider, E.L. (1977). A practical chromogenic procedure for the diagnosis of Krabbe's disease. *Clin. Chim. Acta.* **77**, 53.

Gal, E.M., Hanson, G., and Sherman, A. (1976*b*). Biopterin. I. Profile and quantitation in rat brain. *Neurochem. Res.* **1**, 511.

—— Nelson, J.M., and Sherman, A.D. (1978). Biopterin. III. Purification and characterization of enzymes involved in the cerebral synthesis of 7,8-dihydro-biopterin. *Neurochem. Res.* **3**, 69.

Galdston, M., Steele, J.M., and Dobriner, K. (1952). Alcaptonuria and ochronosis with a report of three patients and metabolic studies in two. *Am. J. Med.* **13**, 432.

Galjaard, H. (1979). Group report on prenatal diagnosis of genetic metabolic diseases in Western Europe. In *Prenatal diagnosis, proc. third European conference, Munich.* (ed. S. Stengel-Rutkowski and E. Schwinger), p.73. Enke Verlag, Stuttgart.

—— (1980*a*). *Genetic metabolic diseases. Early diagnosis and prenatal analysis*, p.114. Elsevier, Amsterdam.

—— (1980*b*). *Genetic metabolic diseases. Early diagnosis and prenatal analysis*, p.641. Elsevier, Amsterdam.

—— Niermeijer, M.F., Hahnemann, N., Mohr, J., and Sorensen, S.A. (1974*b*). An example of rapid prenatal diagnosis of Fabry's disease using micro-techniques. *Clin. Genet.* **5**, 368.

—— Hoogeveen, A., De Wit-Verbeek, H.A., Reuser, A.J.J., Keijzer, W., Westerveld, A., and Bootsma, D. (1974*a*). Tay–Sachs and Sandhoff's disease, intergenic complementation after somatic cell hybridisation. *Exp. Cell. Res.* **87**, 444.

—— —— Keijzer, W., De Wit-Verbeek, H.A., Reuser, A.J.J., Ho, M.W., and Robinson, D. (1975). Genetic heterogeneity in GM$_1$-gangliosidosis. *Nature, Lond.* **257**, 60.

Gall, J.C. jun., Brewer, G.J., and Dern, R.J. (1965). Studies of glucose-6-phosphate dehydrogenase activity in individual erythrocytes: the methemo-globin-elution test for identification of females heterozygous for G6PD deficiency. *Am. J. hum. Genet.* **17**, 359.

Gamble, J.G. and Lehninger, A.L. (1973). Transport of ornithine and citrulline across the mitochondrial membrane. *J. biol. Chem.* **248**, 610.

Gandy, H.M., Keutmann, E.H., and Izzo, A.J. (1960). Characterization of urinary steroids in adrenal hyperplasia: isolation of the metabolites of cortisol,

Compound S and desoxycorticosterone from normotensive patient with adreno-genital syndrome. *J. clin. Invest.* **39**, 364.

Ganschow, R. and Paigen, K. (1967). Separate genes determining the structure and intracellular location of hepatic glucuronidase. *Proc. natl. Acad. Sci. USA* **58**, 938.

Garancis, J.C. (1968). Type II glycogenosis. *Am. J. Med.* **44**, 289.

Gardner, R.J.M. and Hay, H.R. (1974). Hurler's syndrome with clear corneas. *Lancet* **ii**, 845.

Gardo, S. and Papp, Z. (1974). Clinical variations of testicular intersexuality in a family. *J. med. Genet.* **11**, 267.

Garibaldi, L.R., Siliato, F., De Martini, I., Scarsi, M.R., and Romano, C. (1977). Oculocutaneous tyrosinosis. Report of two cases in the same family. *Helv. Paediatr. Acta.* **32**, 173.

Garnica, A.D., Frias, J.L. and Rennert, O.M. (1977). Menkes kinky hair syndrome: is it a treatable disorder? *Clin. Genet.* **11**, 154.

Garland, P.B. and Randle, P.J. (1964). Control of pyruvate dehydrogenase in perfused heart by the intracellular concentration of acetyl-CoA. *Biochem. J.* **91**, 6.

Garrod, A.E. (1901). About alkaptonuria. *Lancet* **ii**, 1484.

—— (1909). *Inborn errors of metabolism*. Frowde, Hodder and Stoughton, London.

Gartler, S.M., Gandini, E., and Ceppellini, R. (1962). Glucose-6-phosphate dehdrogenase deficient mutant in human cell culture. *Nature, Lond.* **193**, 602.

—— Scott, R.C., Goldstein, J.L., Campbell, B., and Sparkes, R. (1971). Lesch–Nyhan syndrome: rapid detection of heterozygotes by use of hair follicles. *Science* **172**, 572.

Gatfield, P.D. and Taller, E. (1971). Accumulation of lysine dipeptides in the brain in hyperpipecolatemia. *Brain. Res.* **29**, 170.

—— Knights, R.M., Devereux, M., and Pozsonyi, J.P. (1969). Histidinemia: report of four new cases in one family and the effect of low-histidine diets. *Can. med. Assoc. J.* **101**, 465.

—— Taller, E., Wolfe, D.M., and Haust, M.D. (1975). Hyperornithinemia, hyper-ammonemia, and homocitrullinuria associated with decreased carbamyl phosphate synthetase I activity. *Ped. Res.* **9**, 488.

—— —— Hinton, G.G., Wallace, A.C., Abdelnour, G.M., and Haust, M.D. (1968). Hyperpipecolatemia: a new metabolic disorder associated with neuropathy and hepatomegaly: a case study. *Can. med. Assoc. J.* **99**, 1215.

Gatt, S. and Rapport, M.M. (1966). Isolation of ß-galactosidase and ß-glucosidase from brain. *Biochim. Biophys. Acta.* **113**, 567.

Gatti, R., Costantino, G., Borrone, C., Salemi, M., Filocamo, M., Romeo, G., and Durand, P. (1980). Detection of carriers for fucosidosis in Calabria. *Helv. Paediatr. Acta. (Suppl.)* **45**, 30.

—— Borrone, C., Durand, P., De Virgilis, S., Sanna, G., Cao, A., Von Figura, K., Kresse, H., and Paschke, E. (1982). Sanfilippo type D disease: clinical findings in two patients with a new variant of mucopolysaccharidosis III. *Eur. J. Ped.* **138**, 168.

Gaull, G.E. (1969). Pathogenesis of maple syrup urine disease: observation during dietary management and treatment of coma by peritoneal dialysis. *Biochem. Med.* **3**, 130.

—— (1977). Deficiency of methionine adenosyltransferase: a new etiology for hypermethioninemia in infancy. In *The biochemistry of adenosylmethionine* (ed.

F. Salvatore, E. Borek, V. Zappia, H.G. Williams-Ashman, and F. Schlenk), p.37. Columbia University Press, New York.

—— and Tallan, H.H. (1974). Methionine adenosyltransferase deficiency: new enzymatic defect associated with hypermethioninemia. *Science* **186**, 59.

—— Rassin, D.K., and Sturman, J.A. (1968). Pyridoxine-dependency in homocystinuria. *Lancet* **ii**, 1302.

—— Sturman, J.A., and Schaffner, F. (1974). Homocystinuria due to cystathionine synthase deficiency: enzymatic and ultrastructural studies. *J. Ped.* **84**, 381.

—— Rassin, D.K., Solomon, G.E., Harric, R.C., and Sturman, J.A. (1970). Biochemical observations on so-called hereditary tyrosinemia. *Ped. Res.* **4**, 337.

—— Tallan, H.H., Londsdale, D., Przyrembel, H., Schaffner, F., and Von Bassewitz, D.B. (1981). Hypermethioninemia associated with methionine adenosyltransferase deficiency: clinical, morphological and biochemical observations on four patients. *J. Ped.* **98**, 734.

Gautier, J.C., Laudat, P.H., Rosa, A., Gray, F., and Lhermitte, F. (1973). Maladie de Refsum: test de charge en phytol chez un descendant. *Nouv. Presse. med.* **2**, 2029.

Geever, R.F., Wilson, L.B., Nallaseth, F.S., Milner, P.F., Bittner, M., and Wilson, J.T. (1981). Direct identification of sickle cell anemia by blot hydridization. *Proc. natl. Acad. Sci. USA* **78**, 5081.

Gehler, J., Cantz, M., Stoeckenius, M., and Spranger, J. (1976). Prenatal diagnosis of mucolipidosis II (I-cell disease). *Eur. J. Ped.* **122**, 201.

—— —— Tolksdorf, M., Spranger, J. Gilbert, E., and Drube, H. (1972). Mucopolysaccharidosis VII (beta-glucuronidase deficiency). *Hum. Genet.* **23**, 149.

Geiger, B. and Arnon, R. (1976). Chemical characterization and subunit structure of human *N*-acetyl-hexosaminidases A and B. *Biochemistry* **15**, 3484.

—— —— and Sandhoff, K. (1977). Immunochemical and biochemical investigation of hexosaminidase S. *Am. J. hum. Genet.* **29**, 508.

Gelehrter, T.D. and Snodgrass, P.J. (1974). Lethal neonatal deficiency of carbamyl phosphate synthetase. *New Engl. J. Med.* **290**, 430.

Gellady, A. and Greenwood, R.D. (1972). G-6-PD hemolytic anemia complicating diabetic ketoacidosis. *J. Ped.* **80**, 1037.

Gemsa, D., Woo, C.H., Fudenberg, H.M., and Schmid, R. (1973). Erythrocyte catabolism by macrophages in vitro: the effect of hydrocortisone on erythrophagocytosis and on the induction of heme oxygenase. *J. clin. Invest.* **52**, 812.

Gentz, J. and Lindblad, B. (1972). *p*-hydroxyphenylpyruvate hydroxylase activity in fine-needle aspiration liver biopsies in hereditary tyrosinemia. *Scand. J. clin. Lab. Invest.* **29**, 115.

—— Jagenburg, R., and Zetterström, R. (1965). An inborn error of tyrosine metabolism with cirrhosis of the liver and multiple renal tubular defects (de Toni–Fanconi syndrome). *J. Ped.* **66**, 670.

—— Heinrich, J., Lindblad, B., Lindstedt, S., and Zetterström, R. (1969*a*). Enzymatic studies in a case of hereditary tyrosinemia with hepatoma. *Acta. Paediatr. Scand.* **58**, 393.

—— Johansson, S., Lindblad, B., Lindstedt, S., and Zetterström, R. (1969*b*). Excretion of δ-aminolevulinic acid in hereditary tyrosinemia. *Clin. Chim. Acta* **23**, 257.

George, J.N., O'Brien, R.L., Pollack, S., and Crosby, W.H. (1966). Studies of in vitro primaquine hemolysis: substrate requirement for erythrocyte membrane damage. *J. clin. Invest.* **45**, 1280.

Gerald, P.S. and Efron, M.L. (1961). Chemical study of several varieties of hemoglobin M. *Proc. natl. Acad. Sci. USA* **47**, 1758.

German, J., Simpson, J.L., Morillo-Cucci, G., Passarge, E., and Demayo, A.P. (1973). Testicular feminization and inguinal hernia. *Lancet* **i**, 891.

Gerritsen, S.M., Akkerman, J.W.N., Nijmeijer, B., Sixma, J.J., Witkop, C.J., and White, J. (1977). The Hermansky–Pudlak syndrome. Evidence for a lowered 5-hydroxytryptamine content in platelets of heterozygotes. *Scand. J. Haematol.* **18**, 249.

Gerritsen, T. (1972). Sarcosine dehydrogenase deficiency, the enzyme defect in hypersarcosinemia. *Helv. Paediatr. Acta* **27**, 33.

—— and Waisman, H.A. (1964). Homocystinuria, an error in the metabolism of methionine. *Pediatrics* **33**, 413.

—— —— (1965). Hypersarcosinemia, a new error of metabolism. *Fed. Proc.* **24**, 470.

—— —— (1966). Hypersarcosinemia: an inborn error of metabolism. *New Engl. J. Med.* **275**, 66.

—— Vaughn, J.G., and Waisman, H.A. (1962). The identification of homocystine in the urine. *Biochem. biophys. Res. Commun.* **8**, 493.

Gershoff, S.N. and Prien, E.L. (1967). Effect of daily MgO and vitamin B_6 administration to patients with recurrent oxalate kidney stones. *Am. J. clin. Nutr.* **20**, 393.

Ghadimi, H.K. (1981). Histidinemia: biochemistry and behaviour. *Am. J. Dis. Child.* **135**, 210.

Ghadimi, H., Binnington, V.I., and Pecora, P. (1965). Hyperlysinemia associated with retardation. *New Engl. J. Med.* **273**, 723.

—— Binnington, V.I., and Pecora, P. (1964). Hyperlysinemia associated with mental retardation (abst). *Proc. Soc. Ped. Res., Seattle.* 41.

—— Partington, M.W., and Hunter, A. (1961). A familial disturbance of histidine metabolism. *New Engl. J. Med.* **265**, 221.

—— —— —— (1962). Inborn error of histidine metabolism. *Pediatrics* **29**, 714.

Giannelli, F. and Pawsey, S.A. (1974). DNA repair synthesis in human heterokaryons. II. A test for heterozygosity in XP and some insight in the structure of the defective enzyme. *J. Cell Sci.* **15**, 163.

—— —— (1976). DNA repair synthesis in human heterokaryons. III. The rapid and slow complementing varieties of Xeroderma pigmentosum. *J. Cell. Sci.* **20**, 207.

Gibberd, F.B., Billimoria, J.D., Page, N.G.R., and Retsas, S. (1979). Heredopathia atactica polyneuritiformis (Refsum's disease) treated by diet and plasma exchange. *Lancet* **i**, 575.

Gibbs, D.A. and Watts, R.W.E. (1966). An investigation of the possible role of xanthine oxidase in the oxidation of glyoxylate to oxalate. *Clin. Sci.* **31**, 285.

—— —— (1969). The variation of urinary oxalate excretion with age. *J. lab. clin. Med.* **73**, 901.

—— —— (1970). The action of pyridoxine in primary hyperoxaluria. *Clin. Sci.* **38**, 277.

—— McFadyen, I.R., Crawford, M. d'A., de Muinck Keizer, E.E., Headhouse-Benson, C.M., Wilson, T.M., and Farrant, P.H. (1984). First-trimester diagnosis of Lesch–Nyhan syndrome. *Lancet* **ii**, 1180.

Giblett, E.R., Ammann, A.J., Wara, D.W., Sandman, R., and Diamond, L.K. (1975). Nucleoside-phosphorylase deficiency in a child with severely defective T-cell immunity and normal B-cell immunity. *Lancet* **i**, 1010.

—— Anderson, J.E., Cohen, F., Pollara, B., and Meuwissen, H.J. (1972). Adenosine-deaminase deficiency in two patients with severely impaired cellular immunity. *Lancet* **ii**, 1067.

Gibson, J.B., Carson, N.A.J., and Neill, D.W. (1964). Pathological findings in homocystinuria. *J. clin. Pathol.* **17**, 427.

Gibson, Q.H. (1948). The reduction of methaemoglobin in red blood cells and studies on the cause of idiopathic methaemoglobinaemia. *Biochem. J.* **42**, 13.

Gilbert, A. and Lereboullet, P. (1901). La cholemie simple familiale. *Sem. Med.* **21**, 241.

—— —— and Herscher, M. (1907). Les trois cholemies congenitales. *Bull. Mem. Soc. med. Hop. Paris* **24**, 1203.

Gilbert, E., Varakis, J., Opitz, J., and Zu Rhein, G.M. (1975*b*). Generalised gangliosidosis type 2. *Z. Kinderheilk.* **120**, 151.

—— Dawson, G., Zurhein, G.M., Opitz, J.M., and Spranger, J.W. (1973). I-cell disease, mucolipidosis II. Pathological, histochemical, ultrastructural and biochemical observation in four cases. *Z. Kinderheilk.* **114**, 259.

Gilbert, F., Kucherlapati, R.P., Creagan, R.P., Murnane, M.J., Darlington, G.J., and Ruddle, F.H. (1975*a*). The assignment of hexosaminidase A and B to individual chromosomes. *Proc. natl Acad. Sci. USA* **72**, 263.

Gilbert-Dreyfus, S., Sebaoun, C.A., and Belaisch, J. (1957). Étude d'un cas familial d'androgynoidisme avec hypospadias grave, gynécomastie et hypero-éstrogénie. *Ann. Endocrinol. (Paris)* **18**, 93.

Gilboa, V., Ber, A., Lewitis, Z., and Hasenfratz, J. (1963), Goitrous myxedema due to iodide trapping defect. *Arch. intern. Med.* **112**, 212.

Gilles, F.H. and Deuel, R.K. (1971). Neuronal cytoplasmic globules in the brain in Morquio's syndrome. *Arch. Neurol. (Chicago)* **25**, 393.

Gillespie, J.B. and Perucca, L.G. (1960). Congenital generalised indifference to pain (congenital analgia). *Am. J. Dis. Child.* **100**, 124.

Ginns, E.I., Brady, R.O., Stowens, D.E., Furbish, F.S., and Barranger, J.A. (1980). A new group of glucocerebrosidase isozymes found in human white blood cells. *Biochem. biophys. Res. Commun.* **97**, 1103.

Ginsberg, L.C., Donnelly, P.V., Di Ferrante, D.T., Di Ferrante, N., and Caskey, T. (1978). *N*-acetylglucosamine 6-sulfate sulfatase in man: deficiency of the enzyme in a new mucopolysaccharidosis. *Ped. Res.* **12**, 805.

Girardin, E.P., de Wolfe, M.J., and Crocker, J.F.S. (1979). Treatment of cystinosis with cysteamine. *J. Ped.* **94**, 838.

Gitlin, P. and Craig, J.M. (1963). The thymus and other lymphoid tissues in congenital agammaglobulinemia. *Pediatrics* **32**, 517.

Gitzelmann, R. (1967). Hereditary galactokinase deficiency, a newly recognized cause of juvenile cataracts. *Ped. Res.* **1**, 14.

—— (1972). Deficiency of uridine diphosphate galactose 4-epimerase in blood cells of an apparently healthy infant. Preliminary communication. *Helv. Paediatr. Acta.* **27**, 125.

—— and Baerlocher, K. (1973). Vorteile und Nachteile der Fructose in der Nahrung. *Pädiat. Fortbildiung. Praxis.* **37**, 40.

—— —— and Prader, A. (1973). Hereditäre Störungen im Fruktose- und Galactose-Stoffwechsel. *Monatsschr. Kinderheilkind.* **121**, 174.

—— and Haigis, E. (1978). Appearance of active UDP-galactose 4'-epimerase in cells cultured from epimerase-deficient persons. *J. inher. metab. Dis.* **1**, 41.

—— and Hansen, R.G. (1974). Biogenesis of galactose: evidence in galactosemic infants (abst.). *Ped. Res.* **8**, 137.

—— —— (1980). Galactose metabolism, hereditary defects and their clinical significance. In *Inherited disorders of carbohydrate metabolism* (ed. D. Burman, J.B. Holton and C.A. Pennock), p.61. MTP Press, Lancaster.

—— and Steinmann, B. (1973). Uridine diphosphate galactose 4'-epimerase deficiency. II. Clinical follow-up biochemical studies and family investigation. *Helv. Paediatr. Acta* **28**, 497.

—— —— Mitchell, B., and Haigis, E. (1976). Uridine diphosphate galactose 4'-epimerase deficiency IV. Report of eight cases in three families. *Helv. Paediatr. Acta* **31**, 441.

—— Wiesman, U.N., Spycher, M.A., Herschkowitz, N., and Giedion, A. (1978). Unusually mild course of ß-glucuronidase deficiency in two brothers (mucopolysaccharidosis VII) *Helv. Paediatr. Acta* **33**, 413.

—— Steinmann, B., Otten, A., Dumermuth, G., Herdan, M., Reubi, J.C., and Cuenod, M. (1977). Nonketotic hyperglycinemia treated with strychnine, a glycine receptor antagonist. *Helv. Paediatr. Acta* **32**, 517.

Givens, J.R., Wiser, W.L., Summitt, R.L., Kerber, I.J., Andersen, R.N., Pittaway, D.E., and Fish, S.A. (1974). Familial male pseudohermaphroditism without gynecomastia due to deficient testicular 17-ketosteroid reductase activity. *New. Engl. J. Med.* **291**, 938.

Gjessing, L.R. (1963). Studies of functional neural tumors. II. Cystathioninuria. *Scand. J. clin. lab. Invest.* **15**, 474.

—— (1964). Cystathioninuria during a load of thyroxine. *Scand. J. clin. lab. Invest.* **16**, 680.

—— and Halvorsen, S. (1965). Hypermethioninaemia in acute tyrosinosis. *Lancet* **ii**, 1132.

Gjone, E. (1974). Familial lecithin:cholesterol acyltransferase deficiency: a clinical survey. *Scand. J. clin. lab. Invest.* **33**, (*suppl.*) 137, 73.

—— and Bergaust, B. (1969). Corneal opacity in familial plasma cholesterol ester deficiency. *Acta. Ophthalmol.* **47**, 222.

—— and Norum, K.R. (1968). Familial serum-cholesterol ester deficiency: clinical study of a patient with a new syndrome. *Acta. Med. Scand.* **183**, 107.

Glaser, J.H. and Sly, W.S. (1973). ß-glucuronidase deficiency mucopolysaccharidosis: methods for enzymatic diagnosis. *J. lab. clin. Med.* **82**, 969.

—— McAlister, W.H., and Sly, W.S. (1974). Genetic heterogeneity in multiple lysosomal enzyme hydrolase deficiency. *J. Ped.* **85**, 192.

Glazebrook, A.J. (1945). Wilson's disease. *Edinburgh med. J.* **52**, 83.

Glenn, B.L., Glenn, H.G., and Omtvedt, I.T. (1968). Congential porphyria in the domestic cat (*Felis catus*): preliminary investigations on inheritance pattern. *Am. J. vet. Res.* **29**, 1653.

Glick, N.R., Snodgrass, P.J., and Schafer, I.A. (1976). Neonatal argininosuccinic aciduria with normal brain and kidney but absent liver argininosuccinate lyase activity. *Am. J. hum. Genet.* **28**, 22.

Glomset, J.A. and Norum, K.R. (1973). The metabolic role of lecithin:cholesterol acyltransferase: perspective from pathology. *Adv. Lipid Res.* **2**, 1.

—— —— and King, W. (1970). Plasma lipoproteins in familial lecithin:cholesterol acyltransferase deficiency; lipid composition and reactivity *in vitro*. *J. clin. Invest.* **49**, 1827.

—— Nichols, A.V., Norum, K.R., King, W., and Forte, T. (1973). Plasma lipoproteins in familial lecithin:cholesterol acyltransferase deficiency: further studies of very low and low density lipoprotein abnormalities. *J. clin. Invest.* **52**, 1078.

—— Applegate, K., Forte, T., King, W.C., Mitchell, C.D., Norum, K.R., and

Gjone, E. (1980*a*). Abnormalities in lipoproteins of *d*<1.006 g/ml in familial lecithin:cholesterol acyltransferase deficiency. *J. Lipid Res.* **21**, 1116.

—— Mitchell, C.D., King, W.C., Applegate, K.R., Forte, T., Norum, K.R., and Gjone, E. (1980*b*). *In vitro* effects of lecithin:cholesterol acyltransferase on apolipoprotein distribution in familial lecithin:cholesterol acyltransferase deficiency. *Ann. N.Y. Acad. Sci.* **348**, 224.

Gloriex, F.H., Scriver, C.S., Delvin, E., and Mohyuddin, F. (1971). Transport and metabolism of sarcosine in hypersarcosinemia and normal phenotypes. *J. clin. Invest.* **50**, 2313.

Glössl, J. and Kresse, H. (1978). A sensitive procedure for the diagnosis of *N*-acetyl-galactosamine 6-sulfate sulfatase deficiency in classical Morquio's disease. *Clin. Chim. Acta.* **88**, 111.

—— —— (1982). Impaired degradation of keratan sulphate by Morquio A fibroblasts. *Biochem. J.* **203**, 335.

—— Truppe, W., and Kresse, H. (1979). Purification and properties of *N*-acetylgalactosamine 6-sulphate sulphatase from human placenta. *Biochem. J.* **181**, 37.

—— Lembeck, K., Gamse, G., and Kresse, H. (1980). Morquio's disease type A. Absence of material crossreacting with antibodies against *N*-acetylgalacto-samine-6-sulfate sulfatase. *Hum. Genet.* **54**, 87.

—— Maroteaux, P., Di Natale, P., and Kresse, H. (1981). Different properties of residual *N*-acetylgalactosamine-6-sulfate sulfatase in fibroblasts from patients with mild and severe forms of Morquio disease type A. *Ped. Res.* **15**, 976.

Glueck, C.J., Levy, R.I., and Fredrickson, D.S. (1968). Acute tendinitis and arthritis. A presenting syndrome of familial type II hyperlipoproteinemia. *J. Am. Med. Ass.* **206**, 2895.

Godwin, J.T., Fowler, M.F., Dempsey, E.F., and Henneman, P.H. (1958). Primary hyperoxaluria and oxalosis: report of a case and review of the literature. *New Engl. J. Med.* **259**, 1099.

Goebelsmann, U., Zachmann, M., Davajan, V., Israel, R., Mestman, J.H., and Mishell, D.R. (1976). Male pseudohermaphroditism consistent with 17,20 desmolase deficiency. *Gynecol. Invest.* **7**, 138.

—— Horton, R., Mestman, J.H., Arce, J.J., Nagata, Y., Nakamura, R.M., Thorneycroft, I.H., and Mishell, D.R. (1973). Male pseudohermaphroditism due to testicular 17ß-hydroxysteroid dehydrogenase deficiency. *J. clin. Endocr. Metab.* **36**, 867.

Goedde, H.W., Gehring, D., and Hofmann, R.A. (1965). On the problem of a 'silent gene' in pseudocholinesterase polymorphism. *Biochim. Biophys. Acta* **107**, 391.

Goka, T.J., Stevenson, R.E., Hefferan, P.M., and Howell, R.R. (1976). Menkes disease: a biochemical abnormality in cultured human fibroblasts. *Proc. natl. Acad. Sci. USA* **73**, 604.

Gold, R.J.M., Maag, U.R., Neal, J.L., and Scriver, C.R. (1974). The use of biochemical data in screening for mutant alleles and in genetic counselling. *Ann. hum. Genet.* **37**, 315.

Goldberg, A. (1959). Acute intermittent porphyria. A study of 50 cases. *Q. J. Med.* **28**, 183.

—— and Rimington, C. (1962). *Diseases of porphyrin metabolism*. C.C. Thomas, Springfield, Illinois.

—— —— and Lockhead, A. (1967). Hereditary coproporphyria. *Lancet* i, 632.

—— Moore, M.R., Beattie, A.D., Hall, P.E., McCallum, J., and Grant, J.K.

(1969). Excessive urinary excretion of certain porphyrinogenic steroids in human acute intermittent porphyria. *Lancet* **i**, 115.

Goldberg, J.D., Truex, J.H., and Desnick, R.J. (1977). Tay–Sachs disease: an improved, fully automated method for heterozyote identification by tear ß-hexosaminidase assay. *Clin. Chim. Acta* **77**, 43.

Golden, M.P., Lippe, B.M., Kaplan, S.A., Lavin, N., and Slavin, J. (1978). Management of adrenal hyperplasia using serum dehydroepiandrosterone sulfate and 17-hydroxyprogesterone concentrations. *Pediatrics* **61**, 67.

Goldfischer, S. (1965). The localisation of copper in the pericanalicular granules (lysosomes) of liver in Wilson's disease. *Am. J. pathol.* **46**, 977.

—— and Sternlieb, I. (1968). Changes in the distribution of hepatic copper in relation to the progression of Wilson's disease (hepatolenticular degeneration). *Am. J. Pathol.* **53**, 883.

—— Coltoff-Schiller, B., Biempica, L., and Wolinski, H. (1975). Lysosomes and the sclerotic arterial lesions in Hurler's syndrome. *Hum. Pathol.* **6**, 633.

Goldman, H., Scriver, C.R., Aaron, K., Delvin, E., and Canlos, Z. (1971). Adolescent cystinosis: comparisons with infantile and adult forms. *Pediatrics* **47**, 979.

—— —— and Pinsky, L. (1970). Use of dithiothreitol to correct cystine storage in cultured cystinotic fibroblasts. *Lancet* **i**, 811.

Goldsmith, L.A. (1976). The ichthyoses. *Prog. med. Genet.* **1**, 185.

—— (1983). Tyrosinemia and related disorders. In *The metabolic basis of inherited disease*, 5th edn. (ed. J.B. Stanbury, J.B. Wyngaarden, D.S. Fredrickson, J.L. Goldstein, and M.S. Brown), p.287. McGraw-Hill, New York.

—— and Reed, J. (1976). Tyrosine-induced eye and skin lesions. *J. Am. med. Ass.* **236**, 382.

—— Thorpe, J.M., and Marsh, R.F. (1981). Tyrosine aminotransferase deficiency in mink (*Mustela vison*): a model for human tyrosinemia II. *Biochem. Genet.* **19**, 687.

—— —— and Roe, C.R. (1979). Hepatic enzymes of tyrosine metabolism in tyrosinemia II. *J. invest. Dermatol.* **73**, 530.

—— Kang, E., Bienfang, D.C., Jimbow, K., Gerald, P., and Baden, H.P. (1973). Tyrosinemia with plantar and palmar keratosis and keratitis. *J. Ped.* **83**, 798.

Goldstein, J.L. (1972). The cardiac manifestations of homozygous and heterozygous forms of familial type II hyperbetalipoproteinemia. *Birth defects: original article series VIII*, p.202. Alan R. Liss, New York.

—— (1973). Genetic aspects of hyperlipidemia in coronary heart disease. *Hosp. Pract.* **8**, 53.

—— and Brown, M.S. (1979). The LDL receptor locus and the genetics of familial hypercholesterolemia. *Ann. Rev. Genet.* **13**, 259.

—— —— and Stone, N.J. (1977). Genetics of the LDL receptor: evidence that the mutations affecting binding and internalization are allelic. *Cell* **12**, 629.

—— Campbell, B.K., and Gartler, S.M. (1973*a*). Homocystinuria: heterozygote detection using phytohemagglutinin-stimulated lymphocytes. *J. clin. Invest.* **52**, 218.

—— Dana, S.E., and Brown, M.S. (1974*b*). Esterification of low density lipoprotein in human fibroblasts and its absence in homozygous familial hypercholesterolemia. *Proc. natl. Acad. Sci. USA* **71**, 4288.

—— Basu, S.K., Brunschede, G.Y., and Brown, M.S. (1976). Release of low density lipoprotein from its cell surface receptor by sulfated glycosaminoglycans. *Cell* **7**, 85.

—— Dana, S.E., Brunschede, G.Y., and Brown, M.S. (1975). Genetic heterogenity in familial hypercholesterolemia: evidence for two different mutations affecting functions of low-density lipoprotein receptor. *Proc. natl. Acad. Sci. USA* **72**, 1092.

—— Schrott, H.G., Hazzard, W.R., Bierman, E.L., and Motulsky, A.G. (1973*b*). Hyperlipidemia in coronary heart disease. II. Genetic analysis of lipid levels in 176 families and delineation of a new inherited disorder, combined hyperlipidemia. *J. clin. Invest.* **52**, 1544.

Goldstein, M.L., Kolodny, E.H., Gascon, G.G., and Gilles, F.H. (1974*a*). Macular cherry-red, myoclonic epilepsy and neurovascular storage in a 17-year-old girl. *Trans. Am. neurol. Assoc.* **99**, 110.

Gollan, J.L., Huang, S.M., Billing, B., and Sherlock, S. (1975). Prolonged survival in three brothers with severe type II Crigler–Najjar syndrome. Ultrastructural and metabolic studies. *Gastroenterology* **68**, 1543.

Gompertz, D. (1975). Organic acidaemias. In *The treatment of inherited metabolic disease* (ed. D.N. Raine), p.191. MTP Press, Lancaster.

—— Goodey, P.A., and Bartlett, K. (1973*b*). Evidence for the enzymatic defect in ß-methylcrotonylglycinuria. *FEBS Lett.* **32**, 13.

—— Bartlett, K., Blair, D., and Stern, C.M.M. (1973*a*). Child with a defect in leucine metabolism associated with ß-hydroxyisovaleric aciduria and ß-methylcrotonylglycinuria. *Arch. Dis. Child.* **48**, 975.

—— Draffan, G.H., Watts, J.L., and Hull, D. (1971). Biotin-responsive ß-methylcrotonylglycinuria. *Lancet* **ii**, 22.

—— Goodey, P.A., Saudubray, J.M., Charpentier, C., and Chignolle, A. (1974*b*). Prenatal diagnosis of methylmalonic acidemia. *Pediatrics* **54**, 511.

—— Stoors, C.N., Bau, D.C.K., Peters, T.J., and Hughes, E.A. (1970). Localization of the enzyme defect in propionicacidemia. *Lancet* **i**, 1140–1143.

—— Saudubray, J.M., Charpentier, C., Bartlett, K., Goodey, P.A., and Draffan, G.H. (1974*a*). A defect in L-isolcucine metabolism associated with α-methyl-ß-hydroxybutyric and α-methylacetoacetic aciduria: quantitative *in vivo* and *in vitro* studies. *Clin. Chim. Acta* **57**, 269.

—— Goodey, P.A., Thom, H., Russell, G., MacLean, M.W., Ferguson-Smith, M.E., and Ferguson-Smith, M.A. (1973*c*). Antenatal diagnosis of propionic acidaemia. *Lancet* **ii**, 1010.

—— —— —— —— Johnston, A.W., Mellor, D.H., MacLean, M.W., Ferguson-Smith, M.E., and Ferguson-Smith, M.A. (1975). Prenatal diagnosis and family studies in a case of propionicacidemia. *Clin. Genet.* **8**, 244.

Gonatas, N.K. and Gonatas, J. (1965). Ultrastructural and biochemical observations in a case of systemic later infantile lipidosis and its relationship to Tay Sachs disease and gargoylism. *J. Neuropath. exp. Neurol.* **24**, 318.

—— Terry, R.D, and Winkler, R. (1963). A case of juvenile lipidosis: the significance of electronmicroscopic and biochemical observations of a cerebral palsy. *J. Neuropathol. exp. Neurol.* **22**, 557.

Goodman, R.M. (1979). *Genetic disorders among Jewish people.* Johns Hopkins University Press, Baltimore and London.

Goodman, S.I. and Browder, J.A. (1970). Hydroxylysinuria in association with trisomy 21. *Lancet* **ii**, 1141.

—— and Kohloff, J.G. (1975). Glutaric aciduria: inherited deficiency of glutaryl-CoA dehydrogenase activity. *Biochem. Med.* **13**, 138.

—— Mace, J.W., and Pollak, S. (1971). Serum γ-glutamyl transpeptidase deficiency. *Lancet* **i**, 234.

—— McIntyre, C.A., and O'Brien, D. (1967). Impaired intestinal transport of proline in a patient with familial iminoaciduria. *J. Ped.* **71**, 246.

—— Browder, J.A., Hiles, R.A., and Miles, B.S. (1972). Hydroxylysinemia; a disorder due to a defect in the metabolism of free hydroxylysine. *Biochem. Med.* **6**, 344.

—— Hambridge, K.M., Mahoney, C.P., and Striker, G.E. (1973*a*). Renal homotransplantation in the treatment of cystinosis. In *Cystinosis* (ed. J.D. Schulman), p.225. US Government Printing Office, Washington DC.

—— Mace, J.W., Turner, B., and Garrett, W.J. (1973*b*). Antenatal diagnosis of argininosuccinic aciduria. *Clin. Genet.* **4**, 236.

—— McCabe, E.R.B., Fennessy, P.V., and Mace, J.W. (1979). Multiple acyl-CoA dehydrogenase deficiency (glutaric aciduria type II). Abs. no. 563. *Ped. Res.* **13**, 419.

—— —— —— —— (1980*b*). Multiple acyl-CoA dehydrogenase deficiency (glutaric aciduria type II) with transient hypersarcosinemia and sarcosinuria; possible inherited deficiency of an electron transfer flavoprotein. *Ped. Res.* **14**, 12.

—— Mace, J.W., Moles, B.S., Teng, C.C., and Brown, S.B. (1974). Defective hydroxyproline metabolism in type II hyperprolinemia. *Biochem. Med.* **10**, 329.

—— Markey, S.P., Moe, P.G., Miles, B.S., and Teng, C.C. (1975). Glutaric aciduria; a 'new' disorder of amino acid metabolism. *Biochem. Med.* **12**, 12.

—— Moe, P.G., Hammond, K.B., Mudd, S.H., and Uhlendorf, B.W. (1970). Homocystinuria with methylmalonic aciduria: two cases in a sibship. *Biochem. Med.* **4**, 500.

—— Norenberg, M.D., Shikes, R.H., Breslich, D.J., and Moe, P.G. (1977). Glutaric aciduria: biochemical and morphologic considerations. *J. Ped.* **90**, 746.

—— Solomans, C.C., Muschenheim, F., McIntyre, C.A., Miles, B. and O'Brien, D. (1968). A syndrome resembling lathyrism associated with iminodipeptiduria. *Am. J. Med.* **45**, 152.

—— Gallegos, D.A., Pullin, C.J., Halpern, B., Truscott, R.J.W., Wise, G., Wilcken, B., Ryan, E.D., and Whelen, D.T. (1980*a*). Antenatal diagnosis of glutaric acidemia. *Am. J. hum. Genet.* **32**, 695.

Gordon, B.A. Carson, R., and Haust, M.D. (1980). Unusual clinical and ultrastructural features in a boy with biochemically typical mannosidosis. *Acta Paediatr. Scand.* **69**, 787.

Gordon, E.R., Shaffer, E.A., and Sass-Kortsak, A. (1976). Bilirubin secretion and conjugation in the Crigler–Najjar syndrome type II. *Gastroenterology* **70**, 761.

Goss, S.J. and Harris, H. (1975). New method for mapping genes in human chromosomes. *Nature, Lond.* **255**, 680.

—— —— (1977). Gene transfer by means of cell fusion. II. The mapping of 8 loci on human chromosome 1 by statistical analysis of gene assortment in somatic cell hybrids. *J. Cell. Sci.* **25**, 39.

Goto-Tomura, R., Takesue, Y., and Takesue, S. (1976). Immunological similarity between NADH-cytochrome b5 reductase of erythrocytes and liver microsomes. *Biochim. Biophys. Acta* **423**, 293.

Gout, J.-P., Serre, J.-C., Dieterlen, M., Antener, I., Frappat, P., Bost, M., and Beaudoing, A. (1977). Une nouvelle cause d'hyperméthioninemie de l'enfant: le déficit en 5-adenosyl-methionine-synthétase. *Arch. Fr. Pediatr.* **34**, 416.

Goutieres, F., Arsenio-Nunes, M.L., and Aicardi, J. (1979). Mucolipidosis IV. *Neuropadiatrie* **10**, 321.

Goyer, R.A., Tsuchiya, K., Leonard, D.L., and Kahyo, H. (1972). Aminoaciduria in Japanese workers in lead and cadmium industries. *Am. J. clin. Path.* **57**, 635.

Grabowski, G.A., Walling, L., and Desnick, R.J. (1980). Human mannosidosis: in vitro and in vivo studies of cofactor supplementation. In *Enzyme therapy in genetic disease: 2* (ed. R.J. Desnick), p.319. Alan R. Liss, New York.

—— —— Ikonne, J.U., Wolfe, L.S., and Desnick, R.J. (1977). Enzyme manipulation. Evaluation of oral $ZnSO_4$ supplementation in mannosidosis type II. *Ped. Res.* **11**, 456.

Grandchamp, B. and Nordmann, Y. (1977). Decreased lymphocyte coproporphyrinogen III oxidase activity in hereditary coproporphyria. *Biochem. biophys. Res. Commun.* **74**, 1089.

Gräsbeck, R., Gordin, R., Kantero, I, and Kuhlbäck, B. (1960). Selective vitamin B_{12} malabsorption and proteinuria in young people. *Acta Med. Scand.* **167**, 289.

Gravel, R.A., Lam, K.F., Scully, K.J., and Hsia, Y.E. (1977). Genetic complementation of propionyl-CoA carboxylase deficiency in cultured human fibroblasts. *Am. J. hum. Genet.* **29**, 378.

—— Mahoney, M.J., Ruddle, F.H., and Rosenberg, L.E. (1975). Genetic complementation in heterokaryons of human fibroblasts defective in cobalamin metabolism. *Proc. natl. Acad. Sci. USA* **72**, 3181.

Gray, C.H., Muir, H., and Neuberger, A. (1950a). Studies in congenital porphyria. 3. The incorporation of ^{15}N into the haem and glycine of haemoglobin. *Biochem. J.* **47**, 542.

—— Neuberger, A., and Sneath, P.H.A. (1950b). Studies in congenital porphyria. 2. Incorporation of ^{15}N in the stercobilin in the normal and in porphyric. *Biochem. J.* **47**, 87.

Gray, G.M. (1975). Carbohydrate digestion and absorption. Role of the small intestine. *New Engl. J. Med.* **292**, 1225.

—— Conklin, K.A., and Townley, R.R.W. (1976). Sucrase-isomaltase deficiency. Absence of an inactive enzyme variant. *New Engl. J. Med.* **294**, 750.

Green, H. and Chan, T. (1973). Pyrimidine starvation induced by adenosine in fibroblasts and lymphoid cells: role of adenosine deaminase. *Science* **182**, 836.

Greenberg, B.H., Blackwelder, W.C., and Levy, R.I. (1977). Primary type V hyperlipoproteinemia. A descriptive study in 32 families. *Ann. intern. Med.* **87**, 526.

Greene, M.L., Boyle, J.A., and Seegmiller, J.E. (1970). Substrate stabilization: genetically controlled reciprocal relationship of two human enzymes. *Science* **167**, 887.

—— Fujimoto, W.Y., and Seegmiller, J.E. (1969b). Urinary xanthine stones: a rare complication of allopurinol therapy. *New Engl. J. Med.* **280**, 426.

—— Hug, G., and Schubert, W.K. (1969a). Metachromatic leukodystrophy: treatment with arylsulfatase A. *Arch. Neurol.* **20**, 147.

—— Stifel, F.B., and Herman, R.H. (1972) 'Ketotic hypoglycaemia' due to hepatic fructose-1,6-diphosphate deficiency. Treatment with folic acid. *Am. J. Dis. Child.* **124**, 415.

Gregersen, N., Lauritzen, R., and Rasmussen, K. (1976). Suberylglycine excretion in the urine of a patient with dicarboxylic aciduria. *Clin. Chim. Acta* **70**, 417.

—— Brandt, N.J., Christensen, E., Grøn, I., Rasmussen, K., and Brandt, S. (1977). Glutaric aciduria: clinical and laboratory findings in two brothers. *J. Ped.* **90**, 740.

—— Kølvraa, S., Rasmussen, K., Christensen, E., Brandt, N.J., Ebbesen, F., and Hansen, F.H. (1980). Biochemical studies in a patient with defects in the metabolism of Acyl-CoA and sarcosine: another possible case of glutaric aciduria type II. *J. inher. metab. Dis.* **3**, 67.

Gregoire, A., Perier, O., and Dustin, P. (1966). Metachromatic leukodystrophy, an electron microscopic study. *J. Neuropathol. exp. Neurol.* **25**, 617.

Greiling, H. (1957). Beitrag zur Entstehung der ochronose bei Alkaptonurie. *Klin. Wschr.* **35**, 889.

Greter, J., Hagberg, B., Steen, G., and Söderhjelm, U. (1978). 3-Methylglutaconic aciduria: report on a sibship with infantile progressive encephalopathy. *Eur. J. Pediatr.* **129**, 231.

Griffin, J.E. (1979). Testicular feminization associated with a thermolabile androgen receptor in cultured human fibroblasts. *J. clin. Invest.* **64**, 1624.

—— and Durrant, J.L. (1981). The frequency of qualitative receptor defects in 32 families with androgen resistance. *Clin. Res.* **29**, 505A.

—— Punyashthiti, K., and Wilson, J.D. (1976). Dihydrotestosterone binding by cultured human fibroblasts: comparison of cells from control subjects and from patients with hereditary male pseudohermaphroditism due to androgen resistance. *J. clin. Invest.* **57**, 1342.

Griffin, J.L., Goren, E., Schaumburg, M., Engel, K., and Loriaux, L. (1977). Adrenomyeloneuropathy: a probable variant of adrenoleukodystrophy. I. Clinical and endocrinological aspects. *Neurology* **27**, 1107.

Griffin, R.F. and Elsas, L.J. (1975). Classic phenylketonuria: diagnosis through heterozygote detection. *J. Ped.* **86**, 512.

Grimes, A.J., Meisler, A., and Dacie, J.V. (1964). Hereditary non-spherocytic haemolytic anaemia. A study of red-cell carbohydrate metabolism in twelve cases of pyruvate-kinase deficiency. *Br. J. Haematol.* **10**, 403.

Groebe, H., Krins, M., Schmidberger, H., Von Figura, K., Harzer, K., Kresse, H., Paschke, E., Sewell, A., and Ullrich, K. (1980). Morquio syndrome (mucopolysaccharidosis IVB) associated with ß-galactosidase deficiency. Report of two cases. *Am. J. hum. Genet.* **32**, 258.

Groen, J.J. (1964). Gaucher's disease: Hereditary transmission and racial distribution. *Arch. Intern. Med.* **113**, 543.

Gross, R.T., Hurwitz, R.E., and Marks, P.A. (1958). An hereditary enzymatic defect in erythrocyte metabolism: glucose-6-phosphate dehydrogenase deficiency. *J. clin. Invest.* **37**, 1176.

Gross, S.R. (1964). Hematologic studies on erythropoietic porphyria: a new case with severe hemolysis, chronic thrombocytopenia and folic acid deficiency. *Blood* **23**, 762.

—— (1975). Animal models of glycogen storage conditions – their relations to human disease. *West. J. Med.* **123**, 194.

—— and Mayer, S.E. (1974). Characterization of the phosphorylase b to a converting activity in skeletal muscle extracts of mice with the phosphorylase b kinase deficiency mutation. *J. biol. Chem.* **249**, 6710.

Grosser, Y. and Eales, L. (1973). Patterns of fecal porphyrin excretion in the hepatocutaneous porphyrias. *S. Afr. med. J.* **27**, 2162.

Groth, C.G., Collste, H., Dreborg, S., Håkansson, G., Lundgren, G., and Svennerholm, L. (1980). Attempt at enzyme replacement in Gaucher disease by renal transplantation. In *Enzyme therapy in genetic disease: 2* (ed. R.J. Desnick) p.475. Alan R. Liss, New York.

—— Hagenfeldt, L., Blomstrand, R., Dreborg, S., Löfström, B., Öckerman, P., Samuelsson, K., Svennerholm, L., Werner, B., and Westberg, G. (1971). Splenic transplantation in a case of Gaucher's disease. *Lancet* i, 1260.

Grover, W.E. and Strutton, M.C. (1975). Copper infusion therapy in trichopoliodystrophy. *J. Ped.* **86**, 216.

Grunfeld, J.P., Leporrier, M., Droz, D., Bensaude, I., Hinglais, N., and Crosnier, J. (1975). Le transplantation renale chez les sujets atteints de maladie de Fabry. *Nouv. Presse med.* **4**, 2081.

Guggenheim, M.A., McCabe, E.R.B., Roig, M., Goodman, S.I., Lum, G.M., Bullen, W.W., and Ringel, S.P. (1980). Glycerol kinase deficiency with neuromuscular, skeletal, and adrenal abnormalities. *Ann. Neurol.* **7**, 441.

Guguen-Guillouzo, C., Szajnert, M.F., Schapira, F., Belanger, L., and Grenier, A. (1979). Liver fetal isozymes in hereditary tyrosinemia. *Europ. J. Cancer.* **15**, 1131.

Guibaud, P., Divry, P., Dubois, Y., Collombel, C., and Larbre, F. (1973). Une observation d'acidemie isovalerique. *Arch. franc. Ped.* **30**, 633.

—— Maire, I., Goddon, R., Teyssier, G., Zabot, M.T., and Mandon, G. (1979). Mucopolysaccharidose type VII par deficit en ß-glucuronidase: étude d'une famille. *J. Genet. hum.* **27**, 29.

Gunn, C.H. (1938). Hereditary acholuric jaundice in a new mutant strain of rats. *J. Hered.* **29**, 137.

Gustavson, K.-H. and Hagberg, B. (1971). The incidence and genetics of metachromatic leucodystrophy in northern Sweden. *Acta Paediatr. Scand.* **60**, 585.

Guthrie, RE. and Susi, A. (1963). A simple phenylalanine method for detecting phenylketonuria in large populations of newborn infants. *Pediatrics* **32**, 338.

Gutman, A.B. and Yü, T.-F. (1957). Protracted uricosuric therapy in tophaceous gout. *Lancet* **ii**, 1258.

Gutsche, B.B., Scott, E.M., and Wright, R.C. (1967). Hereditary deficiency of pseudocholinesterase in Eskimos. *Nature (Lond.)* **215**, 322.

Güttler, F. and Hansen, G. (1977a). Different phenotypes for phenylalanine hydroxylase deficiency. *Ann. clin. Biochem.* **14**, 124.

—— —— (1977b). Heterozygote detection in phenylketonuria. *Clin. Genet.* **11**, 137.

—— and Wamberg, E. (1977). On indications for treatment of the hyperphenylalaninemic neonate. *Acta Paediat. Scand.* **66**, 339.

—— Kaufman, S., and Milstien, S. (1977). Phenylalanine has no effect on dihydropteridine reductase activity in phenylketonuria fibroblasts. *Lancet* **ii**, 1139.

Gwinup, G., Wieland, R.G., Besch, P.K., and Hamwi, G.J. (1966). Studies on the mechanism of the production of the testicular feminization syndrome. *Am. J. Med.* **41**, 448.

Habib, R., Bargeton, E., Brissaud, H.-H., Raynaud, J., and Le Ball, J.-C. (1962). Constations anatomique chez en enfant d'un syndrome de Lowe. *Arch. franc. Ped.* **19**, 945.

Habig, W.H., Pabst, M.J., Fleishchner, G., Gatmaitan, Z., Arias, I.M., and Jakoby, W.B. (1974). The identity of glutathione *S*-transferase B with ligandin, a major binding protein of liver. *Proc. natl. Acad. Sci. USA* **10**, 3879.

Haddad, J.G., Chyu, K.J., Hahn, T.J., and Stamp, T.C.B., (1973). Serum concentrations of 24-hydroxyvitamin D in sex-linked hypophosphatemic vitamin D-resistant rickets. *J. lab. clin. Med.* **81**, 22.

Hadorn, B., Tarlow, M.J., Lloyd, J.K., and Wolff, O.H. (1969). Intestinal enterokinase deficiency. *Lancet* **i**, 812.

Haeger-Aronson, B. (1963). Erythropoietic porphyria. A new type of inborn error of metabolism. *Am. J. Med.* **35**, 450.

—— and Krook, G. (1966). Erythropoietic protoporphyria. A study of known cases in Sweden. *Acta Med. Scand.* **245**, (suppl.) 48.

—— Stathers, G., and Swahn, G. (1968). Hereditary coproporphyria: Study of a Swedish family. *Ann. intern. Med.* **69**, 221.

Hagberg, B. (1963a). The clinical diagnosis of Krabbe's infantile leucodystrophy. *Acta Paediatr. Scand.* **52**, 213.

—— (1963b). Clinical symptoms, signs and tests in metachromatic leukodystrophy. In *Brain lipids and lipoproteins and the leukodystrophies* (ed. J. Folch-Pi and H. Bauer), p.134. Elsevier, Amsterdam.

—— Kollberg, H., Sourander, P., and Åkesson, H.O. (1970). Infantile globoid cell leucodystrophy (Krabbe's disease): a clinical and genetic study of 32 Swedish cases. *Neuropäediatrie* **1**, 74.

Hagen, G.A., Niepomniszcze, H., Haibach, H., Bigazzi, M., Hati, R., Rapoport, B., Jimeniz, C., De Groot, L.J., and Frawley, T.E. (1971). Peroxidase deficiency in familial goiter with iodine organification defect. *New Engl. J. Med.* **285**, 1394.

Hagenfeldt, L., Larsson, A., and Andersson, R. (1978). The γ-Glutamyl cycle and amino acid transport. *New Engl. J. Med.* **299**, 587.

—— —— and Zetterstrom, R. (1974). Pyroglutamic aciduria. *Acta Paediatr. Scand.* **63**, 1.

Haggard, M.E. and Lockhart, L. (1967). Megaloblastic anaemia and orotic aciduria. A hereditary disorder of pyrimidine metabolism responsive to uridine. *Am. J. Dis. Child.* **113**, 733.

Hagge, W., Brodehl. J., and Gellissen, K. (1967). Hypersarcosinemia. *Ped. Res.* **1**, 409.

Hahn, L.C., Ben-Yoseph, Y., and Nadler, H.L. (1980). Glycoprotein and ganglioside α-N-acetylneuraminidases in sialidosis and mucolipidoses. *Am. J. hum. Genet.* **32**, 41A.

Hähnel, R., Hähnel, E., Wysocki, S.J., Wilkinson, S.P., and Hockey, A. (1982). Prenatal diagnosis of X-linked ichthyosis. *Clin. Chim. Acta* **120**, 143.

Hakami, N., Neiman, P.E., Canellos, G.P., and Lazerson, J. (1971). Neonatal megaloblastic anaemia due to inherited transcobalamin II deficiency in two siblings. *New Engl. J. Med.* **285**, 1163.

Håkansson, G. and Svennerholm, L. (1979). Diagnosis of Gaucher disease with natural labeled and artificial substrates. In *Prenatal diagnosis* (ed. J.D. Murken, S. Stengel-Rutkowski, and E.Schwinger), p.301. Enke, Stuttgart.

Haldane, J.B.S. (1937–1938). A hitherto unexpected complication in the genetics of human recessives. *Ann. Eugen.* **8**, 263.

Hall, C.A. and Finkler, A.E. (1967). The dynamics of transcobalamin II. A vitamin B_{12} binding substance in plasma. *J. lab. clin. Med.* **65**, 459.

Hall, C.W., Cantz, M., and Neufeld, E.F. (1973). A ß-glucurosidase deficiency mucopolysaccharidosis: studies in cultured fibroblasts. *Arch. Biochem. Biophys.* **155**, 32.

Hällén, J. and Krook, H. (1963). Follow-up studies on an unselected ten-year material of 360 patients with liver cirrhosis in one community. *Acta Med. Scand.* **173**, 479.

Halley, D.J.J., Winchester, B.G., Burditt, L.J., D'Azzo, A., Robinson, D., and Galjaard, H. (1980). Comparison of the α-mannosidases in fibroblast cultures from patients with mannosidosis and mucolipidosis II and from controls. *Biochem. J.* **187**, 541.

—— Keijzer, W., Jaspers, N.G.J., Niermeijer, M.F., Kleijer, W.J., Boué, J. Boué, A., and Bootsma, D. (1979). Prenatal diagnosis of xeroderma pigmentosum (Group C) using assays of unscheduled DNA synthesis and postreplication repair. *Clin. Genet.* **16**, 137.

Halverstadt, D.B. and Wenzl, J.E. (1974). Primary hyperoxaluria and renal transplantation. *J. Urol.* **111**, 398.

Halvorsen, S. (1980). Screening for disorders of tyrosine metabolism. In *Neonatal screening for inborn errors of metabolism* (ed. H. Bickel, R. Guthrie, and G. Hammersen) p.45. Springer, New York.

—— and Gjessing, L. (1970). Studies on the amino acid levels in serum and urine of infants and children with 'metabolic liver diseases'. *Ped. Res.* **4**, 216.

—— Stokke, O., and Jellum, E.A. (1979). A variant form of 2-methyl-3-hydroxybutyric and 2-methylacetoacetic aciduria. *Acta Paediatr. Scand.* **68**, 123.

—— Hygstedt, O., Jagenburg, R., and Sjaasted, O. (1969). Cellular transport of L-histidine in Hartnup disease. *J. clin. Invest.* **48**, 1552.

—— Pande, H., Løken, A.C., and Gjessing, L.R. (1966). Tyrosinosis. A study of 6 cases. *Arch. Dis. Child.* **41**, 238.

Hambraeus, L. and Broberger, O. (1967). Penicillamine treatment of cystinosis. *Acta Paediatr. Scand.* **56**, 243.

—— Hardell, L.I., Westphal, O., Lorentsen, R., and Hjorth, G. (1974). Argininosuccinic aciduria: Report of three cases and the effect of high and reduced protein intake on the clinical state. *Acta Paediatr. Scand.* **63**, 525.

Hamers, M.N. (1978). *Immunochemical studies on Fabry's disease*. Thesis, University of Amsterdam.

—— Westerveld, A., Meera Khan, P., and Tager, J.M. (1977). Characterization of α-galactosidase isoenzymes in normal and Fabry human–Chinese hamster somatic cell hybrids. *Hum. Genet.* **36**, 289.

Hamilton, H.B. and Neel, J.V. (1963). Genetic heterogeneity in human acatalasia. *Am. J. hum. Genet.* **15**, 408.

Hammersen, G., Houghton, S., and Levy, H.L. (1975). Rennes-like variant of galactosemia: clinical and biochemical studies. *J. Ped.* **87**, 50.

—— Willie, L., Schmitt, H., Lutz, P., and Bickel, H. (1978). Maple syrup urine disease: Treatment of the acutely ill newborn. *Eur. J. Pediatr.* **129**, 157.

Hanley, W.B., Linsao, L., Davidson, W., and Moes, C.A.F. (1970). Malnutrition with early treatment of phenylketonuria. *Ped. Res.* **4**, 318.

Harcourt, B. and Ashton, N. (1973). Ultrastructure of the optic nerve in Krabbe's leukodystrophy. *Br. J. Ophthalmol.* **96**, 864.

Hardisty, R.M. and Mills, D.C.B. (1972). The platelet defect associated with albinism. *Ann. N.Y. Acad. Sci.* **201**, 429.

—— —— and Ketsa-Ard, K. (1972). The platelet defect associated with albinism. *Br. J. Haematol.* **23**, 679.

Harker, L.A., Slichter, S.J., Scott, C.R., and Ross, R. (1974). Homocystinemia: Vascular injury and arterial thrombosis. *New Engl. J. Med.* **291**, 537.

Harlan, W.R. Jun., Winesett, P.S., and Wasserman, A.J. (1967). Tissue lipoprotein lipase in normal individuals and in individuals with exogenous hyper-triclyceridemia and the relationship of this enzyme to assimilation of fat. *J. clin. Invest.* **46**, 239.

Harmeyer, J. and Plonait, H. (1967). Generalisierte hyperaminoaciduria mit erblicher rachitis bei schweinen. *Helv. Paediatr. Acta* **22**, 216.

Harper, P.S., Laurence, K.M., Parkes, A., Wusteman, F.S., Kresse, H., Von Figura, K., Ferguson-Smith, M.A., Duncan, D.M., Logan, R.W., Hall, F., and Whiteman, P. (1974). Sanfilippo A disease in the fetus. *J. med. Genet.* **11**, 123.

Harpey, J.-P., Rosenblatt, D.S., Cooper, B.A., Le Moël, G., Roy, C., and Lafourcade, J. (1981). Homocystinuria caused by 5,10-methylene-tetrahydro-folate reductase deficiency: a case in an infant responding to methionine, folinic acid, pyridoxine and vitamin B_{12} therapy. *J. Ped.* **98**, 275.

Harries, J.T., Piesowicz, A.T., Seakins, J.W.T., Francis, D.E.M., and Wolff,

O.H., (1971). Low proline diet in type-1 hyperprolinaemia. *Arch. Dis. Child.* **46**, 72.

Harris, D.J., Yang, B.I., Thompson, R.M., and Wolf, B. (1981). Propionyl-CoA carboxylase deficiency presenting as non-ketotic hyperglycinemia. *J. med. Genet.* **18**, 156.

—— —— Wolf, B., and Snodgrass, P.J. (1980). Dysautonomia in an infant with secondary hyperammonemia due to propionyl coenzyme A carboxylase deficiency. *Pediatrics* **65**, 107.

Harris, E.D. Jun. and Sjoerdsma, A. (1966*a*). Collagen profile in various clinical conditions. *Lancet* **ii**, 707.

—— —— (1966*b*). Effect of penicillamine on human collagen and its possible application to treatment of scleroderma. *Lancet* **ii**, 996.

Harris, H. (1980). *The principles of human biochemical genetics*, 3rd edn. p.353. Elsevier, Amsterdam.

—— and Robson, E.B. (1959). A genetical study of ethanolamine phosphate excretion in hypophosphatasia. *Ann hum. Genet.* **23**, 421.

—— —— (1963). Fractionation of human serum cholinesterase components by gel filtration. *Biochim. Biophys. Acta* **73**, 649.

—— and Whittaker, M. (1962). The serum cholinesterase variants: a study of twenty-two families selected via the 'intermediate' phenotype. *Ann. hum. Genet.* **26**, 73.

—— Penrose, L.S., and Thomas, D.H.H. (1959). Cystathioninuria. *Ann. hum. Genet.* **23**, 442.

—— Mittwoch, U., Robson, E.B., and Warren, F.L. (1955*a*). Phenotypes and genotypes in cystinuria. *Ann. hum. Genet.* **20**, 57.

—— —— —— —— (1955*b*). The pattern of amino-acid excretion in cystinuria. *Ann. hum. Genet.* **19**, 196.

—— Whittaker, M., Lehmann, H., and Silk, E. (1960). The pseudocholinesterase variants. Esterase levels and dibucaine numbers in families selected through suxamethonium sensitive individuals. *Acta Genet. Stat. Med.* **10**, 1.

Harris, R.C. (1961). Mucopolysaccharide disorder: A possible new geneotype of Hurler's syndrome (abstract). *Am. J. Dis. Child.* **102**, 741.

Harrison, H.E., Harrison, H.C., Lipshitz, F., and Johnson, A.D., (1966). Growth disturbance in hereditary hypophosphatemia. *Am. J. Dis. Child.* **112**, 290.

Hartlage, P.L., Coryell, M.E., Hall, W.K., and Hahn, D.A. (1974). Argininosuccinic aciduria: Perinatal diagnosis and early dietary management. *J. Ped.* **85**, 86.

Harzer, K. (1979). Prenatal enzymic diagnosis of sphingolipidoses using natural lipid substrates. In *Proceedings of the third European conference on prenatal diagnosis* (ed. J.D. Murken, S. Stengel-Rutkowski, and E. Schwinger) p.288. Enke, Stuttgart.

—— (1980). Enzymic diagnosis in 27 cases with Gaucher's disease. *Clin. Chim. Acta* **106**, 9.

—— and Benz, H.U. (1974). Deficiency of lactosyl sulfatide sulfatase in metachromatic leukodystrophy. *Hoppe-Seyler's Z. Physiol. Chem.* **355**, 744.

—— Zahn, V., Stengel-Rutkowski, S., and Gley, E.O. (1975*b*). Pränatale diagnose der metachromatische leucodystrophie. *Dtsch. med. Wschr.* **100**, 951.

—— Schlote, W., Peiffer, J., Benz, H.U., and Anzil, A.P. (1978). Neurovisceral lipidosis compatible with Niemann–Pick disease Type C: Morphological and biochemical studies of a late infantile case and enzyme and lipid assays in a prenatal case of the same family. *Acta Neuropathol.* **43**, 97.

—— Stengel-Rutkowski, S., Gley, E.O., Albert, A., Murken, J.D., Zahn, V., and

Henkel, K.P. (1975*a*). Pränatale diagnose der GM_2-gangliosidose typ 2. *Dtsch. med. Wschr.* **100**, 106.

Hashem, N., Bootsma, D., Keijzer, W., Greene, A.E., Coriell, L., Thomas, G.H., and Cleaver, J.E. (1980). Clinical characteristics, DNA repair, and complementation groups in xeroderma pigmentosum patients from Egypt. *Cancer Res.* **40**, 13.

Hashim, S.A. and Van Itallie, T.B. (1965). Cholestyramine resin therapy for hypercholesterolemia. *J. am. med. Assoc.* **192**, 289.

Hashimoto, K., Gross, B.G., and Lever, W.F. (1965). Angiokeratoma corporis diffusum. Histochemical and electron microscopic studies of the skin. *J. Invest. dermatol.* **44**, 119.

Hashimoto, T., Minato, H., Kuroda, Y., Toshima, K., Ohara, K., and Miyao, M. (1978). Monozygotic twins with presumed metachromatic leukodystrophy: Activity of arylsulfatase A in serum of patients and family. *Arch. Neurol.* **35**, 689.

Hasilik, A. and Neufeld, E.F. (1980*a*). Biosynthesis of lysosomal enzymes in fibroblasts; phosphorylation of mannose residues. *J. biol. Chem.* **255**, 4946.

—— —— (1980*b*). Biosynthesis of lysosomal enzymes in fibroblasts; synthesis as precursors of higher molecular weight. *J. biol. Chem.* **255**, 4937.

—— Waheed, A., and Von Figura, K. (1981). Enzymatic phosphorylation of lysosomal enzymes in the presence of UDP-*N*-acetylglucosamine. Absence of the activity in I-cell fibroblasts. *Biochem. biophys. Res. Commun.* **98**, 761.

—— Klein, U., Waheed, A., Strecker, G., and Von Figura, K. (1980). Phosphorylated oligosaccharides in lysosomal enzymes: Identification of α-*N*-acetylglucosamine(1)phospho(6)mannose diester groups. *Proc. natl. Acad. Sci. USA* **77**, 7074.

Haskins, M.E., Aguirre, G.D., Jezyk, P.F., and Patterson, D.F. (1980). The pathology of feline arylsulfatase B deficient mucopolysaccharidosis. *Am. J. Pathol.* **101**, 657.

—— Jezyk, P.F., Desnick, R.J., McDonough, S.K., and Patterson, D.F. (1979*a*). Mucopolysaccharidosis in a domestic short-haired cat: A disease distinct from that seen in the Siamese cat. *J. Am. Vet. Med. Assoc.* **175**, 384.

—— —— —— —— (1979*b*). Alpha-L-iduronidase deficiency in a cat. A model of mucopolysaccharidosis I. *Ped. Res.* **13**, 1294.

—— —— —— and Patterson, D.F. (1981). Animal model of human disease. Mucopolysaccharidosis I (Hurler, Scheie, Hurler/Scheie syndrome). *Comp. Pathol. Bull.* **VIII**, 3.

Hatano, H., Ohkido, M., Matsuo, I., Arai, R., and Mamiya, G. (1968). Aminoaciduria in xeroderma pigmentosum. *Acta Dermatovener.* **48**, 571.

Hauser, G.A. (1965). Testicular feminization. In *Intersexuality* (ed. C. Overzier), p.255. Academic Press Ltd., London.

Haworth, J.C., Perry, T.L., Blass, J.P., Hansen, S., and Urquhart, N. (1976). Lactic acidosis in three sibs due to defects in both pyruvate dehydrogenase and α-ketoglutarate dehydrogenase complexes. *Pediatrics* **58**, 564.

Hayashi, A., Suzuki, T., and Shimizu, A. (1968). Properties of hemoglobin M: Unequivalent nature of the α and ß subunits in the hemoglobin molecule. *Biochim. Biophys. Acta* **168**, 262.

Haymond, M.W., Ben-Galim, E., and Strobel, K.S. (1978). Glucose and alanine metabolism in children with maple syrup urine disease. *J. clin. Invest.* **62**, 398.

—— Karl, I.E., Feigin, R.D., Devivo, D., and Pagliara, A.S. (1973). Hypoglycemia and maple syrup urine disease — defective gluconeogenesis. *Ped. Res.* **7**, 500.

552   *References*

Haynes, M.D., Carter, R.F., Pollard, A.C., and Carey, W.F. (1980). Light and electron microscopy of infantile anf foetal tissues in cystinosis. *Micron* **11**, 443.

Haynes, M.E. and Robertson, E. (1981). Can oculocutaneous albinism be diagnosed prenatally? *Prenatal Diagnosis* **1**, 85.

Healy, L.A. and Hall, A.P. (1970). The epidemiology of hyperuricaemia. *Bull. rheum. Dis.* **20**, 600.

Hearing, V.J., Ekel, T.M., and Montague, P.M. (1981). Mammalian tyrosinase: Isozymic forms of the enzyme. *Int. J. Biochem.* **13**, 99.

Hegesh, E. and Avron, M. (1967). The enzymatic reduction of ferrihemoglobin. 1. The reduction of ferrihemoglobin in red blood cells and hemolysates. *Biochim. Biophys. Acta* **146**, 91.

—— Calmanovici, N., and Avron, M. (1968). New method for determining ferrihemoglobin reductase (NADH-methemoglobin reductase) in erythrocytes. *J. lab. clin. Med.* **72**, 339.

Heilmeyer, L. and Clotten, R. (1964). The erythropoietic porphyria. *Acta Haematol.* **31**, 137.

—— —— and Heilmeyer, L. jun. (1966). *Disturbances in heme synthesis* (trans. M. Steiner). Charles C. Thomas, Springfield, Illinois.

Hellier, M.D., Perrett, D., and Holdsworth, C.D. (1970). Dipeptide absorption in cystinuria. *Br. med. J.* **iv**, 782.

—— Chir, B., Holdsworth, C.D., and Perrett, D. (1973). Dibasic amino acid absorption in man. *Gastroenterology*, **65**, 613.

Henderson, J.F., Fraser, J.H., and McCoy, E.E. (1974). Methods for the study of purine metabolism in human cells *in vitro*. *Clin. Biochem.* **7**, 339.

—— Rosenbloom, F.M., Kelley, W.N., and Seegmiller, J.E. (1968). Variations in purine metabolism of cultured skin fibroblasts from patients with gout. *J. clin. Invest.* **47**, 1511.

Henkind, P. and Ashton, N. (1965). Ocular pathology in homocystinuria. *Trans. Ophthalmol. Soc. U.K.* **85**, 21.

Henry, C.G., Strauss, A.W., Keating, J.P., and Hillman, R.E. (1981). Congestive cardiomyopathy associated with ß-ketothiolase deficiency. *J. Ped.* **99**, 754.

Henslee, J.G. and Jones, M.E. (1981). Ornithine synthesis from glutamate in mitochondria from rat small intestinal mucosa. *Fed. Proc.* **40**, 1683.

Herb, J.K., Subramanian, S., and Robinson, H. (1973). Type III mucopolysaccharidosis: Report of a case with severe mitral valve involvement. *J. Ped.* **82**, 101.

Herman, R.H., Rosensweig, N.S., Stifel, F.B., and Herman, Y.F. (1969). Adult formiminotransferase deficiency: a new entity. *Clin. Res.* **17**, 304.

Herndon, J.H., Steinberg, D., Uhlendorf, B.W., and Fales, H.M. (1969). Refsum's disease: characterization of the enzyme defect in cell culture. *J. clin. Invest.* **48**, 1017.

Hers, H.G. (1959). Etude enzymatique sur fragments hépatiques. Applications à la classification des glycogénoses. *Rev. Int. Hépat.* **9**, 35.

—— (1963). α-Glucosidase deficiency in generalised glygogen storage disease (Pompe's disease). *Biochem. J.* **86**, 11.

—— (1980). Carbohydrate metabolism and its regulation. In *Inherited disorders of carbohydrate metabolism* (ed. D. Burman, J.B. Holton, and C.A. Pennock), p.3. MTP Press, Lancaster.

—— and Joassin, G. (1961). Anomalie de l'aldolase hépatique dans l'intolérance au fructose. *Enzymol. Biol. clin.* **1**, 4.

—— and Van Hoof, F. (1968). Glycogen storage disease: Type VI glycogenosis. In

Carbohydrate metabolism and its disorders (ed. F. Dickens, P.J. Randle, and W.J. Whelan), p.166. Academic Press, New York.

Herz, F., Kaplan, E., and Scheye, E.S. (1970). Diagnosis of erythrocyte glucose-6-phosphate dehydrogenase deficiency in a Negro male despite hemolytic crisis. *Blood* **35**, 90.

Heston, W.E., Hoffman, H.A., and Rechcigle, M. (1965). Genetic analysis of liver catalase activity in two substrains of C 57 BL mice. *Genet. Res.* **6**, 387.

Heukels-Dully, M.J. and Niermeijer, M.F. (1976). Variation in lysosomol enzyme activity during growth in culture of human fibroblasts and amniotic fluid cells. *Exp. Cell. Res.* **97**, 304.

Heydorn, K., Damsgaard, E., Horn, N., Mikkelsen, M., Tygstrup, I., Vestermark, S., and Weber, J. (1975). Extra-hepatic storage of copper. A male fetus suspected of Menkes' disease. *Hum. Genet.* **29**, 171.

Hickman, S. and Neufeld, E.F. (1972). A hypothesis for I-cell disease: defective hydrolases that do not enter lysosomes. *Biochem. biophys. Res. Commun.* **49**, 992.

—— Shapiro, L.J., and Neufeld, E.F. (1974). A recognition marker required for uptake of a lysosomal enzyme by cultured fibroblasts. *Biochem. biophys. Res. Commun.* **57**, 55.

Higginbottom, M.C., Sweetman, L., and Nyhan, W.L. (1978). A syndrome of methylmalonic aciduria, homocystinuria, megaloblastic anaemia and neurologic abnormalities in a vitamin B_{12}-deficient breast-fed infant of a strict vegetarian. *New Engl. J. Med.* **299**, 317.

Hill, A. and Zaleski, W.A. (1971). Tyrosinosis: biochemical studies of an unusual case. *Clin. Biochem.* **4**, 263.

Hillman, R.E. and Keating, J.P. (1974). ß-Ketothiolase deficiency as a cause of the 'Ketotic hyperglycinemia syndrome'. *Pediatrics* **53**, 221.

Hindfelt, B. (1975). On mechanisms of hyperammonemic coma — with particular reference to hepatic encephalopathy. *Ann. N.Y. Acad. Sci.* **252**, 116.

Hirani, S., Winchester, B.G., and Patrick, A.D. (1977). Measurement of the mannosidase activities in human plasma by a differential assay. *Clin. Chim. Acta* **81**, 135.

Hirschfeld, A.G. and Fleshman, J.K. (1969). An unusually high incidence of congenital adrenal hyperplasia in the Alaskan Eskimo. *J. Ped.* **75**, 492.

Hirschhorn, R. (1980). Treatment of genetic disease by allo-transplantation. *Birth defects: original article series* **XVI**, vol. 1, p.429. Alan R. Liss, New York.

—— and Martin, D.W. jun. (1978). Enzyme defects in immunodeficiency diseases. *Springer Seminars in Immuno-pathology* **1**, 299.

—— Beratis, N., and Rosen, F.S. (1976). Characterisation of residual enzyme activity in fibroblasts from patients with adenosine deaminase deficiency and combined immuno-deficiency: evidence for a mutant enzyme. *Proc. natl. Acad. Sci. USA* **73**, 213.

—— —— —— Parkman, R., Stern, R. and Polmar, S. (1975). Adenosine-deaminase deficiency in a child diagnosed prenatally. *Lancet* **i**, 73.

Hirsch-Kauffmann, M. and Doppler, W. (1981). Biochemical studies on a patient with complete APRT-deficiency. *Proceedings of the 6th international congress of human genetics* **P5**, 53.

Hitzig, W.H. and Willi, H. (1961). Hereditäre lymphoplasmocytäre dysgenesie. *Schweiz. med. Wschr.* **91**, 1625.

—— Dohmann, U., Pluss, H.J., and Vischer, D. (1974). Hereditary trans-cobalamin II deficiency: Clinical findings in a new family. *J. Ped.* **85**, 622.

Ho, M.W. (1973). Hydrolysis of ceramide trihexoside by a specific α-galactosidase from human liver. *Biochem. J.* **133**, 1.

—— and Light, N.D. (1973). Glucocerebrosidase: reconstitution from macromolecular components depends on acidic phospholipids. *Biochem. J.* **136**, 821.

—— and O'Brien, J.S. (1969). Hurler's syndrome: deficiency of a specific beta-galactosidase isoenzyme. *Science* **165**, 611.

—— —— (1970). Stimulation of acid beta-galactosidase activity by chloride ions. *Clin. Chim. Acta* **30**, 531.

—— —— (1971). Gaucher's disease: deficiency of acid ß-glucosidase and reconstitution of enzyme activity *in vitro. Proc. natl. Acad. Sci USA* **68**, 2810.

—— Beutler, S., Tennant, L., and O'Brien, J.S. (1972*b*). Fabry's disease: evidence for a physically altered α-galactosidase. *Am. J. hum. Genet.* **24**, 256.

—— Seck, J., Schmidt, D., Veath, M.L., Johnson, W., Brady, R.O., and O'Brien, J.S. (1972*a*). Adult Gaucher's disease: kindred studies and determination of acid ß-glucosidase in cultured fibroblasts. *Am. J. hum. Genet.* **24**, 37.

Hoar, D.I. and Rudd, N.L. (1976). Prenatal diagnosis of hypophosphatasia. *Lancet* **i**, 1194.

Hobbs, J.R., Hugh-Jones, K., Barrett, A.J., Byrom, N., Chambers, D., Henry, K., James, D.C.O., Lucas, C.F., Rogers, T.R., Benson, P.F., Tansley, L.R., Patrick, A.D., Mossman, J., and Young, E.P. (1981). Reversal of clinical features of Hurler's disease and biochemical improvement after treatment by bone marrow transplantation. *Lancet* **ii**, 709.

Hobolth, N. and Pedersen, C. (1978). Six cases of the mild form of the Hunter syndrome in five generations: three affected males with progeny. (Abstract) *Clin. Genet.* **13**, 121.

Hochberg, R.B., McDonald, P.D., Feldman, M. and Lieberman, S. (1974). Studies on the biosynthetic conversion of cholesterol into pregnenolone. *J. biol. Chem.* **249**, 1277.

Hockaday, T.D.R., Clayton, J.E., Frederick, E.W., and Smith, L.H. jun. (1964). Primary hyperoxaluria. *Medicine* **43**, 315.

—— Frederick, E.W., Clayton, J.E., and Smith, L.H. jun. (1965). Studies on primary hyperoxaluria. II. Urinary oxalate, glycolate, and glyoxylate measurement by isotope dilution method. *J. lab. clin. Med.* **65**, 677.

Hocking, D.J., Jolly, R.D., and Batt, R.D. (1972). Deficiency of α-mannosidase in Angus cattle. An inherited lysosomal storage disease. *Biochem. J.* **128**, 69.

Hockwald, R.S., Clayman, C.B., and Alving, A.S. (1952). Toxicity of primaquine in Negroes. *J. Am. Med. Assoc.* **149**, 1568.

Hodgkin, W.E., Giblett, E.R., Levine, H., Bauer, W., and Motulsky, A.G. (1965). Complete pseudocholinesterase deficiency: genetic and immunologic characterization. *J. clin. Invest.* **44**, 486.

Hodson, P., Goldblatt, J., and Beighton, P. (1979). Non-neuronopathic Gaucher's disease presenting in infancy. *Arch. Dis. Child.* **54**, 707.

Hoefnagel, D. and Pomeroy, J. (1970). Hydroxylysinuria. *Lancet* **i**, 1342.

—— Andrew, E.D., Mireault, N.G., and Berndt, W.O. (1965). Hereditary choreoathetosis, self mutilation and hyperuricaemia in young males. *New Engl. J. Med.* **273**, 130.

Hoeksema, H.L., Reuser, A.J.J., Hoogeveen, A.T., Westerveld, A., Braidman, I., and Robinson, D. (1977). Characterization of hexosaminidase isoenzymes in man–Chinese hamster somatic cell hybrids. *Am. J. hum. Genet.* **29**, 14.

Hoffbauer, R.W. and Schrempf, G. (1976). Phenylalanine hydroxylation in cultured fibroblasts from patients with phenylketonuria. *Lancet* **ii**, 194.

—— —— and Monch, E. (1976). Phenylalanine hydroxylation in phenylketonuria. *Lancet* ii, 1031.

Hofstad, F., Seip, M., and Eriksen, L. (1973). Congenital erythropoietic porphyria with a hitherto undescribed porphyrin pattern. *Acta Paed. Scand.* **62**, 380.

Hogben, L., Worrall, R.L., and Zieve, I. (1932). The genetic basis of alcaptonuria. *Proc. R. Soc. Edinburgh.* **52**, 264.

Hokanson, J.T., O'Brien, W.E., Idemoto, J., and Schafer, I.A. (1978). Carrier detection in ornithine transcarbamylase deficiency. *J. Ped.* **93**, 75.

Hollister, D.W., Cohen, A.H., Rimoin, D.L., and Silberberg, R. (1975). The Morquio syndrome (mucopolysaccharidosis IV): Morphologic and biochemical studies. *Johns Hopkins med. J.* **137**, 176.

Hollowell, J.G. Jun., Coryell, M.E., Hall. W.K., Findley, J.K., and Thevaos, T.G. (1968). Homocystinuria as affected by pyridoxine, folic acid and vitamin B_{12}. *Proc. Soc. exp. Biol. Med.* **129**, 327.

—— Hall, W.K., Coryell, M.E., McPherson, J. Jun., and Hahn, D.A. (1969). Homocystinuria and organic aciduria in a patient with vitamin B_{12} deficiency. *Lancet* ii, 1428.

Holmes, E.W. and O'Brien, J.S. (1978). Feline GM_1-gangliosidosis: characterisation of the residual liver acid ß-galactosidase. *Am. J. hum. Genet.* **30**, 505.

—— —— (1979). Separation of glycoprotein-derived oligosaccharides by thin-layer chromatography. *Analyt. Biochem.* **93**, 167.

—— Wyngaarden, J.B., and Kelley, W.N. (1973). Human glutamine phosphoribosylpyrophosphate amidotransferase. Two molecular forms interconvertible by purine ribo nucleotides and phosphoribosylpyrophosphate. *J. biol. Chem.* **248**, 6035.

—— Mason, D.H., Goldstein, L.I., Blount, R.E., and Kelley, W.N. (1974). Xanthine oxidase deficiency: Studies in a previously unreported case. *Clin. Chem.* **20**, 1076.

Holmes, L.B., McGowan, B.L., and Efron, M.L. (1969). Lowe's syndrome: A search for the carrier state. *Pediatrics* **44**, 358.

Holstein, T.J., Burnett, J.B. and Quevedo, W.C. jun. (1967). Genetic regulation of multiple forms of tyrosinase in mice. Action of a and b loci. *Proc. Soc. exp. Biol. Med.* **126**, 415.

Holton, J.B., Lewis, F.J.W., and Moore, G.R. (1964). Biochemical investigation of histidinaemia. *J. clin. Path.* **17**, 671.

Holtzaple, P.G., Rea, C., Bovee, K., and Segal, S. (1971). Characteristics of cystine and lysine transport in renal and jejunal tissue from cystinuric dogs. *Metabolism* **20**, 1016.

Holtzman, N.A. (1976). Menkes' kinky hair syndrome. A genetic disease involving copper. *Fed. Proc.* **35**, 2276.

—— Mellits, E.D., and Kallman, C.H. (1974). Neonatal screening for phenylketonuria: II. Age dependence of initial phenylalanine in infants with PKU. *Pediatrics* **53**, 353.

Holzel, A., Schwarz, V., and Sutcliffe, K.W. (1959). Defective lactose absorption causing malnutrition in infancy. *Lancet* i, 1126.

Hom, B.L. (1967). Plasma turnover of [57]colbalt–vitamin B_{12} bound to transcobalamin I and II. *Scand. J. Haematol.* **4**, 321.

Hommes, F.A., Polman, H.A., and Reerink, J.D. (1968). Leigh's encephalopathy: an inborn error of gluconeogenesis. *Arch. Dis. Child.* **43**, 423.

—— Bendien, K., Elema, J.D., Bremer, H.J., and Lombeck, I. (1976). Two cases of phosphoenolpyruvate carboxykinase deficiency. *Acta Paediatr. Scand.* **65**, 233.

—— Kuipers, J.R.G., Elema, J.D., Jansen, J.F., and Jonxis J.H.P. (1968). Propionic acidemia, a new inborn error of metabolism. *Ped. Res.* **2**, 519.

Hooft, C., Carton, D., and De Schrijver, F. (1969). Cystathioninemia in three siblings. In *Enzymopenic anaemias, lysosomes and other papers* (ed. J.D. Allen, K.S. Holt, J.T. Ireland, and R.J. Pollitt), p.200. Churchill Livingstone, London.

—— Valcke, R., and Herpol, J. (1964). Étude clinique du syndrome de Lowe. *Acta Paediatr. Belgium* **18**, 197.

—— Carton, D., De Schrijver, F., Delbeke, M.J., Samijn, W., and Kint, J. (1971). Juvenile cystinosis in two siblings. In *Inherited disorders of sulphur metabolism* (ed. N.A.J. Carson and D.N. Raine), p.141. Churchill Livingstone, Edinburgh.

—— Timmermans, J., Snoek, J., Antener, I., Oyaert, W., and Van den Hende, C. (1965). Methionine malabsorption syndrome. *Ann. Paediatr.* (*Basel*) **205**, 73.

—— Carton, D., Snoek, J., Timmermans, J., Antener, I., Van den Hende, C., and Oyaert, W. (1968). Further investigations in the methionine syndrome. *Helv. Paediatr. Acta* **23**, 334.

Hoogenraad, N.J., Mitchell, J.D., Don, N.A., Sutherland, T.M., and McLeay, A.C. (1980). Detection of carbamyl phosphate synthetase I deficiency using duodenal biopsy samples. *Arch. Dis. Child.* **55**, 292.

Hoogeveen, A., D'Azzo, A., Brossmer, R., and Galjaard, H. (1981). Correction of combined ß-galactosidase/neuraminidase deficiency in human fibroblasts. *Biochem. biophys. Res. Commun.* **103**, 292.

—— Verheijen, F.W., D'Azzo, A., and Galjaard, H. (1980). Genetic heterogeneity in human neuraminidase deficiency. *Nature, Lond.* **285**, 500.

Hopkins, I.A., Townley, R.R.W., and Shipman, R.T. (1969a). Cerebral thrombosis in a patient with homocystinuria. *J. Ped.* **75**, 1082.

—— Connelly, J.F., Dawson, A.G., Hird, F.J.R., and Maddison, T.G. (1969b). Hyperammonenia due to ornithine transcarbamylase deficiency. *Arch. Dis. Child.* **44**, 143.

Hopkins, J. and Tudhope, G.R. (1973). Glutathione peroxidase in human red cells in health and disease. *Br. J. haematol.* **25**, 563.

Hopkinson, D.A., Cook, P.J.L., and Harris, H. (1969), Further data on the adenosine deaminase polymorphism and a report of a new phenotype. *Ann. hum. Genet.* **32**, 361.

—— Corney, G., Cook, P.J.L., Robson, E.B., and Harris, H. (1970). Genetically determined electrophoretic variants of human red-cell NADH diaphorase. *Ann. hum. Genet.* **34**, 1.

Hoppel. C.L. and Tomec, R.J. (1972). Carnitine palmityltransferase. Location of two enzymatic activities in rat liver mitochrondria. *J. biol. Chem.* **247**, 832.

Hopwood, J.J. (1979). α-L-iduronidase, ß-D-glucuronidase and 2-sulf-L-iduronate-2-sulfatase: preparation and characterisation of radioactive substrates from heparin. *Carbohyd. Res.* **69**, 203.

—— and Elliot, H. (1981). The diagnosis of the Sanfilippo C syndrome using monosaccharide and oligosaccharide substrates to assay acetyl-CoA: 2-amino-2-deoxy-α-glucoside N- acetyltransferase activity. *Clin. Chim. Acta* **112**, 67.

—— and Muller, V. (1979). Biochemical discrimination of Hurler and Scheie syndromes. *Clin. Sci.* **57**, 265.

—— Holzman, I., and Drash, A.L. (1977). Fructose-1,6-diphosphatase deficiency. *Am. J. Dis. Child.* **131**, 418.

Horn, N. (1976). Copper incorporation studies on cultured cells for prenatal diagnosis of Menkes' disease. *Lancet* **i**, 1156.

—— (1980). Menkes X-linked disease: heterozygous phenotype in uncloned fibroblast cultures. *J. med. Genet.* **17**, 257.

—— (1981). Menkes X-linked disease: prenatal diagnosis of hemizygous males and heterozygous females. *Prenatal Diagnosis* **1**, 107.

—— Mooy, P., and McGuire, V.M. (1980). Menkes X-linked disease: two clonal cell populations in heterozygotes. *J. med. Genet.* **17**, 262.

Horwitz, A.L. and Dorfman, A. (1978). The enzymatic defect in Morquio's disease: the specificity of *N*-acetylhexosamine sulfatases. *Biochem. biophys. Res. Commun* **80**, 819

Hoskins, L.C., Winawer, S.J., Broitman, S.A., Gottlieb, L.S., and Zamchick, N. (1967). Clinical giardiasis and intestinal malabsorption. *Gastroenterology* **53**, 265.

Hovig, T. and Gjone, E. (1974). Familial lecithin:cholesterol acyltransferase deficiency: Ultrastructural studies on lipid deposition and tissue reactions. *Scand. J. clin. lab. Invest.* **33**, (suppl.) 135.

Howell, R.R. (1968). Hyperuricemia in childhood. *Fed. Proc.* **27**, 1078.

—— Ashton, D.M., and Wyngaarden, J.B. (1962). Glucose-6-phosphatase deficiency glycogen storage disease. Studies on the interrelationships of carbohydrate, lipid, and purine abnormalities. *Pediatrics* **29**, 553.

—— Kaback, M.M., and Brown, B.I. (1971). Type IV glycogen storage disease: Branching enzyme deficiency in skin fibroblasts and possible heterozygote detection. *J. Ped.* **78**, 638.

—— Stevenson, R.E., Philiky, R.L., and Berry, D.H. (1976). Hepatic adenomata in patients with Type I glycogen storage disease (Von Gierke's). *J. Am. med. Assoc.* **236**, 1481.

Hsia, D.Y.-Y., Driscoll, K.W., Troll, W., and Knox, W.E. (1956). Detection by phenylalanine tolerance tests of heterozygous carriers of phenylketonuria. *Nature, Lond.* **178**, 1239.

Hsia, Y.E., Scully, K.J., and Rosenberg, L.E. (1969). Defective propionate carboxylation in ketotic hyperglycinaemia. *Lancet* **i**, 757.

—— —— —— (1971). Inherited propionyl-CoA carboxylase deficiency in 'ketotic hyperglycinaemia'. *J. clin. Invest.* **50**, 127.

Hsieh, E. and Jaffé, E.R. (1971). Electrophoretic and functional variants of NADH-methaemoglobin reductase in hereditary methemoglobinemia. *J. clin. Invest.* **50**, 196.

Huang, I.-Y., Fujii, H., and Yoshida, A. (1980). Structure and function of normal and variant human phosphoglycerate kinase. *Hemoglobin* **4**, 601.

Huang, P.W.H., Rozdilsky, B., Gerrard, J.W., Goluboff, N., and Holman, C.H. (1970). Crigler–Najjar syndrome in four of five siblings with post-mortem findings in one. *Arch. Pathol.* **90**, 536.

Hudson, F.P., Mordaunt, V.L., and Leahy, I. (1970). Evaluation of treatment begun in first three months of life in 184 cases of phenylketonuria. *Arch. Dis. Child.* **45**, 5.

Hudson, L.D.S., Fiddler, M.B., and Desnick, R.J. (1980). Immunologic aspects of enzyme replacement therapy. An evaluation of the immune response to unentrapped, erythrocyte- and liposome-entrapped enzyme in C3H/HeJ Gus[h] mice. In *Enzyme therapy in genetic disease: 2* (ed. R.J. Desnick), p.163. Alan R. Liss, New York.

Hue, L. and Hers, H.G. (1972). The conversion of [4-^{3}H] fructose and of [4-^{3}H] glucose to liver glycogen in the mouse. An investigation of the glyceraldehyde crossroads. *Eur. J. Biochem.* **29**, 268.

Hug, G. (1962). Glucagon tolerance test in glycogen storage disease. *J. Ped.* **60**, 545.

—— (1980). Pre- and postnatal diagnosis of glycogen storage disease. In *Inherited disorders of carbohydrate metabolism* (ed. D. Burman, J.B. Holton, and C.A. Pennock), p.327. MTP Press, Lancaster.

—— Schubert, W.K., and Shwachman, H. (1965). Imbalance of liver phosphorylase and accumulation of hepatic glycogen in a girl with progressive disease of the brain. *J. Ped.* **67**, 741.

—— Chuck, G., Walling, L., and Schubert, W.K. (1974). Liver phosphorylase deficiency in glycogenosis type VI: Documentation by biochemical analysis of hepatic biopsy specimens. *J. lab. clin. med.* **84**, 26.

—— Schubert, W.K., Chuck, G.,.and Garancis, J.C. (1967). Liver phosphorylase: deactivation in child with progressive brain disease, increased hepatic glycogen and increased urinary catecholamines. *Am. J. Med.* **42**, 139.

Hughes, I.A. and Laurence, K.M. (1979). Antenatal diagnosis of congenital adrenal hyperplasia. *Lancet* **ii**, 7.

—— and Winter, J.S.D. (1976). The application of a serum 17OH- progesterone radioimmunoassay to the diagnosis and management of congenital adrenal hyperplasia. *J. Ped.* **88**, 766.

Hughes, W.S. and Senior, J.R. (1975). The glucose-galactose malabsorption syndrome in a 23 year old woman. *Gastroenterology* **68**, 142.

Huguley, C.M., Bain, J.A., Rivers, S.L., and Scoggins, R.B. (1959). Refractory megaloblastic anaemia associated with excretion of orotic acid. *Blood* **14**, 615.

Huijing, F. (1964). Amylo-1,6-glucosidase activity in normal leucocytes and in leucocytes of patients with glycogen-storage disease. *Clin. Chim. Acta* **9**, 269.

—— (1975). Glycogen metabolism and glycogen-storage diseases. *Physiol. Rev.* **55**, 609.

—— and Fernandes, J. (1969). X-chromosomal inheritance of liver glycogenosis with phosphorylase kinase deficiency. *Am. J. hum. Genet.* **21**, 275.

—— Warren, R.J., and McLeod, A.G.W. (1973). Elevated activity of lysosomal enzymes in amniotic fluid of a fetus with mucolipidosis II. *Clin. Chim. Acta* **44**, 453.

—— Klein Obbink, H.J., and von Creveld, S. (1968). The activity of the debranching enzyme system in leucocytes. *Acta Genet. (Basel)* **18**, 128.

Hülsmann, W.C. and Fernandes, J. (1971). A child with lactacidemia and fructose-diphosphatase deficiency in the liver. *Ped. Res.* **5**, 633.

—— Oei, T.L., and Creveld, S. (1961). Phosphorylase activity in leukocytes from patients with glycogen storage disease. *Lancet* **ii**, 581.

Hultberg, B. and Masson, P.K. (1975). Activation of residual acidic α-mannosidase activity in mannosidosis tissues by metal ions. *Biochem. biophys. Res. Commun.* **67**, 1473.

—— and Öckerman, P.A. (1970). ß-glucosidase activities in human tissues; findings in Gaucher's disease. *Clin. Chim. Acta* **28**, 169.

—— Lindsten, J., and Sjöblad, S. (1976). Molecular forms and activities of glycosidases in culture of amniotic fluid cells. *Biochem. J.* **155**, 599.

—— Sjöblad, S., and Öckerman, P.A. (1973). 4-methylumbelliferyl-ß-glucosidase in cultured human fibroblasts from controls and patients with Gaucher's disease. *Clin. Chim. Acta* **49**, 93.

Hultquist, D.E. and Passon, P.G. (1971). Catalysis of methemoglobin reduction by erythrocyte cytochrome b5 and cytochrome b5 reductase. *Nature new Biol.* **229**, 252.

Humbel, R. (1976). Rapid method for measuring arylsulfatases A and B in leucocytes as a diagnosis for sulfatidosis, mucosulfatidosis and mucopolysaccharidosis VI. *Clin. Chim. Acta* **68**, 339.

Hunt, D.M. (1974). Primary defect in copper transport underlies mottled mutants in the mouse. *Nature, Lond.* **249**, 852.

—— (1976). A study of copper treatment and tissue copper levels in the murine congenital copper deficiency, mottled. *Life Sci.* **19**, 1913.

Hunter, C. (1917). A rare disease in two brothers. *Proc. R. Soc. Med.* **10**, 104.

Hunter, J., Thompson, R.P.H., Rake, M.O., and Williams, R. (1971). Controlled trial of phetharbital, a non-hypnotic barbiturate in unconjugated hyperbilirubinaemia. *Br. med. J.* **ii**, 497.

Hunziker, N. (1980). Richer–Hanhart syndrome and tyrosinemia type II. *Dermatologica* **160**, 180.

Hurler, G. (1920). Ueber einen Typ multiplier Abartungen, vorwiegend am Skelettsystem. *Z. Kinderheilkd.* **24**, 220.

Hutchinson, J.H. and McGirr, E.M. (1954). Hypothyroidism as an inborn error of metabolism. *J. clin. Endocrinol. Metab.* **14**, 869.

—— —— (1956). Sporadic nonendemic goitrous cretinism. *Lancet* **i**, 1035.

Hutton, R.A., MacNab, A.J., and Rivers, P.A. (1976). Defect of platelet function associated with chronic hypoglycaemia. *Arch. Dis. Child.* **51**, 49.

Huttunen, J.K., Ehnholm, C., Kinnunen, P.K.J., and Nikkilä, E.A. (1975). An immunochemical method for selective measurement of two triglyceride lipases in human postheparin plasma. *Clin. Chim. Acta* **63**, 335.

Hyanek, J., Bremer, H.J., and Slavik, M. (1969). 'Homocystinuria' and [urinary] excretion of ß-amino acids in patients treated with 6-azauridine. *Clin. Chim. Acta* **25**, 288.

Iber, F.L., Chalmers, T.C., and Uzman, L.L. (1957b). Studies of protein metabolism in hepatolenticular degeneration. *Metabolism* **6**, 388.

—— Rosen, H., Levenson, S.M., and Chalmers, T.C. (1957a). The plasma amino acids in patients with liver failure. *J. lab. clin. med.* **50**, 417.

Idelson, L.I., Rustamov, R.S., and Kolesnikova, A.C. (1973). Hereditary hemolytic anemia with deficiency of 6-phosphogluconic dehydrogenase (6-PGH) and drug induced hemoglobinuria. *Klin. Med. (Moscow)* **6**, 130.

Igarashi, M., Belchis, D., and Suzuki, K. (1976b). Brain gangliosides in adrenoleukodystrophy. *J. Neurochem.* **27**, 327.

—— Schaumburg, H.H., Powers, J., Kishimoto, Y., Kolodny, E., and Suzuki, K. (1976a). Fatty acid abnormality in adrenoleukodystrophy. *J. Neurochem.* **76**, 851.

Ikeda, Y., Noda, C., and Tanaka, K. (1981). Purification and characterization of isovaleryl-CoA dehydrogenase from rat liver mitochondria. In *Metabolism and clinical implications of branched chain amino and ketoacids* (ed. M. Walser and J.R. Williamson), p.41. Elsevier/North-Holland, New York.

Illingworth, B. and Brown, D.H. (1964). Glycogen storage diseases, types III, IV and VI. In *Ciba foundation symposium: control of glycogen metabolism* (ed. W.J. Whelan and M.P. Cameron), p.336. Elsevier Publishing Co., Amsterdam.

—— and Cori, G.T. (1952). Structure of glycogens and amylopectins. III. Normal and abnormal human glycogen. *J. biol. Chem.* **199**, 653.

Imerslund, O. and Bjornsted, P. (1963). Familial vitamin B_{12} malabsorption. *Acta Haematol.* **30**, 1.

Imperato-McGinley, J., Peterson, R.E., Leshin, M., Griffin, J.E., Cooper, G., Draghi, S., Berenyi, M., and Wilson J.D. (1980). Steroid 5 α-reductase deficiency in a 65 year old male pseudohermaphrodite: the natural history, ultrastructure of the testes and evidence for inherited enzyme heterogeneity. *J. clin. endocr. Metab.* **50**, 15.

—— —— Gautier, T., and Sturla, E. (1979). Male pseudohermaphroditism secondary to 5 α-reductase deficiency — a model for the role of androgens in both the development of the male phenotype and the evolution of a male gender identity. *J. Steroid Biochem.* **11**, 637.

Irreverre, F., Mudd, S.H., Heizer, W.D., and Laster, L. (1967). Sulfite oxidase deficiency: studies of a patient with mental retardation, dislocated ocular lenses, and abnormal urinary excretion of 5-sulfo-L-cysteine, sulfite, and thiosulfate. *Biochem. Med.* **1**, 187.

Isemura, M., Hanyo, T., Gejyo, F., Nakazawa, R., Igarashi, R., Matsuo, S., Ikeda, K., and Sato, Y. (1979). Prolidase deficiency with imidopeptiduria. A familial case with and without clinical symptoms. *Clin. Chim. Acta* **93**, 401.

Isenberg, J.N. and Sharp, H.L. (1975). Aspartylglucosaminuria: psychomotor retardation masquerading as a mucopolysaccharidosis. *J. Ped.* **86**, 713.

Ivemark, B.I., Svennerholm, L., Thoren, C., and Tunnell, R. (1963). Niemann–Pick disease in infancy. *Acta Pediatr.* **52**, 391.

Iwamoto, A., Naito, C., Teramoto, T., Katu, H., Kaka, M., Kariya, T., Shimuzu, T., Oka, H., and Oda, T. (1978). Familial lecithin:cholesterol acyltransferase deficiency complicated with unconjugated hyperbilirubinemia and peripheral neuropathy. The first reported case in the Far East. *Acta Med. Scand.* **204**, 219.

Jackson, A.H., Sancovich, H.A., Ferramola, A.M., Evans, N., Games, D.E., Matlin, S.A., Elder, G.H., and Smith, S.G. (1975*b*). Macrocyclic intermediates in the biosynthesis of porphyrins. *Phil. Trans. R. Soc. Lond. B.* **273**, 119.

Jackson, J.D., Smith, F.G., Litmann, N.N., Yuile, C.L., and Latta, H. (1962). The Fanconi syndrome with cystinosis. Electron microscopy of renal biopsy specimens from five patients. *Am. J. Med.* **33**, 893.

Jackson, S.H., Dennis, A.W., and Greenberg, M. (1975*a*). Iminopeptiduria: a genetic defect in recycling collagen. A method for determining prolidase in red blood cells. *Can. med. Assoc. J.* **113**, 759.

Jacobsen, C.D., Gjone, E., and Hovig, T. (1972). Sea-blue histiocytes in familial lecithin:cholesterol acyltransferase deficiency. *Scand. J. Haematol.* **9**, 106.

Jacoby, L.B., Littlefield, J.W., Milunsky, A., Shih, V.E., and Welroy, R.G. (1972). A microassay for argininosuccinase in cultured cells. *Am. J. hum. Genet.* **24**, 321.

—— Shih, V.E., Struckmeyer, C., Niermeijer, M.F., and Boué, J. (1981). Variation in argininosuccinate synthetase activity in amniotic fluid cell cultures: implications for prenatal diagnosis of citrullinemia. *Clin. Chim. Acta* **116**, 1.

Jaeger, A., Tempe, J.D., Geisler, F., Nordmann, Y., and Mantz, J.M. (1975). La coproporphyrie héréditaire. *Nouv. Presse Med.* **4**, 2783.

Jaffé, E.R. (1966). Hereditary methemoglobinemias associated with abnormalities in the metabolism of erythrocytes. *Am. J. Med.* **41**, 786.

Jagiello, G. and Atwell, J.D. (1962). Prevalence of testicular feminization. *Lancet* **i**, 329.

Jänne, O., Perheentupa, J., and Vikho, R. (1970). Plasma and urinary steroids in an eight-year-old boy with 3ß-hydroxysteroid dehydrogenase deficiency. *J. clin. Endocr.* **531**, 162.

Jansen, A.J.M. and Trijbels, F.J.M. (1982). A new radiochemical assay for fructose-1,6-diphosphatase in human leucocytes. *Clin. Chim. Acta* **119**, 143.

Jatzkewitz, H. (1958). Zwei Typen von Cerebrosid-schwelfelsaureestern als Sog. 'Pralipoide' und Speichersubstanzen bei der Leukodystrophie, Typ Scholz (metachromatische Form der diffusen Sklerose). *Z. Physiol. Chem.* **311**, 279.

—— and Mehl, E. (1969). Cerebroside sulfatase and arylsulfatase A deficiency in metachromatic leucodystrophy. *J. Neurochem.* **16**, 19.

—— and Stinshoff, K. (1973). An activator of cerebroside sulfatase in human normal liver and in cases of congenital metachromatic leucodystrophy. *FEBS Lett.* **32**, 129.

Jay, B., Carruthers, J., Treplin, M.C.W., and Winder, A.F. (1976). Human albinism. In *The eye and inborn errors of metabolism* (ed. D. Bergsma, A.J. Bron, and E. Cotlier). *Birth defects: original article series XII*, Vol. 3, p.415. Alan R. Liss, New York.

Jeandet, J. and Lestradet, H. (1961). L'hyperlactacidémic cause probable de l'hyperuricémie dans la glycogenose hépatique. *Rev. fr. Etudes. clin. Biol.* **6**, 71.

Jeffcoate, S.L., Brooks, R.V., and Prunty, F.T.G. (1968). Secretion of androgens and oestrogens in testicular feminization: studies *in vivo* and *in vitro* in two cases. *Br. med. J.* **i**, 208.

Jeffcoate, T.N.A., Fliegner, J.R.H., Russell, S.H., Davis, J.C., and Wade, A.P. (1965). Diagnosis of the adrenogenital syndrome before birth. *Lancet* **ii**, 553.

Jellum, E., Kluge, T., Börresen, H.C., Stokke, O., and Eldjarn, L. (1970). Pyroglutamic aciduria — a new inborn error of metabolism. *Scand. J. clin. lab. Invest.* **26**, 327.

Jenner, F.A. and Pollitt, R.J. (1967). Large quantities of 2-acetamido-1(ß-L-aspartamido)-1,2-dideoxyglucose in the urine of mentally retarded siblings. *Biochem. J.* **103**, 48P.

Jensen, J., Blankenhorn, D.H., and Kornerup. V. (1967). Coronary disease in familial hypercholesterolemia. *Circulation* **36**, 77.

Jensen, O.A., Pedersen, C., Schwartz, M., Vestermark, S., and Warburg, M. (1978). Hurler–Scheie phenotype: report of an inbred sibship with tapeto-retinal degeneration and electron-microscopic examination of the conjunctiva. *Ophthalmologica* **176**, 194.

Jepson, J.B. (1971). Hartnup disease. In *Cellular organelles and membranes in mental retardation* (ed. P.F. Benson), p.55. Churchill Livingstone, Edinburgh.

—— (1978). Hartnup disease. In *The metabolic basis of inherited disease*, 4th edn. (ed. J.B. Stanbury, J.B. Wyngaarden, and D.S. Fredrickson), p.1563. McGraw-Hill, New York.

—— Smith, A.J., and Strang, L.B. (1958). An inborn error of metabolism with urinary excretion of hydroxyacids, ketoacids, and amino acids. *Lancet* **ii**, 1334.

Jervis, G.A. (1947). Studies on the phenylpyruvic oligophrenia: the position of the metabolic error. *J. biol. Chem.* **169**, 651.

—— (1953). Phenylpyruvic oligophrenia deficiency of phenylalanine-oxidizing system. *Proc. Soc. exp. Biol. Med.* **82**, 514.

Jeune, M., Francois, R., and Jarlot, B. (1959). Contribution à l'étude des polycories glycogéniques du foie. *Rev. int. Hépatol.* **9**, 1.

—— Collombel, C., Michel, M., David, M., Guibaud, P., Guerrier, G., and Albert, J. (1970). Hyperleucinisoleucinémie par défaut partiel de transamination associée á une hyperprolinémie de type 2. Observation familiale d'une double aminoacidopathie. *Ann. Pediatr.* **17**, 349.

Jim, R.T.S. and Chu, F.K. (1963). Hyperbilirubinemia due to glucose-6-phosphate dehydrogenase deficiency in a newborn Chinese infant. *Pediatrics* **31**, 1046.

Jimbow, K., Fitzpatrick, T.B., Szabo, G., and Hori, Y. (1975). Congenital circumscribed hypomelanosis: a characterization based on electron microscopic study of tuberous sclerosis, nevus depigmentosis and piebaldism. *J. Invest. Dermatol.* **64**, 50.

Jøergensen, S.K. and With, T.K. (1955). Congenital porphyria in swine and cattle in Denmark. *Nature Lond.* **176**, 156.

—— —— (1963). Porphyria in domestic animals. Danish observations in pigs and cattle and comparison with human porphyria. *Ann. N.Y. Acad. Sci.* **104**, 701.

John, E.G., Bhat, R., and Vidyasagar, D. (1980). Neonatal survival after early diagnosis and treatment of argininosuccinic aciduria. *J. Ped.* **97**, 867.

Johnson, D.L. and Desnick, R.J. (1978). Molecular pathology of Fabry's disease: physical and kinetic properties of α-galactosidase A in cultured human endothelial cells. *Biochim. Biophys. Acta* **538**, 195.

—— Del Monte, M.A., Cotlier, E., and Desnick, R.J. (1975). Fabry disease: diagnosis of hemizygotes and heterozygotes by α-galactosidase A activity in tears. *Clin. Chim. Acta* **63**, 81.

Johnson, F. (1956). A method for demonstrating calcium oxalate in tissue sections. *J. Histochem. Cytochem.* **4**, 404.

Johnson, J.D., Kretchmer, N., and Simoons, F.J. (1974). Lactose malabsorption: its biology and history. *Adv. Pediatr.* **31**, 197.

Johnson, J.L., Hainline, B.E., and Rajagopalan, K.V. (1980c). Characterization of the molybdenum cofactor of sulfite oxidase, xanthine oxidase, and nitrate reductase. *J. biol. Chem.* **255**, 1783.

—— Waud, W.R., Rajagopalan, K.V., Duran, M., Beemer, F.A., and Wadman, S.K. (1980d). Inborn errors of molybdenum metabolism: combined deficiencies of sulfite oxidase and xanthine dehydrogenase in a patient lacking the molybdenum cofactor. *Proc. natl. Acad. Sci. USA* **77**, 3715.

Johnson, W.G. and Brady, R.O. (1972). Ceramide trihexosidase from human placenta. In *Methods in enzymology* Vol. 28 (ed. S.P. Colowick and N.O. Kaplan), p.849. Academic Press, New York.

—— Gal, A.E., Miranda, A.F., and Pentchev, P.G. (1980b). Diagnosis of adult Gaucher disease: use of a new chromogenic substrate, 2-hexadecanoylamino-4-nitrophenyl-ß-D-glucopyranoside, in cultured skin fibroblasts. *Clin. Chim. Acta* **102**, 91.

—— Thomas, G.H., Miranda, A.F., Driscoll, J.M., Wigger, J.H., Yeh, M.N., Schwartz, R.C., Cohen, C.S., Berdon, W.E., and Koenigsberger, M.R. (1980a). Congenital sialidosis: biochemical studies: clinical spectrum in four sibs; two successful prenatal diagnoses. *Am. J. hum. Genet.* **32**, 43A.

Johnston, A.W., Warland, B.J., and Weller, S.D.V. (1966). Genetic aspects of angiokeratoma corporis diffusum. *Ann. hum. Genet.* (*Lond.*) **30**, 25.

—— Weller, S.D., and Warland, B.J. (1968). Angiokeratoma corporis diffusum. Some clinical aspects. *Arch. Dis. Child.* **43**, 73.

—— Frost, P., Spaeth, G.L., and Renwich, J.H. (1969). Linkage relationships of the angiokeratoma (Fabry) locus. *Ann. hum. Genet.* (*Lond.*) **32**, 369.

Johnston, S.S. (1978). Homocystinuria. *Ophthalmologica* **176**, 282.

Jolly, R.D. and Desnick, R.J. (1979). Inborn errors of lysosomal catabolism. Principles of heterozygote detection. *Am. J. med. Genet.* **4**, 293.

—— Janmaat, A., and Van de Water, N.S. (1977). Heterozygote detection: a comparative study using neutrophils, lymphocytes, and two reference parameters in the bovine mannosidosis model. *Biochem. Med.* **18**, 402.

—— Van de Water, N.S., Janmaat, A., Slack, P.M., and McKenzie, R.G. (1980). Zinc therapy in the bovine mannosidosis model. In *Enzyme therapy in genetic disease:* 2 (ed. R.J. Desnick), p. 305. Alan R. Liss, New York.

—— Thompson, K.G., Murphy, C.E., Mankelow, B.W., Bruere, A.N., and Winchester, B.G. (1976). Enzyme replacement therapy. An experiment of nature in a chimeric mannosidosis calf. *Ped. Res.* **10**, 219.

Jonas, A. and Schneider, J.A. (1983). Cited by Schneider, J.A. and Schulman, J.D. (1983). In *The metabolic basis of inherited disease*, 5th edn. (ed. J.B. Stanbury, J.B. Wyngaarden, D.S. Fredrickson, J.L. Goldstein, and M.S. Brown), p.1848. McGraw-Hill, New York.

Jones, H.W. and Scott, W.W. (1971). *Hermaphroditism, genital anomalies and related endocrine disorders*, 2nd edn. Williams and Wilkins, Baltimore.

Jones, M.E. (1980). Pyrimidine nucleotide biosynthesis in animals: genes, enzymes and regulation of UMP biosynthesis. *Ann. Rev. Biochem.* **49**, 253.

—— Anderson, A.D., Anderson, C., and Hodes, S. (1961). Citrulline synthesis in rat tissues. *Arch. Biochem. Biophys.* **95**, 499.

Jones, M.Z. and Dawson, G. (1981). Caprine ß-mannosidosis. *J. biol. Chem.* **256**, 5185.

—— and Laine, R.A. (1981). Caprine obligosaccharide storage disease. *J. biol. Chem.* **256**, 5181.

Jongsma, A.P.M., Hagmeijer, J., and Mieralhan, P. (1974). Regional mapping of TPI, DLH-B and PEP-B on chromosome 12 of man. In *Second workshop on human gene mapping, Rotterdam conference*, p.189. National Foundation, Basel.

Jordan, S.W. (1964). Electron microscopy of Gaucher cells. *Exptl. mol. Pathol.* **3**, 76.

Josso, N. (1971). Interspecific character of the Mullerian-inhibiting substance: action of the human fetal testis, ovary and adrenal on the fetal rat mullerian duct in organ culture. *J. clin. Endocr. Metab.* **32**, 404.

—— (1981). Differentiation of the genital tract: stimulators and inhibitors. In *Mechanisms of sex differentiation in animals and man* (ed. C.R. Austin, and R.G. Edwards), p.165. Academic Press, London.

Jost, A. (1953). Problems of fetal endocrinology: the gonadal and hypophyseal hormones. *Recent Progr. Horm. Res.* **8**, 379.

—— (1961). The role of fetal hormones in prenatal development. *Harvey Lect.* **55**, 201.

—— (1970). Hormonal factors in the sex differentiation of the mammalian foetus. *Phil. Trans. R. Soc. Lond. B.* **259**, 119.

Junien, C., Vibert, M., Weil, D., Van-Cong, N., and Kaplan, J.C. (1978). Assignment of NADH cytochrome b_5 reductase (DIA_1 locus) to human chromosome 22. *Hum. Genet.* **42**, 233.

—— Leroux, A., Lostanlen, D., Reghis, A., Boué, J., Nicholas, H., Boué, A., and Kaplan, J.C. (1981). Prenatal diagnosis of congenital enzymopenic methaemoglobinaemia with mental retardation due to generalized cytochrome b_5 reductase deficiency: first report of two cases. *Prenatal Diagnosis.* **1**, 17.

Justice, P., Ryan, C., Hsia, D. Y.-Y., and Krmpotik, E. (1970). Amylo-1,6-glucosidase in human fibroblasts: studies in type III glycogen storage disease. *Biochem. biophys. Res. Commun.* **39**, 301.

—— Wenger, D.A., Naidu, S., and Rosenthal, I.M. (1977). Enzymatic studies in a new variant of GM_1-gangliosidosis in an older child. *Ped. Res.* **11**, 407.

Kaback, M. (1981). *Summary of worldwide Tay–Sachs disease screening and detection*. University of California, Los Angeles.

—— Percy, A.K., and Kasselberg, A.G. (1972). *In vitro* studies in sulfatide lipidosis. In *Sphingolipids, sphingolipidoses and allied disorders* (*Adv. exp. med. Biol.* **19**) (ed. B.W. Volk and S.M. Aronson), p.451. Plenum Press, New York.

—— Zeiger, R.S., Reynolds, L.W., and Sonneborn, M. (1973a). Tay–Sachs disease. A model for the control of recessive genetic disorders. In *Proceedings of the 4th international conference on birth defects* (ed. A.G. Motulsky and W. Lenz), p.248. Excerpta Medica, Amsterdam.

—— —— —— —— (1974). Approaches to the control and prevention of Tay–Sachs disease. In *Progress in medical genetics* (ed. A.G. Steinberg and A.G. Bearn), p.103. Grune and Stratton, New York.

—— Sloan, H.R., Sonneborn, M., Herndon, R.M., and Percy, A.K. (1973b).

GM₁-gangliosidosis type I: in utero detection and fetal manifestations. *J. Ped.* **82**, 1037.

—— Miles, J., Yaffe, M., Itibashi, H., McIntyre, H., Goldberg, M., and Mohandas, T. (1978). Hexosaminidase A (HexA) deficiency in early adulthood: a new type of GM₂-gangliosidosis. *Am. J. hum. Genet.* **30**, 31A.

Kaebling, R., Craig, J.B., and Pasamanick, B. (1961). Urinary porphobilinogen: results of screening 2500 psychiatric patients. *Arch. Gen. Psychiatry* **5**, 494.

Kahlke, W. and Richterich, R. (1965). Refsum's disease (heredopathia atactica polyneuritiformis), an inborn error of lipid metabolism with storage of 3,7,11,15-tetramethyl hexadecanoic acid. II. Isolation and identification of the storage product. *Am. J. Med.* **39**, 237.

Kahn, A., Buc, H.A., Girot, R., Cottreau, D., and Griscelli, C. (1978). Molecular and functional anomalies in two new mutant glucose-phosphate-isomerase variants with enzyme deficiency and chronic hemolysis. *Hum. Genet.* **40**, 293.

Kaibara, H., Eguchi, M., Shibata, K., and Takagishi, K. (1979). Hurler–Scheie phenotype: a report of two pairs of inbred sibs. *Hum. Genet.* **53**, 37.

Kaiser-Kupfer, M.I., Valle, D., and Bron, A.J. (1980*a*). Clinical and biochemical heterogeneity in gyrate atrophy. *Am. J. Ophthalmol.* **89**, 219.

—— De Monasterio, F.M., Valle, D., Walser, M., and Brusilow, S. (1980*b*). Gyrate atrophy of the choroid and retina: improved visual function following reduction of plasma ornithine by diet. *Science* **210**, 1128.

—— —— —— —— —— (1981). Visual results of a long-term trial of a low-arginine diet in gyrate atrophy of the choroid and retina. *Ophthalmology* **88**, 307.

Kajii, T., Matsuda, I., Ohsawa, T., Katsunuma, H., Ichida, T., and Arashima, S. (1974). Hurler–Scheie genetic compound (mucopolysaccharidosis IH-1S) in Japanese brothers. *Clin. Genet.* **6**, 394.

Kakimoto, Y., Taniguchi, K., and Sano, I. (1969). D-ß-aminoisobutyrate: pyruvate aminotransferase in mammalian liver and excretion of ß-aminoisobutyrate by man. *J. biol. Chem.* **244**, 335.

Kalckar, H.M., Anderson, E.P., and Isselbacher, K.J. (1956). Galactosemia, a congenital defect in a nucleotide transferase. *Biochim. Biophys. Acta* **20**, 262.

Kaloud, H., Sitzmann, F.C., Ayer, R., and Palinuf, F. (1973). Klinische und biochemische Befunde bei einem Kleinkind mit hereditaren Galakto-kimasdefekt. *Klin. Padiatr.* **185**, 18.

Kalow, W. and Staron, N. (1957). On the distribution and inheritance of atypical forms of human serum cholinesterase as indicated by dibucaine numbers. *Can. J. Biochem. Physiol.* **35**, 1305.

Kamaryt, J., Mrskos, A., Podhradska, O., Kolcova, V., Cabalska, B., Duczynska, N., and Borzymowska, J. (1978). PKU locus: genetic linkage with human amylase (Amy) loci and assignment to linkage Group 1. *Hum. Genet.* **43**, 205.

Kamisaka, K., Listowsky, I., Gatmaitan, Z., and Arias, I.M. (1975). Interactions of bilirubin and other ligands with ligandin. *Biochemistry* **14**, 2175.

Kampine, J.P., Brady, R.O., Kanfer, J.N., Feld, M., and Shapiro, D. (1967). Diagnosis of Gaucher's disease and Niemann–Pick disease with small samples of venous blood. *Science* **155**, 86.

Kan, Y.W. and Dozy, A.M. (1978). Polymorphism of DNA sequence adjacent to human α-globin structural gene: relationship to sickle mutation. *Proc. natl. Acad. Sci. USA* **75**, 5631.

—— Lee, K.Y., Furbetta, M., Angius, A., and Cao, A. (1980). Polymorphism of DNA sequence in the ß-globin gene region. *New Engl. J. Med.* **302**, 185.

Kang, A.H. and Trelstad, R.L. (1973). A collagen defect in homocystinuria. *J. clin. Invest.* **52**, 2571.

Kang, E.S., Snodgrass, P.J., and Gerald, P.S. (1972). Methylmalonyl coenzyme A racemase defect: another cause of methylmalonic aciduria. *Ped. Res.* **6**, 875.

—— —— —— (1973). Ornithine transcarbamylase deficiency in the newborn infant. *J. Ped.* **82**, 642.

Kanno, T., Sudo, K., Takeuchi, I., Kanda, S., Honda, N., Nishimura, Y., and Oyama, K. (1980). Hereditary deficiency of lactate dehydrogenase M-subunit. *Clin. Chim. Acta* **108**, 267.

Kano, I. and Yamakawa, T. (1974). The properties of α-galactosidase remaining in kidney and liver of patients with Fabry's disease. *Chem. Phys. Lipids* **13**, 283.

Kanwar, Y.S., Manaligod, J.G., and Wong, P.W.K. (1976). Morphologic studies in a patient with homocystinuria due to 5,10-methylenetetrahydrofolate reductase deficiency. *Ped. Res.* **10**, 598.

Kaplan, A., Achord, D.T., and Sly, W.S. (1977). Phosphohexosyl components of a lysosomol enzyme are recognized by pinocytosis receptors on human fibroblasts. *Proc. natl. Acad. Sci. USA* **74**, 2026.

Kaplan, B.H. (1971). δ-aminolevulinic acid synthetase from the particulate fraction of liver of porphyric rats. *Biochim. Biophys. Acta.* **235**, 381.

Kaplan, D. (1969). Classification of the mucopolysaccharidoses based on the pattern of mucopolysacchariduria. *Am. J. Med.* **47**, 729.

—— McKusick, V., Trebach, S., and Lazarus, R. (1968). Keratosulfate-chondroitin sulfate peptide from normal urine and from urine of patients with Morquio syndrome (mucopolysaccharidosis IV). *J. lab. clin. Med.* **71**, 48.

Kaplan, J.C. and Beutler, E. (1967). Electrophoresis of red-cell NADH- and NADPH-diaphorases in normal subjects and patients with congenital methemoglobinemia. *Biochem. biophys. Res. Commun.* **29**, 605.

—— Leroux, A., and Beauvais, P. (1979). Formes cliniques et biologiques du déficit en cytochrome b₅ reductase. *C. r. Séanc. Soc. Biol.* **173**, 368.

Kaplowitz, N., Javitt, N., and Harber, L.C. (1968). Isolation of erythrocytes with normal protoporphyrin levels in erythropoietic protoporphyria. *New Engl. J. Med.* **278**, 1077.

Kark, R.A.P. and Rodriguez-Budelli, M. (1977). The spectrum of ataxia syndromes due to lipoamide dehydrogenase deficiency. *Neurology* **27**, 359.

—— —— (1979). Pyruvate dehydrogenase deficiencies in six of fourteen unselected patients with spinocerebellar degenerations. *Neurology* **29**, 126.

—— Blass, J.P., and Engel, W.K. (1974). Pyruvate oxidation in neuromuscular disease: evidence of a genetic defect in two families with the clinical syndrome of Friedreich's ataxia. *Neurology* **24**, 964.

—— Engel, W.K., Blass, J.P., Steinberg, D., and Walsh, G.O. (1971). Heredopathia atactica polyneuritiformis (Refsum's disease): a second trial of dietary therapy in two patients. *Birth defects: Original article series VII*, **1**, 53.

Karpati, G., Carpenter, S., Eisan, A.A., Wolfe, L.S., and Feindel, W. (1974). Multiple peripheral nerve entrapments: an unusual phenotypic variant of the Hunter syndrome (mucopolysaccharidosis II) in a family. *Arch. Neurol.* **31**, 418.

—— —— Engel, A.G., Watters, G., Allen, J., Rothman, S., Klassen, G., and Mamer, O.A. (1975). The syndrome of systemic carnitine deficiency. *Neurology* **25**, 16.

Kasbekar, D.K., Nagabhushanam, A., and Greenberg, D.M. (1964). Purification and properties of orotic acid-decarboxylating enzymes from calf thymus. *J. biol. Chem.* **239**, 4245.

Käser, H., Cottier, P., and Antener, I. (1962). Glycoglycinuria, a new familial syndrome. *J. Ped.* **61**, 386.

Kato, T. (1977). Renal handling of dibasic amino acids and cystine in cystinuria. *Clin. Sci. mol. Med.* **53**, 9.

Kattamis, C.A., Kyriazakou, M., and Chaidas, S. (1969). Favism. Clinical and biochemical data. *J. med. Genet.* **6**, 34.

Kaufman, J.M., Greene, M.L., and Seegmiller, J.E. (1968). Urine uric acid to creatinine ratio — a screening test for inherited disorders of purine metabolism. *J. Ped.* **73**, 583.

Kaufman, S. (1957). The enzymatic conversion of phenylalanine to tyrosine. *J. biol. Chem.* **226**, 511.

—— (1959). Studies on the mechanisim of the enzymatic conversion of phenylalanine to tyrosine. *J. biol. Chem.* **234**, 2677.

—— (1971). The phenylalanine hydroxylating system from mammalian liver. *Adv. Enzymol.* **35**, 245.

—— (1976). Phenylketonuria: biochemical mechanisms. In *Advances in neurochemistry* Vol. 2 (ed. B.W. Agranoff and M.H. Aprison), p.1. Plenum Press, New York and London.

—— Holtzman, N.A., Milstien, S., Butler, I.J., and Krumholz, A. (1975). Phenylketonuria due to a deficiency of dihydropteridine reductase. *New Engl. J. Med.* **293**, 785.

—— Berlow, S., Summer, G.K., Milstien, S., Schulman, J.D., Orloff, S., Spielberg, S., and Pueschel, S. (1978). Hyperphenylalaninemia due to a deficiency of biopterin. A variant form of phenylketonuria. *New Engl. J. Med.* **299**, 673.

Kawamura, N., Moser, A.B., Moser, H.W., Ogino, T., Suzuki, K., Schaumburg, H., Milunsky, A., Murphy, J., and Kishimoto, Y. (1978). High concentration of hexacosanoate in cultured skin fibroblasts of lipids from adrenoleukodystrophy patients. *Biochem. biophys. Res. Commun.* **82**, 114.

Kaye, C.I. and Nadler, H.L. (1976). Transport of L-cystine by cultivated skin fibroblasts of normal subjects and patients with cystinosis. *Ped. Res.* **10**, 637.

Kaziro, Y. and Ochoa, S. (1964). The metabolism of propionic acid. *Adv. Enzymol.* **26**, 283.

Kazy, Z., Rozovsky, I.S., and Bakharev, V.A. (1982). Chorion biopsy in early pregnancy: a method of early prenatal diagnosis for inherited disorders. *Prenatal Diagnosis.* **2**, 39.

Keating, J.P., Feigin, R.D., Tenenbaum, S.M., and Hillman, R.E. (1972). Hyperglycinemia with ketosis due to a defect in isoleucine metabolism: a preliminary report. *Pediatrics* **50**, 890.

Keeler, C.E. (1953). The Caribe Cuna moon-child and its heredity. *J. Hered.* **44**, 163.

Keenan, B.S., Meyer, W.J. III, Hadjian, A.J., Jones, H.W., and Migeon, C.J. (1974). Syndrome of androgen insensitivity in man: absence of α-dihydrotestosterone binding protein of skin fibroblasts. *J. clin. Endocr. Metab.* **38**, 1143.

Keeton, B.R. and Moosa, A. (1976). Organic aciduria. Treatable cause of floppy infant syndrome. *Arch. Dis. Child.* **51**, 636.

Keitt, A.S. (1966). Pyruvate kinase deficiency and related disorders of red cell glycolysis. *Am. J. Med.* **41**, 762.

—— (1969). Hemolytic anemia with impaired hexokinase activity. *J. clin. Invest.* **48**, 1997.

—— Smith, T.W., and Jandl, J.H. (1966). Red cell 'pseudomosaicism' in congenital methemoglobinemia. *New Engl. J. Med.* **275**, 398.

Kekomäki, M., Visakorpi, J.K., Perheentupa, J., and Saxen, L. (1967). Familial protein intolerance with deficient transport of basic amino acids: an analysis of 10 patients. *Acta Paediatr. Scand.* **56**, 617.

Kellerman, G. and Walter, H. (1972). On the population genetics of the ceruloplasmin polymorphism. *Hum. Genet.* **15**, 84.

Kelley, W.N. and Meade, J.C. (1971). Studies on hypoxanthine-guanine phosphoribosyltransferase in fibroblasts from patients with the Lesch–Nyhan syndrome. Evidence for genetic heterogeneity. *J. biol. Chem.* **246**, 2953.

—— Fox, I.H., and Wyngaarden, J.B. (1970). Essential role of phosphoribosyl pyrophosphate (PRPP) in regulation of purine biosynthesis in cultured human fibroblasts. *Clin. Res.* **18**, 457.

—— Greene, M.L., Rosenbloom, F.M., Henderson, J.F., and Seegmiller, J.E. (1969). Hypoxanthine-guanine phosphoribosyltransferase deficiency in gout. *Ann. intern. Med.* **70**, 155.

—— Rosenbloom, F.M., Henderson, J.F., and Seegmiller, J.E. (1967). A specific enzyme defect in gout associated with overproduction of uric acid. *Proc. natl. Acad. Sci. USA* **57**, 1735.

—— —— Seegmiller, J.E., and Howell, R.R. (1968). Excessive production of uric acid in type I glycogen storage disease. *J. Ped.* **72**, 488.

Kelly, S. and Bakhru-Kishore, R. (1979). Fluorimetric assay of acid lipase in human leucocytes. *Clin. Chim. Acta* **97**, 239.

Kelly, T.E. and Graetz, G. (1977). Isolated acid neuraminidase deficiency: a distinct lysosomal storage disease. *Am. J. med. Genet.* **1**, 31.

—— Reynolds, L.W., and O'Brien, J.S. (1976). Segregation within a family of two mutant alleles for hexosaminidase A. *Clin. Genet.* **9**, 540.

—— Thomas, G.H., Taylor, H.A., McKusick, V.A., Sly, W., Glaser, J.H., Robinow, M., Luzzatti, L., Espiritu, C., Feingold, M., Bull, M.J., Ashenhurst, E.M., and Ives, E.J. (1975). Mucolipidosis III (Pseudo-Hurler polydystrophy): clinical and laboratory studies in a series of 12 patients. *Johns Hopkins med. J.* **137**, 156.

Kennaway, N.G. and Buist, N.R.M. (1971). Metabolic studies in a patient with hepatic cytosol tyrosine aminotransferase deficiency. *Ped. Res.* **5**, 287.

—— and Curtis, H.C. (1981). Complementation analysis in fibroblasts from eight patients with clinically different forms of citrullinaemia. *J. inher. metab. Dis.* **4**, 23.

—— Weleber, R.G., and Buist, N.R.M. (1980). Gyrate atrophy of the choroid and retina with hyperornithinemia: biochemical and histologic studies and response to vitamin B_6. *Am. J. hum. Genet.* **32**, 529.

—— Harwood, P.J., Ramberg, D.A., Koler, R.D., and Buist, N.R.M. (1975). Citrullinemia: enzymatic evidence for genetic heterogeneity. *Ped. Res.* **9**, 554.

Kennedy, P., Swash, M., and Dean, M. (1973). Cervical cord compression in mucopolysaccharidosis. *Dev. med. Child. Neurol.* **15**, 194.

Kenyon, K.R. and Sensenbrenner, J.A. (1971). Mucolipidosis II (I-cell disease): ultrastructural observations of conjunctiva and skin. *Invest. Ophthalmol.* **10**, 555

—— —— (1974). Electron microscopy of cornea and conjunctiva in childhood cystinosis. *Am. J. Ophthalmol.* **78**, 68.

Kerr, C.B. and Wells, R.S. (1965). Sex-linked ichthyosis. *Ann. Hum. Genet.* **29**, 33.

Kerry, K.R. and Townleuy, R.R.W. (1965). Genetic aspects of intestinal sucrase-isomaltase deficiency. *Aust. Paediatr. J.* **1**, 223.

Khachadurian, A.K. (1967). Persistent elevation of the erythrocyte sedimentation rate (ESR) in familial hypercholesterolemia. *J. Med. Liban* **20**, 31.

—— (1968). Migratory polyarthritis in familial hypercholesterolemia (type II hyperlipoproteinemia). *Arthr. Rheum.* **11**, 385.

—— (1971). A general view of clinical and laboratory features of familial hypercholesterolemia (type II hyperbetalipoproteinemia). *Protides biol. Fluids* **19**, 315.

—— and Abu Feisal, K. (1958). Alkaptonuria: report of a family with seven cases appearing in four successive generations with metabolic studies in one patient. *J. chron. Dis.* **7**, 455.

—— and Uthman, S.M. (1973). Experiences with the homozygous cases of familial hypercholesterolemia. A report of 52 patients. *Nutr. Metab.* **15**, 132.

Kieffer, N.M., Burns, S.J., and Judge, N.G. (1976). Male pseudohermaphroditism of the testicular feminizing type in a horse. *Equine. vet. J.* **8**, 38.

Kihara, H., Valente, M., Porter, M.T., and Fluharty, A.L. (1973*b*). Hyperdibasic-aminoaciduria in a mentally retarded homozygote with a peculiar response to phenothiazines. *Pediatrics* **5**, 223.

—— Ho, C.K., Fluharty, A.L., Tsay, K.K., and Hartlage, P.L. (1980). Prenatal diagnosis of metachromatic leucodystrophy in a family with pseudo arylsulfatase A deficiency by the cerebroside sulfate loading test. *Ped. Res.* **14**, 224.

—— Porter, M.T., Fluharty, A.L., Scott, M.L., De la Flor, S.D., Trammell, J.L., and Nakamura, R.N. (1973*a*). Metachromatic leukodystrophy: ambiguity of heterozygote identification. *Am. J. ment. Defic.* **77**, 389.

Kiil, R. and Rokkones, T. (1964). Late manifesting variant of branched-chain ketoaciduria (maple syrup urine disease). *Acta Paediatr. Scand.* **53**, 356.

Kim, C.W., Okada, A., Konishi, H., and Nakajima, T. (1978). Urinary excretion of amino acids in patients receiving intravenous hyperalimentation. *Clin. Chim. Acta* **83**, 151.

Kim, Y.J. and Rosenberg, C.E. (1974). On the mechanism of pyridoxine responsive homocystinuria. II. Properties of normal and mutant cystathionine ß-synthase from cultured fibroblasts. *Proc. natl. Acad. Sci. USA* **71**, 4821.

King, R.A. and Witkop, C.J. jun. (1976). Hairbulb tyrosinase activity in oculocutaneous albinism. *Nature, Lond.* **263**, 69.

—— and Witkop, C.J. (1977). Detection of heterozygotes for tyrosinase-negative oculocutaneous albinism by hairbulb tyrosinase assay. *Am. J. hum. Genet.* **29**, 164.

—— Creel, D., Cervenka, J., Okoro, A.N., and Witcop, C.J (1980). Albinism in Nigeria with delineation of new recessive oculocutaneous type. *Clin. Genet.* **17**, 259.

Kinoshita, J.H. (1965). Cataracts in galactosemia. *Invest. Ophthalmol.* **4**, 786.

—— Dvornik, D., Kraml, M., and Gabbay, K. (1968). The effect of an aldose reductase inhibitor on the galactose-exposed rabbit lens. *Biochim. Biophys. Acta* **158**, 472.

Kint, J.A. (1970). Fabry's disease, α-galactosidase deficiency. *Science* **167**, 1268.

—— (1971). On the existence and the enzymic interconversion of the isoenzymes of α-galactosidase in human organs. *Arch. int. Physiol.* **79**, 633.

—— (1973). Antagonistic action of chondroitin sulphate and cetylpyridinium chloride on human ß-galactosidase. *FEBS Lett.* **36**, 53.

—— (1974). *In vitro* restoration of deficient ß-galactosidase activity in liver of patients with Hurler and Hunter disease. *Nature, Lond.* **250**, 424.

—— and Carton, D. (1968). Deficient argininosuccinase activity in brain in argininosuccinate aciduria. *Lancet* **ii**, 635.

—— —— (1971). New evidence for the identity of homoserine deaminase and cystathionase in human liver. *Arch. int. Physiol. Biochem.* **79**, 202.

—— Dacremont, G., Carton, D., Orye, E., and Hooft, C. (1973). Mucopoly-saccharidosis: secondarily induced abnormal distribution of lysosomal iso-enzymes. *Science* **181**, 352.

Kirkland, R.T., Kirkland, J.L., Johnson, C., Horning, M., Librick, L., and Clayton, G.W. (1973). Congenital lipoid adrenal hyperplasia in an eight-year-old phenotypic female. *J. clin. Endocr. Metab.* **36**, 488.

Kirkman, H.N., Simon, E.R., and Pickard, B.M. (1965). Seattle variant of glucose-6-phosphate dehydrogenase. *J. lab. clin. Med.* **66**, 834.

Kistler, J.P., Lott, I.T., Kolodny, E.H., Friedman, R.B., Nersasian, R., Schnur, J., Mihm, M.C., Dvorak, A.M., and Dickersin, R. (1977). Mannosidosis: new clinical presentation, enzyme studies and carbohydrate analysis. *Arch. Neurol.* **34**, 45.

Kitagawa, T., Owada, N., Sakiyama, T., Aoki, K., Kamoshita, S., Amenomori, T., and Kobayashi, T. (1978). *In utero* diagnosis of Gaucher disease. *Am. J. hum. Genet.* **30**, 322.

Kitamura, M., Iijima, N., Hasimoto, F., and Hiratsuka, A. (1971). Hereditary deficiency of subunit H of lactate dehydrogenase. *Clin. Chim. Acta* **34**, 419.

Kitao, T., Sugita, Y., Yoneyama, Y., and Hattori, K. (1974). Methemoglobin reductase (cytochrome b_5 reductase) deficiency in congenital methemo-globinemia. *Blood* **44**, 879.

Kivirikko, K.I. and Myllyla, R. (1979). Collagen glycosyltransferases. In *International review of connective tissue research* Vol. 8 (ed. D.A. Hall and D.S. Jackson), p.23. Academic Press, New York.

Kizaki, H. and Sakurada, T. (1976). A micro-assay method for hypoxanthine–guanine and adenine phosphoribosyltransferases. *Anal. Biochem.* **72**, 49.

Kjellman, B., Gamstorp, J., Brun, A., Öckerman, P.A., and Palmgren, B. (1969). Mannosidosis: a clinical and histopathologic study. *J. Ped.* **75**, 366.

Klebanoff, S.J. (1968). Myeloperoxidase-halide-hydrogen peroxide antibacterial system. *J. Bacteriol.* **95**, 2131.

Kleijer, W.J., Niermeijer, M.F., and Galjaard, H. (1978). Prenatal diagnosis of genetic metabolic diseases in 118 pregnancies at risk. *Monogr. hum. Genet.* **9**, 217.

—— Van der Veer, E., and Niermeijer, M.F. (1976b). Rapid prenatal diagnosis of GM_1-gangliosidosis using microchemical methods. *Hum. Genet.* **33**, 299.

—— Wolffers, G.M., Hoogeveen, A., and Niermeijer, M.F. (1976a). Prenatal diagnosis of Maroteaux-Lamy disease. *Lancet* ii, 50.

—— Thoomes, R., Galjaard, H., Wendel, U., and Fowler, B. (1984b). First trimester (chorion biopsy) diagnosis of citrullinaemia and methylmalonicaciduria. *Lancet* ii, 1340.

—— van Diggelen, O.P., Janse, H.C., Galjaard, H., Dumez, Y., and Boué, J. (1984a). First trimester diagnosis of Hunter syndrome on chorionic villi. *Lancet* ii, 472.

—— Hoogeveen, A., Verheyen, F.W., Niermeijer, M.F., Galjaard, H., O'Brien, J.S., and Warner, T.S. (1979). Prenatal diagnosis of sialidosis with combined neuraminidase and ß-galactosidase deficiency. *Clin. Genet.* **16**, 60.

Klein, U., Kresse, H., and Von Figura, K. (1978). Sanfilippo syndrome type C: deficiency of acetyl-CoA: α-glucodaminide *N*-acetyltransferase in skin fibro-blasts. *Proc. natl. Acad. Sci. USA*. **75**, 5185.

Kleinman, P., Winchester, P., and Volbert, F. (1976). Sulfatide cholecystosis. *Gastrointestinal Radiology* **1**, 99.

Klenk, E. and Kahlke, W. (1963). Über das Vorkommen der 3,7,11,15-tetramethylhexadecansäure (Phytansäure) in den Cholesterinestern und andern

Lipoidfraktionen der Organe bei einem Krankheitsfall unbekannter Genese (verdacht auf Refsum Syndrom). *Hoppe-Seylers' Z. Physiol. Chem.* **333**, 133.

Klujber, L., Cholnoky, P., and Méhes, K. (1969). Urinary excretion of beta-aminoisobutyric acid in Down's syndrome and in idiopathic retardation. *Hum. Hered.* **19**, 567.

Knapp, A. (1960). Über eine neue, hereditäre, von Vitamin-B$_6$ abhängige Störung im Tryptophan-Stoffwechsel. *Clin. Chim. Acta* **5**, 6.

Knorr, D., Bidlingmaier, F., and Engelhardt, D. (1973). Reifenstein's syndrome, a 17ß-hydroxysteroid-oxyreductase deficiency? *Acta Endocrinol. (Kbh.) suppl.* **173**, 37.

—— —— Butenandt, O., Schnakenburg, K.V., and Wagner, W. (1977). Test for heterozygosity of congenital adrenal hyperplasia. In *Congenital adreal hyperplasia* (ed. P.A. Lee, L.P. Plotnick, A.A. Kowarski, and C.J. Migeon), p.495. University Park Press, Baltimore.

Knox, W.E. (1958). Sir Archibald Garrod's 'inborn errors of metabolism'. II. Alkaptonuria. *Am. J. hum. Genet.* **10**, 95.

—— and Lemay-Knox, M. (1951). The oxidation in liver of L-tyrosine to acetoacetate through *p*-hydroxyphenylpyruvate and homogentisic acid. *Biochem. J.* **49**, 686.

Knudson, A.G. jun., Di Ferrante, N., and Curtis, J.E. (1971). Effect of leucocyte transfusion in a child with type II mucopolysaccharidosis. *Proc. natl. Acad. Sci. USA* **68**, 1738.

Knudson, K.B., Sparberg, M., and Lecocq, F. (1977). Porphyria precipitated by fasting. *New Engl. J. Med.* **277**, 350.

—— Welsh, J.D., Kronenberg, R.S., Vanderveen, J.E., and Heidelbaugh, N.D. (1968). Effect of a non-lactose diet on human intestinal disaccharidase activity. *Am. J. dig. Dis.* **13**, 593.

Kobayashi, T., Scaravilli, F., and Suzuki, F. (1980*a*). Biochemistry of twitcher mouse: an authentic murine model of human globoid cell leukodystrophy. In *Neurological mutations affecting myelination* (ed. N. Baumann), p.253. Elsevier, Amsterdam.

—— Nagara, H., Suzuki, K., and Suzuki, K. (1982). The twitcher mouse: determination of genetic status by galactosylceramidase assays on clipped tail. *Biochem. Med.* **27**, 8.

—— Ohta, M., Goto, I., Tanaka, Y., and Kuroiwa, Y. (1979). Adult type mucolipidosis with ß-galactosidase and sialidase deficiency: histological and biochemical studies. *J. Neurol.* **221**, 137.

—— Yamanaka, T., Jacobs, J., Teixeira, F., and Suzuki, K. (1980*b*). The twitcher mouse: an enzymatically authentic model of human globoid cell leukodystrophy (Krabbe disease). *Brain. Res.* **202**, 479.

Koch, J., Stokstad, E.L.R., Williams, H.E., and Smith, L.H. jun. (1967). Deficiency of 2-oxoglutarate: glyoxylate carboligase activity in primary hyperoxaluria. *Proc. natl. Acad. Sci. USA* **57**, 1123.

Koch, R., Blaskovics, M., Wenz, E., Fishler, K., and Schaeffler, G. (1974). Phenylalaninemia and phenylketonuria. In *Heritable disorders of amino acid metabolism* (ed. W.L. Nyhan), p.109. John Wiley and Sons, New York.

Kodama, H., Ishimoto, Y., Shimomura, M., Hirota, T., and Ohmori, S. (1975). Isolation of two new sulfur-containing amino acids from the urine of a cystathioninuric patient. *Physiol. Chem. Phys.* **7**, 147.

—— Umemura, S., Shimomura, M., Mizuhara, S., Arata, J., Yamamoto, Y., Yasutake, A., and Izumiya, N. (1976*a*). Studies on a patient with iminopeptiduria. I. Identification of urinary iminopeptides. *Physiol. Chem. Phys.* **8**, 463.

Kodama, J. (1971). Unusual sulfur-containing amino acids in the urine of homocytinuric patients. II. *S*-(ß-carboxyethyl) homocysteine, *S*-(ß-carboxyethylthio) homocysteine and *S*-(ß-carboxyethylthio) cysteine. *Physiol Chem. Phys.* **3**, 159.

Kodama, S., Seki, A., Hanabusa, M., Morisita, Y., Sakurai, T., and Matsuo, T. (1976*b*). Mild variant of maple syrup urine disease. *Eur. J. Pediatr.* **124**, 31.

Koepp, P. and Hoffman, B. (1975). Detection of heterozygotes for phenylketonuria and hyperphenylalaninemia by gas-chromatographic analysis of aromatic acid excretion in urine. *Clin. Chim. Acta* **58**, 215.

Koff, E., Kammerer, B., Boule, P., and Pueschel, S.M. (1979). Intelligence and phenylketonuria: effects of diet termination. *J. Ped.* **94**, 534.

Köhlin, P. and Mehlin, K. (1968). Hereditary fructose intolerance in four Swedish families. *Acta. Paediatr. Scand.* **57**, 24.

Kohn, G., Livini, N., Ornoy, A., Sekeles, E., Beyth, Y., Legum, C., Bach, G., and Cohen, M.M. (1977). Prenatal diagnosis of mucolipidosis IV by electron microscopy. *J. Ped.* **90**, 62.

Kolhouse, J.F. and Allen, R.H. (1977). Recognition of two intracellular cobalamin binding proteins and their identification as methylmalonyl-CoA mutase and methionine synthetase. *Proc. natl. Acad. Sci. USA* **74**, 921.

Kolodny, E.H. and Mumford, R.A. (1976). Arylsulfatases A and B in metachromatic leukodystrophy and Maroteaux-Lamy syndrome: studies with 4-methylumbelliferyl sulfate. In *Current trends in sphingolipidosis and allied disorders (Adv. exp. med. Biol.* **68**) (ed. B.W. Volk and L. Schneck), p.239. Plenum Press, New York.

—— Adams, R.D., Haller J.S., Joseph, J., Crumbrine, P.K., and Raghavan, S.S. (1980). Late onset globoid cell leukodystrophy. *Ann. Neurol.* **8**, 219.

—— Wald, I., Moser, H.W., Cogan, D.G., and Kubuwara, T. (1973). GM$_2$-gangliosidosis without deficiency in the artificial substrate cleaving activity of hexosaminidase A and B. *Neurology* **23**, 427.

Kølvraa, S. (1979). Inhibition of the glycine cleavage system by branched-chain amino acid metabolites. *Ped. Res.* **13**, 889.

—— Rasmussen, K., and Brandt, N.J. (1976). D-glyceric acidemia: biochemical studies of a new syndrome. *Ped. Res,* **10**, 825.

Komrower, G.M. (1973). Treatment of galactosaemia. In *Treatment of inborn errors of metabolism* (ed. J.W.T. Seakins, R.A. Saunders, and C. Toothill), p.113. Churchill Livingstone, Edinburgh.

—— and Lee, D.H. (1970). Long-term follow-up of galactosaemia. *Arch. Dis Child.* **45**, 367.

—— and Robins, A.J. (1969). Plasma amino acid disturbance in infancy. I. Hypermethioninaemia and transient tyrosinaemia. *Arch. Dis. Child.* **44**, 418.

—— and Sardharwalla, I.B. (1971). The dietary treatment of homocystinuria. In *Inherited disorders of sulphur metabolism* (ed. N.A.J. Carson and D.N. Raine), p.254. Churchill Livingstone, Edinburgh.

—— and Westall, R. (1967). Hydroxykynureninuria. *Am. J. Dis. Child.* **113**, 77.

—— and Wilson, V.K. (1963). Homocystinuria. *Proc. R. Soc. Med.* **56**, 996.

—— Sardharwalla, I.B., Coutts, J.M.J., and Ingham, D. (1979). Management of maternal phenylketonuria: an emerging clinical problem. *Br. med. J.* **i**, 1383.

—— Schwarz, V., Holzel, A., and Goldberg, L. (1956). A clinical and biochemical study of galactosaemia. A possible explanation of the nature of the biochemical lesion. *Arch. Dis. Child.* **31**, 254.

—— Wilson, V., Clamp, J.R., and Westall, R.G. (1964). Hydroxykynureninuria.

A case of abnormal tryptophan metabolism probably due to a deficiency of kynureninase. *Arch. Dis. Child.* **39**, 250.

Konrad, P.N., Richards, F. II, Valentine, W.N., and Paglia, D.E. (1972). γ-Glutamyl-cysteine synthetase deficiency. A cause of hereditary hemolytic anemia. *New Engl. J. Med.* **286**, 557.

—— McCarthy, D.J., Mauer, A.M., Valentine, W.N., and Paglia, D.E. (1973). Erythrocyte and leukocyte phosphoglycerate kinase deficiency with neurologic disease. *J. Ped.* **82**, 456.

Kopelman, H., Asatoor, A.M., and Milne, M.D. (1964). Hyperprolinaemia and hereditary nephritis. *Lancet* **ii**, 1075.

Kopits, S.E. (1976). Orthopedic complications of dwarfism. *Clin. Orthop.* **114**, 153.

—— Perovic, M.N., McKusick, V.A., Robinson, R.A., and Bailey, J.A. (1972). Congenital atlantoaxial dislocations in various forms of dwarfism. *J. Bone Joint Surg.* **54A**, 1349.

Koppe, J.G., Marinkovic-Ilsen, A., Rijken, Y., De Groot, W.P., and Jobsis, A.C. (1978). X-linked ichthyosis. A sulfatase deficiency. *Arch. Dis. Child.* **53**, 803.

Kornfeld, M., Snyder, R.D., and Wenger, D.A. (1977). Fucosidosis with angiokeratoma. Electron microscopic changes in the skin. *Arch. Pathol. lab Med.* **101**, 478.

Kornfeld, R. and Kornfeld, S. (1973). Comparative aspects of glycoprotein structure. *Ann. Rev. Biochem.* **45**, 217.

Korth-Schutz, S., Virdis, R., Saenger, P., Chow, D.M., Levine, L.S., and New, M.I. (1978). Serum androgens as a continuing index of adequacy of treatment of congenital adrenal hyperplasia. *J. clin. Endocr. Metab.* **46**, 452.

Kosower, N.D. (1968). Discussion of 'Glutathione deficiency' by H.K. Prins, J.A. Loos, and C. Zürcher. In *Hereditary disorders of erythrocyte metabolism* (ed. E. Beutler) (City of Hope Symposia Series Vol. 1), p.176. Grune and Stratton, New York.

Koster, J.F., Slee, R.G., and Hülsmann, W.C. (1974). The use of leucocytes as an aid in the diagnosis of glycogen storage disease type II (Pompe's disease). *Clin. Chim. Acta* **51**, 319.

—— Fernades, J., Slee, R.G., Van Berkel, T.J.C., and Hulsmann, W.C. (1973). Hepatic phosphorylase deficiency: a biochemical study. *Biochem. biophys. Res. Commun.* **53**, 282.

Kowarski, A., Russell, A., and Migeon, C.J. (1968). Aldosterone secretion rate in the hypertensive form of congenital adrenal hyperplasia. *J. clin. Endocr.* **28**, 1445.

Kozinn, P.J., Wiener, H., and Cohen, P. (1957). Infantile familial amaurotic idiocy. *J. Ped.* **51**, 58.

Krabbe, K. (1916). A new familial, infantile form of diffuse brain sclerosis. *Brain* **39**, 74.

Kraemer, K.H., Coon, H.G., Pettiga, R.A., Barrett, S.F., Rahe, A.E., and Robbins, J.H. (1975). Genetic heterogeneity in xeroderma pigmentosum. Complementation groups and their relationship to DNA repair rates. *Proc. natl. Acad. Sci. USA* **72**, 59.

Kramer, J.W., Davis, W.C., and Prieur, D.J. (1977). Chediak-Higashi syndrome of cats. *Lab. Invest.* **36**, 554.

Kramer, S., Viljoen, E., Mayer, A.M., and Metz, J. (1965). The anemia of erythropoietic porphyria with the first description of the disease in an elderly patient. *Br. J. Haematol.* **11**, 666.

Krane, S.M., Pinnell, S.R., and Erbe, R.W. (1972). Lysyl-protocollagen hydroxylase deficiency in fibroblasts from siblings with hydroxylysine-deficient collagen. *Proc. natl. Acad. Sci. USA* **69**, 2899.

Kranhold, J.F., Loh, D., and Morris, R.C. jun. (1969). Renal fructose metabolizing enzymes: significance in hereditary fructose intolerance. *Science* **165**, 402.

Krauss, R.M., Levy, R.I., and Fredrickson, D.S. (1974). Selective measurement of two lipase activities in postheparin plasma from normal subjects and patients with hyperlipoproteinemia. *J. clin. Invest.* **54**, 1107.

Kredich, N.M. and Hershfield, M.S. (1983). Immunodeficiency diseases caused by adenosine deaminase deficiency and purine nucleotide phosphorylase deficiency. In *The metabolic basis of inherited disease*, 5th ed. (ed. J.B. Stanbury, J.B. Wyngaarden, D.S Fredrickson, J.L. Goldstein, and M.S. Brown), p.1157. McGraw-Hill, New York.

Kresse, H. (1973). Mucopolysaccharidosis III A (Sanfilippo disease): deficiency of heparin sulfamidase in skin fibroblasts and leucocytes. *Biochem. biophys. Res. Commun.* **54**, 1111.

—— and Holtfrederich, D. (1980). Thiosulfate-mediated increase of arylsulfatase activities in multiple sulfatase deficiency disorder fibroblasts. *Biochem. bipohys. Res. Commun.* **97**, 41.

—— Fuchs, W., Glössl, J., Holtfrederich, D., and Gilberg, W. (1981). Liberation of N-acetylglucosamine-6-sulfate by human ß-N-acetylhexosaminidase A. *J. biol. Chem.* **256**,, 12926.

—— Paschke, E., Von Figura, K., Gilberg, W., and Fuchs, W. (1980). Sanfilippo disease type D; deficiency of N-acetylglucosamine-6-sulfate sulfatase required for heparan sulfate degradation. *Proc. natl. Acad. Sci. USA* **77**,6822.

—— Von Figura, K., Klein, U., Glössl, J., Paschke, E., and Pohlmann, R. (1982). Enzymic diagnosis of the genetic mucopolysaccharide storage disorders. In *Methods in enzymology* Vol. 83 (ed. V. Ginsburg), p.559. Academic Press, New York and London.

—— Wiesmann, U., Cantz, M., Hall, C.W., and Neufeld, E.F. (1971). Biochemical heterogeneity of the Sanfilippo syndrome: preliminary characterization of two deficient factors. *Biochem. biophys. Res. Commun.* **42**, 892.

Kretchmer, N., Stone, M., and Bauer, C. (1958). Hereditary enzymatic effects as illustrated by hypophosphatasia. *Ann. N.Y. Acad. Sci.* **75**, 279.

—— Levine, S.Z., McNamar, H., and Barnett, H.L. (1956). Certain aspects of tyrosine metabolism in the young. I. The development of the tyrosine oxidizing system in human liver. *J. clin. Invest.* **35**, 236.

Krieg, T., Kirsch, E., Matzen, K., and Muller, P.K. (1981). Osteogenesis imperfecta: biochemical and clinical evaluation of 13 cases. *Klin. Wochenschr.* **59**, 91.

Krieger, I. and Hart, Z.H. (1974). Valine-sensitive nonketotic hyperglycinemia. *J. Ped.* **85**, 43.

—— and Tanaka, K. (1976). Therapeutic effects of glycine in isovaleric acidemia. *Ped. Res.* **10**, 25.

Krivit, W., Desnick, R.J., Lee, J., Moller, J., Wright, F., Sweeley, C.C., Snijder, P.D., and Sharp, H.L. (1972). Generalized accumulation of neutral glycosphingolipids with GM_2-ganglioside accumulation in the brain. *Am. J. Med.* **52**, 763.

Krnjević, K. (1965). Actions of drugs on single neurones in the cerebral cortex. *Br. med. Bull.* **21**, 10.

Kroll, W.A. and Lichte, K.-H. (1973). Cystinosis: a review of the different forms and of recent advances. *Humangenetik* **20**, 75.

—— and Schneider, J.A. (1974). Decrease in free-cystine content of cultured cystinotic fibroblasts by ascorbic acid. *Science* **186**, 1040.

—— Becker, F.L.A., and Schneider, J.A. (1974). Measurement of intracellular aminoacids in cultured skin fibroblasts. *Biochem. Med.* **10**, 368.

Krooth, R.S. (1965). Properties of diploid cell strains developed from patients with an inherited abnormality of uridine biosynthesis. Cold Spring Harbor Symposium on quantitative biology, *Hum. Gen.* **29**, 189.

—— (1967). Some properties of diploid cell strains developed from the tissues of patients with inherited biochemical disorders. *In Vitro* **2**, 82.

—— and Weinberg, A.N. (1961). Studies on cell lines developed from the tissues of patients with galactosemia. *J. exp. Med.* **113**, 1155.

—— Howell, R.R., and Hamilton, H.B. (1962). Properties of acatalasic cells growing *in vitro*. *J. exp. Med.* **115**, 313.

Krovetz, L.J., Lorincz, A.E., and Schiebler, G.L. (1965). Cardiovascular manifestations of the Hurler syndrome: hemodynamic and angiocardiographic observations in 15 patients. *Circulation* **31**, 132.

Kubilus, J., Tarcio, A.J., and Baden, H.P. (1979). Steriod sulfatase deficiency in sex-linked ichthyosis. *Am. J. hum. Genet.* **31**, 50.

Kuhl, P., Olek, K., Wardenbach, P., and Grzeschik, K.-H. (1979). Assignment of a gene for human quinoid-dihydropteridine reductase (QDPR,EC 1.6.5.1) to chromosome 4. *Hum. Genet.* **53**, 47.

Kuhnle, U., Chow, D., Rapaport, R., Pang, S., Levine, L.S., and New, M.I. (1981). The 21-hydroxylase activity in the glomerulosa and fasciculata of the adrenal cortex in congenital adrenal hyperplasia. *J. clin. Endocr. Metab.* **52**, 534.

Kumamoto, C.A. and Fessler, J.H. (1980). Biosynthesis of A_1B procollagen. *Proc. natl. Acad. Sci. USA* **77**, 6434.

Kuo, P.T., Hayase, K., Kostis, J.B., and Moreyra, A.E. (1979). Use of combined diet and cholestipol in long-term (7-7½ years) treatment of patients with type II hyperlipoproteinemia. *Circulation* **59**, 199.

Kurczynski, T.W., Muir, W.A., Fleisher, L.D., Palomaki, J.F., Gaull, G.E., Rassin, D.K., and Abramowsky, C. (1980). Maternal homocystinuria: studies of an untreated mother and fetus. *Arch. Dis. Child.* **55**, 721.

Kuroda, Y., Osawa, T., Ito, M., Watanabe, T., Takeda, E., Toshima, K., and Miyao, M. (1980). Relationship between skin histidase activity and blood histidine response to histidine intake in patients with histidinemia. *J. Ped.* **97**, 269.

Kusakabe, T. and Miyake, T. (1963). Defective deiodination of ^{131}I-labeled L-diiodotyrosine in patients with simple goitre. *J. clin. Endocr. Metab.* **23**, 132.

—— —— (1964). Thyroidal deiodination defect in three sisters with simple goiter. *J. clin. Endocr. Metab.* **24**, 456.

Kushner, J.P. and Barbuto, A.J. (1975). An inherited defect in porphyria cutanea tarda (PCT): decreased uroporphyrinogen decarboxylase activity (urodecarb). *Clin. Res.* **23**, 403A.

—— —— and Lee, G.R. (1976). An inherited enzymatic defect in porphyria cutanea tarda. Decreased uroporphyrinogen decarboxylase activity. *J. clin. Invest.* **58**, 1089.

Kuzell, W.C., Glover, R.P., Gibbs, J.O., and Blau, R. (1964). Effect of sulfinpyrazone on serum uric acid in gout. A long-term study. *Geriatrics* **19**, 894.

Kvittingen, E.A., Halvorsen, S., and Jellum, E. (1983). Deficient fumarylaceto-acetase fumarylhydrolase activity in lymphocytes and fibroblasts from patients with hereditary tyrosinaemia. *Ped. Res.* **17**, 541.

Kwiterovich, P.O. jun., Fredrickson, D.S. and Levy, R.I. (1974). Familial hypercholesterolemia (one form of familial type II hyperlipoproteinemia). A study of its biochemical, genetic, and clinical presentation in childhood. *J. clin. Invest.* **53**, 1237.

—— Sloan, H.R., and Fredrickson, D.S. (1970). Glycolipids and other lipid constituents of normal human liver. *J. Lipid. Res.* **11**, 322.

Kyllerman, M. and Steen, G. (1977). Intermittently progressive dyskinetic syndrome in glutaric aciduria. *Neuropädiatre* **8**, 397.

Kyriakides, E.C., Paul, B., and Balint, J.A. (1972). Lipid accumulation and acid lipase deficiency in fibroblasts from a family with Wolman's disease, and their apparent correction *in vitro*. *J. lab. clin. Med.* **80**, 810.

LaBadie, G.U., Beratis, N.G., Price, P.M., and Hirschhorn, K. (1981). Studies of the copper-binding proteins in Menkes and normal cultured skin fibroblasts lysates. *J. Cell. Physiol.* **106**, 173.

La Du, B.N. (1967*a*). The enzymatic deficiency in tyrosinemia. *Am. J. Dis. Child.* **113**, 54.

—— (1967*b*). Histidinemia. *Am. J. Dis. Child.* **113**, 88.

—— (1978). Histidinemia. In *The metabolic basis of inherited disease*, 4th end. (ed. J.B. Stanbury, J.B. Wyngaarden, and D.S. Fredrickson), p.317. McGraw-Hill, New York.

—— and Gjessing, L.R. (1978). Tyrosinosis and tyrosinemia. In *The metabolic basis of inherited disease*, 4th edn. (ed. J.B. Stanbury, J.B. Wyngaarden, and D.S. Fredrickson), p.256. McGraw-Hill, New York.

—— Zannoni, V.G., Laster, L., and Seegmiller, J.E. (1958). The nature of the defect in tyrosine metabolism in alcaptonuria. *J. biol. Chem.* **230**, 251.

—— Howell, R.R., Jacoby, G.A., Seegmiller, J.E., and Zannoni, V.G. (1962). The enzymatic defect in histidinemia. *Biochem. biophys. Res. Commun.* **7**, 398.

—— —— —— —— Sober, E.K., Zannoni, V.G., Canby, J.P., and Ziegler, L.K. (1963). Clinical and biochemical studies on two cases of histidinemia. *Pediatrics* **32**, 216.

Lake, B.D. (1965). A reliable rapid screening test for sulfatide lipidosis. *Arch. Dis. Child.* **40**, 284.

—— and Patrick, A.D. (1970). Wolman's disease: deficiency of Eboo resistant acid esterase activity with storage of lipids in lysosomes. *J. Ped.* **76**, 262.

Lam, S.T.S., Fensom, A.H., Coleman, D., Heaton, D., Morsman, J. Nicolaides, K., and Rodeck, C.H. (1984). Steroid sulphatase activity in trophoblast samples. *J. Obstet. Gynaecol.* **5**, 24.

Lamberg, B.A. (1973). Congenital euthyroid goitre and partial peripheral resistance to thyroid hormones. *Lancet* i, 854.

Lamont, N.M., Hathorn, M., and Joubert, S.M. (1961). Porphyria in the African. A study of 100 cases. *Q. J. Med.*, new series, **30**, 373.

La Motta, R.V.,Woronick, C.L., and Reinfrank, R.F. (1970). Multiple forms of serum cholinesterase: molecular weights of isoenzymes. *Arch. Biochem. Biophys.* **136**, 448.

Lampert, I.A. and Lewis, P.D. (1975). Staining of sulfatides in metachromatic leukodystrophy with alcian blue at high salt concentrations. *Histochemistry* **43**, 269.

Lamy, M., Dubois, R., Rossier, A., Frezal, J., Loeb, H., and Blancher, G. (1960). La glycogénose par déficience en phosphorylase hépatique. *Arch. Fr. Pediatr.* **17**, 14.

Lancaster, G., Mamer, O.A., and Scriver, C.R. (1974). Branched-chain alpha-keto acids isolated as oxime derivatives: relationship to the corresponding hydroxy acids and amino acids in maple syrup urine disease. *Metabolism* **23**, 257.

Lancet (1974). Histidinaemia: to treat or not to treat? *Lancet* i, 719.

Landing, B.H., Donnel, G.N., Alfi, O.S., Neustein, H.B., Lee, F.A., Ng, W.G.,

Bergren, W.R., and Surgeon, P. (1976). Fucosidosis: clinical, pathologic, and biochemical studies of five patients. In *Current trends in sphingolipidoses and allied disorders* (ed. B.W. Volk and L. Schneck), p.147. Plenum Press, New York.

Lange, A.J., Arion, W.J., and Beaudet, A.L. (1980). Type Ib Glycogen storage disease is caused by a defect in the glucose-6-phosphate translocase of the microsomal glucose-6-phosphatase system. *J. biol. Chem.* **255**, 8381.

Langenbeck, U., Rudiger, H.W., Schulze-Schencking, M., Keller, W., Brackertz, D., and Goedde, H.W. (1971). Evaluation of a heterozygote test for maple syrup urine disease in leucocytes and cultured fibroblasts. *Humangenetik* **11**, 304.

—— Wendel, U., Mench-Hoinowski, A., Kuschel, D., Becker, K., Przyrembel, H., and Bremer, H.J. (1978). Correlations between branched-chain amino acids and branched-chain α-keto acids in blood in maple syrup urine disease. *Clin. Chim. Acta* **88**, 283.

Langer, T., Strober, W., and Levy, R.I. (1972). The metabolism of low density lipoprotein in familial type II hyperlipoproteinemia. *J. clin. Invest.* **51**, 1528.

Lanzkowsky, P. (1970). Congenital malabsorption of folate. *Am. J. Med.* **48**, 580.

Lapiere, C.M., Lenaers, A., and Kohn, L.D. (1971). Procollagen peptidase: an enzyme excising the coordination peptides of procollagen. *Proc. natl. Acad. Sci. USA* **68**, 3054.

Larbrisseau, A., Brochu, P., Ng, Y.K., Jashmin, G., Potier, M., Vanasse, M., and Hausser, C. (1978). Première observation Canadienne d'une fucosidose. *Union méd Can.* **107**, 968.

Larmande, A. and Timsit, E. (1955). Á propos de 20 cas de xeroderma pigmentosum. *Algéire Med.* **59**, 557.

Larochelle, J., Mortezai, A., Belanger, M., Tremblay, M., Claveau, J.C., and Aubin, G. (1967). Experience with 37 infants with tyrosinemia. *Can. med. Assoc. J.* **97**, 1051.

Larsson, A. and Mattsson, B. (1976). On the mechanism of 5-oxoproline over-production in 5-oxoprolinuria. *Clin. Chim. Acta* **67**, 245.

—— Zetterstrom, R., Hörnell, H., and Porath, U. (1976). Erythrocyte glutathione synthetase in 5-oxoprolinuria: kinetic studies of the mutant enzyme and detection of heterozygotes. *Clin. Chim. Acta* **73**, 19.

Lasker, M. (1941). Essential frustosuria. *Hum. Biol.* **13**, 51.

—— (1950). The question of arabinosuria. *Am. J. clin. Path.* **20**, 485.

—— (1955). Mortality of persons with xyloketosuria. *Hum. Biol.* **27**, 294.

Laster, L.G., Mudd, S.H., Finkelstein, J.D., and Irreverre, F. (1965b). Homocystinuria due to cystathionine-synthase deficiency: the metabolism of L-methionine. *J. clin. Invest.* **44**, 1708.

—— Spaeth, G.L., Mudd, S.H., and Finkelstein, J.D. (1965a). Homocystinuria due to cystathionine synthase deficiency. *Ann. intern. Med.* **63**, 1117.

Lauseker, C., Heidt, P., Fischer, D., Hartley, B.H., and Löhr, G.W. (1965). Anémie hémolytique constitutionnelle avec déficit en 6-phospho-gluconate-deshydrogénase. *Arch. fr. Pediatr.* **22**, 789.

Laver, M. and Fairley, K.F. (1971). D-penicillamine in pregnancy. *Lancet* **i**, 1019.

Law, E.A. and Fowler, B. (1976). ß-mercaptolactate-cysteine disulfiduria in a mentally retarded Scottish male. *J. ment. Dis. Res.* **20**, 99.

Layzer, R.B., Rowland, L.P., and Ranney, H.M. (1967). Muscle phosphofructo-kinase deficiency. *Arch. Neurol.* **17**, 512.

Lazarus, S.S., Vethamany, V.G., Schneck, L., and Volk, B.W. (1967). Fine structure and histochemistry of peripheral blood cells in Niemann–Pick disease. *Lab. Invest.* **17**, 155.

Leaf, G. and Neuberger, A. (1948). The preparation of homogentisic acid and of 2:5-dihydroxyphenylethylamine. *Biochem. J.* **43**, 606.

Leclerc, J.L., Hould, F., Lelievre, M., and Gagne, F. (1971). Maladie de Wolman: etude anatomo-clinique d'une nouvelle observation avec absence de calcifications radiologiques et macroscopiques des surrenals. *Lavel Med.* **42**, 461.

Lederer, B. and Stalmans, W. (1976). Human liver glycogen phosphorylase. Kinetic properties and assay in biopsy specimens. *Biochem. J.* **159**, 689.

—— Van Hoof, F., Van Den Berghe, G., and Hers, H.G. (1975). Glycogen phosphorylase and its converter enzymes in haemolysates of normal human subjects and patients with Type VI glycogen storage disease. *Biochem. J.* **147**, 23.

Lee, F.A., Donnell, G.N., and Gwinn, J.L. (1977). Radiographic features of fucosidosis. *Pediatr. Radiol.* **5**, 204.

Lee, J.E. and Yoshida, A. (1976). Purification and chemical characterization of human hexosaminidases A and B. *Biochem. J.* **159**, 535.

Lee, P.A. and Gareis, F.J. (1975). Evidence for partial 21-hydroxylase deficiency among heterozygote carriers for congenital adrenal hyperplasia. *J. clin. Endocr. Metab.* **41**, 415.

Lee, R.E., Worthington, C.R., and Glew, R.H. (1973). The bilayer nature of deposits occurring in Gaucher's disease. *Arch. Biochem. Biophys.* **159**, 259.

Leeming, R.J., Blair, J.A., and Rey, F. (1976). Biopterin derivatives in atypical phenylketonuria. *Lancet* **i**, 99.

Lehmann, A.R., Kirk-Bell, S., Arlett, C.F., Paterson, M.C., Lohman, P.H.M., De Weerd-Kastelein, E.A., and Bootsma, D. (1975). Xeroderma pigmentosum cells with normal levels of excision repair have a defect in DNA synthesis after UV irradiation. *Proc. natl. Acad. Sci. USA* **72**, 219.

—— —— —— Harcourt, S.A., De Weerd-Kastelein, E.A., Keijzer, W., and Hall-Smith, P. (1977). Repair of ultraviolet light damage in a variety of human fibroblast cell strains. *Cancer. Res.* **37**, 904.

Lehmann, H. and Huntsman, R.G. (1972). *The hemoglobinopathies,* p.1401. McGraw-Hill, New York.

—— and Ryan, E. (1956). The familial incidence of low pseudocholinesterase level. *Lancet* **ii**, 124.

Lehnert, W., Niederhoff, H., Junker, A., Saule, H., and Frasch, W. (1979). A case of biotin-responsive 3-methylcrotonylglycine- and 3-hydroxyisovaleric aciduria. *Eur. J. Pediatr.* **132**, 107.

Lehrach, H., Frischauf, A.M., Hanahan, D., Wozney, J., Fuller, F., Crkvenjakov, R., Boedtker, H., and Doty, P. (1978). Construction and characterisation of a 2.5-kilobase procollagen clone. *Proc. natl. Acad. Sci. USA* **75**, 5417.

Lehrer, R.I. and Cline, M.J. (1969). Leukocyte myeloperoxidase deficiency and disseminated candidiasis: the role of myeloperoxidase in resistance to Candida infection. *J. clin. Invest.* **48**, 1478.

Leigh, D. (1951). Subacute necrotizing encephalomyelopathy in an infant. *J. Neurol. Neurosurg. Psychiatry* **14**, 216.

Lemaitre, L., Remy, J., Farriaux, J.P., Dhondt, J.L., and Walbaum, R. (1979). Radiologic signs of mucolipidosis II or I-cell disease. *Pediatr. Radiol.* **7**, 97.

Lemonnier, A., Pousset, J.L., Charpentier, C., and Moatti, N. (1971). Isolement et identification d'un nouveau métabolite urinaire d'oxydation de l'homocystine. *Clin. Chim. Acta* **33**, 359.

Lemonnier, F., Charpentier, C., Odièvre, M., Larrègue, M., and Lemonnier, A. (1979). Tyrosine aminotransferase isoenzyme deficiency. *J. Ped.* **94**, 931.

Lenke, R.R. and Levy, H.L. (1980). Maternal phenylketonuria and hyperphenylalaninaemia: an international survey of untreated and treated pregnancies. *New Engl. J. Med.* **303**, 1202.

Lenoir, G., Rivron, M., Gubler, M.-C., Dufier, J.-L., Tome, F.S.M., and Guivarch, M. (1977). La maladie de Fabry. Traitement du syndrome acrodyniforme par la carbamazepine. *Arch. franc. Ped.* **34**, 704.

Lenz, W. (1959). Quelques remarques au sujet du travail de W. Taillard et A. Prader: étude génétique du syundrome de feminisation testiculaire totale et partielle. *J. Genet. hum.* **8**, 199.

Leonard, J.V., Seakins, J.W.T., and Griffin, N.K. (1979a). ß-hydroxy-ß-methylglutaric aciduria presenting as Reye's syndrome. *Lancet* **i**, 680.

—— —— —— and Marshall, W.C. (1979b). ß-hydroxy-ß-methylglutaric aciduria, Reye's syndrome, and Echovirus II. *Lancet* **i**, 1147.

—— —— Bartlett, K., Hyde, J., Wilson, J., and Clayton, B. (1981). Inherited disorders of 3-methylcrotonyl-CoA carboxylation. *Arch. Dis. Child.* **56**, 53.

Leonidas, J.C. (1965). Essential fructosuria. *N.Y. State. J. Med.* **65**, 2257.

Leroux, A., Junien, C., Kaplan, J.C., and Bamberger, J. (1975). Generalized deficiency of cytochrome b5 reductase in congenital methemoglobinemia with mental retardation. *Nature, Lond.* **258**, 619.

Leroy, J.G. and Demars, R.I. (1967). Mutant enzymatic and cytological phenotypes in cultured human fibroblasts. *Science* **157**, 804.

—— Spranger, J.W., Feingold, M., Opitz, J.M., and Crocker, A.C. (1971). I-cell disease: a clinical picture. *J. Ped.* **79**, 360.

—— Ho, M.W., McBrinn, M.C., Zielke, K., Jacob, I., and O'Brien, J.S. (1972). I-cell disease: biochemical studies. *Ped. Res.* **6**, 752.

—— Van Elsen, A., Martin, J.J., Dumon, J.E., Hulet, A.E., Okada, S., and Navarro, C. (1973). Infantile metachromatic leukodystrophy: confirmation of a prenatal diagnosis. *New Engl. J. Med.* **288**, 1365.

Lesch, M. and Nyhan, W.L. (1964). A familial disorder of uric acid metabolism and central nervous system function. *Amer. J. Med.* **36**, 561.

Leshin, M., Griffin, J.E., and Wilson, J.D. (1978). Hereditary male pseudohermaphroditism associated with an unstable form of 5 α-reductase. *J. clin. Invest.* **62**, 685.

Lester, F.T. and Cusworth, D.C. (1973). Lysine infusion in cystinuria: theoretical renal thresholds for lysine. *Clin. Sci. mol. Med.* **44**, 99.

Leung, M.K.K., Fessler, L.I., Greenberg, D.B., and Fessler, J.H. (1979). Separate amino and carboxyl procollagen peptidases in chick embryo tendon. *J. biol. Chem.* **254**, 224.

Levene, C.I. and Gross, J. (1959). Alterations in state of molecular aggregation of collagen induced in chick embryos by ß-aminopropionitrile (laythyrus factor). *J. exp. Med.* **110**, 771.

Levene, P.A. and La Forge, F.B. (1914). Note on a case of pentosuria. *J. biol. Chem.* **18**, 319.

Levi, A.J., Sherlock, S., Scheuer, P.J., and Cumings, J.N. (1967). Presymptomatic Wilson's disease. *Lancet* **ii**, 575.

Levin, B., Oberholzer, V.G., and Sinclair, L. (1969b). Biochemical investigations of hyperammonaemia. *Lancet* **ii**, 170.

—— Dobbs, R.H., Burgess, E.A., and Palmer, T. (1969a). Hyperammonaemia. A variant type of deficiency of liver ornithine transcarbamylase. *Arch. Dis. Child.* **44**, 162.

Levin, E.Y. (1968). Uroporphyrinogen III cosynthetase in bovine erythropoeitic porphyria. *Science* **161**, 907.

—— (1971). Enzymatic properties of uroporphyrinogen III cosynthetase. *Biochemistry* **10**, 4669.

Levin, S., Moses, S.W., Chayoth, R., Jagoda, N., and Steinitz, K. (1967). Glycogen storage disease in Israel. A clinical, biochemical and genetic study. *Isr. J. med. Sci.* **3**, 397.

Levine, L.S., Lieber, E., Pang, S., and New, M.I. (1980). Male pseudohermaphroditism due to 17-ketosteroid reductase deficiency diagnosed in the newborn period. *Ped. Res.* **14**, 480.

—— New, M.I., Pollack, M., and Dupont, B. (1979). Prenatal diagnosis of congenital adrenal hyperplasia. *Lancet*, **ii**, 637.

—— Zachmann, M., New, M.I., Prader, A., Pollack, M.S., O'Neill, G.J., Yang, S.Y., Oberfield, S.E., and Dupont, B. (1978). Genetic mapping of the 21-hydroxylase deficiency gene within the HLA linkage group. *New Engl. J. Med.* **299**, 911.

Levine, S.Z. and Hoenig, E.M. (1972). Astrocytic gliosis of vascular adventitia and arachnoid membrane in infantile Gaucher's disease. *J. Neuropath, exp. Neurol.* **31**, 147.

—— Gordon, H.H., and Marples, E. (1941). A defect in the metabolism of tyrosine in premature infants II. Spontaneous occurrence and eradication by vitamin C. *J. clin. Invest.* **20**, 209.

Levy, H.L. (1973). Newborn screening for metabolic disorders. *New Engl. J. Med.* **288**, 1299.

—— (1980*a*). Screening for galactosaemia. In *Inherited disorders of carbohydrate metabolism* (ed. D. Burman, J.B. Holton, and C.A. Pennock), p.133. MTP Press, Lancaster.

—— and Mudd, S.H. (1973). Homocystinuria due to bacterial contamination in pyridoxine-unresponsive cystathioninemia. *Ped. Res.* **7**, 162.

—— Coulombe, J.T., and Shih, V.E. (1980). Newborn urine screening. In *Neonatal screening for inborn errors of metabolism* (ed. H. Bickel, R. Guthrie, and G. Hammersen), p.89. Springer, Berlin.

—— Shih, V.E., and Madigan, P.M. (1974). Routine newborn screening for histidinemia. Clinical and biochemical results. *New Engl. J. Med.* **291**, 1214.

—— Karolkewicz, V., Houghton, S.A., and MacReady, R.A. (1970*b*). Screening the 'normal' population in Massachusetts for phenylketonuria. *New Engl. J. Med.* **282**, 1455.

—— Erickson, A.M., Lott, I.T., and Kurtz, D.J. (1973). Isovaleric acidemia: results of family study and dietary treatment. *Pediatrics* **52**, 83.

—— Shih, V.E., Madigan, P.M., and MacCready, R.A. (1969*b*). Transient tyrosinemia in full-term infants. *J. Am. med. Ass.* **209**, 249.

—— Mudd, S.H., Schulman, J.D., Dreyfus, P.M., and Abeles, R.H. (1970*a*). A derangement in B_{12} metabolism associated with homocystinuria, cystathioninemia, hypomethioninemia and methylmalonic aciduria. *Am. J. Med.* **48**, 390.

—— Shih, V.E., Madigan, P.M., Karolkewicz, V., Carr, J.R., Lum, A., Richards, A.S., Crawford, J.D., and MacCready, R.A. (1969*a*). Hypermethioninemia with other hyperaminoacidemias. *Am. J. Dis. Child.* **117**, 96.

Levy, R.I. (1980*b*). Drugs used in the treatment of hypolipoproteinemias. In *The pharmacological basis of therapeutics*, 6th end. (ed. A.G. Gilman, L.S. Goodman, and A. Gilman), p.834. Macmillan, New York.

Lewis, B., Chait, A., February, A.W., and Mattock, M. (1973). Functional overlap between 'chylomicra' and 'very low density lipoproteins' of human plasma during alimentary lipaemia. *Atherosclerosis* **17**, 455.

Lewis, G.M., Spencer-Peet, J., and Stewart, K.M. (1963). Infantile hypoglycaemia due to inherited deficiency of glycogen synthetase in liver. *Arch. Dis. Child.* **38**, 40.

Li, S.-C., Hirabayashi, Y., and Li, Y.-T. (1981). A protein activator for the enzyme hydrolysis of GM_2 ganglioside. *J. biol. Chem.* **256**, 6234.

Li, Y.-T. and Li, S.-C. (1971). Anomeric configuration of galactose residues in ceramide trihexosides. *J. biol. Chem.* **246**, 3769.

Libert, J. Van Hoff, F., Tousaint, D., Roozitalab, H., Kenyon, K.R., and Green, W.R. (1979). Ocular findings in metachromatic leucodystrophy: an electron microscopic and enzyme study in different clinical and genetic variants. *Arch. Opthalmol.* **97**, 1495.

Lichtenstein, J.R., Martin, J.R., Kohn, L.D., Byers, P.H., and McKusick, V.A. (1973). Defect in conversion of procollagen to collagen in a form of the Ehlers–Danlos syndrome. *Science* **182**, 298.

Lichtenstein, L. and Kaplan, L. (1954). Hereditary ochronosis: Pathological changes observed in two necropsied cases. *Am. J. Path.* **30**, 99.

Liddell, J., Lehmann, H., and Davies, D. (1963). Harris and Whittaker's pseudocholinesterase variant with increased resistance to fluoride. A study of four families and the identification of the homozygote. *Acta. Genet.* **13**, 95.

–––– –––– and Silk, E. (1962). A 'silent' pseudocholinesterase gene. *Nature, Lond.* **193**, 561.

Lidsky, A.S., Güttler, F., and Woo, S.L.C. (1985). Prenatal diagnosis of classic phenylketonuria by DNA analysis. Lancet i, 549.

Liebaers, I. and Neufeld, E.F. (1976). Iduronate sulfatase activity in serum, lymphocytes, and fibroblasts — simplified diagnosis of the Hunter syndrome. *Ped. Res.* **10**, 733.

–––– Di Natale, P., and Neufeld, E.F. (1977). Iduronate sulfatase in amniotic fluid: an aid in the prenatal diagnosis of the Hunter syndrome. *J. Ped.* **90**, 423.

Lieberman, E., Shaw, K.N.F., and Donnell, G.N. (1967). Cystathioninuria in galactosemia and certain types of liver disease. *Pediatrics* **40**, 828.

Lieberman, J.S., Oshtory, M., Taylor, R.G., and Dreyfus, P.M. (1980). Perinatal neuropathy as an early manifestation of Krabbe's disease. *Arch. Neurol.* **37**, 446.

Lieberman, T.W., Podos, S.M., and Hartstein, J. (1966). Acute glaucoma, ectopia lentis and homocystinuria. *Am. J. ophthalmol.* **61**, 252.

Liem, K.O. and Hooghwinkel, G.J. (1975). The use of α-L-iduronidase activity in leucocytes for the detection of Hurler and Scheie syndromes. *Clin. Chim. Acta* **60**, 259.

–––– Giesberts, M.A.H., Van de Kamp, J.J.P., Van Pelt, J.F., and Hooghwinkel, G.J.M. (1976). Sanfilippo B disease in two related sibships. Biochemical studies in patients, parents and sibs. *Clin. Genet.* **10**, 273.

Light, I.J., Berry, H.K., and Sutherland, J.M. (1966). Amino-acidemia of prematurity. *Am. J. Dis. Child.* **112**, 229.

–––– Sutherland, J.M., and Berry, H.K. (1973). Clinical significance of tyrosinemia of prematurity. *Am. J. Dis. Child.* **125**, 243.

Lin, C.Y. and Kumar, S. (1972). Pathway for the synthesis of fatty acids in mammalian tissues. *J. biol. Chem.* **247**, 604.

Lin, Y.N. and Radin, N.S. (1973). Alternate pathways of cerebroside catabolism. *Lipids* **8**, 732.

Lindblad, B., Lindstedt, S., and Steen, G. (1977). On the enzymic defects in hereditary tyrosinemia. *Proc. natl. Acad. Sci. USA.* **74**, 4641.

–––– Lindstrand, K., Svanberg, B., and Zetterström, R. (1969). The effect of cobamide coenzyme in methylmalonic acidemia. *Acta Paediatr. Scand.* **58**, 178.

Linder, D. and Gartler, S.M. (1965). Glucose-6-phosphate dehydrogenase mosaicism: utilization as cell marker in the study of leiomyomas. *Science* **150**, 67.

Lindstedt, S., Norberg, K., Steen, G., and Wahl, E. (1976). Structure of some aliphatic dicarboxylic acids found in the urine of an infant with congenital lactic acidosis. *Clin. Chem.* **22**, 1330.

Linker, A., Evans, L.R., and Langer, L.O. (1970). Morquio's disease and mucopolysaccharide excretion, *J. Ped.* **77**, 1039.

Linn, T.C., Pettit, F.H., and Reed, L.J. (1969*b*). α-keto acid dehydrogenase complexes. X. Regulation of the activity of the pyruvate dehydrogenase complex from beef kidney mitochondria by phosphorylation and dephosphorylation. *Proc. natl. Acad. Sci. USA* **62**, 234.

—— Hucho, F., and Reed, L.J. (1969*a*). α-keto acid dehydrogenase complexes. XI. Comparative studies of regulatory properties of the pyruvate dehydrogenase complexes from kidney, heart and liver mitochondria. *Proc. natl. Acad. Sci. USA* **64**, 227.

Linnell, J.C., Quadros, E.V., Elliott, P.G., and Malleson, P. (1980). Defective adenosylcobalamin synthesis in a case of transcobalamin II deficiency. *J. inher. metab. Dis.* **3**, 95.

—— Matthews, D.M., Mudd, S.H., Uhlendorf, B.W., and Wise, I.J. (1976). Cobalamins in fibroblasts cultured from normal control subjects and patients with methylmalonic aciduria. *Ped. Res.* **10**, 179.

Linneweh, F., Schaumlöffel, E., and Vetrella, M. (1970). Galaktokinase-Defekt bei einem Neugeborenen. *Klin. Wochenschr.* **48**, 31.

Lipid Research Clinics, The (1980). Population study data book, Vol I, visit II. Cited by Herbert, P.N., Assmann, G., Gotto, A.M., and Fredrickson, D.S. Familial lipoprotein deficiency (Tables 29–3, 29–4). In *The metabolic basis of inherited diseases*, 5th edn. (ed. J.B. Stanbury, J.B. Wyngaarden, D.S. Fredrickson, J.S. Goldstein, and M.S. Brown), p.592. McGraw-Hill, New York.

Lissitzky, S., Bismuth, J., Codaccioni, J.L., and Cartouzou, G. (1968). Congenital goiter with iodoalbumin replacing thyroglobulin and defect of deiodination of iodotyrosines: serum origin of the thyroid iodoalbumin. *J. clin. Endocr. Metab.* **28**, 1797.

—— Torresani, J., Burrow, G.N., Bouchilloux, S., and Chabaud, O. (1975). Defective thyroglobulin export as a cause of congenital goiter. *Clin. Endocrinol.* **4**, 363.

Litman, N., Kanter, A.I., and Finberg, L. (1975). Galactokinase deficiency presenting as pseudotumor cerebri. *J. Ped.* **86**, 410.

Little, J.A., Whayne, T.F., Bhagwat, A.G., Buckley, G.C., and Kallos, A. (1970). A case of type I hyperlipoproteinemia unusually sensitive to dietary alcohol and fat with induction of lipemia. *Clin Res.* **18**, 736.

Little, P.F.R., Annison, G., Darling, S., Williamson, R., Camba, L., and Modell, B. (1980). Model for antenatal diagnosis of beta-thalassemia and other monogenic disorders by molecular analysis of linked DNA polymorphisms. *Nature, Lond.* **285**, 144.

Ljungren, J.-G., Lindstrom, H., and Hjern, B. (1973). The concentration of peroxidase in normal and adenomatous human thyroid tissue with special reference to patients with Pendred's syndrome. *Acta Endocrinol.* **72**, 272.

Lockman, L.A., Hunninghake, D.B., Krivit, W., and Desnick, R.J. (1973). Relief of pain of Fabry's disease by diphenylhydantoin. *Neurology* **23**, 871.

—— Kennedy, W.R., and White, J.G. (1967). The Chediak–Higashi syndrome: electrophysiologic and electron microscopic observations on the peripheral neuropathy. *J. Ped.* **70**, 942.

Lockwood, D.H., Merimee, T.J., Edgar, P.J., Greene, M.L., Fujimoto, W.Y., Seegmiller, J.E., and Howell, R.R. (1969). Insulin secretion in type I glycogen storage disease. *Diabetes* **18**, 755.

Loder, P.B. and De Gruchy, G.C. (1965). Red-cell enzymes and co-enzymes in non-spherocytic congenital haemolytic anaemias. *Br. J. Haematol.* **11**, 21.

Loeb, H., Tondeur, M., Jonniaux, G., Mockel-Pohl, S., and Vamos-Hurwitz, E. (1969). Biochemical and ultrastructural studies in a case of mucopoly-saccharidosis 'F' (fucosidosis). *Helv. Paediatr. Acta* **24**, 519.

Löhr, G.W., Blume, K.G., Rüdiger, H.W., Sokal, G., and Gulbis, E. (1968). A new type of pyruvate-kinase deficiency of human erythrocytes. *Lancet* **i**, 753.

Lomholt, J.C. and With, T.K. (1969). Hereditary coproporphyria. A family with unusually few and mild symptoms. *Acta. Med. Scand.* **186**, 83.

Long, W.K., Kirkman, H.N., and Sutton, H.E. (1965). Electrophoretically slow variants of glucose-6-phosphate dehydrogenase from red cells of Negroes. *J. lab. clin. Med.* **65**, 81.

Longhi, R.C., Fleisher, L.D., Tallan, H.H., and Gaull, G.E. (1977). Cysta-thionine ß-synthase deficiency: a qualitative abnormality of the deficient enzyme modified by vitamin B_6 therapy. *Ped. Res.* **11**, 100.

Lonsdale, D., Faulkner, W.R., Price, J.W., and Smeby, R.R. (1969). Intermittent cerebellar ataxia associated with hyperpyruvic acidemia and hyperalaninuria. *Pediatrics* **43**, 1025.

Loonen, M.C.B., Udlugt. L., and Franke, C.L. (1974). Angiokeratoma corporis diffusum and lysosomal enzyme deficiency. *Lancet* **ii**, 785.

—— Schram, A.W., Koster, J.F., Niermeijer, M.F., Busch, H.F.M., Martin, J.J., Brouwer-Kelder, B., Mekes, W., Slee, R.G., and Tager, J.M. (1981*a*). Identification of heterozygotes for glycogenosis 2 (acid maltase deficiency). *Clin. Genet.* **19**, 55.

—— Busch, H.F.M., Koster, J.F., Martin, J.J., Niermeijer, M.F., Schram, A.W., Brouwer-Kelder, B., Mekes, W., Slee, R.G., and Tager, J.M. (1981*b*), A family with different clinical forms of acid maltase deficiency (glycogenosis type II): biochemical and genetic studies. *Neurology* **31**, 1209.

Lorenzen, F., Pang, S., New, M.I., Pollack, M.S., Oberfield, S., Dupont, B., Chow, D., and Levine, D.S. (1980). Studies of the C-21 and C-19 steroids and HLA genotyping in siblings and parents of patients with congenital adrenal hyperplasia due to 21-hydroxylase deficiency. *J. clin. Endocr. Metab.* **50**, 572.

Lott, I.T., Dulaney, J.T., Milunsky, A., Hoefnagel, D., and Moser, H.W. (1976). Apparent biochemical homozygosity in two obligatory heterozygotes for metachromatic leukodstrophy. *J. Ped.* **89**, 438.

—— Erickson, A.M., and Levy, H. (1972). Dietary treatment of an infant with isovaleric acidemia. *Pediatrics* **49**, 617.

Lough, J., Fawcett, J., and Wiegensberg, B. (1970). Wolman's disease: an electron, histochemical and biochemical study. *Arch. Pathol. (Chicago)* **89**, 103.

Lowden, J.A. and LaRamee, M.-A. (1972). Problems in prenatal diagnosis using sphingolipid hydrolase assays. In *Sphingolipids, sphingolipidoses and allied disorders* (ed. B.W. Volk and S.M. Aronson), p.257. Plenum Press, New York.

—— and O'Brien, J.S. (1979). Sialidosis: a review of human neuraminidase deficiency. *Am. J. hum. Genet.* **31**, 1.

—— Cutz, E., Conen, P.E., Rudd, N., and Doran, T. (1973). Prenatal diagnosis of GM_1-gangliosidosis. *New Engl. J. Med.* **288**, 225.

Lowe, C.U., Terrey, M., and MacLaughlan, E.A. (1952). Organic-aciduria, decreased renal ammonia production, hydrophthalmos, and mental retardation; clinical entity. *Am. J. Dis. Child.* **83**, 164.

—— Sokal, J.E., Mosovich, L.L., Sarcione, E.J., and Doray, B.H. (1962). Studies in liver glycogen disease. Effects of glucagon and other agents on metabolic pattern and clinical status. *Am. J. Med.* **33**, 4.

Lowe, M.D. and Tapp, E. (1966). Renal damage caused by anhydro-4-epi-tetracycline. *Arch. Path.* **81**, 362.

Lowry, R.B. and Renwick, D.H.G. (1971). The relative frequency of the Hurler and Hunter syndromes. *New Engl. J. Med.* **284**, 221.

Lubs, H.A. jun., Vilar, O., and Bergenstal, D.M. (1959). Familial male pseudohermaphroditism with labial testes and partial feminization: endocrine studies and genetic aspects. *J. clin. Endocr. Metab.* **19**, 1110.

Luhby, A.L., Eagle, F.J., Roth, E., and Cooperman, J.M. (1961). Relapsing megaloblastic anemia in an infant due to a specific defect in gastro-intestinal absorption of folic acid. *Am. J. Dis. Child.* **102**, 482.

Luijten, J.A.F.M., Van der Heijden, M.C.M., Rijksen, G., Willemse, J., and Staal, G.E.J. (1978). Characterization of the arylsulfatase A of three cases of metachromatic leukodystrophy: one of the late infantile, one of the juvenile and one of the adult variant. *J. mol. Med.* **3**, 227.

Lustberg, T.J., Schulman, J.D., and Seegmiller, J.E. (1969). Metabolic fate of homogentisic acid-I-^{14}C (HGA) in alcaptonuria and effectiveness of ascorbic acid in preventing experimental ochronosis. *Arthritis Rheum.* **12**, 678.

Lustmann, J., Bimstein, E., and Yatziv, S. (1975). Dentigenous cysts and radiolucent lesions of the jaw associated with Hunter's syndrome. *J. oral Surg.* **33**, 679.

Lusty, C.J. (1978). Carbamoylphosphate synthetase I of rat-liver mitochondia. *Eur. J. Biochem.* **85**, 373.

Lutzner, M.A., Tierney, J.H., and Benditt, E.P. (1965). Giant granules and widespread cytoplasmic inclusions in a genetic syndrome of Aleutian mink. *Lab. Invest.* **14**, 2063.

Luzzatto, L., Usanga, E.A., and Reddy, S. (1969). Glucose-6-phosphate dehydrogenase deficient red cells: resistance to infection by malarial parasites. *Science* **164**, 839.

Lynch, H.T., Anderson, D.E., Smith, J.L., Howell, J.B., and Krush, A.J. (1967). Xeroderma pigmentosum, maligant melanoma and congenital ichthyosis. *Arch. Dermatol.* **96**, 625.

Lyon, I.C.T., Procopis, P.G., and Turner, B. (1971). Cystathioninuria in a well baby population. *Acta. Paediatr. Scand.* **60**, 324.

Lyon, M.F. (1962). Sex chromatin and gene action in the mammalian X-chromosome. *Am. J. hum. Genet.* **14**, 135.

—— and Hawkes, S.G. (1970). X-linked gene for testicular feminization in the mouse. *Nature, Lond.* **227**, 1217.

Lyons, L.B., Cox, R.P., and Dancis, J. (1973). Complementation analysis of maple syrup urine disease in heterokaryons derived from cultured human fibroblasts. *Nature, Lond.* **243**, 533.

McArdle, B. (1951). Myopathy due to a defect in muscle glycogen breakdown. *Clin. Sci.* **10**, 13.

—— (1969). Skeletal muscle glycogenoses other than type II. In *Some inherited disorders of muscle and brain* (ed. J.D. Allen and D.N. Raine), p.46. E & S Livingstone Ltd., Edinburgh and London.

McBride, O.W. and Ozer, H.L. (1973). Transfer of genetic information by purified metaphase chromosomes. *Proc. natl. Acad. Sci. USA* **70**, 1258.

McBrinn, M.C., Okada, S., Ho, M.W., Hu, C.C., and O'Brien, J.S. (1969). Generalized gangliosidosis: impaired cleavage of galactose from a mucopolysaccharide and a glycoprotein. *Science* **163**, 946.

McCabe, E.R.B., Sadava, D., Bullen, W.W., McKelvey, H.A., and Seltzer, W.K. (1982). Human glycerol kinase deficiency: enzyme kinetics and fibroblast hybridization. *J. inher. metab. Dis.* **5**, 177.

—— Fennessey, P.V., Guggenheim, M.A., Miles, B.S., Bullen, W.W., Sceats, D.J., and Goodman, S.I. (1977). Human glycerol kinase deficiency with hyperglycerolemia and glyceroluria. *Biochem. biophys. Res. Commun.* **78**, 1327.

—— Melvin, T.-R., O'Brien, D., Montgomery, R.R., Robinson, W.A., Bhasker, C., and Brown, R.I. (1980). Neutropenia in a patient with type 1B glycogen storage disease: in vitro response to lithium chloride. *J. Ped.* **97**, 944.

McCance, R.A., Morrison, A.B., and Dent, C.E. (1955). The excretion of phosphoethanolamine and hypophosphatasia. *Lancet* **i**, 131.

McCord, J.M. and Fridovich, I. (1969). Superoxide dismutase. An enzymic function for erythrocuprein (hemocuprein). *J. biol. Chem.* **244**, 6049.

McCulloch, J.C. and Marliss, E.B. (1975). Gyrate atrophy of the choroid and retina with hyperornithinemia. *Am. J. Ophthalmol.* **80**, 1047.

—— Arshinoff, S.A., Marliss, E.B., and Parker, J.A. (1978). Hyperornithinemia and gyrate atrophy of the choroid and retina. *Ophthalmology* **85**, 918.

McCully, K.S. (1969). Vascular pathology of homocysteinemia: implications for the pathogenesis of arteriosclerosis. *Am. J. Path.* **56**, 111.

McCurdy, P.R., Kirkman, H.N., Naiman, J.L., Jim, R.T.S., and Pickard, B.M. (1966). A Chinese variant of glucose-6-phosphate dehydrogenase. *J. lab. clin. Med.* **67**, 374.

MacDermot, K.D., Nelson, W., Reichert, C.M., and Schulman, J.D. (1980). Attempts at use of strychnine sulfate in the treatment of nonketotic hyperglycinemia. *Pediatrics* **65**, 61.

McDonald, L., Bray, C., Field, C., Love, F., and Davies, B. (1964). Homocystinuria, thrombosis, and the blood-platelets. *Lancet* **i**, 745.

McDonald, J.A. and Kelley, W.N. (1971). Lesch–Nyhan syndrome: altered kinetic properties of mutant enzyme. *Science* **171**, 689.

McDonald, J.E. and Henneman, P.H. (1965). Stone dissolution in vivo and control of cystinuria with D-penicillamine. *New Engl. J. Med.* **273**, 578.

MacFaul, R., Cavanagh, N., Lake, B.D., Stephens, R., and Whitfield, A.E. (1982). Metachromatic leucodystrophy: review of 38 cases. *Arch. Dis. Child.* **57**, 168.

McGirr, E.M. and Hutchinson, J.H. (1953). Radioactive iodine studies in non-endemic goitrous cretins. *Lancet* **i**, 1117.

—— Clement, W.E., Currie, A.R., and Kennedy, J.S. (1959). Impaired dehalogenase activity as a cause of goitre with malignant changes. *Scott. Med. J.* **4**, 232.

McInnes, R.R., Arshinoff, S.A., Bell, L., Marliss, E.B., and McCulloch, J.C. (1981). Hyperornithinaemia and gyrate atrophy of the retina: improvement of vision during treatment with a low arginine diet. *Lancet* **i**, 513.

MacKenzie, D.Y. and Woolf, L.I. (1959). Maple syrup urine disease, an inborn error of the metabolism of valine, leucine and isoleucine associated with gross mental deficiency. *Br. med. J.* **i**, 90.

MacKenzie, R.E. (1979). Formiminotransferase-cyclodeaminase, a bifunctional protein from pig liver. In *Chemistry and biology of pteridines* (ed. R.L. Kisliuk and G.M. Brown), p.443. Elsevier/North-Holland, New York.

McKusick, V.A. (1972). *Heritable disorders of connective tissue.* C.V. Mosby, Saint Louis.

—— (1978). *Mendelian inheritance in man,* 5th end. Johns Hopkins University Press, Baltimore.

—— (1980. The anatomy of the human genome. *Am. J. Med.* **69**, 267.

—— and Neufeld, E.F. (1983). The mucopolysaccharide storage diseases. In *The metabolic basis of inherited disease*, 5th ed. (ed. J.B. Stanbury, J.B. Wyngaarden, D.S. Fredrickson, J.L. Goldstein, and M.S. Brown), p.751. McGraw-Hill, New York.

—— Hall, J.G., and Char, F. (1971). The clinical and genetic characteristics of homocystinuria. In *Inherited disorders of sulphur metabolism* (ed. N.A.J. Carson and D.N. Raine), p.179. Churchill Livingstone, London.

—— Howell, R.R., Hussels, I.E., Neufeld, E.F., and Stevenson, R. (1972). Allelism, non allelism and genetic compounds among the mucopolysaccharidoses: hypothesis. *Lancet* **i**, 993.

McLaughlin, J., Askanas, V., and Engel, W.K. (1980). Adrenomyeloneuropathy: increased accumulation of very long chain fatty acids in cultured skeletal muscle. *Biochem. biophys. Res. Commun.* **92**, 1202.

McLaurin, A.W., Beisel, W.R., McCormick, G.J., Scalettar, R., and Herman, R.H. (1961). Primary hyperoxaluria. *Ann. Intern. Med.* **55**, 70.

McMaster, K.R., Powers, J.M., Hennigar, G.R., Wohltmann, H.J., and Farr, G.H. (1979). Nervous system involvement in type IV glycogenosis. *Arch. Pathol. Lab. Med.* **103**, 105.

McMichael, H.B., Webb, J., and Dawson, A.M. (1965). Lactase deficiency in adults: a cause of 'functional' diarrhoea. *Lancet* **i**, 717.

MacNab, G.H. (1955). Maldescent of the testicle. *J. R. Coll. Surg. Edinb.* **1**, 126.

McNair, A., Gudmand-Høyer, E., Jarnum, S., and Orrild, L. (1972). Sucrose malabsorption in Greenland. *Br. med. J.* **ii**, 19.

McNary, W. and Lowenstein, L.M. (1965). A morphological study of the renal lesion in angiokeratoma corporis diffusum universale (Fabry's disease). *J. Urol.* **93**, 641.

McPhee, G.B., Logan, R.W., and Primrose, D.A.A. (1975). Fucosidosis: how many cases undetected? *Lancet* **ii**, 462.

McReynolds, J.W., Mantagos, S., Brusilow, S., and Rosenberg, L.E. (1978). Treatment of complete ornithine transcarbamylase deficiency with nitrogen-free analogues of essential amino acids. *J. Ped.* **93**, 421.

Madan, K. and Shoemaker, J. (1980). XY females with enzyme deficiencies of steroid metabolism: a brief review. *Hum. Genet.* **53**, 291.

Madden, J.D., Walsh, P.C., MacDonald, P.C., and Wilson, J.D. (1975). Clinical and endocrinological characterization of a patient with the syndrome of incomplete testicular feminization. *J. clin. Endocr. Metab.* **40**, 751.

Madzarovová-Nohejlova, J. (1973). Trehalase deficiency in a family. *Gastroenterology* **65**, 130.

Magnus, I.A., Jarrett, A., Prankerd, T.A.J., and Rimington, C. (1961). Erythropoietic protoporphyria: a new syndrome with solar urticaria due to protoporphyrinaemia. *Lancet* **ii**, 448.

Maher, J.A., Espstein, F.H., and Hand, E.A. (1958). Xanthomatosis and coronary heart disease: necropsy study of two affected siblings. *Arch. Intern. Med.* **102**, 437.

Mahoney, M.J. and Rosenberg, L.E. (1970). Inherited defects of B_{12} metabolism. *Am. J. Med.* **48**, 584.

—— Hart, A.C., Steen, V.D., and Rosenberg, L.E. (1975a). Methylmalonic acidemia: biochemical heterogenity in defets of 5′-deoxyadenosylcobalamin synthesis. *Proc. natl. Acad. Sci. USA* **72**, 2799.

—— Rosenberg, L.E., Lindblad, B., Waldenström, J., and Zetterström, R.

(1975*b*). Prenatal diagnosis of methylmalonic aciduria. *Acta. Paediatr. Scand.* **64**, 44.

—— —— Mudd, S.H., and Uhlendorf, B.W. (1971). Defective metabolism of vitamin B_{12} in fibroblasts from patients with methylmalonic aciduria. *Biochem. biophys. Res. Commun.* **44**, 375.

Mahuran, D. and Lowden, J.A. (1980). The subunit and polypeptide structure of hexosaminidases from human placenta. *Can. J. Biochem.* **58**, 287.

Maire, I., Zabot, M.T., Mathieu, M., Cotte, J., and Hermier, M. (1978). Mannosidosis: tissue culture studies in relation to prenatal diagnosis. *J. inher. metab. Dis.* **1**, 19.

—— Mandon, G., Zabot, M.T., Mathieu, M., and Guibaud, P. (1979). ß-glucuronidase deficiency: enzyme studies in an affected family and prenatal diagnosis. *J. inher. metab. Dis.* **2**, 29.

Malamud, N. (1966). Neuropathology of phenylketonuria. *J. Neuropathol. exp. Neurol.* **25**, 254.

Malan, C., Neethling, A.C., Shanley, B.C., Gompertz, D., Bartlett, K., and Schraader, E.B. (1977). Isovaleric acidemia in two South African children. *S. Afr. med. J.* **52**, 980.

Malekzadeh, M.H., Neustein, H.B., Schneider, J.A., Pennisi, A.J., Ettenger, R.B., Uittenbogaart, C.H., Kogut, M.D., and Fine, R.N. (1977). Cadaver renal transplantation in children with cystinosis. *Am. J. Med.* **63**, 525.

Malone, M.J. and Stoffyn, P. (1966). A comparative study of brain and kidney glycolipids in metachromatic leukodystrophy. *J. Neurochem.* **13**, 1037.

—— Szoke, M.C., and Looney, G.L. (1975). Globoid leukodystrophy. I. Clinical and enzymatic studies. *Arch. Neurol.* **32**, 606.

Mamelle, J.C., Vanier, M.T., Baraton, G., Gilly, J., Carrier, H., Guichard, Y., Richard, A., and Gilly, R. (1975). Etude clinique, ultrastructural et biochimique d'un cas de gangliosidose GM_1 de type 2. *Arch. franc. Péd.* **32**, 925.

Mann, T.P., Wilson, K.M., and Clayton, B.E. (1965). A deficiency state arising in infants on synthetic foods. *Arch. Dis. Child.* **40**, 364.

Manowitz, P., Shapiro, S.S., and Goldstein, L. (1977). Ascorbate-2-sulfate levels in metachromatic leukodystrophy patients. *Biochem. Med.* **18**, 274.

Mantagos, S., Genel, M., and Tanaka, K. (1979). Ethylmalonic-adipic aciduria: in vivo and in vitro studies indicating deficiency of activities of multiple acyl-CoA dehydrogenases. *J. clin. Invest.* **64**, 1580.

—— Tsagaraki, S., Burgess, E.A., Oberholzer, V., Palmer, T., Sacks, J., Baibas, S., and Valaes, T. (1978). Neonatal hyperammonemia with complete absence of liver carbamyl phosphate synthetase activity. *Arch. Dis. Child.* **53**, 230.

Manz, H.J., Schuelein, M., McCullough, D.C., Kishimoto, Y., and Eiben, R.M. (1980). New phenotypic variant of adrenoleucodystrophy. Pathologic, ultrastructural and biochemical study in two brothers. *J. neurol. Sci.* **45**, 245.

Mapes, C.A., Anderson, R.L., Sweeley, C.C., Desnick, R.J., and Krivit, W. (1970). Enzyme replacement in Fabry's disease, an inborn error of metabolism. *Science* **169**, 987.

Marescau, B., Pintens, J., Lowenthal, A., and Terheggen, H.G. (1979*b*). Excretion of alpha-keto-gamma-guanidinovaleric acid and its cyclic form in patients with hyperargininemia. *Clin. Chim. Acta* **98**, 35.

—— —— —— —— and Adrianssens, K. (1979*a*). Arginase and free amino acids in hyperargininemia: leucocyte arginine as a diagnostic parameter for heterozygotes. *J. clin. Chem. clin. Biochem.* **17**, 211.

Maret, A., Salvayre, R., Negre, A., and Dousteblazy, L. (1980). Electrofocusing separation of molecular forms of splenic beta-glucosidase in normal subject and Gaucher's disease. *Biomedicine* **33**, 80.

Marimo, B. and Giannelli, F. (1975). Gene dosage effect in human trisomy 16. *Nature, Lond.* **256**, 204.

Marmion, L.H., O'Neill, B.P., Schafer, R.A., Shafer, N.J., Grekin, R.J., and Feringa, E.R. (1979). The adrenoleucomyeloneuropathy complex: a clinical evaluation of disease expression in four generations of a kindred (abstract). *Neurology* **29**, 560.

Maroteaux, P., Levêgue, B., Marie, J., and Lamy, M. (1963). Une nouvelle dysostose avec élimination urinaire de chondroitin-sulfate B. *Presse méd.* **71**, 1849.

Marsac, C., Augereau, C., Feldman, G., Wolf, B., Hansen, T.L., and Berger, R. (1982). Prenatal diagnosis of pyruvate carboxylase deficiency. *Clin. Chim. Acta* **119**, 121.

Marsh, C.A. and Gourlay, G.C. (1971). Evidence for a non lysosomal α-mannosidase in rat liver homogenates. *Biochim. Biophys. Acta* **235**, 142.

Marshall, W.C., Ockenden, B.C, Fosbrooke, A.S., and Cumings, J.N. (1968). Wolman's disease: a rare lipidosis with adrenal calcification. *Arch. Dis. Child.* **44**, 331.

Marstein, S., Jellum, E., Halpern, B., Eldjarn, L., and Perry, T.L. (1976). Biochemical studies of erythrocytes in a patient with pyroglutamic acidemia (5-oxoprolinemia). *New Engl. J. Med.* **295**, 406.

Martensson, E. (1966). Neutral glycolipids of human kidney: isolation, identification and fatty acid composition. *Biochim. Biophys. Acta* **116**, 196.

Martin, J.J., Leroy, J.G., Farriaux, J.R., Fontaine, G., Desnick, R.J. and Cabello, A. (1975). I-cell disease (mucolipidosis II). *Acta Neuropath.* **33**, 285.

Mason, P.A., Fraser, R., Semple, P.F., and Morton, J.J. (1979). The interaction of ACTH and angiotensin II in the control of corticosteroid plasma concentration in man. *J. steroid Biochem.* **10**, 235.

Massie, R.W. and Hartman, R.C. (1957). Albinism and sicklemia in a Negro family. *Am. J. hum. Genet.* **9**, 127.

Masson, P.R., Lundblad, A., and Autio, S. (1974). Mannosidosis: detection of the disease and of heterozygotes using serum and leukocytes. *Biochem. biophys. Res. Commun.* **56**, 296.

Matalon, R., Arbogast, B., and Dorfman, A. (1974*b*). Deficiency of chondroitin sulfate *N*-acetyl-galactosamine 4-sulfate sulfatase in Maroteaux–Lamy syndrome. *Biochem. biophys. Res. Commun.* **61**, 1450.

Matalon, R. and Deanching, M. (1977). The enzymic basis for the phenotypic variation of Hurler and Scheie syndromes. *Ped. Res.* **11**, 519.

—— and Dorfman, A. (1969). Acid mucopolysaccharides in cultured human fibroblasts. *Lancet* **ii**, 838.

—— —— (1972). Hurler's syndrome; an α-L-iduronidase deficiency. *Biochem. biophys. Res. Commun.* **47**, 959.

—— —— (1974). Sanfilippo A syndrome: sulfamidase deficiency in cultured fibroblasts and liver. *J. clin. Invest.* **54**, 907.

—— —— and Nadler, H.L. (1972). A chemical method for the antenatal diagnosis of mucopolysaccharidoses. *Lancet* **i**, 798.

—— Justice, P., and Deanching, M.N. (1977*a*). Phenylalanine hydroxylase in human placenta: novel system for study of phenylketonuria. *Lancet* **i**, 853.

—— Michals, K., Justice, P., and Deanching, M.N. (1977*b*). Glucose-6-phosphatase activity in human placenta: possible detection of heterozygote for glycogen-storage disease type I. *Lancet* **i**, 1360.

—— Arbogast, B., Justice, P., Brandt, E.K., and Dorfman, A. (1974*a*). Morquio's syndrome: deficiency of chondroitin sulfate N-acetylhexosamine sulfate sulfatase. *Biochem. biophys. Res. Commun.* **61**, 759.

—— Horwitz, A., Wappner, R., Deanching, M., and Brandt, I.K. (1978). Keratan and heparan sulfaturia — a mucopolysaccharidosis with an enzyme defect not previously identified (abstract). *Ped. Res.* **12**, 453.

—— Wappner, R., Deanching, M., Brandt, I.K., and Horwitz, A. (1982). Keratan and heparan sulphaturia: a mucopolysaccharidosis with an enzyme defect not previously identified. *J. inher. metab. Dis.* **5**, suppl. **1**, 57.

Mathoth, Y., Zaizov, R., Hoffman, J., and Klibansky, C. (1974). Clinical and biochemical aspects of chronic Gaucher's disease. *Isr. J. med. Sci.* **10**, 1523.

Martensson, E. (1966). Neutral glycolipids of human kidney: isolation, identification and fatty acid composition. *Biochim. Biophys. Acta* **116**, 196.

Matin, M.A. and Sylvester, P.E. (1980). Clinicopathological studies of oculo cerebro renal syndrome of Lowe, Terry, and MacLachlan. *J. ment. Defic. Res.* **24**, 1.

Matsuda, I., Pearson, T., and Holtzman, M.A. (1974). Determination of apoceruloplasmin by radio-immunoassay in nutritional copper deficiency, Menkes kinky hair syndrome, Wilson's disease and umbilical cord blood. *Ped. Res.* **8**, 821.

—— Sugai, M., and Kajii, T. (1970). Ornithine loading test in Lowe's syndrome. *J. Ped.* **77**, 127.

—— Anakura, M., Arashima, S., Saito, Y., and Oka, Y. (1976). Variant form of citrullinemia. *J. Ped.* **88**, 824.

—— Arashima, S., Nambru, H., Takekoshi, Y., and Anakura, M. (1971). Hyperammonemia due to a mutant enzyme of ornithine transcarbamylase. *Pediatrics* **48**, 595.

—— —— Oka, Y., Mitsuyama, T., Ariga, S., Ikeuchi, T., and Ichida, T. (1975). Prenatal diagnosis of fucosidosis. *Clin. Chim. Acta* **63**, 55.

Matsui, S.M., Mahoney, M.J., and Rosenberg, L.E. (1983). The natural history of the inherited methylmalonic acidemias. *New Engl. J. Med.* **308**, 857.

Matsumoto, N., Ishihara, T., Oda, E., Miwa, S., Nakashima, K., Uchino, F., and Fukumoto, Y. (1973). Fine structure of the spleen and liver in glucosephosphate isomerase (GPI) deficiency hereditary nonspherocytic hemolytic anemia — selctive reticulocyte destruction as a mechanism of hemolysis. *Acta Haematol. Jap.* **36**, 46.

Matsushima, A. and Orii, T. (1981). The activity of carbamylphosphate synthetase I (CPS I) and ornithine transcarbamylase (OTC) in the intestine and the screening of OTC deficiency in the rectal mucosa. *J. inher. metab. Dis.* **4**, 83.

—— Taga, T., Orii, T., Matsuda, Y., Tsuji, A., and Katsunuma, N. (1979). Hyperammonemia due to ornithine transcarbamylase deficiency accompanied by decreased activity of carbamylphosphate synthetase. *Brain. Dev. (Japan)* **11**, 343.

Matsuura, F., Nunez, H.A., Grabowski, A.G., and Sweeley, C.C. (1981). Structural studies of urinary oligosaccharides from patients with mannosidosis. *Arch. Biochem. Biophys.* **207**, 337.

Matz, D., Enzenauer, J., and Menne, F. (1975). Ueber einen Fall von atypischer Galaktosämie. *Humangenetik* **27**, 309.

Maurer, H.M., Wolff, J.A., Buckingham, S., and Spielvogel, A.R. (1972). 'Impotent' platelets with prolonged bleeding times. *Blood* **39**, 490.

Maury, C.P.J. (1980). Accumulation of glycoprotein-derived metabolites in neural and visceral tissues in aspartylglycosaminuria. *J. lab. clin. Med.* **96**, 838.

Maury, P. (1979) Accumulation of two glycoasparagines in the liver in aspartyl-glycosaminuria. *J. biol. Chem.* **254**, 1513.

—— and Palo, J. (1980). Characterisation of the storage material of peripheral lymphocytes in aspartylglycosaminuria. *Clin. Sci.* **58**, 165.

Mauvais-Jarvis, P., Bercovici, J.P., Crepy, O., and Gauthier, F. (1970). Studies on testosterone metabolism in subjects with testicular feminization syndrome. *J. clin. Invest.* **49**, 31.

Max, S.R., Nelson, P.G., and Brady, R.O. (1970). The effect of denervation on the composition of muscle gangliosides. *J. Neurochem.* **17**, 1516.

—— MacLaren, N.K., Brady, R.O., Bradley, R.M., Rennels, M.B., Tanaka, J., Garcia, J.H., and Cornblath, G.M. (1974). GM$_3$ (Hematoside) sphingolipo-dystrophy. *New Engl. J. Med.* **291**, 929.

Mayes, J.S. and Beutler, E. (1977). Alpha-galactosidase A from human placenta. Stability and subunit size. *Biochim. Biophys. Acta* **484**, 408.

—— and Guthrie, R. (1968). Detection of heterozygotes for galackokinase deficiency in a human population. *Biochem. Genet.* **2**, 219.

Maytum, W.J., Goldstein, N.P., McGuckin, W.F., and Owen, C.A. jun. (1961). Copper metabolism in Wilson's disease, Laennec's cirrhosis and hemachromatosis: studies with radiocopper (^{64}Cu). *Proc. Mayo Clin.* **36**, 641.

Medeiros-Neto, G.A., Bloise, W., and Ulhoa-Cintra, A.B. (1972). Partial defect of iodide trapping mechanism in two siblings with congenital goiter and hypothyroidism. *J. clin. Endocr. Metab.* **35**, 370.

Medes, G. (1972). A new error of tyrosine metabolism: tyrosinosis. The intermediary metabolism of tyrosine and phenylalanine. *Biochem. J.* **26**, 917.

Medical Research Council (1963). Report to the MRC of the conference on phenylketonuria. Treatment of phenylketonuria. *Br. med. J.* **i**, 1691.

Meuwisse, G.W. and Dahlqvist, A. (1968). Glucose-galactose malabsorption. A study with biopsy of the small intestinal mucosa. *Acta Paediatr. Scand.* **57**, 273.

Méhes, K., Klujber, L., Lassu, G., and Kajtár, P. (1972). Hypophosphatasia: screening and family investigations in an endogamous Hungarian village. *Clin. Genet.* **3**, 60.

Mehl, E. and Jatzkewitz, H. (1965). Evidence for the genetic block in meta-chromatic leucodystrophy. *Biochem. biophys. Res Commun.* **19**, 407.

—— —— (1968). Cerebroside-3-sulfate as a physiological substrate of arylsulfatase A. *Biochim. Biophys. Acta* **151**, 619.

Meienhofer, M.-C., Lagrange, J.-L., Cottreau, D., Lenoir, G., Dreyfus, J.-C., and Kahn, A. (1979). Phosphofructokinase in human blood cells. *Blood* **54**, 389.

Meisler, M. and Rattazzi, M.C. (1974). Immunological studies of ß-galactosidase in normal human liver and in GM$_1$ gangliosidosis. *Am. J. Hum. Genet.* **26**, 683.

Meister, A. (1973). On the enzymology of amino acid transport. Transport in kidney and probably other tissues is mediated by a cycle of enzymic reactions involving glutathione. *Science* **180**, 33.

—— (1981). On the cycles of glutathione metabolism and transport. In *Symposium on biological cycles (Honoring Sir Hans A. Krebs). Current topics in cellular regulation*, Vol. 18 (ed. B. Horecker and E. Stadtman), p.21.

Melançon, S.B., Khachardurian, A.K., Nadler, H.L., and Brown, B.I. (1973). Metabolic and biochemical studies in fructose-1,6-diphosphatase deficiency. *J. Ped.* **82**, 650.

—— Lee, S.Y., and Nadler, H.L. (1971). Histidase activity in cultivated human amniotic fluid cells. *Science* **173**, 627.

Melchior, J.C. and Clausen, J. (1968). Metachromatic leukodystrophy in early childhood: treatment with a diet deficient in vitamin A. *Acta Pediatr. Scand.* **57**, 2.

Mellman, I.S., Lin, P.-F., Ruddle, F.H., and Rosenberg, L.E. (1979*a*). Genetic control of cobalamin binding in normal and mutant cells: assignment of the gene for 5-methyltetrahydrofolate:L-homocysteine 5-methyltransferase to human chromosome 1. *Proc. natl. Acad. Sci. USA* **76**, 405.

—— Willard, H.F., Youngdahl-Turner, P., and Rosenberg, L.E. (1979*b*). Cobalamin coenzyme synthesis in normal and mutant human fibroblasts: evidence of a processing enzyme activity deficient in cbl C cells. *J. biol. Chem.* **254**, 11847.

—— Youngdahl-Turner, P., Willard, H.F., and Rosenberg, L.E. (1977). Intracellular binding of radioactive hydroxocobalamin to cobalamin-dependent apoenzymes in rat liver. *Proc. natl. Acad. Sci. USA* **74**, 916.

Menkes, J.H. (1959). Maple syrup urine disease: isolation and identification of organic acids in the urine. *Pediatrics* **23**, 348.

—— (1966). Idiopathic hyperglycinemia: isolation and identification of three previously undescribed urinary ketones. *J. Ped.* **69**, 413.

—— (1968). Cerebral proteolipids in phenylketonuria. *Neurology* **18**, 1003.

—— and Solcher, H. (1967). Maple syrup urine disease: effects of diet therapy on cerebral lipids. *Arch. Neurol.* **16**, 486.

—— Hurst, P.L., and Craig, J.M. (1954). New syndrome: progressive familial infantile cerebral dysfunction associated with unusual urinary substance. *Pediatrics* **14**, 462.

—— Philippart, M., and Fiol, R.E. (1965). Cerebral lipids in maple syrup disease. *J. Ped.* **66**, 584.

—— Alter, M., Steigleder, G.K., Weakley, D.R., and Sung, J.H. (1962). A sex-linked recessive disorder with retardation of growth, peculiar hair and cerebral and cerebellar degeneration. *Pediatrics* **29**, 764.

—— O'Brien, J.S., Okada, S., Grippo, J., Andrews, J.M., and Cancilla, P.A. (1971). Juvenile GM$_2$-gangliosidosis. *Arch. Neurol.* **25**, 14.

Mentzer, W.C., jun., Baehner, R.L., Schmidt-Schonbein, H., Robinson, S.H., and Nathan, D.G. (1971). Selective reticulocyte destruction in erythrocyte pyruvate kinase deficiency. *J. clin. Invest.* **50**, 688.

Menzies, I.S. and Seakins, J.W.T. (1976). Sugars. In *Chromatographic and electrophoretic techniques*, Vol. 1: *Paper and thin layer chromatography*, 4th edn. (ed. I. Smith and J.W.T. Seakins), p.183. William Heinemann, London.

Mercier, C. and Whelan, W.J. (1973). Further characterization of glycogen from type IV glycogen storage disease. *Eur. J. Biochem.* **40**, 221.

Merin, S., Livni, N., Berman, E.R., and Yatziv, S. (1975). Mucolipidosis IV. Ocular, systemic and ultrastructural findings. *Invest. Ophthal.* **14**, 437.

Merkatz, I.R., New, M.I., Peterson, R.E., and Seaman, M.P. (1969). Prenatal diagnosis of adrenogenital syndrome by amniocentesis. *J. Ped.* **75**, 977.

Mersmann, G. and Buddecke, E. (1977). Evidence for material from mannosidosis fibroblasts cross reacting with anti-acidic and α-mannosidase antibodies. *FEBS Lett.* **73**, 123.

—— Von Figura, K., and Buddecke, E. (1976). Mannosidosis: storage of mannose-containing material in cultured human mannosidosis cells and metabolic correction by pig kidney α-mannosidase. *Hoppe-Seyler's Z. Physiol. Chem.* **357**, 641.

Meuwissen, H.J., Pollara, B., and Pickering, R.J. (1975). Combined immuno-deficiency disease associated with adenosine deaminase deficiency. *J. Ped.* **86**, 169.

Meyer, U.A. (1973). Intermittent acute porphyria. Clinical and biochemical studies of disordered heme biosynthesis. *Enzyme* **16**, 366.

—— and Schmid, R. (1978). The porphyrias. In *The metabolic basis of inherited disease*, 5th edn. (ed. J.B. Stanbury, J.B. Wyngaarden, D.S. Fredrickson, J.L. Goldstein, and M.S. Brown), p.1166. McGraw-Hill, New York.

—— Strand, L.J., Doss, M., Rees, A.C., and Marver, H.S. (1972). Intermittent acute porphyria — demonstration of a genetic defect in porphobilinogen metabolism. *New Engl. J. Med.* **286**, 1277.

Meyer, W.J. III, Migeon, B.R., and Migeon, C.J. (1975). Locus on human X chromosome for dihydrotestosterone receptor and androgen insensitivity. *Proc. natl. Acad. Sci. USA* **72**, 1469.

Michals, K., Matalon, R., and Wong, P.W.K. (1978). Importance of methionine restriction. Dietary treatment of tyrosinemia type 1. *Am. Diet. Assoc.* **73**, 508.

Michels, V.V. and Beaudet, A.L. (1978). Arginase deficiency in multiple tissues of argininemia. *Clin. Genet.* **13**, 61.

—— —— (1980). Cholesteryl lignocerate. Hydrolysis in adrenoleucodystrophy. *Ped. Res.* **14**, 21.

—— Driscoll, D.J., Ferry, G.D., Duff, D.F., and Beaudet, A.L. (1979). Pulmonary vascular obstruction associated with cholesteryl ester storage disease. *J. Ped.* **94**, 621.

Migeon, B.R. and Huijing, F. (1974). Glycogen storage disease associated with phosphorylase kinase deficiency: evidence of inactivation. *Am. J. hum. Genet.* **26**, 360.

—— Der Kaloustian, V.M., Nyhan, W.L., Young, W.J., and Childs, B. (1968). X-linked hypoxanthine-guanine phosphoribosyl transferase deficiency: heterozygote has two clonal populations. *Science* **160**, 425.

—— Sprenkle, J.A., Liebaers, I., Scott, J.F., and Neufeld, E.F. (1977). X-linked Hunter syndrome: the heterozygous phenotype in cell culture. *Am. J. hum. Genet.* **29**, 448.

—— Moser, H.W., Moser, A.B., Axelman, J., Sillence, D., and Norum, R.A. (1981). Adrenoleukodystrophy: evidence for X linkage, inactivation and selection favouring the mutant allele in heterozygous cells. *Proc. natl. Acad. Sci. USA* **78**, 5066.

Migeon, C.J. and Kenny, F.M. (1966). Cortisol production rate. V. Congenital virilizing adrenal hyperplasia. *J. Ped.* **69**, 779.

Milch, R.A. (1957). Inheritance of alcaptonuria. *Bull. Hosp. Joint Dis.* **18**, 103.

Millán, J.L., Whyte, M.P., Avioli, L.V., and Fishman, W.H. (1980). Hypophosphatasia (adult form): quantitation of serum alkaline phosphatase isoenzyme activity in a large kindred. *Clin. Chem.* **26**, 840.

Miller, A.L. (1978). I-cell disease: isoelectric focusing, concanavalin A-sepharose 4B binding and kinetic properties of human liver ß-D-galactosidase. *Biochim. Biophys. Acta* **522**, 174.

—— Freeze, H.H., and Kress, B.C. (1981). I-cell disease. In *Lysosomes and lysosomal storage diseases* (ed. J.W. Callahan and J.A. Lowden), p.271. Raven Press, New York.

Miller, J.D., McCluer, R., and Kanfer, J.N. (1973). Gaucher's disease: Neurologic disorder in two siblings. *Ann. Inst. Med.* **78**, 883.

Miller, W.L. and Reimann, B.E.F. (1972). Childhood variant of Niemann–Pick disease. *Am. J. clin. Pathol.* **58**, 450.

Milne, M.D., Lewis, A.A.G., and Lyle, W.H., eds (1968). *Penicillamine*. In Proceedings of conference, the Royal Society of Medicine, London, 27 November 1967. *Postgrad. med. J.* suppl., October 1968.

—— London, D.R., and Asatoor, A.M. (1962). Citrullinuria in cases of cystinuria. *Lancet* **ii**, 49.

—— Asatoor, A.M., Edwards, K.D.G., and Loughridge, L.W. (1961). The intestinal absorption defect in cystinuria. *Gut* **2**, 323.

—— Crawford, M.A., Girao, C.B., and Loughridge, L.W. (1960). The metabolic disorder in Hartnup disease. *Q. J. Med.* **29**, 407.

Milstien, S. and Kaufman, S. (1975). Studies on the phenylalanine hydroxylase system *in vivo*. *J. biol. Chem.* **250**, 4782.

Milunsky, A. (1973). *The prenatal diagnosis of hereditary disorders*. C.C. Thomas Publishers, Springfield, Illinois.

—— and Neufeld, E.F. (1973). The Hunter syndrome in a 46XX girl. *New Engl. J. Med.* **288**, 106.

—— and Tulchinsky, D. (1977). Prenatal diagnosis of congenital adrenal hyperplasia due to 21-hydroxylase deficiency. *Pediatrics* **59**, 768.

Minio-Paluello, F., Gautier, A., and Magnenat, P. (1968). L'ultrastructure du foie humaine dans un cas de Crigler–Najjar. *Acta Hepatosplenol.* **15**, 65.

Minno, A.M. and Rogers, J.A. (1957). Ochronosis: report of a case. *Ann. intern. Med.* **46**, 179.

Minor, R.R. (1980). Collagen metabolism. A comparison of diseases of collagen and diseases affecting collagen. *Am. J. Pathol.* **98**, 227.

Miranda, A.F., Nette, E.G., Hartlage, P.L., and Di Mauro, S. (1979). Phosphorylase isoenzymes in normal myophosphorylase-deficient human heart. *Neurology* **29**, 1538.

Miras, J.C., Mantzos, D.J., and Levis, M.G. (1966). The isolation and partial characterization of glycolipids of normal human leucoyctes. *Biochem. J.* **98**, 782.

Mishkel, M., Nazir, D.J., and Crother, S. (1975). A longitudinal assessment of lipid ratios in the diagnosis of type III hyperlipoproteinemia. *Clin. Chim. Acta* **58**, 121.

Mitch, W.E. and Walser, M. (1977). Nitrogen balance of uremic patients receiving branched chain ketoacids and the hydroxy analogue of methionine as substitutes for the respective amino acids. *Clin. Nephrol.* **8**, 341.

Mitchell, B., Haigis, E., Steinmann, B., and Gitzelmann, R. (1975). Reversal of UDP-galactose 4-epimerase deficiency of human leukocytes in culture. *Proc. natl. Acad. Sci. USA* **72**, 5026.

Mitchener, W.M. (1967). Hyperuricaemia and mental retardation with athetosis and self-mutilation. *Am. J. Dis. Child.* **113**, 195.

Mitsumoto, H. (1979). McArdle disease: phosphorylase activity in regenerating muscle fibers. *Neurolgy* **29**, 258.

Mittwoch, U. (1963). The demonstration of mucopolysaccharide inclusions in the lymphocytes of patients with gargoylism. *Acta Haematol.* **29**, 202.

Miwa, S. and Nishina, T. (1974). Studies on pyruvate kinase (PK) deficiency. I. Clinical, hematological and erythrocyte enzyme studies. *Acta Haematol. Jap.* **37**, 1.

—— Sato, T., and Murao, H. (1972). A new type of phosphofructokinase deficiency: hereditary non-spherocytic hemolytic anemia. *Acta Haematol. Jap.* **35**, 113.

—— Nishina, T., Kakehashi, Y., Kitamura, M., Hiratsuka, A., and Shizume, K. (1971). Brief notes. Studies on erythrocyte metabolism in a case with hereditary deficiency of H-subunit of lactate dehydrogenase. *Acta Haematol. Jap.* **34**, 228.

Miyatake, T. (1969). A study on glycolipid in Fabry's disease. *Jap. J. exp. Med.* **39**, 35.

—— and Suzuki, K. (1972). Globoid cell leukodystrophy: additional deficiency of psychosine galactosidase. *Biochem. biophys. Res. Commun.* **48**, 538.

—— Atsumi, T., Obayashi, T., Mizuno, Y., Ando, S., Ariga, T., Matsui-Nakamura, K., and Yamada, T. (1979). Adult type neuronal storage disease with neuraminidase deficiency. *Ann. Neurol.* **6**, 232.

Miyazaki, K. and Ohtaki, N. (1975). Tyrosinase as glycoprotein. *Arch. Dermatol. Forsch.* **252**, 211.

Miyazaki, M., Fukuda, J., Aki, M., Ezawa, T., Kitamura, M., and Tamura, Z. (1971). A case of hepatic encephalomyelopathy associated with citrullinemia. *Brain Nerve* **23**, 19.

Mize, C.E., Herndon, J.H., Blass, J.P., Milne, G.W.A., Follansbee, C., Laudat, P., and Steinberg, D. (1969). Localisation of the oxidative defect in phytanic acid degradation in patients with Refsum's disease. *J. clin. Invest.* **48**, 1033.

Mjølnerod, O.K., Rasmussen, K., Dommerud, S.A., and Gjeruldsen, S.T. (1971). Congenital connective-tissue defect probably due to D-penicillamine treatment in pregnancy. *Lancet* **i**, 673.

Mohamed, S.D., McKay, E., and Galloway, W.H. (1965). Juvenile familial megaloblastic anaemia due to selective malabsorption of vitamin B_{12}. *Q. J. Med.* **35**, 433.

Mollica, F., Pavone, L., and Antener, I. (1971). Pure familial hyperprolinemia: isolated inborn error of amino acid metabolism without other anomalies in a Sicilian family. *Pediatrics* **48**, 225.

Mommaerts, W.F.H.M., Illingworth, B., Pearson, C.M., Guillory, P.J., and Seraydarian, K. (1959). A functional disorder of muscle associated with the absence of phosphorylase. *Proc. natl. Acad. Sci. USA* **45**, 791.

Money, J. and Ehrhardt, A.A. (1972). Gender dysmorphic behaviour and fetal sex hormones. *Rec. Progr. Horm. Res.* **28**, 735.

Mongeau, J.-G., Hilgartner, M., Worthen, H.G., and Frimpter, G.W. (1966). Cystathioninuria: study of an infant with normal mentality, thrombocytopenia, and renal calculi. *J. Ped.* **69**, 1113.

Monteleone, J.A., Beutler, E., Monteleone, P.L., Utz, C.L., and Casey, E.C. (1971). Cataracts, galactosuria and hypergalactosemia due to galactokinase deficiency in a child. *Am. J. Med.* **50**, 403.

Moore, M.R., Thompson, G.G., Goldberg, A., Ippen, H., Seubert, A., and Seubert, S. (1978). The biosynthesis of haem in congenital (erythropoietic) porphyria. *Int. J. Biochem.* **9**, 933.

Moorhead, P.J., Cooper, D.J., and Timperley, W.R. (1975). Progressive peripheral neuropathy in patient with primary hyperoxaluria. *Br. med. J.* **ii**, 312.

Moosa, A. and Dubowitz, V. (1971). Late infantile metachromatic leukodystrophy: effect of low vitamin A diet. *Arch. Dis. Child.* **46**, 381.

—— and Hughes, E.A. (1974). L-glutamine therapy in Leigh's encephalomyelopathy. *Arch. Dis. Child.* **43**, 246.

Moosman, K. and Bojanowsky, A. (1975). Rezidivierende candidiosis bec myeloperoxydasemangel. *Monatsschr. Kinderheilk.* **123**, 408.

Moran, T.J. and Yunis, E. J. (1962). Studies on ochronosis. 2. Effects of injection of homogentisic acid and ochronotic pigment in experimental animals. *Am. J. Path.* **40**, 359.

Morgan, H.D., Stewart, W.K., Lowe, K.G., Stowers, J.M., and Johnstone, J.H. (1962). Wilson's disease and the Fanconi syndrome. *Q. J. Med.* **31**, 361.

Morgan, J.M., Hartley, M.W., Miller, A.C., and Diethelm, A.G. (1974). Successful renal transplantation in hyperoxaluria. *Arch. Surg.* **109**, 430.

Morin, C.L., Thompson, M.W., Jackson, S.H., and Sass-Kortsak, A. (1971).

Biochemical and genetic studies in cystinuria. Observation on double heterozygotes of genotype I/II. *J. clin. Invest.* **50**, 1961.

Morquio, L. (1929). Sur une forme de dystrophie osseuse familiale. *Bull. Soc. Pédiat. Paris.* **27**, 145.

Morreels, C.L. jun., Fletcher, B.D., Weilbaecher, R.G., and Dorst, J.P. (1968). The roentgenographic features of homocystinuria. *Radiology* **90**, 1150.

Morris, J.M. (1953). The syndrome of testicular feminization in male pseudohermaphrodites. *Am. J. Obstet. Gynecol.* **65**, 1192.

—— (1964). Defective coupling of iodotryosine in familial goiters. *Arch. Int. Med.* **114**, 417.

Morris, M.D. and Fisher, D.A. (1967). Trypsinogen deficiency disease. *Am. J. Dis. Child.* **114**, 203.

—— Lewis, B.D., Dooland, P.D., and Harper, H.A. (1961). Clinical and biochemical observations on an apparently nonfatal variant of branched-chain ketoaciduria (maple syrup urine disease). *Pediatrics* **28**, 918.

Morrison, W.J., Christopher, N.L., Bayless, T.M., and Dana, E.A. (1974). Low lactase levels: evaluation of the radiologic diagnosis. *Radiology* **111**, 513.

Morrow, G. III (1967). Citrullinemia. A preliminary case report. *Am. J. Dis. Child.* **113**, 157.

—— and Barness, L.A. (1972). Combined vitamin responsiveness in homocystinuria. *J. Ped.* **81**, 946.

—— Mahoney, M.J., Mathews, C., and Lebowitz, J. (1975). Studies of methylmalonic coenzyme A carbonylmutase activity in methylmalonic acidemia. I. Correction of clinical, hepatic and fibroblast data. *Ped. Res.* **9**, 641.

—— Revsin, B., Mathews, C., and Giles, H. (1976). A simple rapid method for prenatal detection of defects in propionate metabolism. *Clin. Genet.* **10**, 218.

—— Schwarz, R.H., Hallock, J.A., and Barness, L.A. (1970). Prenatal detection of methylmalonic acidemia. *J. Ped.* **77**, 120.

—— Barness, L.A., Cardinale, G.J., Abeles, R.G., and Flaks, J.G. (1969). Congenital methylmalonic acidemia: enzymatic evidence for two forms of the disease. *Proc. natl. Acad. Sci. USA* **63**, 191.

—— Revsin, B., Clark, R., Lebowitz, J., and Whelan, D.T. (1978). A new variant of methylmalonic acidemia: defective coenzyme-apoenzyme binding in cultured fibroblasts. *Clin. Chim. Acta* **85**, 67.

Mortelmans, K., Friedberg, E.C., Slor, H., Thomas, G., and Cleaver, J.E. (1976). Defective thymine dimer excision by cell-free extracts of xeroderma pigmentosum cells. *Proc. natl. Acad. Sci. USA* **73**, 2757.

Moser, H.W. and Chen, W.W. (1983). Ceramidase deficiency: Farber's lipogranulomatosis. In *The metabolic basis of inherited disease*, 5th edn. (ed. J.B. Stanbury, J.B. Wyngaarden, D.S. Fredrickson, J.L. Goldstein, and M.S. Brown), p.820. McGraw-Hill, New York.

—— Moser, A.B., and McKhann, G.M. (1967). The dynamics of a lipidosis: turnover of sulfatide, steroid sulfate, and polysaccharide sulfate in metachromatic leukodystrophy. *Arch. Neurol.* **17**, 494.

—— Prenskey, A.L., Wolfe, H.J., and Rosman, N.P. (1969). Farber's lipogranulomatosis: report of a case and demonstration of an excess of free ceramide and ganglioside. *Am. J. Med.* **47**, 869.

—— Braine, H., Pyeritz, R.E., Ullman, D., Murray, C., and Asbury, A.K. (1980*a*). Therapeutic trial of plasmapheresis in Refsum disease and in Fabry disease. In *Enzyme therapy in genetic diseases: 2* (ed. R.J. Desnick), p.491. Alan R. Liss, New York.

—— Moser, A.B., Frayer, K.K., Chen, W., Schulman, J.D., O'Neill, B.P., and

Kishimoto, Y. (1981). Adrenoleukodystrophy: increased plasma content of saturated very long chain fatty acids. *Neurology.* **31**, 1241.

—— —— Kawamura, N., Migeon, B., O'Neill, B.P., Fenselau, C., and Kishimoto, Y. (1980*b*). Adrenoleukodystrophy: studies of the phenotype, genetics and biochemistry. *Johns Hopkins med. J.* **147**, 217.

—— —— Murphy, J., Suzuki, K., Schaumburg, H., and Kishimoto, Y. (1980*c*). Adrenoleukodystrophy: elevated C_{26} fatty acid in cultured skin fibroblasts. *Ann. Neurol.* **7**, 542.

—— —— Powers, J.M., Nitowsky, H.M., Schaumburg, H.H., Norum, R.A., and Migeon, B.R. (1982). The prenatal diagnosis of adrenoleukodystophy: demonstration of increased hexacosanoic acid levels in cultured amniocytes and fetal adrenal gland. *Ped. Res.* **16**, 172.

—— O'Brien, J.S., Atkins, L., Fuller, T.C., Klinman, A., Janowska, S., Russell, P.G., Bartsocas, C.S., Cosini, B., and Dulaney, J.T. (1974). Infusion of normal HL-A identical leucocytes in Sanfilippo disease type B. *Arch. Neurol.* **31**, 329.

Mossman, J. and Patrick, A.D. (1982). Prenatal diagnosis of mucopolysaccharidosis by two-dimensional electrophoresis of amniotic fluid glycosaminoglycans. *Prenatal Diagnosis* **2**, 169.

—— Patrick, A.D., Fensom, A.H., Tansley, L.R., Benson, P.F., Der Kaloustian, V.M., and Dudin, G. (1981). Correct prenatal diagnosis of a Hurler fetus where amniotic fluid cell cultures were of maternal origin. *Prenatal Diagnosis* **1**, 121.

—— Young, E.P., Patrick, A.D., Fensom, A.H., Ellis, M., Benson, P.F., and Der Kaloustian, V.M. (1983). Prenatal tests for Sanfilippo disease type B in four pregnancies. *Prenatal Diagnosis* **3**, 347.

Mudd, S.H. and Levy, H.L. (1978). Disorders of transsulfuration. In *The metabolic basis of inherited disease*, 4th end. (ed. J.B. Stanbury, J.B. Wyngaarden, and D.S. Fredrickson), p. 458. McGraw-Hill, New York.

—— Irreverre, F., and Laster, L. (1967). Sulfite oxidase deficiency in man: demonstration of the enzymatic defect. *Science* **156**, 1599.

—— Levy, H.L., and Abeles, R.H. (1969). A derangement in B_{12} metabolism leading to homocystinemia, cystathioninemia and methylmalonicaciduria. *Biochem. biophys. Res. Commun.* **35**, 121.

—— Finkelstein, J.D., Irreverre, F., and Laster. L. (1964). Homocystinuria: an enzymatic defect. *Science* **143**, 1443.

—— Laster, L., Finkelstein, J.D., and Irreverre, F. (1966). Studies on homocystinuria. In *Amines and schizophrenia* (ed. H.E. Himwich, S.S. Kety, and J.R. Smithies), p.247. Pergamon Press, New York.

—— Uhlendorf, B.W., Freeman, J.M., Finkelstein, J.D., and Shih, V.E. (1972). Homocystinuria associated with decreased methylenetetrahydrofolate reductase activity. *Biochem. biophys. Res. Commun.* **46**, 905.

—— —— Hinds, K.R., and Levy, H.L. (1970*b*). Deranged B_{12} metabolism: studies of fibroblasts grown in tissue culture. *Biochem. Med.* **4**, 215.

—— Edwards, W.A., Loeb, P.M., Brown, M.S., and Laster, L. (1970*a*). Homocystinuria due to cystathionine synthase deficiency: the effect of pyridoxine. *J. clin. Invest.* **49**, 1762.

Muensch, H., Yoshida, A., Altland, K., Jensen, W., and Goedde, H.-W. (1978). Structural difference at the active site of dibucaine resistant variant of human plasma cholinesterase. *Am. J. hum. Genet.* **30**, 302.

Muir, H. and Jacobs, S. (1967). Protein-polysaccharides of pig laryngeal cartilage. *Biochem. J.* **103**, 367.

—— Mittwoch, U., and Bitter, T. (1963). The diagnostic value of isolated urinary mucopolysaccharides and of lymphocyte inclusions in gargoylism. *Arch. Dis. Child.* **38**, 358.

Mulivor, R.A., Mennoti, M., Zackai, E.H., and Harris, H. (1978). Prenatal diagnosis of hypophosphatasia, genetics, biochemical, and clinical studies. *Am. J. hum. Genet.* **30**, 271.

Müller, C.R., Westerveld, A., Migl, B., Franke, W., and Ropers, H.H. (1980). Regional assignment of the gene locus for steroid sulfatase. *Hum. Genet.* **54**, 201.

Müller, D., Pilz, H., and Ter Meulen, V. (1969). Studies on adult metachromatic leucodystrophy. *J. Neurol. Sci.* **9**, 567.

Muller, H. and Harzer, K. (1980). Partial purification of acid sphingomyelinase from normal and pathological (M. Niemann–Pick Type C) human brain. *J. Neurochem.* **34**, 446.

Munro, G.F. and Miller, D.R. (1970). Mechanism of fructose diphosphate activation of a mutant pyruvate kinase (PK) from human red cells. *Biochim. Biophys. Acta* **206**, 87.

Murad, S. and Kishimoto, Y. (1978). Chain elongation of fatty acid in brain: a comparison of mitochondrial and microsomal enzyme activities. *Arch. Biochem. Biophys.* **185**, 300.

Murphey, W.H., Lindmark, D.G., Patchen, L.I., Housler, M.E., Harrod, E.K., and Mosovich, L. (1973). Serum carnosinase deficiency concomitant with mental retardation. *Ped. Res.* **7**, 601.

Murphy, J.V., Wolfe, H.J., Balazs, E.A., and Moser, H.W. (1971). A patient with deficiency of arysulfatases, A, B, C and steroid sulfatase. In *Lipid storage diseases: Enzymatic defects and clinical implications* (ed. J. Bernsohn and H.J. Grossman), p.67. Academic Press, New York.

Murray, G.J., Furbish, F.S., Doebber, T.W., Wu, M.S., Bugianesi, R.L., Brady, R.O., and Barranger, J.A. (1983). Chemical and enzymatic modifications result in improved targeting of ß-glucuronidase to rat liver non-parenchymal cells. (Abstract) *7th International symposium on glycoconjugates, Lund-Ronneby, Sweden, 17–23 July.* Rahms i, Lund.

Murray, J.C., Lindberg, K.A., and Pinnell, S.R. (1976). Inhibition of lysyl hydroxylase by homogentisic acid: a proposed connective tissue defect in alcaptonuria. *Clin. Res.* **24**, A15.

Mustajoki, P. and Koskelo, P. (1976). Hereditary hepatic porphyrias in Finland. *Acta Med. Scand.* **200**, 171.

Myerowitz, R. and Neufeld, E.F. (1981). Maturation of α-L-iduronidase in cultured human fibroblasts. *J. biol. Chem.* **256**, 3044.

Myers, J.P., Sung, J.H., Cowan, D., and Wolff, A. (1963). Pathological findings in the central and peripheral nervous systems in Chediak–Higashi's disease and the findings of cytoplasmic neuronal inclusions. *J. Neuropathol. exp. Neurol.* **22**, 357.

Myrianthopoulos, N.C. and Aronson, S.M. (1966). Population dynamics of Tay–Sachs disease. I. Reproductive fitness and selection. *Am. J. hum. Genet.* **18**, 313.

Nadler, H.L. and Gerbie, A.B. (1969). Enzymes in non cultured amniotic fluid cells. *Am. J. Obstet. Gynecol.* **103**, 710.

—— Inouye, T., and Hsia, D.Y.-Y. (1969). Classical galactosemia: a study of fifty-five cases. In *Galactosaemia* (ed. D.Y.-Y. Hsia), p.127. C.C. Thomas, Springfield, Illinios.

Nakagawa, S., Kumin, S., and Nitowsky, H.M. (1980). Studies on the activities and properties of lysosomal hydrolases in fractional populations of human peripheral blood cells. *Clin. Chim. Acta* **101**, 33.

Nagamani, M., McDonough, P.G., Ellegood, J.O., and Mahesh, V.B. (1978).

Maternal and amniotic fluid 17 α-hydroxyprogesterone levels during pregnancy: diagnosis of congenital adrenal hyperplasia in utero. *Am. J. Obstet. Gynecol.* **130**, 791.

Nakamura, E., Rosenberg, L.E., and Tanaka, K. (1976). Microdetermination of methylmalonic acid and other short chain dicarboxylic acids by gas chromatography: use in prenatal diagnosis of methylmalonic acidemia and in studies of isovaleric acidemia. *Clin. Chim. Acta* **68**, 127.

Nakashima, K. (1973). Further evidence of molecular alteration and abberation of erythrocyte pyruvate kinase. *Clin. Chim. Acta* **55**, 245.

Nance, W.E., Jackson, C.E., and Witkop, C.J. jun. (1970). Amish albinism: a distinctive autosomal recessive phenotype. *Am. J. hum. Genet.* **22**, 579.

Narayanan, N.S., Siegel, R.C., and Martin, G.R. (1972). On the inhibition of lysyl oxidase by ß-aminopropionitrile. *Biochem. biophys. Res. Commun.* **46**, 745.

Narisawa, K., Igarashi, Y., Otomo, H., and Tada, K. (1978). A new variant of glycogen storage disease type I probably due to a defect in the glucose-6-phosphatase transport system. *Biochem. biophys. Res. Commun.* **83**, 1360.

—— Wada, Y., Saito, T., Suzuki, H., Kudo, M., Arakawa, T.S., Katsushima, N., and Tsuboi, R. (1977). Infantile type of homocytinuria with $N^{5,10}$-methylenetetrahydrofolate reductase defect. *Tohoku J. exp. Med.* **121**, 185.

Natale, P.J. and Tremblay, G.C. (1969). On the availability of intramitochondrial carbamoylphosphate for the extramitochondrial biosynthesis of pyrimidines. *Biochem. biophys. Res. Commun.* **37**, 512.

Nathan, D.G., Oski, F.A., Sidel, V.W., and Diamond, L.K. (1965). Extreme hemolysis and red cell distortion in erythrocyte pyruvate kinase deficiency. II. Measurements of erythrocyte glucose consumption, potassium flux and adenosine triphosphate stability. *New Engl. J. Med.* **272**, 118.

Natowizc, M.R., Chi, M.M.-Y., Lowry, O.H., and Sly, W.S. (1979). Enzymatic identification of mannose 6-phosphate on the recognition marker for receptor-mediated pinocytosis of ß-glucuronidase by human fibroblasts. *Proc. natl. Acad. Sci. USA* **76**, 4322.

Navon, R., Padeh, B., and Adam, A. (1973). Apparent deficiency of hexosaminidase A in healthy members of a family with Tay–Sachs disease. *Am. J. hum. Genet.* **25**, 287.

Naylor, E.W. (1980). Newborn screening for maple syrup urine disease (branched chain ketoaciduria). In *Neonatal screening for inborn errors of metabolism* (ed. H. Bickel, R. Guthrie, and G. Hammersen), p.19. Springer, Berlin.

—— and Cederbaum, S.D. (1981). Urinary pyrimidine excretion in arginase deficiency. *J inher. metab. Dis.* **4**, 207.

Naylor, S.L., Klebe, R.J., and Shows, T.B. (1978). Argininosuccinic aciduria: assignment of the argininosuccinate lyase gene to the pter→q22 region of human chromosome 7 by bioautography. *Proc. natl. Acad. Sci. USA* **75**, 6159.

—— Lalley, P.A., Elliott, R.W., Brown, J.A., and Shows, T.B. (1980). Evidence for homologous regions of human chromosome 3 and mouse chromosome 9 predicts location of human genes. *Am. J. hum. Genet.* **32**, 159a.

Neale, G., Clark, M., and Levin, B. (1965). Intestinal sucrase deficiency presenting as sucrose intolerance in adult life. *Br. med. J.* **ii**, 1223.

Necheles, T.F., Rai, U.S., and Cameron, D. (1970). Congenital nonspherocytic hemolytic anemia associated with an unusual erythrocyte hexokinase abnormality. *J. lab. clin. Med.* **76**, 593.

Negishi, H. and Benke, P.J. (1977). Epithelial cells and Von Gierke's disease. *Ped. Res.* **11**, 936.

—— Morishita, Y., Kodama, S., and Matsuo, T. (1974). Platelet glucose-6-

phosphatase activity in patients with Von Gierke's disease and their parents. *Clin. Chim. Acta* **53**, 175.

Neil, J.F., Merikangas, J.R., and Glew, R.H. (1979). EEG findings in adult neuropathic Gaucher's disease. *Clin. Electroencephalo.* **10**, 198.

Nes, N. (1966). Testikulaer feminisering hos storfe. *Nord. Vet. Med.* **18**, 19.

Neuberger, A., Rimington, C., and Wilson, J.M.G. (1947). Studies on alcaptonuria. II. Investigations on a case of human alcaptonuria. *Biochem. J.* **41**, 438.

Neufeld, E.F., Liebaers, I., Epstein, C.J., Yatziv, S., Milunsky, A., and Migeon, B.R. (1977). The Hunter syndrome in females: is there an autosomal recessive form of iduronate sulfatase deficiency? *Am. J. hum. Genet.* **29**, 455.

Neustein, H.R., Stevenson, S.S., and Krainer, L. (1955). Oxalosis with renal calcinosis due to calcium oxalate. *J. Ped.* **47**, 624.

Neuwelt, E.A., Stumpf, D., Austin, J., and Kohler, P. (1971). A monospecific antibody to human sulfatase A. Preparation, characterization and significance. *Biochim. Biophys. Acta* **236**, 333.

—— Barranger, J.A., Brady, R.O., Pagel, M., Furbish, F.S., Quirk, J.M., Moor, G.E., and Frenkel, E. (1981). Delivery of hexosaminidase A to the cerebrum following osmotic modification of the blood–brain barrier. *Proc. natl. Acad. Sci. USA* **78**, 5838.

Neville, B.G.R., Harris, R.F., Stern, D.J., and Stern, J. (1971). Maternal histidinemia. *Arch. Dis. Child.* **46**, 119.

—— Lake, B.D., Stephens, R., and Sanders, M.D. (1973). A neurovisceral storage disease with vertical supranuclear ophthalmoplegia and its relationship to Niemann–Pick disease. *Brain* **96**, 97.

—— Bentovim, A., Clayton, B.E., and Shepherd, J. (1972). Histidinemia. Study of relations between clinical and biological findings in 7 subjects. *Arch. Dis. Child.* **47**, 190.

Nevin, N.C., Cumings, J.N., and McKeown, F. (1967). Refsum's syndrome, heredopathia atactica polyneuritiforms. *Brain* **90**, 419.

New, M.I. (1970). Male pseudohermaphroditism due to 17 α-hydroxylase deficiency. *J. Clin. Invest.* **49**, 1930.

—— (1977). Present status of prenatal diagnosis of congenital adrenal hyperplasia. In *Congenital adrenal hyperplasia* (ed. P.A. Lee, L.P. Plotnick, A.A. Kowalski, and C.J. Migeon), p.511. University Park Press, Baltimore.

—— and Seaman, M.P. (1970). Secretion rates of cortisol and aldosterone precursors in various forms of congenital adrenal hyperplasia. *J. clin. Endocr. Metab.* **30**, 361.

—— Miller, B., and Peterson, R.E. (1966). Aldosterone secretion in normal children and in children with adrenal hyperplasia. *J. clin. Invest.* **45**, 412.

Newcombe, D.S. and Willard, J.M. (1971). A spectrophotometric assay for HGPRT. *Analyt. Biochem.* **43**, 451.

Newcomer, A.D. and McGill, D.B. (1967). Disaccharidase activity in the small intestine: prevalence of lactase deficiency in 100 healthy subjects. *Gastroenterology* **53**, 881.

Newell, F.W., Matalon, R., and Meyer, S. (1975). A new mucolipidosis with psychomotor retardation, corneal clouding and retinal degeneration. *Am. J. Ophthalmol.* **80**, 440.

New England congenital hypothyroidism collaborative (1981). Effects of neonatal screening for hypothyroidism: prevention of mental retardation by treatment before clinical manifestations. *Lancet* **ii**, 1095.

New England regional screening program and the New England congenital hypothyroidism collaborative (1982). Pitfalls in screening for hypothyroidism. *Pediatrics* **70**, 16.

Ng, W.G., Bergren, W.R., and Donnell, G.N. (1973). A new variant of galactose-1-phosphate uridyltransferase in man: the Los Angeles variant. *Ann. hum. Genet.* **37**, 1.

—— Donnell, G.N., Koch, R., and Bergren, W.R. (1976). Biochemical and genetic studies of plasma and leucocyte α-L-fucosidase. *Am. J. hum. Genet.* **28**, 42.

Ng, Y.K. and Wolfe, L.S. (1975). Characterisation of oligosaccharides and glycopeptides excreted in the urine of GM_1-gangliosidosis patients. *Biochem. biophys. Res. Commun.* **66**, 123.

Niazi, M., Coleman, D.V., Mowbray, J.F., and Blunt, S. (1979). Tissue typing amniotic fluid cells — potential use for detection of contaminating maternal cells. *J. med. Genet.* **16**, 21.

Nicholls, A.C., Pope, F.M., and Schloon, H. (1979). Biochemical heterogeneity of osteogenesis imperfecta: new variant. *Lancet* **i**, 1193.

Nichols, J. (1970). Antenatal diagnosis and treatment of the adrenogenital syndrome. *Lancet* **i**, 83.

—— and Gibson, G.G. (1969). Antenatal diagnosis of the adrenogenital syndrome. *Lancet* **ii**, 1068.

Niederwieser, A., Giliberti, P., and Baerlocher, K. (1973). ß-mercaptolactate cysteine disulfiduria in two normal sisters. Isolation and characterization of ß-mercaptolactate cysteine disulfide. *Clin. Chim. Acta* **43**, 405.

—— Curtius, H.-C., Wang, M., and Leupold, D. (1982*b*). Atypical phenylketonuria with defective biopterin metabolism. Monotherapy with tetrahydrobiopterin or sepiapterin, screening and study of biosynthesis in man. *Eur. J. Pediatr.* **138**, 110.

—— Matasovic, A., Steinmann, B., Baerlocher, K., and Kempken, B. (1976). Hydantoin-5-propionic aciduria in folic acid nondependent formimino glutamic aciduria observed in two siblings. *Ped. Res.* **10**, 215.

—— Giliberti, P., Matasovic, A., Pluznik, S., Steinmann, B., and Baerlocher, K. (1974). Folic acid non-dependent formiminoglutamicaciduria in two siblings. *Clin. Chim. Acta* **54**, 293.

—— Staudenmann, W., Wang, M., Curtius, H.-C., Atares, M., and Cardesa-Garcia, J. (1982*a*). Hyperphenylalaninemia with neopterin deficiency — a new enzyme defect presumably of GTP cyclohydrolase. *Eur. J. Pediatr.* **138**, 97.

—— Curtius, H.-C., Gitzelmann, R., Otten, A., Baerlocher, K., Blehova, B., Berlow, S., Gröbe, H., Rey, F., Schaub, J., Scheibenreiter, S., Schmidt, H., and Viscontini, M. (1980). Excretion of pterins in phenylketonuria and phenylketonuria variants. *Helv. Paediatr. Acta* **35**, 335.

Nielsen, K.B., Wamberg, E., and Weber, J. (1979). Successful outcome of pregnancy in a phenylketonuric woman after low-phenylalanine diet introduced before conception. *Lancet* **i**, 1245.

Niemann, A. (1914). Ein unbekanntes Krankheitsbild. *Jahrb. Kinderheilk.* **79**, 1.

Niepomniszcze, H., Castells, S., De Groot, L.J., Refetoff, S., Kim, O.S., Rapoport, B., and Hati, R. (1973). Peroxidase defect in congenital goiter with complete organification block. *J. clin. Endocr. Metab.* **36**, 347.

Nikkila, E.A. and Aro, A. (1973). Family study of serum lipids and lipoproteins in coronary heart disease. *Lancet* **i**, 954.

Nilsson, L.R., Borgfors, N., Gamstrop, I., Holst, H.E., and Liden, G. (1964). Nonendemic goitre and deafness. *Acta. Paediatr.* **53**, 117.

Nilsson, O. and Svennerholm, L. (1982). Accumulations of glucosylceramide and glucosylsphingosine (psychosine) in cerebrum and cerebellum in infantile and juvenile Gaucher disease. *J. Neurochem.* **39**, 709.

Nishikawa, M., Ito, S., Sano, K., Nishioeda, Y., Nasako, Y., Fujisawa, T., Mori, T., and Seo, K. (1970). A case of familial cystathioninuria with goiter and some anomalies. *Endocrinol. Jap.* **17**, 57.

Nishimura, R.N. and Barranger, J.A. (1980). Neurologic complications of Gaucher's disease Type 3. *Arch. Neurol.* **37**, 92.

Nitowsky, H.M. and Grunfield, A. (1967). Lysosomal α-glucosidase in type II glycogenosis: activity in leucocytes and cell cultures in relation to genotype. *J. lab. clin. Med.* **69**, 472.

—— Sassa, S., Nakagawa, M., and Jagani, N. (1978). Prenatal diagnosis of congenital erythropoietic porphyria (abs.). *Ped. Res.* **12**, 455.

Nixon, A.D. and Buchanan, J.C. (1967). Haemolytic anaemia due to pyruvate kinase deficiency. *N. Z. med. J.* **66**, 859.

Norden, A.G.W. and O'Brien, J.S. (1973). Ganglioside GM₁ ß-galactosidase: studies in human liver and brain. *Arch. Biochem. Biophys.* **159**, 383.

—— —— (1975). An electrophoretic variant of ß-galactosidase with altered catalytic properties in a patient with GM₁-gangliosidosis. *Proc. natl. Acad. Sci. USA* **72**, 240.

—— Tennant, L.L., and O'Brien, J.S. (1974). GM₁ ganglosidase ß-galactosidase A: purification and studies of the enzyme from human liver. *J. biol. Chem.* **249**, 7969.

Norden, N.E., Lundblad, A., Svensson, S., Ockerman, P.-A., and Autio, S. (1973). A mannose-containing trisaccharide isolated from urines of three patients with mannosidosis. *J. biol. Chem.* **248**, 6210.

Norton, P.M., Roitman, E., Snyderman, S.E., and Holt, L.E. jun. (1962). A new finding in maple syrup urine disease. *Lancet* **i**, 26.

Norum, K.R. and Gjone, E. (1967). Familial plasma lecithin:cholesterol acyltransferase deficiency. Biochemical study of a new inborn error of metabolism. *Scand. J. clin. Lab. Invest.* **20**, 231.

—— —— (1968). The effect of plasma transfusion on the plasma cholesteryl esters in patients with familial plasma lecithin:cholesterol acyltransferase deficiency. *Scand. J. clin. Lab. Invest.* **22**, 339.

Nuki, G., Astrin, K., Brenton, D., Cruikshank, M., Lever, J., and Seegmiller, J. (1973). Purine and pyrimidine nucleotide pools in azaguanine resistant human lymphoblasts deficient in hypoxanthine-guanine phosphoribosyltransferase. *Am. J. hum. Genet.* **25**, 56A.

Nwokoro, N. and Neufeld, E.F. (1979). Detection of Hunter heterozygotes by enzymatic analysis of hair roots. *Am. J. hum. Genet.* **31**, 42.

Nyhan, W.L. (1983). Nonketotic hyperglycinemia. In *The metabolic basis of inherited disease*, 5th edn. (ed. J.B. Stanbury, J.B. Wyngaarden, D.S. Fredrickson, J.L. Goldstein, and M.S. Brown), p.561. McGraw-Hill, New York.

—— Borden, M., and Childs, B. (1961). Idiopathic hyperglycinemia, a new disorder of amino acid metabolism. The concentrations of other amino acids in the plasma and their modification by the administration of leucine. *Pediatrics* **27**, 539.

—— Oliver, W.J., and Lesch, M. (1965). A familial disorder of uric acid metabolism and central nervous system function. II. *J. Ped.* **67**, 257.

—— Bakay, B., Connor, J.D., Marks, J.F., and Keele, D.K. (1970). Hemizygous expression of glucose-6-phosphate dehydrogenase in erythrocytes of heterozygotes for the Lesch–Nyhan syndrome. *Proc. natl. Acad. Sci. USA* **65**, 214.

—— James, J.A., Teberg, A.J., Sweetman, L., and Nelson, L.G. (1969). A new

disorder of purine metabolism with behavioural manifestations. *J. Ped.* **74**, 20.
—— Pesek, J., Sweetman, L., Carpenter, D.G., and Carter, C.H. (1967). Genetics of an X-linked disorder of uric acid metabolism and cerebral function. *Ped. Res.* **1**, 5.
Oakes, B.W., Danks, D.M., and Campbell, P.E. (1976). Human copper deficiency: ultrastructural studies of the aorta and skin in a child with Menkes' syndrome. *Exp. mol. Path.* **25**, 82.
Oakey, R.E., Cawood, M.T., and MacDonald, P.R. (1974). Biochemical and clinical observations in a pregnancy with placental sulphatase and other enzyme deficiencies. *Clin. Endocrinol.* **3**, 131.
Oates, J.A., Nirenberg, P.Z., Jepson, J.B., Sjoerdsma, A., and Udenfriend, S. (1963). Conversion of phenylalanine to phenylethylamine in patients with phenylketonuria. *Proc. Soc. exp. Biol. Med.* **112**, 1078.
Oberholzer, V.G. and Briddon, A. (1978). 3-amino-2-piperidone in the urine of patients with hyperornithinemia. *Clin. Chim. Acta* **87**, 411.
—— and Palmer, T. (1976). Increased excretion of *N*-carbamoyl compounds in patients with urea cycle defects. *Clin. Chim. Acta* **68**, 73.
—— Levin, B., Burgess, E.A., and Young, W.F. (1967). Methylmalonic aciduria: An inborn error of metabolism leading to chronic metabolic acidosis. *Arch. Dis. Child.* **42**, 492.
O'Brien, D. and Butterfield, L.J. (1963). Further studies on renal tubular conservation of free amino acids in early infancy. *Arch. Dis. Child.* **38**, 437.
—— and Jensen, C.B. (1963). Pyridoxine dependency in two mentally retarded subjects. *Clin. Sci. mol. Med.* **24**, 179.
O'Brien, J.F., Cantz, M., and Spranger, J. (1974). Maroteaux–Lamy disease (mucopolysaccharidosis VI) subtype A. Deficiency of *N*-acetyl-galactosamine-4-sulphatase. *Biochem. biophys. Res. Commun.* **60**, 1170.
O'Brien, J.S. (1972*a*). Sanfilippo syndrome: Profound deficiency of alpha-acetylglucosaminidase activity in organs and skin fibroblasts from type B patients. *Proc. natl. Acad. Sci. USA* **69**, 1720.
—— (1972*b*). Ganglioside storage diseases. In *Advances in human genetics*, Vol. 3 (ed. K. Hirschhorn and H. Harris), p.39. Plenum Press, New York.
—— (1977). Pitfalls in the prenatal diagnosis of Tay–Sachs disease. In *Tay–Sachs disease: Screening and prevention* (ed. M. Kaback, D. Rimoin, and J. O'Brien), p.283. Alan R. Liss, New York.
—— (1978*a*). The cherry-red spot – myoclonus syndrome: a newly recognised inherited lysosomal storage disease due to acid neuraminidase deficiency. *Clin. Genet.* **14**, 55.
—— (1978*b*). GM$_1$-gangliosidosis. In *The metabolic basis of inherited disease*, 4th edn. (ed. J.B. Stanbury, J.B. Wyngaarden, and D.S. Fredrickson), p.841. McGraw-Hill, New York.
—— (1981). Enzymology of the sialidoses. In *Lysosomes and lysosomal storage diseases* (ed. J.W. Callahan and J.A. Lowden), p.263. Raven Press, New York.
—— (1983). The gangliosidoses. In *The metabolic basis of inherited disease*, 5th edn. (ed. J.B. Stanbury, J.B. Wyngaarden, D.S. Fredrickson, J.L. Goldstein, and M.S. Brown), p.945. McGraw-Hill, New York.
—— and Geiger, B. (1979). Normal adult with absent HEX A: immunoreactive HEX A is present. *Am. J. hum. Genet.* **31**, 642.
—— and Norden, A.G.W. (1977). Nature of the mutation in adult ß-galactosidase deficient patients. *Am. J. hum. Genet.* **29**, 184.
—— and Sampson, E.L. (1965). Myelin membrane: a molecular abnormality. *Science* **150**, 1613.

—— and Warner, T.G. (1980). Sialidosis: delineation of subtypes by neuraminidase assay. *Clin. Genet.* **17**, 35.

—— Miller, A.L., Loverde, A.W., and Veath, M.L. (1973). Sanfilippo disease type B: enzyme replacement and metabolic correction in cultured fibroblasts. *Science* **181**, 753.

—— Okada, S., Chen, A., and Fillerup, D.L. (1970). Tay–Sachs disease; detection of heterozygotes and homozygotes by serum hexosaminidase assay. *New Engl. J. Med.* **283**, 15.

—— Tennant, L.L., Veath, M.L., Scott, C.R., and Bucknall, W.E. (1978). Characterisation of unusual hexosaminidase A-deficient human mutants. *Am. J. hum. Genet.* **30**, 602.

—— Gugler, E., Giedon, A., Wiessmann, U., Hershkowitz, N., Meier, C., and Leroy, J. (1976). Spondyloepiphyseal dysplasia, corneal clouding, normal intelligence and ß-galactosidase deficiency. *Clin. Genet.* **9**, 495.

—— Okada, S., Fillerup, D., Veath, M.L., Adornata, B., Brenner, P.H., and Leroy, J.G. (1971). Tay–Sachs disease: prenatal diagnosis. *Science* **172**, 61.

Öckerman, P.A. (1965). Assay by a spectrofluorimetric method of glucose-6-phosphate in the liver in glycogen storage disease type 1. *Clin. Chim. Acta* **12**, 445.

—— (1967*a*). A generalised storage disorder resembling Hurler's syndrome. *Lancet* **ii**, 239.

—— (1967*b*). Deficiency of beta-galactosidase and alpha-mannosidase; primary enzyme defects in gargoylism and a new generalised disease. *Acta. Paediatr. Scand.* **177**, 35.

—— (1968). Lysosomal acid hydrolases in the liver in gargoylism. Deficiency of 4-methylumbelliferyl-ß-galactosidase. *Scand. J. clin. Lab. Invest.* **22**, 142.

—— (1969). Mannosidosis: isolation of oligosaccharide storage material from brain. *J. Ped.* **75**, 360.

—— (1973). Mannosidosis. In *Lysosomes and storage diseases* (ed. H.G. Hers and F. van Hoof), p.292. Academic Press, New York.

—— and Kohlin, P. (1968). Tissue acid hydrolase activities in Gaucher's disease. *Scand. J. clin. Lab. Invest.* **22**, 62.

O'Daly, S. (1968). An abnormal sulphydryl compound in urine. *Ir. J. med. Sci.* **1**, 578.

Odell, G.B. (1959). The dissociation of bilirubin from albumin and its clinical implications. *J. Ped.* **55**, 268.

Odievre, M. (1976). L'avenir des enfants atteints d'intolérance héréditaire au fructose. *Rev. Ped.* **12**, 449.

—— Gentil, C., Gautier, M., and Alagille, D. (1978). Hereditary fructose intolerance in childhood. Diagnosis, management and course in 55 patients. *Am. J. Dis. Child.* **132**, 605.

—— Charpentier, C., Cathelineau, L., Vedrenne, J., Delacoux, F., and Mercie, C. (1973). Hyperammoniémia constitutionelle avec déficit en carbamyl-phosphate-synthetase. *Arch. fr. Ped.* **30**, 5.

Oehmichen, M. (1980). Enzyme-histochemical differentiation of neuroglia and microglia: a contribution to the cytogenesis of microglia and globoid cells. *Path. Res. Pract.* **168**, 244.

—— and Grüninger, H. (1974). The origin of multinucleated giant cells in experimentally induced and spontaneous Krabbe's disease (globoid cell leuko-dystrophy). *Beitr. Path. Bd.* **153**, 111.

—— Wietholter, H., and Gencic, M. (1980). Cytochemical markers for mononuclear phagocytes as demonstrated in reactive microglia and globoid cells. *Acta Histochem.* **66**, 243.

Ogino, T. and Suzuki, K. (1981). Specificities of human and rat brain enzymes of cholesterol ester metabolism towards very long chain fatty acids: implications for biochemical pathogenesis for adrenoleukodystrophy. *J. Neurochem.* **36**, 776.

—— Schaumburg, H.H., Suzuki, K., Kishimoto, Y., and Moser, A.B. (1978). Metabolic studies of adrenoleukodystophy. In *Myelination and demyelination* (ed. J. Palo), p.601. Plenum Press, New York.

Ohmori, S., Kodama, H., Ikegami, T., Mizuhara, S., Aura, T., Isshiki, G., and Uremura, I. (1972). Unusual sulfur-containing amino acids in the urine of homocystinuric patients. III. Homocysteic acid, homocysteine sulfinic acid, *S*-(carboxymethylthio) homocysteine and *S*-(3-hydroxy-3-carboxy- η-propyl) homocysteine. *Physiol. Chem. Phys.* **4**, 286.

Ohnishi, A. and Dyck, P.J. (1974). Loss of small peripheral sensory neurons in Fabry disease. Histologic and morphometric evaluation of cutaneous nerves, spinal ganglia and posterior columns. *Arch. Neurol.* **31**, 120.

Ohno, S. (1978). The role of H-Y antigen in primary sex determination. *J. Am. med. Assoc.* **239**, 217.

Ohtaki, N. and Miyazaki, K. (1973). Immunologic homogeneity and electrophoretic heterogeneity of mouse melanoma tyrosinase. *J. Invest. Dermatol.* **61**, 339.

Oka, Y., Matsuda, I., Arashima, S., Anakura, M., Mitsuyama, T., and Nagamatsu, I. (1976). Citrate treatment in a patient with pyruvate decarboxylase deficiency. *Tohoku J. exp. Med.* **118**, 131.

Okada, S. and O'Brien, J.S. (1968). Generalised gangliosidosis: beta-galactosidase deficiency. *Science* **160**, 1002.

—— —— (1969). Tay–Sachs disease. Generalised absence of ß-D-*N*-acetyl-hexosaminidase component. *Science* **165**, 698.

—— McCrae, N., and O'Brien, J.S. (1972). Sandhoff's disease (GM$_2$-gangliosidosis type 2). Clinical, chemical and enzyme studies in five patients. *Ped. Res.* **6**, 606.

—— Veath, M.L., and O'Brien J.S. (1970). Juvenile GM$_2$-gangliosidosis, partial deficiency of hexosaminidase A. *J. Ped.* **77**, 1063.

—— —— Leroy, J., and O'Brien, J.S. (1971). Ganglioside GM$_2$-storage diseases: hexosaminidase deficiencies in cultured fibroblasts. *Am. J. hum. Genet.* **23**, 55.

Okeda, R., Suzuki, Y., Horiguchi, S., and Fujii, T. (1979). Fetal globoid cell leukodystrophy in one of twins. *Acta Neuropathol.* **47**, 151.

Okken, A., Van der Blij, J.F., and Hommes, F.A. (1973). Citrullinemia and brain damage. *Ped. Res.* **7**, 52.

Okolicsanyi, L., Orlando, R., Venuti, M., Dalbrun, G., Cobelli, C., Ruggeri, A., and Salvan, A. (1981). A modelling study of the effect of fasting on bilirubin kinetics in Gilbert's syndrome. *Am. J. Physiol.* **240**, 266.

Okoro, A.N. (1975). Albinism in Nigeria. *Br. J. Dermatol.* **92**, 485.

Old, J.M., Ward, R.H.T., Petrou, M., Karagozlu, F., Modell, B., and Weatherall, D.J. (1982). First-trimester fetal diagnosis for haemoglobinopathies: three cases. *Lancet* **ii**, 1413.

O'Leary, J.A. (1965). Comparative studies of the gonad in testicular feminization and cryptorchidism. *Fertil. Steril.* **16**, 813.

Oldfield, J.E., Allen, P.H., and Adair, J. (1956). Identification of cystine calculi in mink. *Proc. Soc. exp. Biol. Med.* **91**, 560.

Olsen, B.R. and Prockop, D.J. (1974). Ferritin-conjugated antibodies used for labeling of organelles involved in the cellular synthesis and transport of procollagen. *Proc. natl. Acad. Sci. USA* **71**, 2033.

Olson, J.S., Ballou, D.P., Palmer, G., and Massey, V. (1974). The mechanism of action of xanthine oxidase. *J. biol. Chem.* **249**, 4363.

Olsson, R., Sourander, P., and Svennerholm, L. (1966). Experimental studies on the pathogenesis of leucodystrophies. I. The effect of intracerebrally injected sphingolipids in the rat brain. *Acta Neuropathol. (Berl.)* **6**, 153.

Omenn, G.S. and Cheung, S.C.Y. (1974). Phosphoglycerate mutase isozyme marker for tissue differentiation in man. *Am. J. hum. Genet.* **26**, 393.

O'Neill, B.P., Moser, H.W., and Marmion, L.C. (1980). The adrenoleukomyelo-neuropathy (ALMN) complex: elevated C_{26} fatty acid in skin fibroblasts and correlation with disease expression in three generations of a kindred. *Neurology* **30**, 352.

Opitz, J.M., Simpson, J.L., Sarto, G.E., Summitt, R.L., New, M., and German, J. (1972). Pseudovaginal perineoscrotal hypospadias. *Clin. Genet.* **3**, 1.

Orci, L., Carpenter, J.-L., Perrelet, A., Anderson, R.G.W., Goldstein, J.L., and Brown, M.S. (1978). Occurrence of low density lipoprotein receptors within large pits on the surface of human fibroblasts as demonstrated by freeze-etching. *Exp. Cell. Res.* **113**, 1.

Orii, T., Nakamura, F., Kudoh, T., Tsuchihaski, K., and Nakao, R. (1975). A profound deficiency of $(CH_3{}^{14}C)$ choline sphingomyelin cleaving enzyme in Niemann–Pick disease type B. *Tohoku J. exp. Med.* **117**, 193.

Orme, R.L.E. (1970). Wolman's disease: an unusual presentation. *Proc. R. Soc. Med.* **63**, 489.

Osang, M., Santer, R., Zahn, V., and Schaub, J. (1979). Prenatal diagnosis of hypophosphatasia. *Eur. J. Pediatr.* **130**, 225.

Oski, F.A. and Bowman, H. (1969). A low Km phosphoenolpyruvate mutant in the Amish with red cell pyruvate kinase deficiency. *Brit. J. Haematol.* **17**, 289.

—— and Diamond, L.K. (1963). Erythrocyte pyruvate kinase deficiency resulting in congenital nonspherocytic hemolytic anemia. *New Engl. J. Med.* **269**, 763.

—— Nathan, D.G., Sidel, V.W., and Diamond, L.K. (1964). Extreme hemolysis and red-cell distortion in erythrocyte pyruvate kinase deficiency. I. Morphology, erythrokinetics and family enzyme studies. *New Engl. J. Med.* **270**, 1023.

Osler, W. (1904). Ochronosis: the pigmentation of cartilages, sclerotics and skin in alkaptonuria. *Lancet* **i**, 10.

Otten, J. and Vis, H.L. (1968). Acute reversible renal tubular dysfunction following intoxication with methyl-3-chromone. *J. Ped.* **73**, 422.

Owada, M. and Neufeld, E. (1982). Is there a mechanism for introducing acid hydrolases into liver lysosomes that is independent of mannose 6-phosphate recognition? Evidence from I-cell disease. *Biochem. biophys. Res. Commun.* **105**, 814.

—— Nishiya, O., Sakiyama, T., and Kitagawa, T. (1980). Prenatal diagnosis of I-cell disease by measuring altered α-mannosidase activity in amniotic fluid. *J. inher. met. Dis.* **3**, 117.

Owen, C.A. jun. and McConahey, W.M. (1956). An unusual iodinated protein of the serum in Hashimoto's thyroiditis. *J. clin. Endocrinol. Metab.* **16**, 1570.

Oyanagi, K., Miura, R., and Yamanouchi, T. (1970). Congenital lysinuria: a new inherited transport disorder of dibasic amino acids. *J. Ped.* **77**, 259.

—— Takagi, N., Kitabatake, M., and Nakao, T. (1967). Hartnup disease. *Tohoku J. exp. Med.* **91**, 383.

Ozaki, H., Mizutani, M., Oka, E., Ohtahara, S., Kimoto, H., Tanaka, T., Hakozaki, H., Takahashi, K., and Suzuki, Y. (1978). Farber's disease (disseminated lipogranulomatosis): the first case reported in Japan. *Acta Med. Okayama* **32**, 69.

Pabico, R.C., Atanacio, B.C., McKenna, B.A., Pamurcoglu, T., and Yodaiken, R. (1973). Renal pathologic lesions and functional alterations in a man with Fabry's disease. *Am. J. Med.* **55**, 415.

Padgett, G.A. (1968). The Chediak-Higashi syndrome. *Adv. vet. Sci. Comp. Med.* **12**, 239.

—— Leader, R.W., Gorham, J.R., and O'Mary, C.C. (1965). The familial occurrence of the Chediak–Higashi syndrome in mink and cattle. *Genetics* **49**, 505.

Paglia, D.E. and Valentine, W.N. (1970). Evidence for molecular alteration of pyruvate kinase as a consequence of erythrocyte aging. *J. lab. clin. Med.* **76**, 202.

—— —— (1971). Additional kinetic distinctions between normal pyruvate kinase and a mutant isozyme from human erythrocytes. Correction of the kinetic anomaly by fructose-1,6-diphosphate. *Blood* **37**, 311.

—— —— and Rucknagel, D.L. (1972). Defective erythrocyte pyruvate kinase with impaired kinetics and reduced optimal activity. *Brit. J. haematol.* **22**, 651.

—— Shende, A., Lanzkowsky, P., and Valentine, W.N. (1981). Hexokinase 'New Hyde Park'. A low activity erythrocyte isozyme in a Chinese kindred. *Am. J. Hematol.* **10**, 107.

—— Paredes, R., Valentine, W.N., Dorantes, S., and Konrad, P.N. (1975). Unique phenotypic expression of glucose phosphate isomerase deficiency. *Am. J. hum. Genet.* **27**, 62.

—— Valentine, W.N., Baughan, M.A., Miller, D.R., Reed, C.F., and McIntyre, O.R. (1968). An inherited molecular lesion of erythrocyte pyruvate kinase. Identification of a kinetically aberrant isozyme associated with premature hemolysis. *J. clin. Invest.* **47**, 1929.

Pagliara, A.S., Karl, I.E., Hammond, M., and Dipnis, D.M. (1973). Hypoglycaemia in infancy and childhood, Parts I and II. *J. Ped.* **82**, 365.

Pahwa, A., Pahwa, S., O'Reilly, R., and Good, R.A. (1978). Treatment of the immunodeficiency diseases — progress toward replacement therapy emphasising cellular and macromolecular engineering. *Springer Semin. Immunopath.* **1**, 355.

Paigen, K., Laborca, C., and Watson, G. (1979). A regulatory locus for mouse beta-glucuronidase induction, *Gur*, controls messenger RNA activity. *Science* **203**, 554.

Pain, R.W., Welch, F.W., Woodroffe, A.J., Handley, D.A., and Lockwood, W.H. (1975). Erythropoietic uroporphyria of Günther first presenting at 58 years with positive family studies. *Br. med. J.* **iii**, 621.

Palmer, T., Oberholzer, V.G., Burgess, E.A., Butler, L.J., and Levin, B. (1974). Hyperammonemia in 20 families: Biochemical and genetic survey, including investigations in 3 new families. *Arch. Dis. Child.* **49**, 443.

Palo, J. and Savolainen, H. (1973). Biochemical diagnosis of aspartylglucosaminuria. *Ann. clin. Res.* **5**, 156.

—— Pollitt, R.J., Pretty, K.M., and Savolainen, H. (1973). Glycoasparagine metabolites in patients with aspartylglycosaminuria: comparison between English and Finnish patients with special reference to storage materials. *Clin. Chim. Acta* **47**, 69.

—— Riekkinen, P., Arstila, A.U., Autio, S., and Kivimaki, T. (1972). Aspartylglucosaminuria. II. Biochemical studies on brain, liver, kidney and spleen. *Acta Neuropath.* **20**, 217.

Pammenter, M., Albrecht, C., Liebenberg, N.V.D.W., and Van Jaarsveld, P.

(1978). Afrikander cattle congenital goiter: characteristics of its morphology and iodoprotein pattern. *Endocrinology* **102**, 954.

Pang, S., Levine, L.S., Chow, D., Faiman, C., and New, M.I. (1979). Serum androgen concentrations in neonates and young infants with virilizing congenital adrenal hyperplasia (VCAH). *Clin. Endocrinol.* **11**, 575.

Papadopoulos, S.N., Vagenakis, A.G., Moschos, A., Koutras, D.A., Matsaniotis, N., and Malamos, B. (1970). A case of a partial defect of the iodide trapping mechanism. *J. clin. Endocrinol. Metab.* **30**, 302.

Pare, C.M.B., Sandler, M., and Stacey, R.C. (1957). 5-hydroxytryptamine deficiency in phenylketonuria. *Lancet* **i**, 511.

Park, Y.K. and Linkswiler, H. (1970). Effect of vitamin B$_6$ depletion in adult man on the excretion of cystathionine and other methionine metabolites. *J. Nutr.* **100**, 110.

Parker, C.E., Shaw, K.N.F., Jacobs, E.E., and Gutenstein, M. (1970*a*). Hydroxylysinuria. *Lancet* **i**, 1119.

Parker, R., Snedden, W., and Watts, D.W.E. (1970*b*). The quantitative determination of hypoxanthine and xanthine ('oxypurines') in skeletal muscle from two patients with congenital xanthine oxidase deficiency (xanthinuria). *Biochem. J.* **116**, 317.

Parkman, R., Gelfand, W.E., Rosen, F.S., Sanderson, A., and Hirschhorn, R. (1975). Severe combined immunodeficiency and adenosine deaminase deficiency. *New Engl. J. Med.* **292**, 714.

Parks, G.A., Bermudez, J.A., Anast, C.S., Bongiovanni, A.M., and New, M.I. (1971). Pubertal boy with 3 ß-hydroxysteroid dehydrogenase defect. *J. clin. Endocrinol. Metab.* **33**, 269.

Parr, C.W., and Fitch, L.I. (1967). Inherited quantitative variations of human phosphogluconate dehydrogenase. *Ann. hum. Genet.* **30**, 339.

Parry, M.F., Root, R.K., Metcalf, J.A., Delaney, K.K., Kaplow, L.S., and Richard, W.J. (1981). Myeloperoxidase deficiency prevalence and clinical significance. *Ann. int. Med.* **95**, 293.

Partington, M.W., Delahaye, D.J., Masotti, R.E., Read, J.H., and Roberts, B. (1968). Neonatal tyrosinaemia: a follow-up study. *Arch. Dis. Child.* **43**, 195.

—— and Hennen, B.K.E. (1967). The Lesch–Nyhan syndrome: self-destructive biting, mental retardation, neurological disorder and hyperuricaemia. *Dev. Med. Child. Neurol.* **9**, 563.

Pascal, T.A., Gaull, G.E., Beratis, N.G., Gillam, B.M., and Tallan, H.H. (1978). Cystathionase deficiency: evidence for genetic heterogeneity in primary cystathioninuria. *Ped. Res.* **12**, 125.

—— —— —— —— —— and Hirschhorn, K. (1975*b*). Vitamin B$_6$-responsive and -unresponsive cystathioninuria: two variant molecular forms. *Science* **190**, 1209.

—— —— —— —— —— —— and Parker, C. (1975*a*). Cystathionase in long term lymphoid cell lines. Evidence for altered enzyme protein in cystathioninuria. *Ped. Res.* **9**, 315.

Passon, P.G. and Hultquist, D.E. (1972). Soluble cytochrome b5 reductase from human erythrocytes. *Biochim. Biophys. Acta* **275**, 62.

Patel, V. and Zeman, W. (1976). Variability of expressivity of α-fucosidase deficiency. In *Current trends in sphingolipidoses and allied disorders* (ed. B.W. Volk and L. Schneck), p.167. Plenum Press, New York and London.

Patrick, A.D. (1962). The degradative metabolism of L-cysteine and L-cystine *in vitro* by liver in cystinosis. *Biochem. J.* **83**, 248.

—— (1965). A deficiency in glucocerebrosidase in Gaucher's disease. *Biochem. J.* **97**, 17C.

—— (1978). Biochemical studies on amniotic fluid and its cells. In *Towards the prevention of fetal malformation* (ed. J.B. Scrimgeour), p.165. Edinburgh University Press, Edinburgh.

—— and Lake, B.D. (1968). Cystinosis: electron microscopic evidence of lysosomal storage of cystine in lymph node. *J. Clin. Pathol.* **21**, 571.

—— —— (1969*a*). Deficiency of an acid lipase in Wolman's disease. *Nature* **222**, 1067.

—— —— (1969*b*). An acid lipase deficiency in Wolman's disease. *Biochem. J.* **112**, 298.

—— (1973). Wolman disease. In *Lysosomes and storage diseases* (ed. H.G. Hers and F. Van Hoof), p.453. Academic Press, New York and London.

—— Berlin, R.D., and Schulman, J.D. (1979). Gamma-glutamyl transferase: studies of normal and cystinotic human leukocytes, rabbit neutrophiles and rat liver. *Ped. Res.* **13**, 1058.

—— Willcox, P., Stephens, R., and Kenyon, V.G. (1976). Prenatal diagnosis of Wolman's disease. *J. med. Genet.* **13**, 49.

—— Young, E., Kleijer, W.J., and Niermeijer, M.F. (1977). Prenatal diagnosis of Niemann–Pick disease type A using a chromogenic substrate. *Lancet* i, 144.

Patriquin, H.B., Kaplan, P., Kind, H.P., and Giedion, A. (1977). Neonatal mucolipidosis II (I-cell disease): clinical and radiologic features in three cases. *Am. J. Roentgenol.* **129**, 37.

Patton, V.M. and Dekaban, A.S. (1971). GM_1-gangliosidosis and juvenile cerebral lipidosis. *Arch. Neurol.* **24**, 529.

Paul, T.D., Brandt, I.K., Elsas, L.J., Jackson, C.E., Mamunes, P., Nance, C.S., and Nance, W.E. (1978). Phenylketonuria heterozygote detection in families with affected children. *Am. J. hum. Genet.* **30**, 293.

—— —— —— —— Nance, C.S., and Nance, W.E. (1979*b*). Linkage analysis using heterozygote detection in phenylketonuria. *Clin. Genet.* **16**, 216.

—— Greco, J. jun., Brandt, I.K., Jackson, C.E., and Nance, W.E. (1979*a*). Is there a heterozygote advantage in the birth weight and number of children born to PKU heterozygotes? *Am. J. hum. Genet.* **31**, 140A.

Pavone, L., Mollica, F., and Levy, H.L. (1975). Asymptomatic type-II hyperprolinaemia associated with hyperglycinaemia in three sibs. *Arch. Dis. Child.* **50**, 637.

—— Moser, H.W., Mollica, F., Reitano, C., and Durand, P. (1980). Farber's lipogranulomatosis: ceramide deficiency and prolonged survival in three relatives. *Johns Hopkins Med. J.* **147**, 193.

Pawsey, S.A., Magnus, I.A., Ramsey, C.A., Benson, P.F., and Giannelli, F. (1979). Clinical, genetic and DNA repair studies on a consecutive series of patients with xeroderma pigmentosum. *Quart. J. Med.* **48**, 179.

Paxton, J.W., Moore, M.R., Beattie, A.D., and Goldberg, A. (1974). 17-oxosteroid conjugates in plasma and urine of patients with acute intermittent porphyria. *Clin. Sci. mol. Med.* **46**, 207.

Pedrini, V., Lenuzi, L., and Zambotti, V. (1962). Isolation and identification of keratosulphate in urine of patients affected by Morquio-Ullrich disease. *Proc. Soc. exp. Biol. Med.* **110**, 847.

Pehlke, D.M., McDonald, J.A., Holmes, E.W., and Kelly, W.N. (1972). Inosinic acid dehydrogenase activity in the Lesch–Nyhan syndrome. *J. clin. Invest.* **51**, 1398.

Peiffer, J. (1959). Uber, die metachromatischen Leukodystrophien (Typ. Scholz). *Arch. Psychiatr. Nervenkr.* **199**, 386.

Pellefigure, F., Butler, J. De B., Spielberg, S.P., Hollenberg, M.D., Goodman,

S.I., and Schulman, J.D. (1976). Normal amino acid uptake by cultured human fibroblasts does not require γ-glutamyl transpeptidase. *Biochem. bipohys. Res. Commun.* **73**, 997.

Pellissier, J.F., Hassoun, J. Gambarelli, D., Byron, P.A. Casanova, P., and Toga, M. (1976). Maladie de Niemann–Pick, type C de Crocker. *Acta Neuropath. (Berl.)* **34**, 65.

Peltonen, L., Palotie, A., and Prockop, D.J. (1980). A defect in the structure of type 1 procollagen in a patient who had osteogenesis imperfecta: Excess mannose in the COOH-terminal propeptide. *Proc. natl. Acad. Sci. USA* **77**, 6178.

Pennock, C.A. (1976). A review and selection of simple laboratory methods used for the study of glycosaminoglycan excretion and the diagnosis of the mucopolysaccharidoses. *J. clin. Pathol.* **29**, 111.

—— Mott, M.G., and Barstone, G.F. (1970). Screening for mucopolysaccharidoses. *Clin. Chim. Acta* **27**, 93.

Penrose, LS. (1935). Inheritance of phenylpyruvic amentia (phenylketonuria). *Lancet* **ii**, 192.

Pentchev, P.G., Brady, R.O., Blair, H.E., Britton, D.E., and Sorrell, S.H. (1978). Gaucher disease; isolation and comparison of normal and mutant glucocerebrosidase from human spleen tissue. *Proc. natl. Acad. Sci. USA* **75**, 3970.

—— Gal, A.E., and Hibbert, S.R. (1975). Replacement therapy for inherited enzyme deficiency. Sustained clearance of accumulated glucocerebroside in Gaucher's disease following infusion of purified glucocerebrosidase. *J. mol. Med.* **1**, 73.

—— Gal, A.E., Booth, A.D., Omodeo-Sale, F., Fouks, J., Neumeyer, B.A., Quirk, J.M., Dawson, G., and Brady, R.O. (1980). A lysosomal storage disorder in mice characterized by a dual deficiency of sphingomyelinase and glucocerebrosidase. *Biochim. Biophys. Acta* **619**, 669.

Penttinen, R.P., Lichtenstein, J.R., Martin, G.R., and McKusick, V.A. (1975). Abnormal collagen metabolism in cultured cells in osteogenesis imperfecta. *Proc. natl. Acad. Sci. USA* **72**, 586.

Percy, A.K. and Brady, R.O. (1968). Metachromatic leucodystrophy: diagnosis with samples of venous blood. *Science* **161**, 594.

—— Farrell, D.F. and Kaback, M.M. (1972). Cerebroside sulphate (sulphatide) sulphohydrolase: an improved assay method. J. Neurochem. **19**, 233.

—— Kaback, M.M., and Herndon, R.M. (1977). Metachromatic leukodystrophy: comparison of early and late-onset forms. *Neurology* **27**, 933.

Pergament, E., Heimler, A., and Shah, P. (1973). Testicular feminisation and inguinal hernia. *Lancet* **ii**, 740.

Perheentupa, J. and Visakorpi, J.K. (1965). Protein intolerance with deficient transport of basic amino acids: Another inborn error of metabolism. *Lancet* **ii**, 813.

—— Raivio, K.O., and Nikkilä, E.A. (1972). Hereditary fructose intolerance. *Acta Med. Scand.* **542**, 65.

Perkoff, G.T., Parker, V.J., and Hahn, R.F. (1962). The effect of glucagon in three forms of glycogen storage disease. *J. clin. Invest.* **41**, 1099.

Pernow, B.B., Havel, R.J., and Jennings, D.B. (1967). The second wind phenomenon in McArdle's syndrome. *Acta Med. Scanc. (Suppl)* **472**, 294.

Perry, T.L. (1974). Homocystinuria. In *Heritable disorders of aminoacid metabolism* (ed. W.L. Nyhan), p.395. John Wiley and Sons, New York.

—— and Hansen, S. (1978). Biochemical effects in man and rat of three drugs which can increase brain GABA content. *J. Neurochem.* **30**, 679.

—— —— and MacDougall, L. (1966). Homolanthionine excretion in homocystinuria. *Science* **152**, 1750.

—— Urquhart, N., and Hansen, S. (1977). Studies of the glycine cleavage enzyme system in brain from infants with gylcine encephalopathy. *Ped. Res.* **12**, 1192.

—— Hansen, S., Love, D., and Finch, C.A. (1968c). N-acetylcystathionine: A new urinary amino-acid in congenital cystathioninuria. *Nature, Lond.* **219**, 178.

—— —— MacDougall, L., and Warrington, P.D. (1967a). Sulphur containing amino acids in the plasma and urine of homocystinurics. *Clin. Chim. Acta* **15**, 409.

—— Robinson, G.C., Teasdale, J.M., and Hansen, S. (1967b). Concurrence of cystathioninuria, nephrogenic diabetes insipidus and severe anemia. *New Engl. J. Med.* **276**, 721.

—— Applegarth, D.E., Evans, M.E., Hansen, S., and Jellum, E. (1975a). Metabolic studies of a family with massive formiminoglutamicaciduria. *Ped. Res.* **9**, 117.

—— Hansen, S., Love, D.L., Crawford, L.E., and Tischler, B. (1968a). Treatment of homocystinuria with a low-methionine diet, supplemental cystine and a methyl donor. *Lancet* **ii**, 474.

—— —— Tischler, B., Bunting, R., and Berry, K. (1967c). Carnosinemia: A new metabolic disorder associated with neurologic disease and mental defect. *New Engl. J. Med.* **277**, 1219.

—— —— —— Richards, F.M., and Sokol, M. (1973). Unrecognised adult phenylketonuria. Implications for obstetrics and psychiatry. *New Engl. J. Med.* **289**, 395.

—— Hardwick, D.F., Dixon, G.H., Dolman, C.L., and Hansen, S. (1965). Hypermethioninemia: a metabolic disorder associated with cirrhosis, islet cell hyperplasia and renal tubular degeneration. *Pediatrics* **36**, 236.

—— —— Hansen, S., Love, D.L., and Isreals, S. (1968b). Cystathioninuria in two healthy siblings. *New Engl. J. Med.* **278**, 590.

—— Wirtz, M.L.K., Kennaway, N.G., Hsia, Y.E., Atienza, F.C., and Uemura, H.S. (1980). Amino acid and enzyme studies of brain and other tissues in an infant with argininosuccinic aciduria. *Clin. Chim. Acta* **105**, 257.

—— Urquhart, N., MacLean, J., Evans, M.E., Hansen, S., Davidson, A.G.F., Applegarth, D.A., MacLeod, P.J., and Lock, J.E. (1975b). Nonketotic hyperglycinemia. Glycine accumulation due to absence of glycine cleavage in brain. *New Engl. J. Med.* **292**, 1269.

Persico, M.G., Toniolo, D., Nobile, C., D'Urso, M., and Luzzatto, L. (1981). cDNA sequence of human glucose 6-phosphate dehydrogenase cloned in pBR322. *Nature, Lond.* **294**, 778.

Peters, S.P., Coyle, P., and Glew, R.H. (1976). Differentiation of ß-glucocerebrosidase from ß-glucosidase in human tissues using sodium taurocholate. *Arch. Biochem. Biophys.* **175**, 569.

—— Glew, R.H., and Lee, R.E. (1977b). Gaucher's disease. In *Practical enzymology of the sphingo-lipidoses* (ed. R.H. Glew and S.P. Peters), p.71. Alan R. Liss, New York.

—— Coyle, P., Coffee, C.J., Glew, R.H., Kuhlenschmidt, M.S., Rosenfeld, L., and Lee, Y.C. (1977a). Purification and properties of a heat-stable glucocerebrosidase activating factor from control and Gaucher spleen. *J. biol. Chem.* **252**, 563.

Peterson, M.L. and Herber, R. (1967). Intestinal sucrase deficiency. *Trans. Ass. Am. Phys.* **80**, 275.

Peterson, D.I., Bacchus, A., Seaich, L., and Kelly, T.E. (1975). Myelopathy associated with Maroteaux–Lamy syndrome. *Arch. Neurol.* **32**, 127.

Peterson, R.E., Imperato-McGinley, J. Gautier, T., and Sturla, E. (1977). Male pseudohermaphroditism due to steroid 5 α-reductase deficiency. *Am. J. Med.* **62**, 170.

Petit, F.H., Yeaman, S.J., and Reed, L.J. (1978). Purification and characterization of branched chain α-ketoacid dehydrogenase complex of bovine kidney. *Proc. natl. Acad. Sci. USA* **75**, 4881.

Petrelli, M. and Blair, J.D. (1975). The liver in GM_1-gangliosidosis types 1 and 2. *Arch. Pathol.* **99**, 111.

Pettersson, G. and Bonnier, G. (1937). Inherited sex-mosaic in man. *Hereditas* **23**, 49.

Pflugshaupt, R., Scherz, R., and Bütler, R. (1976). Polymorphism of human red cell adenosine deaminase, esterase D, glutamate pyruvate transaminase and galactose-1-phosphate-uridyltransferase in the Swiss population. *Hum. Hered.* **26**, 161.

Philippart, M. and Menkes, J.H. (1967). Isolation and characterization of the principal cerebral glycolipids in the infantile and adult forms of Gaucher's disease. In *Inborn disorders of sphingolipid metabolism* (ed. S.M. Aronson and B.W. Volk), p.389. Pergamon Press, New York.

—— Franklin, S.S., and Gordon, A. (1972). Reversal of an inborn sphingolipidosis (Fabry's disease) by kidney transplantation. *Ann. intern. Med.* **77**, 195.

—— Martin, L., Martin, J.J., and Menkes, J.H. (1969). Niemann–Pick disease. *Arch. Neurol. (Chicago)* **20**, 227.

—— Sarlieve, L., Meurant, C., and Mechler, L. (1971). Human urinary sulfatides in patients with sulfatidosis (metachromatic leukodystrophy). *J. Lipid Res.* **12**, 434.

Phillips, J.A. III, Panny, S.R., Kazazian, H.H. jun., Boehm, C.D., Scott, A.F., and Smith, K.D. (1980). Prenatal diagnosis of sickle cell anemia by restriction endonuclease analysis: Hind III polymorphisms in γ-globin genes extend test applicability. *Proc. natl. Acad. Sci. USA* **77**, 2853.

Phillips, N.C., Robinson, D., and Winchester, B.G. (1974). Human liver α-D-mannosidase activity. *Clin. Chim. Acta* **55**, 11.

Phillips, S.F. and McGill, D.B. (1973). Glucose-galactose malabsorption in an adult: perfusion studies of sugar, electrolyte and water transport. *Am. J. dig. Dis.* **18**, 1017.

Picard, J.Y., Tran, D., and Josso, N. (1978). Biosynthesis of labelled anti-mullerian hormone by fetal testes: evidence for the glycoprotein nature of the hormone and for its disulfide-bonded structure. *Mol. Cell. Endocrin.* **12**, 17.

Pick, L. (1927). Uber die lipoidzellige splenohepatomegalie Typus Niemann–Pick als Stoffwechselerkrankung. *Med. Klin.* **23**, 1483.

Pickering, W.R. and Howell, R.R. (1972). Galactokinase deficiency: clinical and biochemical findings in a new kindred. *J. Ped.* **81**, 50.

Piesowicz, A.T. (1968). Hyperprolinaemia. *Arch. Dis. Child.* **43**, 748.

Pikula, B., Plamenac, P., Ćurčić, B., and Nikulin, A. (1973). Myocarditis caused by primary oxalosis in a 4-year-old child. *Vichows Arch. (Pathol. Anat.)* **358**, 99.

Pilz, H. and Heipertz, R. (1974). The fatty acid composition of cerebrosides and sulfatides in a case of adult metachromatic leukodystrophy. *Z. Neurol.* **206**, 203.

—— Müller, D., and Linke, I. (1973). Histochemical and biochemical studies of urinary lipids in metachromatic leukodystrophy and Fabry's disease. *J. lab. clin. Med.* **81**, 7.

—— Duensing, I., Heipertz, R., Seidel, D., Lowitzsch, K., Hopf, H.C., and

Goebel, H.H. (1977). Adult metachromatic leucodystrophy. *Eur. Neurol.* **15**, 301.

Pimstone, N.R. (1975). The hepatic aspects of the porphyrias. In *Modern trends in gastroenterology* (ed. A.E. Read), p.373. Butterworths, London.

Pinnell, S.R., Krane, S.M., Kenzora, J.E., and Glimcher, M.J. (1972). A heritable disorder of connective tissue: hydroxylysine-deficient collagen disease. *New Engl. J. Med.* **286**, 1013.

Piñol Aguadé, J., Castells, A., Indacochea, A., and Rodes, J. (1969). A case of biochemically unclassifiable hepatic porphyria. *Br. J. Dermatol.* **81**, 270.

—— Herrero, C., Almeida, J., Castells, A., Ferrando, J., De Asprer, J., Palou, A., and Gimenez, A. (1975). Porphyrie hépato-érythrocytaire. Une nouvelle forme de porphyrie. *Ann. Derm. Syphiligr.* **102**, 125.

Pinsky, L. and Krooth, R.S. (1967*a*). Studies on the control of pyrimidine biosynthesis in human diploid cell strains. I. Effect of 6-azauridine on cellular phenotype. *Proc. natl. Acad. Sci. (Wash.)* **57**, 925.

—— —— (1967*b*). Studies on the control of pyrimidine biosynthesis in human diploid cell strains. II. Effects of 5-azaorotic acid, barbituric acid and pyrimidine precursors on cellular phenotype. *Proc. natl. Acad. Sci. (Wash.)* **57**, 1267.

—— Kaufman, M., and Summitt, R.L. (1981). Congenital androgen insensitivity due to a qualitatively abnormal androgen receptor. *Am. J. med. Genet.* **10**, 91.

—— —— Straisfeld, C., Zilahi, B., and Hall, C. St.-G. (1978). 5 α-reductase activity of genital and nongenital skin fibroblasts from patients with 5 α-reductase deficiency, androgen insensitivity, or unknown forms of male pseudohermaphroditism. *Am. J. med. Genet.* **1**, 407.

—— Miller, J., Shanfield, B., Walters, G., and Wolfe, L.S. (1974). GM$_1$-gangliosidosis in skin fibroblast cultures: Enzymatic differences between types 1 and 2 and observations in a third variant. *Am. J. hum. Genet.* **26**, 563.

Piomelli, S., Lamola, A.A., Poh-Fitzpatrick, M.B., Seaman, C., and Harber, L.C. (1975). Erythropoietic protoporphyria and lead intoxication: The molecular basis for difference in cutaneous photosensitivity. I. Different rates of disappearance of protoporphyrin from the erythrocytes, both *in vivo* and *in vitro*. *J. clin. Invest.* **56**, 1519.

Pitt, D. (1967). Phenylalanine maintenance in phenylketonuria. *Aust. Pediat. J.* **3**, 161.

Pittelkow, R.B., Kierland, R.R., and Montgomery, H. (1957). Polariscopic and histochemical studies in angiokeratoma corporis diffusum. *Arch. Dermatol.* **76**, 59.

Pittman, C.S. and Pittman, J.A. jun. (1966). A study of the thyroglobulin, thyroidal protease and iodoproteins in two congenital goitrous cretins. *Am. J. Med.* **40**, 49.

Pittman, G., Deodhar, S., Schulman, J.B., and Lando, J.E. (1971). Nephropathic cystinosis in a young adult — report of a case. *Lab. Invest.* **24**, 442.

Plotkin, G.R. and Isselbacher, K.J. (1964). Secondary disaccharidase deficiency in adult celiac disease (non-tropical sprue) and other malabsorption states. *New Engl. J. Med.* **271**, 1033.

Poduslo, S.E., Tennekoon, G., Price, D., Miller, K., McKhann, G.M. (1976). Fetal metachromatic leukodystrophy: Pathology, biochemistry and a study of in vitro enzyme replacement in CNS tissue. *J. Neuropathol. exp. Neurol.* **35**, 622.

Poenaru, L. and Dreyfus, J.-C. (1973). Electrophoretic heterogeneity of human α-mannosidase. *Biochim. Biophys. Acta* **303**, 171.

—— —— Boué, J., Nicolesco, H., Ravise, N., and Bamberger, J. (1976). Prenatal diagnosis of fucosidosis. *Clin. Genet.* **10**, 260.

—— Girard, S., Thepot, F., Madelenat, P., Huraux-Rendu, C., Vinet, M.-C., and Dreyfus, J.-C. (1979). Antenatal diagnosis in three pregnancies at risk for mannosidosis. *Clin. Genet.* **16**, 428.

Pollack, M.S., Maurer, D., Levine, L.S., New, M.I., Pang, S., Duchon, M., Owens, R.P., Merkatz, I.R., Nitowsky, H.M., Sachs, G., and Dupont, B. (1979). Prenatal diagnosis of congenital adrenal hyperplasia (21-hydroxylase deficiency) by HLA typing. *Lancet* **i**, 1107.

Pollitt, R.J. (1973). Argininosuccinate lyase levels in blood, liver and cultured fibroblasts of a patient with argininosuccinic aciduria. *Clin. Chim. Acta* **46**, 33.

—— and Jenner, F.A. (1969). Enzymatic cleavage of 2-acetamido-1(ß-L-aspartamido)-1,2-dideoxy-ß-D-glucose by human plasma and seminal fluid. Failure to detect the heterozygous state for aspartylglucosaminuria. *Clin. Chim. Acta* **25**, 413.

—— —— and Mersky, H. (1968). Aspartylglucosaminuria. An inborn error of metabolism associated with mental defect. *Lancet*, **ii**, 253.

Polmar, S.H., Wetzler, E., Stern, R.C., and Hirschhorn, R. (1975). Restoration of in vitro lymphocyte responses with exogenous ADA in a patient with severe CID. *Lancet* **ii**, 743.

—— Stern, R.C., Schwartz, A.L., Wetzler, E.M., Chase, P.A., and Hirschhorn, R. (1976). Enzyme replacement therapy for adenosine deaminase deficiency and severe combined immunodeficiency. *New Engl. J. Med.* **295**, 1337.

Polonowski, C. and Bier, H. (1970). Pseudo-trypsinogen deficiency due to lack of intestinal enterokinase. *Acta Paed. Scand.* **59**, 458.

Pomeranz, M.M., Friedman, L.J., and Tunick, I.S. (1941). Roentgen findings in alcaptonuric ochronosis. *Radiology* **37**, 295.

Pomeroy, J., Efron, M.L., Dayma, J., and Hoefnagel, D. (1968). Hartnup disorder in a New England family. *New Engl. J. Med.* **278**, 1214.

Pommier, J. Tourniaire, J., Deme, D., Chalendar, D., Bornet, H., and Nunez, J. (1974). A defective thyroid peroxidase solubilized from a familial goitre with iodine organification defect. *J. clin. Endocrin. Metab.* **39**, 69.

—— —— Rahmoun, B., Deme, D., Pallo, D., Bornet, H., and Nunez, J. (1976). Thyroid iodine organification defects: A case with lack of thyroglobulin iodination and a case without any peroxidase activity. *J. clin. Endocrin. Metab.* **42**, 319.

Pope, F.M., Martin, G.R., Lichtenstein, J.R., Penttinen, R., Gerson, B., Rowe, D.W., and McKusick, V.A. (1975). Patients with Ehlers–Danlos syndrome type IV lack type III collagen. *Proc. natl. Acad. Sci. USA* **72**, 1314.

Popp, R.A. (1962). Studies of the mouse hemoglobin locus. II. Position of the hemoglobin locus with respect to albinism and shaker-l loci. *J. Hered.* **53**, 73.

Port, A.E. and Hunt, D.M. (1979). A study of the copper-binding proteins in liver and kidney tissue of neonatal normal and mottled mutant mice. *Biochem. J.* **183**, 721.

Porter, I.H., Schulze, J., and McKusick, V.A. (1962). Genetical linkage between the loci for glucose-6-phosphate dehydrogenase deficiency and colour blindness in American Negroes. *Ann. hum. Genet.* **26**, 107.

Porter, M.T., Fluharty, A.L., and Kihara, H. (1969). Metachromatic leuco-dystrophy: arylsulphatase A deficiency in skin fibroblast cultures. *Proc. natl. Acad. Sci. (Wash.)* **62**, 887.

—— —— —— (1971b). Correction of abnormal sulfate metabolism in cultured metachromatic leukodystrophy fibroblasts. *Science* **172**, 1263.

—— —— Harris, S.E., and Kihara, H. (1970). The accumulation of cerebroside sulfates by fibroblasts in culture from patients with late infantile metachromatic leucodystrophy. *Arch. Biochem.* **138**, 646.

—— —— Trammell, J., and Kihara, H. (1971a). A correlation of intracellular cerebroside sulfatase activity in fibroblasts with latency in metachromatic leukodystrophy. *Biochem. biophys. Res. Commun.* **44**, 660.

Porter, S. and Lowe, B.A. (1963). Congenital erythropoietic protoporphyria: 1. Case reports, clinical studies and porphyrin analyses in two brothers. *Blood* **22**, 521.

Poulos, A. (1981). Diagnosis of Refsum's disease using $(1-^{14}C)$phytanic acid as substrate. *Clin. Genet.* **20**, 247.

Poulson, R. (1976). The enzymic conversion of protoporphyrinogen IX to protoporphyrin IX in mammalian mitochondria. *J. biol. Chem.* **251**, 3730.

Powell, G.F. and Maniscalco, R.M. (1976). Bound hydroxyproline excretion following gelatin loading in prolidase deficiency. *Metabolism* **25**, 503.

—— Kurosky, A., and Maniscalco, R.M. (1977). Prolidase deficiency: Report of a second case with quantitation of the excessively excreted amino acids. *J. Ped.* **91**, 242.

—— Rasco, M.A., and Maniscalco, R.M. (1974). A prolidase deficiency in man with iminopeptiduria. *Metabolism* **23**, 505.

Powell, L.W., Hemingway, E., Billing, B.H., and Sherlock, S. (1967). Idiopathic unconjugated hyperbilirubinemia (Gilbert's syndrome): A study of 42 families. *New Engl. J. Med.* **277**, 1108.

Powers, S.G. (1981). Regulation of rat liver carbamyl phosphate synthetase. I. Inhibition by metal ions and activation by amino acids and other chelating agents. *J. biol. Chem.* **256**, 11160.

Prader, A. and Gurtner, H.P. (1955). Das Syndrom des Pseudohermaphroditismus masculinus bei Kongenitaler Nebennierenrinden Hyperplasie ohne Androgen-Uberproduktion. *Helv. Pediatr. Acta* **10**, 397.

—— Kind, H.P., and De Luca, H.F. (1976). Pseudovitamin D deficiency (vitamin D dependency). In *Inborn errors of calcium and bone metabolism* (ed. H. Bickel and J. Stern), p.115. University Park Press, Baltimore.

Prensky, A.L., Carr, S., and Moser, H.W. (1968). Development of myelin in inherited disorders of amino acid metabolism. Arch. Neurol. **19**, 552.

—— and Moser, H.W. (1966). Brain lipids, proteolipids and free amino acids in maple syrup urine disease. *J. Neurochem.* **13**, 863.

—— Ferreira, G., Carr, S., and Moser, H.W. (1967). Ceramide and ganglioside accumulation in Farber's lipogranulomatosis. *Proc. Soc. exp. Biol. Med. (N.Y.)* **126**, 725.

Price, R.G. and Dance, N. (1972). The demonstration of multiple heat stable forms of *N*-acetyl-ß-glucosaminidase in normal human serum. *Biochim. Biophys. Acta* **271**, 145.

Pritchard, D.H., Napthine, D.V., and Sinclair, A.J. (1980). Globoid cell leukodystrophy in polled Dorset sheep. *Vet. Path.* **17**, 399.

Prockop, D.J., Kivirikko, K.I., Tuderman, L., and Guzman, N.A. (1979). The biosynthesis of collagen and its disorders. *New Engl. J. Med.* **301**, 13 and 77.

Procopis, P.G. (1979). Computerised tomography in the leukodystrophies. *Clin. exp. Neurol.* **16**, 309.

—— Camakaris, J., and Danks, D.M. (1981). A mild form of Menkes' steely hair syndrome. *J. Ped.* **98**, 97.

Prohaska, J.R. and Wells, W.W. (1974). Copper deficiency in the developing rat brain: a possible model for Menkes' steely-hair disease. *J. Neurochem.* **23**, 91.

Proia, R.L. and Neufeld, E.F. (1982). Synthesis of ß-hexosaminidase in cell-free translation and in intact fibroblasts: An insoluble precursor α-chain in a rare form of Tay–Sachs disease. *Proc. natl. Acad. Sci. USA* **79**, 6360.

Prota, G. (1980). Recent advances in the chemistry of melanogenesis in mammals. *J. Invest. Dermatol.* **75**, 122.

—— and Thomson, R.H. (1976). Melanin pigmentation in mammals. *Endeavour* **35**, 32.

Przyrembel, H., Leupold, D., Tosberg, P., and Bremer, H.J. (1973). Amino acid excretion of premature infants receiving different amounts of protein. *Clin. Chim. Acta* **49**, 27.

—— Bachmann, D., Lombeck, I., Becker, K., Wendel, U. Wadman, S.K., and Bremer, H.J. (1975). Alpha-ketoadipic aciduria, a new inborn error of lysine metabolism: Biochemical studies. *Clin. Chim. Acta* **58**, 257.

—— Wendel, U., Becker, K., Bremer, H.J., Bruinvis, L., Ketting, D., and Wadman, S.K. (1976). Glutaric aciduria type II: Report on a previously undescribed metabolic disorder. *Clin. Chim. Acta* **66**, 227.

Pucholt, V., Fitzsimmons, J.S., Gelsthorpe, K., Reynolds, MA., and Milner, R.D.G. (1980). Location of the gene for 21-hydroxylase deficiency. *J. med. Genet.* **17**, 447.

Pueschel, S.M., Bresnan, M.J., Shih, V.E., and Levy, S.M. (1979). Thiamine-responsive intermittent branched chain ketoaciduria. *J. Ped.* **94**, 628.

Pullon, D.H.H. (1980). Homocystinuria and other methioninemias. In *Neonatal screening for inborn errors of metabolism* (ed. H. Bickel, R. Guthrie, and G. Hammersen), p.29. Springer, Berlin.

Qazi, Q.H., Hill, J.G., and Thompson, M.W. (1971). Steroid studies in parents of patients with congenital virilizing adrenal hyperplasia. *J. clin. Endocrinol. Metab.* **33**, 23.

Querido, A., Stanbury, J.B., Kassenaar, A.A.H., and Meijer, J.W.A. (1956). The metabolism of iodotyrosines. III. Di-iodotyrosine dehalogenating activity of human thyroid tissue. *J. clin. Endocrinol. Metal.* **16**, 1096.

Quigley, H.A. and Kenyon, K.R. (1974). Ultrastructural and histochemical studies of a newly recognised form of systemic mucopolysaccharidosis (Maroteaux–Lamy syndrome, mild phenotype). *Am. J. Ophthalmol.* **77**, 809.

Quinlan, C.D. and Martin, E.A. (1970). Refsum's syndrome: Report of three cases. *J. Neurol. Neurosurg. Psychiat.* **33**, 817.

Quisling, R.W., Moore, G.R., Jahrsdoerfer, R.A., and Canrell, R.W. (1979). Osteogenesis imperfecta. *Arch. Otolaryngol.* **105**, 207.

Qureshi, I.A., Letarte, J., Ouellet, R., and Lelievre, M. (1981*b*). Sodium benzoate therapy and dietary control in hyperargininemia. *Ped. Res.* **15**, 638.

—— —— —— and Lemieux, B. (1978). Enzymologic and metabolic studies in two families affected by argininosuccinic aciduria. *Ped. Res.* **12**, 256.

—— —— —— Lelievre, M., and Laberge, C. (1981*a*). Ammonia metabolism in a family affected by hyperargininemia. *Diabet. Metab.* **7**, 5.

Raabe, W.A. (1981). Ammonia and disinhibition in cat motor cortex by ammonium acetate, monofluoroacetate and insulin-induced hypoglycemia. *Brain Res.* **210**, 311.

Raafat, R., Hashemian, M.P., and Abrishami, M.A. (1973). Wolman's disease: Report of two new cases with a review of the literature. *Am. J. clin. Path.* **59**, 490.

Rabier, D., Cathelineau, L., and Kamoun, P. (1979). Letter to the Editor: Lack of mitochondrial enzymes of the urea cycle in human white blood cells. *Ped. Res.* **13**, 207.

Rac, R., Hill, G.N., Pain, R.W., and Mulhearn, C.J. (1968). Congenital goitre in merino sheep due to an inherited defect in the biosynthesis of thyroid hormone. *Vet. Sci.* **9**, 209.

Raghavan, S.S., Gajewski, A., and Kolodny, E.H. (1981). Leukocyte sulfatidase for the reliable diagnosis of metachromatic leukodystrophy. *J. Neurochem.* **36**, 724.

—— Topol, J. and Kolodny, E.H. (1980). Leukocyte ß-glucosidase in homozygotes and heterozygotes for Gaucher's disease. *Am. J. hum. Genet.* **32**, 158.

Rahman, A.N. and Lindenberg, R. (1963). The neuropathology of hereditary dystopic lipidosis. *Arch. Neurol. (Chicago)* **9**, 373.

Raijman, L. (1979). Double deficiencies of urea cycle enzymes in human liver. *Biochem. Med.* **21**, 226.

Raivio, K., Perheentupa, J., and Nikkilä, E.A. (1967). Aldolase activities in the liver in parents of patients with hereditary fructose intolerance. *Clin. Chim. Acta* **17**, 275.

Ramage, I. and Cunningham, W.L. (1975). The occurrence of low α-L-fucosidase activities in normal human serum. *Biochim. Biophys. Acta* **403**, 473.

Ramot, B. (1976). A family with pyrimidine 5'nucleotidase deficiency. *Blood* **47**, 919.

Rampini, S.U. (1973). Die kongenitalen Störungen des phenylalaninstoffwechsels. *Schweiz. Med. Wschr.* **103**, 537.

—— (1976). Cited by Van de Kamp (1979), p.33.

Ramsay, C.A. and Giannelli, F. (1975). The erythemal action spectrum and DNA repair synthesis in xeroderma pigmentosum. *Brit. J. Dermatol.* **92**, 49.

—— Coltart, T.M., Blunt, S., Pawsey, S.A., and Giannelli, F. (1974). Prenatal diagnosis of xeroderma pigmentosum. Report of the first successful case. *Lancet* **ii**, 1109.

Ramsey, M.S. and Dickson, D.H. (1975). Lens fringes in homocystinuria. Br. J. Ophthalmol. **59**, 338.

—— Yanoff, M., and Fine, B.S. (1972). The ocular histopathology of homocystinuria: A light and electronmicroscopic study. *Am. J. Ophthalmol.* **74**, 377.

Ramsey, R.B., Banik, W.L., and Davison, A.N. (1979). Adrenoleukodystrophy: Brain cholesterol esters and other neutral lipids. *J. Neurol. Sci.* **40**, 189.

Randle, P.J., Sugden, P.H., Kerbey, A.L., Radclife, P.M., and Hutson, N.J. (1979). Regulation of pyruvate oxidation and the conservation of glucose. *Biochem. Soc. Symp.* **43**, 67.

Ransome-Kuti, O., Kretchmer, N., Johnson, J.D., and Gribble, J.T. (1975). A genetic study of lactose digestion in Nigerian families. *Gastroenterology* **68**, 431.

Rao, D.C., Keats, B.J., Laloud, J.M., Morton, N.E., and Yee, S. (1979). A maximum likelihood map of chromosome 1. *Am. J. hum. Genet.* **31**, 680.

Rao, G.H.R., Gerrard, J.M., Witkop, C.J., and White, J.G. (1981). Platelet aggregation independent of ADP release or prostaglandin synthesis in patients with Hermansky–Pudlak syndrome. *Prostaglandins Med.* **6**, 459.

—— White, J.G., Jachimowicz, A.A., and Witkop, C.J. jun. (1974). Nucleotide profiles of normal and abnormal platelets by high-pressure liquid chromatography. *J. lab. clin. Med.* **84**, 839.

Rapin, I., Suzuki, K., Suzuki, K., and Valsamis, M. (1976). Adult (chronic) GM$_2$-gangliosidosis. Atypical spinocerebellar degeneration in a Jewish sibship. *Arch. Neurol.* **33**, 120.

—— Goldfischer, S., Katzman, R., Engel, J., and O'Brien, J.S. (1978). The cherry-red spot — myoclonus syndrome. *Ann. Neurol.* **3**, 234.

Rasmussen, K., Ando, T., Nyhan, W.L., Hull, D., Cottom, D., Donnell, G.,

Wadlington, W., and Kilroy, A.W. (1972). Excretion of propionylglycine in propionic acidemia. *Clin. Sci.* **42**, 665.

Rasmussen, H. and Bordier, P. (1974). *The physiological and cellular basis of metabolic bone disease.* Williams and Wilkins, Baltimore.

—— Pechet, M., Anast, C., Mazur, A., Gertner, J., and Broadus, A.E. (1980). Long term treatment of familial hypophosphatemic rickets with oral phosphate and 1 α-hydroxyvitamin D₃. *J. Ped.* **99**, 16.

Rathbun, J.C. (1948). 'Hypophosphatasia'. A new developmental anomaly. *Am. J. Dis. Child.* **75**, 822.

Ratner, S. (1973). Enzymes of arginine and urea synthesis. *Adv. Enzymol.* **39**, 3.

—— (1976). Enzymes of arginine and urea synthesis. In *The urea cycle* (ed. S. Grisolia, R. Báguena, and F. Mayor), p.181. John Wiley and Sons, New York.

Rattazzi, M.C. and Davidson, R.G. (1972). Prenatal detection of Tay–Sachs disease. In *Antenatal diagnosis* (ed. A. Dorfman). p.208. University of Chicago Press, Chicago.

—— —— (1977). Prenatal diagnosis of metachromatic leukodystrophy by electrophoretic and immunologic techniques. *Ped. Res.* **11**, 1030.

—— Marks, J., and Davidson, R.G. (1973). Electrophoresis of arylsulfatase from normal individuals and patients with metachromatic leucodystrophy. *Am. J. hum. Genet.* **25**, 310.

—— Brown, J., Davidson, R.G., and Shows, T.B. (1976). Studies on complementation of ß-hexosaminidase deficiencies in human GM₂-gangliosidosis. *Am. J. hum. Genet.* **28**, 143.

Rattenbury, J.M., Blau, K., Sandler, M., Pryse-Davies, J., Clark, P.J., and Pooley, S.S.F. (1976). Prenatal diagnosis of hypophosphatasia. *Lancet* i, 306.

Raynaud, E.J., Escourolle, R., Baumaur, H., Turpin, J.C., Dubois, G., Malpuech, G., and Lagarde, R. (1975). Metachromatic leukodystrophy. Ultrastructural and enzymatic study of a casc of variant O form. *Arch. Neurol* **32**, 834.

Read, D.H., Harrington, D.D., Keenan, T.W., and Hinsman, E.J. (1976). Neuronal visceral GM₁-gangliosidosis in a dog with ß-galactosidase deficiency. *Science* **194**, 442.

Reddi, O.S., Reddy, S.V., and Reddy, K.R.S. (1977). A sibship with hypervalinemia. *Hum. Genet.* **39**, 139.

—— Reddy, M.V.R., and Reddy, K.R.S. (1978). Familial hydroxykynureninuria. *Hum. Hered.* **28**, 238.

Redeker, A.G., and Bronow, R.S. (1964). Erythropoietic protoporphyria presenting as hydroa aestivale. *Arch. Dermatol.* **89**, 104.

Reed, G.B. jun., Dixon, J.F.P., Neustein, H.B., Donnell, G.N., and Landing, B.H. (1968). Type IV glycogenosis: patient with absence of a branching enzyme α-1,4-glucan α-1-4-glucan 6-glucosyl transferase. *Lab. Invest.* **19**, 546.

Reed, L.J., Pettit, F.H., Yeaman, S.J., Teague, W.M., and Bleile, D.M. (1980). Structure, function and regulation of the mammalian pyruvate dehydrogenase complex. *Proc. Eur. Biochem. Soc.* **60**, 47.

Reem, G.H. (1975). Phosphoribosylpyrophosphate overproduction: a new metabolic abnormality in the Lesch–Nyhan syndrome. *Science* **190**, 1098.

—— and Friend, C. (1975). Purine metabolism in murine virus-induced erythroleukemic cells during differentiation *in vitro. Proc. natl. Acad. Sci. USA* **72**, 1630.

Reeves, G., Rigby, P.G., Rosen, H., Friedell, G.H., and Emerson, C.P. (1963). Hemochromatosis and congenital nonspherocytic hemolytic anemia in siblings. *J. Am. Med. Assoc.* **186**, 123.

Refetoff, S., De Groot, L.J., and Barsano, C.P. (1980). Defective thyroid hormone feedback regulation in the syndrome of peripheral resistance to thyroid hormone. *J. clin. Endocrinol. Metab.* **51**, 41.

—— De Wind, L.T., and De Groot, L.J. (1967). Familial syndrome combining deaf-mutism, stippled epiphyses, goiter and abnormally high PBI: Possible target organ refratoriness to thyroid hormone. *J. clin. Endocrinol.* **27**, 279.

Refsum, S. (1945). Heredo-ataxia hemeralogica polyneuritiformis et tidligere ikke beskrevet familiaert syndrome? En foreløbig meddelelse. *Nord. Med.* **28**, 2682.

—— (1946). Heredopathia atactica polyneuritiformis. *Acta Psychiatr. Scand.* (*Suppl.*) **38**, 9.

Reifenstein, E.C. jun. (1947). Hereditary familial hypogonadism. *Clin. Res.* **3**, 86.

Reitman, M.L. and Kornfeld, S. (1981). UDP-*N*-acetylglucosamine:glycoprotein *N*-acetyl-glucosamine-1-phosphotransferase. Proposed enzyme for the phosphorylation of the high mannose oligosaccharide units of lysosomal enzymes. *J. biol. Chem.* **256**, 4275.

—— Varki, A., and Kornfeld, S. (1981). Fibroblasts from patients with I-cell disease and pseudo-Hurler polydystrophy are deficient in uridine 5′-diphosphate-*N*-acetylglucosamine: glycoprotein *N*-acetylglucosaminylphosphotransferase activity. *J. clin. Invest.* **67**, 1574.

Renal Transplant Registry, Advisory Committee to the (1975). Renal transplantation in congenital and metabolic diseases. A report from the ASC/NIH Renal transplant registry. *J. Am. med. Ass.* **232**, 148.

Renlund, M., Chester, M.A., Lundblad, A., Aula, P., Raivio, K.O., Autio, S., and Koskela, S.-L. (1979). Increased urinary excretion of free *N*-acetylneuraminic acid in thirteen patients with Salla disease. *Eur. J. Biochem.* **101**, 245.

Renteria, V.G. and Ferrans, V.J. (1976). Intracellular collagen fibrils in cardiac valves of patients with Hurler syndrome. *Lab. Invest.* **34**, 263.

—— —— and Roberts, W.C. (1976). The heart in Hurler syndrome. Gross histologic and ultrastructural observations in five necropsy cases. *Am. J. Cardiol.* **38**, 487.

Resibois, A. (1969). Electron microscopic study of metachromatic leucodystrophy. III. Lysosomal nature of inclusions. *Acta Neuropath.* (*Berl.*) **13**, 149.

—— (1971). Electron microscopic studies of metachromatic leucodystrophy. *Pathol. Europ.* **6**, 278.

Reuser, A.J.J. and Galjaard, H. (1976). Characterization of hexosaminidase C and S in fibroblasts from control individuals and patients with Tay–Sachs disease. *FEBS Lett.* **71**, 1.

—— Koster, J.F., Hoogeveen, A., and Galjaard, H. (1977). Biochemical, immunological and cell genetic studies in glycogenosis II. *Am. J. hum. Genet.* **30**, 132.

Rey, J., Frézal, J., Royer, P., and Lamy, M. (1966). L'absence congénitale de lipase pancréatique. *Arch. franc. Pediat.* **23**, 5.

—— Harpey, J.-P., Leeming, R.-J., Blair, J.-A., Aircardi, J., and Rey, J. (1977). Les hyperphenylalaninémies avec activité normale de la phenylalanine-hydroxylase. *Arch. franc. Pediat.* **34**, 109.

Rhead, W.J., Hall, C.L., Tanaka, K. (1981). Novel tritium release assays for isovaleryl-CoA dehydrogenases. *J. biol. Chem.* **256**, 1616.

Rhead, W.R. and Tanaka, K. (1980). Demonstration of a specific mitochondrial isovaleryl-CoA dehydrogenase deficiency in fibroblasts from patients with isovaleric acidemia. *Proc. natl. Acad. Sci. USA* **77**, 580.

Ricciuti, F.C., Gelehrter, T.D., and Rosenberg, L.E. (1976). X-chromosome inactivation in human liver: confirmation of X-linkage of ornithine transcarbamylase. *Am. J. hum. Genet.* **28**, 332.

Richards, F. II, Cooper, M.R., Pearce, L.A., Cowan, R.J., and Spurr, C.L. (1974). Familial spinocerebellar degeneration, hemolytic anemia and glutathione deficiency. *Arch. intern. Med.* **124**, 534.

Richardson, K.E. and Tolbert, N.E. (1961). Oxidation of glyoxylic acid to oxalic acid by glycolic acid oxidase. *J. biol. Chem.* **236**, 1280.

Richman, P.G. and Meister, A. (1975). Regulation of γ-glutamyl-cysteine-synthetase by nonallosteric feedback inhibition by glutathione. *J. biol. Chem.* **250**, 1422.

Riddick, F.A. jun, Desai, K.B., Stanbury, J.B., and Murison, P.J. (1969). Familial goiter with diminished synthesis of thryoglobulin. *Z. Exp. Med.* **150**, 203.

Rietra, P.J.G.M., Brouwer-Kelder, E.M., De Groot, W.P., and Tager, J.M. (1976). The use of biochemical parameters for the detection of carriers of Fabry's disease. *J. mol. Med.* **1**, 237.

—— Molenaar, J.L., Hamers, M.N., Tager, J.M., and Borst, P. (1974). Investigation of the α-galactosidase deficiency in Fabry's disease using antibodies against the purified enzyme. *Eur. J. Biochem.* **46**, 89.

Rifkind, A.B. (1976). Drug-induced exacerbations of porphyria. *Primary Care* **3**, 665.

Riley, J.D. (1960). Gout and cerebral palsy in a three-year-old boy. *Arch. Dis. Child.* **35**, 293.

Rimington, C. and With, T.K. (1973). Porphyrin studies in erythropoietic porphyria. *Danish med. Bull.* **20**, 5.

Ritchie, J.W.K. and Carson, N.A.J. (1973). Pregnancy and homocystinuria. *J. Obstet. Gynaecol. Br. Commonw.* **80**, 664.

Rittman, L.S., Tennant, L.L., and O'Brien, J.S. (1980). Dog GM_1-gangliosidosis: characterization of the residual liver acid ß-galactosidase. *Am. J. hum. Genet.* **32**, 880.

Robbins, J.H., Kraemer, K.H., Lutzner, M.A., Festoff, B.W., and Coon, H.G. (1974). Xeroderma pigmentosum: an inherited disease with sun sensitivity, multiple cutaneous neoplasms and abnormal DNA repair. *Ann intern. Med.* **80**, 221.

Robert, M.F., Schultz, D.J., Wolf, B., Cochran, W.D., and Schwartz, A.L. (1972). Successful treatment of a neonate with propionic acidemia and severe hyperammonemia by peritoneal dialysis. *Arch. Dis. Child.* **54**, 962.

Robertson, D.H. (1961). Nitrofurazone-induced haemolytic anaemia in a refractory case of Trypanosoma rhodesiense sleeping sickness: the haemolytic trait and self-limiting haemolytic anaemia. *Ann. Trop. Med. Parasit.* **55**, 49.

Robillard, J.E., Mangeaux, J.-G., Morin, C.L., Dallaire, L., and Massicotte, P. (1973). Syndrome de Lowe: défauts multisystémique de transport. *Union med. Canad.* **102**, 1496.

Robins, E., Hirsch, H.E., and Emmous, S.S. (1968). Glycosidases in the nervous system. I. Assay, some properties and distribution of beta-galactosidase, beta-glucuronidase and beta-glucosidase. *J. biol. Chem.* **243**, 4246.

Robinson, B.H. and Sherwood, W.G. (1975). Pyruvate dehydrogenase phosphatase deficiency. Cause of congenital chronic lactic acidosis in infancy. *Ped. Res.* **9**, 935.

—— Taylor, J., and Sherwood, W.G. (1977). Deficiency of dihydro-lipoyl dehydrogenase (a component of the pyruvate and α-keto-glutarate dehydrogenase complex): A cause of congenital chronic lactic acidosis in infancy. *Ped. Res.* **11**, 1198.

—— —— —— (1980). The genetic heterogeneity of lactic acidosis: occurrence of recognizable inborn errors of metabolism in a pediatric population with lactic acidosis. *Ped. Res.* **14**, 956.

—— Sherwood, W.G., Lampty, M., and Lowden, J.A. (1976). ß-methylglutaconic aciduria: A new disorder of leucine metabolism. *Ped. Res.* **10**, 371.

—— Taylor, J., Kahler, S.G., and Kirkman, H.N. (1981). Lactic acidemia, neurologic deterioration and carbohydrate dependence in a girl with dihydrolipoyl dehydrogenase deficiency. *Eur. J. Pediatr.* **136**, 35.

—— Sherwood, W.G., Taylor, J., Balfe, J.W., and Mamer, O.A. (1979). Acetoacetyl CoA thiolase deficiency: A cause of severe ketoacidosis in infancy simulating salicylism. *J. Ped.* **95**, 228.

Robinson, D. (1974). Multiple forms of glycosidases in the normal and pathological states. *Enzyme* **18**, 114.

—— and Stirling, J.L. (1968). *N*-acetyl-ß-D-glucosaminidases in human spleen. *Biochem. J.* **107**, 321.

Robinson, W.G., Nagle, R., Bachnawat, B.K., Kupiecki, E.P., and Coon, M.J. (1957). Coenzyme A thiol esters of isobutyric methacrylic and ß-hydroxyisobutyric acids as intermediates in the enzymatic degradation of valine. *J. biol. Chem.* **224**, 1.

Robson, E.B. and Rose, G.A. (1957). The effect of intravenous lysine on the renal clearances of cystine, arginine and ornithine in normal subjects, in patients with cystinuria and Fanconi syndrome and in their relatives. *Clin. Sci.* **16**, 75.

Rochiccioli, P. and Dutau, G. (1974). Trouble de l'hormonosynthése thyroidienne par déficit en iodotyrosine-déshalogenase. *Arch. fr. Ped.* **31**, 25.

Rockson, S., Stone, R., Van der Weyden, M., and Kelley, W.M. (1974). Lesch–Nyhan syndrome: evidence for abnormal adrenergic function. *Science* **186**, 934.

Rodeck, C.H. (1980). Fetoscopy guided by real-time ultrasound for pure fetal blood samples, fetal skin samples and examination of the fetus *in utero*. *Br. J. Obstet. Gynaec.* **87**, 449.

—— and Campbell, S. (1978). Sampling pure fetal blood by fetoscopy in the second trimester of pregnancy. *Br. med. J.* **2**, 728.

—— Fensom, A.H., Benson, P.F., and Ellis, M. (1983*b*). Prenatal exclusion of late infantile metachromatic leucodystrophy in a late-presenting pregnancy by assay of fetal leucocytes. *Prenatal Diagnosis* **3**, 257.

—— Patrick, A.D., Pembrey, M.E., Tzannatos, C., and Whitfield, A.E. (1982). Fetal liver biopsy for prenatal diagnosis of ornithine carbamyl transferase deficiency. *Lancet* **ii**, 297.

—— Tansley, L.R., Benson, P.F., Fensom, A.H., and Ellis, M. (1983*a*). Prenatal exlusion of Hurler's disease by leucocyte α-L-iduronidase assay. *Prenatal Diagnosis* **3**, 61.

Roden, L. (1980). Structure and metabolism of connective tissue proteoglycans. In *The biochemistry of glycoproteins and proteoglycans* (ed. W.J. Lennarz), p.267. Plenum Press, New York.

Rodriguez-Budelli, M.M., Kark, R.A.P., Blass, J.P., and Spence, M.A. (1978). Action of physostigmine on inherited ataxias. *Adv. Neurol.* **21**, 195.

Roe, T.G., Ng, W.G., Bergren, W.R., and Donnell, G.N. (1973). Urinary galactitol in galactosemic patients. *Biochem. Med.* **7**, 266.

Roelofs, R.I., Engel, W.K., and Chauvin, P.B. (1972). Histochemical phosphorylase activity in regenerating muscle fibers from myophosphorylase-deficiency patients. *Science* **177**, 795.

Roels, H., Qantacher, J., Kint, A., Van der Ecken, H., and Vrints, L. (1970). Generalised gangliosidosis — GM₁. Morphological study. *Eur. Neurol.* **3**, 129.

Roerdink, F.H., Gouw, W.L.M., Okken, A., Van der Blij, J.F., Luit-De Haan, G., Hommes, F.A., and Huisjes, H.J. (1973). Citrullinemia. Report of a case with studies on antenatal diagnosis. *Ped. Res.* **7**, 863.

Rogers, L.E., Warford, L.R., Patterson, R.B., and Porter, F.S. (1968). Hereditary orotic aciduria. 1. A new case with family studies. Pediatrics. **42**, 415.

Rogers, P.A., Fisher, R.A., and Harris, H. (1975). An examination of the age-related patterns of decay of the hexokinases of human red cells. *Clin. Chim. Acta* **65**, 291.

Rome, L.H., Garvin, A.J., and Neufeld, E.F. (1978). Human kidney α-L-iduronidase: purification and characterisation. *Arch. Biochem. Biophys.* **189**, 344.

Romeo, G. and Levin, E.Y. (1969). Uroporphyrinogen III cosynthetase in human congenital erythropoietic porphyria. *Proc. natl. Acad. Sci. USA* **63**, 856.

—— and Migeon, B.R. (1970). Genetic inactivation of the α-galactosidase locus in carriers of Fabry's disease. *Science* **170**, 180.

—— Childs, B., and Migeon, B.R. (1972). Genetic heterogeneity of α-galactosidase in Fabry's disease. *FEBS Lett.* **27**, 161.

—— Borrone, C., Gatti, R., and Durant, P. (1977). Fucosidosis in Calabria: founder effect or high gene frequency? *Lancet* **i**, 368.

—— Urso, M., Pisacane, A., Blum, E., De Falco, A., and Ruffilli, A. (1975). Residual activity of α-galactosidase A in Fabry's disease. *Biochem. Genet.* **13**, 615.

Root, R.K. and Metcalf, J.A. (1977). H₂O₂ release from human granulocytes during phagocytosis: relationship to superoxide anion formation and cellular catabolism of H₂O₂: studies with normal and cytochalasin-B treated cells. *J. clin. Invest.* **60**, 1266.

Rose, C.I. and Haines, D. (1978). Familial hyperglycerolemia. *J. clin. Invest.* **61**, 163.

Rose, H.G., Kranz, P., Weinstock, M., Juliano, J., and Haft, J.I. (1973). Inheritance of combined hyperlipoproteinemia: Evidence for a new lipoprotein phenotype. *Am. J. Med.* **148**, 160.

Rosemann, U. and Friede, R.L. (1967). Entry of labelled donor cells from the bloodstream into the CNS. *J. Neuropathol. exp. Neurol.* **26**, 144.

Rosen, H.M. and Klebanoff, S.J. (1979). Bactericidal activity of a superoxide anion generating system: a model for the polymorphonuclear leukocyte. *J. exp. Med.* **149**, 27.

—— Yoshimura, N., Hodgman, J.M., and Fischer, J.E. (1977). Plasma amino acid patterns in hepatic encephalopathy of differing etiology. *Gastroenterology.* **72**, 483.

Rosen, J.F., Fleischman, A.R., Finberg, L., Hamstra, A., and de Luca, H.F. (1979). Rickets with alopecia: an inborn error of vitamin D metabolism. *J. Ped.* **94**, 729.

Rosenberg, L.E. (1978). Disorders of propionate, methylmalonate and cobalamin metabolism. In *The metabolic basis of inherited disease* (ed. J.B. Stanbury, J.B. Wyngaarden, and D.S. Fredrickson), 4th ed., p.411. McGraw-Hill, New York.

—— Crawhall, J.C., and Segal, S. (1967). Intestinal transport of cystine and cysteine in man: evidence for separate mechanisms. *J. clin. Invest.* **46**, 30.

—— Durant, J.L., and Holland, J.M. (1965). Intestinal absorption and renal extraction of cystine and cysteine in cystinuria. *New Engl. J. Med.* **273**, 1239.

—— Lilljeqvist, A.-C., and Hsia, Y.E. (1968). Methylmalonic aciduria: Metabolic block localisation and vitamin B_{12} dependency. *Science* **162**, 805.

—— Downing, S.J., Durant, J.L., and Segal, S. (1966). Cystinuria: Biochemical evidence for three biochemically distinct diseases. *J. clin. Invest.* **45**, 365.

Rosenblatt, D.S. and Erbe, R.W. (1977). Methylene tetrahydrofolate reductase in cultured human cells. II. Genetic and biochemical studies of methylenetetra-hydrofolate reductase deficiency. *Pediatrics* **11**, 1141.

—— and Scriver, C.R. (1968). Heterogeneity in genetic control of phenylalanine metabolism in man. *Nature, Lond.* **218**, 677.

—— Cooper, B.A., Lue-Shing, S., Wong, P.W.K., Berlow, S., Narisawa, K., and Baumgartner, R. (1979). Folate distribution in cultured human cells. *J. clin. Invest.* **63**, 1019.

Rosenbloom, F.M., Kelley, W.N., Henderson, J.F., and Seegmiller, J.E. (1967). Lyon hypothesis and X-linked disease. *Lancet* **ii**, 305.

—— Henderson, J.F., Caldwell, I.A., Kelley, W.N., and Seegmiller, J.E. (1968). Biochemical bases of accelerated purine biosynthesis *de novo* in human fibroblasts lacking hypoxanthine-guanine phosphoribosyltransferase. *J. biol. Chem.* **243**, 1166.

Rosenfield, R.L., Laurence, A.M., Liao, S., and Landau, R.L., (1971). Androgens and androgen responsiveness in the feminizing testis syndrome: Comparison of complete and 'incomplete' forms. *J. clin. Endocrinol. Metab.* **32**, 625.

Rosewater, S., Gwinup, G., and Hamwi, G.J. (1965). Familial gynecomastia. *Ann. intern. Med.* **63**, 377.

Rösler, A., Leiberman, E., Rosenmann, A., Ben-Uzilio, R., and Weidemfeld, J. (1979). Prenatal diagnosis of 11 ß-hydroxylase deficiency congenital adrenal hyperplasia. *J. clin. Endocrinol. Metab.* **49**, 546.

Ross, G., Dunn, D., and Jones, M.E. (1978). Ornithine synthesis from glutamate in rat intestinal mucosa homogenates: Evidence for the reduction of glutamate to gamma-glutamyl semialdehyde. *Biochem. biophys. Res. Commun.* **85**, 140.

Roth, K., Cohn, R., Yandrasitz, J., Preti, G., Dodd, P., and Segal, S. (1976). Beta-methylcrotonic aciduria associated with lactic acidosis. *J. Ped.* **88**, 229.

Roth, M. and Felgenhauer, W.-R. (1968). Recherche de l'excrétion d'acide homogentisique urinaire chez des hétérozygotes pour l'alcaptonurie. *Enzymol. Biol. clin.* **9**, 53.

Rothstein, M. and Miller, L.L. (1954). The conversion of lysine to pipecolic acid in the rat. *J. biol. Chem.* **211**, 851.

Rowe, D.W., McGoodwin, E.B., Martin, G.R., Sussman, M.D., Grahn, D., Faris, B., and Franzblau, C. (1974). A sex-linked defect in the cross-linking of collagen and elastin associated with the mottled locus in mice. *J. exp. Med.* **139**, 180.

Rowland, L.P., Lovelace, R.E., Schotland, D.L., Araki, S., and Carmel, P. (1966). The clinical diagnosis of McArdle's disease: Identification of another family with deficiency of muscle phosphorylase. *Neurology* **16**, 93.

Roy, L.P. and Pollard, A.C. (1978). Cysteamine therapy for cystinosis. *Lancet* **ii**, 729.

Rubino, A., Zimbalatti, F., and Auricchio, S. (1964). Intestinal disaccharidase activities in adult and suckling rats. *Biochim. Biophys. Acta* **92**, 305.

Rudd, N.L., Miskin, M., Hoar, D.I., Benzie, R., and Doran, T.A. (1976). Prenatal diagnosis of hypophosphatasia. *New Engl. J. Med.* **295**, 146.

Rudiger, H.W., Langenbeck, U., Schulze-Schenking, M., Goedde H.W., and Schuchmann, L. (1972). Defective decarboxylase in branched chain ketoacid oxidase multienzyme complex in classic type of maple syrup urine disease. *Humangenetik* **14**, 257.

Rundles, R.W., Metz, E.N., and Silberman, H.R. (1966). Allopurinol in the treatment of gout. *Ann. intern. Med.* **64**, 229.

Russell, A., Statter, M., and Abzug, S. (1977). Methionine-dependent formimino glutamic acid transferase deficiency: Human and experimental studies in its therapy. *Hum. Hered.* **27**, 205.

—— Levin, B., Oberholzer, V.G., and Sinclair, L. (1962). Hyperammonemia: a new instance of an inborn enzymatic defect of the biosynthesis of urea. *Lancet* **i**, 699.

Russell, G., Thom, H., Tarlow, M.J., and Gompertz, D. (1974). Reduction of plasma propionate by peritoneal dialysis. *Pediatrics* **53**, 281.

Russell, R.G.G. (1965). Excretion of inorganic pyrophosphate in hypophosphatasia. *Lancet* **ii**, 461.

—— Bisaz, S., Donath, A., Morgan, D.B., and Fleisch, H. (1971). Inorganic pyrophosphate in plasma in normal persons and in patients with hypophosphatasia, osteogenesis imperfecta and other disorders of bone. *J. clin. Invest.* **50**, 961.

Russell, S.B., Russell, J.D. and Littlefield, J.W. (1971). ß-glucuronidase activity in fibroblasts cultured from patients with and without cystic fibrosis. *J. med. Genet.* **8**, 441.

Rutsaert, J., Menn, R., and Resibois, A. (1973). Ultrastructure of sulfatide storage in normal and sulfatase-deficient fibroblasts in vitro. *Lab Invest.* **29**, 527.

—— Tondeur, M., Vamos-Hurwitz, E., and Dustin, P. (1977). The cellular lesions of Farber's disease and their experimental reproduction in tissue culture. *Lab. Invest.* **36**, 474.

Ryman, B.E. (1980). Recent advances and problems in the glycogen storage diseases. In *Inherited disorders of carbohydrate metabolism* (ed. D. Burman, J.B. Holton, and C.A. Pennock), p.289. MTP Press, Lancaster.

Sabina, R.L., Swain, J.L., Patten, B.M., Ashizawa, T., O'Brien, W.E., and Holmes, E.W. (1980). Disruption of the purine nucleotide cycle: A potential explanation for muscle dysfunction in myoadenylate deaminase deficiency. *J. clin. Invest.* **66**, 1419.

Sachs, B. (1896). A family form of idiocy, generally fatal, associated with early blindness. *J. nerv. ment. Dis.* **21**, 475.

Sack, G. (1980). Clinical diversity in Gaucher's disease. *Johns Hopkins med. J.* **146**, 166.

Sadamoto, M. (1966). Nature of cultured cells of the skin from acatalasemic individuals with Takahara's disease. *Acta Med. Okayama* **20**, 193.

Saenger, P., Goldman, A.S., Levine, L.S., Korthschutz, S. Muecke, E.C., Katsumata, M., Doberne, Y., and New, M.I. (1978). Prepubertal diagnosis of steroid 5α-reductase deficiency. *J. clin. Endocrinol. Metab.* **46**, 627.

Saez, J.M., De Peretti, E., Morera, A.M., David, M., and Bertrand, J. (1971). Familial male pseudohermaphroditism with gynecomastia due to a testicular 17-ketosteroid reductase defect. I. Studies in vivo. *J. clin. Endocrinol. Metab.* **32**, 604.

Sagebiel, R.W. and Parker, F. (1968). Cutaneous lesions of Fabry's disease: Glycoplipid lipidosis — light and electron microscopic findings. *J. Invest. Dermatol.* **50**, 208.

Saheki, T., Tsuda, M., Takada, S., Kusumi, K., and Katsunuma, T. (1980). Role of argininosuccinate synthetase in the regulation of urea synthesis in the rat and argininosuccinate synthetase-associated metabolic disorder in man. *Adv. Enzymol.* **18**, 221.

—— Ueda, A., Hosoya, M., Kusumi, K., Takada, S., Tsuda, M., and Katsunuma, T. (1981). Qualitative and quantitative abnormalities of argininosuccinate synthetase in citrullinemia. *Clin. Chim. Acta* **109**, 325.

Salafsky, I.S. and Nadler, H.L. (1971). Alpha-1,4-glucosidase activity in Pompe's disease. *J. Ped.* **79**, 794.

—— —— (1973). A fluorometric assay of α-glucosidase and its application in the study of Pompe's disease. *J. lab. clin. Med.* **81**, 450.

Salter, R.H. (1968). The muscle glycogenoses. *Lancet* **i**, 1301.

Samuelsson, K. and Zetterstrom, R. (1971). Ceramides in a patient with lipogranulomatosis (Farber's disease) with chronic course. *Scand. J. clin. Lab. Invest.* **27**, 393.

Sandhoff, K. and Harzer, K. (1973). Total hexosaminidase deficiency in Tay–Sachs disease (variant 0). In *Lysosomes and storage diseases* (ed. H.G. Hers and F. Van Hoof), p.345. Academic Press, New York.

Sando, G.N. and Neufeld, E.F. (1977). Recognition and receptor mediated uptake of a lysosomal enzyme, α-L-iduronidase, by cultured human fibroblasts. *Cell* **12**, 619.

Sanfilippo, S.J., Podosin, R., Langer, L.O. jun., and Good, R.A. (1963). Mental retardation associated with acid mucopolysacchariduria (heparitin sulphate type). *J. Ped.* **63**, 837.

Sann, L., Mathieu, M., Bourgeois, J., Bienvenu, J., and Bethenod, M. (1980). In vivo evidence for defective activity of glucose-6-phosphatase in type 1B glycogenosis. *J. Ped.* **96**, 691.

Sannes, L.J. and Hultquist, D.E. (1978). Effects of hemolysate concentration, ionic strength and cytochrome b5 concentration on the rate of methemoglobin reduction in hemolysates of human erythrocytes. *Biochim. Biophys. Acta* **544**, 547.

Sano, S. and Granick, S. (1961). Mitochondrial coproporphyrinogen oxidase and protoporphyrin formation. *J. biol. Chem.* **236**, 1173.

Santiago-Borrero, P.J., Santini, R. jun., Perez-Santiago, E., and Maldonado, N. (1973). Congenital isolated defect of folic acid absorption. *J. Ped.* **82**, 450.

Sardharwalla, I.B., Komrower, G.M., and Schwarz, V. (1980). Pregnancy in classical galactosaemia. In *Inherited disorders of carbohydrate metabolism* (ed. D. Burman, J.B. Holton, and C.A. Pennock), p.125. MTP Press, Lancaster.

—— Fowler, B., Robins, A.J., and Komrower, G.M. (1974). Detection of heterozygotes for homocystinuria: Study of sulphur-containing amino acids in plasma and urine after L-methionine loading. *Arch. Dis. Child.* **49**, 553.

—— Jackson, S.H., Hawke, H.D., and Sass-Kortsak, A. (1968). Homocystinuria: a study with low-methionine diet in three patients. *Can. med. Ass. J.* **99**, 731.

Sass, M.D. (1968). Observations on the role of TPNH dehydrogenase in human red cells. *Clin. Chim. Acta* **21**, 101.

—— Caruso, C.J., and Farhangi, M. (1967). TPNH-methemoglobin reductase deficiency: A new red-cell enzyme defect. *J. lab. clin. Med.* **70**, 760.

Sassa, S., Bradlow, H.L., and Kappas, A. (1979a). Steriod induction of δ-aminolevulinic acid synthase and porphyrins in liver. Structure-activity studies and the permissive effects of hormones on the induction process. *J. biol. Chem.* **254**, 10011.

—— Levere, R.D., Solish, G., and Kappas, A. (1974b). Studies on the porphyrin-heme biosynthetic pathway in cultured human amniotic cells. *J. clin. Invest.* **53**, 70a.

—— Solish, G., Levere, R.D., and Kappas, A. (1975). Studies in porphyria. IV. Expression of the gene defect of acute intermittent porphyria in cultured human skin fibroblasts and amniotic cells: Prenatal diagnosis of the porphyric trait. *J. exp. Med.* **142**, 722.

—— Zalar, G.L., Poh-Fitzpatrick, M.B., and Kappas, A. (1979*b*). Studies in porphyria. IX. Detection of the gene defect of erythropoietic protoporphyria in mitogen-stimulated human lymphocytes. *Trans. Ass. Am. Phys.,* **92**, 268.

—— Granick, S., Bickers, D.R., Bradlow, H.L., and Kappas, A. (1974*a*). A microassay for uroporphyrinogen I synthase, one of three abnormal enzyme activities in acute intermittent porphyria, and its application to the study of the genetics of this disease. *Proc. natl. Acad. Sci. USA* **71**, 732.

Sass-Kortsak, A. (1965). Copper metabolism. *Adv. clin. Chem.* **8**, 1.

—— (1974). Hepatolenticular degeneration (Kinnier Wilson's disease). In *Handbuch der inneren Medizin*, Vol. 7 (ed. H. Schwiegk), 5th ed., p.622. Springer, Berlin.

Sato, K., Imai, F., Hatayama, I., and Roelofs, R.I. (1977). Characterization of glycogen phosphorylase isoenzymes present in cultured skeletal muscle from patients with McArdle's disease. *Biochem. biophys. Res. Commun.* **78**, 663.

Saudubray, J.M., Dreyfus, J.C., Cepanec, C., Le Lo'ch, H., Trung, P.H., and Mozziconacci, P. (1973). Acidose lactigue, hypoglycemie et hepatomegalie par deficit héréditaire en fructose-1,6-diphosphatase hépatique. *Arch. fr. Pediatr.* **30**, 609.

Saugstad, L.F. (1973). Increased 'reproductive casualty' in heterozygotes for phenylketonuria. *Clin. Genet.* **4**, 105.

—— (1977). Heterozygote advantage for the phenylketonuria allele. *J. med. Gen.* **14**, 20.

Saunders, M., Sweetman, L., Robinson, B., Roth, K., Cohn, R., and Gravel, R.A. (1979). Biotin-response organicaciduria: multiple carboxylase defects and complementation studies with propionic acidemia in cultured fibroblasts. *J. clin. Invest.* **64**, 1695.

Sawaki, S., Hattori, N., Morikawa, N., and Yamada, K. (1967). Oxidation and reduction of glyoxylate by lactate dehydrogenase. *J. Vit. (Kyoto)* **13**, 93.

Saxon, A. (1973. Hemodialysis for oxaluric renal failure. *New Engl. J. Med.* **288**, 526.

Schaap, T. and Bach. G. (1980). Incidence of mucopolysaccharidoses in Israel: is Hunter disease a 'Jewish disease'? *Hum. Genet.* **56**, 221.

Schafer, I.A., Scriver, C.R., and Efron, M.L. (1962). Familial hyperprolinemia, cerebral dysfunction and renal anomalies occurring in a family with hereditary nephritis and deafness. *New Engl. J. Med.* **267**, 51.

Schapira, F., Schapira, G., and Dreyfuss, J.C. (1961–1962). La lésion enzymatique de la fructosurie bénigne. *Enzym. Biol. clin.* **1**, 170.

Schaub, J., Janka, G.E., Christomanou, H., Sandhoff, K., Permanetter, W., Hübner, G., and Meister, P. (1980). Wolman's disease: Clinical, biochemical and ultrastructural studies in an unusual case without striking adrenal calcification. *Eur. J. Ped.* **135**, 45.

—— Daumling, S. Curtius, H.-C., Niederwieser, A., Bartholomé, K., Viscontini, M., Schircks, B., and Bieri, J.H. (1978). Tetrahydrobiopterin therapy of atypical phenylketonuria due to defective dihydrobiopterin biosynthesis. *Arch. Dis. Child.* **53**, 674.

Schaumburg, H.H., Powers, J.M., Raine, C.S., Suzuki, K., and Richardson, E.P. jun. (1975). Adrenoleucodystrophy. A clinical and pathological study of 17 cases. *Arch. Neurol.* **32**, 577.

————— Spencer, P.S., Griffin, J.W., Prineas, J.W., and Boehme, D.M. (1977). Adrenomyeloneuropathy: a probable variant adrenoleucodystrophy. II. General pathologic, neuropathologic and biochemical aspects. *Neurology* **27**, 1114.

Schedewie, H., Willich, E., Grobe, H., Schmidt, H., and Muller, K.M. (1973). Skeletal findings in homocystinuria. A collaborative study. *Ped. Radiol.* **1**, 12.

Scheie, H.G., Hambrick, G.W. jun. and Barness, L.A. (1962). A newly recognised forme fruste of Hurler's disease (gargoylism). *Am. J. Ophthalmol.* **53**, 753.

Scheinberg, I.H. and Gitlin, D. (1952). Deficiency of ceruloplasmin in patients with hepatolenticular degeneration (Wilson's disease). *Science* **116**, 484.

Scherz, R., Pflugshaupt, R., and Bütler, R. (1976). A new genetic variant of galactose 1-phosphate uridyl transferase. *Hum. Genet.* **35**, 51.

Scherzer, L.S. and Ilson, J.B. (1969). Normal intelligence in the Lesch–Nyhan syndrome. *Pediatrics* **44**, 116.

Schettler, F.G. (1969). Essential familial hypercholesterolemia. In *Artherosclerosis* (ed. F.G. Schettler and G.S. Boyd), p.543. Elsevier, Amsterdam.

Schibanoff, J.M., Kamoshita, S., and O'Brien, J.S. (1969). Tissue distribution of glycosphingolipids in a case of Fabry's disease. *J. Lipid Res.* **10**, 515.

Schiff, L., Schubert, W.K., McAdams, A.J., Spiegel, E.L., and O'Donnell, J.F. (1968). Hepatic cholesterol ester storage disease, a familial disorder. I. Clinical aspects. *Am. J. Med.* **44**, 538.

Schimke, R.N., McKusick, V.A., Huang, T., and Pollack, A.P. (1965). Homocystinuria. *J. Am. med. Assoc.* **193**, 711.

————— Zakheim, R.M., Corder, R.C., and Hug, G. (1973). Glycogen storage disease type IX: Benign glycogenosis of liver and hepatic phosphorylase kinase deficiency. *J. Ped.* **83**, 1031.

Schlesinger, P.H., Cotton, R.G.H., and Danks, D.M. (1976). Phenylketonuria phenotype detectable in fibroblasts. *Lancet* **ii**, 1245.

————— Watson, B., Cotton, R.G.H., and Danks, D.M. (1979). Urinary dihydroxanthopterin in the diagnosis of malignant hyperphenylalaninaemia and phenylketonuria. *Clin. Chim. Acta* **92**, 187.

Schmickel, R.D., Distler, J.J., and Jourdian, G.W. (1975). Accumulation of sulfate-containing acid mucopolysaccharides in I-cell fibroblasts. *J. lab. clin. Med.* **86**, 672.

Schmid, R., Axelrod, J., Hammaker, L., and Swarm, R.L. (1958). Congenital jaundice in rats due to a defective glucuronide formation. *J. clin. Invest.* **37**, 1123.

————— and Mahler, R. (1959). Chronic progressive myopathy with myoglobinuria: Demonstration of a glycogenolytic defect in the muscle. *J. clin. Invest.* **38**, 2044.

————— Robbins, P.W., and Traut, R.R. (1959). Glycogen synthesis in muscle lacking phosphorylase. *Proc. natl. Acad. Sci, USA* **45**, 1236.

————— Schwartz, S., and Sundberg, D. (1955). Erythropoietic (congenital) porphyria: A rare abnormality of the normoblasts. *Blood* **10**, 416.

————— ————— and Watson, C.J. (1954). Porphyria content of bone marrow and liver in various forms of porphyria. *Arch. intern. Med.* **93**, 167.

Schmidt, L. (1974). The biochemical detection of metabolic disease: Screening tests and a systematic approach to screening. In *Heritable disorders of amino acid metabolism: patterns of clinical expression and genetic variation* (ed. W.L. Nyhan), p.675, John Wiley and Sons, New York.

Schmidt, R., Von Figura, K., Paschke, E., and Kresse, H. (1977). Sanfilippo's disease type A: sulfamidase activity in peripheral leukocytes of normal, heterozygous and homozygous individuals. *Clin. Chim. Acta* **80**, 7.

Schmoeckel, C. and Hohlfed, M. (1979). A specific ultrastructural marker for disseminated lipogranulomatosis (Farber). *Arch. Dermatol. Res.* **266**, 187.

Schneider, A.J., Kinter, W.B., and Stirling, C.E. (1966). Glucose-galactose malabsorption. *New Engl. J. Med.* **274**, 305.

Schneider, A.S., Valentine, W.N., Hattori, M., and Heins, H.L. jun. (1965). Hereditary hemolytic anemia with triosephosphate isomerase deficiency. *New Engl. J. Med.* **272**, 229.

—— —— Baughan, M.A., Paglia, D.E., Shore, N.A., and Heins, H.L. jun. (1968*c*). Triosephosphate isomerase deficiency. A multi-system inherited enzyme disorder. Clinical and genetic aspects. In *Hereditary disorders of erythrocyte metabolism* (ed. E. Beutler). p.265. Grune and Stratton, New York.

Schneider, E.L., Ellis, W.G., Brady, R.O., McCulloch, J.R., and Epstein, C.J. (1972*a*). Prenatal Niemann–Pick disease: biochemical and histologic examination of a 19-gestational week fetus. *Ped. Res.* **6**, 720.

—— —— —— —— —— (1972*b*). Infantile Gaucher's disease: *in utero* diagnosis and fetal pathology. *J. Ped.* **81**, 1134.

—— Pentchev, P.G., Hibbert, S.R., Sawitsky, A., and Brady, R.O. (1978). A new form of Niemann–Pick disease characterized by temperature-labile sphingomyelinase. *J. med. Genet.* **15**, 370.

Schneider, G., Genel, M., Bongiovanni, A.M., Goldman, A.S., and Rosenfeld, R.L. (1975). Persistent testicular $\triangle^5$-isomerase 3ß-hydroxysteroid dehydrogenase ($\triangle^5$-3ß-HSD) deficiency in the $\triangle^5$-3ß-HSD form of congenital adrenal hyperplasia. *J. clin. Invest.* **55**, 681.

Schneider, J.A., and Seegmiller, J.E. (1972). Cystinosis and the Fanconi syndrome. in *The metabolic basis of inherited disease*, 3rd ed. (ed. J.B. Stanbury, J.B. Wyngaarden, and D.S. Fredrickson), p.1581. McGraw-Hill, New York.

—— Bradley, K., and Seegmiller, J.E. (1967*b*). Increased cystine in leukocytes from individuals homozygous and heterozygous for cystinosis. *Science* **157**, 1321.

—— —— —— (1968*b*). Transport and intracellular fate of cysteine-^{35}S in leukocytes from normal subjects and patients with cystinosis. *Ped. Res.* **2**, 441.

—— Schulman, J.D., and Thoene, J.G. (1981). Cysteamine therapy in cystinosis. *New Engl. J. Med.* **304**, 1172.

—— Wong, V., and Seegmiller, J.E. (1969). The early diagnosis of cystinosis. *J. Ped.* **74**, 114.

—— Rosenbloom, F.M., Bradley, K.H., and Seegmiller, J.E. (1967*a*). Increased free-cystine content of fibroblasts cultured from patients with cystinosis. *Biochem. biophys. Res. Commun.* **29**, 527.

—— Wong, V. Bradley, K.H., and Seegmiller, J.E. (1968*a*). Biochemical comparisons of the adult and childhood forms of cystinosis. *New Engl. J. Med.* **279**, 1253.

—— Schlesselman, J.J., Mendoza, S.A., Orloff, S., Thoene, J.G., Kroll, W.A., Godfrey, A.D., and Schulman, J.D. (1979). Ineffectiveness of ascorbic acid therapy in nephropathic cystinosis. *New Engl. J. Med.* **300**, 756.

—— Verronst, P.M., Kroll, W.A., Garvin, A.J., Horger, E.O., Wong, V.G., Spear, G.S., Jacobson, C., Pellett, O.L., and Becker, P.L.A. (1974). Prenatal diagnosis of cystinosis. *New Engl. J. Med.* **290**, 878.

Schneider, P.B. and Kennedy, E.P. (1967). Sphingomyelinase in normal human spleen and in spleens from subjects with Niemann–Pick disease. *J. Lipid. Res.* **8**, 202.

Schnyder, U.W., Konrad, B.B., Schreier, K., Nerz, P., and Crefeld, W. (1968). Über Ichthyosen. *Dtsch. Med. Wschr.* **98**, 423.

Scholnick, P.L., Hammaker, E., and Marver, H.S. (1972). Soluble δ-aminolevulinic acid synthetase of rat liver. II. Studies related to the mechanism of enzyme action and heme inhibition. *J. biol. Chem.* **247**, 4132.

Scholz, W. (1925). Klinische, pathologisch-anatomische und erbbiologische Untersuchungen bei familiarer diffuser Hirnsklerose in kindesalter. *Z. Ges. Neurol. Psych.* **99**, 42.

Schonenberg, H. and Lindenfelser, R. (1974). Farber-syndrom (disseminierte lipogranulomatose). *Monatsschr. Kinderheilk.* **122**, 153.

Schram, A.W. and Tager, J.M. (1981). The specificity of lysosomal hydrolases: human α-galactosidase isoenzymes. *Trends Biochem. Sci.* **6**, 328.

—— Hamers, M.N., Brouwer-Kelder, B., Donker-Koopman, W.E., and Tager, J.M. (1977). Enzymological properties and immunological characterization of α-galactosidase isoenzymes from normal and Fabry human liver. *Biochim. Biophys. Acta* **482**, 125.

—— De Groot, P.G., Hamers, M.N., Brouwer-Kelder, B., Donker-Koopman, W.E., and Tager, J.M. (1978). Further characterization of two forms of *N*-acetyl- α -galactosaminidase from human liver. *Biochim. Biophys. Acta* **525**, 410.

Schreier, K. and Flaig, H. (1956). Urinary excretion of indolepyruvic acid in normal conditions and Folling's disease. *Klin. Wschr.* **34**, 1213.

Schrieken, R.M., Kerber, R., Jonasescu, V.V., and Zellweger, H. (1975). Cardiac manifestations of the mucopolysaccharidoses. *Circulation* **52**, 700.

Schrijvers, J. and Hommes, F.A. (1975). Activity of fructose-1,6-diphosphatase in human leucocytes. *New Engl. J. Med.* **292**, 1298.

Schröter, W. (1970). 2,3-Diphosphoglyceratstoffwechsel und 2,3-diphospho-gluceratmutase-mangel in erythrozyten. *Blut* **20**, 311.

—— Koch, H.H., Wonneberger, B., Kalinowsky, W., Arnold, A., Blume, K.G., and Hüther, W. (1974). Glucose phosphate isomerase deficiency with congenital nonspherocytic hemolytic anemia: A new variant (Type Nordhorn). 1. Clinical and genetic studies. *Ped. Res.* **8**, 18.

Schuchmann, L., Colombo, J.P., and Fischer, H. (1980). Hyperammonemia due to ornithine transcarbamylase deficiency – a cause of lethal metabolic crisis during the newborn period and infancy. *Klin. Paed.* **192**, 281.

Schulman, J.D. (1973). Sulfur metabolism. In *Cystinosis* (ed. J.D. Schulman), p.67. US Government Printing Office, Washington DC.

—— Fujimoto, W.Y., Bradley, K.H., and Seegmiller, J.E. (1970*b*). Identification of heterozygous genotype for cystinosis *in utero* by a new pulse labeling technique: preliminary report. *J. Ped.* **77**, 468.

—— Lustberg, T.J., Kennedy, J.L., Muscles, M., and Seegmiller, J.E. (1970*c*). A new variant of maple syrup urine disease (branched chain ketoaciduria). *Am. J. Med.* **49**, 118.

—— Wong, V.G., Kuwabara, T., Bradley, K.H., and Seegmiller, J.E. (1970*a*). Intracellular cystine content of leukocyte populations in cystinosis. *Arch. intern. Med.* **125**, 660.

—— Goodman, S.I., Mace, J.W., Patrick, A.D., Tietze, F., and Butler, E.J. (1975). Glutathionuria: Inborn error of metabolism due to tissue deficiency of γ-glutamyl transpeptidase, *Biochem. biophys. Res. Commun.* **65**, 68.

Schulte, M.-J. and Lenz, W. (1977). Fatal sorbitol infusion in patient with fructose-sorbitol intolerance. *Lancet* **ii**, 188.

Schultz, M.G. (1962). Male pseudohermaphroditism diagnosed with aid of sex chromatin technique. *J. Am. vet. Med. Assoc.* **140**, 241.

Schumate, J.B., Kaiser, K.K., Carroll, J.E., and Brooke, M.H. (1980). Adenylate deaminase deficiency in a hypotonic infant. *J. Ped.* **96**, 885.

Schumert, Z., Rosenmann, A., Landau, H., and Rösler, A. (1980). 11-deoxycortisol in amniotic fluid: prenatal diagnosis of congenital adrenal hyperplasia due to 11 ß-hydroxylase deficiency. *Clin. Endocrinol.* **12**, 257.

Schutgens, R.B.H., Beemer, F.A., Tegelaers, W.H.H., and De Groot, W.P. (1979*a*). Mild variant of argininosuccinic aciduria. *J. inher. metab. Dis.* **2**, 13.

——— Heymans, H., Ketel, A., Veder, H.A., Duran, M., Ketting, D., and Wadman, S.K. (1979*b*). Lethal hypoglycaemia in a child with a deficiency of 3-hydroxy-3-methylglutaryl coenzyme A lyase. *J. Ped.* **94**, 89.

——— Middleton, B., Van der Blij, J.F., Oorthuys, J.W.E., Veder, H.A., Vulsma, T., and Tegelaers, W.H.H. (1982). Beta-ketothiolase deficiency in a family confirmed by *in vitro* enzymatic assays in fibroblasts. *Eur. J. Ped.* **139**, 39.

Schutta, H.S. and Johnson, L. (1969). Clinical signs and morphologic abnormalities in Gunn rats treated with sulfadiethoxine. *J. Ped.* **75**, 1070.

——— Pratt, R.T.C., Metz, H., Evans, K.A., and Carter, C.O. (1966). A family study of the late infantile and juvenile forms of metachromatic leucodystrophy. *J. med. Genet.* **3**, 86.

Schwartz, J.M. and Reiss, A.L. (1974). Erythrocyte diaphorases: their identity and role. *Program of the American society of hematology, 17th annual meeting.* (Abstract 192).

——— Paress, P.S., Ross, J.M., Dipillo, F., and Rizek, R. (1972). Unstable variant of NADH methemoglobin reductase in Puerto Ricans with hereditary methemoglobinemia. *J. clin. Invest.* **51**, 1594.

Schwartz, R., Ashmore, J., and Renold, A.E. (1957). Galactose tolerance test in glycogen storage disease. *Pediatrics* **19**, 585.

Schwartz, T.M. and Schackelford, R.M. (1973). Pseudodistemper is apparently new ailment of mink. *U.S. Fur Rancher* **52**, 6.

Schwartz, V., Golberg, L., Komrower, G.M., and Holzel, A. (1956). Some disturbances of erythrocyte metabolism in galactosaemia. *Biochem. J.* **62**, 34.

Schweizer, W. (1947). Studies on the effect of l-tyrosine on the white rat. *J. Physiol.* **106**, 167.

Scialom, C., Najean, Y., and Bernard, J. (1966). Anémie hémolytique congénital non sphérocytaire avec déficit incomplet en 6-phosphogluconate deshydro-génase. *Nouv. Rev. fr. Hematol.* **69**, 452.

Scott, C.R., Lagunoff, D., and Pritzl, P. (1973). A mucopolysaccharide storage disease with involvement of the renal glomerular epithelium. *Am. J. Med.* **54**, 549.

——— Clark, S.H., Teng, C.C., and Svedberg, K.R. (1970*a*). Clinical and cellular studies on sarcosinemia. *J. Ped.* **77**, 805.

——— Hakami, N., Teng, C.C., and Sagerson, R.N. (1972). Hereditary transcobalamin II deficiency: The role of transcobalamin II in vitamin B_{12}-mediated reactions. *J. Ped.* **81**, 1106.

Scott, E.M. (1960). The relation of diaphorase of human erythrocytes to inheritance of methemoglobinemia. *J. clin. Invest.* **39**, 1176.

——— (1968). Congenital methemoglobinemia due to DPNH-diaphorase deficiency. In *Hereditary disorders of erythrocyte metabolism* (ed. E. Beutler), p.102. Grune and Stratton, New York.

——— (1973). Inheritance of two types of deficiency of human serum cholinesterase. *Ann. hum. Genet., Lond.* **37**, 139.

——— and Powers, R.F. (1974). Properties of the C_5 variant form of human serum cholinesterase. *Am. J. hum. Genet.* **26**, 189.

——— Duncan, I.W., and Ekstrand, V. (1965). The reduced pyridine nucleotide dehydrogenase of human erythrocytes. *J. biol. Chem.* **240**, 481.

—— Weaver, D.D., and Wright, R.C. (1970*b*). Discrimination of phenotypes in human serum cholinesterase deficiency. *Am. J. hum. Genet.* **22**, 363.

Scott-Emuakpor, A., Higgins, J.V., and Kohrman, A.F. (1972). Citrullinemia: a new case with implications concerning adaptation to defective urea synthesis. *Ped. Res.* **6**, 626.

Scowen, E.F., Stansfield, A.G., and Watts, R.W.E. (1959). Oxalosis and primary hyperoxaluria. *J. Path. Bact.* **77**, 195.

Scriver, C.R. (1965). Hartnup disease. *New Engl. J. Med.* **273**, 530.

—— (1968). Renal tubular transport of proline, hydroxyproline and glycine. III. Genetic basis for more than one mode of transport in human kidney. *J. clin. Invest.* **47**, 823.

—— (1969). Inborn errors of amino acid metabolism. *Brit. med. Bull.* **25**, 35.

—— (1970). Vitamin D dependency. *Pediatrics* **45**, 361.

—— (1978). Familial iminoglycinuria. In *The metabolic basis of inherited disease* (ed. J.B. Stanbury, J.B. Wyngaarden, and D.S. Fredrickson), 4th edn., p.1593. McGraw-Hill, New York.

—— (1983). Familial iminoglycinuria. In *The metabolic basis of inherited disease* (ed. J.B. Stanbury, J.B. Wyngaarden, D.S. Fredrickson, J.L. Goldstein, and M.S. Brown), 5th edn., p. 1800. McGraw-Hill, New York.

—— and Cameron, D. (1969). Pseudohypophosphatasia. *New Engl. J. Med.* **281**, 604.

—— and Hutchison, J.H. (1963). The vitamin B_6 deficiency syndrome in human infancy: Biochemical and clinical observations. *Pediatrics* **31**, 240.

—— and Rosenberg, L.E. (1973). *Amino acid metabolism and its disorders.* W.B. Saunders, Philadelphia.

—— Clow, C.L., and Silverberg, M. (1966*a*). Hypermethioninaemia in acute tyrosinosis. *Lancet* **i**, 153.

—— Davies, E., and Cullen, A.M. (1964). Application of a simple micromethod to the screening of plasma for a variety of aminoacidopathies. *Lancet* **ii**, 230.

—— Pueschel, S., and Davies, E. (1966*b*). Hyper-ß-alaninemia associated with ß-aminoaciduria and γ-aminobutyricaciduria, somnolence and seizures. *New Engl. J. Med.* **274**, 635.

—— Schafer, I.A., and Efron, M.L. (1961). New renal tubular amino acid transport system and a new hereditary disorder of amino acid metabolism. *Nature, Lond.* **192**, 672.

—— Mackenzie, S., Clow, C.L., and Delvin, F. (1971). Thiamine-responsive maple syrup urine disease. *Lancet* **i**, 310.

—— Whelan, D.T., Clow, C.L., and Dallaire, L. (1970). Cystinuria: Increased prevalence in patients with mental disease. *New Engl. J. Med.* **287**, 783.

—— MacDonald, W., Reade, T., Glorieux, F.H., and Nogrady, B.L. (1977*b*). Hypophosphatemic nonrachitic bone disease: an entity distinct from X-linked hypophosphatemia in renal defect, bone involvement, and inheritance. *Am. J. med. Genet.* **1**, 101.

—— Perry, J.R.T., Lasley, L., Clow, C.L., Coulter, D., and Laberge, C. (1977*a*). Neonatal tyrosinemia (NT) in the Eskimo. Result of protein polymorphism. *Ped. Res.* **11**, 411.

Seakins, J.W.T. and Ersser, R.S. (1967). Effects of amino acid loads on a healthy infant with the biochemical features of Hartnup disease. *Arch. Dis. Child.* **42**, 682.

Sealock, R.R., Gladston, M., and Steele, J.M. (1940). Administration of ascorbic acid to an alkaptonuric patient. *Proc. Soc. exp. Biol. Med.* **44**, 580.

Seaman, C., Wyss, S., and Piomelli, S. (1980). The decline in energetic metabolism with aging of the erythrocyte and its relationship to cell death. *Am. J. Haematol.* **8**, 31.

Searle, A.G. (1968). *Comparative genetics of coat colour in mammals.* Academic Press, London.

Seashore, M.R., Durant, J.L., and Rosenberg, L.E. (1972). Studies on the mechanism of pyridoxine-responsive homocystinuria. *Ped. Res.* **6**, 187.

Seegmiller, J.E. (1968). Proceedings of the seminars on the Lesch–Nyhan syndrome: Pathology and pathological physiology. *Fed. Proc.* **27**, 1042.

—— (1976). Inherited deficiency of hypoxanthine-guanine phosphoribosyl-transferase in X-linked uric aciduria (the Lesch–Nyhan syndrome and its variants). In *Advances in human genetics*, Vol. 6, (ed. H. Harris and K. Hirschhorn), p.75. Plenum Press, New York.

—— Rosenbloom, F.W., and Kelley, W.N. (1967). Enzyme defect associated with a sex-linked human neurological disorder and excessive purine synthesis. *Science* **155**, 1682.

—— Zannoni, V.G., Laster, L., and La Du, B.N. (1961). An enzymatic spectro-photometric method for the determination of homogentisic acid in plasma and urine. *J. biol. Chem.* **236**, 774.

—— Friedmann, T., Harrison, H.E., Wong, V., and Schneider, J.A. (1968). Cystinosis. *Ann. intern. Med.* **68**, 883.

Seftel, H.C., Baker, S.G., Sandler, M.P., Forman, M.B., Joffe, B.I., Mendelsohn, D., Jenkins, T., and Mieny, C.J. (1980). A host of hypercholesterolaemic homozygotes in South Africa. *Br. med. J.* **281**, 633.

Segal, S., Blair, A., and Roth, H. (1965). The metabolism of galactose by patients with congenital galactosemia. *Am. J. Med.* **38**, 62.

Seiji, M. and Kikuchi, A. (1969). Acid phosphatase activity in melanosomes. *J. Invest. Dermat.* **52**, 212.

—— Fitzpatrick, T.B., Simpson, R.T., and Birbeck, M.S.C. (1963). Chemical compositon and terminology of specialized organelles (melanosomes and melanin granules) in mammalian melanocytes. *Nature, Lond.* **197**, 1082.

Sekiguchi, T. and Sekiguchi, F. (1973). Interallelic complementation in hybrid cells derived from Chinese hamster diploid clones deficient in hypoxanthine-guanine phosphoribosyl-transferase activity. *Exp. Cell Res.* **77**, 391.

Seligman, P.A., La Donna, L., Steiner, B.S., and Allen, R.A. (1980). Studies of a patient with megaloblastic anaemia and an abnormal transcobalamin II. *New Engl. J. Med.* **303**, 1209.

Selkoe, D.J. (1969). Familial hyperprolinemia and mental retardation. A second metabolic type. *Neurology* **19**, 494.

Serratrice, G., Monges, A., Roux, H., Aquaron, R., and Gambarelli, D. (1969). Forme myopathique du deficit en phosphofructokinase. *Rev. Neurol.* **120**, 271.

Setlow, R.B. (1968). The photochemistry, photobiology and repair of poly-nucleotides. In *Progress in nucleic acid research and molecular biology* Vol. 8, (ed. J.N. Davidson and W.E. Cohn), p.257. Academic Press, New York.

—— Regan, J.D., German, J., and Carrier, W.L. (1969). Evidence that xeroderma pigmentosum cells do not perform the first step in the repair of ultraviolet damage to their DNA. *Proc. natl. Acad. Sci. USA* **64**, 1035.

Shafai, T., Sweetman, L., Weyler, W., Goodman, S.I., Fennessey, P.V., and Nyhan, W.L. (1978). Propionic acidemia with severe hyperammonemia and defective glycine metabolism. *Pediatrics* **92**, 84.

Shah, S.N., Peterson, N.A., and McKean, C.M. (1972). Lipid composition of human cerebral white matter and myelin in phenylketonuria. *J. Neurochem.* **19**, 2369.

Shapira, E. and Nadler, H.L. (1975). The nature of the residual arylsulfatase activity in metachromatic leukodystrophy. *J. Ped.* **86**, 881.

Shapira, J.E., Phillips, J.A., Byers, P.H., Sanders, R., Holbrook, K.A., Levin, L.S., Dorst, J., Barsh, G.S., Peterson, K.E., and Goldstein, P. (1982). Prenatal diagnosis of lethal osteogenesis imperfecta (OI type II). *J. Ped.* **100**, 127.

Shapiro, L.J. and Mohandas, T. (1980). Molecular genetics of X-linked ichthyosis. *Ped. Res.* **14**, 527.

—— Hall, C.W., Leder, I.G., and Neufeld, E.F. (1976a). The relationship of α -L-iduronidase and Hurler corrective factor. *Arch. Biochem. Biophys.* **172**, 156.

—— Mohandas, T., Weiss, R., and Romeo, G. (1979b). Non-inactivation of an X-chromosome locus in man. *Science* **204**, 1224.

—— Weiss, R., Webster, D., and France, J.T. (1978). X-linked ichthyosis due to steroid-sulphatase deficiency. *Lancet* **i**, 70.

—— Cousins, L., Fluharty, A.L., Stevens, R.L., and Kihara, H. (1976b). Steroid sulfatase deficiency. *Ped. Res.* **11**, 894.

—— Aleck, K.A., Kaback, M.M., Itabashi, H., Desnick, R.J., Brand, N., Stevens, R.L., Fluharty, A.L., and Kihara, H. (1979a). Metachromatic leukodystrophy without arylsulphatase A deficiency. *Ped. Res.* **13**, 1179.

Shapiro, S.L., Sheppard, G.L., Dreifuss, F.E., and Newcombe, D.S. (1966). X-linked recessive inheritance of a syndrome of mental retardation with hyper-uricaemia. *Proc. Soc. exp. Biol. (NY)* **122**, 609.

Shaw, K.N.F., Boder, F., Gutenstein, M., and Jacobs, E.E. (1963). Histidinemia. *J. Ped.* **63**, 720.

Sheffield, L.J., Schlesinger, P., Faull, K., Halpern, B.J., Schier, G.M., Cotton, R.G.H., Hammond, J., and Danks, D.M. (1977). Iminopeptiduria, skin ulcerations and edema in a boy with prolidase deficiency. *J. Ped.* **91**, 578.

Sheldon, W. (1964). Congenital pancreatic lipase deficiency. *Arch. Dis. Child.* **39**, 268.

Sherwood, W.G., Saunders, M., Robinson, B.H., Brewster, T., and Gravel, R.A. (1982). Lactic acidosis in biotin-responsive multiple carboxylase deficiency caused by holocarboxylase synthetase deficiency of early and late onset. *J. Ped.* **101**, 546.

Shibata, Y., Higashi, T., Hirai, H., and Hamilton, H.B. (1967). Immunochemical studies on catalase. II. An anticatalase reacting component in normal, hypocatalasic and acatalasic human erythrocytes. *Arch. Biochem. Biophys.* **118**, 200.

Shih, V.E. and Efron, M.L. (1972). Urea cycle disorders. In *The metabolic basis of inherited disease* (ed. J.B. Stanbury, J.B. Wyngaarden, and D.S. Fredrickson) 3rd edn., p.370. McGraw-Hill, New York.

—— and Mandell, R. (1974). Metabolic defect in hyperornithinemia. *Lancet* **ii**, 1522.

—— Efron, M.L., and Moser, H.W. (1969). Hyperornithinemia, hyper-ammonemia and homocitrullinuria. A new disorder of amino acid metabolism associated with myoclonic seizures and mental retardation. *Am. J. Dis. Child.* **117**, 83.

—— Mandell, R., and Scholl, M.L. (1974). Historical observation in maple syrup urine disease. *J. Ped.* **85**, 868.

—— Berson, E.L., Mandell, R., and Schmidt, S.Y. (1978). Ornithine ketoacid transaminase deficiency in gyrate atrophy of the choroid and retina. *Am. J. hum. Genet.* **30**, 174.

—— Mandell, R., Jacoby, L.B., and Berson, E.L. (1981). Genetic complementation analysis in fibroblasts from gyrate atrophy (GA) and the syndrome of hyperornithinemia, hyperammonemia and homocitrullinuria. *Red. Res.* **15**, 569.

—— Salam, M.Z., Mudd, S.H., Uhlendorf, B.W., and Adams, R.D. (1972). A new form of homocystinuria due to $N^{5,10}$-methylenetetrahydrofolate reductase deficiency. *Ped. Res.* **6**, 135.

—— Abroms, I.F., Johnson, J.L., Carney, M., Mandell, R., Robb, R.M., Cloherty, J.P., and Rajagopalan, K.V. (1977). Sulfite oxidase deficiency: Biochemical and clinical investigations of a hereditary metabolic disorder in sulfur metabolism. *New Engl. J. Med.* **297**, 1022.

Shimada, K., Gill, P.J., Silbert, J.E., Douglas, W.H.J., and Fanburg, B.L. (1981). Involvement of cell surface heparin sulfate in the binding of lipoprotein lipase to cultured bovine endothelial cells. *J. clin. Invest.* **68**, 995.

Shokeir, M.H.K. and Shreffler, D.C. (1969). Cytochrome oxidase deficiency in Wilson's disease: A suggested ceruloplasmin function. *Proc. natl. Acad. Sci. USA* **62**, 867.

Short, E.M., Conn, H.O., Snodgrass, P.J., Campbell, A.G.M., and Rosenberg, L.E. (1973). Evidence for X-linked dominant inheritance of ornithine transcarbamylase deficiency. *New Engl. J. Med.* **288**, 7.

Shows, T.B. and Brown, J.A. (1975). Human X-linked genes regionally mapped utilizing X-autosome translocations and somatic cell hybrids. *Proc. natl. Acad. Sci. USA* **72**, 2125.

—— —— Halley, L.L., Goggin, A.P., Eddy, R.L., and Byers, M.G. (1978). Assignment of a α-galactosidase (α-GAL) gene to the q22 → qter region of the X chromosome in man. *Cytogenet. Cell Genet.* **22**, 541.

—— Scrofford-Wolf, L., Brown, J.A., and Meisler, M. (1979). GM_1-gangliosidosis: chromosome 3 assignment of the ß-galactosidase A gene (ß-GAL_A). *Somat. Cell Genet.* **5**, 147.

Shreffler, D.C., Brewer, G.J., Gall, J.C., and Honeyman, M.S. (1967). Electrophoretic variation in human serum ceruloplasmin: A new genetic polymorphism. *Biochem. Genet.* **1**, 101.

Shull, R.M., Munger, R.J., Spellacy, E., Hall, C.W., Constantopoulos, G., and Neufeld, E.F. (1982). Animal model of human disease: canine α-L-iduronidase deficiency, a model of mucopolysaccharidosis I. *Am. J. Pathol.* **109**, 244.

Sibert, J.R. (1979). Antenatal detection of congenital adrenal hyperplasia. *Lancet* **ii**, 37.

Siciliano, M.J., Bordelon, M.R., and Kohler, P.O. (1978). Expression of human adenosine deaminase after fusion of adenosine deaminase-deficient cells with mouse fibroblasts. *Proc. natl. Acad. Sci. USA* **75**, 936.

Sidbury, J.B. jun., Cornblath, M., Fisher, J., and House, E. (1961). Glycogen in erythrocytes of patient with glycogen storage disease. *Pediatrics* **27**, 103.

—— Mason, J., Burns, W.B. jun., and Ruebner, B.H. (1962). Type IV glycogenosis: Report of a case proven by characterization of glycogen and studied at necropsy. *Bull. Johns Hopkins Hosp.* **111**, 157.

Siegel, R.C. (1975). The connective tissue defect in homocystinuria (HS). *Clin. Res.* **23**, 263a.

—— Black, C.M., and Bailey, A.J. (1979). Cross-linking of collagen in the X-linked Ehlers–Danlos type V. *Biochem. Biophys. Res. Commun.* **88**, 281.

Siegenbeek van Heukelom, L.H., Akkerman, J.W.N., Staal, G.E.J., Stoop, J.W., and Zegers, B.J.M. (1976). An abnormal form of purine defective T-cell and normal B-cell immunity. *Clin. Chim. Acta* **72**, 117.

—— —— —— De Bruyn, C.H.M.M., Stoop, J.W., Zegers, B.J.M., De Bree,

P.K., and Wadman, S.K. (1977). A patient with purine nucleoside phosphorylase deficiency: enzymological and metabolic aspects. *Clin. Chim. Acta* **74**, 271.

Siemerling, E. and Creutzfeld, H.G. (1923). Bronzekranheit und sklerosierende encephalomyelitis (Diffuse Sklerose). *Arch. Psychiatr. Nervenkr.* **68**, 217.

Siimes, M.A., Rihiala, E.-L., and Leisti, J. (1979). Hexokinase deficiency in erythrocytes: A new variant in 5 members of a Finnish family. *Scan. J. Haematol.* **22**, 214.

Siiteri, P.K, and Wilson, J.B. (1974). Testosterone formation and metabolism during male sexual differentiation in the human embryo. *J. clin. Endocrinol. Metab.* **38**, 113.

Silk, D.B.A., Perrett, D., Clark, M.L., Stephens, A.D., and Scowen, E.F. (1975). A study of the renal handling and intestinal absorption of dibasic amino acids in a patient with genotype +/II heterozygous cystinuria and idiopathic hypercalcuria. *Clin. Chim. Acta* **59**, 195.

Sillence, D.O., Rimoin, D.L., and Danks, D.M. (1979*b*). Clinical variability in osteogenesis imperfecta – variable expressivity or genetic heterogeneity. *Birth Defects* **15**, 113.

—— Senn, A., and Danks, D.M. (1979*a*). Genetic heterogeneity in osteogenesis imperfecta. *J. med. Genet.* **16**, 101.

Sillero, M.A.G., Sillero, A., and Sols, A. (1969). Enzymes involved in fructose metabolism in liver and the glyceraldehyde metabolic crossroads. *Eur. J. Biochem.* **10**, 345.

Silverman, J.L. (1962). Apparent dominant inheritance of hypophosphatasia. *Arch. intern. Med.* **110**, 191.

Silvers, W.K. (1979). *The coat colors of mice. A model for mammalian gene action and interaction.* Springer-Verlag, New York.

Simard, H., Barry, A., Villeneuve, B., Petitclerc, C., Garneaux, R., and Delâge, J.-M. (1972) Porphyrie erythropoiétique congénitale. *Can. Med. Ass. J.* **106**, 1002.

Simell, O. and Takki, K. (1973). Raised plasma-ornithine and gyrate atrophy of the choroid and retina. *Lancet* **i**, 1031.

—— Johansson, T., and Aula, P. (1973). Enzyme defect in saccharopinuria. *J. Ped.* **82**, 54.

Similä, S. (1979). Hydroxyproline metabolism in Type II hyperprolinaemia. *Ann. clin. Biochem.* **16**, 177.

Simmonds, H.A., Sahota, A., Potter, C.F., and Cameron, J.S. (1978). Purine metabolism and immunodeficiency: urinary purine excretion as a diagnostic screening test in adenosine deaminase and purine nucleoside phosphorylase deficiency. *Clin. Sci. molec. Med.* **54**, 579.

—— Van Acker, K.J., Cameron, J.S., and Snedden, W. (1976). The identification of 2,8-dihydroxyadenine, a new component of urinary stones. *Biochem. J.* **157**, 485.

—— Watson, J.G., Hugh-Jones, K., Perrett, D., Sahota, A., and Potter, C.F. (1979). Deoxynucleoside excretion in adenosine deaminase deficiency and purine nucleoside phosphorylase deficiency. In *Inborn errors of specific immunity* (ed. B. Pollara, R.J. Pickering, H.J. Meuwissen, and I.H. Porter), p.377. Academic Press, New York.

Simoni, G., Brambati, B., Danesino, C., Rossella, F., Terzoli, G.L., Ferrari, M., and Fraccaro, M. (1983). Efficient direct chromosome analysis and enzyme determinations from chorionic villi samples in the first trimester of pregnancy. *Hum. Genet.* **63**, 349.

Simons, L.A., Reichl, D., Myant, N.B., and Mancini, M. (1975). The metabolism of the apoprotein of plasma low density lipoprotein in familial hyperbetalipo-proteinaemia in the homozygous form. *Atherosclerosis* **21**, 283.

Simoons, F.J. (1970). Primary adult lactose intolerance and the milking habit: a problem in biological and cultural interrelations. II. A culture historical hypothesis. *Am. J. dig. Dis.* **15**, 695.

Simopoulos, A.P., Marshall, J.R., Delea, C.S., and Bartter, F.C. (1971). Studies on the deficiency of 21-hydroxylation in patients with congenital adrenal hyperplasia. *J. clin. Endocrinol. Metab.* **32**, 438.

Singer, J.D., Cotlier, E., and Krimmer, R. (1973). Hexosaminidase A in tears and saliva for rapid identification of Tay–Sachs disease and its carriers. *Lancet* **ii**, 1116.

Singer, H.S., and Schafer, I.A. (1970). White cell ß-galactosidase activity. *New Engl. J. Med.* **285**, 571.

Singh, I., Tavella, D., and Di Ferrante, N. (1975). Measurements of aryl-sulphatases A and B in human serum. *J. Ped.* **86**, 574.

—— Moser, H.W., Moser, A.B., and Kishimoto, Y. (1981). Adrenoleuko-dystrophy: impaired oxidation of long chain fatty acids in cultured skin fibroblasts and adrenal cortex. *Biochem. biophys. Res. Commun.* **102**, 1223.

Singh, J., Di Ferrante, N.M., Niebes, P., and Tavella, D. (1976). *N*-acetyl-galactosamine 6-sulfate sulfatase in man: absence of the enzyme in Morquio disease. *J. clin. Invest.* **57**, 1036.

Singh, R.R., Lawrence, W.H., and Autian, J. (1972). Embryonic-fetal toxicity and teratogenic effects of a group of methacrylate esters in rats. *J. dent. Res.* **51**, 1632.

Singh, S., Willers, I., and Goedde, H.W. (1977). Heterogeneity in maple syrup urine disease: aspects of co-factor requirement and complementation in cultured fibroblasts. *Clin. Genet.* **11** 277.

Sipila, I. (1980). Inhibition of arginine-glycine amidinotransferase by ornithine. A possible mechanism for the muscular and chorioretinal atrophies in gyrate atrophy of the choroid and retina with hyperornithinemia. *Biochim. Biophys. Acta* **613**, 79.

—— Rapola, J., Simell, O., and Vannas, A. (1981). Supplementary creatinine as a treatment for gyrate atrophy of the choroid and retina. *New Engl. J. Med.* **304**, 867.

—— Simell, O., Rapola, J., Sainio, K., and Tuuteri, L. (1979). Gyrate atrophy of the choroid and retina with hyperornithinemia: Tubular aggregates and type 2 fiber atrophy in muscle. *Neurology* **29**, 996.

Sjaastad, O., Berstad, J., Gjesdahl, P., and Gjessing, L. (1976). Homo-carnosinosis. 2. A familial metabolic disorder associated with spastic paraplegia, progressive mental deficiency and retinal pigmentation. *Acta Neurol. Scand.* **53**, 275.

Slack, J. (1969). Risks of ischaemic heart disease in familial hyperlipoprotein-aemic states. *Lancet* **ii**, 1380.

Slavick, M., Lovenberg, W., and Keiser, H.R. (1973). Changes in serum and urine amino acids in patients with progressive systemic sclerosis treated with 6-azauridine triacetate. *Biochem. Pharmacol.* **22**, 1295.

Sloan, H.R., Uhlendorf, B.W., Jacobson, C.B., and Fredrickson, D.S. (1969*a*). ß-galactosidase in tissue culture derived from human skin and bone marrow: Enzyme defect in GM₁-gangliosidosis. *Ped. Res.* **3**, 532.

—— —— Kanfer, J.N., Brady, R.O., and Fredrickson, D.S. (1969*b*). Deficiency of sphingomyelin-cleaving enzyme activity in tissue cultures derived from patients with Niemann–Pick disease. *Biochem. biophys. Res. Commun.* **34**, 582.

Sloan, W.R. and Walsh, P.C. (1976). Familial persistent mullerian duct syndrome. *J. Urol.* **115**, 459.

Slördahl, S., Lie, S.O., Jellum, E. and Stokke, O. (1979). Increased need for L-cysteine in hereditary tyrosinemia. *Ped. Res.* **13**, 74.

Sly, W.S., Quinton, B.A., McAlister, W.H., and Rimoin, D.L. (1973). Beta-glucuronidase deficiency. Report of clinical, neurologic and biochemical features of a new mucopolysaccharidosis. *J. Ped.* **82**, 249.

Smith, A.J. and Strang, L.B. (1958). An inborn error of metabolism with the urinary excretion of α-hydroxybutyric acid and phenylpyruvic acid. *Arch. Dis. Child.* **33**, 109.

Smith, I., Lobascher, M., and Wolff, O.H. (1973). Factors influencing outcome in early treated phenylketonuria. In *Treatment of inborn errors of metabolism* (ed. J.W.T. Seakins, R.A. Saunders, and C. Toothill), p.41. Churchill Livingstone, London..

—— Francis, D.E.M., Clayton, B.E., and Wolff, O.H. (1975). Comparison of an amino acid mixture and protein hydrolysates in treatment of infants with phenylketonuria. *Arch. Dis. Child.* **50**, 864.

—— Lobascher, M.E., Stevenson, J.E., Wolff, O.H., Schmidt, H., Grubel-Kaiser, S., and Bickel, H. (1978). Effect of stopping low-phenylalanine diet on intellectual progress of children with phenylketonuria. *Br. med. J.* **2**, 723.

Smith, J., Zellweger, H., and Afifi, A.K. (1967). Muscular form of glycogenesis type II (Pompe). *Neurology* **17**, 537.

Smith, L.H. jun. and Williams, H.E. (1967). Treatment of primary hyper-oxaluria. *Mod. Treat.* **4**, 522.

—— Jones, J.D., and Keating, F.R. jun. (1969). Primary hyperoxaluria. In *Renal stone research symposium* (ed. A. Hodgkinson and B.E.C. Nordin), p.297. Churchill Livingstone, London..

Smith, M.E. (1967). The metabolism of myelin lipids. *Adv. Lipid Res.* **5**, 241.

Smolin, L.A., Benevenga, J., and Berlow, S. (1981). The use of betaine for the treatment of homocystinuria. *J. Ped.* **99**, 467.

Snyder, P.D. jun., Desnick, R.J., and Krivit, W. (1972). The glycosphingolipids and glycosyl hydrolases of human blood platelets. *Biochem. biophys. Res. Commun.* **46**, 1857.

Snyder, R.D., Carlow, T.J., Ledman, J., and Wenger, D.A. (1976). Ocular findings in fucosidosis. In *The eye and inborn errors of metabolism* (ed. D. Bergsma, A.J. Bron, and E. Cotlier), p.241. Alan R. Liss, New York.

Snyderman, S.E. (1965). An eczematoid dermatitis in histidine deficiency. *J. Ped.* **66**, 212.

—— (1974). Maple syrup urine disease. In *Heritable disorders of aminoacid metabolism: Patterns of clinical expression and genetic variation* (ed. W.L. Nyhan), p.17. John Wiley, New York.

—— (1975). Maple syrup urine disease. In *The treatment of inherited metabolic disease* (ed. D.N. Raine), p.71. MTP Press, Lancaster.

—— Norton, P.M., Roitman, E., and Holt, J.E. jun. (1964). Maple syrup urine disease with particular reference to dietotherapy. *Pediatrics* **34**, 454.

—— Sansaricq, C., Chen, W.J., Norton, P.M., and Phansalkar, S.V. (1977). Argininemia. *J. Ped.* **90**, 563.

—— Boyer, A., Roitman, E., Holt, L.E., and Prose, P.H. (1963). The histidine requirement of the infant. *Pediatrics* **31**, 786.

—— Sansaricq, C., Phansalkar, S.V., Schacht, R.G., and Norton, P.M. (1975). The therapy of hyperammonemia due to ornithine transcarbamylase deficiency in a male neonate. *Pediatrics* **56**, 65.

Sobel, D.O., Gutai, J.P., Jones, J.C., Wagener, D.K., and Smith, W. (1980). Detection of heterozygote of 21-hydroxylase deficiency. *Lancet* **i**, 47.

Sobel, E.H., Clark, L.C., Fox, R.P., and Robinow, M. (1953). Rickets, deficiency of 'alkaline' phosphatase activity and premature loss of teeth in childhood. *Pediatrics* **11**, 309.

Sobel, M.E., Yamamoto, T., Adams, S.L., Dilauro, R., Avvedimento, V.E., De Crombrugghe, B., and Pastan, I. (1978). Construction of a recombinant bacterial plasmid containing a chick pro-alpha 2 collagen gene sequence. *Proc. natl. Acad. Sci. USA* **75**, 5846.

Sockalosky, J.J., Ulstrom, R.A., De Luca, H.F., and Brown, D.M. (1980). Vitamin D-resistant rickets: end-organ unresponsiveness to 1,25 $(OH)_2D_3$. *J. Ped.* **96**, 701.

Soeparto, P., Stobo, E.A., and Walker-Smith, J.A. (1972). Role of chemical examination of the stool in the diagnosis of sugar malabsorption in children. *Arch. Dis. Child.* **47**, 56.

Soffer, D., Yamanaka, T., Wenger, D.A., Suzuki, K., and Suzuki, K. (1980). Central nervous system involvement in adult-onset Gaucher's disease. *Acta Neuropathol.* **49**, 1.

Sokal, J.E., Lowe, C.U., Sacione, E.J., Mosovich, L.L., and Doray, B.H. (1961). Studies of glycogen metabolism in liver glycogen disease (Von Gierkes disease); six cases with similar metabolic abnormalities and responses to glucagon. *J. clin. Invest.* **40**, 364.

Solem, E. (1974). The absolute configuration of ß-aminoisobutyric acid formed by degradation of thymine in man. *Clin. Chim. Acta* **53**, 183.

—— Agarwal, D.P., and Goedde, H.W. (1975). The determination of ß-aminoisobutyric acid in human serum by ion-exchange chromatography. *Clin. Chim. Acta* **59**, 203.

—— Jellum, E., and Eldjarn, L. (1974). The absolute configuration of ß-aminoisobutyric acid in human serum and urine. *Clin. Chim. Acta* **50**, 393.

Solitaire, G.B., Shih, V.E., Nelligan, D.J., and Dolan, T.F. (1969). Argininosuccinic aciduria: clinical, biochemical, anatomical and neuropathological observations. *J. ment. Defic. Res.* **13**, 153.

Solomon, E., Swallow, D., Burgess, S., and Evans, L. (1979). Assignment of the human acid α-glucosidase gene (α GLU) to chromosome 17 using somatic cell hybrids. *Ann. hum. Genet.* **42**, 273.

Solomons, C.C., Goodman, S.I., and Riley, C.M. (1967a). Calcium carbimide in the treatment of primary hyperoxaluria. *New Engl. J. Med.* **276**, 207.

—— —— —— (1967b). Treatment of hyperoxaluria. *New Engl. J. Med.* **277**, 1425.

Sorensen, L.B., Tesar, J.T., Ellman, M.H., and Colwell, N. (1972). A new case of xanthinuria. *Am. J. Med.* **53**, 690.

Spaeth, G.L. and Barber, G.W. (1967). Prevalence of homocystinuria among the mentally retarded. Evaluation of a specific screening test. *Pediatrics* **40**, 586.

—— and Frost, P. (1965). Fabry's disease: its ocular manifestations. *Arch. Ophthalmol. (Chicago)* **74**, 760.

Spear, G.S. (1973). The pathology of the kidney. In *Cystinosis* (ed. J.D. Schulman), p.37. US Government Printing Office, Washington.

—— (1974). Pathology of the kidney in cystinosis. In *Pathology annual* (ed. S.C. Sommers), p.81. Appleton-Century-Crofts, New York.

—— Scusser, R.J., Schulman, J.D., and Alexander, F. (1971). Polykaryocytosis in the visceral glomerular epithelium in cystinosis with description of an unusual clinical variant. *Johns Hopkins Med. J.* **129**, 83.

Spector, E.B., Lochridge, O., and Bloom, A. (1975*a*). Citrulline metabolism in normal and citrullinemic human lymphocyte lines. *Biochem. Genet.* **13**, 471.

—— Cederbaum, S.D., Bernard, B., and Ballard, C.A. (1975*b*). Properties of human adult and fetal red blood cell arginase: a possible prenatal, diagnostic test for arginase deficiency. *Proceedings of the international symposium on inborn errors of metabolism in man*, p.214. Karger, Basel.

—— Kiernan, M., Bernard, B., and Cederbaum, S.D. (1980). Properties of fetal and adult red blood cell arginase deficiency. *Am. J. hum. Genet.* **32**, 79.

Spencer, A.G. and Franglen, G.T. (1952). Gross amino-aciduria following a lysol burn. *Lancet* **i**, 190.

Spencer, N., Hopkinson, D.A., and Harris, H. (1968). Adenosine deaminase polymorphism in man. *Ann. hum. Genet. Lond.* **32**, 9.

Spense, M.W., Goldbloom, A.L., Burgess, J.K., D'Entremont, D., Ripley, B.A., and Weldon, K.L. (1977). Heterozygote detection in angiokeratoma corporis diffusum (Anderson-Fabry disease). *J. med. Genet.* **14**, 91.

—— Mackinnon, K.E., Burgess, J.K., D'Entremont, D.M., Belitsky, P., Lannon, S.G., and Macdonald, A.S. (1976). Failure to correct the metabolic defect by renal allotransplantation in Fabry's disease. *Ann. intern. Med.* **84**, 13.

Sperling, O., Boer, P., Persky-Brosh, S., Kanare, K.E., and De Vries, A. (1972). Altered kinetic property of erythrocyte phosphoribosylpyrophosphate synthase in excessive purine production. *Rev. Eur. Etud. clin. biol.* **XVII**, 703.

Spielberg, S.P., Kramer, L.I., Goodman, S.I., Butler, J., Tietze, F., Quinn, P., and Schulman, J.D. (1977). 5-oxoprolinuria: Biochemical observations and case report. *J. Ped.* **91**, 237.

Spranger, J. (1972). The systemic mucopolysaccharidoses. *Ergebn Inn. Med. Kinderheilk. MF* **32**, 166.

—— and Wiedemann, H.R. (1970). The genetic mucolipidoses; diagnosis and differential diagnosis. *Humangenetik* **9**, 113.

—— Gehler, J., and Cantz, M. (1977). Mucolipidosis I: a sialidosis. *Am. J. med. Genet.* **1**, 21.

—— Langer, L.O. jun., and Wiedemann, H.R. (1974*b*). *Bone dysplasias, an atlas of constitutional disorders of skeletal development.* Saunders, Philadelphia.

—— Gehler, J., O'Brien, J.F., and Cantz, M. (1974*a*). Chondroitinsulphaturia with α-L-iduronidase deficiency. *Lancet* **ii**, 1082.

—— Cantz, M., Gehler, J., Liebaers, I., and Theiss, W. (1978). Mucopolysaccharidosis II (Hunter disease) with corneal opacities: Report of two patients at the extremes of a wide clinical spectrum. *Eur. J. Ped.* **129**, 11.

Srivastava, S.K. and Beutler, E. (1974). Studies on human ß-D-N-acetyl-hexosaminidase. *J. biol. Chem.* **249**, 2054.

—— Awasthi, Y.C., Yoshida, A., and Beutler, E. (1974). Studies in the human hexosaminidases. *J. biol. Chem.* **249**, 2043.

Staal, G.E., Koster, J.F., and Nijessen, J.G. (1972). A new variant of red blood cell pyruvate kinase deficiency. *Biochim. Biophys. Acta* **258**, 685.

—— —— Kamp, H., Van Milligen-Boersma, L., and Veeger, C. (1971). Human erythrocyte pyruvate kinase. Its purification and some properties. *Biochim. Biophys. Acta* **227**, 86.

Stacey, T.E., MacNab, A., and Strang, L.B. (1980). Recent work on treatment of Type 1 glycogen storage disease. In *Inherited disorders of carbohydrate metabolism* (ed. D. Burman, J.B. Holton, and C.A. Pennock), p.315. MTP Press, Lancaster.

Ställberg-Stenhagen, S. and Svennerholm, L. (1965). Fatty acid composition of human brain sphingomyelins: normal variation with age and changes during myelin disorders. *J. Lipid. Res.* **6**, 146.

Stanbury, J.B. (1978). Familial goiter. In *The metabolic basis of inherited disease* (ed. J.B. Stanbury, J.B. Wyngaarden, and D.S. Fredrickson), 4th edn., p.206. McGraw-Hill, New York.

—— and Chapman, E.M. (1960). Congenital hypothyroidism with goitre: Absence of an iodide-concentrating mechanism. *Lancet* **i**, 1162.

—— and Janssen, M.-A. (1963). Labeled iodoalbumin in the plasma in thryotoxicosis after I^{125} and I^{131}. *J. clin. Endocrinol. Metab.* **23**, 1056.

—— Kassenaar, A.A.H., and Meijer, J.W.A. (1956*b*). The metabolism of iodotyrosines. I. The fate of mono- and di-iodotyrosine in normal subjects and in patients with various diseases. *J. clin. Endocrinol. Metab.* **16**, 735.

—— Meijer, J.W.A., and Kassenaar, A.A.H. (1956*a*). The metabolism of iodotyrosines. II. The metabolism of mono- and di-iodotyrosine in certain patients with familial goiter. *J. clin. Endocrinol. Metab.* **16**, 848.

—— Kassenaar, A.A.H., Meijer, J.W.A., and Terpstra, J. (1955). The occurrence of mono- and di-iodotyrosine in the blood of a patient with congenital goiter. *J. clin. Endocrinol. Metab.* **15**, 1216.

Starzl, T.E., Marchioro, T.L., Sexton, A.W., Illingworth, B., Waddell, W.R., Faris, T.D., and Herrmann, T.J. (1965). The effect of portacaval transposition on carbohydrate metabolism: Experimental and clinical observations. *Surgery* **57**, 687.

—— Putnam, C.W., Porter, K.A., Halgrimson, E.G., Corman, J., Brown, B.I., Gotlin, R.W., Rodgerson, D.O., and Greene, H.L. (1973). Portal diversion for the treatment of glycogen storage disease in humans. *Ann. Surg.* **178**, 525.

States, B. and Segal, S. (1969). Thin-layer chromatrographic separation of cystine and the *N* ethylmaleimide adducts of cysteine and glutathione. *Anal. Biochem.* **27**, 323.

—— Harris, D., and Segal, S. (1974). Uptake and utilization of exogenous cystine by cystinotic and normal fibroblasts. *J. clin. Invest.* **53**, 1003.

—— Blazer, B., Harris, D., and Segal, S. (1975). Prenatal diagnosis of cystinosis. *J. Ped.* **87**, 558.

Steffes, M.M. and Wong, E.T. (1979). Prenatal diagnosis of congenital adrenal hyperplasia. *Lancet* **ii**, 303.

Stein, W.H. and Moore, S. (1954). The free amino acids of human blood plasma. *J. biol. Chem.* **211**, 915.

Stein, H., Berman, E.R., Lioni, N., Merin, S., Sleskin, J., and Cohen, T. (1974). Pseudo-Hurler polydystrophy (mucolipidosis III): A clinical, biochemical and ultrastructural study. *Isr. J. med. sci.* **10**, 463.

Steinberg, D. (1965). Remarks on the biochemical basis of Refsum's disease. *Nord. Med.* **73**, 570.

—— (1983). Phytanic acid storage disease (Refsum's disease). In *The metabolic basis of inherited disease* (ed. J.B. Stanbury, J.B. Wyngaarden, D.S. Fredrickson, J.L. Goldstein, and M.S. Brown), 5th edn., p. 735. McGraw-Hill, New York.

—— Avigan, J., Mize, C., Eldjarn, L., Try, K., and Refsum, S. (1965). Conversion of $U-C^{14}$-phytol to phytanic acid and its oxidation in heredopathia atactica polyneuritiformis. *Biochem. biophys. Res. Commun.* **19**, 783.

—— Mize, C.E., Herndon, J.H. jun., Fales, H.M., Engel, W.K., and Vroom, F.Q (1970). Phytanic acid in patients with Refsum's syndrome and response to dietary treatment. *Arch. intern. Med.* **125**, 70.

—— —— Avigan, J., Fales, H.M., Eldjarn, L., Stokke, O., and Refsum, S.

(1967). Studies on the metabolic error in Refsum's disease. *J. clin. Invest.* **46**, 313.

Steinherz, R., Tietze, F., Railford, D., Gahl, W., and Schulman, J.D. (1982*b*). Patterns of amino acid efflux from isolated normal and cystinotic human leucocyte lysosomes. *J. biol. Chem.* **257**, 6041.

—— —— Gahl, W., Triche, T., Chiang, H., Modesti, A., and Schulman, J.D. (1982*a*). Cystine accumulation and clearance by normal and cystinotic leucocytes exposed to cystine dimethyl ester. *Proc. natl. Acad. Sci. USA* **79**, 4446.

Steinman, L., Tharp, B.R., Dorfman, L.J., Forno, L.S., Sogg, R.L., Kelts, K.A., and O'Brien, J.S. (1980). Peripheral neuropathy in the cherry-red spot myoclonus syndrome (Sialidosis type I). *Ann. Neurol.* **7**, 450.

Steinmann, B. and Gitzelmann, R. (1976). Fruktose und Sorbitol in Infusions-flüssigkeiten sind nicht immer harmlos. In *Monosaccharides and polyalcohols in nutrition, therapy and dietetics* (ed. G. Ritzel and G. Brubacher). *Int. J. Vit. Nutr. Res. (Suppl.* **15**), 289.

—— Baerlocher, K., and Gitzelmann, R. (1975*b*). Hereditäre Störungen des Fruktosestoffwechsels. Belastungsproben mit Fruktose, Sorbitol und Dihydroxy-aceton. *J. nutr. Metab.* **18**, 115.

—— Martin, G.R., Baum, B.I., and Crystal, R.G. (1979). Synthesis and degradation of collagen by skin fibroblasts from controls and from patients with osteogenesis imperfecta. *FEBS Lett* **101**, 269.

—— Gitzelmann, R., Vogel, A., Grant, M.E., Harwood, R., and Sear, C.H.J. (1975*a*). Ehlers–Danlos syndrome in two siblings with deficient lysyl hydroxylase activity in cultured skin fibroblasts but only mild hydroxylysine deficit in skin. *Helv. Paediatr. Acta* **30**, 255.

—— Tuderman, L., Peltonen, L., Martin, G.R., McKusick, V.A., and Prockop, D.J. (1980). Evidence for a structural mutation of procollagen type 1 in a patient with the Ehlers–Danlos syndrome type VII. *J. biol. Chem.* **255**, 8887.

Sternlieb, I. (1966). The Kayser–Fleischer ring, a histochemical and electron microscopic study. *Med. Radiogr. Photogr.* **42**, 14.

—— and Scheinberg, I.H. (1968). Prevention of Wilson's disease in asymptomatic patients. *New Engl. J. Med.* **278**, 352.

—— Van den Hamer, C.J.A., Morell, A.G., Alpert, S., Gregoriadis, G., and Scheinberg, I.H. (1973). Lysosomal defect of hepatic copper excretion in Wilson's disease (hepatolenticular degeneration). *Gastroenterology* **64**, 99.

Sternowsky, H.J., Robon, J., Hutterer, F., and Gaull, G. (1973). Determination of α-ketoacids as silylated oximes in urine and serum by combined gas chromatography-mass spectrometry. *Clin. Chim. Acta* **47**, 371.

Stetten, D.W. jun. and Stetten, M.R. (1960). Glycogen metabolism. *Physiol. Rev.* **40**, 505.

Stevens, R.L., Fluharty, A.L., Kihara, H., Kaback, M.M., Shapiro, L.J., Marsh, B., Sandhoff, K., and Fischer, G. (1981). Cerebroside sulfatase activator deficiency induced metachromatic leukodystrophy. *Am. J. hum. Genet.* **33**, 900.

Stevenson, R.E., Taylor, H.A., and Parks, S.E. (1978). ß-galactosidase deficiency: prolonged survival in three patients following early central nervous deteriora-tion. *Clin. Genet.* **13**, 305.

—— Howell, R.R., McKusick, V.A., Suskind, R., Hanson, J.W., Eliot, D.E., and Neufeld, E.F. (1976). The iduronidase-deficient mucopolysacharidoses: clinical and roentgenographic features. *Pediatrics* **57**, 111.

Stich, H.F. (1975). Response to homozygous and heterozygous xeroderma pigmentosum cells to several chemical and viral carcinogens. In *Molecular mechanisms for repair of DNA* (ed. P.C. Hanawalt and R.B. Setlow), p.773. Plenum Press, New York.

Stickler, G.B. (1969). Familial hypophosphatemic vitamin-D resistant rickets. *Acta. Paediatr. Scand.* **58**, 213.

—— Jowsey, J., Phil, D., and Bianco, A.J. (1971). Possible detrimental effect of large doses of vitamin D in familial hypophosphatemic vitamin D-resistant rickets. *J. Ped.* **79**, 68.

Stirling, J.L. (1972). Separation and characterization of *N*-acetyl-ß-glucosaminidases A and P from maternal serum. *Biochim. Biophys. Acta* **271**, 154.

—— Robinson, D., Fensom, A.H., Benson, P.F., and Baker, J.E. (1978). Fluorimetric assay for prenatal detection of Hurler and Scheie homozygotes or heterozygotes. *Lancet* **i**, 147.

—— —— —— —— —— and Button, L.R. (1979). Prenatal diagnosis of two Hurler fetuses using an improved assay for methylumbelliferyl-α-L-iduronidase. *Lancet* **ii**, 37.

Stokke, K.T. and Norum, K.R. (1971). Determination of lecithin:cholesterol acyltransfer in human blood plasma. *Scand. J. clin. lab. Invest.* **27**, 21.

Stokke, O., Goodman, S.I., and Moe, P.G. (1976). Inhibition of brain glutamate decarboxylase by glutamate, glutaconate and ß-hydroxyglutarate: explanation of the symptoms in glutaric aciduria. *Clin. Chim. Acta* **66**, 411.

—— Eldjarn, L., Jellum, E., Pande, H., and Waaler, P.E. (1972). ß-methylcrotonyl-CoA carboxylase deficiency: A new metabolic error in leucine degradation. *Pediatrics* **49**, 726.

—— —— Norum, K.R., Steen-Johnsen, J., and Halvorsen, S. (1967). Methylmalonic aciduria: A new inborn error of metabolism which may cause fatal acidosis in the neonatal period. *Scand. J. clin. lab. Invest.* **20**, 313.

Stone, N.J., Levy, R.I., Fredrickson, D.S., and Verter, J. (1974). Coronary artery disease in 116 kindred with familial type II hyperlipoproteinemia *Circulation* **49**, 476.

Stoop, J.W., Eijsvoogel, V.P., Zegers, B., Blok-Schut, B., Van Bekkum, D.W., and Ballieux, R.E. (1976). Selective severe cellular immunodeficiency – effect of thymus transplantation and transfer factor administration. *Clin. Immunol. Immunopathol.* **6**, 289.

—— Zegers, B.J.M., Hendricks, G.F.M., Van Heukelom, L.H.S., Staal, G.E.J., De Bree, P.K., Wadman, S.K., and Ballieux, R.E. (1977). Purine nucleoside phosphorylase deficiency associated with selective cellular immunodeficiency. *New Engl. J. Med.* **296**, 651.

Strand, L.J., Felsher, B., Redeker, A.G., and Marver, H.S. (1970). Enzymatic abnormalities in heme biosynthesis in intermittent acute porphyria: Decreased hepatic conversion of porphobilinogen to porphyrins and increased δ-aminolevulinic acid synthetase activity. *Proc. natl. Acad. Sci. USA* **67**, 1315.

—— Meyer, U.A., Felsher, B.F., Redeker, A.G., and Marver, H.S. (1972). Decreased red cell uroporphyrinogen I synthetase activity in intermittent acute porphyria. *J. clin. Invest.* **51**, 2530.

Strecker, G. and Michalski, J.C. (1978). Biochemical basis of six different types of sialidosis. *FEBS Lett.* **85**, 20.

—— and Montreuil, J. (1979). Glycoproteins et glycoproteinoses. *Biochemie* **61**, 1199.

—— Michalski, J.C., Montreuil, J., and Farriaux, J.-P. (1976*b*). Deficit in neuraminidase associated with mucolipidosis II (I-cell disease). *Biomedicine* **25**, 238.

—— Fournet, B., Bouquelet, S., Montreuil, J., Dhondt, J.L., and Farriaux, J.-P.

(1976*a*). Etude chimique des mannosides urinaires excrétés au cours de la mannosidose. *Biochemie* **58**, 579.

—— Peers, M.-C., Michalski, J.C., Hondi-Assah, T., Fournet, B., Spik, G., Montreuil, J., Farriaux, J.-P., Maroteaux, P., and Durand, P. (1977). Structure of nine sialyl-oligosaccharides accumulated in urine of eleven patients with three different types of sialidosis. *Eur. J. Biochem.* **75**, 391.

Strecker, H.J. (1965). Purification and properties of rat liver ornithine-δ-transaminase. *J. biol. Chem.* **240**, 1225.

Strickland, G.T. (1972). Febrile penicillamine eruption. *Arch. Neurol.* **26**, 474.

—— Frommer, D., Leu, M.-L., Pollard, R., Sherlock, S., and Cumings, J.N. (1973). Wilson's disease in the United Kingdom and Taiwan. I. General characteristics of 142 cases and prognosis. II. A genetic analysis of 88 cases. *Q. J. Med.* **42**, 619.

Strife, C.E., Zuroweste, E.L., Emmett, E.A., Finelli, V.N., Petering, H.G., and Berry, H.K. (1977). Tyrosinemia with acute intermittent porphyria: amino-levulinic acid dehydratase deficiency related to elevated urinary aminolevulinic acid levels. *J. Ped.* **90**, 400.

Strömme, J.H., Borud, O., and Moe, P.J. (1976). Fatal lactic acidosis in a newborn attributable to a congenital defect of pyruvate dehydrogenase. *Ped. Res.* **10**, 60.

Stumpf, D.A., Neuwelt, E., Austin, J., and Kohler, P. (1971). Metachromatic leukodystrophy (MLD). X. Immunological studies of the abnormal sulfatase A. *Arch. Neurol.* **25**, 427.

—— Austin, J.H., Crocker, A.C., and La France, M. (1973). Mucopolysaccharidosis type VI (Maroteaux–Lamy syndrome): Arylsulphatase B deficiency in tissues. *Am. J. Dis. Child.* **126**, 747.

Su, T.-S., Beaudet, A.L., and O'Brien, W.E. (1983). Abnormal mRNA for argininosuccinate synthetase in citrullinaemia. *Nature, Lond.* **301**, 533.

Suda, M. and Takeda, Y. (1950*a*). Metabolism of tyrosine. I. Application of successive adaptation of bacteria for the analysis of the enzymatic breakdown of tyrosine. *J. Biochem. (Tokyo)* **37**, 375.

—— —— (1950*b*). Metabolism of tyrosine. II. Homogentisicase. *J. Biochem. (Tokyo)* **37**, 381.

Sugita, M., Connolly, P., Dulaney, J.T., and Moser, H.W. (1973). Fatty acid composition of free ceramides of kidney and cerebellum from a patient with Farber's disease. *Lipids* **8**, 401.

—— Dulaney, J.T., and Moser, H.W. (1972). Ceramidase deficiency in Farber's disease (Lipogranulomatosis). *Science* **178**, 1100.

—— Williams, M., Dulaney, J.T., and Moser, H.W. (1975). Ceramidase and ceramide synthesis in human kidney and cerebellum. *Biochim. Biophys. Acta* **398**, 125.

—— Iwamori, M., Evans, J., McCleur, R.H., Dulaney, J.T., and Moser, H.W. (1974). High performance liquid chromatography of ceramides: Application to analysis in human tissues and demonstration of ceramide excess in Farber's disease. *J. Lipid Res.* **15** 223.

Sulcova, J., Jirasek, J.E., and Starka, L. (1973). Transformation of testosterone into dihydrotestosterone by the primordia of human genitalia and by the fetal suprascapular skin. *Steroids Lipid Res.* **4**, 129.

Sung, J.H. (1979). Autonomic neurons affected by lipid storage in the spinal cord of Fabry's disease. Distribution of autonomic neurons in the sacral cord. *J. Neuropathol. exp. Neurol.* **38**, 87.

—— Hayano, M., and Desnick, R.J. (1977). Mannosidosis: pathology of the nervous system. *J. Neuropathol. exp. Neurol.* **36**, 807.

Sussman, M., Lichtenstein, J.R., Nigra, T.P., Martin, G.R., and McKusick, V.A. (1974). Hydroxylysine-deficient skin collagen in a patient with a form of the Ehlers–Danlos syndrome. *J. Bone. Joint. Surg.* **56A**, 1128.

Suzuki, K. (1968). Cerebral GM_1-gangliosidosis: chemical pathology of visceral organs. *Science* **159**, 1471.

—— (1970). Ultrastructural study of experimental globoid cells. *Lab. Invest.* **23**, 612.

—— (1972). Neurochemical aspects of mucopolysaccharidoses. In *Handbook of neurochemistry* (ed. A. Lajtha), Vol. VII, p.17. Plenum Press, New York.

—— (1977). Krabbe disease and GM_1-gangliosidosis. In *Practical enzymology of the sphingolipidoses* (ed. R.H. Glew and S.P. Peters), p.101. Alan R. Liss Inc. New York.

—— and Suzuki, Y. (1970). Globoid cell leucodystrophy (Krabbe's disease). Deficiency of galactocerebroside-ß-galactosidase. *Proc. natl. Acad. Sci. USA* **66**, 302.

—— —— (1973). Krabbe's disease. In *Lysosomes and storage diseases* (ed. H.G. Hers and F. Van Hoof), p.395. Academic Press, New York.

—— —— (1983). Galactosylceramide lipidosis: Globoid cell leukodystrophy (Krabbe's disease). In *The metabolic basis of inherited disease* (ed. J.B. Stanbury, J.B. Wyngaarden, D.S. Fredrickson, J.L. Goldstein, and M.S. Brown), 5th edn., p.858. McGraw-Hill, New York.

—— Schneider, E.L., and Epstein, C.J. (1971*b*). In utero diagnosis of globoid cell leukodystrophy (Krabbe's disease). *Biochem. biophys. Res. Commun.* **45**, 1363.

—— Suzuki, K., and Chen, G.C. (1967). Isolation and chemical characterization of metachromatic granules from a brain with metachromatic leucodystrophy. *J. Neuropathol. exp. Neurol.* **26**, 537.

—— Suzuki, Y., and Fletcher, T.F. (1972). Further studies of galactocerebroside ß-galactosidase in globoid cell leukodystrophy. In *Sphingolipids, sphingo-lipidoses and allied disorders* (ed. B.W. Volk and S.M. Aronson), p. 487. Plenum Press, New York.

—— Tanaka, H., and Suzuki, K. (1976). Studies on the pathogenesis of Krabbe's leukodystrophy: Cellular reaction of the brain to exogenous galactosylsphingo-sine, monogalactosyl diglyceride and lactosylceramide. In *Current trends in sphingolipidoses and allied disorders* (ed. B.W. Volk and L. Schneck), p.99. Plenum Press, New York.

—— Suzuki, K., Rapin, I. Suzuki, Y., and Ishii, N. (1970*a*). Juvenile GM_2-gangliosidosis: Clinical variant of Tay–Sachs disease or a new disease. *Neurology* **20**, 190.

Suzuki, Y. and Suzuki, K. (1971). Krabbe's globoid cell leucodystrophy: deficiency of galatocerebrosidase in serum, leucocytes and fibroblasts. *Science* **171**, 73.

—— Crocker, A.C., and Suzuki, K. (1971*c*). GM_1-gangliosidosis. *Arch. Neurol.* **24**, 58.

—— Miyatake, T., Fletcher, T.F., and Suzuki, K. (1974). Glycosphingolipid ß-galactosidases. III. Canine form of globoid cell leukodystrophy: Comparison with the human disease. *J. biol. Chem.* **249**, 2109.

—— Jacob, J., Suzuki, K., Kutty, K.M., and Suzuki, K. (1971*a*). GM_2-gangliosidosis with total hexosaminidase deficiency. *Neurology* **21**, 313.

—— Nakamura, N., Fukuoka, K., Shimada, Y., and Wono, M. (1977). ß-galactosidase deficiency in juvenile and adult patients. *Hum. Genet.* **36**, 219.

—— Austin, J., Suzuki, K., Armstrong, D., Schlenker, J., and Fletcher, T. (1970*b*). Studies in globoid leukodystrophy: Enzymatic and lipid findings in the canine form. *Exp. Neurol.* **29**, 65.

Svennerholm, L. (1962). The chemical nature of normal human brain and Tay–Sachs gangliosides. *Biochem. biophys. Res. Commun.* **9**, 436.

—— (1967). Metabolism of gangliosides in cerebral lipidoses. In *Inborn disorders of sphingolipid metabolism* (ed. S.M. Aronson and B.W. Volk), p.169. Pergamon Press, Oxford.

—— Håkansson, G., and Vanier, M.-T. (1975). Clinical pathology of Krabbe's disease. IV. *Acta. Pediatr. Scand.* **64**, 649.

—— Vanier, M.-T., and Månsson, J.-E. (1980). Krabbe disease: A galactosyl-sphingosine (psychosine) lipidosis. *J. Lipid Res.* **21**, 53.

—— Håkansson, G., Månsson, J.-E., and Vanier, M.-T. (1979). The assay of sphingolipid hydrolases in white blood cells with labeled natural substrates. *Clin. Chim. Acta* **92**, 53.

—— Dreborg, S., Erikson, A., Groth, C.G., Hillborg, P.O., Håkansson, G., Nilsson, O., and Tibblin, E. (1982). Gaucher disease of the Norbottinian type (Type III). Phenotypic manifestation. In *Gaucher disease: A century of delineation and research* (ed. R.J. Desnick, S. Gatt, and G.A. Grabowski), p.67. Alan R. Liss, New York.

Swallow, D.M., Evans, L., Saha, L.Y., and Harris, H. (1976). Characterization and tissue distribution of *N*-acetylhexosaminidase C: suggetive evidence for a separate hexosaminidase locus. *Ann. hum. Genet.* **40**, 55.

—— O'Brien, J.S., Hoogeveen, A.T., and Bucks, D.W. (1981). Electrophoretic analysis of glycoprotein enzymes in the sialidoses and mucolipidoses. *Ann. hum. Genet.* **45**, 29.

—— Stokes, D.C., Corney, G., and Harris, H. (1974). Differences between the hexosaminidase isoenzymes in serum and tissues. *Ann. hum. Genet. Lond.* **37**, 287.

Sweeley, C.C. (1980) Cell surface glycolipids. *Am. chem. Soc. Symp. Series* **128**, Washington DC.

—— and Klionsky, B. (1963). Fabry's disease: classification as a sphingolipidosis and partial characterization of a novel glycolipid. *J. biol. Chem.* **238**, 3148.

—— —— (1964). Fabry's disease: The isolation and characterization of a ceramide trihexoside from kidney. *6th Int. Cong. Biochem. Abstr.* New York.

Sweetman, L. and Nyhan, W.L. (1974). Propionyl-CoA carboxylase deficiency in ß-methylcrotonylglycinuria. *Clin. Res.* **22**, 237A.

—— —— Trauner, D.A., Merritt, A., and Singh, M. (1980). Glutaric aciduria type II. *J. Ped.* **96**, 1020.

—— Weyler, W., Shafai, T., Young, P.F., and Nyhan, W.L. (1979). Prenatal diagnosis of propionic acidemia. *J. Am. med. Ass.* **242**, 1048.

Swift, P.N., Benson, P.F., and Studdy, J.D. (1973). Genetic studies in mental subnormality. II. The application of genetic principles to screening for metabolic disorders: The Leybourne Grange survey of mentally subnormal siblings. *Brit. J. Psychiat.*, Special Publication No. 8, p.61.

Swift, T.R. and McDonald, T.F. (1976). Peripheral nerve involvement in Hunter syndrome (mucopolysaccharidosis II). *Arch. Neurol.* **33**, 845.

Szeinberg, A., Sheba, C., Ramot, B., and Adam, A. (1960). Differences in hemolytic susceptibility among subjects with glucose-6-phosphate dehydro-genase (G.6-P.D.) deficiency. *Clin. Res.* **8**, 18.

Tabas, I. and Kornfeld, S. (1980). Biosynthetic intermediates of ß-glucuronidase contain high mannose oligosaccharides with blocked phosphate residues. *J. biol. Chem.* **255**, 6633.

Tabei, T. and Heinrichs, W.L. (1976). Diagnosis of placental sulfatase deficiency. *Am. J. Obstet. Gynecol.* **124**, 409.

Tabira, T., Goto, I., Kuroiwa, Y., and Kikuchi, M. (1974). Neuropathological and biochemical studies in Fabry's disease. *Acta Neuropath.* **30**, 345.

Tada, K., Morikawa, T., and Arakawa, T. (1966). Prolinuria: Transport of proline by leukocytes. *Tohoku J. exp. Med.* **90**, 189.

—— Wada, Y., and Arakawa, T. (1967*b*). Hypervalinemia: Its metabolic lesion and therapeutic approach. *Am. J. Dis. Child.* **113**, 64.

—— Takada, G., Omura, K., and Itokawa, Y. (1978). Congenital lactic acidosis due to pyruvate carboxylase deficiency: absence of an inhibitor of TPP-ATP phosphoryl transferase. *Eur. J. Ped.* **127**, 141.

—— Yokoyama, Y., Nakagawa, H., and Arakawa, T. (1968). Vitamin B$_6$ dependent xanthurenic aciduria. The second report. *Tohoku J. exp. Med.* **95**, 107.

—— —— —— Yoshida, T., and Arakawa, T. (1967*a*). Vitamin B$_6$ dependent xanthurenic aciduria. *Tohoku J. exp. Med.* **93**, 115.

—— Sugita, K., Fujikawi, K., Kesaki, T., Takada, G., and Omura, K. (1973). Hyperalaninemia with pyruvicemia in a patient suggestive of Leigh's encephalo-myelopathy. *Tohoku J. exp. Med.* **109**, 13.

—— Tateda, H., Arashima, S., Sakay, K., Kitagawa, T., Aoki, K., Suha, S., Kawanura, M., Oura, T., Takesada, M., Kuroda, Y., Yamashita, F., Matsuda, I., and Naruse, H. (1982). Intellectual development in patients with untreated histidinemia. A collaborative study group of neonatal screening for inborn errors of metabolism in Japan. *J. Ped.* **101**, 562.

Taddeini, L. and Watson, C.J. (1968). The clinical porphyrias. *Sem. Hematol* **5**, 335.

Tagliavini, F., Pietrini, V., Pilleri, G., Trabattoni, G., and Lechi, A. (1979). Case report. Adult metachromatic leukodystrophy: Clinicopathological report of two familial cases with slow course. *Neuropath. appl. Neurobiol.* **5**, 233.

Tahmoush, A.J., Alpers, D.H., Feigin, R.D., Armbrustmacher, V., and Prensky, A.L. (1976). Hartnup disease. Clinical, pathological and biochemical observations. *Arch. Neurol.* **33**, 797.

Taillard, W. and Prader, A. (1957). Étude génétique du syndrome de feminisation testiculaire totale et partielle. *J. Genet. hum.* **6**, 13.

Takahara, S. (1967). Acatalasemia. *Asian med. J.* **10**, 46.

—— (1968). Acatalasemia in Japan. In *Hereditary disorders of erythrocyte metabolism* (ed. E. Beutler), Vol. 1, p.21. Grune and Stratton, New York.

—— Hamilton, H.B., Neel, J.V., Kobara, T.Y., Ogura, Y., and Nishimura, E.T. (1960). Hypocatalasemia: a new genetic carrier state. *J. clin. Invest.* **39**, 610.

Takamizawa, M., Toru, M., Kojima, T., Watanabe, A., and Hirokawa, K. (1973). An autopsy case of juvenile hepatocerebral degeneration (non-Wilsonian I nose type) with mental retardation with special reference to amino acids metabolism. *Psych. Neurol. Jap.* **75**, 370.

Takebe, H. (1978). Relationship between DNA defects and skin cancer in xeroderma pigmentosum. *J. supramol. Struct. Supp.* **2**, 30.

—— Miki, Y., Kozuka, T., Furuyama, J.-I., Tanaka, K., Sasaki, M.S., Fujiwara, Y., and Akiba, H. (1977). DNA repair characteristics and skin cancers of xeroderma pigmentosum patients in Japan. *Cancer Res.* **37**, 490.

Takeuchi, F., Hanaoka, F., Yano, E., Yamada, M., Horiuchi, Y., and Akaoka, I. (1981). The mode of genetic transmission of a gouty family with increased phosphoribosylpyrophosphate synthetase activity. *Hum. Genet.* **58**, 322.

Takki, K. (1974). Gyrate atrophy of the choroid and retina associated with hyperornithinemia. *Br. J. Ophthalmol.* **58**, 3.

—— and Milton, R.C. (1981). The natural history of gyrate atrophy of the choroid and retina. *Ophthalmology* **88**, 292.

—— and Simell, O. (1976). Gyrate atrophy of the choroid and retina with hyperornithinemia. *Birth Defects: Original Article Series XII*, **3** 373.

Tallman, J.F., Brady, R.O., Navon, R., and Padeh, B. (1974). Ganglioside catabolism in hexosaminidase A-deficient adults. *Nature, Lond.* **252**, 254.

Tamaya, T., Nioka, S., Furuta, N., Boku, S., Motoyama, T., Ohono, Y., and Okada, H. (1978). Preliminary studies on steroid-binding proteins in human testes of testicular feminization syndrome. *Fertil. Steril.* **30**, 170.

Tanaka, H. and Suzuki, K. (1975). Lactosylceramide ß-galactosidase in human sphingolipidoses: Evidence for two genetically distinct enzymes. *J. biol. Chem.* **250**, 2324.

Tanaka, J., Garcia, J.H., Max, S.R., Viloria, J.E., Kamijo, Y., Maclaren, N.K., Cornblath, M., and Brady, R.O. (1975). Cerebral sponginess and GM₃-gangliosidosis: Ultrastructure and probable pathogenesis. *J. Neuropathol. exp. Neurol.* **34**, 249.

Tanaka, K. (1975). Disorders of organic acid metabolism. In *Biology of brain dysfunction* (ed. G.E. Gaull). Vol. 3, p.145. Plenum Press, New York.

—— Mandell, R., and Shih, V.E. (1976). Metabolism of (1-¹⁴C) and (2-¹⁴C) leucine in cultured skin fibroblasts from patients with isovaleric acidemia. Characterization of metabolic defects. *J. clin. Invest.* **58**, 164.

—— Budd, M.A., Efron, M.E., and Isselbacher, K.J. (1966). Isovaleric acidemia: a new genetic defect of leucine metabolism. *Proc. natl. Acad. Sci. USA* **56**, 236.

—— Hine, D.G., West-Dull, A., and Lynn, T.B. (1980*b*). A gas-chromatographic method for analysis of urinary organic acids: Retention indices of 155 metabolically important compounds. *Clin. Chem.* **26**, 1839.

—— West-Dull, A., Hine, D.G., and Lynn, T.B. (1980*a*). A gas chromatographic method for analysis of urinary organic acids. II. Description of procedure and its application for diagnosis of patients with organic acidurias. *Clin. Chem.* **26**, 1847.

—— Akino, M., Hagi, Y., Doi, M., and Shiota, T. (1981). The enzymatic synthesis of sepiapterin by chicken kidney preparations. *J. biol. Chem.* **256**, 2963.

Tanaka, K.R. (1969). Pyruvate kinase. In *Biochemical methods in red cell genetics* (ed. J.J. Yunis), p.167. Academic Press, New York.

—— and Paglia, D.E. (1971). Pyruvate kinase deficiency. *Semin. Hematol.* **8**, 367.

Tancredi, F., Gauzzi, G., and Auricchio, S. (1970). Renal iminoglycinuria without intestinal malabsorption of glycine and imino acids. *J. Ped.* **76**, 386.

Tang, T.T., Good, T.A., Dyken, P.R., Johnson, S.D., McCready, S.R., Sy, S.T., Lardy, H.A., and Rudolph, F.B. (1972). Pathogenesis of Leigh's encephalomyelopathy. *J. Ped.* **81**, 189.

Tarlow, M.J., Hadorn, B., Arthurton, M.W., and Lloyd, J.K. (1970). Intestinal enterokinase deficiency. *Arch. Dis. Child.* **45**, 651.

—— Seakins, J.W., Lloyd, J.K., Matthews, D.M., Cheng, B., and Thomas, A.J., (1972). Absorption of amino acids and peptides in a child with a variant of Hartnup disease and coexistent coeliac disease. *Arch. Dis. Child.* **47**, 798.

Tarui, S., Kono, N., Nasu, T., and Nishikawa, M. (1969). Enzymatic basis for the coexistence of myopathy and hemolytic disease in inherited muscle phosphofructokinase deficiency. *Biochem. biophys. Res. Commun.* **34**, 77.

—— Okuno, G., Ikura, Y., Tanaka, T., Suda, M., and Nishikawa, M. (1965). Phosphofructokinase deficiency in skeletal muscle: a new type of glycogenosis. *Biochem. biophys. Res. Commun.* **19**, 517.

Taskinen, M.-R., Nikkilä, E.A., Huttunen, J.K., and Hilden, H. (1980). A micromethod for assay of lipoprotein lipase activity in needle biopsy samples of human adipose tissue and skeletal muscle. *Clin. Chim. Acta.* **104**, 107.

Taunton, O.D., Greene, H.L., Stifel, F.B., Hofeldt, F.D., Lufkin, E.G., Hagler, L., Herman, Y., and Herman, R.H. (1978). Fructose-1,6-diphosphatase deficiency, hypoglycemia and response to folate therapy in a mother and her daughter. *Biochem. Med.* **19**, 260.

Tauro, G.P., Danks, D.M., Rowe, P.B., Van der Weyden, M.B., Schwarz, M.A., Collins, V.L., and Neal, B.W. (1976). Dihydrofolate reductase deficiency causing megaloblastic anemia in two families. *New Engl. J. Med.* **294** 466.

Tay, W. (1881). Symmetrical changes in the region of the yellow spot in each eye of an infant. *Trans. ophthalmol. Soc, UK.* **1**, 55.

Taylor, C.J. and Green, S.H. (1981). Menkes' syndrome (trichopoliodystrophy): use of scanning electron-microscope in diagnosis and carrier identification. *Dev. Med. Child. Neurol.* **23**, 361.

Taylor, H.A., Thomas, G.H., Aylsworth, A., Stevenson, R.E., and Reynolds, L.W. (1975). Mannosidosis: deficiency of a specific α-mannosidase component in cultured fibroblasts. *Clin. Chim. Acta* **59**, 93.

—— —— Miller, C.S., Kelly, T.E., and Siggers, D. (1973). Mucolipidosis III (pseudo-Hurler polydystrophy): Cytological and ultrastructural observations of cultured fibroblast cells. *Clin. Genet.* **4**, 388.

Taylor, J., Robinson, B.H., and Sherwood, G. (1978). A defect in branched chain amino acid metabolism in a patient with congenital lactic acidosis due to dihydrolipoyl dehydrogenase deficiency. *Ped. Res.* **12**, 60.

Taylor, R.E. and Farrell, R.K. (1973). Light and electon microscopy of peripheral blood neutrophiles in a killer whale affected with Chediak–Higashi syndrome. *Fed. Proc.* **32**, 822.

Taylor, R.T. and Jenkins, W.T. (1966). Leucine aminotransferase: I. Colorimetric assays. II. Purification and characterization. III. Activation of ß-mercaptoethanol. *J. biol. Chem.* **241**, 4391, 4396, 4406.

Taylor, W.O.G. (1978). Eldridge-Green Lecture, 1978. Visual disabilities of oculocutaneous albinism and their alleviation. *Trans. Ophthalmol. Soc. UK* **98**, 423.

Tedesco, T.A. and Mellman, W.J. (1967). Argininosuccinate synthetase activity and citrulline metabolism in cells cultured from a citrullinemic subject. *Proc. natl. Acad. Sci. USA* **57**, 829.

—— Nonow, R., Miller, K., and Mellman, W.J. (1972). Galactokinase: evidence for a new racial polymorphism. *Science* **178**, 176.

—— Miller, K.L., Rawnsley, B.E., Adams, M.C., Markus, H.B., Orkwiszewski, K.G., and Mellman, W.J. (1977). The Philadelphia variant of galactokinase. *Am. J. hum. Genet.* **29**, 240.

Teisberg, P. and Gjone, E. (1974). The lecithin:cholesterol acyltransferase deficiency locus in man: Probable linkage to the alpha-haptoglobin locus on chromosome no 16. *Nature, Lond.* **249**, 550.

—— —— and Olaisen, B. (1975). Genetics of lecithin:cholesterol acyltransferase deficiency. *Ann. hum. Genet.* **38**, 327.

Tellez-Nagel, I., Rapin, I., Iwamoto, T., Johnson, A.B., Norton, W.T., and Nitowsky, H. (1976). Mucolipidosis IV. Clinical, ultrastructural, histochemical and chemical studies of a case including a brain biopsy. *Arch. Neurol.* **33**, 828.

Tenenhouse, H.S. and Scriver, C.R. (1979). Renal brush border membrane adaptation to phosphorus deprivation in the *Hyp/Y* mouse. *Nature, Lond.* **281**, 225.

Teree, T.M., Friedman, A.B., Kest, L.M., and Fetterman, G.H. (1970). Cystinosis and proximal tubular nephropathy in siblings. Progressive development of the physiological and anatomical lesion. *Am. J. Dis. Child.* **119**, 481.

Terheggen, H.G., Lowenthal, A., Lavinka, F., and Colombo, J.P. (1975*a*). Familial hyperargininemia. *Arch. Dis. Child.* **50**, 57.

—— —— —— —— and Rogers, S. (1975*b*). Unsuccessful trial of gene replacement in arginase deficiency. *Z. Kinderheilk.* **119**, 1.

—— Schwenk, A., Lowenthal, A., Van Sande, M., and Colombo, J.P. (1969). Argininemia with arginase deficiency. *Lancet* **ii**, 748.

Terry, K. and Linker, A. (1964). Distinction among four forms of Hurler's syndrome. *Proc. Soc. biol. Med.* **115**, 394.

Thalhammer, O. (1973). Histidinamie: Biochemische anomalie oder Krankheit? *Wschr. Kinderheilk.* **121**, 201.

—— (1980). Neonatal screening for histidinemia. In *Neonatal screening for inborn errors of metabolism* (ed. V. Bickel, R. Guthrie, and G. Hammerson), p.59. Springer, Berlin, Heidelberg, New York.

—— Gitzelmann, R., and Pantlitschko, M. (1968). Hypergalactosemia and galactosuria due to galactokinase deficiency in a newborn. *Pediatrics* **42**, 441.

—— Havelec, L., Knoll, E., and Wehle, E. (1977). Intellectual level (IQ) in heterozyotes for phenylketonuria (PKU). *Hum. Genet.* **38**, 285.

Theron, C.N. and Van Jaarseveld, P.P. (1975). The thyroidal serum iodoproteins in a congenital bovine goitre. *S. Afr. med. J.* **42**, 756.

Thier, S.O., Fox, M., Segal, S., and Rosenberg, I.E. (1964). Cystinuria: in vitro demonstration of an intestinal transport defect. *Science* **143**, 482.

—— Segal, S., Fox, M., Blair, A., and Rosenberg, I.E. (1965). Cystinuria: defective intestinal transport of dibasic amino acids and cystine. *J. clin. Invest.* **44**, 442.

Thoene, J.G. and Lemons, R. (1980). Cystine depletion of cystinotic tissues by phosphocysteamine (WR 638). *J. Ped.* **96**, 1043.

—— Baker, H., Yoshino, M., and Sweetman, L. (1981). Biotin responsive carboxylase deficiency associated with sub-normal plasma and urinary biotin. *New Engl. J. Med.* **304**, 817.

—— Oshima, R.G., Olson, D.L., and Schneider, J.A. (1976). Cystinosis: Intracellular cystine depletion by aminothiols *in vitro* and *in vivo*. *J. clin. Invest.* **58**, 180.

—— —— Ritchie, D.G., and Schneider, J.A. (1977*a*). Cystinotic fibroblasts accumulate cystine from intracellular protein degradation. *Proc. natl. Acad. Sci. USA* **74**, 4505.

—— Batshaw, M., Spector, E., Kulovich, S., Brusilow, S., Walser, M., and Nyhan, W. (1977*b*). Neonatal citrullinemia: Treatment with keto analogues of essential amino acids. *J. Ped.* **90**, 218.

Thomas, G.H. and Howell, R.R. (1972). Arylsulphatase A activity in human urine: quantitative studies on patients with lysosomal disorders including MLD. *Clin. Chim. Acta* **36**, 99.

—— Taylor, H.A., Reynolds, L.W., and Miller, C.S. (1973). Mucolipidosis III. Multiple lysosomal enzyme abnormalities in serum and cultured fibroblast cells. *Ped. Res.* **7**, 751.

—— —— Miller, C.S., Axelman, J., and Migeon, B.R. (1974). Genetic complementation after fusion of Tay–Sachs and Sandhoff's cells. *Nature, Lond.* **250**, 580.

—— Tiller, G.E., Reynolds, L.W., Miller, C.S., and Bace, J.W. (1976). Increased levels of sialic acid associated with a sialidase deficiency in I-cell disease (mucolipidosis II) fibroblasts. *Biochem. Biophys. Res. Commun.* **71**, 188.

—— Tipton, R.E., Ch'ien, L.T., Reynolds, L.W., and Miller, C.S. (1978). Sialidase (α -*N*-acetyl neuraminidase) deficiency: The enzyme defect in an adult with macular cherry red spots and myoclonus without dementia. *Clin. Genet.* **13**, 369.

—— Haslam, R.H.A., Batshaw, M.L., Capute, A.J., Neidengard, L., and Ransom, J.L. (1975). Hyperpipecolic acidemia associated with hepatomegaly, mental retardation, optic nerve dysplasia and progressive neurological disease. *Clin. Genet.* **8**, 376.

—— Raghavan, S., Kolodny, E.H., Frisch, A., Neufeld, E.F., O'Brien, J.S., Reynolds, L.W., Miller, C.S., Shapiro, J., Kazazian, H.H. jun., and Heller, R.H. (1982). Nonuniform deficiency of hexosaminidase A in tissues and fluids of two unrelated individuals. *Ped. Res.* **16**, 232.

Thomas, P.K., King, R.H.M., Kocen, R.S., and Brett, E.M. (1977). Comparative ultrastructural observations on peripheral nerve abnormalities in the late infantile, juvenile and late onset forms of metachromatic leukodystrophy. *Acta Neuropath. (Berlin)* **39**, 237.

Thompson, R.P.H., Nicholson, D.C., Farnan, M.B., Whitemore, D.N., and Williams, R. (1970). Cutaneous porphyria due to a malignant primary hepatoma. *Gastroenterology* **59**, 779.

Thorpe, R. and Robinson, D. (1978). Purification and serological studies of human α -L-fucosidase in the normal and fucosidosis states. *Clin. Chim. Acta* **86**, 21.

Tiepolo, L., Zuffardi, O., Fraccaro, M., Di Natale, D., Gargantini, L., Muller, C.R., and Ropers, H.H. (1980). Assignment by deletion mapping of the steroid sulfatase X-linked ichthyosis locus to Xp223. *Hum. Genet.* **54**, 205.

Tietze, F. (1973). Enzymatic reduction of cystine and other disulfides. In *Cystinosis* (ed. J.D. Schulmann), p.67. DHEW Publication No (NIH) 72–249. US Government Printing Office, Washington DC.

—— Bradley, K.H. and Schulman, J.D. (1972). Enzymatic reduction of cystine by subcellular fractions of cultured and peripheral leukocytes from normal and cystinotic individuals. *Ped. Res.* **6**, 649.

Timme, A.H. (1971). The ultrastructure of the liver in human symptomatic porphyria: A preliminary communication. *S. Afr. J. Lab. clin. Med.* **17** (Special issue), 58.

Tio, T.H., Leijnse, B., Jarrett, A. and Rimington, C. (1957). Acquired porphyria from a liver tumor. *Clin. Sci.* **16**, 517.

Tippett, P. and Danks, D.M. (1974). The clinical and biochemical findings in three cases of hypersarcosinaemia and one case of transient hypersarcosinuria associated with folic acid deficiency. *Helv. Paediatr. Acta* **29**, 261.

Tischfield, J.A., Creagan, R.P., and Ruddle, F.H. (1974*a*). Assignment of a selectable gene: Adenine phosphoribosyl-transferase to chromosome 16. *Cytogenet. Cell Genet.* **13**, 167.

—— Nichols, E. and Ruddle, F.H. (1974*b*). Assignment of adenosine deaminase to chromosome 20. *Cytogenet. Cell Genet.* **13**, 1.

Tizianello, A., De Ferranti, G., Garibotto, G., Gurreri, G., and Robaudo, C. (1980). Renal metabolism of amino acids and ammonia in subjects with normal renal function and in patients with chronic renal insufficiency. *J. clin. Invest.* **65**, 1162.

Tobes, M. and Mason, M. (1975). L-kynurenine aminotransferase and dl- α - aminoadipate aminotransferase. I. Evidence for identity. *Biochem. biophys. Res. Commun.* **62**, 390.

Tobin, W.E., Huijing, F., Porro, R.S. and Salzman, R.T. (1973). Muscle phosphofructokinase deficiency. *Arch. Neurol.* **28**, 128.

Tome, F.M.S., Fardeau, M., and Leoir, G. (1977). Ultrastructure of muscle and sensory nerve in Fabry disease. *Acta Neuropathol.* **38**, 187.

Tomer, K.B., Rothman, R., Yudkoff, M., and Segal, S. (1977). Unusual pattern of metabolites in the urine of a child with tyrosinemia: Glyceraldehyde. *Clin. Chim. Acta* **81**, 109.

Tomlinson, A. and Westall, R.G. (1964). Argininosuccinic aciduria: argininosuccinase and arginase in human blood cells. *Clin. Sci.* **26**, 261.

Tondeur, M. and Loeb, H. (1969). Etude ultrastructurelle du foie dans la maladie de Morquio. *Ped. Res.* **3**, 19.

—— Neufeld, E.F. (1974). The mucopolysaccharidoses. Biochemistry and ultrastructure. In *Molecular pathology* (ed. R.A. Good, S.B. Day, and J.J. Yunis), p.600. Charles C. Thomas, Springfield, Illinois.

—— Vamos-Hurwitz, E., Mockel-Pohl, S., Dereume, J.P., Cremer, N., and Loeb, H. (1971). Clinical, biochemical and ultrastructural studies in a case of chondrodystrophy presenting the I-cell phenotype in culture. *J. Ped.* **79**, 366.

Toppet, M., Vamos-Hurwitz, E., Jonniaux, G., Cremer, N., Tondeur, M., and Pelc, S. (1978). Farber's disease as a ceramidosis: Clinical, radiological and biochemical aspects. *Acta. Paediatr. Scand.* **67**, 113.

Tornheim, K. and Lowenstein, J.M. (1972). The purine nucleotide cycle: The production of ammonia from aspartate by extracts of rat skeletal muscle. *J. biol. Chem.* **247**, 162.

—— —— (1974). The purine nucleotide cycle: Interactions with oscillations of the glycolytic pathway in muscle extracts. *J. biol. Chem.* **249**, 3241.

—— —— (1975). The purine nucleotide cycle: Control of phosphofructokinase and glycolytic oscillations in muscle extracts. *J. biol. Chem.* **250**, 6304.

Torsvik, H. (1972). Studies on the protein moiety of serum high density lipoprotein from patients with familial lecithin:cholesterol acyltransferase deficiency. *Clin. Genet.* **3**, 188.

Touraine, J.L. and Malik, M.C. (1980). Fetal liver transplantation in Fabry's disease. *Proc. EBMT Meeting*, Sils Maria.

—— —— Perrot, H., Maire, I., and Revillard, J.P. (1979b). Maladie de Fabry: Deux maladies ameliores par la greffe de cellules de foie foetal. *Nouv. Presse Med.* **8**, 1499.

—— —— Traeger, J., Perrot, H., and Maire, I. (1979a). Attempt at enzyme replacement by fetal liver transplantation in Fabry's disease. *Lancet* **i**, 1094.

Touster, O., Mayberry, R.H., and McCormick, D.B. (1957). The conversion of $1-^{13}C$-D-glucuronolactone to $5-^{13}C$-xylulose in a pentosuric human. *Biochim. Biophys. Acta* **24**, 196.

Townes, P.L., Bryson, M.C., and Miller, G. (1967). Further observations on trypsinogen deficiency disease: report of a second case. *J. Ped.* **71**, 220.

Townley, R.R.W., Khaw, K.-T., and Shwachman, H. (1965). Quantitative assay of disaccharidase activities of small intestinal mucosal biopsy specimens in infancy and childhood. *Pediatrics* **36**. 911.

Treacher, R.J. (1965). Intestinal absorption of lysine in cystinuric dogs. *J. Path.* **75**, 309.

Trefz, F.K., Bartholomé, K., Bickel, H., Lutz, P., and Schmidt, H. (1978). *In vivo* determination of phenylalanine hydroxylase activity using heptadeuterophenylalanine and comparison to the *in vitro* assay values. *Monogr. hum. Genet.* **9**, 108.

Trelstad, R.L., Rubin, D., and Gross, J. (1977). Osteogenesis imperfecta congenita. Evidence for a generalised molecular disorder of collagen. *Lab. Invest.* **36**, 501.

Tremblay, R.R., Foley, T.P. jun., Corvol, P., Park, I.J., Kowarski, A., Blizzard, R.M., Jones, H.W. jun. and Migeon, C.J. (1972). Plasma concentration of testosterone, dihydrotestosterone, testosterone-oestradiol binding globulin and putuitary gonadotrophins in the syndrome of male pseudohermaphroditism with testicular feminization. *Acta. Endocrinol.* **70**, 331.

Trijbels, J.M.F., Sengers, R.C.A., Bakkeren, J.A.J.M., De Kort, A.F.M., and Deutman, A.F. (1977). L-ornithine-ketoacid transferase deficiency in cultured fibroblasts of a patient with hyperornithinemia and gyrate atrophy of the choroid and retina. *Clin. Chim. Acta* **79**, 371.

Trojak, J.E., Ho, C.H., Roesel, R.A., Levin, L.S., Koptis, S.E., Thomas, G.H., and Toma, S. (1980). Morquio-like syndrome (MPS IVB) associated with deficiency of a ß-galactosidase. *Johns Hopkins Med. J.* **146**, 75.

Trotter, W.R. (1960). The association of deafness with thyroid dysfunction. *Brit. med. Bull.* **16**, 92.

Truscott, R.J.W., Pullin, C.J., Halpern, B., Hammond, J., Haan, E., and Danks, D.M. (1979*b*). The identification of 3-keto-2-methylvaleric acid and 3-hydroxy-2-methylvaleric acid in a patient with propionic acidaemia. *Biomed. Mass Spectrom.* **6**, 294.

—— Hick, L., Pullin, C., Halpern, B., Wilcken, B., Griffiths, H., Silink, M., Kilham, H., and Grunseit, F. (1979*a*). Dicarboxylic aciduria: The response to fasting. *Clin. Chim. Acta* **94**, 31.

Try, K. (1969). Heredopathia atactica polyneuritiformis (Refsum's disease): The diagnostic value of phytanic acid determination in serum lipids. *Eur. Neurol.* **2**, 296.

Tryggvason, K., Robey, P.G., and Martin, G.R. (1980). Biosynthesis of type IV procollagens. *Biochemistry* **19**, 1284.

Tsai, C.S., Burgett, M.W., and Reed, L.J. (1973). α-keto acid dehydrogenase complexes. XX. A kinetic study of the pyruvate dehydrogenase complex from bovine kidney. *J. biol. Chem.* **248**, 8348.

Tsay, G.C., and Dawson, G. (1975). Glycopeptide storage in fibroblasts from patients with inborn errors of glycoprotein and glycosphingolipid catabolism. *Biochem. biophys. Res. Commun.* **63**, 807.

Tsujii, T., Morita, T., Matsuyama, Y., Matsui, T., Tamura, M., and Matsuoka, Y. (1976). Sibling cases of chronic recurrent hepatocerebral disease with hyper-citrullinemia. *Gastroenterol. J.* **11**, 328.

Tsukamoto, I., Yoshinaga, T., and Sano, S. (1979). The role of zinc with special reference to the essential thiol groups in δ-aminolevulinic acid dehydratase of bovine liver. *Biochim. Biophys. Acta* **570**, 167.

Tsung-Sheng, S., Beaudet, A.L., and O'Brien, W.E. (1983). Abnormal mRNA for argininosuccinate synthetase in citrullinaemia. *Nature, Lond.* **301**, 533.

Tubergen, D.G., Krooth, R.S., and Heyn, R.M. (1969). Hereditary orotic aciduria with normal growth and development. *Am. J. Dis. Child.* **118**, 864.

Tugwell, P. (1973). Glucose-6-phosphate dehydrogenase deficiency in Nigerians with jaundice associated with lobar pneumonia. *Lancet* **i**, 968.

Turakainen, H., Larjava, H., Saarni, H., and Penttinen, R. (1980). Synthesis of hyaluronic acid and collagen in skin fibroblasts cultured from patients with osteogenesis imperfecta. *Biochim. Biophys. Acta* **628**, 388.

Turnbull, A., Baker, H., Vernon-Roberts, B., and Magnus, J.A. (1973). Iron metabolism in porphyria cutanea tarda and in erythropoietic protoporphyria. *Qt. J. Med.* **42**, 341.

Turner, B.M. and Brown, D.A. (1972). Amino acid excretion in infancy and early childhood: A survey of 200 000 infants. *Med. J. Aust.* **1**, 62.

—— and Hirschhorn, K. (1977). Gaucher's disease: evidence for biochemical heterogeneity. *Am. J. hum. Genet.* **29**, 6.

—— —— (1978). Properties of ß-glucosidase in cultured skin fibroblasts from controls and patients with Gaucher's disease. *Am. J. hum. Genet.* **30**, 346.

—— Turner, V.S., and Hirschhorn, K. (1979). Metabolic correction of fucosidosis fibroblasts by human α-L-fucosidase. *J. Cell Physiol.* **98**, 225.

—— —— Beratis, N.G., and Hirschhorn, K. (1975). Polymorphism of human α-fucosidase. *Am. J. hum. Genet.* **27**, 651.

—— Smith, M., Turner, V.S., Kucherlapati, R.S., Ruddle, F.H., and Hirschhorn, K. (1978). Assignment of gene locus for human alpha-fucosidase to chromosome 1 by analysis of somatic cell hybrids. *Somat. Cell. Genet.* **4**, 45.

Turner, G., Dey, J., and Turner, B. (1967). Homocystinuria: A report of two Australian families. *Aust. Ped. J.* **3**, 48.

Tytgat, G.N., Rubin, C.E., and Saunders, D.R. (1971). Synthesis and transport of lipoprotein particles by intestinal absorptive cells in man. *J. clin. Invest.* **50**, 2065.

Udem, L., Ranney, H.M., Dunn, H.F., and Pisciotta, A. (1970). Some observations on the properties of hemoglobin $M_{milwaukee-1}$. *J. mol. Biol.* **48**, 489.

Udenfriend, S. and Cooper J.R. (1952). The enzymatic conversion of phenylalanine to tyrosine. *J. biol. Chem.* **194**, 503.

Uhlendorf, B.W. and Mudd, S.H. (1968). Cystathionine synthase in tissue culture derived from human skin: enzymatic defect in homocystinuria. *Science* **160**, 1007.

—— Conerly, E.B., and Mudd, S.H. (1973). Homocystinuria: studies in tissue culture. *Ped. Res.* **7**, 645.

Ulick, S. (1976). Diagnosis and nomenclature of the disorders of the aldosterone biosynthetic pathway. *J. clin. Endocrin. Metab.* **43**, 92.

—— Gautier, E., Vetter, K.K., Markello, J.R., Yaffe, S., and Lowe, C.U., (1964). An aldosterone biosynthetic defect in a salt-losing disorder. *J. clin. Endocrin. Metab.* **24**, 669.

Ullrich, K., Mersmann, G., Weber, E., and von Figura, K. (1978). Evidence for lysosomal enzyme recognition by human fibroblasts via a phosphorylated carbohydrate moiety. *Biochem. J.* **170**, 643.

Ulrich, J., Herschkowitz, N., Heitz, P., Sigrist, Th., and Baelocher, P. (1978). Adrenoleukodystrophy. Preliminary report of a connatal case. Light and electron microscopical, immunohistochemical and biochemical findings. *Acta Neuropath. (Berlin)* **43**, 77.

Umemura, S. (1978). Studies on a patient with iminodipeptiduria. II. Lack of prolidase activity in blood cells. *Physiol. Chem. Phys.* **10**, 279.

Upchurch, K.S., Leyva, A., Arnold, W., Holmes, E.W., and Kelley, W. (1975). Hypoxanthine phosphoribosyl transferase deficiency: association of reduced catalytic activity with reduced levels of immunologically detectable enzyme protein. *Proc. natl. Acad. Sci. (Wash.)* **72**, 4142.

Utermann, G., Hees, M., and Steinmetz, A. (1977). Polymorphism of apolipoprotein E and occurrence of dysbetalipoproteinaemia in man. *Nature, Lond.* **269**, 604.

Uys, C.J. and Eales, L. (1963). The histopathology of the liver in acquired porphyria. *S. Afr. J. lab. clin. Med.* **9**, 190.

Uzman, L.L., Iber, F.L., Chalmers, T.C., and Knowlton, M. (1956). The mechanism of copper deposition in the liver in hepatolenticular degeneration (Wilson's disease). *Am. J. med. Sci.* **231**, 511.

Vagenakis, A.G., Hamilton, C., Maloof, F., Braverman, L.E., and Ingbar, S.H. (1972). The concentration and binding of thyroxine in the serum of patients with the testicular feminization syndrome: observations on the effects of ethinyl estradiol and norethandrolone. *J. clin. Endocrin. Metab.* **34**, 327.

Valenta, L.J., Bode, H., Vickery, A.L., Caulfield, J.B., and Maloof, F. (1973). Lack of thyroid peroxidase activity as the cause of congenital goitrous hypothyroidism. *J. clin. Endocrin. Metab.* **36**, 830.

Valentine, W.N., Anderson, H.M., and Paglia, D.E. (1972). Studies on human erythrocyte metabolism. II. Nonspherocytic haemolytic anaemia, high red cell ATP and ribosephosphate pyrophosphokinase (RPK E.C. 2.7.6.1) deficiency. *Blood* **39**, 674.

—— Bennett, J.M., and Krivit, W. (1973). Nonspherocytic haemolytic anaemia with increased red cell adenine dinucleotides, glutathione and basophilic stippling and ribosephosphate pyrophosphokinase (RPK) deficiency: studies on two new kindreds. *Br. J. Haematol.* **24**, 157.

—— Fink, K., and Paglia, D.E. (1976*a*). Hereditary haemolytic anaemia with human erythrocyte pyrimidine 5′-nucleotidase deficiency. *J. clin. Invest.* **54**, 866.

—— Paglia, D.E., and Fink, K. (1976*b*). Lead poisoning. Association with hemolytic anemia, basophilic stippling, erythrocyte pyrimidine 5′-nucleotidase deficiency, and intraerythrocytic accumulation of pyrimidines. *J. clin. Invest.* **58**, 926.

—— Schneider, A.S., Baughan, M.A., Paglia, D.E., and Heins, H.L. jun. (1966). Hereditary hemolytic anemia with triosephosphate isomerase deficiency. Studies in kindreds with co-existent sickle cell trait and erythrocyte glucose-6-phosphate dehydrogenase deficiency. *Am. J. Med.* **41**, 27.

—— Oski, F.A., Paglia, D.E., Baughan, M.A., Schneider, A.S., and Naiman, J.L. (1967). Hereditary hemolytic anemia with hexokinase deficiency. Role of hexokinase in erythrocyte aging. *New Engl. J. Med.* **276**, 1.

—— Hsieh, H.-S., Paglia, D.E., Anderson, H.M., Baughan, M.A., Jaffé, E.R., and Garson, O.M. (1968). Hereditary hemolytic anemia: association with phosphoglycerate kinase deficiency in erythrocytes and leukocytes. *Trans. Assoc. Am. Physicians.* **81**, 49.

—— Hsieh, H., Paglia, D.E., Anderson, H.M., Baughan, M.A., Jaffé, E.R., and Garson, O.M. (1969). Hereditary hemolytic anemia associated with phosphoglycerate kinase deficiency in erythrocytes and leucocytes. A probable X-chromosome-linked syndrome. *New Engl. J. Med.* **280**, 528.

Valle, D., Boison, A.P., and Kaiser-Kupfer, M.I. (1979*a*). Complementation analysis of gyrate atrophy of the choroid and retina. *Ped. Res.* **13**, 427.

—— Kaiser-Kupfer, M.I., and Del Valle, L.A. (1977). Gyrate atrophy of the choroid and retina: deficiency of ornithine aminotransferase in transformed lymphocytes. *Proc. natl. Acad. Sci. USA* **74**, 5159.

—— Pai, G.S., Thomas, G.H., and Pyeritz, R.E. (1980*a*). Homocystinuria due to cystathionine ß-synthase deficiency: clinical manifestations and therapy. *Johns Hopkins Med. J.* **146**, 110.

—— Phang, J.M., and Goodman, S.I. (1974). Type II hyperprolinemia: absence of $\triangle'$-pyrroline-5-carboxylic acid dehydrogenase activity. *Science* **185**, 1053.

—— Goodman, S.I., Harris, S.C., and Phang, J.M. (1979*b*). Genetic evidence for a common enzyme catalysing the second step in the degradation of proline and hydroxyproline. *J. clin. Invest.* **64**, 1365.

—— Walʾer, M., Brusilow, S.W., and Kaiser-Kupfer, M.I. (1980*b*). Gyrate atrophy of choroid and retina: amino acid metabolism and correction of hyperornithinemia with an arginine deficient diet. *J. clin. Invest.* **65**, 371.

—— Goodman, S.I., Applegarth, D.A., Shih, V.E., and Phang, J.M. (1976). Type II hyperprolinemia. *J. clin. Invest.* **58**, 598.

Valman, H.B., Patrick, A.D., Seakins, J.W.T., Platt, J.W., and Gompertz, D. (1973). Family with intermittent maple syrup urine disease. *Arch. Dis. Child.* **48**, 255.

Vamos, E., Liebaers, I., Bousard, N., Libert, J., and Perlmutter, N. (1981). Multiple sulphatase deficiency with early onset. *J. inher. metab. Dis.* **4**, 103.

Van Acker, K.J., Simmonds, H.A., Potter, C., and Cameron, J.S. (1977). Complete deficiency of adenine phosphoribosyltransferase. Report of a family. *New Engl. J. Med.* **297**, 127.

Van Biervliet, J.P.G.M., Bruinvis, L., Van der Heiden, C., Ketting, D., Wadman, S.K., Willemse, J.L., and Monnens, L.A.H., (1977). Report of a patient with severe, chronic lactic acidaemia and pyruvate carboxylase deficiency. *Dev. Med. Child Neurol.* **19**, 392.

Van Bogaert, L., Seitelberger, F., and Edgar, G.W.F. (1963). Etudes neuropathologiques et neurochimiques sur un cas de Niemann–Pick chez un jeune enfant. *Acta. Neuropath.* **3**, 57.

Vance, D.E., Krivit, W., and Sweeley, C.C. (1969). Concentrations of glycosyl ceramides in plasma and red cells in Fabry's disease: a glycolipid lipidosis. *J. Lipid Res.* **10**, 188.

Van Creveld, S. (1961). Clinical course of glycogen storage disease. *Chem. Weekblad.* **57**, 445.

—— and Huijing., F. (1965). Glycogen storage disease. *Am. J. Med.* **38**, 554.

Van de Kamp, J.J.P. (1979). The Sanfilippo syndrome: a clinical and genetical study of 75 patients in the Netherlands (Doctoral thesis, University of Leiden).

—— Niermeijer, M.F., Von Figura, K., and Giesberts, M.A.H. (1981). Genetic heterogeneity and clinical variability in the Sanfilippo syndrome (types A, B and C). *Clin. Genet.* **20**, 152.

Van Den Bergh, A.A.H., and Muller, P. (1916). Ueber eine direkte und eine indirekte Diazoreaktion auf Bilirubin. *Biochem. Z.* **77**, 90.

Van der Bergh, F.A.J.J.M., Rietra, P.J.G.M., Kolk-Vegter, A.J., Bosch, E., and Tager, J.M. (1976). Therapeutic implications of renal transplantation in a patient with Fabry's disease. *Acta Med. Scand.* **200**, 249.

Van Den Berghe, G. (1978). Metabolic effects of fructose in the liver. *Curr. Top. Cell Regul.* **13**, 97.

Van der Hagen, C.B., Borresen, A.L., Molne, K., Oftedal, G., Bjoro, K., and Berg, K. (1973). Metachromatic leukodystrophy. 1. Prenatal detection of arylsulphatase A deficiency. *Clin. Genet.* **4**, 256.

Van der Heiden, C., Gerards, L.J., Van Biervliet, J.P.G.M., Desplanque, J., De Bree, P.K., Van Sprang, F.J., and Wadman, S.K. (1976). Lethal neonatal argininosuccinate lyase deficiency in four children from one sibship. *Helv. Paediatr. Acta* **31**, 407.

Van der Zee, S.P.M., Schretlen, E.D.A.M., and Monnens, L.A.H. (1968). Megaloblastic anaemia in Lesch–Nyhan syndrome. *Lancet* i, 1427.

—— Trijbels, J.M.F., Monnens, L.A.H., Hommes, F.A., and Schretlen, E.D.A.M. (1971). Citrullinaemia with rapidly fatal neonatal course. *Arch. Dis. Child.* **46**, 847.

Van Diggelen, O.P., Hoogeveen, A.T., Smith, P.J., Reuser, A.J.J., and Galjaard, H. (1982). Enhanced proteolytic degradation of normal ß-galactosidase in the lysosomal storage disease with combined ß-galactosidase and neuraminidase deficiency. *Biochim. Biophys. Acta* **703**, 69.

—— Schram, A.W., Sinnott, M.L., Smith, P.J., Robinson, D. and Galjaard, H. (1981). Turnover of ß-galactosidase in fibroblasts from patients with genetically different types of ß-galactosidase deficiency. *Biochem. J.* **200**, 143.

Vandwater, N.S., Jolly, R.D., and Farrow, B.R.H. (1979). Canine Gaucher disease. *Austr. J. Exp. Biol. Med. sci.* **57**, 551.

Van Dyke, D.L., Fluharty, A.L., Schafer, I.A., Shapiro, L.J., Kihara, H., and Weiss, L. (1981). Prenatal diagnosis of Maroteaux–Lamy syndrome. *Am. J. med. Genet.* **8**, 235.

Van Elsen, A.F. and Leroy, J.G. (1975). Arginase isoenzymes in human diploid fibroblasts. *Biochem. biophys. Res. Commun.* **62**, 191.

—— —— (1978). Response to Cederbaum and Spector letter. *Am. J. hum. Genet.* **30**, 91.

—— —— Vanneuville, F.J., and Vercruyssen, A.L. (1976). Isoenzymes of serum *N*-acetyl-ß- D-glucosaminidase in the I-cell disease heterozygote. *Hum. Genet.* **31**, 75.

Van Eys, J. and Garms, P. (1971). Pyruvate kinase deficiency hemolytic anaemia: A model for correlation of clinical syndrome and biochemical anomalies. *Ad. Ped.* **18**, 203.

Van Gelderen, H.H. and Dooren, L.J. (1963). Studies in oligophrenia. II. Amino-aciduria in mentally deficient children. *Acta. Paed. (Uppsala)* **52**, 41.

—— and Teijema, H.L. (1973). Hyperlysinaemia: Harmless inborn error of metabolism? *Arch. Dis. Child.* **48**, 892.

Van Gennip, A.H., Van Bree-Blom, E.J., Grift, J., De Bree, P.K., and Wadman, S.K. (1980). Urinary purines and pyrimidines in patients with hyperammonemia of various origins. *Clin. Chim. Acta* **104**, 227.

Van Halbeck, H., Dorland, L., Veldink, G.A., Vliegenhart, J.F.G., Strecker, G., Michalski, J.-C., Montreuil, J. and Hull, W.E. (1980). A 500 Mhz′H NMR study of urinary oligosaccharides from patients with mannosidosis. *FEBS Lett.* **121**, 71.

Van Heeswijk, P.J., Blank, C.H., Seegmiller, J.E., and Jacobson, C.B. (1972). Preventive control of the Lesch–Nyhan syndrome. *Obstet. Gynec.* **40**, 109.

—— Trijbels, J.M.F., Schretlen, E.D.A.M., Van Munster, P.J.J., and Monnens, L.A.H. (1969). A patient with a deficiency of serum carnosinase activity. *Acta Paediatr. Scand.* **58**, 584.

Van Hoof, F. (1973). Mucopolysaccharidoses. In *Lysosomes and storage diseases* (ed. H.G. Hers and F. Van Hoof), p.217. Academic Press, New York.

—— and Hers, H.G. (1964). L'ultrastructure des cellules hepatiques dans la maladie de Hurler (gargoylisme). *C.R. Acad. Sci.* **259**, 1281.

—— —— (1967). The subgroups of type III glycogenosis. *Eur. J. Biochem.* **2**, 265.

—— —— (1968). Mucopolysaccharidosis by absence of α-fucosidase. *Lancet* **i**, 1198.

—— Hue, L., De Barsy, T., Jacquemin, P., Devos, P., and Hers, H.G. (1972). Glycogen storage diseases. *Biochimie* **54**, 745.

Vanier, M.-T. (1978). Biochemical studies in sphingomyelin storage disorders. In *Enzymes and lipid metabolism* (ed. S. Gatt, L. Freysz, and P. Mandel). Plenum Press, New York.

—— Svennerholm, L. (1975). Chemical pathology of Krabbe's disease. III. Ceramide hexosides and gangliosides of brain. *Acta. Paediatr. Scand.* **64**, 641.

—— —— (1976). Chemical pathology of Krabbe's disease: The occurrence of psychosine and other neutral sphingoglycolipids. *Adv. exp. Med. Biol.* **68**, 115.

—— Revol, A., and Boué, A. (1979). Semimicrotechniques for the assay of sphingohydrolase activities measured with natural labelled substrates. In *Proceedings of the Third European Conference on Prenatal Diagnosis* (ed. J.D. Murken, S. Stengel-Rutkowski, and E. Schwinger) p.292. Enke, Stuttgart.

—— Svennerholm, L., Månsson, J.-E., Håkansson, G., Boué, A., and Lindsten, J. (1981). Prenatal diagnosis of Krabbe disease. *Clin. Genet.* **20**, 79.

Van Mullem, P.J., and Ruiter, M. (1970). Fine structure of the skin in angiokeratoma corporis diffusum (Fabry's disease). *J. Pathol.* **101**, 221.

Van Slyke, D.D., Hiller, A., Weisiger, J.R., and Cruz, W.O. (1946). Determination of carbon monoxide in blood and of total and active hemoglobin by carbon monoxide capacity: inactive hemoglobin and methemoglobin contents of normal blood. *J. biol. Chem.* **166**, 121.

Van Voorthuizen, W.F., Dinsart, C., Flavell, R.A., Devijlder, J.J., and Vassart, G. (1978). Abnormal cellular localization of thyroglobulin mRNA associated with hereditary congenital goiter and thyroglobulin deficiency. *Proc. natl. Acad. Sci. USA* **75**, 74.

Varadi, S. (1958). Haematological aspects in a case of erythropoietic porphyria. *Br. J. Haematol.* **4**, 270.

Varki, A. and Kornfeld, S. (1980). Structural studies of phosphorylated high mannose-type oligosaccharides. *J. biol. Chem.* **255**, 10847.

—— Reitman, M.L., Vannier, M., Kornfeld, S., Grubb, J.H., and Sly, W.S. (1982). Demonstration of the heterozygous state for I-cell disease and pseudo-Hurler polydystrophy by assay of *N*-acetylglucosaminylphosphotransferase in white blood cells and fibroblasts. *Am. J. hum. Genet.* **34**, 717.

Veale, A.M.O. (1980) Screening for phenylketonuria. In *Neonatal screening for inborn errors of metabolism* (ed. H. Bickel, R. Guthrie, and G. Hammersen), p.7. Springer, Berlin.

—— Lyon, I.C.T., and Houston, J.B. (1972). Multiple screening for inborn errors of metabolism in New Zealand and other countries in the Pacific basin. *Abst. Eur. Soc. Ped. Res.*, Germany, 11–14 September.

Vecchio, F., Carnevale, F. and Di Bitonto, G. (1976). Galactokinase deficiency in an Italian infant. In *Inborn errors of calcium and bone metabolism* (ed. H. Bickel and J. Stern), p.317. MTP Press Ltd, Lancaster.

Verhoeven, G. and Wilson, J.D. (1976). Cytosol androgen receptor in sub-mandibular gland and kidney of the normal mouse and the mouse with testicular feminization. *Endocrinology* **99**, 79.

Verhue, W. and Hers, H.G. (1966). A study of the reaction catalysed by the liver branching enzyme. *Biochem J.* **99**, 222.

Verity, M.A. and Montasir, M. (1976). Infantile Gaucher's disease: neuropathology, acid hydrolase activities and negative staining observations. *Neuropädiatrie* **8**, 89.

Verma, I.C., Sud, N., and Manerikar, S. (1974). Homocystinuria: Report of two cases in siblings. *Ind. Ped.* **11**, 753.

Vermorken, A.J.M., Van Bennekom, C.A., De Bruyn, C.H.M.M., and Dei, T.L. (1980). Heterozygote detection in Fabry's disease using mailed hair roots. *Br. J. Dermatol.* **103**, 101.

Vethamany, V.G., Welch, J.P., and Vethamany, S.K. (1972). Type D Niemann–Pick disease (Nova Scotia variant). *Arch. Path.* **93**, 537.

Vidgoff, J. and Buist, N.R.M. (1977). Serum hexosaminidase activity in I-cell disease carriers. *Hum. Genet.* **36**, 307.

—— —— and O'Brien, J.S. (1973). Absence of ß-*N*-acetyl-D-hexosaminidase A activity in a healthy woman. *Am. J. hum. Genet.* **25**, 372.

Vigneron, C., Marchal, C., Deifts, C., Vidailhet, M., Pierson, M., and Neimann, N. (1970). Déficit partiel et transitoire en galactokinase érythrocytaire chez un nouveau-né. *Arch. fr. Péd.* **27**, 523.

Vimal, C.M., Fensom, A.H., Heaton, D., Ward, R.H.T., Garrod, P., and Penketh, R.J.A. (1984). Prenatal diagnosis of argininosuccinic aciduria by analysis of cultured chorionic villi. *Lancet* **ii**, 521.

Virchow, R. (1866). Ein Fall von allgemeiner Ochronose der Knorpel und Knorpelähnlichen Theile. *Arch. Path. Anat.* **37**, 212.

Visser, H.K.A. and Cost, W.S. (1964). A new hereditary defect in the biosynthesis of aldosterone: urinary C21-corticosteroid pattern in three related patients with a salt-losing syndrome, suggesting an 18-oxidation defect. *Acta Endocrinol.* **47**, 589.

Vital, C., Battin, J., Rivel, J., and Heheunstre, J.P. (1976). Aspects ultrastructuraux des lesions du nerf peripherique dans un cas de maladie de Farber. *Rev. Neurol.* **132**, 419.

Vives-Corrons, J.L., Marie, J., Pujades, M.A., and Kahn, A. (1980). Hereditary erythrocyte pyruvate-kinase (PK) deficiency and chronic hemolytic anemia: Clinical, genetic and molecular studies on six new Spanish patients. *Hum. Genet.* **53**, 401.

Vladutiu, G.D. and Rattazzi, M.C. (1975). Abnormal lysosomal hydrolases excreted by cultured fibroblasts in I-cell disease (mucolipidosis II). *Biochem. biophys. Res. Commun.* **67**, 956.

—— —— (1978). Desialylation of ß-hexosaminidase and its effect on uptake by fibroblasts. *Biochim. Biophys. Acta* **539**, 31.

Volpe, R., Knowlton, T.G., Foster, A.D., and Conen, P.E. (1968). Testicular feminization: A study of two cases, one with a seminoma. *Can. Med. Assoc. J.* **98**, 438.

Volpintesta, E.J. (1974). Menkes' kinky hair syndrome in a black infant. *Am. J. Dis. Child.* **128**, 244.

Von Figura, K. and Klein, U. (1979). Isolation and characterization of phosphorylated oligosaccharides from α-*N*-acetylglucosaminidase that are recognized by cell surface receptors. *Eur. J. Biochem.* **94**, 347.

—— and Kresse, H. (1974). Quantitative aspects of pinocytosis and the intracellular fate of *N*-acetyl-α-D-glucosaminidase in Sanfilippo B fibroblasts. *J. clin. Invest.* **53**, 85.

—— —— (1976). Sanfilippo disease type B: presence of material cross reacting with antibodies against α-*N*-acetylglucosaminidase. *Eur. J. Biochem.* **61**, 581.

—— Lögering, M., Mersmann, G., and Kresse, H. (1973). Sanfilippo B disease: Serum assay for detection of homozygous and heterozygous individuals in three families. *J. Ped.* **83**, 607.

—— Van de Kamp, J.J., and Niermeijer, M.F. (1982). Prenatal diagnosis of Morquio's disease type A (*N*-acetylgalactosamine 6-sulphate sulphatase deficiency). *Prenatal Diagnosis* **2**, 67.

Von Frerichs, F.T. (1861). Pathologisch-anatomischer Atlas Klinik der Leberkrankheiten, Vol. 2, p.62. Viewig, Brunswick, Germany.

Vora, S., Corash, L., Engel, W.K., Durham, S., Seaman, C., and Piomelli, S. (1980*b*). The molecular mechanism of the inherited phosphofructokinase deficiency associated with hemolysis and myopathy. *Blood* **55**, 629.

—— Seaman, C., Durham, S., and Piomelli, S. (1980*a*). Isozymes of human phosphofructokinase: Identification and subunit structural characterization of a new system. *Proc. natl. Acad. Sci. USA* **77**, 62.

Voute, P.A. jun. and Wadman, S.K. (1968). Cystathioninuria in hepato-blastoma. *Clin. Chim. Acta* **22**, 373.

Wabrek, A.J., Millard, P.R., Wilson, W.B. jun., and Pion, R.J. (1971). Creation of a neovagina by the Frank non-operative method. *Obstet. Gynecol.* **37**, 408.

Wachtel, S.S. (1979). Immunogenetic aspects of abnormal sexual differentiation. *Cell* **16**, 691.

—— and Koo, G.C. (1981). H-Y antigen in gonadal differentiation. In *Mechanisms of sex differentiation in animals and man* (ed. C.R. Austin and R.G. Edwards), p.255. Academic Press, London.

Wada, Y. (1965). Idiopathic hypervalinemia: valine and α-keto-acids in blood following an oral dose of valine. *Tohoku. J. Exp. Med.* **87**, 322.

—— Tada, K., Minagawa, A., Yoshida, T., Morikawa, T., and Okamura, T. (1963). Idiopathic hypervalinemia. *Tohoku J. exp. Med* **81**, 46.

Waddington, R.T. (1972). A case of primary liver tumour associated with porphyria. *Br. J. Surg.* **59**, 653.

Waheed, A., Hasilik, A., and Von Figura, K. (1981*b*). Processing of the phosphorylated recognition marker in lysosomal enzymes: Characterization and partial purification of a microsomal α-*N*-acetylglucosamine phosphodiesterase. *J. biol. Chem.* **256**, 5717.

—— Pohlmann, R., Hasilik, A., and Von Figura, K. (1981*a*). Subcellular location of two enzymes involved in the synthesis of phosphorylated recognition markers in lysosomal enzymes. *J. biol. Chem.* **256**, 4150.

—— —— —— —— Van Elsen, A., and Leroy, J.G. (1982). Deficiency of UDP-*N*-acetylglucosamine:lysosomal enzyme *N*-acetylglucosamine-1-phospho-transferase in organs of I-cell patients. *Biochem biophys. Res. Commun.* **105**, 1052.

Waisbren, S.E., Schnell, R.R., and Levy, H.L. (1980). Diet termination in children with phenylketonuria: a review of psychological assessments used to determine outcome. *J. inher. metab. Dis.* **3**, 149.

Waldenstrom, J. (1957). The porphyrias as inborn errors of metabolism. *Am. J. Med.* **22**, 758.

Waldman, T.A., Mogielnicki, R.P., and Strober, W. (1973). The proteinuria of cystinosis: Its patterns and pathogenesis. In *Cystinosis* (ed. J.D. Schulman), p.55. DHEW Publication no (NIH) 72–249. US Government Printing Office, Washington, DC.

Walker, A.C., Stack, E.M., and Horsfall, W.A. (1970). Familial male pseudo-hermaphroditism. *Med. J. Aust.* **1**, 156.

Walker-Smith, J.A., Turner, B., Blomfield, J., and Wise, G. (1973). Therapeutic implications of copper deficiency in Menkes' steely-hair syndrome. *Arch. Dis. Child.* **48**, 958.

Walkey, S.U., Blakemore, W.F., and Purpura, D.P. (1981). Alterations in neurone morphology in feline mannosidosis. *Acta Neuropathol.* **53**, 75.

Wallace, B.J., Aronson, S.M., and Volk, B.W. (1963). Histochemical and bio-chemical studies of globoid cell leucodystrophy (Krabbe's disease). *J. Neurochem.* **11**, 367.

—— Volk, B.W., and Lazarus, S.S. (1964). Fine structural localization of acid phosphatase activity in neurones of Tay–Sachs disease. *J. Neuropath. exp. Neurol.* **23**, 676.

—— Kaplan, D., Adachi, M., Schneck, L., and Volk, B.W. (1966). Muco-polysacharidosis type III: morphologic and biochemical studies of two siblings with Sanfilippo syndrome. *Arch. Path.* **82**, 462.

Waller, H.D. (1968). Glutathione reductase deficiency. In *Hereditary disorders of erythrocyte metabolism* (ed. E. Beutler), p.185. Grune and Stratton, New York.

Wallis, K., Gross, M., Kohn, R., and Zaidman, J. (1971). A case of Wolman's disease. *Helv. Paediatr. Acta* **26**, 98.

Walsh, J.R. (1971). A distinctive pigment of the skin in New Guinea indigenes. *Ann. hum. Genet. (Lond.)* **34**, 379.

Walsh, P.C., Madden, J.D., Harrod, M.J., Goldstein, J.L., MacDonald, P.C., and Wilson, J.D. (1974). Familial incomplete male pseudohermaphroditism type 2. Decreased dihydrotestosterone formation in pseudovaginal perineoscrotal hypospadias. *New Engl. J. Med.* **291**, 944.

Walshe, J.M. (1956). Penicillamine, a new oral therapy for Wilson's disease. *Am. J. Med.* **21**, 487.

—— (1967). The physiology of copper in man and its relation to Wilson's disease. *Brain* **90**, 149.

—— (1969). Management of penicillamine nephropathy in Wilson's disease: a new chelating agent. *Lancet* **ii**, 1401.

—— (1970). Wilson's disease: its diagnosis and management. *Brit. J. Hosp. Med.* **4**, 91.

—— (1975). Wilson's disease (hepatolenticular degeneration). In *The treatment of inherited metabolic disease* (ed. D.N. Raine), p.171. MTP Press, Lancaster.

—— (1982). Treatment of Wilson's disease with trientine (triethylene tetramine) dihydrochloride. *Lancet* **i**, 643.

Walters, T.R. (1967). Congenital megaloblastic anemia responsive to N^5-formyl-tetrahydrofolic acid administration. (abstract) *J. Ped.* **70**, 686.

Wang, Y.M. and Van Eys, J. (1970). The enzymatic defect in essential pentosuria. *New Engl. J. Med.* **282**, 892.

Warner, J.O. (1975). Juvenile onset metachromatic leucodystrophy: Failure of response on a low vitamin A diet. *Arch. Dis. Child.* **50**, 735.

Warner, T.G. and O'Brien, J.S. (1979). Synthesis of 2'(4-methylumbelliferyl)-α-D-N-acetylneuraminic acid and detection of skin fibroblast neuraminidase in normal humans and in sialidosis. *Biochemistry* **13**, 2783.

—— —— (1982). GM_1-gangliosidosis and Sandhoff disease. In *Genetic errors of glycoprotein metabolism* (ed. P. Durand and J.S. O'Brien), p.179. Edi-ermes, Milano; Springer, Berlin.

Warren, R.J., Condron, C.J., Hollister, D., Huijing, F., Neufeld, E.F., Hall, C.W., McLeod, A.G.W., and Lorincz, A.E. (1973). Antenatal diagnosis of mucolipidosis II (I-cell disease). *Ped. Res.* **7**, 343.

Waterbury, L. and Frankel, E.P. (1972). Hereditary nonspherocytic hemolysis with erythrocyte phosphofructokinase deficiency. *Blood* **39**, 4115.

Waterson, J.R., Winter, W.P., and Schmickel, R.D. (1974). Cysteine activation in cultured cystinotic cells. The specific activity of cysteinyl-tRNA synthetase and $tRNA_{cys}$ and the determination of the Michaelis-Menten constants for cysteinyl-tRNA synthetase. *J. clin. Invest.* **54**, 182.

Watson, B., Gormley, I.P., Gardiner, S.E., Evans, H.J., and Harris, H. (1972). Reappearance of murine hypoxanthine guanine phosphoribosyl transferase activity in mouse Ag cells after attempted hybridisation with human lines. *Exp. Cell Res.* **75**, 401.

Watson, C.J., (1960). The problem of porphyria – some facts and questions. *New Engl. J. Med.* **263**, 1205.

—— and Schwartz, S. (1941). A simple test for urinary porphobilinogen. *Proc. Soc. exp. Biol. Med.* **47**, 393.

—— Perman, V., Surrell, F.A., Hoyt, H.H., and Schwartz, S. (1958). Some studies of the comparative biology of human and bovine porphyria erythropoietica. *Trans. Assoc. Am. Phys.* **71**, 196.

—— Runge, W., Taddeini, L., Bossenmaier, I., and Cardinal, R. (1964). A suggested control gene mechanism for the excessive production of types I and III porphyrins in congenital erythropoietic porphyria. *Proc. natl. Acad. Sci. USA* **52**, 478.

Watts, D.W.E., Watts, J.E.M., and Seegmiller, L.E. (1965). Xanthine oxidase activity in human tissues and its inhibition by allopurinol (4-hydroxypyrazolo [3,4-d] pyrimidine). *J. lab. clin. Med.* **66**, 688.

—— Engleman, K., Klinenberg, J.R., Seegmiller, J.E., and Sjoerdsma, A. (1964). The enzyme defect in a case of xanthinuria. *Biochem. J.* **90**, 4P.

—— McKeran, R.O., Brown, E., Andrews, T.M., and Griffiths, M.I. (1974). Clinical and biochemical studies on treatment of Lesch–Nyhan syndrome. *Arch. Dis. Child.* **49**, 693.

—— Chalmers, R.A., Gibbs, D.A., Lawson, A.M., Purkiss, P., and Spellacy, E. (1979). Studies on some possible biochemical treatments of primary hyperoxaluria. *Qt. J. Med.* **48**, 259.

Weatherall, D.J. (1960). Enzyme deficiency in haemolytic disease of the newborn. *Lancet* **ii**, 835.

Weber, H.-P., Harms, E., and Knöpfle, G. (1979). Adoleszenten-Zystinose Literatürbersicht mit eigener Beobachtung. *Klin. Pädiatr.* **191**, 8.

Webster, D.R., Simmonds, H.A., Barry, D.M.J., and Becroft, D.M.O. (1981). Pyrimidine and purine metabolites in ornithine carbamoyl transferase deficiency. *J. inher. metab. Dis.* **4**, 27.

Webster, H. de F. (1962). Schwann cell alterations in metachromatic leucodystrophy. *J. Neuropath. exp. Neurol.* **21**, 534.

Wehinger, H., Witt, I., Lösel, I., Denz-Seibert, G., and Sander, C. (1975). Intravenous copper in Menkes' kinky hair syndrome. *Lancet* **i**, 1143.

Weil, D., Van Cong, N., Rebourcet, R., Gross, M.S., and Frezal, J. (1979). Regional mapping of enzyme loci on human chromosomes 2, 17, 5 and X by use of somatic cell hybridization. *Cytogenet. Cell Genet.* **45**, 155.

Weinberg, A.G., Mize, C.E., and Worthen, H.G. (1976). The occurrence of hepatoma in the chronic form of hereditary tyrosinemia. *J. Ped.* **88**, 434.

Weinberger, A., Sperling, O., Rabinovitz, M., Brosh, S., Adam, A., and De Vries, A. (1974). High frequency of cystinuria among Jews of Libyan origin. *Hum. Hered.* **24**, 568.

Weingeist, T.A. and Blodi, F.C. (1973). Fabry's disease: Ocular findings in a female carrier. A light and electron microscopic study. *Arch. Ophthalmol.* **85**, 169.

Weinhouse, S. and Friedmann, B. (1951). Metabolism of labeled 2-carbon acids in the intact rat. *J. biol. Chem.* **191**, 707.

Weleber, R.G. and Kennaway, N.G. (1981). Clinical trial of vitamin B_6 for gyrate atrophy of the choroid and retina. *Ophthalmology* **88**, 316.

Wellner, V.P., Sekura, R., Meister, A., and Larsson, A. (1974). Glutathione synthetase deficiency, an inborn error of metabolism involving the γ-glutamyl cycle in patients with 5-oxoprolinuria (pyroglutamic aciduria). *Proc. natl. Acad. Sci. USA* **71**, 2505.

Wells, R.S. and Kerr, C.B. (1965). Genetic classification of ichthyosis. *Arch. Dermatol.* **92**, 1.

—— —— (1966). Clinical features of autosomal dominant and sex-linked ichthyosis in an English population. *Br. med. J.* **i**, 947.

Wells, M.A. and Dittmer, J.C. (1967). A comprehensive study of the postnatal changes in the concentration of the lipids of developing rat brain. *Biochemistry* **6**, 3169.

Welsh, C., Gay, S., Rhode, R.K., Pfister, R., and Miller, E.J. (1980). Collagen heterogeneity in normal rabbit cornea. I. Isolation and biochemical characterization of the genetically-distinct collagens. *Biochim. Biophys. Acta* **625**, 78.

Wendel, U. and Claussen, U. (1979). Antenatal diagnosis of maple syrup urine disease. *Lancet* **i**, 161.

—— —— and Langenbeck, U. (1980*b*). Pattern of branched-chain α-keto acids in amniotic fluid. *Clin. Chim. Acta* **120**, 267.

—— Rudiger, H.W., Passarge, E., and Mikkelsen, M. (1973). Maple syrup urine disease: rapid prenatal diagnosis by enzyme assay. *Humangenetik* **19**, 127.

—— —— Przyrembel, H., and Bremer, H.J. (1975). Alpha-ketoadipic aciduria: degradation studies with fibroblasts. *Clin. Chim. Acta* **58**, 271.

—— Becker, K., Przyrembel, H., Bulla, M., Manegold, C., Menchhionowski, A., and Langenbeck, U. (1980*a*). Peritoneal dialysis in maple-syrup urine disease: Studies on branched-chain amino and ketoacids. *Eur. J. Ped.* **134**, 57.

Wenger, D.A. (1977). Niemann–Pick disease. In *Practical enzymology of the sphingolipidoses* (ed. R.H. Glew and S.P. Peters), p.39. Alan R. Liss, New York.

—— (1978). Assay of ß-glucosidase and sphingomyelinase for identification of patients and carriers of Gaucher's disease and Niemann–Pick disease. In *Enzymes of lipid metabolism* (ed. S. Gatt, L. Freysz, and P. Mandel), p.707. Plenum Press, New York.

—— Barth, G., and Gitchens, J.H. (1977). Nine cases of sphingomyelin lipidosis, a new variant in Spanish–American children. Juvenile variant of Niemann–Pick disease with foamy and sea-blue histiocyte. *Am. J. Dis. Child.* **131**, 955.

—— Sattler, M., and Hiatt, W. (1974*a*). Globoid cell leukodystrophy: deficiency of lactosyl ceramide ß-galactosidase. *Proc. natl. Acad. Sci. USA* **71**, 854.

—— Roth, S., and Sattler, M. (1982). Acute neuronopathic (infantile) and chronic non-neuronopathic (adult) Gaucher disease in full siblings. *J. Ped.* **100**, 252.

—— Sattler, M., and Clark, C. (1975). Partial purification of galactosyl- and lactosylceramide ß-galactosidase from human brain. *Trans. Am. Soc. Neurochem.* **6**, 151.

—— —— and Markey, S.P. (1973). Deficiency of monogalactosyl diglyceride ß-galactosidase activity in Krabbe's disease. *Biochem. biophys. Res. Commun.* **53**, 680.

—— Tarby, T.J. and Wharton, C. (1978*a*). Macular cherry-red spots and myoclonus with dementia: Coexistent neuraminidase and ß-galactosidae deficiencies. *Biochem. biophys. Res. Commun.* **82**, 589.

—— Clark, C., Sattler, M., and Wharton, C. (1978*b*). Synthetic substrate ß-glucosidase activity in leukocytes: A reproducible method for the identification of patients and carriers of Gaucher's disease. *Clin. Genet.* **13**, 145.

—— Sattler, M., Clark, C., and McKelvey, H. (1974*b*). An improved method for the identification of patients and carriers of Krabbe's disease. *Clin. Chim. Acta* **56**, 199.

—— —— Kudoh, T., Snyder, S.P., and Kingston, R.S. (1980*b*). Niemann–Pick disease: A genetic model in Siamese cats. *Science* **208**, 1471.

—— —— Mueller, O.T., Myers, G.G., Schneider, R.S., and Nixon, G.W. (1980*a*). Adult GM$_1$-gangliosidosis: Clinical and biochemical studies on two patients and comparison to other patients called variant or adult GM$_1$-gangliosidosis. *Clin. Genet.* **17**, 323.

Werder, E.A., Siebenmann, R.E., Knorr-Murset, G., Zimmermann, A., Sizonenko, P.C., Theintz, P., Girard, J., Zachmann, M. and Prader, A. (1980). The incidence of congenital adrenal hyperplasia in Switzerland: a survey of patients born in 1960 to 1974. *Helv. Paediatr. Acta* **35**, 5.

West, C.D. (1971). Toward better therapy for resistant rickets. *J. ped.* **79**, 181.

West, J.C., Goodman, S.I., Schroter, G.P., Bloustein, P.A., Hambridge, K.M. and Weil, R. (1977). Pediatric kidney transplantation for cystinosis. *J. Ped. Surg.* **12**, 651.

Westall, R.G. (1963). Dietary treatment of a child with maple syrup urine disease (branched chain keto-aciduria). *Arch. Dis. Child.* **38**, 485.

—— Dancis, J., and Miller, S. (1957). Maple syrup urine disease – a new molecular disease. *Am. J. Dis. Child.* **94**, 571.

Wetterberg, L. (1975). Report on an international survey of safe and unsafe drugs in acute intermittent porphyria. In *Supplement to the proceedings of the first international porphyria meeting – Porphyrins in human diseases* (ed. M. Doss), p.191. Freiburg, Germany.

Weyers, H. and Bickel, H. (1958). Photodermatose mit Aminoacidurie, Indolaceturie und cerebralen Manifestationen (Hartnup–Syndrom). *Klin. Wschr.* **36**, 893.

Weyler, W., Sweetman, L., Maggio, D.C., and Nyhan, W.L. (1977). Deficiency of propionyl-CoA carboxylase and methylcrotonyl-CoA carboxylase in a patient with methylcrotonylglycinuria. *Clin. Chim. Acta* **76**, 321.

Whelan, D.T. and Scriver, C.R. (1968a). Cystathioninuria and renal imino-glycinuria in a pedigree. *New Engl. J. Med.* **278**, 924.

—— —— (1968b). Hyperdibasicaminoaciduria: An inherited disorder of amino acid transport. *Ped. Res.* **2**, 525.

—— Brusso, T., and Spate, M. (1976). Citrullinemia: Phenotypic variations. *Pediatrics* **57**, 935.

—— Hill, R., Ryan, E.D., and Spate, M. (1979). L-glutaric acidemia – Investigation of a patient and his family. *Pediatrics* **63**, 88.

Wherret, J.R. and Hakomori, S. (1973). Characterization of a blood group B glycolipid, accumulating in the pancreas of a patient with Fabry's disease. *J. biol. Chem.* **218**, 3046.

White, J.G. (1966). The Chediak–Higashi syndrome: A possible lysosomal disease. *Blood* **28**, 143.

—— and Witkop, C.J. jun. (1972). Effects of normal and aspirin platelets on defective secondary aggregation in the Hermansky–Pudlak syndrome: A test for storage pool deficient platelets. *Am. J. Pathol.* **68**, 57.

—— —— and Gerritsen, S.M. (1973). The Hermansky–Pudlak syndrome: inclusions in circulating leucocytes. *Br. J. Haem.* **24**, 761.

—— Edson, J.R., Desnick, S.J. and Witkop, C.J. jun. (1971). Studies of platelets in a variant of the Hermansky–Pudlak syndrome. *Am. J. Pathol.* **63**, 319.

White, H.H., Rowland, L.P., Araki, S., Thompson, H.L., and Cowen, D. (1965). Homocystinuria. *Arch. Neurol.* **13**, 455.

Whiteman, P. (1973). The quantitative determination of glycosaminoglycans in urine with Alcian Blue 8 GX. *Biochem. J.* **131**, 351.

—— and Henderson, H. (1977). A method for the determination of amniotic fluid glycosaminoglycans and its application to the prenatal diagnosis of Hurler and Sanfilippo disease. *Clin. Chim. Acta* **79**, 99.

—— and Young, E. (1977). The laboratory diagnosis of Sanfilippo disease. *Clin. Chim. Acta* **76**, 139.

Whiting, M.J. and Elliott, W.H. (1972). Purification and properties of solubilized mitochondrial δ-aminolevulinic acid synthetase and comparison with the cytosol enzyme. *J. biol. Chem.* **247**, 6818.

Whittaker, M. (1967). The pseudocholinesterase variants. A study of fourteen families selected via the fluoride resistant phenotype. *Acta Genet.* **17**, 1.

Wick, H., Brechbühler, T., and Gerard, J. (1970). Citrullinemia: elevated serum citrulline levels in healthy siblings. *Experientia* **26**, 823.

—— Schweizer, K. and Baumgartner, R. (1977). Thiamine dependency in a patient with congenital lacticacidemia due to pyruvate dehydrogenase deficiency. *Agents Actions* **7**, 405.

—— Backmann, C., Baumgartner, R., Brechbühler, T., Colombo, J.P., Wiesmann, U., Mihatsch, M.J., and Ohnacker, H. (1973). Variants of citrullinema. *Arch. Dis. Child.* **48**, 636.

Wiesmann, U.N. and Herschkowitz, N.N. (1974). Studies on the pathogenetic mechanism of I-cell disease in cultured fibroblasts. *Ped. Res.* **8**, 865.

—— and Neufeld, E.F. (1970). Scheie and Hurler syndromes: Apparent identity of the biochemical defect. *Science* **169**, 72.

—— Rossi, E.E., and Herschkowitz, N.N. (1971*b*). Treatment of metachromatic leukodystrophy in fibroblasts by enzyme replacement. *New Engl. J. Med.* **284**, 672.

—— Vassella, F., and Herschkowitz, N.N. (1971*a*). I-cell disease: leakage of lysosomal enzymes into extracellular fluids. *New Engl. J. Med.* **285**, 1090.

—— Meier, C., Spycher, M.A., Schmid, W., Bischoff, A., Gautier, E., and Herschkowitz, N. (1975). Prenatal metachromatic leukodystrophy. *Helv. Paediatr. Acta* **30**, 31.

Wiestner, M., Krieg, T., Horlein, D., Glanville, R.W., Fietzek, P., and Muller, P.K. (1979). Inhibiting effect of procollagen peptides on collagen biosynthesis in fibroblast cultures. *J. biol. Chem.* **254**, 7016.

Wilcken, B. (1978). Cited by Levy *et al.* in Bickel *et al.* (1980), p.93.

—— and Turner, G. (1978). Homocystinuria in New South Wales. *Arch. Dis. Child.* **53**, 242.

Willard, H.F. and Rosenberg, L.E. (1977). Inherited deficiencies of human methylmalonyl-CoA mutase activity: Reduced affinity of mutant apoenzyme for adenosylcobalamin. *Biochem. Biophys. Res. Commun.* **78**, 927.

—— —— (1979*a*). Inborn errors of cobalamin metabolism: Effect of cobalamin supplementation in culture on methylmalonyl-CoA mutase activity in normal and mutant human fibroblasts. *Biochem. Genet.* **17**, 57.

—— —— (1979*b*). Inherited deficiencies of methylmalonyl-CoA mutase activity: biochemical and genetic studies in cultured skin fibroblasts. In *Models for the study of inborn errors of metabolism* (ed. F.A. Hommes), p.297. Elsevier/North Holland Biomedical Press, Amsterdam.

—— —— (1980). Inherited methylmalonyl-CoA mutase apoenzyme deficiency in human fibroblasts: Evidence for allelic heterogeneity, genetic compounds, and codominant expression. *J. clin. Invest.* **65**, 690.

—— Mellman, I.S., and Rosenberg, L.E. (1978). Genetic complementation among inherited deficiencies of methylmalonyl-CoA-mutase activity: Evidence for a new class of human cobalamin mutant. *Am. J. hum. Genet.* **30**, 1.

—— Ambani, L.M., Hart, A.C., Mahoney, M.J., and Rosenberg, L.E. (1976). Rapid prenatal and postnatal detection of inborn errors of propionate, methylmalonate and cobalamin metabolism: A sensitive assay using cultured cells. *Hum. Genet.* **34**, 277.

Willcox, P. and Patrick, A.D. (1974). Biochemical diagnosis of cystinosis using cultured cells. *Arch. Dis. Child.* **49**, 209.

Willems, C., Heusden, A., Hainaut, A., and Chapelle, P. (1971). Hyper-sarcosinemie avec sarcosinurie. *J. Genet. hum.* **19**, 101.

—— Monnens, L.A.H., Trijbels, J.M.F., Sengers, R.A.C., and Veerkamp, J.H. (1974). Pyruvate decarboxylase deficiency in liver. *New Engl. J. Med.* **290**, 406.

Williams, C. and Field, J.B. (1968). Studies in glycogen storage disease. III. Limit dextrinosis: A genetic study. *J. Ped.* **72**, 214.

Williams, H.E. and Field, J.B. (1961). Low leucocyte phosphorylase in hepatic phosphorylase deficient glycogen storage disease. *J. clin. Invest.* **40**, 1841.

—— and Smith, L.H. jun. (1968a). Disorders of oxalate metabolism. *Am. J. Med.* **45**, 715.

—— —— (1968b). L-glyceric aciduria: a new genetic variant of primary oxaluria. *New Engl. J. Med.* **278**, 233.

—— Kendig, E.M., and Field, J.B. (1963). Leukocyte debranching enzyme in glycogen storage disease. *J. clin. Invest.* **42**, 656.

Williams, J.C. and Murray, A.K. (1980). Enzyme replacement in Pompe's disease with an α-glucosidase-low density lipoprotein complex. *Birth defects: original articles Series* XVI, Vol. 1, p.415. Alan R. Liss, New York.

Williamson, M., Koch, R., and Berlow, S. (1979). Diet discontinuation in phenylketonuria. *Pediatrics* **63**, 823.

Williamson, R., Eskdale, J., Coleman, D.V., Niazi, M., Loeffler, E.F., and Modell, B.M. (1981). Direct gene analysis of chorionic villi: a possible technique for first-trimester antenatal diagnosis of haemoglobinopathies. *Lancet* **ii**, 1125.

Wilson, C.S., Mankin, H.T., and Pluth, J.R. (1980). Aortic stenosis and mucopolysaccharidosis. *Ann. intern. Med.* **92**, 496.

Wilson, D.M. and Tapia, H.K. (1974). Xanthinuria in a large kindred. In *Purine metabolism in man* (ed. D. Sperling, A. de Vries, and J.B. Wyngaarden), p.343. Plenum Press, New York.

Wilson, J.D. (1975). Dihydrotestosterone formation in cultured human fibroblasts: Comparison of cells from normal subjects and patients with familial incomplete male pseudohermaphroditism, type 2. *J. biol. Chem.* **250**, 3498.

—— and Laznitzki, I. (1971). Dihydrotestosterone formation in fetal tissues of the rabbit and rat. *Endocrinology* **89**, 659.

—— Harrod, M.G., Goldstein, J.L., Hemsell, D.L., and MacDonald, P.C. (1974). Familial incomplete male pseudohermaphroditism, type 1. Evidence for androgen resistance and variable clinical manifestations in a family with the Reifenstein syndrome. *New Engl. J. Med.* **290**, 1097.

Wilson, R.G. and Masters, P.L. (1977). Neonatal death due to carbamyl phosphate synthetase deficiency. *Aust. Paed. J.* **13**, 119.

Wilson, R.S. and Ruiz, R.S. (1969). Bilateral central retinal artery occlusion in homocystinuria. *Arch. Ophthalmol.* **82**, 267.

Wilson, R.W., Wilson, C.M., Gates, S.C., and Higgins, J.V. (1975). Ketoadipic aciduria: A description of a new metabolic error in lysine tryptophan degradation. *Ped. Res.* **9**, 522.

Wilson, S.A.K. (1912). Progressive lenticular degeneration: a familial nervous disease associated with cirrhosis of the liver. *Brain* **34**, 295.

Wilson-Cox, D.W., Fraser, F.C., and Sass-Kortsak, A. (1972). A genetic study of Wilson's disease; evidence for heterogeneity. *Am. J. hum. Genet.* **24**, 646.

Wimberley, P.D., Harries, J.T., and Burgess, E.A. (1974). Congenital glucose–galactose malabsorption. *Proc. R. Soc. Med.* **67**, 755.

Windhorst, D.B., Zelickson, A.S. and Good, R.A. (1968). A human pigmentary dilution based on a heritable subcellular structural defect – the Chediak–Higashi syndrome. *J. Invest. Dermatol.* **50**, 9.

Winter, J.S.D. (1980). Current approaches to the treatment of congenital adrenal hyperplasia. *J. Ped.* **97**, 81.

Winter, R.M., Swallow, D.M., Baraitser, M., and Purkiss, P. (1980). Sialidosis type 2 (acid neuraminidase deficiency): clinical and biochemical features of a further case. *Clin. Genet.* **18**, 203.

Winterborn, M.H., France, N.E., and Raiti, S. (1970). Incomplete testicular feminization. *Arch. Dis. Child.* **45**, 811.

Winters, R.W., Graham, J.B., Williams, T.F., McFalls, V.W., and Burnett, C.H. (1958) A genetic study of familial hypophosphatemia and vitamin D-resistant rickets with a review of the literature. *Medicine* (Baltimore) **37**, 97.

Wise, D., Wallace, H.J., and Jellinek, E.H. (1962). Angiokeratoma corporis diffusum. A clinical study of eight affected families. *Quart. J. Med.* **31**, 177.

Witkop, C.J. jun. (1971). Albinism. In *Advances in human genetics* (ed. H. Harris and K. Hirschhorn), Vol. 2, p.61. Plenum Press, New York.

—— White, J.G., and King, R.A. (1974). Oculocutaneous albinism. In *Heritable disorders of amino acid metabolism* (ed. W.L. Nyhan), p.177. John Wiley and Sons, New York.

—— MacLean, C.J., Schmidt, P.J., and Henry, J.L. (1966). Medical and dental findings in the Brandywine isolate. *Ala. J. Med. Sci.* **3**, 382.

—— Nance, W.E., Rawls, R.F., and White, J.G. (1970). Autosomal recessive oculocutaneous albinism in man: Evidence for genetic heterogeniety. *Am. J. hum. Genet.* **22**, 55.

—— White, J.G., Nance, W.E., Jackson, C.E., and Desnick, S. (1971). Classification of albinism in man. In *Skin, hair and nails* (ed. D. Bergsma), Vol. 7, p.13. *Birth defects: original articles series*. Williams and Wilkins, The National Foundation, Baltimore.

—— Hill, C.W., Desnick, S.J., Thies, J.K., Thorn, H.L., Jenkins, M. and White J.G. (1973). Ophthalmologic, biochemical, platelet, and ultrastructural defects in the various types of oculocutaneous albinism. *J. Invest. Dermat.* **60**, 443.

Witschel, H. and Meyer, W. (1968). Fabry's disease: Clinical and pathologic studies of a clinical case. *Klin. Wschr.* **46**, 305.

Wolf, B. (1979) Evaluation of biotin responsiveness in cultured fibroblasts from patients with propionic acidemia: Absence of response by structurally altered carboxylases. *Biochem. Genet.* **17**, 709.

—— (1980). Reassessment of biotin-responsiveness in 'unresponsive' propionyl-CoA carboxylase deficiency. *J. Ped.* **97**, 964.

—— and Feldman, G.L. (1982). The biotin-dependent carboxylase deficiencies *Am. J. hum. Genet.* **34**, 699.

—— and Hsia, Y.E. (1978). Biotin-responsiveness in propionic acidemia. *Lancet* **ii**, 901.

—— Willard, H.F., and Rosenberg, L.E. (1980). Kinetic analysis of genetic complementation in heterokaryons of propionyl-CoA carboxylase-deficient human fibroblasts. *Am. J. hum. Genet.* **32**, 16.

—— Hsia, Y.E., Tanaka, K., and Rosenberg, L.E. (1978). Correlation between serum propionate and blood ammonia concentrations in propionic acidemia. *J. Ped.* **93**, 471.

—— Grier, R.E., Allen, R.J., Goodman, S.I., and Kien, C.L. (1983). Biotinidase deficiency: the enzymatic defect in late-onset multiple carboxylase deficiency. *Clin. Chim. Acta* **131**, 273.

—— —— Sweetman, L., Gravel, R., Harris, D.J., and Nyhan, W.L. (1981).

Propionic acidemias: a clinical update. *J. Ped.* **99**, 835.

Wolfe, D.M. and Gatfield, P.D. (1975). Leukocyte urea cycle enzyme in hyperammonia. *Ped. Res.* **9**, 531.

Wolfe, H.J. and Pietra, G.G. (1964). The visceral lesions of metachromatic leuko-dystrophy. *Am. J. Pathol.* **44**, 921.

Wolfe, L.S., Senior, R.G., Ng, Y., and Kin, N.M.K. (1974). The structures of oligosaccharides accumulating in the liver of GM_1-gangliosidosis, type 1. *J. biol. Chem.* **249**, 1828.

—— Callahan, J., Fawcett, J.S., Andermann, P., and Scriver, C. (1970). GM_1-gangliosidosis without chondrostrophy or visceromegaly: ß-galactosidase deficiency with gangliosidosis and the excessive excretion of a keratan sulfate. *Neurology* **20**, 23.

Wolff, J., Thompson, R.H., and Robbins, J. (1964). Congenital goitrous cretinism due to absence of iodide-concentrating ability. *J. clin. Endocrinol. Metab.* **24**, 699.

Wolff, S.M., Dale, D.C., Clark, R.A., Root, R.K., and Kimball, H.R. (1972). The Chediak–Higashi syndrome: Studies of host defenses. *Ann. intern. Med.* **76**, 293.

Wolkoff, A.W., Chowdhury, J.R., Gartner, L.A., Rose, A.L., Biempica, L., Giblin, D.R., Fink, D., and Arias, I.M. (1979). Crigler–Najjar syndrome (type 1) in an adult male. *Gastroenterology* **76**, 3380.

Wolkow, M. and Baumann, E. (1891). Über das Wesen der Alkaptonurie. *Z. Physiol. Chem.* **15**, 228.

Wolman, M. (1964). Histochemistry of lipids in pathology. In *Handbuch der Histochemie* (ed. W. Grauman and K. Neumann), Vol. 5, p.228. Fischer-Verlag, Stuttgart.

—— Sterk, V.V., Gatt, S., and Frenkel, M. (1961). Primary familial scantho-matosis with involvement and calcification of the adrenals. Report of two more cases in siblings of a previously described infant. *Pediatrics* **28**, 742.

Wong, P.W.K. and Pillai, P.M. (1966). Clinical and biochemical observations in two cases of Hartnup disease. *Arch. Dis. Child.* **41**, 383.

—— Justice, P. and Berow, S. (1977*b*). Detection of homozygotes and heterozygotes with methylene tetrahydrofolate reductase deficiency. *J. lab. clin. Med.* **90**, 283.

—— Lambert, A.M., and Komrower, G.M. (1967*b*). Tyrosinaemia and tyrosyluria in infancy. *Dev. Med. Child. Neurol.* **9**, 551.

—— Justice, D. Hruby, M., Weiss, E.B., and Diamond, E. (1977*a*). Folic acid non-responsive homocystinuria due to methylenetetrahydrofolate reductase deficiency. *Pediatrics* **59**, 749.

Wong, V.G. (1973). The eye and cystinosis. In *Cystinosis* (ed. J.D. Schulman), p. 23. DHEW Publication No. (NIH) 72–249. US Government Printing Office, Washington, DC.

—— Lietman, P.S., and Seegmiller, J.E. (1967*a*). Alterations of pigment epithelium in cystinosis. *Arch. Ophthalmol. (Chicago)* **77**, 361.

—— Kewabara, T., Brubaker, R., Olson, W., Schulman, J., and Seegmiller, J.E. (1970). Intralysosomal cystine crystals in cystinosis. *Invest. Ophthalmol.* **9**, 83.

Woo, S.L.C., Gillam, S.S., and Woolf, L.I. (1974). The isolation and properties of phenylalanine hydroxylase from human liver. *Bichem. J.* **139**, 741.

Wood, S. and Nadler, H.L. (1972). Fabry's disease: absence of an α-galactosidase isoenzyme. *Am. J. hum. Genet.* **24**, 250.

Woody, N.C. (1964). Hyperlysinemia. *Am. J. Dis. Child.* **108**, 543.

——— Hutzler, J., and Dancis, J. (1966). Further studies of hyperlysinemia. *Am. J. Dis. Child.* **112**, 577.

——— Snyder, C.H., and Harris, J.A. (1965). Histidinemia. *Am. J. Dis. Child.* **110**, 606.

Woolf, L.I. (1951). Excretion of conjugated phenylacetic acid in phenylketonuria. *Biochem. J.* **49**, ix.

——— (1967). Large scale screening for metabolic disease in the newborn in Great Britain. In *Phenylketonuria and allied metabolic disorders* (ed. J.A. Anderson and K.F. Swaiman), p.50. US Department of Health, Education and Welfare (Children's Bureau), Washington.

——— (1978). The high frequency of phenylketonuria in Ireland and Western Scotland. *J. inher. metab. Dis.* **1**, 101.

Worthy, T.E., Grobner, W., and Kelley, W.N. (1974). Hereditary orotic aciduria: Evidence for a structural gene mutation. *Proc. natl. Acad. Sci. USA* **71**, 3031.

Wosilait, W.D. and Sutherland, E.W. (1956). The relationship of epinephrine and glucagon to liver phosphorylase. II. Enzymatic inactivation of liver phosphorylase. *J. biol. Chem.* **218**, 469.

Wray, S.H., Kuwabara, T., and Sanderson, P. (1976). Menkes' kinky hair disease: a light and electron microscopic study of the eye. *Invest. Ophthalmol.* **15**, 128.

Wright, E.C., Stern, J., Ersser, R., and Patrick, A.D. (1979). Glutathionuria: γ-glutamyl transpeptidase deficiency. *J. inher. metab. Dis.* **2**, 3.

Wright, T. and Pollitt, R. (1973). Psychomotor retardation, epileptic and stuporous attacks, irritability and ataxia associated with ammonia intoxication, high blood ornithine levels and increased homocitrulline in the urine. *Proc. R. Soc. Med.* **66**, 221.

Wysocki, S.J. and Hähnel, R. (1976*a*). 3-hydroxy-3-methylglutaric aciduria: deficiency of 3-hydroxy-3-methylglutaryl-coenzyme A lyase. *Clin. Chim. Acta* **71**, 349.

——— ——— (1976*b*). 3-hydroxy-3-methylglutaric aciduria: 3-hydroxy-3-methyl-glutaryl-coenzyme A lyase levels in leucocytes. *Clin. Chim. Acta* **73**, 373.

——— Wilkinson, S.P., Hähnel, R., Wong, C.Y.B., and Panegyres, P.K. (1976). 3-hydroxy-3-methylglutaric aciduria combined with 3-methylglutaconic aciduria. *Clin. Chim. Acta* **70**, 399.

Yaffe, S.J., Levy, G., Matsuzawa, T., and Baliah, T. (1966). Enhancement of glucuronide-conjugating capacity in a hyperbilirubinemic infant due to apparent enzyme induction by phenobarbital. *New Engl. J. Med.* **275**, 1461.

Yahara, S., Moser, H.W., Kolodny, E.H., and Kishimoto, Y. (1980). Reverse phase high performance liquid chromatography of cerebrosides, sulfatides and ceramides: microanalysis of homolog composition without hydrolysis and application to cerebroside analysis in peripheral nerves of adrenoleukodystrophy patients. *J. Neurochem.* **34**, 694.

Yamada, K., Adachibara, A., Nakazawa, S., Shinkai, A., Nishina, T., and Miwa, S. (1974). Erythrocyte pyruvate kinase deficiency associated with kinetically aberrant isozyme. *Acta Haematol. Jap.* **37**, 17.

Yamakawa, T. and Nagai, Y. (1978). Glycolipids at the cell surface and their biological functions. *Trends Biochem. Sci.* **3**, 128.

Yamamoto, Y., Toyoshima, R., and Muramatsu, K. (1979). Effects of additional protein or methionine and threonine on tyrosine catabolism in rats fed diets high in tyrosine. *Agric. biol. chem.* **43**, 2585.

Yamashita, K., Tachibana, Y., Mihara, K., Okada, S., Yabuuchi, H., and

Kobata, A. (1980). Urinary oligosaccharides of mannosidosis. *J. biol. Chem.* **255**, 5126.

Yamauchi, M., Kitahara, T., Fujisawa, K., Kameda, H., Takasaki, S., Komori, R., Saheki, T., Katsunuma, T., and Katuruma, N., (1980). An autopsied case of hypercitrullinemia in an adult caused by partial deficiency of liver argininosuccinate synthetase. *Acta Hepatol. Jap.* **21**, 326.

Yatziv, S., Statter, M., and Merin, S. (1979). Metabolic studies in two families with hyperornithinemia and gyrate atrophy of choroid and retina. *J. lab. clin. Med.* **93**, 749.

Yeaman, S.J., Hutcheson, E.T., Roche, T.E., Pettit, F.H., Brown, J.R., Reed, L.J., Watson, D.C., and Dixon, G.H. (1978). Sites of phosphorylation on pyruvate dehydrogenase from bovine kidney and heart. *Biochemistry* **17**, 2364.

Yoshida, A. (1967). A single amino acid substitution (asparagine to aspartic acid) between normal (B+) and the common Negro variant (A+) of human glucose-6-phosphate dehydrogenase. *Proc. natl. Acad. Sci. USA* **57**, 835.

—— (1968). The structure of normal and variant human glucose-6-phosphate dehydrogenase. In *Hereditary disorders of erythrocyte metabolism* (City of Hope Symp. Series Vol. 1) (ed. E. Beutler), p.146. Grune and Stratton, New York.

—— (1970). Amino acid substitution (histidine to tyrosine) in a glucose-6-phosphate dehydrogenase variant (G6PD Hektoen) associated with overproduction. *J. mol. Biol.* **52**, 483.

—— and Beutler, E. (1978). Human glucose-6-phosphate dehydrogenase variants: a supplementary tabulation. *Ann. hum. Genet.* **41**, 347.

—— Stamatoyannopoulos, G., and Motulsky, A. (1967). Negro variant of glucose-6-phosphate dehydrogenase deficiency (A−) in man. *Science* **155**, 97.

Yoshida, T., Tada, K., Yokoyama, Y., and Arakawa, T. (1968). Homocystinuria of vitamin B_6 dependent type. *Tohoku J. exp. Med.* **96**, 235.

Yoshinaga, T. and Sano, S. (1980). Coproporphyrinogen oxidase. I. Purification, properties, reaction mechanism and activation by phospholipid. *J. biol. Chem.* **255**, 4722.

Young, E.P. and Patrick, A.D. (1970). Deficiency of acid esterase activity in Wolman's disease. *Arch. Dis. Child.* **45**, 664.

—— Wilson, J., Patrick, A.D., and Crome, L. (1972). Galactocerebrosidase deficiency in globoid cell leucodystrophy of late onset. *Arch. Dis. Child.* **47**, 449.

Young, I.D., Harper, P.S., Newcombe, R.G., and Archer, I.M. (1982). A clinical and genetic study of Hunter's syndrome. 2. Differences between the mild and severe forms. *J. med. Genet.* **19**, 408.

Youngdahl-Turner, P., Mellman, I.S., Allen, R.H., and Rosenberg, L.E. (1979). Protein mediated vitamin uptake: absorptive endocytosis of the transcobalamin II-cobalamin complex by cultured human fibroblasts. *Exp. Cell Res.* **118**, 127.

—— Rosenberg, L.E., and Allen, R.H. (1978). Binding and uptake of transcobalamin II by human fibroblasts. *J. clin. Invest* **61**, 133.

Yü, T.-F. (1965). Secondary gout associated with myeloproliferative diseases. *Arthritis Rheum.* **8**, 765.

—— and Gutman, A.B. (1967). Uric acid nephrolithiasis in gout. Predisposing factors. *Ann. intern. Med.* **67**, 1133.

Yudkoff, M., Foreman, J.W., and Segal, S. (1981). Effects of cysteamine therapy in nephropathic cystinosis. *New Engl. J. Med.* **304**, 141.

—— Yang, W., Snodgrass, P.J., and Segal, S. (1980). Ornithine transcarbamylase deficiency in a boy with normal development. *J. Ped.* **96**, 441.

—— Cohn, R.M., Puschak, R., Rothman, R., and Segal, S. (1978). Glycine therapy in isovaleric acidemia. *J. Ped.* **92**, 813.

Yunis, E.J. and Lee, R.E. (1970). Tubules of globoid leukodystrophy: a right-handed helix. *Science* **169**, 64.

Yutaka, T., Fluharty, A.L., Stevens, R.L., and Kihara, H. (1978). Iduronate sulfatase analysis of hair roots for identification of Hunter syndrome hetero-zygotes. *Am. J. hum. Genet.* **30**, 575.

Zachello, F., Benson, P.F., Brown, S., Croll, P., and Giannelli, F. (1972). Induction of galactokinase in fibroblasts from heterozygous and homozygous subjects. *Nature (New Biol.)* **239**, 95.

Zachmann, M. and Prader, A. (1978). Unusual heterozygotes of congenital adrenal hyperplasia due to 21-hydroxylase deficiency. *Acta Endocrinol.* **87**, 557.

—— —— (1979). Unusual heterozygotes of congenital adrenal hyperplasia due to 21-hydroxylase deficiency confirmed by HLA tissue typing. *Acta Endocrinol.* **95**, 542.

—— Vollmin, J.A., Hamilton, W., and Prader, A. (1972). Steroid 17,20-desmolase deficiency: a new cause of male pseudohermaphroditism. *Clin. Endocrinol.* **1**, 369.

—— —— Murset, G., Curtius, H.C., and Prader, A. (1970). Unusual type of adrenal hyperplasia probably due to deficiency of 3ß-hydroxysteroid de-hydrogenase. *J. clin. Endocr.* **30**, 719.

Zaill, S.S., Charlton, R.W., and Bothwell, T.H. (1962). The haemolytic effect of certain drugs in Bantu subjects with a deficiency of glucose-6-phosphate dehydrogenase. *S. Afr. J. med. Sci.* **27**, 95.

Zaleski, L.A., Dancis, J., Cox, R.P., Hutzler, J., Zaleski, W.A., and Hill, A. (1973). Variant maple syrup urine disease in mother and daughter. *Can. med. Ass. J.* **109**, 299.

Zanella, A., Izzo, C., Rebulla, P., Perroni, L., Mariani, M., Canestri, G., Sansone, G., and Sirchia, G. (1980). The first stable variant of erythrocyte glucose-phosphate isomerase associated with severe hemolytic anemia. *Am. J. Hematol.* **9**, 1.

Zannis, V.I., Just, P.W., and Breslow, J.L. (1981). Human apolipoprotein E isoprotein subclasses are genetically determined. *Am. J. hum. Genet.* **33**, 11.

Zannoni, V.G., Lomtevas, N., and Goldfinger, S. (1969). Oxidation of homo-gentisic acid to ochronotic pigment in connective tissue. *Biochim. Biophys. Acta* **177**, 94.

—— Seegmiller, J.E., and La Du, B.N. (1962), Nature of the defect in alcaptonuria. *Nature, Lond.* **193**, 952.

Zarembski, P.M., Hodgkinson, A., and Cochran, M. (1967). Treatment of primary hyperoxaluria with calcium carbimide. *New Engl. J. Med.* **277**, 1000.

—— Rosen, S.M., and Hodgkinson, A. (1969). Dialysis in the treatment of primary hyperoxaluria. *Br. J. Urol.* **41**, 530.

Zeller, E.A. (1943). Isolierung von Phenylmilchsäure und Phenyltraubensäure aus Harn bei Imbecillitas Phenylpyruvica. *Helv. Chim. Acta* **26**, 1614.

Zellweger, H., Mueller, S., Ionasescu, V., Shochet, S.S., and McCormick, W.F. (1972). Glycogenosis IV: a new cause of infantile hypotonia. *J. Ped.* **80**, 842.

Zerfowski, J. and Sandhoff, K. (1974). Juvenile GM_2-gangliosidose mit veränderter Substratspezifität der Hexosaminidase A. *Acta Neuropath. (Berlin)* **27**, 225.

Zerwekh, J.E., Glass, K., Jowsey, J., and Charles, Y.C. (1979). An unique form of osteomalacia associated with end organ refractoriness to 1,25-dihydroxyvitamin D and apparent defective synthesis of 25-hydroxyvitamin D. *J. Clin. Endocr. Metab.* **49**, 171.

Ziegler, J.B., Lee, C.H., Van der Weyden, M.B., Bagnara, A.S., and Beveridge, J., (1980). Severe combined immunodeficiency and adenine deaminase deficiency: failure of enzyme replacement therapy. *Arch. Dis. Child.* **55**, 452.

—— Van der Weyden, M.B., Lee, C.H., and Daniel, A. (1981). Prenatal diagnosis for adenosine deaminase deficiency. *J. med. Genet.* **18**, 154.

Zipkin, I., Hawkins, G.R., and Mazzarella, M. (1964). The tyrosine, tryptophan, and protein content of human parotid saliva in oral and systemic disease: use of ultraviolet absorption techniques. *Int. Ser. Monog. Oral Biol.* **3**, 331.

Zitman, D., Chazan, S., and Klibansky, C. (1978). Sphingomyelinase activity levels in human peripheral blood leucocytes using (^{3}H) sphingomyelin as substrate: study of heterozygotes and homozygotes for Niemann–Pick disease variants. *Clin. Chim. Acta* **86**, 37.

Zlotogora, J., Bach, G., Barak, V., and Elian, E. (1980). Metachromatic leukodystrophy in the Habbanite Jews: high frequency in a genetic isolate and screening for heterozygotes. *Am. J. hum. Genet.* **32**, 663.

—— Costeff, J. and Elian, E. (1981). Early development in metachromatic leucodystrophy. *Arch. Dis. Child.* **56**, 309.

Zoref, E., De Vries, A., and Sperling, O. (1975). Mutant feedback-resistant phosphoribosylpyrophosphate synthetase associated with purine overproduction and gout; phosphoribosylpyrophosphate and purine metabolism in cultured fibroblasts. *J. clin. Invest.* **56**, 1093.

—— —— —— (1977). Evidence for X-linkage of phosphoribosylpyrophosphate synthetase in man. Studies with cultured fibroblasts from a gouty family with mutant feedback-resistant enzyme. *Hum. Hered.* **27**, 73.

Index